T0344919

SIGNAL PROCESSING FOR 5G

SIGNAL PROCESSING FOR 5G

ALGORITHMS AND IMPLEMENTATIONS

Edited by

Fa-Long Luo, Ph.D., IEEE Fellow
Charlie (Jianzhong) Zhang, Ph.D., IEEE Fellow

WILEY

This edition first published 2016

Registered office
John Wiley & Sons Ltd, The Atrium, Southern Gate, Chichester, West Sussex, PO19 8SQ, United Kingdom

For details of our global editorial offices, for customer services and for information about how to apply for permission to reuse the copyright material in this book please see our website at www.wiley.com.

Library of Congress Cataloging-in-Publication Data

Names: Luo, Fa-Long, editor. | Zhang, Charlie, editor.
Title: Signal processing for 5G : algorithms and implementations / edited by Fa-Long Luo, Charlie Zhang.
Description: Chichester, West Sussex, United Kingdom : John Wiley & Sons Inc., [2016] | Includes bibliographical references and index.
Identifiers: LCCN 2016010334| ISBN 9781119116462 (cloth) | ISBN 9781119116486 (epub) | ISBN 9781119116479 (Adobe PDF)
Subjects: LCSH: Signal processing–Digital techniques–Mathematics. | Mobile communication systems–Standards. | Wireless communication systems–Standards. | Computer algorithms.
Classification: LCC TK5102.9 .S5423 2016 | DDC 621.3845/6–dc23 LC record available at https://lccn.loc.gov/2016010334

A catalogue record for this book is available from the British Library.

Set in 10/12pt, TimesLTStd by SPi Global, Chennai, India.

1 2016

Contents

Part V REFERENCE DESIGN AND 5G STANDARD DEVELOPMENT

Preface

5G wireless technology is developing at an explosive rate and is one of the biggest areas of research within academia and industry. In this rapid development, signal processing techniques are playing the most important role. In 2G, 3G and 4G, the peak service rate was the dominant metric for performance. Each of these previous generations was defined by a standout signal processing technology that represented the most important advance made. In 2G, this technology was time-division multiple access (TDMA); in 3G, it was code-division multiple access (CDMA); and in 4G, it was orthogonal frequency-division multiple access (OFDMA). However, this will not be the case for 5G systems – there will be no dominant performance metric that defines requirements for 5G technologies. Instead, a number of new signal processing techniques will be used to continuously increase peak service rates, and there will be a new emphasis on greatly increasing capacity, coverage, efficiency (power, spectrum, and other resources), flexibility, compatibility, reliability and convergence. In this way, 5G systems will be able to handle the explosion in demand arising from emerging applications such as big data, cloud services, and machine-to-machine communication.

A number of new signal processing techniques have been proposed for 5G systems and are being considered for international standards development and deployment. These new signal processing techniques for 5G can be categorized into four groups:

1. new modulation and coding schemes
2. new spatial processing techniques
3. new spectrum opportunities
4. new system-level enabling techniques.

The successful development and implementation of these technologies for 5G will be challenging and will require huge effort from industry, academia, standardization organizations and regulatory authorities.

From an algorithm and implementation perspective, this book aims to be the first single volume to provide a comprehensive and highly coherent treatment of all the signal processing techniques that enable 5G, covering system architecture, physical (PHY)-layer (down link and up link), protocols, air interface, cell acquisition, scheduling and rate adaption, access procedures, relaying and spectrum allocation. This book is organized into twenty-three chapters in five parts.

Part 1: Modulation, Coding and Waveform for 5G

The first part, consisting of eight chapters, will present and compare the detailed algorithms and implementations of all major candidate modulation and coding schemes for 5G, including generalized frequency division multiplexing (GFDM), filter-bank multi-carrier (FBMC) transmission, universal filtered multi-carrier (UFMC) transmission, bi-orthogonal frequency division multiplexing (BFDM), spectrally efficient frequency division multiplexing (SEFDM), the faster-than-Nyquist signaling (FTN) based time-frequency packing (TFP), sparse code multiple access (SCMA), multi-user shared access (MUSA) and non-orthogonal multiple access (NOMA).

With a focus on FBMC, GFDM, UFMC, BFDM and TFP, Chapter 1 presents a comprehensive introduction to these waveform generation and modulation schemes by covering the basic principles, mathematical models, step-by-step algorithms, implementation complexities, schematic processing flows and the corresponding application scenarios involved.

Chapter 2 is devoted to the FTN data transmission method, with the emphasis on applications that are important for future 5G systems. What is explored in this chapter mainly includes time-FTN methods with non-binary modulation and multi-subcarrier methods that are similar in structure to OFDM. In either, there is an acceleration processing in time or compacting in frequency that makes signal streams no longer orthogonal. FTN can be combined with error-correcting coding structures to form true waveform coding schemes that work at high-bit rates per Hertz and second. As a matter of fact, FTN based systems can potentially double data transmission rates.

The technical evolution from OFDM to FBMC is addressed in Chapter 3, covering the principles, algorithms, designs and implementations of these two schemes. This chapter first presents the details of OFDM-based schemes and the major shortcomings that prevent them from being employed in 5G. Through introduction of synthesis and analysis filter banks, prototype filter design and the corresponding polyphase implementation, Chapter 3 then extensively deals with the working principles of FBMC and compares it with OFDM in terms of performance – power spectral density and out of band power radiation – and complexity – number of fast Fourier transforms and filter banks. One can also see from this chapter that OFDM is a special case of FBMC.

Easy and effective integration with massive multiple-input and multiple-output (MIMO) technology is a key requirement for a modulation and waveform generation scheme in 5G. Chapter 4 demonstrates that FBMC can serve as a viable candidate waveform in the application of massive MIMO. The chapter outlines the system model, algorithm formulation, self-equalization property and pilot contamination of FBMC for massive MIMO channels, and also shows that while FBMC offers the same processing gain as OFDM, it offers the advantages of: more flexible carrier aggregation (CA), higher bandwidth efficiency – because of the absence of cyclic prefix (CP) – blind channel equalization and larger subcarrier spacing, and hence less sensitivity to carrier frequency offset and lower peak-to-average power ratio (PAPR).

Chapter 5 presents a non-orthogonal multicarrier system, namely, spectrally efficient frequency division multiplexing (SEFDM), which packs subcarriers at a frequency separation less than the symbol rate while maintaining the same transmission rate per individual subcarrier. Thus spectral efficiency is improved in comparison with the OFDM system. By transmitting the same amount of data, the SEFDM system can conceptually save up to 45% bandwidth.

This chapter also describes a practical experiment in which the SEFDM concept is evaluated in a CA scenario considering a realistic fading channel. On the other hand, SEFDM involves higher computation complexity and longer processing delays, mainly due to the requirement for complex signal detection. This suggests that advanced hardware implementation is still highly desirable, so as to make SEFDM a better fit to 5G.

As pointed out in Chapter 6, non-orthogonal multi-user superposition and shared access is a promising technology that can increase the system throughput and simultaneously serve massive connections. Non-orthogonal access allows multiple users to share time and frequency resources in the same spatial layer via simple linear superposition or code-domain multiplexing. This chapter overviews all major non-orthogonal access schemes, categorizing them into two groups:

- the non-spreading methods, where modulation symbols are one-to-one mapped to the time/frequency resource elements
- the spreading methods, where symbols are first spread and then mapped to time/frequency resources.

Their design principles, key features, advantages and disadvantages are extensively discussed in this chapter.

Chapter 7 is devoted to a new multiple access scheme, termed NOMA, which introduces power-domain user multiplexing and exploits channel differences among users to improve spectrum efficiency. This chapter also explains the interface design aspects of NOMA, for example multi-user scheduling and multi-user power control, and its combination with MIMO. The performance evaluation and ongoing experimental trials of downlink and uplink NOMA are reported. The simulation results and the measurements obtained from the testbed show that under multiple configurations the cell throughput achieved by NOMA is 30% higher than that of OFDMA.

With a tutorial style, Chapter 8 presents an overview of all the major multicarrier modulation (MCM) candidates for 5G, categorizing them into three groups:

- subcarrier filtered MCM using linear convolution
- subcarrier filtered MCM using circular convolution
- subband windowed MCM.

General comparisons of these candidate algorithms are made in this chapter, covering PAPR, OOB emission, processing and implementation complexity, spectrum efficiency, the requirement of CP, intercarrier interference, intersymbol interference, multipath distortion, orthogonality and the related effects of frequency offset and phase noise, synchronization requirements in both the time domain and the frequency domain, latency, mobility, compatibility and integration with other processing such as massive MIMO.

Part 2: New Spatial Signal Processing for 5G

The five chapters in Part 2 focus on new spatial signal processing technologies for 5G, mainly addressing massive MIMO, full-dimensional MIMO (FD-MIMO), three-dimensional MIMO

(3D-MIMO), adaptive 3D beamforming and diversity, continuous aperture phased MIMO (CAP-MIMO) and orbital angular momentum (OAM) based multiplexing. Chapter 9 mainly deals with the principle, theory, algorithm, design, testing, implementation and prototyping on advanced computing and processing platforms for the massive MIMO technique, which will certainly be employed in 5G standards. Core processing blocks, such as downlink precoding, uplink detection and channel estimation, are reviewed first, after which the emphasis is put on the various hardware implementation issues of massive MIMO, covering radio frequency (RF) front-end calibration, baseband processing, synchronization analyses, testbed design and system prototyping, as well as the corresponding deployment scenarios.

Design and implementation of massive MIMO transmission and reception, which uses millimeter wave (mmWave) bands, is presented in Chapter 10. More specifically, this chapter proposes a framework for the design, analysis, testing and practical implementation of a new MIMO transceiver architecture: CAP-MIMO. Using the concept of beam-space MIMO communication – multiplexing data into multiple orthogonal spatial beams in order to optimally exploit the spatial antenna dimension – CAP-MIMO combines the directivity gains of traditional antennas, the beam-steering capability of phased arrays, and the spatial multiplexing gains of MIMO systems to realize the multi-GBps capacity potential of mmWave technology, as well as the unprecedented operational functionality of dynamic multibeam steering and data multiplexing.

Chapter 11 mainly deals with the modeling and measurement of 3D propagation channels, which play very important roles in designing and implementing an FD-MIMO and 3D-beamforming system. This chapter first presents the fundamental channel descriptions and then provides advanced measurement and modeling techniques for 3D propagation channels. The related measurement results and theoretical analyses for those propagation effects that significantly influence 3D channel behaviors are also outlined. This chapter can serve as a good start for modeling and measuring many other propagation channels arising in application scenarios such as the outdoor-to-indoor scenario and high-density-user scenario.

From theory to practice, all technical aspects of the massive-antenna-based 3D-MIMO techniques are addressed in Chapter 12, with the emphasis on performance evaluation. More specifically, this chapter evaluates the performance of 3D-MIMO with massive antennas by system-level simulation, using practical assumptions and a channel model, and by field trials, with a commercial terminal and networks. In addition, extensive comparisons and analyses of the system-level simulation results and the field-trial test measurements are provided. It is shown that an active antenna system (AAS) can make a good compromise between cost and performance by integrating the active transceivers and the passive antenna array into one unit. This suggests that the AAS can be considered key to commercialization of 3D-MIMO with massive antennas in future 5G systems.

Chapter 13 presents a comprehensive introduction to the basic concept of the OAM of electromagnetic (EM) waves and its applications in wireless communication. It covers the generation, detection of multiplexing and demultiplexing of OAM beams, as well as analyses of the propagation effects in OAM channels. As reported in this chapter, OAM-based multiplexing can increase the system capacity and spectral efficiency of wireless communication links by transmitting multiple coaxial data streams. Moreover, OAM multiplexing can also be combined with the polarization multiplexing and the traditional spatial multiplexing to further improve system performance in terms of the capacity and spectral efficiency.

Part 3: New Spectrum Opportunities for 5G

Organized into four chapters, Part 3 is devoted to signal processing algorithms and their implementation for 5G, taking advantage of new spectrum opportunities, such as the mmWave band and full-duplex (FD) transmission. Chapter 14 provides an overview of the building of a mmWave proof of concept (PoC) system for 5G, covering the RF front end, real-time control, analog-to-digital and digital-to-analog converters, distributed multiprocessor control and baseband processing implementation. Some important requirements of a flexible prototyping platform are discussed in this chapter, along with the software and hardware system architecture needed to enable high-throughput, high-bandwidth applications such as mmWave radio access technology for 5G. For the purpose of showing how to handle design and implementation challenges, a case study of the design of a mmWave PoC system on the basis of a commercial off-the-shelf platform is provided in this chapter as well.

Chapter 15 focuses on mmWave channel modeling and also discusses other signal processing problems for mmWave communication in 5G. Two approaches to meet the requirements of the 5G mmWave channel model are presented in this chapter, namely:

- an enhanced 3GPP-spatial channel model
- a ray-propagation-based statistical model.

Using understanding and analyses of the mmWave channel characteristics, this chapter provides system-design considerations for 5G mmWave band radio access technology and key signal processing technologies related to 5G mmWave communications, including beam acquisition, channel estimation and interference handling.

The general principles and basic algorithms of FD transmission are given in Chapter 16, which explains FD system requirements, self-interference cancellation (SIC) techniques, implementation challenges, impairment mitigation and hardware integration with MIMO. FD operation offers not only the potential to double spectral efficiency (bits/second/Hz) but also improvement of the reliability and flexibility of dynamic spectrum allocation. Meanwhile, SIC is the key to making FD a reality. With the emphasis on signal processing aspects of SIC, this chapter outlines four SIC techniques:

- passive self-interference (SI) suppression in the propagation domain
- active SIC in the analog domain
- active SIC in the digital domain
- auxiliary chain SIC.

Chapter 17 provides an overview of state-of-the-art SI mitigation and cancellation techniques for multi-antenna in-band FD communication, including bidirectional and relay transmission. Design and implementation of FD transceivers is described through concrete examples, notably passive isolation, RF cancellation and nonlinear and adaptive digital cancellation. In the final part of Chapter 17, a demonstration of the in-band full-duplex transceiver is given. The demonstration combines the antenna design with RF and digital cancellation in a relay case, showing that overall SI suppression of nearly 100-dB – down to the noise level – can be achieved, even when using regular low-cost components.

Part 4: New System-level Enabling Technologies for 5G

Part 4 consists of four chapters, which address all the new system-level enabling technologies for 5G, including cloud radio access network (C-RAN), device-to-device (D2D) communication and ultradense networks (UDN). In Chapter 18, major signal processing issues for C-RAN are first reviewed and then the emphasis is moved to two key baseband signal processing steps, namely channel estimation in the uplink and channel encoding/linear precoding in the downlink. Together with theoretical analyses and numerical simulations, the chapter outlines the corresponding algorithms for joint optimization of baseband fronthaul compression and baseband signal processing under different PHY functional splits, whereby uplink channel estimation and downlink channel encoding/linear precoding are carried out either at remote radio heads or at the baseband unit.

Motivated by the consideration that energy efficiency is one of the drivers of 5G networks, Chapter 19 addresses the problem of power allocation for energy efficiency in wireless interference networks. This is formulated as the maximization of the network global energy efficiency with respect to all of the user equipment's transmit power, and a solution to the problem using sequential fractional programming algorithms is outlined. As pointed out at the beginning of this chapter, D2D communication is being considered as one of the key ingredients of 5G wireless networks. Therefore, the use of the sequential fractional programming algorithms in a 5G cellular system with D2D communication is described, including algorithm details, theoretical analyses and numerical simulations.

Chapter 20 is devoted to ultradense networks (UDNs), which are considered to be one of the paramount and dominant approaches to meet the ultra-high traffic volume, density and capacity required for 5G. All of the major technology challenges for deployment and operation of UDN are addressed in this chapter, including site acquisition and expenditure, network operation and management, interference management, mobility management and backhaul resources. Key technologies presented include network coordination, interference mitigation or cancellation-based receivers, dual connectivity, virtual cell, virtual layer, mobility anchor and handover command diversity, as well as joint time and frequency synchronization.

The scope of Chapter 21 is to provide a thorough analysis and discussion of the radio resources management (RRM) aspects of UDNs, with the emphasis on centralized optimization problem modeling and solving. By first presenting a series of mathematical models and programming algorithms for optimal RRM decisions and then applying these algorithms to potential UDN system setups, the chapter explores rate-performance trends as a function of infrastructure densification, as well as the impact of individual RRM dimension optimization on overall performance. It is shown that optimal RRM serves as a key enabler for getting the most of the resource reuse and proximity benefits offered by UDNs.

Part 5: Reference Design and 5G Standard Development

Serving as a practical implementation reference design example and a proof of concept, the real-time prototyping of an FD communication system for 5G is described in Chapter 22, which first reviews major self-interference cancellation schemes and then presents the details of prototyping in hardware architectures, processing flows, programming tools and testing setups. The prototyping system presented in this chapter consists of four main components:

- a dual-polarized antenna
- controller
- field-programmable gate array modules
- the corresponding RF front ends.

All of the key technology issues in converging FD concepts to real silicon are extensively addressed in this chapter: analog and digital SIC, synchronization, reference symbol allocation and channel estimation, cancellation measurement and throughput testing.

Chapter 23 is the last chapter of this book. The standards roadmap from 4G to 5G is first reviewed, and then the major enabling technologies and a more detailed roadmap of the 5G standard development are discussed. As summarized in this chapter, the technologies to be employed in 5G standards should not only enable efficient support of enhanced mobile broadband, which has been a major focus of all the previous generations, but should also enable new services, such as massive-machine-type communications, ultra-reliable communications and ultralow-latency communications. From a standards development and a regulatory-authority point of view, this chapter also shows that new frequency bands above 6 GHz (up to 100 GHz) are expected to play a very important role in 5G networks.

For whom is this book written?

It is hoped that this book will serve not only as a complete and invaluable reference for professional engineers, researchers, manufacturers, network operators, software developers, content providers, service providers, broadcasters, and regulatory bodies aiming at development, standardization, deployment and applications development of 5G systems and beyond, but also as a textbook for graduate students in circuits, signal processing, wireless communications, microwave technology, information theory, antennas and propagation, and system-on-chip implementation.

Fa-Long Luo, Ph.D., IEEE Fellow
Charlie (Jianzhong) Zhang, Ph.D., IEEE Fellow

List of Contributors

Nisar Ahmed University of Southern California, USA, 3740 McClintock Ave., EEB 500, Los Angeles, CA 90089-2565, USA

Angeliki Alexiou University of Piraeus, Greece, 150 Androutsou Street, Office 303, Piraeus-Greece

John B. Anderson Lund University Sweden, Electrical and Information Technology (EIT), Lund University, Box 118, 22100 Lund, SWEDEN

Emilio Antonio-Rodriguez Aalto University, Finland, P.O. Box 13000, FI-00076 AALTO, Finland

Ahsan Aziz National Instruments USA, 11024 Steelton Cove, Austin, TX 78717, USA

Anass Benjebbour NTT DOCOMO, INC. Japan, NTT DOCOMO R&D Center, 3-6 Hikari-no-oka, Yokosuka-shi, Kanagawa, 239-8536, Japan

John H. Brady University of Wisconsin-Madison, USA, Electrical and Computer Engineering, University of Wisconsin-Madison, 1415 Engineering Drive, Madison, WI 53706, USA

Stefano Buzzi University of Cassino and Southern Lazio, Italy, Via G. Di Biasio, 43 - 03043 Cassino (FR) – Italy

Chan-Byoung Chae Yonsei University Korea, Veritas Hall C309, 85 Songdogwahak-ro, Yeonsu-gu, Incheon, Korea

Peng Chen China Telecom Technology Innovation Center, China, Room 1116, China Telecom Beijing Information Science & Technology Innovation Park, Southern Zone of Future Science & Technology City, Beiqijia Town, Changping District, Beijing, 102209, P.R. China

MinKeun Chung Yonsei University Korea, Veritas Hall C325, 85 Songdogwahak-ro, Yeonsu-gu, Incheon, Korea

Giulio Colavolpe University of Parma Italy, viale delle Scienze 181/A, 43124 Parma, Italy

Izzat Darwazeh University College London UK, Department of Electronic & Electrical Engineering, University College London, Torrington Place, London WC1E 7JE, UK

Francesco Di Stasio University of Cassino and Southern Lazio, Italy, via le conche 7, Sessa Aurunca (CE), 8137, Italy

Ove Edfors Lund University Sweden, Dept. of Electrical and Information Technology, P.O. Box 118, SE-221 00 LUND, Sweden

Arman Farhang Trinity College Dublin, Ireland, Electronic & Electrical Engineering, Printing House, Trinity College, Dublin 2, Dublin, Ireland

Behrouz Farhang-Boroujeny University of Utah, USA, Electrical and Computer Engineering Department, University of Utah, Salt Lake City, UT 84112, USA

Antonis Gotsis National Technical University of Athens, Greece, 9 Heroon Polytechneiou Street, Zographou Campus, GR15773, Attica-Greece

Ryan C Grammenos University College London UK, Department of Electronic & Electrical Engineering, University College London, Torrington Place, London WC1E 7JE, UK

Malik Gul National Instruments USA, House # 231, Street # 4, Askari 11, Rawalpindi, Pakistan, 46000

Katsuyuki Haneda Aalto University, Finland, P.O. Box 13000, FI-00076 AALTO, Finland

Xueying Hou China Mobile Research Institute, China, No.32 Xuanwumen West, Xicheng District, Beijing, China, 100091

Wei Jiang University of Duisburg-Essen, Germany, Universität Duisburg-Essen, Digitale Signalverarbeitung, Bismarckstrasse 81, Gebäude BB 1011, D–47057 Duisburg, Germany

Jing Jin China Mobile Research Institute, China, No.32 Xuanwumen West, Xicheng District, Beijing, China, 100091

Eduard Jorswieck Dresden University of Technology, Germany, TU Dresden, Institut für Nachrichtentechnik, Lehrstuhl Theoretische Nachrichtentechnik, 01062 Dresden, Germany

Hyejung Jung Intel Corporation USA, 574 W. Parkside Drive, Palatine, IL, 60067, USA

Thomas Kaiser University of Duisburg-Essen, Germany, Universität Duisburg-Essen, Digitale Signalverarbeitung, Bismarckstrasse 81, Gebäude BB 1011, D–47057 Duisburg, Germany

Jinkyu Kang Korea Advanced Institute of Science and Technology (KAIST), South Korea, EE714, N1 IT Covergence Building, 291 Daehak-ro, Yuseong-gu, Daejeon, 305-701, South Korea

Joonkhyuk Kang Korea Advanced Institute of Science and Technology (KAIST), South Korea, EE714, N1 IT Covergence Building, 291 Daehak-ro, Yuseong-gu, Daejeon, 305-701, South Korea

DongKu Kim Yonsei University Korea, C222, 3rd Engineering Building, School of EEE, Yonsei University, 50 Yonsei-Ro, Seadaemoon-Gu, Seoul, Korea

Jaeweon Kim National Instruments USA, 11500 N. Mopac Expwy, Austin, TX 78759, USA

Yoshihisa Kishiyama NTT DOCOMO, INC. Japan, NTT DOCOMO R&D Center, 3-6 Hikari-no-oka, Yokosuka-shi, Kanagawa, 239-8536, Japan

Dani Korpi Tampere University of Technology, Finland, P.O. Box 692, FI-33101 Tampere, Finland

Nikhil Kundargi National Instruments USA, 3300 Wells Branch Parkway, Apt 2306, Austin, TX 78728, USA

Yongjun Kwak Samsung Electronics Co., Ltd., Korean, 90 Jinsan-ro, APT510-804, Yongin, Gyeonggi-do, South Korea

Juho Lee Samsung Electronics Co., Ltd., Korean, Hyundai Apartment 728-1701, 366, Maeyeong-ro, Yeongtong-gu, Suwon-si, Gyeonggi-do 16701, Korea

Anxin Li NTT DOCOMO, INC. Japan, DOCOMO Beijing Communications Laboratories, 7/F, Raycom Infotech Park Tower A, No. 2 Kexueyuan South Road, Haidian District Beijing, 100190, China

Qian (Clara) Li Intel Corporation USA, 3041 NW Ashford Circle, Hillsboro, OR, 97124, USA

Hao Lin Orange Labs, France, 4 rue du clos courtel, 35512, Cesson Sévigné, France.

Guangyi Liu China Mobile Research Institute, China, No.32 Xuanwumen West, Xicheng District, Beijing, China, 100091

Liang Liu Lund University Sweden, Dept. of Electrical and Information Technology, P.O. Box 118, SE-221 00 LUND, Sweden

Nicola Marchetti Trinity College Dublin, Ireland, Electronic & Electrical Engineering, Printing House, Trinity College, Dublin 2, Dublin, Ireland

Wes McCoy National Instruments USA, 2909 Rabbits Tail Drive, Leander, TX 78641, USA

Andreas F. Molisch University of Southern California, USA, 3740 McClintock Ave, Los Angeles, CA 90089, USA

Takehiro Nakamura NTT DOCOMO, INC. Japan, NTT DOCOMO R&D Center, 3-6 Hikari-no-oka, Yokosuka-shi, Kanagawa, 239-8536, Japan

Karl Nieman National Instruments USA, 1404 Stonethrow Way Austin, TX 78748, USA

Eckhard Ohlmer National Instruments USA, Am Waldschloesschen 2, 01099 Dresden, Germany

Athanasios Panagopoulos National Technical University of Athens, Greece, 9 Heroon Polytechneiou Street, Zographou Campus, GR15773, Attica-Greece

Yong Rao National Instruments USA, 10140 Tularosa Pass, Austin TX 78726, USA

Yongxiong Ren University of Southern California, USA, 3740 McClintock Ave., EEB 500, Los Angeles, CA 90089-2565, USA

Taneli Riihonen Aalto University, Finland, P.O. Box 13000, FI-00076 AALTO, Finland

Akbar M. Sayeed University of Wisconsin-Madison, USA, Electrical and Computer Engineering, University of Wisconsin-Madison, 1415 Engineering Drive, Madison, WI 53706, USA

Keisuke Saito NTT DOCOMO, INC. Japan, NTT DOCOMO R&D Center, 3-6 Hikari-no-oka, Yokosuka-shi, Kanagawa, 239-8536, Japan

Shlomo Shamai (Shitz) Technion–Israel Institute of Technology, Israel, 773, Meyer Bldg., Haifa 32000, Israel

Xiaoming She China Telecom Technology Innovation Center, China, Room 1116, China Telecom Beijing Information Science & Technology Innovation Park, Southern Zone of Future Science & Technology City, Beiqijia Town, Changping District, Beijing, 102209, P.R. China

Min Soo Sim Yonsei University Korea, Veritas Hall C325, 85 Songdogwahak-ro, Yeonsu-gu, Incheon, Korea

Pierre Siohan Orange Labs, France, 4 rue du clos courtel, 35512, Cesson Sévigné, France

Osvaldo Simeone New Jersey Institute of Technology, USA, ECE Department, New Jersey Institute of Technology (NJIT), University Heights Newark NJ 07102, USA

Stelios Stefanatos University of Piraeus, Greece, 150 Androutsou Street, Office 303, Piraeus-Greece

Hui Tong China Mobile Research Institute, China, No.32 Xuanwumen West, Xicheng District, Beijing, China, 100091

Fredrik Tufvesson Lund University Sweden, Dept. of Electrical and Information Technology, P.O. Box 118, SE-221 00 LUND, Sweden

Alessandro Ugolini University of Parma Italy, viale delle Scienze 181/A, 43124 Parma, Italy

Mikko Valkama Tampere University of Technology, Finland, P.O. Box 692, FI-33101 Tampere, Finland

Fei Wang China Mobile Research Institute, China, No.32 Xuanwumen West, Xicheng District, Beijing, China, 100091

Alan E. Willner University of Southern California, USA, 3740 McClintock Ave., EEB 538, Los Angeles, CA 90089-2565, USA

Geng Wu Intel Corporation USA, 3401 Spring Mountain Drive, PLANO, TX, 75025, USA

Guodong Xie University of Southern California, USA, 3740 McClintock Ave., EEB 500, Los Angeles, CA 90089-2565, USA

Tongyang Xu University College London UK, Department of Electronic & Electrical Engineering, University College London, Torrington Place, London WC1E 7JE, UK

Yan Yan University of Southern California, USA, 3740 McClintock Ave., EEB 500, Los Angeles, CA 90089-2565, USA

Yifei Yuan ZTE Corporation China, 9 Royal Avenue, Livingston New Jersey, 07039, USA

Alessio Zappone Dresden University of Technology, Germany, TU Dresden, Institut für Nachrichtentechnik, Lehrstuhl Theoretische Nachrichtentechnik, 01062 Dresden, Germany

Nidal Zarifeh University of Duisburg-Essen, Germany, Universität Duisburg-Essen, Digitale Signalverarbeitung, Bismarckstrasse 81, Gebäude BB 1011, D–47057 Duisburg, Germany

Jianchi Zhu China Telecom Technology Innovation Center, China, Room 1116, China Telecom Beijing Information Science & Technology Innovation Park, Southern Zone of Future Science & Technology City, Beiqijia Town, Changping District, Beijing, 102209, P.R. China

Pingping Zong Intel Corporation USA, 1 Crest Drive, Randolph, NJ, 07869, USA

Part One

Modulation, Coding and Waveform for 5G

1

An Introduction to Modulations and Waveforms for 5G Networks

Stefano Buzzi, Alessandro Ugolini, Alessio Zappone and Giulio Colavolpe

1.1 Motivation and Background

Historically, the evolution of wireless cellular systems has been fueled by the need for increased throughput. Indeed, the need for larger data-rates has been the main driver of the path that has led us from 2G systems[1] to 4G systems, with data-rates evolving from tens of kbit/s up to the current state-of-the-art tens of Mbit/s. Focusing on the physical (PHY) layer, and in particular on the adopted modulation schemes, the transition has been from

[1] Indeed analog 1G cellular systems had no data transmission capability; they just offered voice services.

Signal Processing for 5G: Algorithms and Implementations, First Edition. Edited by Fa-Long Luo and Charlie Zhang.

binary modulations such as the Gaussian minimum shift keying (GMSK), used in the 2G GSM system, to quadrature-amplitude-modulation (QAM) schemes with adaptively chosen cardinality, currently used in 4G systems.

Unlike previous generations of cellular networks, 5G systems will have to accommodate a variety of services and of emerging new applications, and, in order to do that, focusing only on the increase of the data throughput is not enough. In particular, according to the classification in Michailow *et al.* [1], the main reference scenarios currently envisioned for 5G networks are as follows.

- **Very large data-rate wireless connectivity**. Users will be able to download large amounts of data in a short time; a typical application corresponding to this scenario is high-definition video streaming, which of course requires a modulation scheme with large spectral and energy efficiency.
- **Internet of Things (IoT)**. Up to one trillion devices are expected to be connected through the 5G network, enabling users to remotely control things such as cars, washing machines, air conditioners, lights, and so on. Likewise, energy, water and gas distribution companies will take advantage of connected smart meters in order to control their networks. These connected things will have quite limited processing capabilities and will have to transmit small amounts of data sporadically, thus requiring a modulation scheme robust to time synchronization errors and performing well for short communications.
- **Tactile Internet** [2]. This scenario refers to real-time cyber-physical tactile control experiments (such as remote control of drones and/or of rescue robots in emergency situations), and requires a communication service that must be reliable and have small latency. The target latency is in the order of 1 ms, more than one order of magnitude smaller than the latency of current 4G systems. In order to achieve such an ambitious target, the PHY latency of future 5G networks should not exceed 200-300 μs. Other applications, such as on-line gaming and car-to-car and car-to-infrastructure communications, although not directly related to the concept of the tactile Internet, also can take advantage of the low latency requirements [3].
- **Wireless Regional Area Networks (WRAN)**. It is expected that the generous throughput of 5G networks will also suit it to bringing internet broadband access to sparsely populated areas that are not yet covered by wired technologies such as ADSL and optical fiber. In this scenario network devices will have very low mobility, so Doppler effects will be negligible, and also latency will not be a key requirement. In order to be able to meet the throughput demands of bandwidth-hungry residential users, the use of so-called "white spaces" – in other words frequency bands licensed to other services but actually not used – seems unavoidable. It is thus anticipated that the available frequency bands will not be contiguous, and cognitive-like opportunistic spectrum access is a viable option. Millimeter wave frequencies (larger than 20 GHz) also will be used. The modulation format of future 5G systems should thus be able to efficiently exploit the available fragmented and heterogeneous spectrum.

Orthogonal frequency division multiplexing (OFDM) and orthogonal frequency division multiple access (OFDMA) are the modulation technique and the multiple access strategy adopted in long term evolution (LTE) 4G cellular network standards, respectively [4]. OFDM and OFDMA are based on a multicarrier approach and succeeded code division multiple access, as employed in 3G networks, and which was mostly based on a single-carrier approach. Among the chief advantages of OFDM and OFDMA are:

- the ease of implementation of both transmitter and receiver thanks to the use of fast Fourier transform (FFT) and inverse fast Fourier transform (IFFT) blocks
- the ability to counteract multi-path distortion
- the orthogonality of subcarriers, which eliminates intercell interference
- their easy coupling with adaptive modulation techniques
- the ease of integration with multi-antenna hardware, both at the transmitter and receiver.

Nonetheless, there are some key characteristics that make OFDM/OFDMA a less-than-perfect match for the above reference scenarios. First of all, OFDM is based on the use of rectangular pulses in the time domain, which leads to a slowly decaying behavior in the frequency domain; this makes OFDM unsuited for use in fragmented spectrum scenarios, where strict constraints on the out-of-band (OOB) levels are to be fulfilled. In 4G systems, OOB emissions are controlled by inserting null tones at the spectrum edges or, alternatively, by filtering the whole OFDM signal with a selective filter (this is usually known as filtered-OFDM). Both solutions unfortunately lead to a loss in spectral efficiency, since in the former case some of the available subcarriers are actually not modulated, while in the latter case we need a longer cyclic prefix to combat the time dispersion induced by the filtering operation. The need for a long cyclic prefix (CP) in heavy multipath environments is then another factor that degrades the system spectral efficiency. Likewise, the need for strict frequency and time synchronization among blocks and subcarriers in order to maintain orthogonality is a requirement that does not match well with the IoT scenario, wherein many devices have to access the channel with short data frames. Synchronization is also a key issue in the uplink of a cellular network wherein different mobile terminals transmit separately [5], and in the downlink, when base station coordination is used [6, 7]. Additionally, OFDM signals may exhibit large peak-to-average-power ratio (PAPR) values [8], and this has a clear impact on the system energy efficiency.

Based on the above considerations, a very active research track in the area of 5G systems has focused on the search for alternative modulation schemes capable of overcoming the disadvantages of OFDM/OFDMA [9],[2] and of supporting in an optimal way the emerging services and reference scenarios that we have discussed here. The main goal of this research activity is to look for modulation formats with low OOB emissions – so as to fully exploit the fragmented spectrum – and which do not not require a strict orthogonality among subcarriers, so as to simplify synchronization and access procedures.

This chapter provides a review of some of the best recently proposed alternatives to OFDM. Due to space constraints, it is not possible to go deep into details about each modulation scheme; nonetheless, we give mathematical models and block-schemes of the transmitter for all the alternatives considered. We also provide a comparative analysis of these modulations, highlighting their pros and cons, and discussing their ability to operate in the 5G reference scenarios discussed above.

The rest of this chapter is organized into the following three sections. Section 1.2 is devoted to all major alternative modulation formats beyond OFDM, including filter-bank multicarrier

[2] The issue of beyond-OFDM modulation has been also extensively addressed in EU-funded research projects such as 5GNOW [10] and METIS2020 [11].

(FBMC), generalized frequency division multiplexing (GFDM), bi-orthogonal frequency division multiplexing (BFDM), universal filtered multicarrier (UFMC) and time-frequency packing (TFP). In Section 1.3, we will deal with the waveform choice issue by providing some shaping pulses that can be considered as alternatives to the rectangular pulse adopted in OFDM. Section 1.4 is the final section and contains further discussion and provides concluding remarks.

1.1.1 The LTE Solution: OFDM and SC-FDMA

The current 4G standard, the LTE system, is based on the use of the OFDM modulation for the downlink and of the single-carrier frequency division multiple access (SC-FDMA) technique for the uplink [4]. OFDM is an orthogonal block transmission scheme which, in ideal conditions, is not affected by intercarrier interference and intersymbol interference (ISI). Figure 1.1 shows a block scheme for the OFDM modem. A block of K QAM symbols $(s(1), s(2), \ldots, s(K))$ is mapped onto the available K subcarriers and then IFFT is performed. After the IFFT, the CP, whose length must be larger than the channel impulse response duration, is included in the data block, which is then sent to a single-carrier modulator for transmission. After propagation through the channel, the CP is removed, and the block of K observables is passed through an FFT transformation. In ideal conditions, it can be shown that the mth data sample at the output of the FFT block can be written as

$$Z(m) = H(m)s(m) + W(m) \tag{1.1}$$

where $H(m)$ and $W(m)$ are the mth FFT coefficients of the channel impulse response and of the additive disturbance, respectively. Based on Eq. (1.1), the soft estimate of the symbol $s(m)$, to be sent to the data decoding block, is obtained through a simple one-tap equalization:

$$\hat{s}(m) = Z(m)/H(m) \tag{1.2}$$

We also note that, in order to highlight the use of the CP, the OFDM technique that we have just described is sometimes referred to as "CP-OFDM". Despite its simplicity and the aforementioned immunity to multipath distortion, OFDM has some key drawbacks and among these one of the most severe is the large PAPR, which requires amplifiers with an extended linearity range. While in the downlink we can usually afford to have expensive amplifiers at the

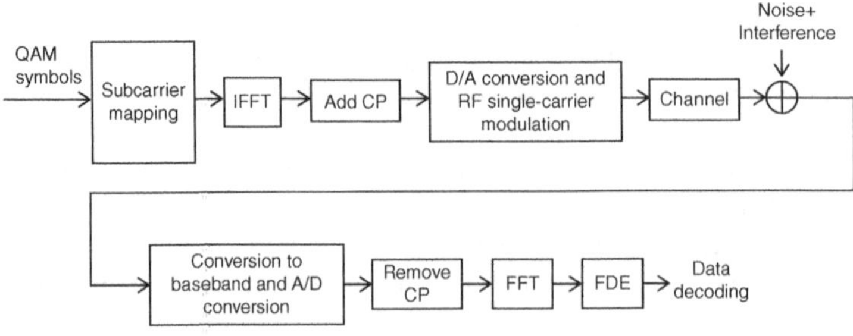

Figure 1.1 Principle of the OFDM modem

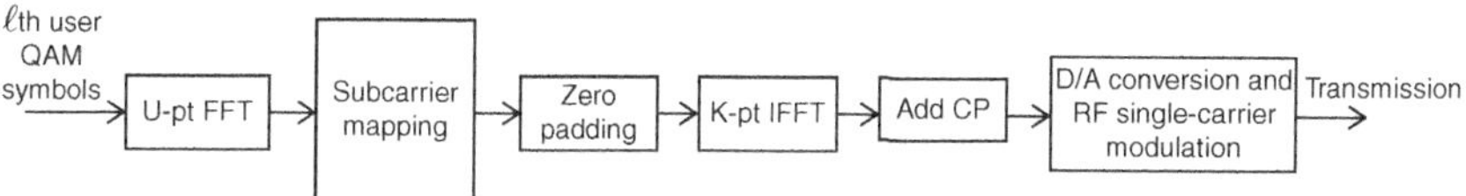

Figure 1.2 Block scheme of the SC-FDMA transmitter used in the uplink of LTE

transmitter (i.e. at the base station), this is not the case in the uplink, where the transmitter is a small mass-market mobile device. Accordingly, the modulation and multiple-access strategy used in the uplink of LTE is the so-called SC-FDMA strategy, a slightly different version of OFDM. Figure 1.2 depicts the typical SC-FDMA transmission scheme implemented in the generic ℓth user mobile device. Letting U denote the number of subcarriers (out of the available K) that have been assigned to the ℓth user in the current resource slot, a block of U QAM symbols is FFTed and mapped onto the assigned subcarriers. At the output of the "subcarrier mapping" block, the "zero padding" block forms a vector of K elements, containing zero values at the positions corresponding to the $K - U$ subcarriers that are not assigned to the ℓth user, and the ℓth user data at the remaining U positions. The K-dimensional block is passed through the IFFT block, then a CP is added and, after upconversion, signal transmission happens. Note that, according to the OFDMA principle, the active users must transmit synchronously so that the base station receiver is able to simultaneously collect the data from the users that are using the K available subcarriers. Due to the U-points FFT operation, the SC-FDMA strategy exhibits a PAPR smaller than that of pure OFDMA, since the transmitted signal is basically equivalent to an oversampled single-carrier signal.

1.2 New Modulation Formats: FBMC, GFDM, BFDM, UFMC and TFP

We now review some of the modulation schemes that are being considered for adoption in future 5G wireless networks.

1.2.1 Filter-bank Multicarrier

As is well-known, in the presence of multi-path channels, plain orthogonal multicarrier modulation formats are not able to maintain orthogonality due to ISI among consecutive multicarrier symbols. The traditional approach in OFDM to counter this issue is to introduce a CP longer than the time spread introduced by the channel. This enables the preservation of traditional transceiver implementations by IFTT and FFT operations, but introduces a time overhead in the communication, resulting into a loss of spectral efficiency.

The approach used by FBMC to overcome this issue is to keep the symbol duration unaltered, thereby avoiding the introduction of any time overhead, and to cope with the overlap among adjacent multicarrier symbols in the time domain by adding an additional filtering at the transmit and receive side, besides the IFFT/FFT blocks. This is done by filtering each output of the FFT by a frequency-shifted version of a lowpass filter $p(t)$, termed a "prototype" filter. This additional filtering, together with the IFFT/FFT operation, forms a synthesis-analysis filter-bank structure, where the prototype filter is designed to significantly suppress ISI.

To begin with, it is worthwhile to observe that the conventional OFDM scheme can also be regarded as an FBMC scheme, with a low-pass FIR prototype filter, with discrete-time impulse response given by

$$p(n) = \begin{cases} \frac{1}{N} & \forall\, n = 1, \ldots, N \\ 0 & \text{elsewhere} \end{cases} \tag{1.3}$$

To see this, let us observe that the input–output relation for the point of index $k = 0$ of an N-point DFT operating on the samples $\{x(n-i)\}_{i=1}^{N}$ can be expressed as:

$$y_0(n) = \frac{1}{N} \sum_{i=1}^{N} x(n-i) \tag{1.4}$$

which can be regarded as the input – output relation of the FIR filter with impulse response in Eq. (1.3). A similar relation can be obtained for the generic nth DFT coefficient, accounting for the frequency shift $e^{-j2\pi ni/N}$. As a consequence, we find that conventional OFDM schemes can be regarded as a particular FBMC scheme with rectangular pulses as prototype filters. However, such a choice of prototype filter does not protect against the ISI caused by multi-path channels. Instead, a prototype filter that guarantees significant ISI suppression is obtained by introducing additional coefficients between the FFT coefficients in the frequency domain. In particular, the number of introduced coefficients between two consecutive DFT coefficients is called the "overlapping factor" of the filter K, which is also equal to the ratio between the filter impulse response duration and the multicarrier symbol period T, thereby determining the number of multicarrier symbols which overlap in the time domain [12]. One prototype filter that is able to ensure a low ISI is that with impulse response:

$$p(t) = 1 + 2 \sum_{k=1}^{K-1} H_k \cos\left(2\pi \frac{kt}{KT}\right) \tag{1.5}$$

where the coefficients H_k are given in Table 1.1, up to $K = 4$.

The samples of the frequency response of $p(t)$ are also reported in Figure 1.3.

From the above discussion, it would seem that a frequency spreading by a factor K is necessary to implement the FBMC scheme. Indeed, one possible implementation is based on a frequency spreading plus a KN-point IFFT at the transmitter, and on a KN-point FFT followed by a despreading at the receiver. This particular implementation has the advantage of requiring only minor modifications with respect to the traditional implementation of OFDM transmissions. However, increasing the FFT size by a factor K poses significant complexity issues. In order to reduce the computational complexity, an alternative implementation has been proposed. This is called polyphase network-FFT (PPN-FFT). PPN-FFT requires no frequency spreading, but at the expense of some additional processing. In the rest of this section this latter approach will be described.

Table 1.1 Frequency-domain prototype filter coefficients

K	H_0	H_1	H_2	H_3
2	1	$\sqrt{2}/2$	-	-
3	1	0.911438	0.411438	-
4	1	0.971960	$\sqrt{2}/2$	0.235147

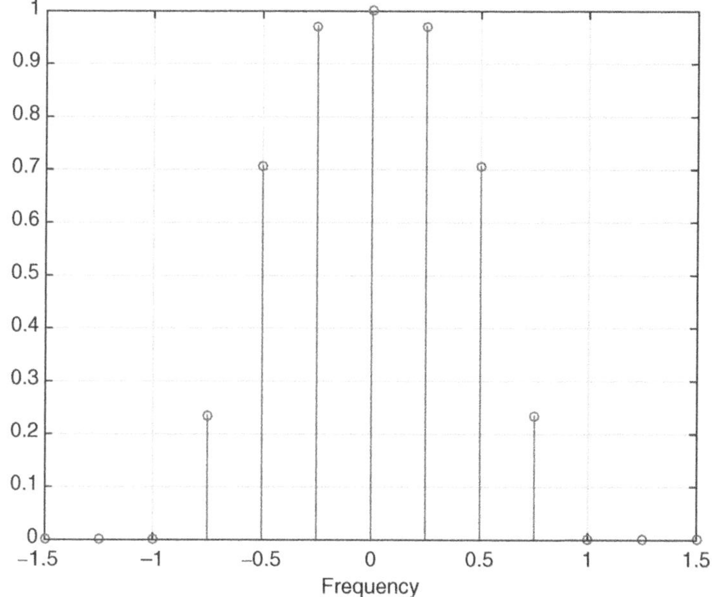

Figure 1.3 Frequency-domain samples of the prototype filter

Let us assume that the time-domain length of the prototype filter can be written as $L = KN$, and let us denote by $\{h_\ell\}_{\ell=1}^{L-1}$ the time-domain filter coefficients. Then, the frequency response of the prototype filter can be written as

$$P_0(f) = \sum_{\ell=0}^{L-1} h_\ell e^{-j2\pi\ell f} = \sum_{p=0}^{N-1} H_p(f) e^{-j2\pi pf} \tag{1.6}$$

where, for all $p = 0, \ldots, N-1$, we have defined the functions

$$H_p(f) = \sum_{k=0}^{K-1} h_{kN+p} e^{-j2\pi fkN} \tag{1.7}$$

We can see that Eq. (1.7) can be regarded as the frequency response of a phase shifter, which gives the name to this implementation of the FBMC modulation scheme. Next, we can obtain the frequency response of the nth filter of the bank by shifting Eq. (1.6) in the frequency domain by a factor n/N. This yields:

$$\begin{aligned} P_n(f) &= \sum_{p=0}^{N-1} H_p(f - n/N) e^{-j2\pi p(f-n/N)} \\ &= \sum_{p=0}^{N-1} \left(\sum_{k=0}^{K-1} h_{kN+p} e^{-j2\pi(f-n/N)kN} \right) e^{-j2\pi p(f-n/N)} \\ &= \sum_{p=0}^{N-1} H_p(f) e^{-j2\pi pf} e^{j2\pi pn/N} \end{aligned} \tag{1.8}$$

where we have exploited the fact that $e^{j2\pi kn} = 1$ for all $k = 0, \ldots, K-1$ and $n = 0, \ldots, N-1$. Then, considering the relations in Eq. (1.6) for all $n = 0, \ldots, N-1$, we can obtain the matrix equation:

$$\begin{pmatrix} P_0(f) \\ P_1(f) \\ \vdots \\ P_{N-1}(f) \end{pmatrix} = \begin{pmatrix} 1 & 1 & 1 & 1 \\ 1 & e^{j2\pi/N} & \cdots & e^{j2\pi(N-1)/N} \\ \vdots & \vdots & \vdots & \vdots \\ 1 & e^{j2\pi(N-1)/N} & \cdots & e^{j2\pi(N-1)^2/N} \end{pmatrix} \begin{pmatrix} H_0(f) \\ e^{-j2\pi f} H_1(f) \\ \vdots \\ e^{-j2\pi(N-1)f} H_{N-1}(f) \end{pmatrix} \tag{1.9}$$

Observing that the square matrix in Eq. (1.9) performs an IDFT operation, and recalling that the final output is obtained by summing the outputs of the individual filters of the bank, we determine that the transmitter of a PPN-FFT system can be implemented as shown in Figure 1.4. A similar scheme is used at the receiver, with the difference that the FFT operation is used in place of the IFFT, and that the frequency shifts are multiples of $-1/N$.

In conclusion, the PPN-FFT scheme can be implemented by adding the phase shifters $e^{-j2\pi pf} H_p(f)$ in series with the usual IFFT/FFT operation performed in conventional OFDM schemes. This entails a slight complexity increase with respect to OFDM, but still results in less complexity than applying a frequency spreading to implement the FBMC scheme.

1.2.2 Generalized Frequency Division Multiplexing

GFDM is a generalized multicarrier modulation that is particularly attractive in scenarios with fragmented spectrum [13, 1]. Indeed one of its main features is its low level of OOB emissions, which makes it well suited for transmission on non-contiguous frequency bands with strict spectral mask constraints. Thanks to the use of the CP, it retains OFDM advantages in

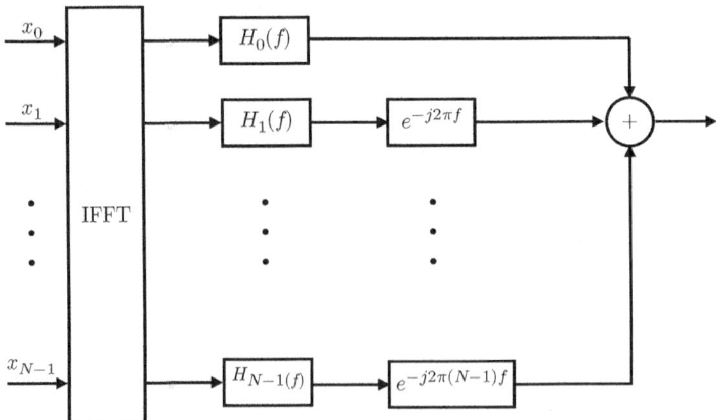

Figure 1.4 PPN-FFT implementation of the transmitter in the FBMC modulation scheme

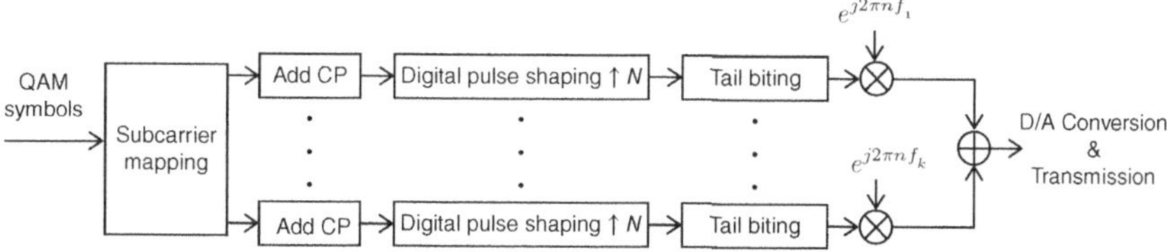

Figure 1.5 The GFDM transmitter

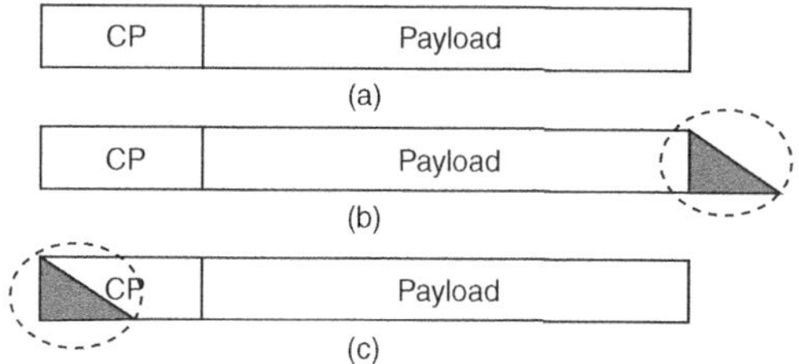

Figure 1.6 The tail-biting operation. (a) the CP is appended to the payload – its length must be set according to the duration of the channel impulse response and of the receive filter impulse response; (b) after passing through the transmit filter, the data packet is longer due to the convolution effect; (c) original length is restored by tail biting and adding the tail to the CP in order to emulate circular convolution

terms of robustness to multipath channels and ease of equalization, and may be efficiently implemented through signal processing in the digital domain. Inspecting Figure 1.5, wherein the block-scheme of a GFDM transmitter is presented, it is seen that GFDM is a pure multicarrier scheme that transmits parallel data streams on carrier frequencies $f_1, f_2, \ldots, f_K$, which are not required to be contiguous. A CP is used to combat time dispersion induced by all the filters, from the transmitter through channel to receiver. In contrast to legacy OFDM, where the CP length is simply required to be larger than the channel impulse response, in GFDM the CP should in principle have a length larger than the sum of the impulse responses of the transmit shaping filter, the channel, and the receive filter. The limitation of the OOB emissions is obtained through the use of pulse shapes; the lower the required OOB emissions, the longer the pulse length in the time domain. An efficient strategy for reducing the length of the CP and, equivalently, the loss in terms of spectral efficiency, is the tail-biting technique [13], the principle of which is shown in Figure 1.6. In this technique, the CP may be chosen to be as long as the sum between the impulse responses of the channel and that of the reception filter – in other words the transmit filter impulse response length is not taken into account – provided that, at the output of the transmit shaping filter, the additional samples that arise from the linear convolution are removed and added at the beginning of the data packet, so as to emulate circular convolution (see Figure 1.6 for an illustration of this operation). Note also that a similar procedure, not described here for the sake of brevity, can be used to reduce the CP length tied to the receive filter.

To provide a mathematical expression for the signal formed by the GFDM transmitter, we use the following notation:

- K denotes the number of available carrier frequencies; the baseband equivalent of the kth frequency band is centered on f_k;
- M denotes the number of QAM symbols forming the data block to be sent on each carrier;
- the M QAM symbols $s(0,k), s(1,k), \ldots, s(M-1,k)$ form the data block to be sent on the kth frequency band, while the $M + M_{CP}$ QAM symbols $\tilde{s}(0,k), \tilde{s}(1,k), \ldots, \tilde{s}(M + M_{CP} - 1, k)$ form the data block at the output of the CP block (see Figure 1.5);
- the transmit pulse is denoted $g_{tx}(n)$ and is a FIR filter of length QN, with N being the number of samples per data interval and Q being the number of signaling intervals that are spanned by the continuous-time version of the transmit pulse; note that the longer the value of Q, the larger the gain in terms of reduction of OOB emissions.

Based on the above notation, the signal at the input of the tail-biting block on the kth branch (carrier) of the transmitter in Fig 1.5 is expressed as

$$x_k(n) = \sum_{m=0}^{M+M_{CP}-1} \tilde{s}(m,k) g_{tx}(n - mM) \tag{1.10}$$

with $n = 0, \ldots, (M + M_{CP} + Q - 1)N - 1$. The subsequent tail-biting procedure reduces the length of this packet of $(Q-1)N$ samples.

One possible receiver architecture is shown in Figure 1.7. Thanks to the use of tail biting and the CP, all linear convolutions are turned into circular convolutions, and one-tap equalization in the frequency domain can be used to remove the ISI introduced by the channel and by the filtering operations. Indeed, after the CP prefix has been removed, we have, on the generic kth branch of the receiver, M data samples which are FFTed in order to obtain the frequency bins $Z(m,k)$:

$$Z(m,k) = S(m,k)H(m,k) + W(m,k) \tag{1.11}$$

where $S(m,k)$ is the mth FFT coefficient of the original QAM data symbols $s(0,k), \ldots, s(M-1,k)$, $W(m,k)$ is the mth FFT coefficient of the rx-filtered overall additive disturbance (i.e. AWGN noise plus adjacent channels interference), and, finally, $H(m,k)$ is the mth FFT coefficient of the impulse response of the composite channel, which is obtained as the convolution of the transmit filer, propagation channel and reception filter.

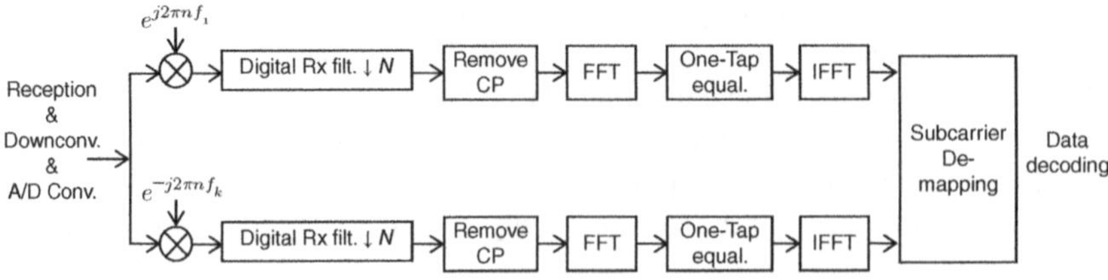

Figure 1.7 Diagram of a possible GFDM receiver

1.2.3 Bi-orthogonal Frequency Division Multiplexing

BFDM is a generalization of the classical CP-OFDM scheme and is able to provide lower intercarrier interference (ICI) and lower ISI. The basic idea is to introduce additional degrees of freedom into the system, which can be designed to obtain the said advantages.

Classical OFDM schemes are based on the orthogonality principle, according to which the prototype filter $g(t)$ should be orthogonal to a suitable time-frequency shifted version of itself, in other words:

$$\left\langle g(t), g(t-\ell T)e^{j2\pi nF(t-\ell T)}\right\rangle = 0, \quad \forall\, \ell, n \neq 0 \tag{1.12}$$

where T is the symbol interval and F is the frequency spacing among adjacent subcarriers. It is known that, due to channel distortions, the orthogonality of the transmissions might be lost unless a CP is used to introduce a guard-time among different symbols. However, this causes an extension of the time duration of the prototype filter, which is suboptimal in doubly dispersive channels because the time and frequency dispersions introduced by the channel are treated differently [14]. A way to overcome this issue is to observe that to obtain perfect demodulation (in the noiseless case) Condition (1.12) is only sufficient but not necessary. Specifically, Condition (1.12) implies that the same filter is used at the transmitter and receiver, but perfect demodulation (in the noiseless case) can also be obtained when the receiver employs a different receive filter, say $\gamma(t)$, provided the following bi-orthogonality condition is met:

$$\left\langle g(t), \gamma(t-\ell T)e^{j2\pi nF(t-\ell T)}\right\rangle = 0, \quad \forall\, \ell, n \neq 0 \tag{1.13}$$

The use of different transmit and receive pulses is precisely the additional degree of freedom enabled by the BFDM modulation scheme. The transmit and receive filters should be designed in order to fulfill Eq. (1.13), while at the same time ensuring low ICI and ISI. A necessary condition for Eq. (1.13) to hold is $TF \geq 1$ [15]. In practice, TF ranges between 1.03 and 1.25, which ensures a good trade-off between spectral efficiency and pulse localization [16, 17].

In doubly dispersive channels, the power of the ICI and ISI depends on the joint time-frequency concentration of the transmit and receive pulses. In more detail, a measure of the power of the ICI and ISI of BFDM is given by the cross-ambiguity function between the transmit and receive pulse, defined as

$$A_{\gamma,g}(\tau,\nu) = \int_t \gamma(t)g^*(t-\tau)e^{-j2\pi\nu t}dt \tag{1.14}$$

Therefore, the transmit pulse $g(t)$ and the receive pulse $\gamma(t)$ should be designed in order to achieve a suitable time-frequency localization. In particular, the following localization properties are desirable [16, 17].

Definition 1.1 *The pulses $g(t)$ and $\gamma(t)$ are said to be polynomially localized of degree $s \geq 0$ if there exists $T_0 > 0$ such that*

$$\int_\tau \int_\nu |A_{\gamma,g}(\tau,\nu)| \left(1 + \left|\frac{\tau}{T_0}\right| + |\nu T_0|\right)^s d\tau d\nu < \infty \tag{1.15}$$

A stronger localization property is the sub-exponential localization.

Definition 1.2 *The pulses $g(t)$ and $\gamma(t)$ are said to be sub-exponentially localized if there exist $T_0 > 0$, $b > 0$, and $\beta \in (0, 1)$ such that*

$$\int_\tau \int_\nu |A_{\gamma,g}(\tau,\nu)| e^{b(|\tau/T_0|+|\nu T_0|)^\beta} d\tau d\nu < \infty \tag{1.16}$$

In practice spline-type pulses are used to obtain a polynomial localization whereas Gaussian pulses enable an exponential localization.

1.2.4 Universal Filtered Multicarrier

Universal Filtered Multicarrier (UFMC) is a multicarrier modulation format that has been proposed by the EU-funded research project 5GNOW [3, 18, 19, 20, 21]. UFMC admits as particular cases the filtered-OFDM and the FBMC modulations. Indeed, while in the former case the whole set of subcarriers is filtered to limit sidelobe effects, and while in FBMC modulations filtering is applied separately to each subcarrier, in UFMC subcarriers are filtered in groups. Denoting again by K the overall number of subcarriers, let us assume that these K subcarriers are divided in B separate groups; although groups are allowed to be composed of different numbers of subcarriers, for the sake of simplicity we assume here that each group is composed of P subcarriers, so that $K = BP$.

Denote now by $\mathbf{s}_1, \mathbf{s}_2, \ldots, \mathbf{s}_B$ the P-dimensional vectors containing the QAM data symbols to be transmitted, and by $\mathbf{V}$ the $(K \times K)$ IFFT matrix; we partition this matrix using the B submatrices $\mathbf{V}_1, \ldots, \mathbf{V}_B$, each of dimension $(K \times P)$:

$$\mathbf{V} = [\mathbf{V}_1 \; \mathbf{V}_2 \; \ldots \; \mathbf{V}_B] \tag{1.17}$$

Equipped with this notation, we can now illustrate the UFMC transmitter operation (see Figure 1.8). The B data vectors $\mathbf{s}_1, \ldots, \mathbf{s}_B$ are processed with the IDFT submatrices $\mathbf{V}_1, \ldots, \mathbf{V}_B$, respectively. Then, they are passed through a pulse shape of length N_g, aimed at attenuating sidelobe levels in the frequency domain, and summed together (see Figure 1.8). In principle, we may use different filters for each branch. Denoting by $\mathbf{F}_i$ the $((K + N_g - 1) \times K)$ Toeplitz matrix describing the convolution operation with the shaping filter, the discrete-time signal to be converted to the analog domain and transmitted at RF is expressed as

$$\mathbf{x} = \sum_{i=1}^{B} \mathbf{F}_i \mathbf{V}_i \mathbf{s}_i \tag{1.18}$$

At the receiver side, denoting by $\mathbf{H}$ the Toeplitz matrix, of dimension $(K + N_g + N_h - 2) \times ((K + N_g - 1))$, where N_h is the length of the propagation channel, and describing linear convolution with the channel impulse response, the discrete-time baseband equivalent of the received signal is given by the following $(K + N_g + N_h - 2)$-dimensional vector:

$$\mathbf{y} = \mathbf{H} \left(\sum_{i=1}^{B} \mathbf{F}_i \mathbf{V}_i \mathbf{s}_i \right) + \mathbf{w} \tag{1.19}$$

with $\mathbf{w}$ being the additive disturbance, made of noise plus possible co-channel interference. It can be seen that Eq. (1.19) describes a classical linear model, and a plethora of well-known

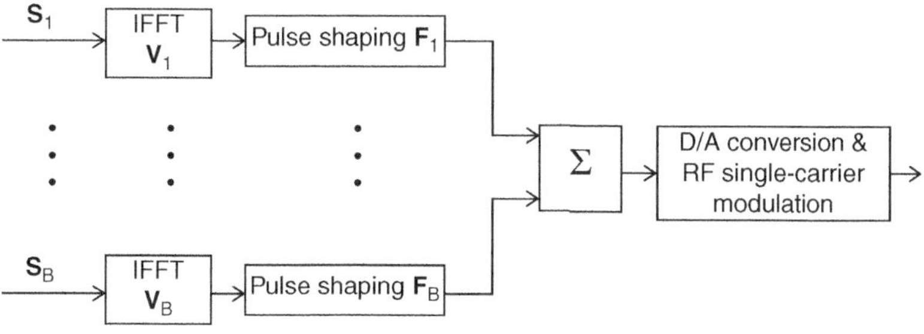

Figure 1.8 The UFMC transmitter

signal processing techniques – matched filtering, linear minimum mean square error estimation, zero-forcing detection and so on – can be used to recover the QAM symbols. Classical FFT-based processing with attendant one-tap equalization in the frequency domain is also possible.

1.2.5 Time-frequency Packing

In traditional digital communications, orthogonal signaling has been often adopted to ensure the absence of ISI and ICI. However, when finite-order constellations are used, it is possible to increase the spectral efficiency of communication systems by giving up the orthogonality condition and by introducing a controlled interference into the signal. This idea was first introduced by Mazo for single-carrier transmissions with the name of faster-than-Nyquist (FTN) signaling [22]. FTN signaling is a linear modulation technique that reduces the time spacing between two adjacent pulses (the symbol time) to well below that ensuring the Nyquist condition, thus introducing controlled ISI [22, 23, 24]. If the receiver can cope with the ISI, the efficiency of the communication system is increased. In the original papers on FTN signaling [22, 23, 24], this optimal time spacing is obtained as the smallest value giving no reduction of the minimum Euclidean distance with respect to the Nyquist case. This ensures that, asymptotically, the ISI-free bit-error rate (BER) performance is reached when optimal detectors are used. More recently, this concept has been extended to multicarrier transmissions by Rusek and Anderson [24]. In this case, intentional ICI is also introduced by reducing the frequency separation among carriers.

A multicarrier FTN signal can be expressed as

$$x(t) = \sqrt{E_s} \sum_n \sum_\ell x_n^{(\ell)} p(t - n\delta_t T) e^{j2\pi\ell\delta_f Ft} \tag{1.20}$$

where E_s is the average energy per symbol, $x_n^{(\ell)}$ is the M-ary symbol transmitted during the nth signaling interval over the ℓth carrier, $p(t)$ is the base pulse, usually a pulse with root raised cosine (RRC) spectrum with roll-off α, and T and F are the symbol time and frequency spacing that ensure orthogonality in the time and frequency domains, respectively.[3] The coefficients $\delta_t \leq 1$ and $\delta_f \leq 1$ are the compression factors for the symbol interval and frequency spacing, respectively. While setting them to 1 results in an orthogonal transmission, they can

[3] As far as F is concerned, its minimum value is $F = \frac{1+\alpha}{T}$.

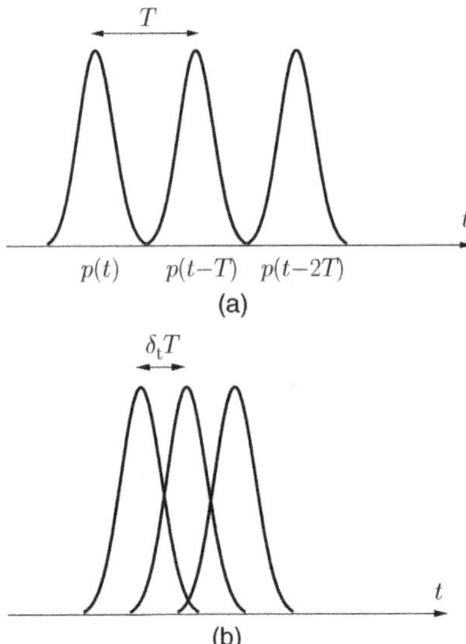

Figure 1.9 Schematic view of (a) orthogonal and (b) FTN signaling in the time domain

be reduced to a given extent without reducing the minimum Euclidean distance. The effects of the application of FTN in the time domain are schematically represented in Figure 1.9, which shows the transmission of a generic pulse $p(t)$ with orthogonal signaling (Figure 1.9(a)) and adopting a coefficient $\delta_{\rm t} < 1$ (Figure 1.9(b)). We can see how interference from adjacent pulses arises in the latter case.

Some scepticism can be raised against this technique. From a practical point of view, FTN may require an optimal detector, the complexity of which easily becomes unmanageable. No hints are provided in the original papers as to how to perform the optimization in the more practical scenario where a reduced-complexity receiver is employed. From a theoretical point of view, although this technique has been proposed to increase the spectral efficiency of a communication system, the uncoded BER is used as figure of merit in place of the spectral efficiency itself.

Before discussing ways to solve these problems, we need to introduce a few definitions. Let us consider the multicarrier transmission in Eq. (1.20), where $\delta_f F$ is the frequency separation between two adjacent carriers and $\delta_t T$ is the symbol time. We will collect in a vector $\mathbf{x}^{(\ell)} = \{x_k^{(\ell)}\}$ the input symbols transmitted over the ℓth carrier. At the receiver side, a discrete-time set of sufficient statistics is extracted using a bank of matched filters and we denote by $\mathbf{y}^{(\ell)} = \{y_k^{(\ell)}\}$ the samples at the output of the matched filter for the ℓth carrier.

Depending on the allowed complexity at the receiver, different strategies can be adopted for detection. For example, the receiver can neglect both ICI and ISI and adopt a symbol-by-symbol detector. In other words, instead of the optimal receiver for the actual channel, we could adopt the optimal receiver for a simplified auxiliary channel, for which the combined effect of ISI and ICI is modeled as a zero-mean Gaussian process independent of the additive thermal noise. Note that the interference is truly Gaussian distributed only if the transmitted symbols are Gaussian distributed as well and this is not the case in practice. Especially when the interference set is small – for example when δ_t and δ_f are close to one – the actual interference distribution may substantially differ from a Gaussian distribution. However, the accuracy of this approximation is not of concern here: assuming Gaussian-distributed interference is required for the auxiliary channel model anyway, to ensure that a symbol-by-symbol receiver is optimal. It is like saying that the Gaussian assumption is a *consequence* of the choice of the symbol-by-symbol receiver. Once the simplified receiver has been selected – suboptimal for the channel at hand but optimal for the considered auxiliary channel – it is possible to compute a lower bound on the information rate for that channel using the technique of Arnold *et al.* [25]. The information rate, also called constrained capacity, is the mutual information when the input symbols are constrained to belong to our finite constellation $\mathcal{X}$. According to mismatched detection [26], this lower bound is *achievable* by that particular suboptimal detector. The achievable spectral efficiency (ASE) is defined as the ratio between the achievable lower bound on the information rate and the product $\delta_f F \delta_t T$

$$\text{ASE} = \frac{I(\mathbf{x}^{(\ell)};\mathbf{y}^{(\ell)})}{\delta_f F \delta_t T}$$

where $\delta_f F$ is a measure of the bandwidth of the given subcarrier.

The most recent extension of the FTN principle is thus time-frequency packing [27], in which it is proposed to optimize δ_f and δ_t in order to maximize the ASE. The idea is very simple: by reducing δ_f and δ_t the achievable information rate $I(\mathbf{x}^{(\ell)};\mathbf{y}^{(\ell)})$ will certainly degrade due to the increased interference. However, the spectral efficiency – in other words $I(\mathbf{x}^{(\ell)};\mathbf{y}^{(\ell)})/\delta_f F \delta_t T$ – can be improved. Hence, the main quantity of interest is not the uncoded BER performance.[4] We may accept a degradation of the information rate provided the spectral efficiency is increased. In other words, instead of keeping the same code, an improvement can be obtained by using a code with a lower rate. Improving the spectral efficiency without increasing the constellation size is convenient since low-order constellations are more robust to impairments such as phase noise and nonlinearities.

In Ref. [27], the main concepts are elucidated with reference to a symbol-by-symbol detector and the additive white Gaussian noise (AWGN) channel, working on the samples at the matched filters output. More sophisticated receiver architectures are considered in Ref. [28], still with reference to the AWGN channel. In general, there are several receiver architectures that have been considered for the detection of TFP signals, that include equalization [29] and filtering, followed by a maximum a posteriori (MAP) symbol detector based, for example, on a BCJR algorithm [30]. One of such advanced filtering techniques is the so-called 'channel shortening' [31], aimed at designing the interference at the MAP detector to properly fit the desired complexity of the detection stage. Further gains can be obtained by using algorithms that detect more than one carrier at a time. In general, the larger the receiver complexity, the higher the gains that this technique can achieve. Its effectiveness has been demonstrated in several scenarios on wireless and optical channels [9, 28, 32, 33], and it appears to be suited for 5G systems as well.

[4] Since there is no need to keep the same Euclidean distance as the Nyquist case, there is no need to employ a base pulse satisfying the Nyquist condition. Thus TFP can be adopted for *any* base pulse.

1.2.6 Single-carrier Schemes

All of the modulation formats considered so far employ multicarrier transmissions. However, while multicarrier formats are compatible with most of the candidate technologies for 5G networks, they might not be the best choice if millimeter waves (mmWaves) are employed.

The use of mmWaves has been proposed as a strong candidate for achieving the spectral efficiency growth required by 5G networks, resorting to the use of the currently unused frequency bands in the range between 20 GHz and 90 GHz. In particular, the E-band, between 70 GHz and 80 GHz, provides 10 GHz of free spectrum, which could be exploited to operate 5G networks. Up until now, the use of mmWaves for cellular communications has been neglected due to the higher atmospheric attenuation that they suffer compared to other frequency bands. However, while this is true for propagation in the macro-cell environments that are typical of past cellular generations, recent measurements suggest that mmWave attenuation is only slightly worse than in other bands, as far as propagation in dense urban environments is concerned [34]. Therefore, mmWaves have recently been reconsidered as a viable technology for cellular communications.

One of the main advantages of multicarrier schemes is their ability to multiplex users in the frequency domain. However, this advantage comes with several disadvantages too. Indeed, this chapter has been concerned with the analysis of possible alternatives to the conventional OFDM scheme, which cope with its shortcomings, but without renouncing the possibility of having a frequency-domain multiplex. However, if mmWaves are used, this feature might not be so crucial, for several reasons.

- As already mentioned, the propagation attenuation of mmWaves make them a viable technology only for small-cell, dense networks, where few users will be associated to any given base station.
- The higher bandwidth would cause low OFDM symbol duration, making it possible to multiplex users in the time domain as efficiently as in the frequency domain.
- mmWaves will be operated together with massive antenna arrays to overcome propagation attenuation. This makes digital beamforming unfeasible, since the energy required for digital-to-analog and analog-to-digital conversion would be huge. Thus, each user will have an own radio-frequency beamforming, which requires users to be separated in time rather than in frequency.

In light of these considerations, one possibility for mmWaves is to dispense with multicarrier transmissions, eliminating its drawbacks, and resorting instead to single-carrier (SC) modulation formats. In Ghosh *et al.* [35], the null cyclic prefix single carrier (NCP-SC) scheme has been proposed for mmWaves. The concept is to transmit a single-carrier signal in which the usual cyclic prefix used by OFDM is replaced by nulls appended at the end of each transmit symbol. The block scheme is shown in Figure 1.10.

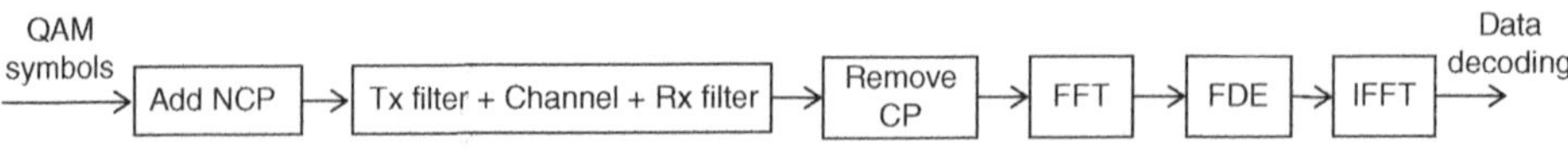

Figure 1.10 Principle of NCP-SC transceiver architecture

The NCP-SC scheme has several advantages over OFDM. In particular:

- The null cyclic prefix is part of the transmit symbol and is fed to the FFT together with the other data samples. This makes it possible to adapt the length of the prefix of each user, without disrupting the frame timing, because the length of each user's transmit symbol is always kept constant to N.
- The NCP-SC has a much lower PAPR and much lower OOB emissions than OFDM. This reduces interference and eases the design and operation of power amplifiers.
- The presence of time intervals in which no useful data are present makes it easier to estimate the interference-plus-noise power at the receiver.

Before concluding this section, it should also be observed that NCP-SC has some drawbacks compared to OFDM too. In particular, it requires a higher computational complexity. As we can see from the NCP-SC scheme in Figure 1.10, both an FFT and IFFT operations are required at the receiver. OFDM, on the other hand, only requires one FFT at the receiver. The resulting complexity increase might become significant, especially for increasing sizes of the FFT.

1.3 Waveform Choice

In this section, we describe some shaping pulses that can be considered as alternatives to the rectangular pulse adopted in OFDM. In practice, we are interested in pulses that achieve a good compromise between their sidelobe levels in the frequency domain, and their extension in the time-domain. The design of discrete-time windows with the discussed properties is a classical topic that arises in many areas of signal processing, such as FFT-based spectrum analysis and the synthesis of finite-impulse-response filters with the window method. Several pulse shapes are thus available in the open literature (see, for example, Proakis and Demetris [36] and Sahin *et al.* [37]). In what follows, we just give three possible examples, namely the evergreen RRC, the pulse proposed in the PHYDYAS research project for use with FBMC [38], and finally the Dolph–Chebyshev (DC) pulse, whose use has been recommended for the UFMC modulation.

- **RRC pulses** are widely used in telecommunication systems to minimize ISI at the receiver. The impulse response of an RRC pulse is

$$p(t) = \begin{cases} \frac{1}{\sqrt{T}}\left(1-\alpha+4\frac{\alpha}{\pi}\right) & t=0 \\ \frac{\alpha}{\sqrt{2T}}\left[\left(1+\frac{2}{\pi}\right)\sin\left(\frac{\pi}{4\alpha}\right)+\left(1-\frac{2}{\pi}\right)\cos\left(\frac{\pi}{4\alpha}\right)\right] & t=\pm\frac{T}{4\alpha} \\ \frac{1}{\sqrt{T}}\frac{\sin\left(\pi\frac{t}{T}(1-\alpha)\right)+4\alpha\frac{t}{T}\cos\left(\pi\frac{t}{T}(1+\alpha)\right)}{\pi\frac{t}{T}\left[1-\left(4\alpha\frac{t}{T}\right)^2\right]} & \text{otherwise} \end{cases}$$

  where T is the symbol interval and α is the roll-off factor, which measures the excess bandwidth of the pulse in the frequency domain.
- **The PHYDYAS pulse** is a discrete-time pulse specifically designed for FBMC systems. Let M be the number of subcarriers. Then the impulse response is

$$p(n) = P_0 + 2\sum_{k=1}^{K-1}(-1)^k P_k \cos\left(\frac{2\pi k}{KM}(n+1)\right)$$

for $n = 0, 1, \ldots, KM - 2$ and $K = 4$, where the coefficients $P_k, k = 0, \ldots, K - 1$ have been selected using the frequency sampling technique [38], and assume the following values:

$$P_0 = 1$$
$$P_1 = 0.97195983$$
$$P_2 = 1/\sqrt{2}$$
$$P_3 = \sqrt{1 - P_1}$$

- **The DC pulse** [39] is significant because, in the frequency domain, it minimizes the main lobe width for a given sidelobe attenuation. Its discrete-time impulse response is [40]:

$$p(n) = \frac{1}{N}\left[10^{-\frac{A}{20}} + 2\sum_{k=1}^{(N-1)/2} T_{N-1}\left(x_0 \cos\left(\frac{k\pi}{N}\right)\right)\cos\left(\frac{2\pi nk}{N}\right)\right]$$

  for $n = 0, \pm 1, \ldots, \pm\frac{N-1}{2}$, where N is the number of coefficients, A is the attenuation of side lobes in dB,

$$x_0 = \cosh\left(\frac{1}{N-1}\cosh^{-1}\left(10^{-\frac{A}{20}}\right)\right)$$

  and

$$T_n(x) = \begin{cases} \cos(n\cos^{-1}(x)) & |x| \leq 1 \\ \cosh(n\cosh^{-1}(x)) & |x| > 1 \end{cases}$$

  is the Chebyshev polynomial of the first kind [41].

In Figure 1.11, we report the spectra of the pulses we have just described. All spectra were computed by performing a 1024 points FFT of pulses of 160 samples in the time domain. The figure compares the rectangular pulse, typical of OFDM, with an RRC pulse having roll-off $\alpha = 0.1$, the PHYDYAS pulse with $M = 1$, and the DC pulse with attenuation $A = -50$ dB. The figure clearly shows that the rectangular pulse is the one with the worst spectral characteristics; on the other hand, the PHYDYAS pulse is the one with the smallest sidelobe levels, while the DC pulse is the one with the smallest width of the main lobe.

1.4 Discussion and Concluding Remarks

This chapter has been devoted to the illustration of some of the most promising modulation schemes for use in forthcoming 5G cellular networks. While legacy OFDM is a robust and mature technology used in several communication systems – indeed, OFDM modulation is the core PHY technology of 4G systems, and is also employed in other systems such as digital audio broadcasting and terrestrial digital video broadcasting – the very stringent requirements of future networks, along with the heterogeneous scenarios that they will have to operate in, has pushed researchers to look for other solutions. One conclusion that can certainly be drawn is that what is the "best" modulation is a question that cannot be easily answered, and indeed

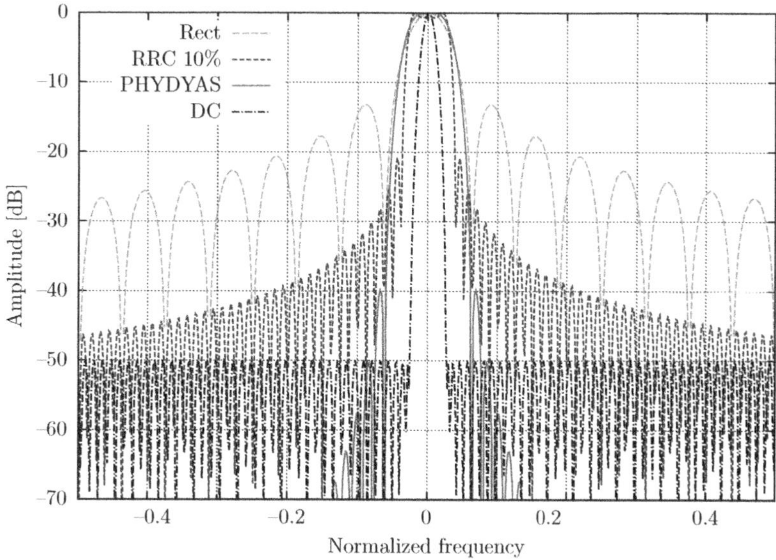

Figure 1.11 Comparison of pulse shapes in the frequency domain

the right answer might be "it depends", in the sense that there is no modulation that performs the best in all possible operating conditions. As an instance, UFMC, by virtue of its low side-lobe levels, is a modulation scheme that has been designed to perform well in scenarios where asynchronous transmissions and carrier frequency offsets may lead to ICI, although this property is retained by FBMC too. Due to its long shaping filters, FBMC unfortunately has a low efficiency in situations where small data packets are to be transmitted, a scenario typical of the IoT. Both UFMC and FBMC do not require the use of a CP, and this is a clear advantage with respect to filtered-OFDM, for instance. On the other hand, when dealing with access to fragmented spectrum, GFDM exhibits great flexibility, since frequency bands can be added and removed in a communication link quite easily and in a flexible way. The latency requirement also plays a key role and in this aspect FBMC again appears a weak choice since the long impulse response of its shaping filters prohibits its use in situations of sporadic traffic and low latency. Considering the issue of pure throughput maximization, it is evident that TFP appears to be the best choice, even though receiver complexity must be carefully taken into account, which makes this modulation clearly unsuited for IoT applications. In a WRAN scenario, on the other hand, in which a vast number of receivers are installed indoors and plugged into the electrical grid, high-complexity receiver are affordable and TFP might be a good option.

Ultimately, the solution to the problem of choosing a new modulation scheme will reside in the so-called software-defined-networking paradigm [42]. Indeed, the trend that we are witnessing in recent years is the increased role of software implementations with respect to hardware implementation of communication services. 5G networks will see a lot of functionality implemented via software as well. In addition, PHY-layer functions will be partly virtualized and implemented in a data-center. A virtualized PHY service will permit tuning of the modulation parameters to the scenario at hand; the modulation scheme itself might be changed according to the operating scenario. In this framework, one might think of a software-defined adaptive PHY, which would certainly be able to cope with the stringent levels of flexibility, scalability, performance and efficiency that 5G networks will require.

References

[1] Michailow, N., Matthé, M., Gaspar, I.S., Caldevilla, A.N., Mendes, L.L., Festag, A., and Fettweis, G. (2014) Generalized frequency division multiplexing for 5th generation cellular networks. *IEEE Trans. Commun.*, **62** (9), 3045–3061.

[2] Fettweis, G. (2014) The tactile internet: applications and challenges. *IEEE Veh. Tech. Mag.*, **9** (1), 64–70.

[3] Schaich, F., Wild, T., and Chen, Y. (2014) Waveform contenders for 5G – suitability for short packet and low latency transmissions, in *79th IEEE Vehic. Tech. Conf. 2014*, IEEE, pp. 1–5.

[4] Ghosh, A., Zhang, J., Andrews, J.G., and Muhamed, R. (2010) *Fundamentals of LTE*, Pearson Education.

[5] Morelli, M. (2004) Timing and frequency synchronization for the uplink of an OFDMA system. *IEEE Trans. Commun.*, **52** (2), 296–306.

[6] Hwang, T., Yang, C., Wu, G., Li, S., and Ye Li, G. (2009) OFDM and its wireless applications: a survey. *IEEE Trans. Veh. Tech.*, **58** (4), 1673–1694.

[7] Irmer, R., Droste, H., Marsch, P., Grieger, M., Fettweis, G., Brueck, S., Mayer, H.P., Thiele, L., and Jungnickel, V. (2011) Coordinated multipoint: Concepts, performance, and field trial results. *IEEE Commun. Mag.*, **49** (2), 102–111.

[8] Ochiai, H. and Imai, H. (2001) On the distribution of the peak-to-average power ratio in OFDM signals. *IEEE Trans. Commun.*, **49** (2), 282–289.

[9] Banelli, P., Buzzi, S., Colavolpe, G., Modenini, A., Rusek, F., and Ugolini, A. (2014) Modulation formats and waveforms for 5G networks: Who will be the heir of OFDM? *IEEE Signal Processing Mag.*, **31** (6), 80–93.

[10] 5GNow website, http://5gnow.eu.

[11] Metis Project website, http://www.metis2020.com.

[12] Bellanger, M. (2010) FBMC physical layer: a primer, *Tech. Rep.*, PHYDYAS.

[13] Fettweis, G., Krondorf, M., and Bittner, S. (2009) GFDM - generalized frequency division multiplexing, in *Proc. Vehicular Tech. Conf.*, Barcelona, Spain, pp. 1–4.

[14] Kozek, W. and Molisch, A.F. (1998) Nonorthogonal pulseshapes for multicarrier communications in doubly dispersive channels. *IEEE J. Select. Areas Commun.*, **16** (8), 1579–1589.

[15] Gröchenig, K. (2001) *Foundations of Time-Frequency Analysis*, Birkhäuser.

[16] Matz, G. and Hlawatsch, F. (eds) (2011) *Fundamentals of Time-varying Communication Channels*, Academic Press.

[17] Matz, G., Bolcskei, H., and Hlawatsch, F. (2013) Time-frequency foundations of communications: concepts and tools. *IEEE Signal Processing Mag.*, **30** (6), 87–96.

[18] Wunder, G., Jung, P., Kasparick, M., Wild, T., Schaich, F., Chen, Y., Brink, S., Gaspar, I., Michailow, N., Festag, A. *et al.* (2014) 5GNOW: non-orthogonal, asynchronous waveforms for future mobile applications. *IEEE Commun. Mag.*, **52** (2), 97–105.

[19] Wunder, G., Kasparick, M., Brink, S., Schaich, F., Wild, T., Gaspar, I., Ohlmer, E., Krone, S., Michailow, N., Navarro, A. *et al.* (2013) 5GNOW: Challenging the LTE design paradigms of orthogonality and synchronicity, in *Proc. Vehicular Tech. Conf.*, Dresden, Germany.

[20] Vakilian, V., Wild, T., Schaich, F., Ten Brink, S., and Frigon, J.F. (2013) Universal-filtered multi-carrier technique for wireless systems beyond LTE, in *Globecom Workshops*, IEEE, pp. 223–228.

[21] Schaich, F. and Wild, T. (2014) Waveform contenders for 5G – OFDM vs. FBMC vs. UFMC, in *6th Int. Symp. Commun. Cont. Sig. Proc. (ISCCSP), 2014*, IEEE, pp. 457–460.

[22] Mazo, J.E. (1975) Faster-than-Nyquist signaling. *Bell System Tech. J.*, **54**, 1450–1462.

[23] Liveris, A. and Georghiades, C.N. (2003) Exploiting faster-than-Nyquist signaling. *IEEE Trans. Commun.*, **47**, 1502–1511.

[24] Rusek, F. and Anderson, J.B. (2005) The two dimensional Mazo limit, in *Proc. IEEE Int. Sympos. Info. Th.*, Adelaide, Australia, pp. 970–974.

[25] Arnold, D.M., Loeliger, H.A., Vontobel, P.O., Kavčić, A., and Zeng, W. (2006) Simulation-based computation of information rates for channels with memory. *IEEE Trans. Inform. Theory*, **52** (8), 3498–3508.

[26] Merhav, N., Kaplan, G., Lapidoth, A., and Shamai, S. (1994) On information rates for mismatched decoders. *IEEE Trans. Inform. Theory*, **40** (6), 1953–1967.
[27] Barbieri, A., Fertonani, D., and Colavolpe, G. (2009) Time-frequency packing for linear modulations: Spectral efficiency and practical detection schemes. *IEEE Trans. Commun.*, **57**, 2951–2959.
[28] Modenini, A., Colavolpe, G., and Alagha, N. (2012) How to significantly improve the spectral efficiency of linear modulations through time-frequency packing and advanced processing, in *Proc. IEEE Intern. Conf. Commun.*, Ottawa, Canada, pp. 3299–3304.
[29] Isam, S., Kanaras, I., and Darwazeh, I. (2011) A truncated SVD approach for fixed complexity spectrally efficient FDM receivers, in *Proc. IEEE Wireless Commun. and Network. Conf.*, Cancun, Mexico, pp. 1584–1589.
[30] Bahl, L.R., Cocke, J., Jelinek, F., and Raviv, J. (1974) Optimal decoding of linear codes for minimizing symbol error rate. *IEEE Trans. Inform. Theory*, **20**, 284–287.
[31] Rusek, F. and Prlja, A. (2012) Optimal channel shortening for MIMO and ISI channels. *IEEE Trans. Wireless Commun.*, **11** (2), 810–818.
[32] Piemontese, A., Modenini, A., Colavolpe, G., and Alagha, N. (2013) Improving the spectral efficiency of non-linear satellite systems through time-frequency packing and advanced processing. *IEEE Trans. Commun.*, **61** (8), 3404–3412.
[33] Colavolpe, G. and Foggi, T. (2014) Time-frequency packing for high capacity coherent optical links. *IEEE Trans. Commun.*, **62**, 2986–2995.
[34] Ghosh, A., Thomas, T.A., Cudak, M., Ratasuk, R., Moorut, P., Vook, F.W., Rappaport, T., G. R, MacCartney, J., Sun, *S*., and Nie, S. (2014) Millimeter wave enhanced local area systems: A high data rate approach for future wireless networks. *IEEE J. Select. Areas Commun.*, **32** (6), 1152–1163.
[35] Cudak, M. *et al.* (2013) Moving towards mmwave-based beyond-4G (b-4G) technology, in *Proc. IEEE 77th Vehic. Tech. Conf. Spring*.
[36] Proakis John, G. and Manolakis Dimitris, G. (1996) *Digital Signal Processing. Principles, Algorithms, and Applications*, Prentice Hall.
[37] Sahin, A., Guvenc, I., and Arslan, H. (2012) A survey on multicarrier communications: Prototype filters, lattice structures, and implementation aspects. *IEEE Commun. Surveys & Tut.*, **16** (3), 1312–1338.
[38] Viholainen, A., Bellanger, M., and Huchard, M. (2008) Prototype filter and structure optimization, *Tech. Rep.*, PHYDYAS.
[39] Dolph, C. (1946) A current distribution for broadside arrays which optimizes the relationship between beam width and side-lobe level. *Proc. IRE*, **34** (6), 335–348.
[40] Antoniou, A. (2000) *Digital Filters: Analysis, Design and Applications*, McGraw-Hill.
[41] Abramowitz, M. and Stegun, I.A. (eds) (1972) *Handbook of Mathematical Functions*, Dover.
[42] Rost, P., Bernardos, C., Domenico, A., Girolamo, M., Lalam, M., Maeder, A., Sabella, D. *et al.* (2014) Cloud technologies for flexible 5G radio access networks. *IEEE Commun. Mag.*, **52** (5), 68–76.

2

Faster-than-Nyquist Signaling for 5G Communication

John B. Anderson

Fifth-generation wireless systems will transmit many more bits than their predecessors in each Hertz, second and square meter of real estate. Many ways to do this are explored in this book, and the focus of this chapter is on more efficient modulation and coding of the signals. One wants more bits per Hertz and second at a given error performance. Fortuitously, the innovations in 5G also raise the signal-to-noise ratio (SNR) by means of smaller cells, MIMO and WiFi-like local methods. This higher SNR is a key to more bits. Proper coding

Signal Processing for 5G: Algorithms and Implementations, First Edition. Edited by Fa-Long Luo and Charlie Zhang.

and modulation can raise the number further, but until now rather few coding methods make good use of higher SNRs. The subject of this chapter is faster-than-Nyquist (FTN) signaling, a leading such method that can potentially double data transmission rates. In addition, it sheds new light on notions of bandwidth, Shannon capacity and complexity that underlie data transmission. FTN is not new, but it is only recently that its implications have become more clear and prototypes have been constructed. The aim of this chapter is to describe two types of FTN: one for single-carrier and one for multi-subcarrier transmission, and to describe what can be hoped for in future FTN technology.

2.1 Introduction to FTN Signaling

FTN began with a 1975 paper by James Mazo [1], in which he made a simple but surprising discovery. He was studying data-bearing orthogonal pulses; that is, pulses that can be detected independently. A result of Nyquist showed that such pulses limited to bandwidth $1/2T$ Hz could not be sent faster than one every T s, and it was widely assumed that sending them faster would raise the error rate of the detector. Mazo found that no such increase had to occur, even if the pulses were sent 25% faster. There was indeed a penalty: the pulses now caused ISI at the detector, and a more complex detection was required. But 25% more data could be sent at the same error rate without consuming more bandwidth.

Presently we will explain this phenomenon in more detail, but first we sketch how the field has evolved since 1975. Mazo's idea of faster pulses lay dormant for 30 years, but other interesting and sometimes troubling questions began to appear and many researchers were concerned about them. Most signaling methods used the orthogonal pulses, but was there a loss from them? How, precisely, were Shannon capacity and signal spectrum related? The entire average power spectral density (PSD) of the signals appeared to play a role, not just a single-number measure of its width. It is possible to attribute carried information to the PSD main lobe and to the stop band of a signal set, and signaling schemes existed that seemed to carry a significant portion in the stop band. Much more was learned about how signal bandwidth, energy and error performance relate, but there were schemes that appeared to violate the traditional Shannon limits that were in use. Finally, all of these puzzles became rather more evident as the signaling carried more bits in each Hertz and second.

Eventually explanations of these issues arose within a certain framework, which came to be called FTN signaling. It now encompasses an entire view of signaling, bandwidth and energy. It proposes coding schemes, but it also clarifies a number of questions. The view begins from the signal PSD: it is fixed, both its pass band and stop band shapes. Data are carried by pulses, but they are free to be non-orthogonal. A revised calculation of the Shannon limit leads to a more liberal limit when the pulses are non-orthogonal; it depends on the PSD and in only a special case is it the traditional textbook limit. FTN can be coded or not, although until now coding has meant that a simple convolutional code precedes the non-orthogonal pulses. Coded FTN is one of the few coding schemes that can work in an effective, natural way at high numbers of bits per Hertz and second. FTN schemes can lie rather close to the Shannon limit; much closer than simple modulation-plus-coding. Those that lie further away can sometimes be trellis-decoded, while those that lie closer require iterative, or "turbo", decoding that simultaneously removes the ISI and decodes the code. The turbo method is more or less identified with FTN.

The rest of this section goes into detail about these concepts and gives a precise explanation of FTN signaling. Section 2.2 introduces coded FTN. FTN based on the usual binary modulation will serve as an introduction and then we will present new work on 4-level modulation. Section 2.3 introduces FTN schemes that are more natural for subcarrier or OFDM schemes. These are based on squeezing together subcarriers in frequency rather than accelerating them in time.

A one-chapter treatment of FTN must leave out many details. Readers wishing to know more about the theory and implementation of FTN are referred to our survey paper and the references therein [2]. Iterative decoders are an essential part of FTN, and these are an offshoot of an older field: turbo equalization. Those schemes simultaneously decode codes and equalize ISI, but they envision a much wider variety of ISIs than those that derive from one fixed, narrowband signal PSD. A survey of turbo equalization with many references is Tüchler and Singer [3]. In its treatment of communication theory this chapter is limited to a few highlights; the details appear in the book by Anderson [4] or the more advanced one by Proakis [5] or in the many other textbooks in the field.

2.1.1 *Definition of FTN: FTN from Detection Theory*

Signal detection theory and Shannon's channel capacity both have interesting things to say about FTN signals. They are free-standing theories and say different things. Detection theory gives a solid physical feel, so it is best to start there.

Most data transmission works by modulating a sequence of pulses with shape $h(t)$ by a sequence of symbols $u_1, u_2, \cdots$ to form a signal of the form $s(t) = \sqrt{E_s} \sum_n u_n h(t - nT)$. This is called linear modulation. A new pulse appears every T seconds, the symbol time. Both $h(t)$ and u_n have unit energy, so that E_s is the symbol energy. If each symbol carries $\log_2 M$ data bits, then $E_b = E_s / \log M$ is the energy per data bit when the signals are uncoded. If the u_n are independent, identically distributed random variables with mean zero, the PSD of $s(t)$ has the same shape as $|H(f)|^2$, where $H(f)$ is the Fourier transform of $h(t)$. In carrier modulation, $h(t)$ and u_n are two-fold, and can be taken as complex or as so-called in-phase and quadrature components. For simplicity in this section, we will take them as a single real function and value, a mode called baseband transmission. In binary signaling $u_n = \pm 1$; a later section will introduce 4-level signaling, where u_n will take values $\sqrt{1/5}\{3, 1, -1, -3\}$. Baseband 2- and 4-ary modulation are called 2PAM and 4PAM; the carrier equivalents take 2 and 4 values in the in-phase and quadrature dimensions—in other words, 4 and 16 two-dimensional values—and are called QPSK and 16QAM, respectively (PAM and QAM here stand for pulse amplitude modulation and quadrature amplitude modulation).

An FTN signal can be defined as follows. Keeping all else the same, accelerate the appearance of the pulses by shortening the symbol time T to τT, $\tau < 1$. This produces

$$s(t) = \sqrt{E_s} \sum_n u_n h(t - n\tau T) \tag{2.1}$$

This form has the same PSD as before but it carries $1/\tau$ more symbols per second. In virtually all linear modulation systems the original $h(t)$ is orthogonal with respect to T, meaning that the integral $\int h(t)h(t - nT)\, dt = 0$ for any integer $n \neq 0$. This greatly simplifies detection; a filter matched to h and sampled at the right time yields an optimal estimate of u_n, with no interference from other u. Nyquist and later researchers studied the properties of such pulses, and an alternate name for them is Nyquist pulses. An important property is the fact that no such pulse has bandwidth less than $1/2T$ Hz; the narrowest such pulse is $h(t) = \sqrt{1/T}\mathrm{sinc}(t/T) = \sqrt{1/T}\sin(\pi t/T)/\pi t/T$, which has unit energy and a "square" spectrum, constant on $[-1/2T, 1/2T]$ Hz and 0 elsewhere. The frequency $1/2T$ Hz is called the Nyquist limit to orthogonal signals. When the acceleration factor τ is less than 1, h is not, in general, orthogonal with respect to τT, and the simple detection is lost. The optimal detector is now a trellis decoder.

FTN signals can begin from any $h(t)$, whether T-orthogonal or not, but one usually begins from an orthogonal h. Another property of such pulses is that they all have 3 dB-down bandwidth $1/2T$ and their power spectrum is antisymmetric about this half-way-down point (if $|H(0)|^2 = K$, $|H(1/2T - \delta)|^2 = K - |H(1/2T + \delta)|^2$, for $0 \leq \delta \leq 1/2T$). A commonly used such pulse is the root raised-cosine (root RC) pulse, the one for which $|H(f)|^2$ has a raised-cosine shape centered at $1/2T$ Hz. The RC shape extends beyond $1/2T$ Hz by a fraction called the excess bandwidth. Pulse shaping is a complex subject, with many implications for FTN, but we will employ only the 30% excess bandwidth root RC pulse in this chapter; this is a favored design in applications.

For the 30% root RC pulse, Figure 2.1 shows two $s(t)$, one simple linear modulation and the other FTN from Eq. (2.1) with $\tau = 0.703$. Both transmit the binary symbols $1, -1, 1, 1, 1$. It can be seen that the signals are rather similar except that one is a fraction 0.297 shorter; as they grow longer both tend to the same PSD.

Something more is going on under the surface, and to see what it is requires some communication theory. First, assume that the channel is linear with additive white Gaussian noise (AWGN) whose PSD is $N_0/2$ watts/Hz. The databit-wise SNR is E_b/N_0. Then in optimal detection the probability of detecting signal $s_2(t)$ given that some other $s_1(t)$ was sent is

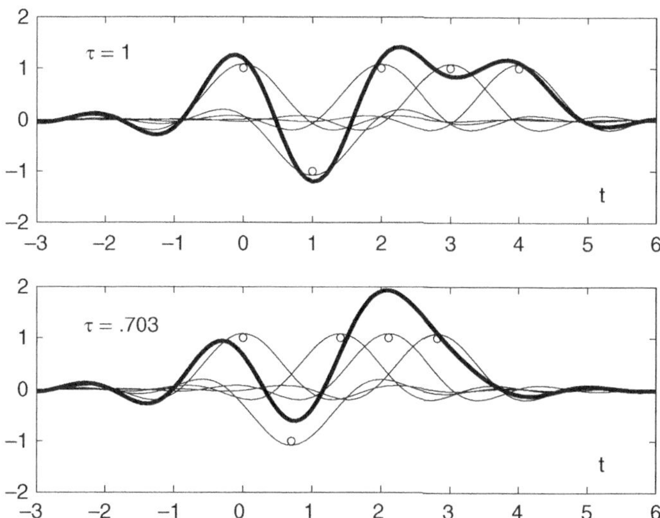

Figure 2.1 A comparison of orthogonal (upper) and $\tau = 0.703$ FTN (lower) transmission of the symbols $1, -1, 1, 1, 1$. Other symbols are 0

$Q(\sqrt{d^2 E_b/N_0})$, where

$$d^2 = (1/2E_b) \int_{-\infty}^{\infty} |s_1(t) - s_2(t)|^2 \, dt \tag{2.2}$$

The quantity d^2 is called the square Euclidean distance between $s_1(t)$ and $s_2(t)$. Here $Q(z)$, $z > 0$, is the integral of the unit Gaussian density over $[z, \infty)$. The worst detection case occurs when d^2 is smallest: if $s_i(t)$ and $s_j(t)$ are signals that are the same up to symbol n' and different afterwards, at least at $n' + 1$, then

$$d^2_{\min} = \min_{i,j} \quad (1/2E_b) \int_{-\infty}^{\infty} |s_i(t) - s_j(t)|^2 \, dt, i \neq j \tag{2.3}$$

is the worst case and is called the *square minimum distance* of the signal set.

We can now investigate the FTN phenomenon in more detail. Notice first from Eq. (2.1) that the difference $s_1(t) - s_2(t)$ depends only on the difference of the respective u_n, not on their actual values. Thus

$$d^2 = (1/2E_b) \int_{-\infty}^{\infty} E_s \left| \sum_{n>n'} \Delta u_n h(t - n\tau T) \right|^2 dt \tag{2.4}$$

where $\Delta u_{n'+1}, \Delta u_{n'+2}, \ldots$ are the respective differences. When s_1 and s_2 differ in only one symbol, $\Delta u_{n'+1} = \pm 2$, the other differences are 0, and it is clear that d^2 is always 2, since the square integral of $h(t)$ is 1.

What is of interest in FTN signaling is the signal difference that produces $d_{\min}$, since this sets the exponential form of the symbol error probability. With orthogonal signals, $d_{\min}$ stems from the "2" case just discussed, since all the distances are simply $2 \times$ [number of nonzero Δu]. What is it with $\tau < 1$? Figure 2.2 compares important error cases for orthogonal signaling ($\tau = 1$) and FTN with $\tau = 0.5$. In the top half, the signal created by $1, -1, 1$ (solid line) appears with the one created by $-1, 1, -1$ (dashed); these lie at square distance 6 apart, compared to $d^2_{\min} = 2$. In the lower half, the same signals, now with $\tau = 0.5$, lie only 1.02 apart. In fact, these signals are a worst-case pair and $d^2_{\min} = 1.02$, which is much less than 2. The distance between signals that differ in a single symbol remains 2, but the FTN acceleration has brought some other pairs closer, so that one of them now sets the minimum distance. Bandwidth reduction has worsened minimum distance! One can ask when one of these other worst cases falls below 2 for the first time. It turns out that this occurs when $\tau = 0.703$, the τ in Fig.2.1. The smallest τ that leads to $d^2_{\min}$ that is still 2 is called the *Mazo limit*.

2.1.1.1 Bit Density

FTN needs a careful definition of time and bandwidth. Any transmission method can send more bits per second by scaling time faster, which scales bandwidth larger by the inverse factor. Time and bandwidth directly trade off, and exist within a single time–bandwidth resource. The proper way to measure signal efficiency is per Hertz and second; that is, to normalize by the product of time and bandwidth consumed. We will call this the *data bit density*, in b/Hz-s. The measure of bandwidth is arbitrary so long as it is consistent; we will measure it as baseband 3 dB bandwidth, for positive frequencies only.

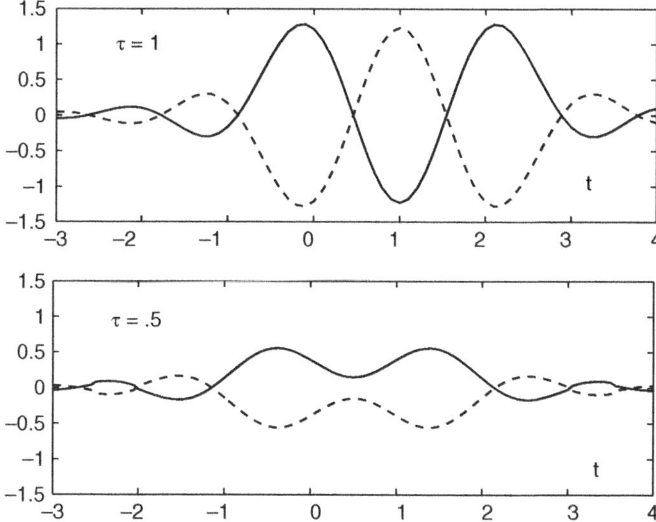

Figure 2.2 Transmission of $1, -1, 1$ (solid) and $-1, 1, -1$ (dashed) by 30% root RC pulses with $\tau = 1$ (upper figure) and $\tau = 0.5$ (lower) pulse spacing. $T = 1$; other symbols are 0. Square distance is much less with $\tau = 0.5$

It is customary to divide signaling methods into two groups:

- those below 2 b/Hz-s: the wideband schemes
- those above: the high energy schemes.

The nature of channel capacity is such that schemes with density above 2 b/Hz-s achieve this only by applying a lot of energy. This is inescapable. We use wideband schemes when we are forced to, in for example space communication; we use high energy schemes when high SNR is available. For reference, simple antipodal orthogonal signaling carries 2 b/Hz-s; that is, symbol time T consumes bandwidth $1/2T$ Hz so that the density is $1/(T/2T) = 2$ b/Hz-s. A rate-1/2 parity-check code reduces this to 1 b/Hz-s. QPSK has 2 b/Hz-s, since the in-phase and quadrature carrier signals have double width but both exist in the same spectrum. 16QAM and 4PAM have density 4 b/Hz-s. The aim of FTN is to improve the densities of these simple schemes, at a given SNR and error performance.

2.1.1.2 Coding, Modulation and Discrete-time Modeling

To have a concrete picture, Figure 2.3 gives a standard picture of coded FTN signaling. Across the top is a rate r convolutional encoder, an interleaver and an ISI mechanism. The last of these models the effect of the FTN pulse transmission via a real-number convolution with a discrete-time channel response sequence $\vec{v}$. The top is standard practice with an ISI channel: a de-interleaver breaks long ISI error events into small pieces that are easily corrected by the convolutional code. What is new in the picture is the bottom half, which is an iterative

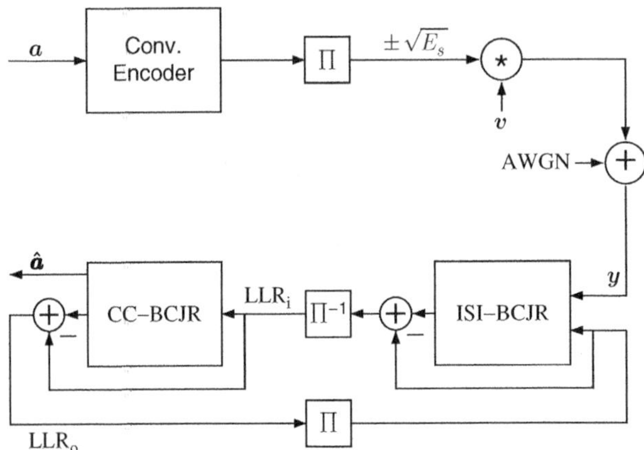

Figure 2.3 A standard picture of convolutionally coded FTN, with iterative decoding. CC denotes convolutional code; BCJR denotes the soft decoder algorithm defined in Section 2.2.1

receiver—in this application called a *turbo equalizer*—first proposed by Douillard *et al.* [8]. There are two soft-output trellis decoders. One, the ISI decoder, attempts to remove the ISI, and the other, a standard soft convolutional decoder, tries to detect which codeword was sent, and coincidentally, the data. Each produces soft estimates of the symbols in the form of a log-likelihood ratio (LLR) of each modulator symbol, defined for the binary case by

$$\text{LLR}(u_n) \triangleq \ln \frac{P[u_n = +1|\vec{y}, \text{LLR}_\text{in}]}{P[u_n = -1|\vec{y}, \text{LLR}_\text{in}]} \tag{2.5}$$

where LLR_in is an observed apriori LLR and $\vec{y}$ is a channel observation. Note that when the conditional probability of u_n is not near $1/2$, the sign of $\text{LLR}(u_n)$ gives the favored symbol and the magnitude gives approximately its log error probability). Each decoder gives its computed LLRs to the other decoder, in a series of iterations that hopefully converges to an accurate estimate of the data.

The figure shows coded FTN. Removing the CC encoder and CC decoder boxes creates an uncoded system, one with only an FTN modulator (the $\vec{v}$ convolution) and demodulator (the ISI decoder). The uncoded system does not need an iterative decoder. It is wise to define carefully here what is meant by "coding" and "modulation". A *modulator* accepts all possible symbol sequences—in uncoded FTN they are directly the data—and produces all possible outputs $s(t)$. A *coded* system selects some but not all of these to form a signal codeword set—this is performed by the convolutional encoder—and the receiver finds the closest signal in the set to what comes in from the channel. It was Shannon's discovery that this strategy can in principle achieve channel capacity.

We note that the distinction between modulator and code is not that modulators have uncorrelated channel outputs. FTN modulator outputs are correlated and the demodulator needs in general to be a trellis decoder.

In this chapter we will assume that a discrete-time model of the channel is available as some $\vec{v}$, subject to the constraint that the continuous-time signals have PSD $|H(f)|^2$. The input to the receiver is $y_n = \{u_n\} * \{v_n\} + \eta_n$, where η_n is AWGN. There are a number of ways to extract such a model, which are reviewed in papers by Anderson et al. and Prlja and Anderson [2, 9]. An all-pass filter $B(z)$ can be applied to y_n, since these have the properties that the noise remains white and the ultimate error performance is unaffected. They have many uses, the most important of which is to make the final model be minimum phase.[1] The final white noise model as a z-transform is thus $V(z)B(z)$, but we will henceforth simply denote it as $V(z)$.

This $V(z)$ defines a certain ISI acting on the sequence u_n. Equalization of such ISI is an ancient subject. Simple ISIs can often be removed without damage to the detector error rate; removal becomes harder as the zeros of $V(z)$ approach the unit circle, eventually requiring a detector working on a trellis ISI model to obtain reasonable performance. Useful FTN adds another order of difficulty because bandwidth is strongly restricted, with not only zeros but *regions* of zeros on the unit circle. Such a model must have infinite response time, which is of course truncated here, but only after considerable length. The ISI removal must be a reduced-trellis method of some kind, as discussed further in Section 2.2.1.

2.1.2 The Shannon Limit for FTN

In 1949 Shannon gave the capacity of the AWGN channel as

$$C_{\text{sq}} = W \log_2[1 + P/(WN_0)] \quad \text{b/s} \tag{2.6}$$

where the total power is P and the noise PSD is $N_0/2$ on a W-size piece of bandwidth $[-f_o - W, -f_o] \cup [f_o, f_o + W]$ Hz. Only signals made from sinc pulses have this square PSD, as denoted by the subscript 'sq'. This formula is the standard one in most textbooks. It can be extended to an arbitrary PSD $|H(f)|^2$ as follows. Consider a small bandwidth of size df. The power in df is $2P|H(f)|^2 df$, where $|H(f)|^2$ is unit-integral on $(-\infty, \infty)$; WN_0 in Eq. (2.6) becomes $N_0\, df$ and the capacity is $df\ \log_2[1 + \frac{2P}{N_0}|H(f)|^2]$. Integrating yields the capacity for the PSD

$$C_{\text{PSD}} = \int_0^\infty \log_2\left[1 + \frac{2P}{N_0}|H(f)|^2\right] df \quad \text{b/s} \tag{2.7}$$

Some analysis shows that if a square PSD and one that only obeys the orthogonal-pulse antisymmetry condition both have the same P and 3-dB bandwidth, then the capacity in Eq. (2.7) is *always larger* than the capacity in Eq. (2.6). This is a statement that more capacity lies in the extension of the spectrum beyond $1/2T$ Hz than what is lost in the reduction before. This "stopband" capacity becomes significant as power grows because the log in Eq. (2.7) magnifies its proportion.

It remains to remove the effect of scaling bandwidth; this will express capacity in b/Hz-s instead of b/s. Let the PSD $P|H(f)|^2$ in Eq. (2.7) be scaled wider by some factor μ. The total power P becomes μP and the integral in Eq. (2.7) becomes μC_{PSD}. The transmitted energy per data bit E_b is power divided by bit rate, which is P/C_{PSD} in either case. That is, we can think of capacity as per unit of bandwidth and E_b will be unaffected, provided that power is scaled along with bandwidth.

With this argument and with the PSD in Eq. (2.7) taken as a standard one with measure 1 Hz, we can take the dimensions in Eq. (2.7) to be b/Hz-s. Technically, this is called a constrained

[1] This is actually done by making it maximum phase and then reversing the sequence at the receiver input.

capacity, the limit to reliable communication constrained to signals with the given PSD shape. To obtain a useful limit in the pictures that follow, we modify the relationship in two more ways:

1. C_{PSD} is set to the b/Hz-s of the practical system under scrutiny, Eq. (2.7) is solved backwards for P/N_0, and E_b/N_0 for C_{PSD} is then $P/N_0 C_{\text{PSD}}$.
2. using a technique of Shannon the calculation is limited to signaling that has a certain data bit error BER.

The result is the limit expressed in a very practical form, as BER versus E_b/N_0. More details of the calculation appear in Anderson *et al.* [2].

A body of work exists on FTN Shannon capacity (see [2, 10] and references therein). From this we can take two striking results. First, codes made up from orthogonal-pulse signals in Eq. (2.1) and detected via this property are *limited to the smaller capacity* of Eq. (2.6). These include almost all codes now in practical application. The proof of this is brief but subtle. If codeword symbols u_n are converted to signals $s(t)$ by sinc pulses in Eq. (2.1) with $\tau = 1$, it is clear that they are subject to capacity per Eq. (2.6). At the receiver, noisy codeword symbols are optimally extracted by a sampler and filter matched to the sinc pulse. If instead wider-band orthogonal pulses are employed, and the codeword symbols extracted in the same way, the decoder performance cannot differ because the noisy symbols are statistically identical. This must be true for any orthogonal pulse $h(t)$. In Section 2.2 we will see that the gap in these two Shannon limits can be wide indeed.

A second interesting result is that codes assembled from FTN signals of the type of Eq. (2.1) with $\tau < 1$ can in principle reach the higher rate of Eq. (2.7) for most h [10]. Furthermore, they can reach most of the way with binary u_n. FTN and non-orthogonal pulses thus open the way to a more favorable Shannon limit. While the higher capacity does not display concrete better codes, it makes reasonable a number of results in the research literature where signals based on non-orthogonal pulses have shown attractive performance.

2.1.3 Summary

FTN can be defined in terms of spacing pulses closer in time so that they are no longer orthogonal. In Section 2.3 we will extend this idea to spacing subcarriers closer in frequency, so that they are again no longer independent of each other. On a deeper level, FTN is about the performance of coded signals made up from non-orthogonal elements. Bandwidth and spectrum shape play a fundamental role in FTN signaling, and they need to be carefully considered. Shannon theory and detection theory have interesting and different things to say about FTN signaling. The latter computes minimum distance and from that estimates the error probability; it allows us to compare practical coded FTN schemes and find a good one. Shannon theory shows that FTN signals usually have a better Shannon limit. In a practical bandwidth-efficient application, the limit is several dB better than Nyquist signaling in terms of SNR.

2.2 Time FTN: Receivers and Performance

In this section we look at standard iterative receivers for coded and uncoded signals of the type defined by Eq. (2.1); that is, signals based on pulses accelerated in time. Frequency FTN, where subcarriers are moved closer in frequency, follows in Section 2.3. First, Section 2.2.1 discusses the basic building block, the BCJR algorithm. Two of these combine to form an iterative receiver for coded FTN. This is applied first to signals with binary symbols u_n in Section 2.2.2 and then to 4-level symbols in Section 2.2.3. Coded FTN systems perform within a few dB of the FTN Shannon limit. Results are given for 4, 5.33 and 6 b/Hz-s. The last has six times the bit density of standard rate 1/2 convolutionally-coded QPSK.

2.2.1 *The BCJR Algorithm and Iterative Decoding*

Named after its four inventors [11], the BCJR algorithm computes log-likelihoods of transitions along paths of a trellis signal structure such as a convolutional code or the ISI created by $\vec{v}$. The algorithm is based on the earlier Baum–Petrie algorithm for identifying Markov models. The BCJR itself does not make decisions, but decisions on data or codeword symbols or on individual trellis transitions can be made by observing the LLRs. In brief, the algorithm observes channel outputs, in our case elements y_n of the sequence $\vec{y}$, and any a priori likelihoods that happen to be available, one set for each trellis stage. By means of simple linear recursions, it computes two working variables, a set of row vectors $\vec{\alpha}_n$ in a forward recursion and a set of column vectors $\vec{\beta}_n$ in a backward recursion. From these the desired likelihoods can finally be computed.

The vectors $\vec{\alpha}_n$ and $\vec{\beta}_n$ list values at stage n for each trellis state. The recursions are vector recursions of the form vector = vector × matrix:

$$\begin{aligned}\vec{\alpha}_n &= \vec{\alpha}_{n-1}\,\vec{\Gamma}_n \\ \vec{\beta}_{n-1} &= \vec{\Gamma}_n\,\vec{\beta}_n\end{aligned} \tag{2.8}$$

where $\vec{\Gamma}_n$ is a different matrix at each stage n, whose (i, j) element has the form

$$\Gamma_n(i,j) = P_{\text{apr}}[\sigma] \times \exp\left(-(w-y)^2\right) \tag{2.9}$$

This is the probability that channel output y_n and the state j at stage n occur, given that the encoder or ISI was in state i at stage $n-1$. Here σ is a modulator or convolutional encoder symbol and from the input LLR comes its a priori probability $P_{\text{apr}}[\sigma]$; w is the trellis branch symbol at n; the entire right-hand factor is the probability of y_n over an AWGN channel; if there is no observation the factor is a constant. The values in the $\vec{\alpha}_n$, $\vec{\beta}_n$ and $\vec{\Gamma}_n$ are combined in various ways to form the desired outputs of the BCJR block.

The algorithm thus handles many variables, but it is otherwise simple and fast. Note that although the algorithm is traditionally shown with LLR inputs and outputs, its recursions use the corresponding probabilities. More details of BCJR operation can be found elsewhere [6, 7].

With reference now to Figure 2.3, the BCJR labeled CC decoder directly decodes the convolutional codewords in coded FTN signaling. It actually performs two functions: likelihoods are produced for the codeword symbols u_n, which are passed to the other BCJR as its a priori information, and in a separate calculation the likelihoods of the data symbols a_n that drive the codeword symbols are produced. The CC BCJR does not see the channel outputs.

The ISI BCJR calculates likelihoods of u_n from observations of the channel outputs and the prior likelihoods from the CC BCJR. Its effect is to remove some of the ISI. The new likelihoods are passed to the CC BCJR, which produces improved likelihoods for the ISI BCJR, and so on in a series of iterations that hopefully converge to a solution. While the ISI BCJR operation is in principle the same as the CC BCJR, bandwidth-efficient FTN signals have a very large ISI trellis model, and some way must be found to simplify the calculation. In the reduced-trellis or channel-shortening approach, the full ISI trellis is approximated by a simpler trellis (a shorter $\vec{v}$) that hopefully is good enough to allow convergence of the iterations. In the reduced-search approach, the full trellis is reachable but the calculations are performed only on the small part that has significant values. This is a kind of sparse matrix calculation of the recursions. In a third, single-BCJR approach called successive interference cancelation, an estimate of the ISI is removed before the CC BCJR rather than passing around likelihoods. Some other approaches are found in Anderson *et al.* [2].

The time-FTN performances in this section were obtained by the two-BCJR structure in Figure 2.3. The ISI BCJR is a reduced-search method called the M-BCJR that works by computing the dominant M elements of the vector $\vec{\alpha}_n$ at each stage n. The outcome is a list of M "alpha paths" through a trellis of alpha values. It is crucial that the channel model be minimum phase, since otherwise it is less clear which alpha paths should be kept or dropped along the way. Some method must be devised to direct the calculation of the $\vec{\beta}_n$, which sees a maximum-phase model and is easily misdirected. A number of M-BCJR procedures are in the literature and are summarized in Prlja and Anderson [9].

The progress of the iterations can be nicely sketched with an EXIT chart that plots the input versus output of the two BCJRs.[2] A canonical example appears in Figure 2.4. There is a line for the CC BCJR (dashed) called the *CC characteristic*; it plots input error rate in u_n versus

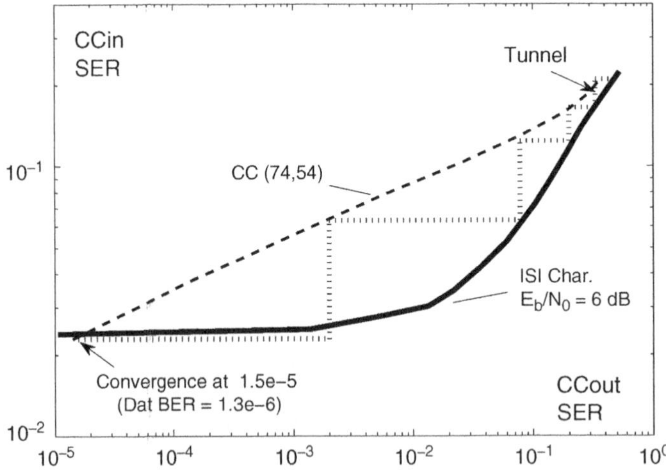

Figure 2.4 A typical codeword letter error rate EXIT plot. This shows error before the soft convolutional decoder plotted against error after it at $E_b/N_0 = 6$ dB. The upper straight line represents code (74,54); the lower curved characteristic is created by $\tau = 0.35$. Ten blocks of 300 k data bits closely follow the dotted trajectory

[2] EXIT, meaning "extrinsic information transfer", is the name given by ten Brink to plots of this type; his original method plotted mutual informations of symbols, whereas we will plot probabilities.

output error rate for the CC BCJR. Another line (solid) is the *ISI characteristic*; it plots output error rate in u_n from the ISI BCJR versus input rate (i.e. at the CC BCJR output). Iterations begin at the upper right in a region near where the two characteristics lie, called the *tunnel*. Iterations stop when the two characteristics touch, which is hopefully at the lower left; if the characteristics touch in the tunnel, decoding cannot get started and will likely fail. If the FTN block is relatively short, the EXIT outcome will vary somewhat from the characteristics, but as the length grows it closely follows them. The heavy lines locate where 10 300 k-long blocks lay in one receiver test.

EXIT charts are an important tool in turbo coding analysis. They originally plotted mutual information, but codeword error rates are better here for several reasons:

- the u_n variable is the means of communication between the BCJRs
- the tunnel is clearly shown
- the two characteristics are defined by straight line asymptotes that are relatively easy to derive.

When the receiver has sufficient resources the CC and ISI characteristics are a function respectively of the convolutional code and the channel model $\vec{v}$ alone. The CC straight line parameters for a number of codes have been published elsewhere [12]. The horizontal asymptote of the ISI characteristic lies approximately at the ISI BCJR output error rate for an AWGN channel with the same E_s/N_0 as the ISI channel. In Figure 2.4, $E_s/N_0 = (1/2)E_b/N_0 = 3$ dB, which yields an error rate $\approx Q(\sqrt{2 \quad 10^{0.3}}) = 0.023$.

Overall, the dynamics of the Figure 2.3 receiver are such that if E_b/N_0 is large enough, the tunnel is open, the iterations converge, and the data error performance is that of the convolutional code at E_b/N_0, since the ISI is effectively removed. Otherwise the tunnel is closed and the receiver is said to be *below threshold*. Clearly, it is important to keep the tunnel open. A good convolutional code for FTN needs two simultaneous properties: a CC characteristic with a small slope, so that it crosses the ISI characteristic far to the left in the figure, and an open tunnel at high error rates, so that the iterations live to achieve the crossing. In the performance plots that follow, the code-alone AWGN performance is plotted as a light dashed line, called the *CC line*. Once the error performance settles to this line, further tests are not shown since they simply follow the line.

If τ lies above the Mazo limit, only one iteration is needed because the ISI BCJR achieves full ISI removal in a single application.[3] If the FTN is not coded, detection of the u_n is performed by just one application of the ISI BCJR. There are no iterations, and error performance will lie in between coded FTN and simple PAM modulation with the same b/Hz-s.

2.2.1.1 Designing for Best Convergence

When the SNR is high, the iterative receiver converges rapidly. There can be compromises that distort the two characteristics in Figure 2.4, but they will remain separate. If the aim is low SNR performance near to the lowest threshold that can be achieved with the code and the ISI, the two BCJRs need to cooperate efficiently. How best to do this is a subtle business that has

[3] The error rate achieved is that of ISI detection, which has the right exponential form but is several times worse than the form for simple 2PAM demodulation. A second iteration achieves the 2PAM form.

been the subject of much research. One principle has been clear from the beginning: With each BCJR there must be "intrinsic subtraction" of the input apriori LLR of u_n from the BCJR's output LLR for the same u_n. This is shown as an outright subtraction of LLRs in Figure 2.3. There are heuristics to justify this simple subtraction, and we use it in Section 2.2.2, although it does fail in the 4-ary BCJR case, as we will take up in Section 2.2.3. Regardless of heuristics, the practical matter is that the two BCJRs must agree on a solution that seems suboptimal to each of them alone. Failure to subtract the input from the output LLR causes one BCJR to dominate. In all but the last iteration, the present BCJRs output for u_n should come more from the prior LLRs for $u_\ell, \ell \neq n$, the so-called "extrinsic information" about u_n.

Aside from intrinsic subtraction, some other principles of BCJR cooperation are described next. If these are not applied a higher threshold will occur and the results to follow will not be achieved.

- Were instrinsic subtraction not applied, the output of each BCJR would be the maximum a posteriori estimate of u_n, an unimpeachable estimate; with subtraction, the two LLRs are strongly damaged, each in a different way, which opens the possibility that mappings or scalings of the outputs will improve them.
- A useful heuristic for scaling is to scale the input likelihoods so that their statistics are those of an AWGN channel. The motivation for this is that AWGN LLRs are Gaussian and the BCJR LLRs after subtraction are nearly perfectly so. The interleavers make successive LLRs also independent, so that each BCJR sees nearly an AWGN channel.
- The LLRs are not very Gaussian in the first iterations. In addition, the CC BCJR performs much more poorly then than the ISI BCJR (in fact it worsens the u_n error rate). The early CC BCJR LLRs should therefore be somewhat ignored (scaled down in absolute value).
- As the quality of the input LLRs into the ISI BCJR improves, the M in the M-BCJR algorithm can be much reduced, eventually to 4–8.
- Precision easily destroys performance. The heart of the difficulty is that probabilities $1-\epsilon$ too near to 1 cannot be well enough expressed in an ordinary number system (for an exposition on this computer science problem see the article by Hayes [13]). One solution is to force the use of logs, the "max log" method, but this has a loss. A better way is to carry along both $1-\epsilon$ and ϵ in calculations, which is equivalent to carrying two likelihoods rather than one likelihood ratio.

2.2.2 *Binary Coded FTN Performance*

We turn now to data bit error versus SNR for convolutionally coded FTN at 2 and 4 bits per Hz-s. For a rate r code and FTN spacing τ, this density is $2r/\tau$. There is thus a τ versus r tradeoff; 4 b/Hz-s may be obtained via $\tau = 1/4, 1/3, 3/8$ with rates $r = 1/2, 2/3, 3/4$ respectively. Convolutional decoding becomes somewhat more difficult as the rate grows, but ISI BCJR implementation becomes much easier as τ grows. The best trade seems to lie near code rate 2/3.

In the following, all convolutional codes are feed-forward; there is no evidence that feedback systematic codes perform better. Rate-1/2 convolutional codes are defined by two left-justified octals (g_1, g_2), where g_1 is the shift register taps that produce the first of each bit pair and g_2 produces the second. Rate-2/3 codes are defined by six octals in the Matlab-style matrix

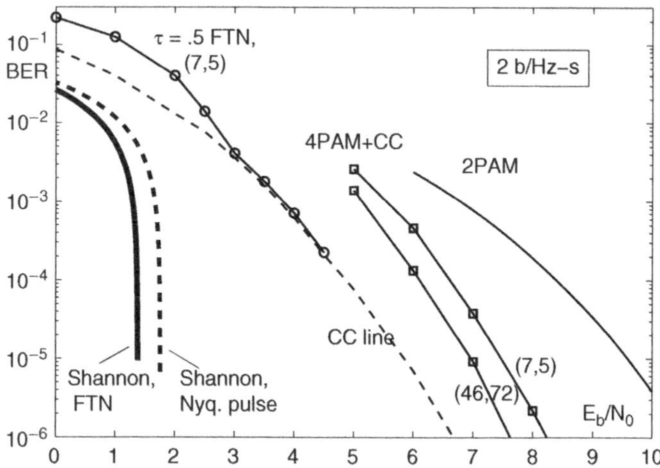

Figure 2.5 BER performance of coded FTN ($r = 1/2, \tau = 1/2$), ordinary $r = 1/2$ convolutional coding with 4PAM modulation, and 2PAM modulation, compared to FTN and Nyquist pulse Shannon limits. All systems run at 2 b/Hz-s with 30% RC spectrum

$[g_{11}g_{12}g_{13}; g_{21}g_{22}g_{23}]$. Here g_{ij} specifies the taps in the shift register that computes the contribution to codeword bit j (of three) from data stream i (of two). For example, octal 54 with memory $m = 3$ means taps 1011, with the present-bit tap leftmost.

Figure 2.5 portrays the overall situation at 2 b/Hz-s. On the left are the Shannon limits for this density, both the one that applies to FTN and to the right of it the traditional limit that applies when the detection uses pulse orthogonality. The rightmost curve is simple 2PAM modulation, lying about 9 dB from the Shannon limit at BER 10^{-6}; a 9–10 dB separation between Shannon and PAM is typical at all densities. About 2–3 dB from Shannon is the behavior of $\tau = 1/2$ combined with the (7,5) rate-1/2 code. Finally, some traditional convolutional coding is shown at about 2 dB gain from simple modulation; to maintain 2 b/Hz-s with rate-1/2 coding, 4PAM modulation is required, together with proper Gray coding.

The two Shannon limits are not much separated at such a low bit density, and Figure 2.5 is shown primarily to compare to older signaling methods. Nonetheless, there are good SNR gains to be had with FTN compared with older methods of similar or less complexity. The M required for $\tau = 1/2$ in the ISI BCJR is about 20, declining at later iterations to about 4. The number of iterations depends on the desired location of the threshold. For the lowest possible threshold it is 20–30; if 1–2 dB more E_b/N_0 can be tolerated, it is only 3–5. In other words, computation rises as we approach capacity, which is no surprise.

Figure 2.6 shows 4 b/Hz-s. The simple modulation is now 4PAM, and it again lies 9–10 dB above the Shannon limit, but there is now 1.2 dB between the FTN and orthogonal-pulse Shannon limits. Uncoded FTN at $\tau = 1/2$ is shown. Test outcomes and CC lines are shown for several coded FTNs with convolutional rate-1/2 and 2/3 (rate-3/4 codes are slightly worse than rate-2/3). Near threshold, the $\tau = 1/4$ case ($r = 1/2$) requires block length above 100 k and M is in the 200–400 range. This is a penalty for too small τ; the ISI has an effective length of 20–25. The $\tau = 1/3$ case ($r = 2/3$) needs M in the 40–100 range, dropping to 4–8 as the

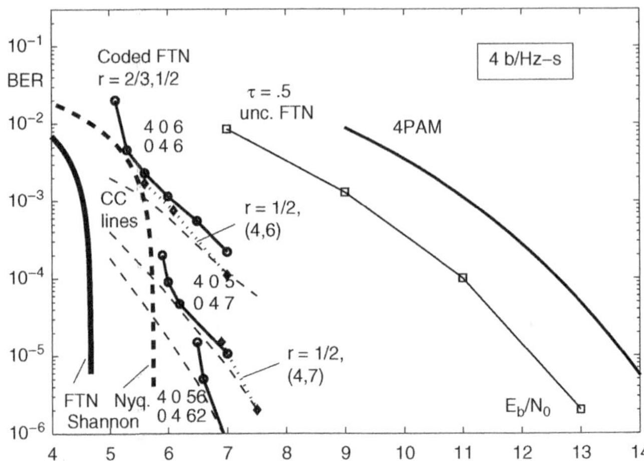

Figure 2.6 BER performance of coded FTN ($r = 1/2, \tau = 1/4 \ldots$ and $r = 2/3, \tau = 1/3$), uncoded FTN ($\tau = 1/2$), and 4PAM modulation, compared to FTN and Nyquist pulse Shannon limits. All systems run at 4 b/Hz-s with 30% RC spectrum

iterations progress. If E_b/N_0 is set to 1 dB above the E_b/N_0 that corresponds to the lowest possible threshold, the early-iteration M is much smaller in both cases.

2.2.3 *Four-level Coded FTN*

We turn now to FTN built upon 4-ary modulation. That is, symbols u_n, encoded or not, feed a 4PAM pulse modulator, whose pulse stream is accelerated by τ as in Eq. (2.1). The u_n alphabet is now the unit-average-energy set $(1/\sqrt{5})\{3, 1, -1, -3\}$. The ISI model is the same $\vec{v}$ as before. The bit density of such a system is $4r/\tau$ b/Hz-s, where the code (if any) has rate r data bits in per codeword bit out.

At 4 data bits/Hz-s, the systems in this section have about the same BER versus E_b/N_0 as the 2-ary modulation systems, but the detection is considerably simpler, primarily because the larger τ means that the ISI BCJR is simpler. Transmission for 5G needs higher bit density, and 4-ary FTN opens the way; the section explores 5.33- and 6-b/Hz-s systems, which are impractical with 2-level modulation but perform quite well with 4-level.

In principle the implementation of 4- and 2-ary FTN are the same, but certain aspects need a significant upgrade. There are now four likelihoods for each n, whose form is

$$\mathrm{LLR}(w, u_n) \triangleq \ln \frac{P[u_n = w|\vec{y}, \mathrm{LLR_{in}}]}{P[u_n \neq w|\vec{y}, \mathrm{LLR_{in}}]} \qquad w = 3, 1, -1, -3 \tag{2.10}$$

The 4-ary ISI BCJR thus takes in and puts out a $4 \times N$ array of LLRs. As before, it actually works with the corresponding probabilities.

With 4-ary LLRs, simple subtraction of LLRs after the CC BCJR and ISI BCJR (as in Figure 2.3) no longer functions. To see how to replace it, we need to consider the step of the BCJR that follows completion of the forward and backward recursions for $\vec{\alpha}_n$ and $\vec{\beta}_n$.

For either the CC or ISI trellis, this step is the calculation of

$$P[i \to j] = \vec{\alpha}_{n-1}(i)\ \Gamma_n(i,j)\ \vec{\beta}_n(j) \tag{2.11}$$

in which $P[i \to j]$ is the probability that trellis state i at trellis stage $n-1$, state j at n, and all the channel outputs $\vec{y}$ simultaneously occur. Here $\Gamma_n(i,j)$ was defined in Eq. (2.9). Individual probabilities of this form are combined to form the outputs of the BCJR block.

To prevent domination by one BCJR—the idea of "intrinsic subtraction"—the P_{apr} factor in Eq. (2.9) needs to be set to a constant; that is, input LLRs are used to find the overall set of $\vec{\alpha}_n$ and $\vec{\beta}_n$ but not in the subsequent individual LLR computations. In the 2-ary case of Section 2.2.2, simple subtraction gives a similar outcome most of the time. In addition, it is equivalent to decorrelating the output and input LLRs. Neither simple subtraction nor decorrelation work well in the 4-ary case.

Another 4-ary subtlety is the nature of the binary code. It remains convolutional in this section. Its input remains the data bits, but the encoder output must reach the modulator as the symbols $\{3, 1, -1, -3\}$. This can be done by a Gray map, but a deeper problem occurs in the CC BCJR decoder: the usual CC BCJR would take in and put out binary LLRs, while the iterative decoder now passes around 4-ary likelihood information. A careful investigation shows that the incoming 4 : 2 and outgoing 2 : 4 LLR conversions lead to a loss, especially when the LLRs are of poor quality. It is better to finesse the problem by expressing the codewords directly in 4-ary symbols. This is easily done. The rate-1/2 code, for example, has bit pairs 00,01,10,11 on its trellis branches, and these can be Gray mapped to symbols $3, 1, -1, -3$. This may seem a drastic move but it is actually not so. The convolutional encoder in either case creates a pseudorandom trellis of codewords whose words have a distance structure and a minimum distance. In either case one optimizes over the shift-register encoder taps, and in the 4-ary case also a Gray map. With rate-2/3 coding, bits come out of the shift registers three at a time, but two applications create six pseudorandom bits (this is actually rate-4/6 coding), which with another fixed map become three pseudorandom 4-ary symbols.

With minimum Euclidean distance as a measure, good error-correcting codes of this type can be found by computer search. Actually, a better method is to evaluate the candidates in short FTN tests, since this can also test for an open tunnel (see Anderson and Zeinali [12] for a 2-ary campaign). In our FTN tests, the resulting codes have about the same good-SNR error performance as traditional binary convolutional codes, but better tunnel characteristics; that is, they better handle the early turbo iterations, with their very poor LLRs.

Some smaller issues also arise in 4-level FTN.

- Four-level modulation implies a much wider LLR range. Energies are typically 3 dB higher, which is in turn exponentiated in the right-hand factor of Eq. (2.9). Consequently, all four symbol probabilities need to be carried along in calculations and care taken with precision.
- The M-BCJR algorithm operation suffers from the LLR range as well: in the reverse (β) recursion the maximum phase encountered there can delete all descendants of the forward recursion before useful calculation can occur in Eq. (2.11). In the tests shown here, the reverse calculation is not "free ranging"; it is performed on just those paths that survive the forward recursion and no others.
- Efficient M-BCJR implementations use look-up tables, but a 4-ary ISI trellis with memory m has 4^m states, which is inconveniently large. It is easier to represent the states in some two-variable manner, employ two tables, and add the results.

2.2.3.1 Test Results

Figure 2.7 shows the outcome of tests at 5.33 and 6 b/Hz-s. The plots share an x-axis so that the behavior as bit density grows is more evident. The convolutional code rate is 2/3, which implies $\tau = 1/2$ and $4/9$; the code used was found by a brief search of short codes and Gray maps. No rate-1/2 codes could be found that function reasonably well at these high densities, because the required $\tau = 2/5$ and $1/3$ lead to ISI characteristics that close the EXIT tunnels. It appears that the higher code rate is essential at these high bit densities.

The energies, CC lines and Shannon limits align in such a way that the decoder either lies below threshold and fails, or it runs at a very low BER: 10^{-8}–10^{-12} in the 6-b/Hz-s case. These are attractive BERs in applications, but they make it difficult to obtain meaningful test results. Instead, the observed E_b/N_0 at threshold is marked with a triangle, and the BER is assumed to follow the CC line (whose lower regions can be estimated from code distance properties). The Shannon FTN and Nyquist pulse limits are now more widely separated (2.2 dB at 6 b/Hz-s) and the FTN limit lies some 12 dB from 8PAM performance at useful BER. Coded FTN gains 8–10 dB of this. At the relatively low FTN decoder complexity, this is an attractive performance.

To perform near threshold these schemes require block length 50 k–100 k data bits and M in the 100–300 range, dropping to 5–20 in later iterations. By giving up a dB in SNR these resources can be much reduced, and only 5–10 iterations are needed.

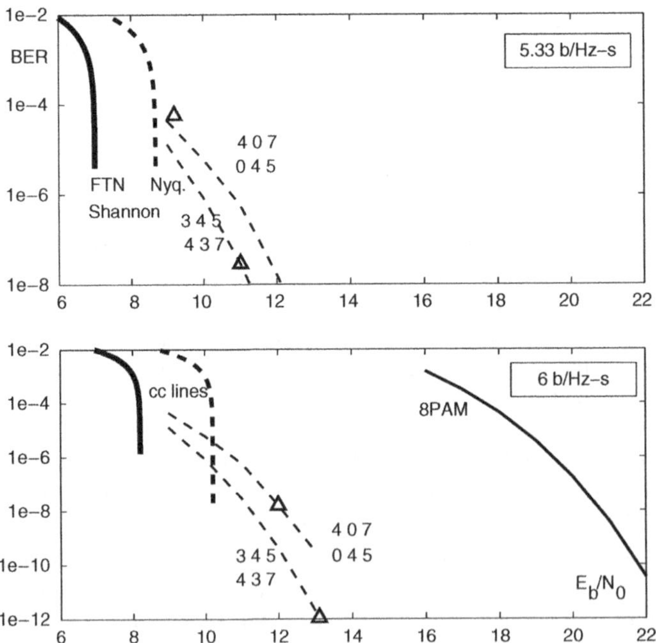

Figure 2.7 BER performance of coded FTN with comparison to 8PAM modulation and FTN and Nyquist pulse Shannon limits. Upper plot: at 5.33 b/Hz-s, $r = 2/3, \tau = 1/2$; lower plot: at 6 b/Hz-s, $r = 2/3, \tau = 4/9$. Triangle symbols are approximate thresholds; above that E_b/N_0 BER follows the CC line

2.2.4 Summary

FTN signaling based on either 2- or 4-ary modulation provides attractive gains over simple modulation or modulation plus binary coding, gains that grow as more storage and computing resources are applied. The gains are more evident as the bit density increases and reach the 6–8 dB range over simple modulation for 5–6-b/Hz-s systems. This high-rate region is of prime interest for 5G. The region is much easier to reach with 4-ary modulation-based FTN, and coding gains are better, so we can conclude that 4-ary modulation is important for 5G.

At the moment no method other than iterative decoding has been applied to coded FTN. The complexity and storage required depends on how close to the lowest possible threshold—and to capacity—one wishes to operate. Lowest-threshold operation requires block length, M-BCJR size and turbo iterations of the order of 100 k, 200 and 30. These rapidly diminish as one moves away from the ultimate SNR performance for the scheme, and the best application of FTN is probably at 1 dB or so above that SNR.

2.3 Frequency FTN Signaling

Until now the focus has been accelerating pulses in time, but it is also possible to squeeze subcarrier transmissions together in frequency. This technique connects naturally to 4G and 5G systems, which emphasize OFDM transmission. Now, however, the ‘O’ no longer holds, and the subcarriers will interfere with each other in a significant way. This *frequency FTN* is similar in principle to time FTN, although the signaling, its analysis and the detection are more complicated. The existence of multiple carriers implies that *phase* among the subcarriers matters; error events are more complex and the same event beginning at different times leads to different distances between signals. The C_{FTN} Shannon-limit bonus in time FTN is no longer significant because the subcarriers occupy approximately a square block of spectrum. Nonetheless, bit density grows for the same E_b/N_0; before less time was consumed, now it is less bandwidth. In addition, research has shown that gains from frequency FTN are *different from those of time FTN*. Applying both techniques at once can lead to a doubling of bit density.

Frequency FTN is not as well explored as time FTN, but chips have been constructed that work with OFDM-like signals (see Section 2.3.2). The approach described here began in papers by Rusek and Anderson [14, 15], but many non-orthogonal frequency approaches have been proposed in recent years and are cited in the survey by Anderson *et al.* [2]. These have in common that something can be gained by abandoning independent subcarriers.

2.3.1 Definition of Frequency FTN

Before proceeding, it will be useful to define a benchmark system. This will make time–frequency FTN clearer and will relate it to OFDM. The symbol time in each subcarrier is τT as before. The time acceleration itself does not change the subcarrier bandwidth. With frequency FTN, we can define a squeeze factor ϕ, $0 < \phi \leq 1$, and take the new subcarrier spacing as $f_\Delta = \phi/T$ Hz. For the time being, let $h(t)$ in Eq. (2.1) be the sinc pulse $\sqrt{1/T}\text{sinc}(t/T)$, and modulate K of these length-N binary pulse trains onto subcarriers whose frequencies are $f_0 + kf_\Delta$, $k = 0, 1, \ldots, K - 1$. The combined FTN can be thought of as a two-dimensional array of points as in Figure 2.8, with a pulse and data symbol

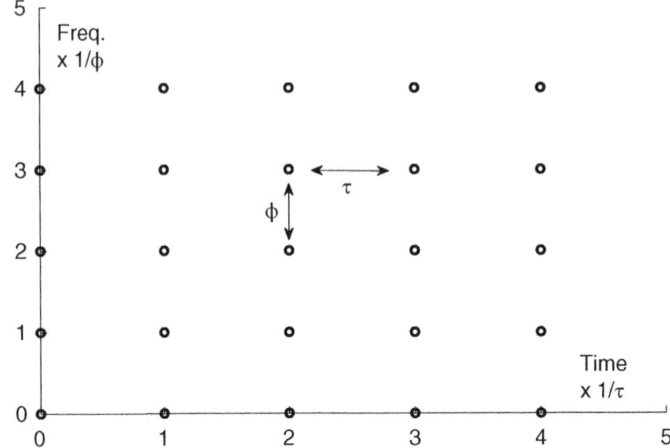

Figure 2.8 Two dimensional FTN framework. One pulse corresponds to each point

corresponding to each point. Each subcarrier has bandwidth $1/T$ Hz, for a total K/T Hz. The in-phase and quadrature signals $I(t)$ and $Q(t)$ carry two transmissions, so that for large N the bits per Hz-s are $\approx 2KN/[(NT)(K/T)] = 2$. The symbol time T cancels, which is a statement that a simple trade of time for bandwidth does not change bit density. We may as well set $T = 1$. Without the FTN squeezes, each pulse can be detected independently. Under time and frequency FTN, they cannot, and the density in a binary system becomes $2/\phi\tau$ b/Hz-s.

If $h(t)$ is orthogonal ($\tau = 1$) but not a sinc, there will be co-channel interference between subcarriers spaced $1/T$ Hz, but this is generally an insignificant effect. It makes sense, therefore, to keep the array pattern in Figure 2.8 as a benchmark.

Turning now to a definition of FTN with K subcarriers, we can define the complete signal as

$$\sqrt{2E_s}[I(t)\cos 2\pi f_0 - Q(t)\sin 2\pi f_0]$$

where

$$I(t) = \sum_{k=0}^{K-1}\sum_{n=0}^{N-1}[a_{k,n}h(t-nT)\cos 2\pi f_k t - b_{k,n}h(t-nT)\sin 2\pi f_k t]$$

$$Q(t) = \sum_{k=0}^{K-1}\sum_{n=0}^{N-1}[b_{k,n}h(t-nT)\cos 2\pi f_k t + a_{k,n}h(t-nT)\sin 2\pi f_k t] \tag{2.12}$$

Here a and b are the symbols associated with the twofold nature of the signal. The signals $\cos 2\pi f_k t$ and $\sin 2\pi f_k t$ are subcarriers at frequencies $f_0 + k\phi/T$, $k = 0, \ldots, K-1$, and f_0 is an overall transmission carrier frequency. The normalized square Euclidean distance between two signals $s^{(1)}(t)$ and $s^{(2)}(t)$ is still $d^2 = (1/2E_b)\int |s^{(1)}(t) - s^{(2)}(t)|^2 dt$. Only the difference in the I and Q matters, and that depends only on the a, b differences. The distance may be written as

$$(1/2)\int[|\Delta I(t)|^2 + |\Delta Q(t)|^2]dt \tag{2.13}$$

as $f_0 \to \infty$; $\Delta I(t)$ and $\Delta Q(t)$ are as I and Q in Eq. (2.12) but with $\Delta a_{k,n}$ and $\Delta b_{k,n}$ instead of $a_{k,n}$ and $b_{k,n}$.

It can be challenging to visualize time–frequency FTN signals. Figure 2.9 shows I and Q for a signal made up of two subcarriers, with $\phi = 0.635$ and $\tau = 1$. Subcarrier 0 at the top carries symbols $a_{0,0} = -1, a_{0,1} = 0, b_{0,0} = 0$ and $b_{0,1} = -1$. The 0s are not meant to be data but are inserted to make the h pulse more clear. Subcarrier 1 at the middle carries $a_{1,0} = 1, a_{1,1} = -1$, $b_{1,0} = 1$ and $b_{1,1} = -1$. Because of phase offsets, the subcarrier 1 symbols are not at all clear in their I and Q contribution. Similarly, the total $I(t)$ and $Q(t)$ signals bear no easy relation to the 8 symbols.

Finding minimum distances for time–frequency FTN is less straightforward than before. As usual, one explores difference events in a, b that begin after some nth interval. An immediate complication is that due to phase effects among the subcarriers the distance of an event now depends on n; that is, on when it starts. There will be a worst case start time. Nonetheless, one can find a Mazo limit, two-dimensional in the symbol time and subcarrier spacing, and the one for 30% root RC pulse shape appears in Figure 2.10. The axes are the τ and ϕ parameters; combinations above and to the right of the line yield square minimum distance 2, with the worst-case event start time. Contours of equal $\phi\tau$ are shown, and it appears that the least product is about 0.6, occurring near $\tau = 0.89, \phi = 0.675$. The value 0.6 can be compared to the time-only Mazo limit, which is set by $\tau = 0.703$. It shows that time and frequency FTN are at least partially independent in their bandwidth-reducing effects.

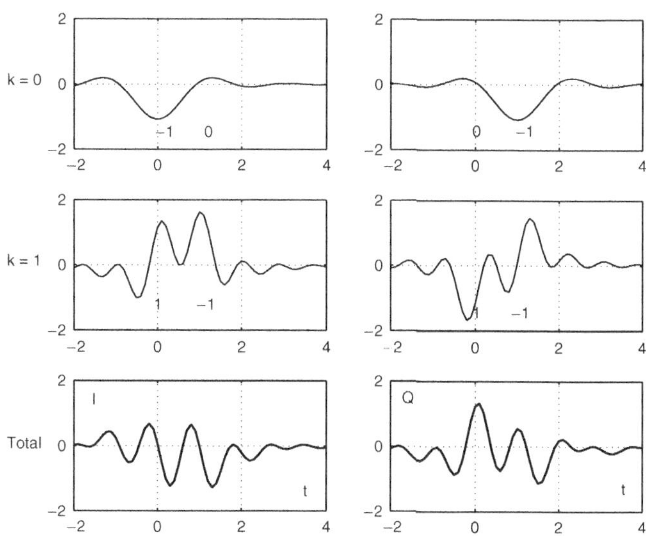

Figure 2.9 I and Q signals for frequency FTN with two subcarriers. 30% root RC pulse, $\tau = 1$, $\phi = 0.635$. Left: I; right: Q. From top to bottom: subcarrier 0, subcarrier 1 and the total signal. Start phase is 0

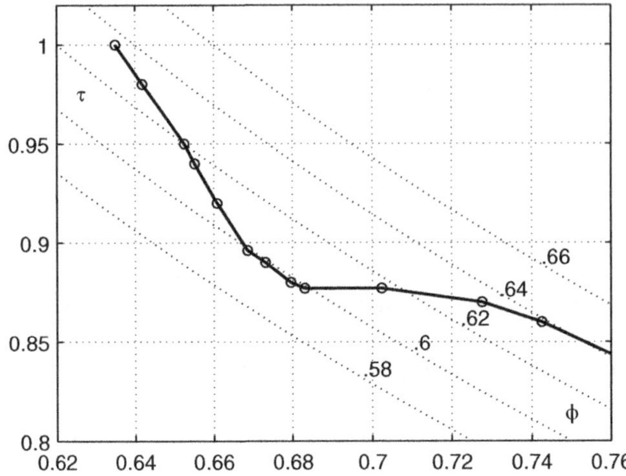

Figure 2.10 Location of 2-dimensional Mazo limit for 30% root RC binary pulse. Dotted lines show constant $\phi\tau$ product. Data per Rusek and Anderson 2009 [15]

The paper by Rusek and Anderson and its references give properties and computational details about time–frequency minimum distances [15]. One worth mentioning now is that a better Mazo limit (a smaller $\phi\tau$ product) occurs when the subcarriers have a synchronous relationship to each other and/or an optimized set of phase offsets among the subcarriers. The Mazo limit so obtained is better than that in Figure 2.10, and $\phi\tau$ approaches 0.5.

2.3.2 Implementation of Frequency FTN

Parts of FTN transmitters and receivers have been implemented in software on numerous occasions, and the outcome is reported in the reviews published on the subject [2, 9, 16] and the references therein. In 2012 a hardware chip transceiver was designed, fabricated and tested by Dasalukunte *et al.*, and this chip implementation and a number of surrounding issues are the subject of a book [16]. The object was to create a competitor for OFDM. Most of the actual hardware work focuses on $\phi = 1$—the value it would have in OFDM—with τ in the range 0.6–1. The iterative detection was a successive interference cancelation design, and the FTN was (7,5)-convolutionally coded. An improvement of 70% in bandwidth efficiency was reached over the ordinary (7,5) code + 2PAM configuration, while achieving the (7,5) CC line above 6 dB E_b/N_0. Figure 2.11 reproduces some important parts of the chip layout.

At time of writing, little is known about receivers for the much more difficult case when both ϕ and τ are in the range 0.5–0.8. It appears that such a receiver must be triply iterative, removing interference from nearby subcarriers, then from neighboring pulses in the trains and then decoding the convolutional code, all in rotation until convergence is obtained.

2.3.3 Summary

The idea of FTN time acceleration has been extended to squeezing subcarriers together in frequency. If there is time FTN but little or no subcarrier squeeze, the system is time FTN

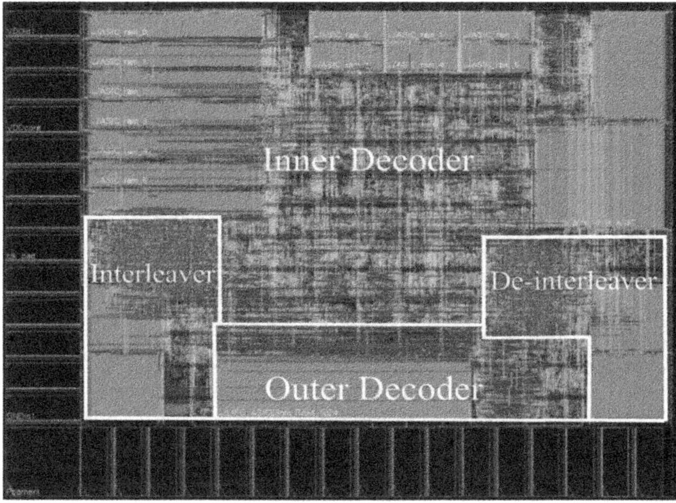

Figure 2.11 Layout of FTN chip fabricated by Dasalukunte. *Source:* Reproduced with permission of D. Dasalukunte, Lund Univ

applied to OFDM. This case is relatively straightforward to implement, and a chip has been constructed. Future challenges include extensions to ϕ significantly below 1 and to nonbinary FTN symbols.

When FTN is applied in both time and frequency to a significant degree, the distance analysis is challenging. Controlled phase shifts applied to subcarriers complicate the analysis, but also lead to better systems; that is, to higher bit density for the same SNR and error rate. The Mazo limit is now the least product $\phi\tau$ that leads to square minimum distance 2. It appears that the least product for time–frequency FTN lies near 0.5, which is a doubling of bit density for the same SNR and error rate.

2.4 Summary of the Chapter

This chapter explored the faster-than-Nyquist data transmission method, with emphasis on applications that are important for future 5G systems. These include time-FTN methods with nonbinary modulation and multi-subcarrier methods that are similar in structure to OFDM. In either, there is an acceleration in time or compacting in frequency that make signal streams no longer orthogonal. Pulses and subcarriers interfere in such a way that a trellis-type detection is required.

FTN can be combined with error-correcting coding structures to form true waveform coding schemes that work at high bits per Hertz and second. These can be far more effective than combining simple coding with modulation. There is some price to pay in complexity and only iterative detection schemes are known at this time, but these are quite efficient so long as the transmission works with an SNR that is 1 dB or so above the iterative detection threshold. We have featured a chip implementation of such a receiver for the multicarrier case.

Although the FTN subject is 40 years old, it is in recent years that many of its implications have been understood. A practical transmission system has spectral sidelobes, and for this case FTN offers a more favorable Shannon limit. The gap between this and the standard textbook limit grows rather large for the SNRs typical of higher-alphabet modulation. The FTN method makes use of non-orthogonal pulses and subcarriers, which contain intentionally created intersymbol and intersubcarrier interference. This is a departure from most earlier methods, which are built on independent pulses and subcarriers.

References

[1] Mazo J. (1975) Faster than Nyquist signaling. *Bell Syst. Tech. J.*, **54**, 1451–1462.

[2] Anderson, J.B., Rusek, F., and Öwall V. (2013) Faster-than-Nyquist signaling. *Proc. IEEE*, **101**, 1817–1830.

[3] Tüchler, M. and Singer, A.C. (2011) Turbo equalization: an overview. *IEEE Trans. Info. Theory*, **57**, 920–952.

[4] Anderson, J.B. (2005) *Digital Transmission Engineering*. IEEE Press–Wiley Interscience.

[5] Proakis, J. (2001) *Digital Communications*, 4th edn. McGraw-Hill.

[6] Anderson, J.B. and Svensson, A. (2003) *Coded Modulation Systems*, Kluwer-Plenum.

[7] Koetter, R., Singer, A.C. and Tüchler, M. (2004) Turbo equalization. *Signal Proc. Mag.*, **21**, 67–80.

[8] Douillard, C. (1995) Iterative correction of intersymbol interference: turbo equalization. *Eur. Trans. Telecomm.*, **6**, 507–511.

[9] Prlja, A. and Anderson, J.B. (2012) Reduced-complexity receivers for strongly narrowband intersymbol interference introduced by faster-than-Nyquist signaling. *IEEE Trans. Commun.*, **60**, 2591–2601.

[10] Rusek, F. and Anderson, J.B. (2009) Constrained capacities for faster than Nyquist signaling. *IEEE Trans. Info. Theory*, **55**, 764–775.

[11] Bahl, L.R., Cocke, J, Jelinek, F and Raviv, J. (1974) Optimal decoding of linear codes for minimizing symbol error rate. *IEEE Trans. Info. Theory*, **20**, 284–287.

[12] Anderson, J.B. and Zeinali, M. (2014) Analysis of best high rate convolutional codes for faster than Nyquist turbo equalization. *Proc. IEEE Int. Symp. Information Theory, Honolulu*.

[13] Hayes, B. (2009) The higher arithmetic, *American Scientist*, **97**, 364–368.

[14] Rusek, F. and Anderson, J.B. (2005) The two dimensional Mazo limit. *Proc. IEEE Int. Symp. Info. Theory, Adelaide*, pp. 970–974.

[15] Rusek, F. and Anderson, J.B. (2009) Multi-stream faster than Nyquist signaling. *IEEE Trans. Commun.*, **57**, 1329–1340.

[16] Dasalukunte, D., Rusek, F., Öwall, V., and Anderson, J.B. (2014). *Faster-than-Nyquist Signaling Transceivers: Algorithms to Silicon*, Springer.

3

From OFDM to FBMC: Principles and Comparisons

Wei Jiang and Thomas Kaiser

3.1 Introduction

With the proliferation of smart phones and tablet computers, the demand on transmission rates of wireless communication systems has grown exponentially in the early years of the twenty-first century. Although wireless local area networks can contribute significantly to

Signal Processing for 5G: Algorithms and Implementations, First Edition. Edited by Fa-Long Luo and Charlie Zhang.

offload wireless traffic, their applications are generally limited to stationary and indoor scenarios. To also provide high-data-rate access in mobile and outdoor environments, cellular systems have to use more and more spectral resources, since the rate at which spectral efficiency is increasing is relatively slow. As a result, the signal transmission bandwidths of cellular systems has become increasingly wide. Global System for Mobile Communications (GSM) [1] adopts the time-division multiple-access (TDMA) technique to support eight voice users over each 200-kHz channel. The spread-spectrum signals of wideband code-division multiple-access (WCDMA) occupy 5 MHz to satisfy the requirements of 3G systems, whose transmission rate is at least 384 kbps. To comply with the International Mobile Telecommunications (IMT) Advanced Standard of the International Telecommunications Union Radiocommunication sector (ITU-R), signal bandwidths of up to 20 MHz and 100 MHz are applied in the 3G Partnership Project (3GPP) long-term evolution (LTE) and LTE-Advanced systems [2], respectively. We can envision 5G systems [3], where transmission rates and signal bandwidths will be further expanded, operating in the higher spectral bands – over 6 GHz – such as millimeter waves (mmWave) [4].

For single-carrier modulation, a higher bandwidth in the frequency domain inevitably corresponds to a narrower symbol period in the time domain. In multipath channels, the delay spread of radio signals causes inter-symbol interference (ISI) if this delay spread cannot be neglected in comparison with the symbol period. Traditionally, a digital filter, referred to as the equalizer, is applied at the receiver to attempt to reverse the distortion incurred in a channel. The number of filter taps required for the equalizer is proportional to the signal bandwidth. In order to effectively mitigate the ISI, the number of filter taps needs to be several hundreds for a signal bandwidth of 20 MHz. That is too complex to implement in a practical system.

Multi-carrier modulation (MCM) is a broadband communication technique, where a wideband signal is split into a number of narrowband signals. The symbol period of a narrowband signal is substantially extended and is far longer than that of a wideband signal. The effect of ISI can be alleviated in an MCM system if the delay spread becomes negligible compared to the extended symbol period. Thanks to the low-complexity implementation of fast Fourier transform (FFT)-based modulator and the application of cyclic prefix (CP), orthogonal frequency-division multiplexing (OFDM) [5] has become the dominant modulation for wired and wireless communication systems. It has been extensively applied, for example in DSL, DVB-T, Wi-Fi, WiMAX, LTE and LTE-Advanced.

Despite its robustness in multipath channels, the rectangularly-pulsed OFDM signal suffers from large sidelobes that can potentially lead to high inter-carrier interference (ICI) and severe adjacent-channel interference (ACI). To flexibly support multi-user scenarios, orthogonal frequency-division multiple access (OFDMA) is applied for OFDM-based systems. In uplinks, however, a synchronization of multiple users that transmit different subsets of subcarriers is difficult to achieve, resulting in unacceptable ICI. Currently, most of the spectral resources below 6 GHz have been allocated, but have been utilized in an inefficient way. This has triggered development of the techniques of cognitive radio (CR) [6] and dynamic spectrum allocation (DSA) [7], through which secondary users are allowed to access holes in the licensed spectra. The OFDM scheme is an attractive modulation technique for CR and DSA due to its flexibility in subcarrier manipulation. However, the large sidelobes of OFDM signals will bring severe ACI, which limits the application of OFDM-based systems in a dynamic spectrum scenario.

To overcome the drawbacks of OFDM, the enhanced schemes, referred to as filtered OFDM, such as the time-domain windowing and active interference cancelation (AIC), have been proposed. On the other hand, the wireless community has begun to explore other advanced MCM techniques for use in 5G systems, with filter bank multi-carrier (FBMC) the most promising one. In the mid-1960s, prior to the emergence of the OFDM technique, Chang [8] contributed pioneering work on FBMC and Saltzberg revealed that a proper design of prototype filter can realize a transmission rate close to the Nyquist rate and achieve a perfect signal reconstruction without ISI and ICI [9]. Relying on the excellent time-frequency localization of advanced prototype filters, FBMC can implement a sidelobe that is as small as possible. This technique drew much attention, from both academia and industry, as demonstrated in the EU's FP7 PHYDYAS project [10] and the IEEE 802.22 standard [11].

This chapter intends to give a concise description of the FBMC technique. In Section 3.2, the rationale of synthesis and analysis filter banks and the design criteria for a prototype filter without ISI and ICI are introduced. Section 3.3 presents the polyphase implementation of a multicarrier system based on the filter bank. As a special case of FBMC, and as a benchmark, the OFDM scheme is depicted in Section 3.4, with the emphasis on its cyclic prefix and guard bands. Section 3.5 describes the FBMC scheme with a prototype filter as used by the PHYDYAS project. In addition, the classical schemes of filtered OFDM are reviewed and compared with the FBMC in terms of the OOB performance achieved and its implementation complexity. Finally, this chapter's conclusions are in Section 3.7.

3.2 The Filter Bank

The filter bank is an array of filters, which are applied to synthesize multicarrier signals at the transmitter and analyze received signals at the receiver. The principle of the synthesis and analysis filters will be depicted in this section.

3.2.1 *The Synthesis Filters*

When a signal $x(t)$ goes through a filter with an impulse response of $h(t)$, its output signal can be expressed as

$$s(t) = h(t) * x(t) \tag{3.1}$$

where $*$ stands for the linear convolution. As illustrated in Figure 3.1, a synthesis filter bank (SFB) consists of an array of filters $h_n(t), n = 1, \ldots, N$. Each filter individually processes its input signal, and the output signals are summed, synthesizing a composite signal $s(t) = \sum_{n=1}^{N} s_n(t)$. The composite signal can be written as

$$s(t) = \sum_{n=1}^{N} h_n(t) * x_n(t) \tag{3.2}$$

Although any filtering is in principle possible, the filter $h_n(t)$ is specifically applied for processing the transmit signal on the nth subcarrier in the field of multicarrier communications. For simplicity, we only take into account baseband signals, ignoring radio-frequency (RF) chains. To be specific, the input signal for each subcarrier can be given by

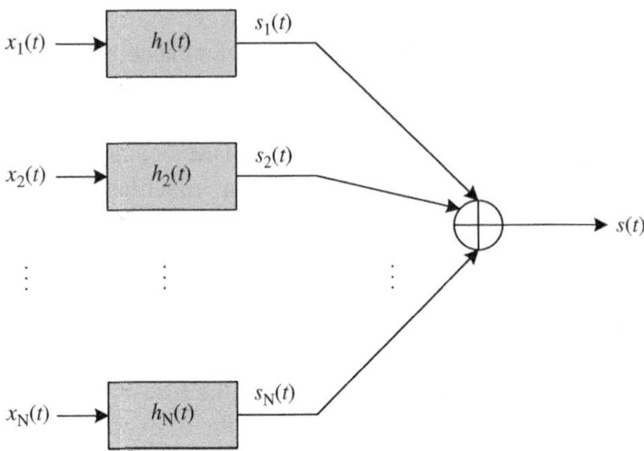

Figure 3.1 Block diagram of the synthesis filter bank

$$x_n(t) = \sum_{m=-\infty}^{\infty} X_{m,n}\delta(t - mT),\ n = 1, \ldots, N \tag{3.3}$$

where $X_{m,n}$ denotes a data symbol transmitted on the nth subcarrier during the mth symbol period, N is the total number of subcarriers, T expresses the length of a symbol period, and $\delta(t)$ denotes the Dirac delta function.

Substituting Eq. (3.3) into Eq. (3.2), the continuous-time baseband transmit signal, which is commonly used in both OFDM and FBMC, can be given by

$$s(t) = \sum_{m=-\infty}^{+\infty}\sum_{n=1}^{N} X_{m,n}h_n(t - mT) \tag{3.4}$$

To achieve orthogonality, the frequency spacing of subcarriers needs to be an integer multiple of the inverse of the symbol period; in other words, $\Delta f = c/T, c = 1, 2,\ldots.$ In general, the frequency spacing is selected as $\Delta f = 1/T$ in order to maximize spectral efficiency. Without loss of generality, we can further denote the carrier frequencies of subcarriers by $f_n = n\Delta f$, $n = 1, 2, \ldots, N$.

The transmit filters are based on a specially designed prototype filter $p_{_T}(t)$, and are modulated by the carrier frequencies f_n, as follows:

$$h_n(t) = p_{_T}(t)e^{2\pi jn\Delta ft+j\phi_n} \tag{3.5}$$

where ϕ_n stands for the phase. Substituting Eq. (3.5) into Eq. (3.4), the baseband transmit signal $s(t)$ can be rewritten as:

$$\begin{aligned} s(t) &= \sum_{m=-\infty}^{\infty}\sum_{n=1}^{N} X_{m,n}p_{_T}(t - mT)e^{2\pi jn\Delta f(t-mT)+j\phi_n} \\ &= \sum_{m=-\infty}^{\infty}\sum_{n=1}^{N} X_{m,n}p_{_T}(t - mT)e^{2\pi jn\Delta ft+j\phi_n} \end{aligned} \tag{3.6}$$

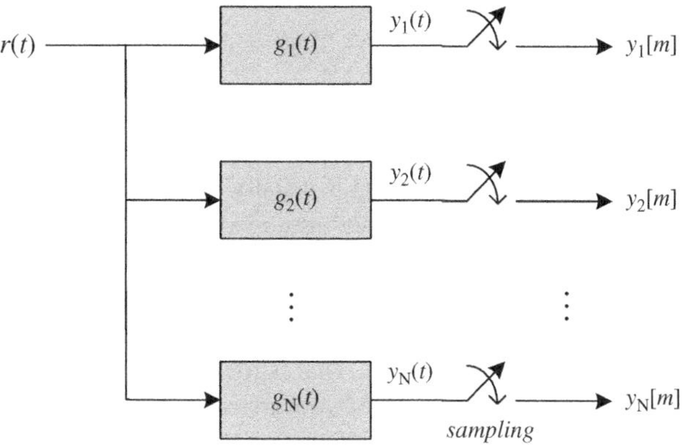

Figure 3.2 Block diagram of the analysis filter bank

3.2.2 *The Analysis Filters*

As shown in Figure 3.2, an analysis filter bank (AFB) consists of an array of filters, which have a common incoming signal $r(t)$. Although any possible filtering can be carried out, in the field of multicarrier communications, each filter analyzes a different subcarrier band of the received signal $r(t)$. Ignoring channel impairments, the input signal at the AFB equals the transmit signal of the SFB; in other words, $r(t) = s(t)$. The analysis filters are based on a specially-designed prototype filter $p_{_R}(t)$. Similar to Eq. (3.5), we can give the impulse response of the analysis filter by

$$g_k(t) = p_{_R}(t)e^{-2\pi jk\Delta ft - j\phi_k}, k = 1, 2, \ldots, N \tag{3.7}$$

For an arbitrary analysis filter $g_k(t)$, $k = 1, 2, \ldots, N$, the continuous-time output signal after filtering can be calculated by

$$\begin{aligned} y_k(t) &= g_k(t) * r(t) \\ &= \sum_{m=-\infty}^{\infty} \sum_{n=1}^{N} X_{m,n} p_{_R}(t) * p_{_T}(t - mT) e^{2\pi jn\Delta ft + j\phi_n} e^{-2\pi jk\Delta ft - j\phi_k} \\ &= \sum_{m=-\infty}^{\infty} \sum_{n=1}^{N} X_{m,n} p_{_R}(t) * p_{_T}(t - mT) e^{2\pi j(n-k)\Delta ft + j(\phi_n - \phi_k)} \end{aligned} \tag{3.8}$$

To properly demodulate the transmit signal at each subcarrier, two conditions need to be satisfied:

1. No ICI in the frequency domain.
2. No ISI in the time domain.

First, the subcarriers need to constitute an orthogonal basis set within a symbol period to avoid the generation of ICI; in other words,

$$\int_0^T e^{2\pi j(n-k)\Delta ft + j(\phi_n - \phi_k)} dt = \begin{cases} T, & k = n \\ 0, & k \neq n \end{cases} \tag{3.9}$$

To achieve orthogonality, the frequency spacing of subcarriers needs to be set as an integer multiple of the inverse of the symbol period; in other words $\Delta f = c/T, c = 1, 2, \ldots$. As mentioned in Section 3.2.1, the frequency spacing is usually selected as $\Delta f = 1/T$ to maximize spectral efficiency. The subcarriers are denoted in an exponential form $e^{2\pi jn\Delta ft}$, $t \in [0, T]$, each of which has two branches: the in-phase (I) and quadrature (Q). The transmit data $X_{m,n}$ is complex-valued; in other words, $X_{m,n} = a_{m,n} + jb_{m,n}$, where $a_{m,n}$ and $b_{m,n}$ are real-valued data. The complex modulated signal $X_{m,n}e^{2\pi jn\Delta ft}$ can be transformed into I- and Q-branches as: $a_{m,n}\cos(2\pi n\Delta ft) - b_{m,n}\sin(2\pi n\Delta ft)$. That is to say, the real part of the transmit data $a_{m,n}$ is modulated on the I-branch signal of the subcarrier $\cos(2\pi n\Delta ft)$, while the imaginary part $b_{m,n}$ is carried by the Q-branch signal $\sin(2\pi n\Delta ft)$. Figure 3.3 gives an example to clarify the question of orthogonality of the subcarriers. The sinusoidal waves $\cos(2\pi n\Delta ft)$ and $\sin(2\pi n\Delta ft)$, $n = 1, 2, 3$ are drawn for one normalized symbol period $t/T \in [0, 1]$. During a symbol period T, all the subcarriers are orthogonal one another, satisfying the requirement of Eq. (3.9).

Second, the selected prototype filter should satisfy the condition that the output signal does not bring ISI to its neighboring symbols in the time domain. The no-ISI condition does not necessarily mean that no overlap should exist among the symbols, but requires at least no interference at the sampling times. As illustrated in Figure 3.4, the composite impulse response should be zero at the sampling times except the original one:

$$p_T(t) * p_R(t)\Big|_{t_s = iT} = \begin{cases} 1, & i = 0 \\ 0, & i \neq 0 \end{cases}$$

where $t_s = iT$ stands for the sampling points on the time axis.

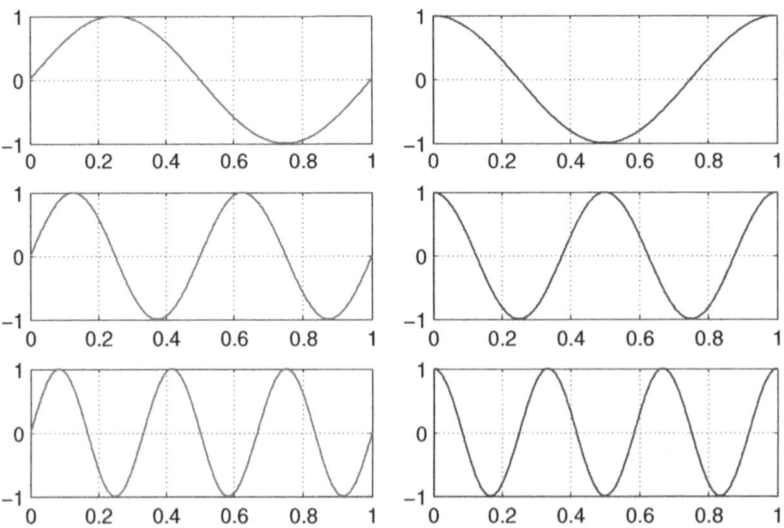

Figure 3.3 The orthogonality of subcarriers

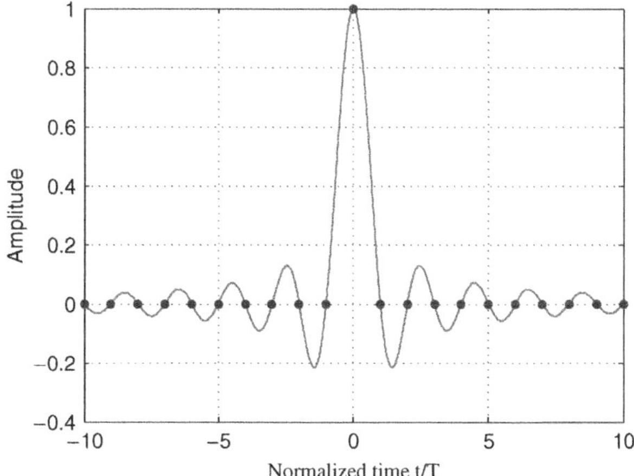

Figure 3.4 An example of signal waveform that does not generate ISI

3.3 Polyphase Implementation

Nowadays, practical communication systems are implemented on digital signal processor or field programmable gate array chipsets in a digital form. So it makes sense to discuss discrete-time implementation of the filter bank.

The SFB has N subcarriers with frequency spacing Δf, resulting in a signal bandwidth of $B_w = N\Delta f$. According to the sampling theorem [12], the sampling interval T_s equals the inverse of the signal bandwidth; in other words $T_s = 1/B_w = T/N$. The length of the symbol period is T, so the corresponding number of signal samples within each symbol period is $T/T_s = N$. The discrete-time prototype filter can be obtained by sampling the continuous-time prototype filter $p_T(t)$ with a sampling rate of T_s, resulting in

$$p_T[l] = p_T(lT_s), l = 0, 1, \ldots, L-1 \tag{3.10}$$

where L is the length of the discrete-time prototype filter $p_T[l]$. It is possible that the length of the prototype filter is larger than the symbol period, and L can be generally selected as an integer multiple of N. Assuming the overlapping factor is K, which means the length of the prototype filter is K times of the symbol period, we have,

$$L = KN \tag{3.11}$$

In signal processing [12], the Z-transform is an important analysis tool, which converts a discrete-time signal into a complex frequency domain. The Z-transform of $p_T[l]$ is calculated by

$$P_T(z) = \sum_{l=0}^{L-1} p_T[l]z^{-l} = \sum_{l=0}^{KN-1} p_T[l]z^{-l} \tag{3.12}$$

Letting $l = k'N + n'$, where $k' = 0, 1, \ldots, (K-1)$ and $n' = 0, 1, \ldots, (N-1)$, Eq. (3.12) can be further transformed into

$$
\begin{aligned}
P_T(z) &= \sum_{n'=0}^{N-1}\sum_{k'=0}^{K-1} p_T[k'N+n']z^{-(k'N+n')} \\
&= \sum_{n'=0}^{N-1} z^{-n'} \sum_{k'=0}^{K-1} p_T[k'N+n']z^{-k'N}
\end{aligned} \tag{3.13}
$$

This constitutes a series of subsequences $p_T^{n'}[k'] = k'N + n'$, $n' = 0, 1, \ldots, (N-1)$, which includes N subsequences. The subsequence $p_T^{n'}[k']$ has a length of K and is called the (n')th polyphase component of the prototype filter $p_T[l]$. The Z-transform of $p_T^{n'}[k']$ is referred to as the (n')th polyphase decomposition of $P_T(z)$, which is defined as

$$
P_T^{n'}(z^N) = \sum_{k'=0}^{K-1} p_T[k'N+n']z^{-k'N} \tag{3.14}
$$

Substituting Eq. (3.14) into Eq. (3.13), we have

$$
P_T(z) = \sum_{n'=0}^{N-1} P_T^{n'}(z^N)z^{-n'} \tag{3.15}
$$

which is the polyphase decomposition of the prototype filter.

Similar to Eq. (3.10), the discrete-time transmit filters can be obtained by sampling $h_n(t)$ with a sampling rate of T_s, as follows:

$$
h_n[l] = h_n(lT_s) = p_T[l]e^{2\pi jn\Delta fl + j\phi_n}, l = 0, 1, \ldots, L-1 \tag{3.16}
$$

The Z-transform of the nth transmit filter can be expressed as

$$
\begin{aligned}
H_n(z) &= \sum_{l=0}^{L-1} h_n[l]z^{-l} \\
&= \sum_{l=0}^{L-1} p_T[l]e^{2\pi jn\Delta fl + j\phi_n} z^{-l} \\
&= \sum_{n'=0}^{N-1}\sum_{k'=0}^{K-1} p_T[k'N+n']e^{2\pi jn(k'N+n')/N + j\phi_n} z^{-(k'N+n')} \\
&= e^{j\phi_n} \sum_{n'=0}^{N-1} e^{2\pi jnn'/N} z^{-n'} \sum_{k'=0}^{K-1} p_T[k'N+n']e^{2\pi jnk'} z^{-k'N} \\
&= e^{j\phi_n} \sum_{n'=0}^{N-1} e^{2\pi jnn'/N} P_T^{n'}(z^N)z^{-n'}
\end{aligned} \tag{3.17}
$$

As mentioned above, the SFB consists of N transmit filters $H_n(z)$, $n = 1, 2, \ldots, N$. Consequently, the Z-transform of SFB can be expressed in matrix form, as follows:

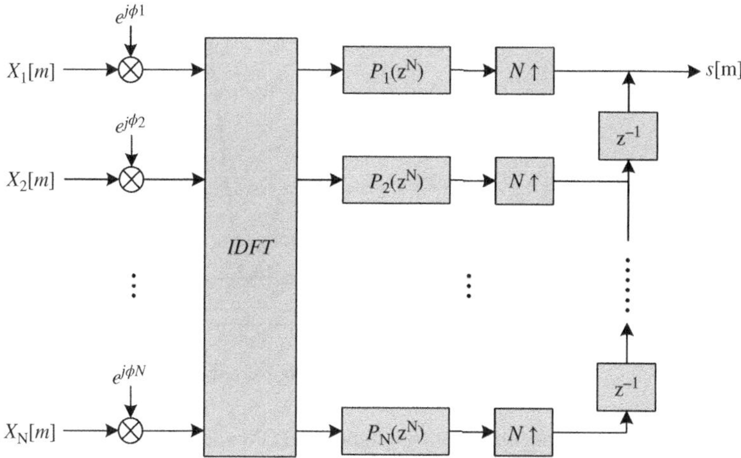

Figure 3.5 A digital implementation of SFB

$$\begin{bmatrix} H_1(z) \\ H_2(z) \\ \vdots \\ H_N(z) \end{bmatrix} = \begin{bmatrix} e^{j\phi_1} \\ e^{j\phi_2} \\ \vdots \\ e^{j\phi_N} \end{bmatrix} \begin{bmatrix} 1 & 1 & \cdots & 1 \\ 1 & W^{-1} & \cdots & W^{-N+1} \\ \vdots & \cdots & \ddots & \vdots \\ 1 & W^{-N+1} & \cdots & W^{(-N+1)^2} \end{bmatrix} \begin{bmatrix} P_T^0(z^N) \\ P_T^1(z^N)z^{-1} \\ \vdots \\ P_T^N(z^N)z^{-N+1} \end{bmatrix} \tag{3.18}$$

where the coefficient $W = e^{j2\pi/N}$. The column vector on the left-hand side stands for the phase rotations, the matrix in the middle is exactly an inverse DFT, and the column vector on the right-hand side is a polyphase decomposition of the prototype filter, which is also called a "polyphase network" (PPN) in the FBMC scheme. Following Eq. (3.18), a structure of the polyphase implementation for the FBMC's modulator is obtained, as shown in Figure 3.5.

3.4 OFDM

The OFDM scheme can be regarded as a special case of FMBC, where a rectangular prototype filter is applied as shown in Figure 3.6.

$$p_0(t) = \begin{cases} 1, & -\frac{T}{2} \le t \le \frac{T}{2} \\ 0, & others \end{cases} \tag{3.19}$$

The discrete-time rectangular prototype filter can be given by:

$$p_0[l] = \begin{cases} 1, & 0 \le l \le N-1 \\ 0, & others \end{cases} \tag{3.20}$$

The OFDM symbols are non-overlapping (the overlapping factor $K = 1$) in the time domain, so the polyphase decomposition is simplified to

$$P_0^{n'}(z^N) = \sum_{k'=0}^{K-1} p_0[k'N + n']z^{-k'N} = \sum_{k'=0}^{0} p_0[n']z^0 = 1, n' = 1, 2, \ldots, N \tag{3.21}$$

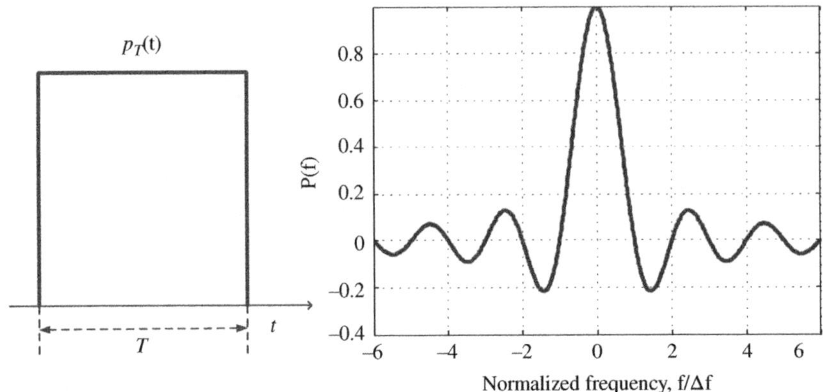

Figure 3.6 The rectangular prototype filter of OFDM and its Fourier transform

Since the phase does not affect the orthogonality of OFDM subcarriers, we can neglect the phase and use $\phi_n = 0, n = 1, 2, \ldots, N$. Thus Eq. (3.18) is simplified into:

$$\mathbf{H}_{OFDM} = \begin{bmatrix} H_1(z) \\ H_2(z) \\ \vdots \\ H_N(z) \end{bmatrix} = \begin{bmatrix} 1 & 1 & \cdots & 1 \\ 1 & W^{-1} & \cdots & W^{-N+1} \\ \vdots & \cdots & \ddots & \vdots \\ 1 & W^{-N+1} & \cdots & W^{(-N+1)^2} \end{bmatrix} \tag{3.22}$$

which is exactly the DFT. That is to say, the OFDM signal can be simply generated by a DFT modulator, which can be further implemented by FFT if the number of subcarriers N is a power of 2. The Fourier transform of the rectangular prototype filter is

$$P_0(f) = \int_{-T/2}^{T/2} e^{-2\pi jft} dt = \frac{\sin \pi fT}{\pi f} = T\text{sinc}\left(\frac{f}{\Delta f}\right) \tag{3.23}$$

where $\text{sinc}(x) = \frac{\sin \pi x}{\pi x}$ is the normalized sinc function. The Fourier transform of the nth subcarrier $p_T(t)e^{2\pi jn\Delta ft}$ can be calculated by:

$$P_n(f) = \int_{-T/2}^{T/2} e^{2\pi jn\Delta ft} e^{-2\pi jft} dt = T\text{sinc}\left(\frac{f - n\Delta f}{\Delta f}\right) \tag{3.24}$$

The Fourier transform of the nth subcarrier can be obtained simply by shifting $P_0(f)$ in the frequency axis with a frequency shift of $n\Delta f$; in other words, $P_n(f) = P_0(f - n\Delta f)$. A set of OFDM subcarriers $\sum_{n=-3}^{n=3} P_n(f)$ in the frequency domain is illustrated in Figure 3.7. It can be observed that ICI does exist at the points of the integer multiple of frequency spacing; in other words, $f_s = n\Delta f$.

3.4.1 Cyclic Prefix

CP refers to the prefixing of an OFDM symbol that is the repetition of the end of the symbol, as illustrated in Figure 3.8. The main objectives are:

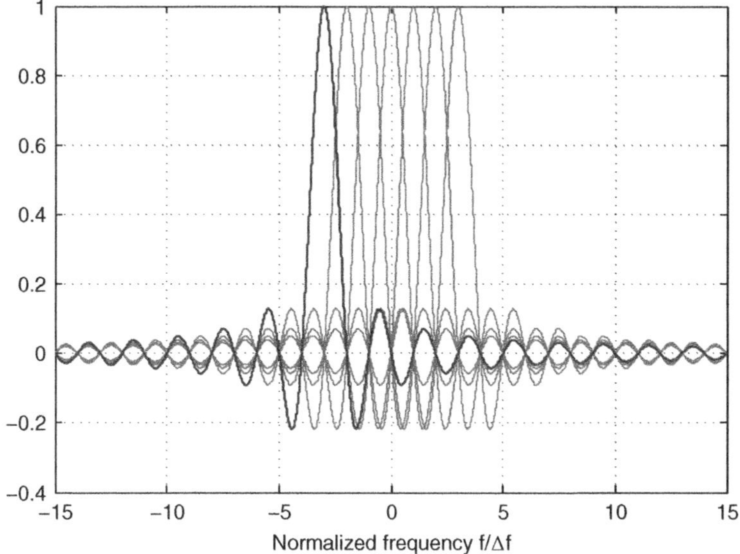

Figure 3.7 A set of OFDM subcarriers

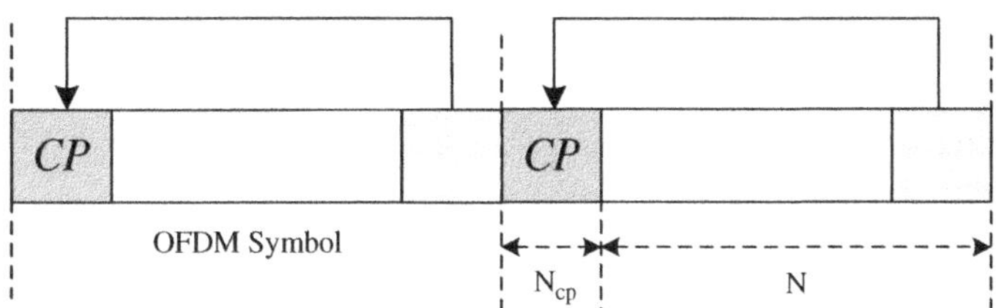

Figure 3.8 The cyclic prefix

- eliminating the ISI from the previous symbol
- converting the linear convolution with a channel filter into a circular convolution, which allows for simple frequency-domain channel estimation and equalization.

Assuming the baseband transmit signal is

$$\boldsymbol{x} = \left[x_1, x_2, \cdots, x_N\right] \tag{3.25}$$

the discrete-time impulse response of the channel, which is assumed to be flat-fading for simplicity, is given by

$$\boldsymbol{h} = \left[h_1, h_2, \cdots, h_{L_h}\right] \tag{3.26}$$

where L_h is the length of discrete-time channel filter. Passing through the channel, an output signal $\boldsymbol{y}$ is obtained by linearly convoluting the input signal with the channel filter; in other words, $\boldsymbol{y} = \boldsymbol{x} * \boldsymbol{h}$, which is calculated as

$$y[s] = \sum_{l=0}^{L_h-1} h[l]x[s-l] \tag{3.27}$$

The length of the output signal is $L_h + N - 1$, which is longer than the input signal if the length of channel filter is more than one; in other words, $L_h > 1$. At the end of each OFDM symbol, the residual part with a length of $L_h - 1$ is overlapped with its following OFDM symbol and causes ISI. Intuitively, an effective method to combat the ISI is inserting between two neighboring OFDM symbols a guard interval that can absorb the residual of the previous OFDM symbol and is removed at the receiver. As a guard interval, the CP can effectively eliminate ISI if its length is selected to be larger than the length of channel filter; in other words, $L_{cp} > L_h$.

The input signal with the insertion of CP can be expressed as

$$\boldsymbol{x}' = \left[x_{(N-L_{cp}+1)}, \cdots, x_{(N-1)}, x_N, x_1, x_2, \cdots, x_N\right] \tag{3.28}$$

The linear convolution of $\boldsymbol{x}'$ and $\boldsymbol{h}$ is

$$y'[s] = \sum_{l=0}^{L_h-1} h[l]x'[s-l] \tag{3.29}$$

The output signal $\boldsymbol{y}' = \boldsymbol{x}' * \boldsymbol{h}$ has a length of $L_h + L_{cp} + N - 1$, and can be denoted

$$\boldsymbol{y}' = \left[y_1, y_2, \cdots, y_{(L_h+L_{cp}+N-1)}\right] \tag{3.30}$$

At the receiver, the CP is discarded and only samples from $L_{cp} + 1$ to $L_{cp} + N$ are extracted and demodulated:

$$\boldsymbol{y} = \left[y_{(L_{cp}+1)}, y_{(L_{cp}+2)}, \cdots, y_{(L_{cp}+N)}\right] \tag{3.31}$$

It can be verified that the output signal $\boldsymbol{y}$ is equal to the result of a cyclic convolution; in other words, $\boldsymbol{y} = \boldsymbol{y}_c = \boldsymbol{x} \otimes \boldsymbol{h}$, where $\otimes$ stands for the cyclic convolution. A numerical example will clarify the conversion of linear and cyclic convolutions. The randomly selected transmit signal and the channel filter are given by:

$$\boldsymbol{x} = \left[0.6294, 0.8116, -0.7460, 0.8268, 0.2647, -0.8049, -0.4430, 0.0938\right] \tag{3.32}$$

$$\boldsymbol{h} = \left[0.9150, 0.9298, -0.6848\right] \tag{3.33}$$

As shown in the upper part of Figure 3.9, the transmit data $\boldsymbol{x}$ is first prefixed with a CP with a length of $N_{cp} = 4$, and then is linearly convoluted with the channel filter $\boldsymbol{h}$. The upper-right figure illustrates the output signal of $\boldsymbol{y}' = \boldsymbol{x}' * \boldsymbol{h}$. The lower parts of Figure 3.9 shows the cyclic convolution between the transmit data and the channel filter; in other words, $\boldsymbol{y}_c = \boldsymbol{x} \otimes \boldsymbol{h}$. Removing the CP part and the residual at the end, we can extract the required signals $\boldsymbol{y}$ for demodulation. It can be observed from Figure 3.9 that $\boldsymbol{y}$ equals $\boldsymbol{y}_{cc}$, which means that a linear convolution can be converted to a cyclic one with the help of a CP.

The motivation for forming a cyclic convolution in the OFDM signals is to simplify channel estimation and equalization. According to signal processing theory, the cyclic convolution of two signals in the time domain, rather than their linear convolution, corresponds to a

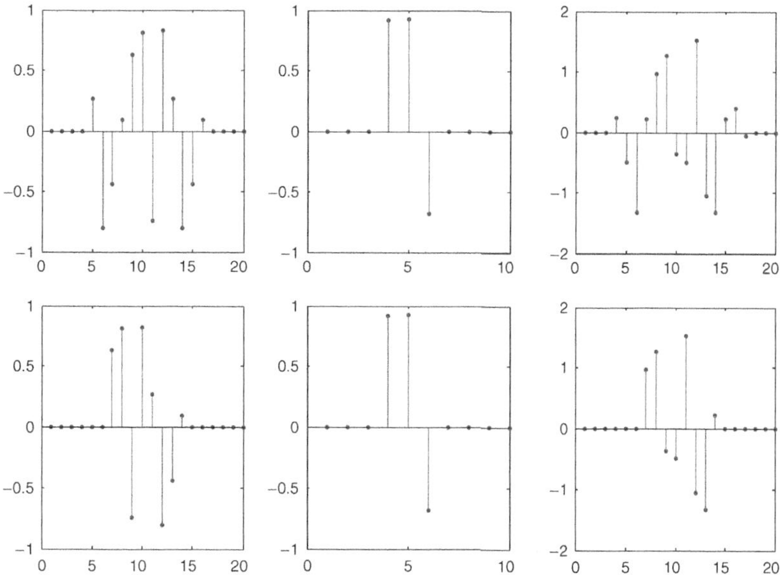

Figure 3.9 The linear and cyclic convolutions

multiplication of their DFT in the frequency domain. The DFT of the output signal $\boldsymbol{y} = \boldsymbol{x} \otimes \boldsymbol{h}$ can be obtained by

$$Y[s] = H[s]X[s], s = 1, \ldots, N \tag{3.34}$$

where $Y[s]$, $H[s]$ and $X[s]$ are DFTs of $\boldsymbol{y}$, $\boldsymbol{h}$ and $\boldsymbol{x}$, respectively. Since the modulation and demodulation are independently carried out on each subcarrier, the channel estimation and equalization become simpler. If $P[s]$ is known transmit data at the receiver, referred to as the pilot, we can estimate the channel response at the insertion point of the pilot as

$$\hat{H}_p[s] = \frac{Y_p[s]}{P[s]} \tag{3.35}$$

where $Y_p[s]$ is the received signal corresponding to $P[s]$. Based on the estimation channel information at the pilots $\hat{H}_p[s]$, the channel responses at all subcarriers $\hat{H}[n]$ can be estimated through an interpolation operation. Thus the recovery of the transmit data can be realized by

$$\hat{X}[s] = \frac{Y[s]}{\hat{H}[s]} \tag{3.36}$$

Although the utilization of CP makes the detection of OFDM signals simpler, the CP inevitably reduces spectral efficiency, which is one of the motivations for seeking a successor to OFDM. As a practical example, the parameters for OFDM modulation in the specification of 3GPP LTE [13] are given in Table 3.1. The DFT length is $N = 2048$ for a subcarrier spacing Δf of 15 kHz and $N = 4096$ for 7.5 kHz. Different OFDM symbols within a slot may have different lengths of CP, ranging from $N_{cp} = 144$ to $N_{cp} = 1024$. The decrease of

Table 3.1 The OFDM parameters in 3GPP LTE

Configuration	Subcarrier spacing Δf (kHz)	Cyclic prefix length N_{CP}	Spectral efficiency loss N_{CP}/N (%)
Normal CP	15	160 or 144	7.8 or 7
Extended CP	15	512	25
Extended CP	7.5	1024	25

Source: 3GPP 2015. Reproduced with permission of 3GPP [13]

spectral efficiency measured by N_{cp}/N can rise to 25%. In comparison, the ISI of the FBMC symbol can be mitigated with a properly-designed polyphase filter and the FMBC symbol does not need a guard interval, boosting spectral efficiency.

3.4.2 Guard Band

Due to the large sidelobes that decay asymptotically as f^{-2} [15], the OOB power leakage of an OFDM signal is unacceptable in practical systems. The interference of OFDM signals on its adjacent channels is around −20 dB, while the so-called adjacent channel interference power ratio (ACIR) specified in 3GPP LTE is −45 dB. Moreover, the ACIR of CR-based systems defined by the FCC is −72 dB. Due to its low-complexity implementation, the insertion of guard bands is often utilized in OFDM systems. Figure 3.10 shows the definition of the channel and transmission bandwidths of 3GPP LTE [14]. Although the whole channel bandwidth is allocated to a dedicated LTE channel, this spectrum cannot be fully used to transmit signals. The guard bands are inserted by deactivating the subcarriers lying at the edges of the spectrum band. The utilization of guard bands somehow alleviates the amount of OOB power leakage but inevitably comes at a cost in spectral efficiency. For instance, as shown in Figure 3.10, the definition of channel and transmission bandwidths for the LTE system are shown. The channel bandwidth is the amount of spectral resource allocated to a dedicated system, while

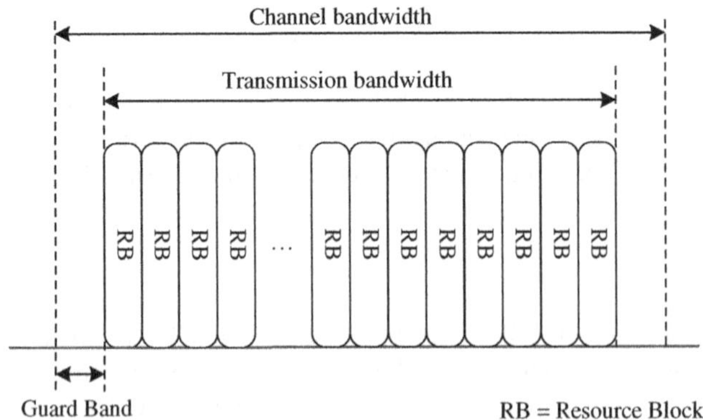

Figure 3.10 Definition of channel bandwidth and transmission bandwidth configuration for LTE carrier. *Source:* 3GPP 2015. Reproduced with permission of 3GPP [14]

Table 3.2 Transmission bandwidth configuration of 3GPP LTE

Channel bandwidth (MHz)	1.4	3	5	10	15	20
Transmission bandwidth N_{RB}	6	15	25	50	75	100
Transmission bandwidth (MHz)	1.08	2.7	4.5	9	13.5	18
Guard band (MHz)	0.32	0.3	0.5	1	1.5	2
Spectral efficiency loss (%)	22.85	10	10	10	10	10

Source: 3GPP 2015. Reproduced with permission of 3GPP [14]

the transmission bandwidth is the width of spectrum that is actually occupied by the transmit signals. Obviously, the transmission bandwidth is not allowed to be larger than the channel bandwidth. In the LTE, the term *resource block* (RB) is defined as a set of OFDM subcarriers, equal to 12 subcarriers, spanning a signal bandwidth of 180 kHz. The transmission bandwidth in the LTE is parameterized by the number of RBs. For example, the 1.4-MHz channel is able to transmit up to six RBs, which is equivalent to a signal bandwidth of 1.08 MHz. The gap between the channel and transmission bandwidth is exactly the width of the guard bands. To give a quantitative evaluation of the loss of spectral efficiency, the parameters related to the guard bands specified in 3GPP LTE are listed in Table 3.2. We can see that the loss of spectral efficiency due to the utilization of guard bands is more than 10% in the LTE system.

3.5 FBMC

In principle, a prototype filter can be designed to achieve a sidelobe as small as possible by means of the filter bank. As an example, we show the Nyquist filter used for generating FBMC signals in EU FP7 PHYDYAS project. According to Bellanger *et al.* [10], the following frequency-domain coefficients can be applied to constitute a desired prototype filter with a overlapping factor of $K = 4$:

$$\boldsymbol{p} = \left[1, 0.97196, 0.707, 0.235147\right] \tag{3.37}$$

Based on these coefficients, the frequency response of the prototype filter is obtained through an interpolation operation, which is expressed as

$$P(f) = \sum_{k=-K+1}^{K-1} p_k \frac{\sin(\pi NK(f - \frac{k}{NK}))}{NK \sin(\pi(f - \frac{k}{NK}))} \tag{3.38}$$

where N is the total number of subcarriers, K is the overlapping factor and p_k is mapped from the aforementioned coefficients, where $p_0 = 1$, $p_{\pm 1} = 0.97196$, $p_{\pm 2} = 0.707$ and $p_{\pm 3} = 0.235147$. Then, its impulse response $p_{_T}(t)$ can be obtained by an inverse Fourier transform, as follows:

$$p_{_T}(t) = 1 + \sum_{k=1}^{K-1} p_k \cos 2\pi \frac{kt}{KT} \tag{3.39}$$

The normalized time and frequency responses of the prototype filter are shown in Figure 3.11. As we can see, this prototype filter spans $K = 4$ FBMC symbols, rather than

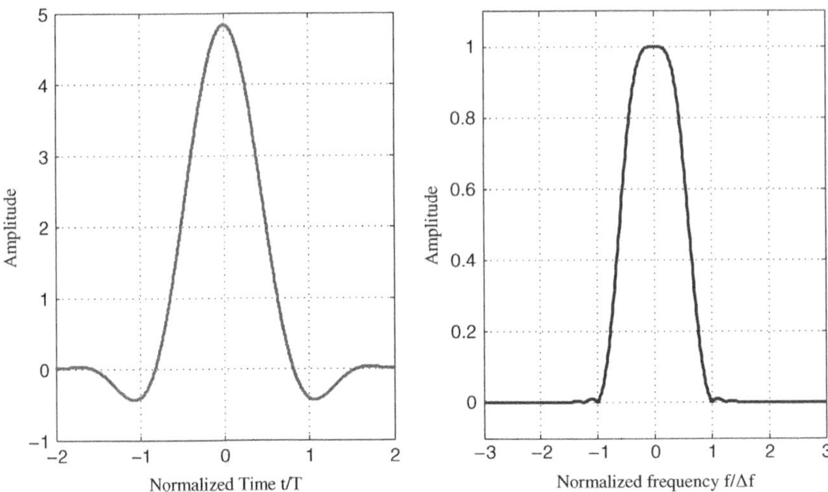

Figure 3.11 The impulse response of the Nyquist filter

the rectangular prototype filter of OFDM, which occupies only a single OFDM symbol. In the frequency domain, the frequency response of the FBMC subcarrier is very compact. The ripples can be neglected, and there is even no ICI between the non-neighboring subcarriers. By sampling $P_T(t)$, a discrete-time prototype filter with a length of KN is obtained, as follows:

$$p_T[s] = p_T(sT_s) = 1 + \sum_{k=1}^{K-1} p_k \cos 2\pi \frac{ks}{KN}, s = 0, 1, \ldots, KN-1 \tag{3.40}$$

Then, the (n')th polyphase decomposition can be obtained as

$$p_T^{n'}[k'] = p_T[k'N + n'], k' = 0, 1, \ldots, K-1 \tag{3.41}$$

The (n')th polyphase decomposition $p_T^{n'}[k']$ is actually an finite impulse response (FIR) filter of length K. This filter is applied for the (n')th FBMC subcarrier, and a number of N FIR filters constitutes the PPN to generate the FBMC signals. Based on the DFT modulator and PPN, the digital implementation of FBMC is illustrated in Figure 3.12, as well as that of OFDM. Thanks to the utilization of DFT, the difference between FBMC and OFDM transmission is only in the implementation of CP and PPN. Hence there should be comparability and a smooth transfer between conventional OFDM and the forthcoming FBMC system, which is very important from the perspective of the deployment of practical systems.

3.6 Comparison of FBMC and Filtered OFDM

The classical sidelobe-suppression methods for OFDM signals – time windowing and AIC – are briefly reviewed in this section. The performance of these filtered OFDM approaches, measured by the OOB power radiation and implementation complexity, is described, together with those of the FBMC scheme.

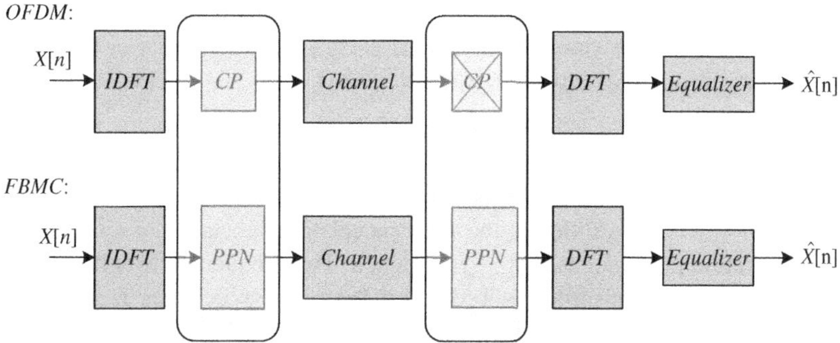

Figure 3.12 The digital implementation of OFDM and FBMC

3.6.1 Classical Approaches to Sidelobe Suppression

3.6.1.1 Time Windowing

According to the Gibbs phenomenon in signal processing [12], a high sidelobe is derived from time-domain discontinuity of a rectangular prototype filter. If the signal's amplitude goes smoothly to zero at the symbol boundaries, its power leakage can be significantly reduced. Hence the time-windowing scheme introduces appropriate windows to reshape the rectangularly-pulsed OFDM symbols. Examples include the half-sine and the Hanning window [15]. Without loss of generality, the window functions can be uniformly formulated as

$$w(t) = R(t) * \eta(t) \tag{3.42}$$

where $R(t)$ is a rectangular pulse:

$$R(t) = \begin{cases} 1, & 0 < t \leq T \\ 0, & others \end{cases} \tag{3.43}$$

For the commonly used half-sine pulse, the generating function $\eta(t)$ can be given by:

$$\eta(t) = \frac{\pi}{2\beta T} \sin(\frac{\pi t}{\beta T})R(t) \tag{3.44}$$

where β denotes the roll-off factor.

The windowing scheme broadens the length of an OFDM symbol, leading to a loss of spectral efficiency. Even if a small value of β is used, the windowing scheme can achieve considerable sidelobe suppression.

3.6.1.2 Active Interference Cancelation

The key idea of AIC is the insertion of several cancelation subcarriers at the edges of OFDM spectrum [16, 17, 18]. The cancelation subcarriers generate a special frequency response, which is exactly the inverse of that of the data subcarriers. That is to say, the cancelation and data subcarriers can be destructively combined and lower OOB power leakage. Suppose the transmit data is $\boldsymbol{d} = [d_1, d_2, \cdots, d_{N_d}]$, where N_d is the number of data subcarriers. A number of N_c cancelation subcarriers carrying $\boldsymbol{g} = [g_1, g_2, \dots, g_{N_c}]$ are inserted at the edges of the signal spectrum, resulting in a composite transmit data $\boldsymbol{x}$:

$$\boldsymbol{x} = [g_1, \dots, g_{N_c/2}, d_1, \dots, d_{N_d}, g_{N_c/2+1}, \dots, g_{N_c}] \tag{3.45}$$

To determine **g**, L specially selected frequency-observation points need to be used. Assuming an unity data $g_{n_c} = 1$ is transmitted at the n_cth cancelation subcarrier, a response at the lth frequency-observation point denoted f_{l,n_c} is obtained. This frequency response f_{l,n_c} is a priori knowledge at both the transmitter and receiver. Then, $L \times N_c$ entries f_{l,n_c} can constitute a precoding matrix **C**. Similarly, the data signal $\boldsymbol{d}$ can form an L-dimensional vector **s**, containing the responses of the data subcarriers at L observation points. Thus the optimization problem can be formulated as a linear least squares problem; in other words,

$$\boldsymbol{g} = \arg \min_{\tilde{\boldsymbol{g}}} ||\mathbf{s} + \mathbf{C}\tilde{\boldsymbol{g}}|| \tag{3.46}$$

3.6.2 *Performance*

In Figure 3.13, the normalized power spectral density of the conventional OFDM signal, the windowed OFDM, AIC and FMBC are given, taking advantage of the classical Welch estimation method [19]. Using the half-sine window given in Eq. (3.44) and a roll-off factor of $\beta = 1/16$, a sidelobe suppression gain of more than 30 dB can be achieved over conventional OFDM, which is considerable. The signal spectrum of AIC is slightly broader than the other techniques, since a total of 20 cancelation subcarriers have been inserted into the guard bands on each side of the signal band. The sidelobe suppression of AIC achieves a gain of about 10 dB over conventional OFDM. It is clear that FMBC is better than the other schemes, not only in its much better sidelobe suppression, but also by providing a much steeper slope at the edges of the signal band. This latter fact allows FBMC to use more subcarriers for signal

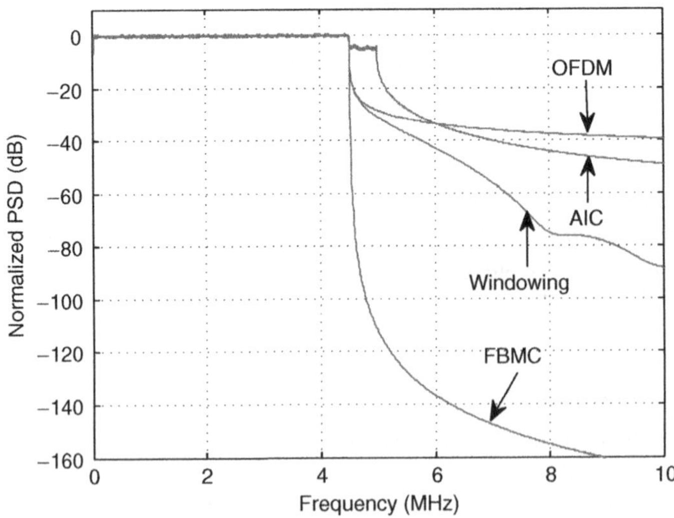

Figure 3.13 The OOB power radiation of the filtered OFDM schemes and FBMC

Table 3.3 Computational complexity

Methods	Complexity expressions
OFDM	$\frac{N}{2}\log_2 N$
Windowing	$\frac{N}{2}\log_2 N + N + N_{cp} + N_r$
AIC	$\frac{N}{2}\log_2 N + {N_c}^3$
FBMC	$\frac{N}{2}\log_2 N + (K+1)N$

transmission and thus to further improve spectral efficiency. More numerical results about the OOB performance of FBMC and the filtered OFDM can be found in the paper by Jiang and Schellmann [20].

3.6.3 Complexity

To clarify the computational complexity comparison, the number of complex multiplications (CM) required for the generation of each OFDM and FBMC symbol are investigated. Although only the complexity of generating the transmit signal at the transmitter is taken into account, the complexity at the receiver can be derived using the same method.

In the case that the number of subcarriers N is a power of two, the FFT can be applied for the modulation of the OFDM signal. Since the insertion of CP does not involve complex multiplication, the complexity of OFDM equals that of the FFT, which needs $\frac{N}{2}\log_2 N$ CMs. For windowed OFDM, the number of samples covered by a window function is $N + N_{cp} + N_r$, where N_{cp} is the length of CP and N_r is the length of the roll-off interval. The windowing is an operation of sample multiplication, rather than the linear convolution, resulting in an additional $N + N_{cp} + N_r$ CMs for each OFDM symbol. For AIC [16], the cancellation subcarriers are generated on a symbol-by-symbol basis by solving the optimization problem in Eq. (3.46). This optimization problem can be efficiently solved by means of *the second-order cone problem* described by Ben-Tal and Nemirovski [21], which requires a complexity amounts to N_c^3 CMs. In the case in which the number of subcarriers N is a power of two, the synthesis filter bank can be efficiently implemented by an N-size FFT and a PPN. At each subcarrier, the output signal of the FFT modulator needs to be processed by a discrete-time filter with a length of K. Hence, the PPN needs KN CMs for the generation of each FBMC symbol. In addition, the phase rotation at each subcarrier needs one CM. The required number of CMs for OFDM, FBMC and filtered OFDM, are presented in Table 3.3. It can be concluded that the modulation of the FBMC signal comes at the cost of reasonable complexity compared to that of OFDM and filtered OFDM.

3.7 Conclusion

In this chapter, the filter bank multicarrier scheme, including the principles of synthesis and analysis filter banks, the design of the prototype filter and its polyphase implementation were presented. As a special case of FBMC and as a benchmark, the details of orthogonal frequency-division multiplexing were described and compared. Despite its robustness against multipath fading and its low-complexity implementation based on the FFT, the OFDM signal suffers from large sidelobes due to the utilization of a rectangular prototype filter. As a result, the OFDM signal is vulnerable to ICI and is highly likely to generate large adjacent channel interference. In addition, the insertion of a CP and guard band could substantially reduce its spectral efficiency. Thanks to the strong capability of signal processing at the filter bank, FBMC can flexibly implement a specially-designed prototype filter that has sidelobes as small as possible. With an excellent frequency–time localization, the utilization of CP and large guard bands can also be avoided in the FBMC scheme, which in turn boost its spectral efficiency. The robustness to ISI, the immunity to ICI and ACI, and the high spectral efficiency of FBMC are achieved at the cost of reasonable implementation complexity. In summary, the FBMC scheme is a potential successor of the dominant OFDM scheme and it is definitely a promising modulation technique for use in forthcoming 5G systems.

References

[1] Hillebrand, F. (2013) The creation of standards for global mobile communication: GSM and UMTS standardization from 1982 to 2000. *IEEE Wireless Commun. Mag.*, **20** (5), 24–33.

[2] Astely, D., Dahlman, E., Furuskar, A., Jading, Y., Lindstrom, M., and Parkvall, S. (2009) LTE: The evolution of mobile broadband. *IEEE Commun. Mag.*, **47** (4), 44–51.

[3] The NGMN Alliance (2015) 5G White Paper, Tech. Rep., Next Generation Mobile Networks.

[4] Pi, Z. and Khan, F. (2011) An introduction to millimeter-wave mobile broadband systems. *IEEE Commun. Mag.*, **49** (6), 101–107.

[5] Weinstein, S.B. (2009) The history of orthogonal frequency-division multiplexing. *IEEE Commun. Mag.*, **47** (11), 26–35.

[6] Haykin, S. (2005) Cognitive radio: brain-empowered wireless commnunications. *IEEE J. Sel. Areas Commun.*, **23** (2), 201–220.

[7] Weiss, T.A. and Jondral, F.K. (2004) Spectrum pooling: an innovative strategy for the enhancement of spectrum efficiency. *IEEE Commun. Mag.*, **42** (3), 8–14.

[8] Chang, R.W. (1966) Synthesis of band-limited orthogonal signals for multichannel data transmission. *Bell Syst. Tech. J.*, **45**, 1775–1796.

[9] Saltzberg, B.R. (1967) Performance of an efficient parallel data transmission system. *IEEE Trans. Commun. Tech.*, **15** (6), 805–811.

[10] Bellanger, M., LeRuyet, D., Roviras, D., and Terré, M. (2010), FBMC physical layer: a primer. http://www.ict-phydyas.org/.

[11] IEEE 802.22 Working Group on Wireless Regional Area Networks, Web page. http://ieee802.org/22/.

[12] Oppenheim, A.V., Willsky, A.S., and Nawab, S.H. (1997) *Signals and Systems*, Pearson.

[13] 3GPP (2015) Technical specifications for evolved universal terrestrial radio access (E-UTRA) - physical channels and modulation, *TS 36.211*.

[14] 3GPP (2015) Technical specifications for evolved universal terrestrial radio access (E-UTRA) - base station (BS) radio transmission and reception, *TS 36.104*.

[15] Farhang-Boroujeny, B. (2011) OFDM versus filter bank multicarrier. *IEEE Signal Process. Mag.*, **28** (3), 92–112.

[16] Yamaguchi, H. (2004) Active interference cancellation technique for MB-OFDM cognitive radio. *Proc. of IEEE Eur. Microw. Conf.*, **2**, 1105–1108.

[17] Brandes, S., Cosovic, I., and Schnell, M. (2006) Reduction of out-of-band radiation in OFDM systems by insertion of cancellation carriers. *IEEE Commun. Lett.*, **10** (6), 420–422.

[18] van de Beek, J. and Berggren, F. (2008) Out-of-band power suppression in OFDM. *IEEE Commun. Lett.*, **12** (9), 609–611.

[19] Welch, P.D. (1967) The use of fast Fourier transform for the estimation of power spectra: a method based on time averaging over short, modified periodograms. *IEEE Trans. Audio Electroacoustics*, **AU-15** (2), 70–73.

[20] Jiang, W. and Schellmann, M. (2012) Suppressing the out-of-band power radiation multi-carrier systems: a comparative study, in *Proc. of IEEE Globecom 2012*, Anaheim, United States, pp. 1477–1482.

[21] Ben-Tal, A. and Nemirovski, A. (2001) *Lectures on Modern Convex Optimization: Analysis, Algorithms, and Engineering Applications*, MPS-SIAM Series on Optimization.

4

Filter Bank Multicarrier for Massive MIMO

Arman Farhang, Nicola Marchetti and Behrouz Farhang-Boroujeny

The recent proposal of massive MIMO as a candidate for the fifth generation of wireless communication networks (5G) has sparked a great deal of interest among researchers. This is due to the fact that such technology can greatly increase the capacity of multiuser networks. Massive MIMO is a multiuser technique with concepts somewhat similar to spread-spectrum systems. The spreading gains of different users come from the corresponding channel gains between each mobile terminal (MT) antenna and the base station (BS) antennas. Therefore, a significant processing gain can be achieved by utilizing a massive number of antenna elements at the BS. In other words, the processing gain can arbitrarily grow by increasing the number of antenna

Signal Processing for 5G: Algorithms and Implementations, First Edition. Edited by Fa-Long Luo and Charlie Zhang.

elements at the BS. As the pioneering work of Marzetta shows, given perfect channel state information (CSI), as the number of BS antennas, in the limit, tends to infinity, the processing gain of the system tends to infinity. Accordingly, the effects of both noise and multiuser interference (MUI) completely fade away [1]. Hence, the network capacity, in theory, can be increased without bound by increasing the number of BS antennas [1].

As suggested by Marzetta [1], orthogonal frequency division multiplexing (OFDM) can be used to convert the frequency-selective channels between each MT and the multiple antennas at the BS to a set of flat fading channels. Thus, the flat gains associated with the set of channels within each subcarrier constitute the spreading gain vector that is used for despreading of the respective data stream.

FBMC methods have their roots in the pioneering work of Chang [2] and Saltzberg [3], who introduced multicarrier techniques over two decades before the introduction and application of OFDM to wireless communication systems. While OFDM relies on cyclic prefix (CP) samples to avoid intersymbol interference (ISI) and to convert the channel to a set of subcarrier channels with flat gains (precisely so when the channel impulse response duration is shorter than the CP). FBMC does not use CP but, adopts a large number of subcarriers, relying on the fact that when each subcarrier band is narrow, it is characterized by an approximately flat gain, and hence may suffer from only a negligible level of ISI.

A new and interesting finding of our research in this area is that in the case of massive MIMO systems, linear combination of the signal components from different channels smooths channel distortion [4, 5]. We call this property of FBMC in massive MIMO channels *self-equalization*. Hence, one may relax the requirement of having approximately flat gains for the subcarriers. This observation, which is confirmed numerically in this chapter, positions FBMC as a strong candidate in the application of massive MIMO. As a result, in a massive MIMO setup, one may significantly reduce the number of subcarriers in an FBMC system. This reduces both system complexity and the latency/delay caused by the synthesis filter bank (at the transmitter) and the analysis filter bank (at the receiver). Also, since linear combination of the signal components equalizes the channel gain across each subcarrier band, one may adopt larger constellation sizes, and hence further improve the system bandwidth efficiency. Moreover, increasing the subcarrier spacing has the obvious benefit of reducing sensitivity to carrier frequency offset.

An additional benefit of FBMC here is that carrier/spectral aggregation – in other words, using non-contiguous bands of spectrum for transmission – becomes a trivial task, since each subcarrier band is confined to an assigned range and has a negligible interference with other bands. This is not the case in OFDM [6].

A major factor in limiting the capacity of non-cooperative multicellular massive MIMO networks working in time-division duplex (TDD) mode is the *pilot contamination* problem [7]. Another contribution of this chapter is to address the pilot contamination problem in massive MIMO networks using cosine modulated multitone (CMT), a particular form of FBMC. This type of FBMC has a blind equalization capability [8]. In this chapter, we extend the blind equalization capability of CMT to massive MIMO applications in order to remove the channel estimation errors caused by contaminated pilots. Based on our observations of numerical results, the proposed blind equalization technique is able to remove pilot contamination effects in multicellular massive MIMO networks and converge towards the optimal linear minimum mean square error (MMSE) performance [9].

The rest of this chapter is laid out as follows. Section 4.1 is dedicated to the system model and formulation of FBMC in massive MIMO channels. The self-equalization property of FBMC in massive MIMO channels is discussed in Section 4.2. A qualitative comparison of FBMC and OFDM in massive MIMO systems is presented in Section 4.3. Section 4.4 addresses the pilot contamination problem in a multicellular massive MIMO network and, finally, conclusions are drawn in Section 4.5.

4.1 System Model and FBMC Formulation in Massive MIMO

Detailed information on different types of FBMC systems can be found in the book by Farhang-Boroujeny [10]. Among the different types of FBMC system, we are more interested in FBMC systems with overlapping subcarriers in the frequency domain, namely staggered multitone (SMT) and CMT, as these provide the highest bandwidth efficiency. Both CMT and SMT can be adopted for massive MIMO, leading to the same performance. However, it turns out that derivation and explanation of the results in the context of CMT are easier to follow. We thus limit our attention in the rest of this chapter to the development of CMT in massive MIMO applications. There are two different implementations of FBMC systems, based on polyphase networks (PPN) and frequency spreading concepts, respectively. In Section 4.1.1, we consider the polyphase implementation of CMT while the frequency spreading (FS) implementation is considered in Section 4.1.2.

In CMT, a set of pulse amplitude modulated (PAM) baseband data streams are vestigial sideband (VSB) modulated and placed at different subcarriers. Moreover, to allow separation of the data symbols (free of ISI and ICI), at the receiver, the carrier phase of the VSB signals is toggled between 0 and $\pi/2$ among adjacent subcarriers. The detailed equations explaining why this approach works can be found in a 1966 paper by Chang [2] and many other publications; a recommended reference is Farhang-Boroujeny's 2010 paper [11]. That author's book [10] provides more details, including the implementation structures and their relevant MATLAB codes.

Demodulation of each subcarrier in CMT is performed in four steps.

1. The received signal is down-converted to base-band using the corresponding carrier frequency to each subcarrier.
2. The demodulated signal is passed through a matched filter that extracts the desired signal at the base-band. Due to the overlap among the adjacent subcarriers, some residuals from adjacent subcarriers remain after matched filtering.
3. A complex-valued single tap equalizer is utilized to equalize the channel effect.[1]
4. As the real part of the equalized signal contains the desired PAM symbol and its imaginary part consists of a mix of ISI components and ICI components from the two adjacent bands, taking the real part of the equalized signal delivers the desired data symbol, free of ISI and ICI.

4.1.1 *Polyphase-based CMT in Massive MIMO*

Consider a multi-user MIMO setup similar to the one discussed in the paper by Marzetta [1]. There are K MTs and a BS in a cell. Each MT is equipped with a single transmit-and-receive

[1] This is based on the assumption that each subcarrier band is sufficiently narrow that it can be approximated by a flat gain. A multi-tap equalizer may be adopted if this approximation is not valid.

antenna, communicating with the cell BS in TDD manner. The BS is equipped with $N \gg K$ transmit/receive antennas, which are used to communicate with the K MTs in the cell *simultaneously*. We also assume, similar to Marzetta, that multicarrier modulation is used for data transmissions. However, we replace OFDM modulation by CMT.

Each MT is distinguished by the BS using the respective subcarrier gains between its antenna and the BS antennas. Ignoring the time and subcarrier indices in our formulation, for simplicity of the equations, a transmit symbol $s(\ell)$ from the ℓth MT arrives at the BS as a vector

$$\mathbf{x}_\ell = (s(\ell) + jq(\ell))\mathbf{h}_\ell \tag{4.1}$$

where $\mathbf{h}_\ell = [h_\ell(0), \ldots, h_\ell(N-1)]^{\mathrm{T}}$ is the channel gain vector whose elements are the gains between the ℓth MT and different antennas at the BS; $q(\ell)$ is the contribution of ISI and ICI. The vector $\mathbf{x}_\ell$ and similar contributions from other MTs, as well as the channel noise vector $\mathbf{v}$, add up and form the BS received signal vector

$$\mathbf{x} = \sum_{\ell=0}^{K-1} \mathbf{x}_\ell + \mathbf{v} \tag{4.2}$$

The BS uses a set of linear estimators that all take $\mathbf{x}$ as their input and provide the estimates of the users' data symbols $s(0), s(1), \cdots, s(K-1)$ at the output. To cast this process in a mathematical formulation and allow introduction of various choices of estimator, we proceed as follows. We define $\tilde{\mathbf{x}} = [\mathbf{x}_{\mathrm{R}}^{\mathrm{T}} \ \ \mathbf{x}_{\mathrm{I}}^{\mathrm{T}}]^{\mathrm{T}}$, $\tilde{\mathbf{v}} = [\mathbf{v}_{\mathrm{R}}^{\mathrm{T}} \ \ \mathbf{v}_{\mathrm{I}}^{\mathrm{T}}]^{\mathrm{T}}$, $\tilde{\mathbf{h}}_\ell = [\mathbf{h}_{\ell,\mathrm{R}}^{\mathrm{T}} \ \ \mathbf{h}_{\ell,\mathrm{I}}^{\mathrm{T}}]^{\mathrm{T}}$, $\breve{\mathbf{h}}_\ell = [-\mathbf{h}_{\ell,\mathrm{I}}^{\mathrm{T}} \ \ \mathbf{h}_{\ell,\mathrm{R}}^{\mathrm{T}}]^{\mathrm{T}}$, $\mathbf{s} = [s(0) \ \ s(1) \cdots s(K-1)]^{\mathrm{T}}$ and $\mathbf{q} = [q(0) \ \ q(1) \cdots q(K-1)]^{\mathrm{T}}$, where the subscripts 'R' and 'I' denote the real and imaginary parts, respectively. Using these definitions, Eq. (4.2) may be rearranged as

$$\tilde{\mathbf{x}} = \mathbf{A} \begin{bmatrix} \mathbf{s} \\ \mathbf{q} \end{bmatrix} + \tilde{\mathbf{v}} \tag{4.3}$$

where $\mathbf{A} = [\tilde{\mathbf{H}} \ \ \breve{\mathbf{H}}]$, and $\tilde{\mathbf{H}}$ and $\breve{\mathbf{H}}$ are $2N \times K$ matrices with columns of $\{\tilde{\mathbf{h}}_\ell, \ \ell = 0, 1, \cdots, K-1\}$ and $\{\breve{\mathbf{h}}_\ell, \ \ell = 0, 1, \cdots, K-1\}$, respectively. Eq. (4.3) has the familiar form that appears in the CDMA literature [12, 13]. Hence, a variety of solutions that have been given for CDMA systems can be immediately applied to the present problem as well. For instance, the matched filter (MF) detector gives an estimate of the vector $\mathbf{s}$, according to the equation

$$\hat{\mathbf{s}}_{\mathrm{MF}} = \mathbf{D}^{-1}\mathbf{\Gamma}\mathbf{A}^{\mathrm{T}}\tilde{\mathbf{x}} \tag{4.4}$$

where $\mathbf{D} = \mathrm{diag}\{\|\tilde{\mathbf{h}}_0\|^2, \ldots, \|\tilde{\mathbf{h}}_{K-1}\|^2\}$, the matrix $\mathbf{\Gamma}$ consists of the first K rows of the identity matrix $\mathbf{I}_{2K}$ of size $2K \times 2K$ and $\hat{\mathbf{s}}_{\mathrm{MF}} = [\hat{s}_{\mathrm{MF}}(0), \ldots, \hat{s}_{\mathrm{MF}}(K-1)]^{\mathrm{T}}$ whose ℓth element, $\hat{s}_{\mathrm{MF}}(\ell)$, is the estimated data symbol of user ℓ. Using Eq. (4.4), each element of $\hat{\mathbf{s}}_{\mathrm{MF}}$ can be expanded as

$$\hat{s}_{\mathrm{MF}}(\ell) = s(\ell) + \sum_{\substack{i=0 \\ i \neq \ell}}^{K-1} \frac{\tilde{\mathbf{h}}_\ell^{\mathrm{T}}}{\|\tilde{\mathbf{h}}_\ell\|^2} (\tilde{\mathbf{h}}_i s(i) + \breve{\mathbf{h}}_i q(i)) + \frac{\tilde{\mathbf{h}}_\ell^{\mathrm{T}}}{\|\tilde{\mathbf{h}}_\ell\|^2} \tilde{\mathbf{v}} \tag{4.5}$$

This leads to a receiver structure similar to that of Marzetta, which shows that when the number of antennas, N, increases to infinity, the multi-user interference and noise effects tend to zero

[1]. Hence $\hat{\mathbf{s}} = \mathbf{s}$, where the vector $\hat{\mathbf{s}}$ is an estimate of $\mathbf{s}$, and the receiver will be optimum. In the context of the CDMA literature, this has the explanation that as N tends to infinity, the processing gain also goes to infinity and accordingly multi-user interference and noise effects fade away.

In realistic situations, when N is finite, the MF estimator is not optimal. A superior estimator is the MMSE estimator

$$\hat{\mathbf{s}} = \mathbf{W}^{\mathrm{T}}\tilde{\mathbf{x}} \tag{4.6}$$

where the coefficient matrix $\mathbf{W}$ is chosen to minimize the cost function

$$\zeta = \mathbb{E}[\|\mathbf{s} - \mathbf{W}^{\mathrm{T}}\tilde{\mathbf{x}}\|^2] \tag{4.7}$$

This solution is optimal in the sense that it maximizes the signal-to-interference-plus-noise ratio (SINR) [13].

Following the standard derivations, the optimal choice of $\mathbf{W}$ is obtained as

$$\mathbf{W}_{\mathrm{o}} = \mathbf{A}(\mathbf{A}^{\mathrm{T}}\mathbf{A} + \sigma_v^2\mathbf{I}_{2K})^{-1}\mathbf{\Gamma}^{\mathrm{T}} \tag{4.8}$$

Here, it is assumed that the elements of the noise vector $\tilde{\mathbf{v}}$ are independent and identically distributed Gaussian random variables with variances of σ_v^2, hence $\mathbb{E}[\tilde{\mathbf{v}}\tilde{\mathbf{v}}^{\mathrm{T}}] = \sigma_v^2\mathbf{I}_{2K}$. The columns of $\mathbf{W}_{\mathrm{o}}$ contain the optimal filter tap weights for different users.

Substitution of Eq. (4.8) into Eq. (4.6) leads to the MMSE solution

$$\begin{aligned}\hat{s}_{\mathrm{MMSE}}(\ell) &= \mathbf{w}_{\mathrm{o},\ell}^{\mathrm{T}}\tilde{\mathbf{h}}_\ell s(\ell) + \sum_{\substack{i=0\\ i\neq\ell}}^{K-1} \mathbf{w}_{\mathrm{o},\ell}^{\mathrm{T}}\tilde{\mathbf{h}}_i s(i)\\ &\quad + \sum_{i=0}^{K-1} \mathbf{w}_{\mathrm{o},\ell}^{\mathrm{T}}\breve{\mathbf{h}}_i q(i) + \mathbf{w}_{\mathrm{o},\ell}^{\mathrm{T}}\tilde{\mathbf{v}}\end{aligned} \tag{4.9}$$

where $\mathbf{w}_{\mathrm{o},\ell}$ is the ℓth column of $\mathbf{W}_{\mathrm{o}}$. Ignoring the off-diagonal elements of $(\mathbf{A}^{\mathrm{T}}\mathbf{A} + \sigma_v^2\mathbf{I}_{2K})$ and also removing the term $\sigma_v^2\mathbf{I}_{2K}$ from Eq. (4.8), one will realize that Eq. (4.9) boils down to the MF tap weights in Eq. (4.5).

The first terms on the right-hand side of Eqs. (4.5) and (4.9) are the desired signal and the rest are the interference-plus-noise terms. We consider $s(\ell)$ and $q(\ell)$ as independent variables with variance of unity. With the assumption of having a flat channel impulse response in each subcarrier band, SINR at the output of the MF and MMSE detectors for user ℓ in a certain subcarrier can be derived, respectively, as

$$\mathrm{SINR}_{\mathrm{MF}}(\ell) = \frac{\|\tilde{\mathbf{h}}_\ell\|^4}{\sum\limits_{\substack{i=0\\ i\neq\ell}}^{K-1} (|\tilde{\mathbf{h}}_\ell^{\mathrm{T}}\tilde{\mathbf{h}}_i|^2 + |\tilde{\mathbf{h}}_\ell^{\mathrm{T}}\breve{\mathbf{h}}_i|^2) + \sigma_v^2\|\tilde{\mathbf{h}}_\ell\|^2} \tag{4.10}$$

and

$$\mathrm{SINR}_{\mathrm{MMSE}}(\ell) = \frac{|\mathbf{w}_{\mathrm{o},\ell}^{\mathrm{T}}\tilde{\mathbf{h}}_\ell|^2}{\sum\limits_{\substack{i=0\\ i\neq\ell}}^{K-1} |\mathbf{w}_{\mathrm{o},\ell}^{\mathrm{T}}\tilde{\mathbf{h}}_i|^2 + \sum\limits_{i=0}^{K-1} |\mathbf{w}_{\mathrm{o},\ell}^{\mathrm{T}}\breve{\mathbf{h}}_i|^2 + \sigma_v^2\|\mathbf{w}_{\mathrm{o},\ell}\|^2} \tag{4.11}$$

4.1.2 FS-based CMT in Massive MIMO

An interesting property of FS-FBMC structure is its high-performance equalization capability. This frequency-spreading equalization (FSE) can be utilized in the context of massive MIMO, as will be discussed next. In the rest of this section, and for our later reference, we review the frequency-spreading structure when applied to CMT, which we hereafter call *FS-CMT*.

The basic idea of the frequency-spreading method is that a discrete-time square-root Nyquist filter, $p(n)$, of size $P = ML$, can be synthesized in the frequency domain using a set of $2M - 1$ distinct tones, as

$$p(n) = \sum_{k=-M+1}^{M-1} c_k e^{j\frac{2\pi kn}{P}} \tag{4.12}$$

Here, c_k is a real-valued coefficient that indicates the frequency response of $p(n)$ at frequency $\omega_k = \frac{2\pi k}{P}$, M is the overlapping factor, which signifies the number of symbols that overlap in the time domain, and L is the sample spacing of the zero-crossings of the Nyquist filter $q(n) = p(n) \star p^*(-n)$.

When the above idea is extended to the multicarrier case, the symbols at each subcarrier should be pulse-shaped in the frequency domain using the filter coefficients c_k. An inverse discrete Fourier transform (IDFT) block of size P can then be used to calculate the time-domain samples from their frequency-domain counterparts. Casting the above discussion into a mathematical form, the FS-CMT symbol can be obtained as

$$\mathbf{x}(n) = \mathbf{F}_P^{\mathrm{H}} \mathbf{\Lambda} \mathbf{\Phi} \mathbf{s}(n) \tag{4.13}$$

Here, $\mathbf{s}(n) = \begin{bmatrix} s_0(n) & s_1(n) & \cdots & s_{L-1}(n) \end{bmatrix}^{\mathrm{T}}$, where $s_k(n)$ indicates the data symbol at the kth subcarrier and nth symbol time, $\mathbf{\Phi}$ indicates the phase-adjustment matrix, which is an $L \times L$ diagonal matrix with the diagonal elements $\{1, e^{j\frac{\pi}{2}}, \ldots\ e^{j\frac{\pi}{2}(L-1)}\}$, $\mathbf{\Lambda}$ is the spreading matrix of size $P \times L$ whose kth column contains the coefficients c_ℓ, which are centered at the center frequency of the kth subcarrier, $\mathbf{F}_P^{\mathrm{H}}$ indicates the IDFT matrix of the size $P \times P$ and $\mathbf{x}(n)$ is a $P \times 1$ vector indicating the nth FS-CMT symbol in the time domain.

Noting that the size of each FS-CMT symbol is P and the symbol spacing is equal to $L/2$, the transmit signal can be constructed by performing the overlap-and-add operations on the symbol vectors as illustrated in Figure 4.1. Assuming an ideal channel and having a perfect

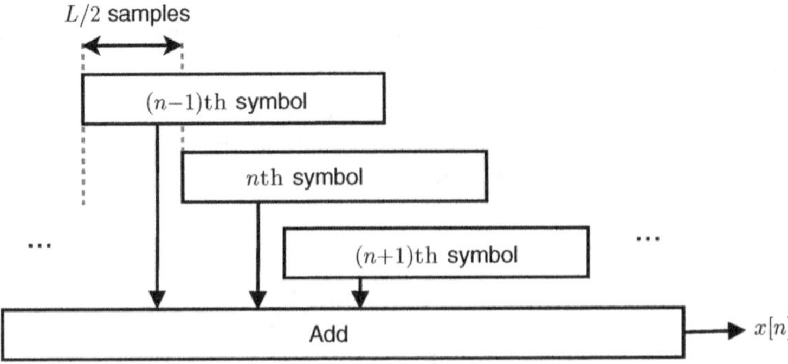

Figure 4.1 Overlap and add operation

synchronization between transmitter and receiver, the P samples that correspond to the nth symbol are collected in the vector $\hat{\mathbf{r}}(n)$ at the receiver side, and the data symbols are recovered according to

$$\hat{\mathbf{s}}(n) = \Re\{\mathbf{\Phi}^{-1}\mathbf{\Lambda}^{\mathrm{T}}\mathbf{F}_P\hat{\mathbf{r}}(n)\} \tag{4.14}$$

Since we assumed an ideal channel, equalization was not considered here. In the presence of the channel, the equalization can be embedded after obtaining the discrete Fourier transform (DFT) output samples [14]. In the following, we extend the equalization scheme proposed by Bellanger for single antenna systems [14] to the context of massive MIMO.

Based on the structure of the CMT receiver, the output at the ith bin of the DFT block from the ℓth user at all the receive antennas can be stacked into an $N \times 1$ vector $\tilde{\mathbf{r}}_i^\ell = r_i^\ell \mathbf{h}_i(\ell)$, where $\mathbf{h}_i(\ell)$ is the channel gain vector whose elements are the gains between the ℓh MT antenna and the BS antennas at the ith frequency bin. The coefficient r_i^ℓ is the received signal at the ith bin of the output of the DFT block at the receiver from user ℓ in an ideal channel. The vectors $\tilde{\mathbf{r}}_i^\ell$ are the received signals from different users that contribute to form the received signal at the BS. Accordingly, the received signal vector can be written as

$$\tilde{\mathbf{r}}_i = \sum_{\ell=0}^{K-1} \tilde{\mathbf{r}}_i^\ell + \mathbf{v}_i = \mathbf{H}_i\mathbf{r}_i + \mathbf{v}_i \tag{4.15}$$

where $\mathbf{H}_i = [\mathbf{h}_i(0), \dots, \mathbf{h}_i(K-1)]$, $\mathbf{r}_i = [r_i^0, \dots, r_i^{K-1}]^{\mathrm{T}}$ and $\mathbf{v}_i$ is an $N \times 1$ vector that contains the FFT samples of the additive white Gaussian noise at the ith bin of the output of the FFT blocks at different receive antennas.

In order to estimate different users' transmitted symbols, a number of equalization techniques can be utilized, for example MF and MMSE. However, as noted above, in realistic situations when the number of BS antennas is finite, MF equalization involves some performance degradation due to the residual multi-user interference. In this section, our discussion will be limited to MMSE-FSE, which is known to be a superior technique.

In MMSE-FSE for massive MIMO systems, similar to single antenna FS-FBMC systems, equalization needs to be performed before despreading the output of the FFT blocks. Therefore, the MMSE linear combination aims at estimating the vector $\mathbf{r}_i$ that contains the equalized samples of different users at the output of their FFT blocks. The MMSE solution of Eq. (4.15) is optimal because it maximizes the SINR [13]. Following the same line of derivation as in Section 4.1.1, the MMSE estimates of the $\mathbf{r}_i$s can be obtained as

$$\hat{\mathbf{r}}_i = \mathbf{W}_i^{\mathrm{H}}\tilde{\mathbf{r}}_i \tag{4.16}$$

where the coefficient matrix $\mathbf{W}_i = \mathbf{H}_i(\mathbf{H}_i^{\mathrm{H}}\mathbf{H}_i + \sigma_v^2\mathbf{I}_K)^{-1}$ contains the optimal MMSE filter tap weights for different users in its columns. It is worth mentioning that the elements of the noise vector $\mathbf{v}_i$ are assumed to be independent and identically distributed Gaussian random variables with variances of σ_v^2 and $\mathbb{E}[\mathbf{v}_i\mathbf{v}_i^{\mathrm{H}}] = \sigma_v^2\mathbf{I}_P$. After the MMSE estimation of the vectors $\hat{\mathbf{r}}_i$ for $i = 0, \dots, P-1$, frequency despreading can be separately applied for each user. To this end, let $\hat{\mathbf{R}} = [\hat{\mathbf{r}}_0, \dots, \hat{\mathbf{r}}_{P-1}]^{\mathrm{T}}$. Consequently, the MMSE estimates of the transmitted data symbols of different users – in other words $\mathbf{s}_\ell$ for $\ell = 0, \dots, K-1$ – can be obtained through frequency despreading

$$\hat{\mathbf{S}} = \Re\{\mathbf{\Phi}^{-1}\mathbf{\Lambda}^{\mathrm{T}}\hat{\mathbf{R}}\} \tag{4.17}$$

where the $L \times K$ matrix $\hat{\mathbf{S}}$ contains the MMSE estimation of different users' transmitted data symbols – the $\hat{\mathbf{s}}_\ell$s – on its columns.

Based on Eqs. (4.15) to (4.17) and the assumption of having a flat fading channel gain over each frequency bin of DFT, the output SINR of the CMT receiver with MMSE channel equalization for the ℓth user at subcarrier i can be calculated as follows

$$\mathrm{SINR}_{\mathrm{MMSE}}^{(\ell,i)} = \frac{P_{\mathrm{S}}^{(\ell,i)}}{P_{\mathrm{I}}^{(\ell,i)}} \tag{4.18}$$

where $P_{\mathrm{S}}^{(\ell,i)}$ and $P_{\mathrm{I}}^{(\ell,i)}$ are signal and interference-plus-noise powers, respectively. $P_{\mathrm{S}}^{(\ell,i)}$ and $P_{\mathrm{I}}^{(\ell,i)}$ can be obtained as

$$P_{\mathrm{S}}^{(\ell,i)} = \sum_{m=-M+1}^{M-1} c_k^2 |\mathbf{w}_{i+m,\Re}^{\mathrm{T}}(\ell)\mathbf{h}_{i+m,\Re}(\ell) + \mathbf{w}_{i+m,\Im}^{\mathrm{T}}(\ell)\mathbf{h}_{i+m,\Im}(\ell)|^2 \tag{4.19}$$

and

$$\begin{aligned} P_{\mathrm{I}}^{(\ell,i)} = & \sum_{m=-M+1}^{M-1} \sum_{\substack{k=0 \\ k\neq\ell}}^{K-1} c_m^2 |\mathbf{w}_{i+m,\Re}^{\mathrm{T}}(\ell)\mathbf{h}_{i+m,\Re}(k) + \mathbf{w}_{i+m,\Im}^{\mathrm{T}}(\ell)\mathbf{h}_{i+m,\Im}(k)|^2 \\ & + \sum_{m=-M+1}^{M-1} \sum_{k=0}^{K-1} c_m^2 |\mathbf{w}_{i+m,\Re}^{\mathrm{T}}(\ell)\mathbf{h}_{i+m,\Im}(k) - \mathbf{w}_{i+m,\Im}^{\mathrm{T}}(\ell)\mathbf{h}_{i+m,\Re}(k)|^2 \\ & + \sum_{m=-M+1}^{M-1} c_m^2 \sigma_v^2 \|\mathbf{w}_{i+m}(\ell)\|^2 \end{aligned} \tag{4.20}$$

where $\mathbf{w}_{i,\Re}(k)$ and $\mathbf{h}_{i,\Re}(k)$ represent the real parts of the kth columns of the matrices $\mathbf{W}_i$ and $\mathbf{H}_i$, respectively. Similarly, the subscript $\Im$ in Eqs. (4.19) and (4.20) indicates the imaginary part of the corresponding vectors. It is worth mentioning that Eq. (4.18) is used as a benchmark to investigate the channel flatness assumption in Section 4.2.2.

4.2 Self-equalization Property of FBMC in Massive MIMO

In conventional (single-input single-output) FBMC systems, in order to reduce channel equalization to single tap per subcarrier, it is often assumed that the number of subcarriers is very large, so each subcarrier band may be approximated by a flat gain. This clearly has the undesirable effect of reducing the symbol rate (per subcarrier), which brings with it:

- the need for longer pilot preambles (equivalently, a reduction in the bandwidth efficiency)
- increase of latency in the channel
- higher sensitivity to carrier frequency offset (CFO)
- higher peak-to-average power ratio (PAPR) due to the large number of subcarriers, which increases the dynamic range of the FBMC signal.

Massive MIMO channels have an interesting property that allows us to resolve the above problems. The MF and MMSE detectors that are used to combine signals from the receive antennas average the distortions from different channels and thus, as the number of BS antennas increases, result in a nearly equalized gain across each subcarrier band. This property of massive MIMO channels, which we call *self-equalization*, is confirmed numerically in Sections 4.2.1 and 4.2.2.

4.2.1 Numerical Study of Polyphase-based CMT in a Massive MIMO Channel

In this section, a broad set of numerical results is presented to confirm the theoretical developments of Section 4.1, as well as the earlier claims on the self-equalization property. It was noted in the previous sections that massive MIMO, through use of a large number of antennas at the BS, can provide a large processing gain. Hence noise and MUI effects can be reduced. In addition, when FBMC is used for signal modulation, linear combination of the signals from multiple antennas at the BS has a flattening effect (i.e. the channel will be equalized) over each subcarrier band. This interesting impact of massive MIMO allows a reduction in the number of subcarriers in FBMC and this, in turn, has the effect of reducing:

- complexity
- the preamble (training) length, thus increasing bandwidth efficiency
- sensitivity to CFO
- PAPR
- system latency.

The first set of results that we present in this section provides evidence of the self-equalization property.

Figure 4.2 presents a set of results that highlights the effect of increasing the number of antennas at the receiver on the signal-to-interference ratio (SIR) for different numbers of subcarriers in the single-user case. The results are presented for a noise-free channel to explore the impact of the number of subcarriers (or equivalently, the width of each subcarrier band) and the number of BS antennas in achieving a flat channel response over each subcarrier band. The results are for a sample set of channel responses generated according to the SUI-4 channel model proposed by the IEEE802.16 broadband wireless access working group [15]. SIRs are evaluated for all subcarrier channels. Note that in each curve, the number of points along the normalized frequency is equal to the number of subcarrier bands, L. For the channel model used here, the total bandwidth, equivalent to the normalized frequency one, is equal to 2.8 MHz. This, in turn, means the subcarrier spacing in each case is equal to $2800/L$ kHz. As an example, when $L = 32$, subcarrier spacing is equal to 87.5 kHz. This, compared to the subcarrier spacing in OFDM-based standards, for example IEEE 802.16 and LTE, is relatively broad; 87.5/15 $\approx$ 6 times larger. As noted above, reducing the number of subcarriers (or, equivalently, increasing the symbol rate in each subcarrier band) reduces transmission latency, increases bandwidth efficiency, and reduces sensitivity to CFO and PAPR.

Next, the channel noise is added to explore similar results to those in Figure 4.2. Since an MF/MMSE receiver in the uplink (or a precoding in the downlink) has a processing gain of N,

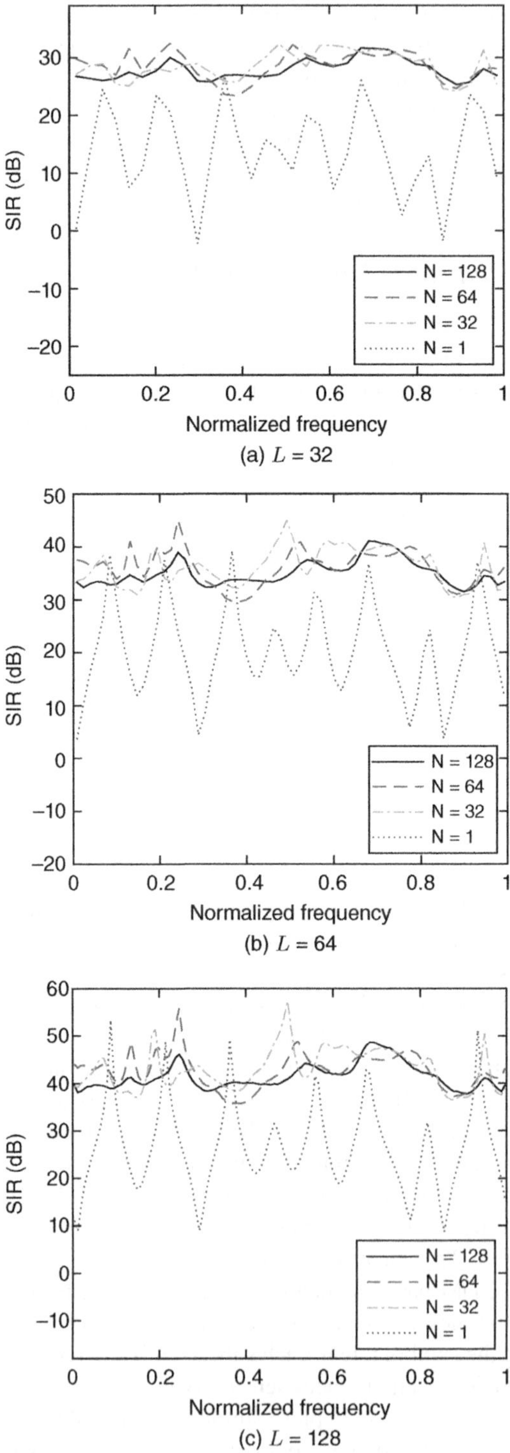

Figure 4.2 SIR comparison of the MF linear combination technique for different numbers of receive antennas, N, and different number of subcarriers, L

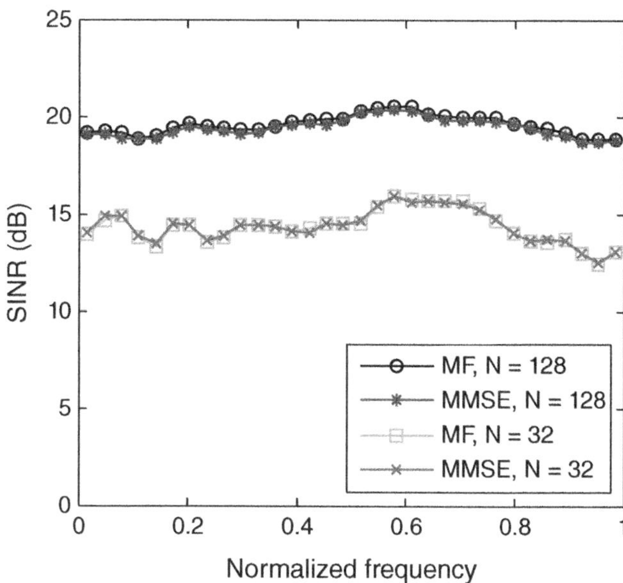

Figure 4.3 SINR comparison between the MMSE and MF linear combination techniques in the single-user case with $L = 32$, when the user's SNR at the receiver input is -1 dB for the two cases of $N = 128$ and $N = 32$

the SINR at the output may be calculated as $\text{SNR}_{\text{in}} + 10\log_{10}N$, where SNR_{in} is the SNR at each BS antenna. The results, presented in Figure 4.3, are for the cases where there are 32 and 128 antennas at the BS, the number of subcarriers is equal to 32 and SNR_{in} is -1 dB. As seen here, the SINR curves in both MMSE and MF receivers coincide. The processing gains for $N = 128$ and $N = 32$ antennas are respectively 21 and 15 dB, and the expected output SINR values 20 and 14 dB are observed.

The above results were presented for the single-user case. The situation changes significantly in the multi-user scenario due to the presence of MUI. As shown in the following results, when multiple MTs simultaneously communicate with a BS, MMSE outperforms MF by a significant margin. This result, which is applicable to both FBMC and OFDM-based MIMO systems, is indeed very interesting and has also recently been reported by Krishnan *et al.* [16].

The analytical SINR relationships derived in Section 4.1 are calculated with the assumption of having a flat channel per subcarrier. Therefore, they can be chosen as benchmarks to evaluate the channel flatness in the subcarrier bands. Figs. 4.4 and 4.5 present the theoretical and simulation results in a multi-user scenario where $K = 6$, $N = 128$, the target SINR is 20 dB (the SNR at each antenna at the BS is selected as $20 - 10\log_{10}N$ dB) and the cases of $L = 64$ and $L = 32$ are examined. As the figures show, the MMSE combining technique is superior to the MF one and its SINR is about the same for all the subcarriers, in other words it has smaller variance across the subcarriers. When $L = 64$ (see Figure (4.4)), for the SINR curves for both MF and MMSE techniques, the simulation results coincide with the theoretical ones almost perfectly, confirming the self-equalization property of linear combination in massive MIMO

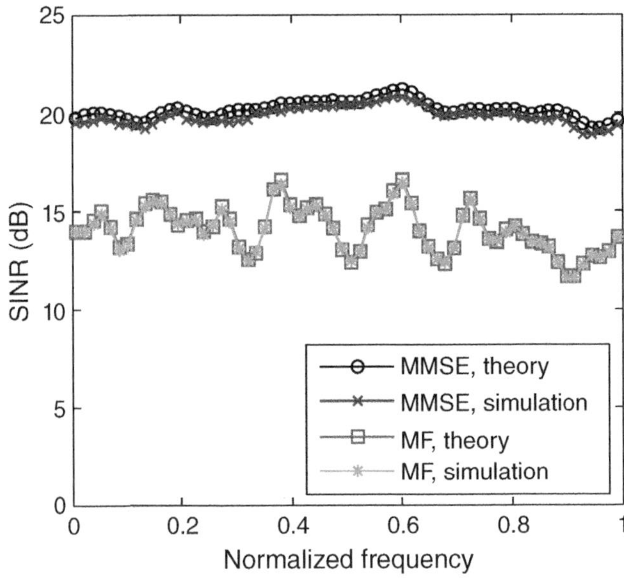

Figure 4.4 SINR comparison between the MMSE and MF linear combination techniques for the case of $K = 6$, $L = 64$ and $N = 128$

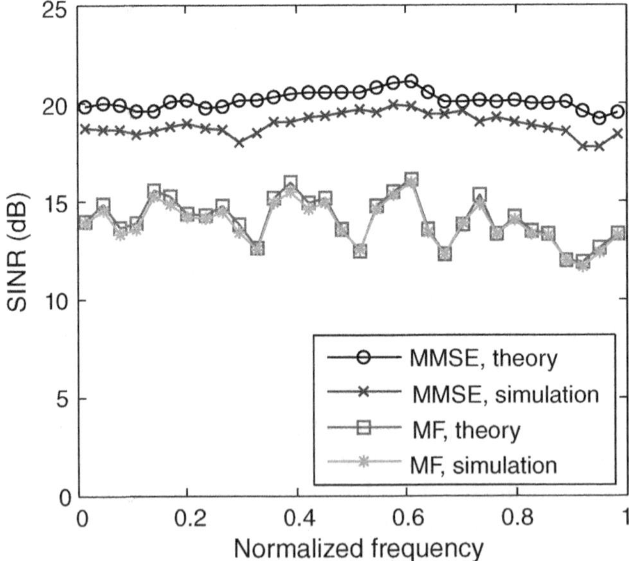

Figure 4.5 SINR comparison between the MMSE and MF linear combination techniques for the case of $K = 6$, $L = 32$ and $N = 128$

FBMC systems. When $L = 32$ (wide-band subcarriers with 87.5 kHz width), the SINR curves from simulations for the MF receiver are still the same as the theoretical curve. However, the SINR simulations for MMSE combining falls 1 dB below the theoretical predictions.

4.2.2 Numerical Study of FS-based CMT in a Massive MIMO Channel

In this section, the theoretical developments of Section 4.1.2 are analyzed and corroborated through numerical results. We use the results of Section 4.2.1 as the basis to evaluate the signal-processing power of the FS method in the context of massive MIMO. It is worth noting that all of our simulations are based on a sample set of channel responses generated according to the SUI-4 channel model proposed by the IEEE802.16 broadband wireless access working group [15]. Additionally, the channels between different users and different antennas are considered to be independent with respect to each other.

In the first set of simulations, a noise-free single-user scenario is considered in order to investigate the self-equalization property in FS-FBMC systems. Figure 4.6 compares the SIR performance of the polyphase network based FBMC (PPN-FBMC) with FS-FBMC systems. A single-tap equalizer per subcarrier is used in the PPN-FBMC structure. SIRs are evaluated at all the subcarrier channels. Note that in each curve, the number of points along the normalized frequency is equal to the number of subcarrier bands, L. From Figure 4.6, it can be understood that a SIR improvement of higher than 30 dB can be achieved through FS-FBMC in comparison with the PPN-FBMC structure. This means that higher-order modulation schemes can be utilized in FS-FBMC-based massive MIMO systems. The total bandwidth is fixed at 2.8 MHz, the number of subcarriers $L = 16$ and an overlapping factor of $M = 4$ is used. This results in a subcarrier spacing of $2800/16 = 175$ kHz. This subcarrier spacing is relatively wide. In

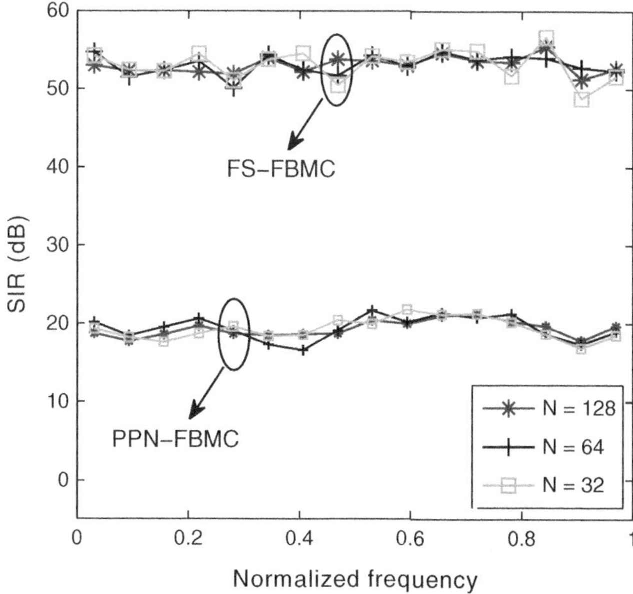

Figure 4.6 SIR for the case $L = 16$, $M = 4$, $K = 1$ and different number of BS antennas

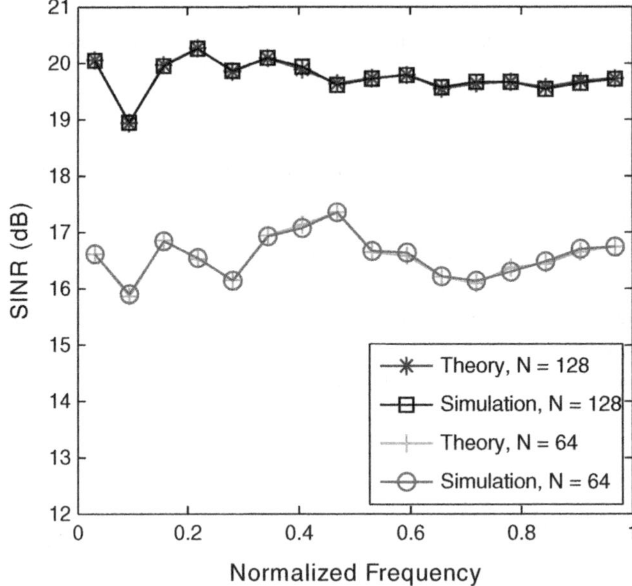

Figure 4.7 SINR evaluation of MMSE linear combination for the case of $L = 16$, $M = 4$, and $K = 6$. The SNR at the receiver input is -1 dB

fact, it is $175/15 \approx 12$ times larger than the subcarrier spacing of OFDM-based systems, such as IEEE 802.16 and LTE.

Next, we consider a multi-user scenario with $K = 6$ users, $L = 16$ subcarriers and $M = 4$. The SNR of -1 dB at the input of BS antennas is considered. The output SINR can be obtained using $\text{SNR}_{\text{in}} + 10\log_{10}N$ in dB. Figure 4.7 presents the theoretical (based on Eq. (4.18)) and simulation-based SINR values at different subcarriers for two cases of $N = 128$ and $N = 64$. As the figure depicts, the theoretical derivations match perfectly with the simulation results. The subcarrier spacing that achieves the same SINR as that of Eq. (4.18) is four times larger than what was suggested in Section 4.2.1. This can be deduced from comparing Figs. 4.7 and 4.4.

4.3 Comparison with OFDM

In the case of OFDM, the multi-user Eq. (4.3) simplifies to

$$\mathbf{x} = \mathbf{H}\mathbf{s} + \mathbf{v}. \tag{4.21}$$

Here, $\mathbf{x}$ is the vector of the received signal samples (over a specified subcarrier), $\mathbf{H}$ is the matrix of channel gains, $\mathbf{s}$ is the vector of data symbols from different users, and $\mathbf{v}$ is the channel noise vector. All these quantities are complex-valued.

The following differences pertain if one compares Eqs. (4.3) and (4.21).

1. While all variables/constants in Eq. (4.3) are real-valued, their counterparts in Eq. (4.21) are complex-valued.
2. The users' data vector **s** in Eq. (4.21) has K elements. This means each user receives multi-user interference from $K-1$ other users. In Eq. (4.3), on the other hand, each user receives interference from $2(K-1)$ users, of which $K-1$ are actual users and the rest we refer to as 'virtual' users. For instance, if the data user of interest is $s(0)$, it may receive interference from $s(1)$, $s(2)$, $\cdots$, $s(K-1)$ (the actual user symbols) and $q(1)$, $q(2)$, $\cdots$, $q(K-1)$ (the virtual user symbols – the contributions from the ISI and ICI components).
3. While the processing gain in the OFDM-based systems is N, equal to the number of elements is each column of channel gain matrix **H**, this number doubles in the CMT-based system.
4. Considering observations 2) and 3), it is readily concluded that both CMT-based and OFDM-based systems suffer from the same level of multi-user interference.

These observations imply that in massive MIMO, signal enhancement through linear combination leads to the same results for both OFDM and CMT-based systems. Nevertheless, CMT offers the following advantages over OFDM.

More flexible carrier aggregation
To make better use of the available spectrum, recent wireless standards put a lot of emphasis on carrier aggregation. A variety of implementations of carrier aggregation has been reported. Apparently, some companies use multiple radios to transmit and receive signals over different portions of the spectrums. Others, such as Fettweis *et al.* [17], suggest modulation and filtering of the aggregated spectra. These solutions are more expensive and less flexible than carrier aggregation in FBMC, which may be one compelling reason to adopt it rather than OFDM.

Lower sensitivity to CFO
As numerically demonstrated in the previous sections, compared to OFDM, FBMC allows an increase in subcarrier spacing. This, in turn, reduces the sensitivity of FBMC to CFO.

Lower PAPR
A reduced number of subcarriers naturally brings a property of low PAPR to the FBMC signal.

Higher bandwidth efficiency
Because of the absence of CP in FBMC, it would be expected to bring higher bandwidth efficiency than OFDM. One point to be noted here is that FBMC usually requires a longer preamble than OFDM. The possibility of reducing the number of subcarriers in FBMC, noted above, can significantly reduce the preamble length in FBMC. Hence, it reduces the overhead of the preamble to a negligible amount.

4.4 Blind Equalization and Pilot Decontamination

The pilot contamination problem in massive MIMO networks operating in TDD mode can limit their expected capacity to a great extent [18]. This section addresses this problem in CMT-based massive MIMO networks, taking advantage of their so-called *blind equalization* property. It is worth mentioning that we consider the polyphase implementation of CMT system here. We extend and apply the blind equalization technique from single antenna case [8] to multicellular massive MIMO systems and show that it can remove channel estimation errors due to the pilot contamination effect without any need for cooperation between different cells or transmission of additional training information. Our numerical results, presented in Section 4.4.1, advocate the efficacy of the proposed blind technique in improving the channel estimation accuracy and removing the residual channel estimation errors caused by users of other cells.

Consider a multicellular massive MIMO network consisting of $C > 1$ cells and K MTs in each cell. Each MT is equipped with a single transmit-and-receive antenna, communicating with the BS in a TDD manner. Each BS is equipped with $N \gg K$ transmit-and-receive antennas that are used to communicate with the K MTs in the cell *simultaneously*.

Each MT is distinguished by the BS using the respective subcarrier gains between its antenna and the BS antennas. Ignoring the time and subcarrier indices in our formulation, and for simplicity of the equations, a transmit symbol $s_c(\ell)$ from the ℓth MT located in the cth cell, arrives at the jth BS as a vector

$$\mathbf{x}_{j\ell} = t_c(\ell)\mathbf{h}_{cj\ell}, \tag{4.22}$$

where $t_c(\ell) = s_c(\ell) + jq_c(\ell)$ and $q_c(\ell)$ is the contribution of ISI and ICI. $\mathbf{h}_{cj\ell} = [h_{cj\ell}(0), \ldots, h_{cj\ell}(N-1)]^{\mathrm{T}}$ indicates the channel gain vector whose elements are the gains between the ℓth MT located in cell c and different antennas at the jth BS. The received signal vector at the jth BS, $\mathbf{x}_j$, contains contributions from its own MTs and the ones located in its neighboring cells apart from the channel noise vector $\mathbf{v}_j$.

$$\mathbf{x}_j = \sum_{c=0}^{C-1}\sum_{\ell=0}^{K-1} \alpha_{cj\ell}\mathbf{x}_{c\ell} + \mathbf{v}_j \tag{4.23}$$

where $\alpha_{cj\ell}$ values are the cross-gain factors between the ℓth user of the cth cell and the BS antennas of the jth cell, which can be thought as path loss coefficients. In general, $\alpha_{cj\ell} \in [0, 1]$. Considering perfect power control for the users of each cell implies that $\alpha_{cj\ell} = 1$ for $c = j$. The vector $\mathbf{x}_j$ is fed into a set of linear estimators at the jth BS to estimate the users' data symbols $s_j(0)$, $s_j(1)$, $\cdots$, $s_j(K-1)$. Eq. (4.23) can be rearranged as

$$\mathbf{x}_j = \mathbf{H}_{jj}\mathbf{t}_j + \sum_{\substack{c=0\\c\neq j}}^{C-1} \mathbf{H}_{cj}\boldsymbol{\alpha}_{cj}\mathbf{t}_c + \mathbf{v}_j \tag{4.24}$$

where the vector $\mathbf{t}_c = [t_c(0), \ldots, t_c(K-1)]^{\mathrm{T}}$, $\boldsymbol{\alpha}_{cj} = \mathrm{diag}\{\alpha_{cj0}, \ldots, \alpha_{cj(K-1)}\}$ and $\mathbf{H}_{cj}$ are $N \times K$ fast-fading channel matrices with the columns $\mathbf{h}_{cj\ell}, \ell = 0, 1, \ldots, K-1$. With the assumption of perfect CSI knowledge, the matched filter tap-weight vector for user ℓ located in the jth cell can be represented as

$$\mathbf{w}_{j\ell} = \frac{\mathbf{h}_{jj\ell}}{\mathbf{h}_{jj\ell}^{\mathrm{H}}\mathbf{h}_{jj\ell}} \tag{4.25}$$

The estimated users' data symbols at the output of the matched filters of the cell j can be mathematically written as

$$\hat{\mathbf{s}}_j = \Re\{\mathbf{D}^{-1}\mathbf{H}_{jj}^{\mathrm{H}}\mathbf{x}_j\} \tag{4.26}$$

where $\mathbf{D} = \mathrm{diag}\{\|\mathbf{h}_{jj0}\|^2, \ldots, \|\mathbf{h}_{jj(K-1)}\|^2\}$ and $\hat{\mathbf{s}}_j$ is the estimation of the vector $\mathbf{s}_j = [s_j(0), \ldots, s_j(K-1)]^{\mathrm{T}}$, which contains the users' transmitted data symbols. As discussed in Marzetta's paper [1], given perfect CSI knowledge at the BS, when the number of antennas, N, tends to infinity, the antenna array gain goes to infinity and hence the MUI and thermal-noise effects vanish. As a result, we have $\hat{\mathbf{s}}_j = \mathbf{s}_j$ and the receiver will be optimum.

The channel gains between the MTs and the BS antennas in each cell are estimated through training pilots transmitted during the uplink phase. The MTs in each cell transmit pilots from a set of mutually orthogonal pilot sequences, which allows the BS to distinguish between the channel impulse responses of different users in the channel estimation stage. As Jose *et al.* argued [18], the channel coherence time does not allow the users of neighboring cells to use orthogonal pilot sequences in the multicellular scenario. In TDD multicellular massive MIMO networks, C base stations use the same set of pilot sequences as well as frequencies. In addition, synchronous transmissions are assumed. Therefore, the same set of pilot sequences being used in neighboring cells will adversely affect the channel estimates at the BS. This effect is called pilot contamination. After correlating the received training symbols with the set of pilot sequences at the jth BS, the estimates of the channel gains between the MTs and massive array antennas of the jth BS can be given as

$$\hat{\mathbf{H}}_{jj} = \mathbf{H}_{jj} + \sum_{\substack{c=0 \\ c\neq j}}^{C-1} \mathbf{H}_{cj}\boldsymbol{\alpha}_{cj} + \tilde{\mathbf{V}}_j \tag{4.27}$$

where the $N \times K$ matrix $\tilde{\mathbf{V}}_j = [\tilde{\mathbf{v}}_j(0), \ldots, \tilde{\mathbf{v}}_j(K-1)]$ contains the channel noise vector $\mathbf{v}_j$ correlated with the pilot sequences on its columns. As one can see from Eq. (4.27), the channel estimates at the jth cell are corrupted by the channel impulse responses of its adjacent cells. Therefore, even with an infinite number of receive antennas at the BS, there will be some MUI from the users of other cells. Figure 4.8 shows this problem in a multicellular massive MIMO network; the dotted arrows show the interference from other cells and the solid one shows the transmitted signal of the desired user in the training phase; in other words, uplink transmission. Pilot contamination can have detrimental effects on the performance of multicellular networks and greatly impair their sum rate [1]. In the next section, we will extend the blind equalization property of CMT to massive MIMO systems in order to purify the channel estimates and tackle the pilot contamination problem without any need for cooperation among the cells or additional training information.

As noted by Farhang-Boroujeny [8], the imaginary part of the CMT symbol at each subcarrier – $q_c(\ell)$ – is formed from a linear combination of a large number of symbols from the

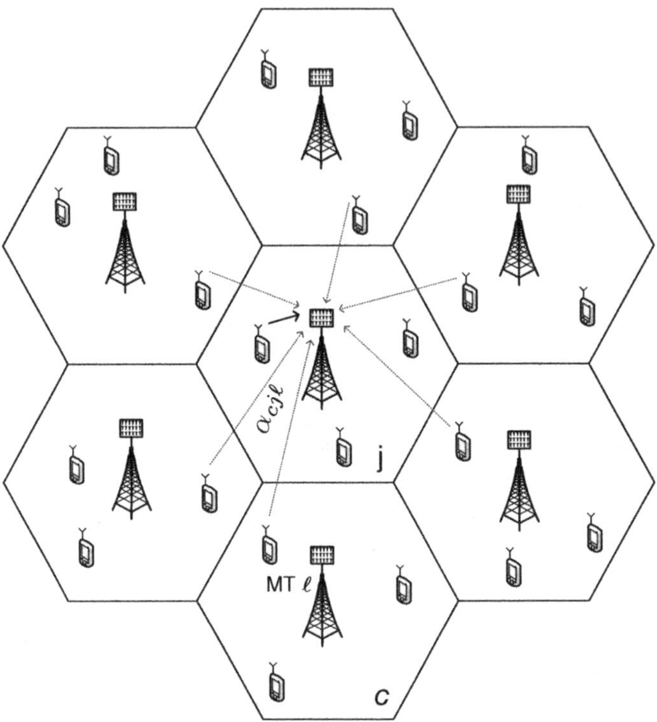

Figure 4.8 The pilot contamination effect in a multicellular massive MIMO network

corresponding and also adjacent subcarriers. Applying the central limit theorem, one can come up with three observations:

1. The favorable real-part of the equalized CMT symbol at each subcarrier is free of ISI and ICI and so its distribution follows that of the respective PAM alphabet.
2. The corresponding imaginary part suffers from ISI and ICI and is distributed in a Gaussian manner.
3. Both the real and imaginary parts of an unequalized symbol at a subcarrier comprise ISI and ICI terms and indeed are distributed in a Gaussian manner.

Based on the aforementioned properties, a blind equalization algorithm similar to the Godard blind equalization algorithm [19] was developed by Farhang-Boroujeny [8] such that the cost function

$$\xi = \mathbb{E}[(|y_k(n)|^p - R)^2] \tag{4.28}$$

is minimized. $y_k(n)$ is the equalizer output (in the case here, the equalizer output of the kth subcarrier channel), p is integer (usually set equal to 1 or 2), n is the iteration index, $R = \mathbb{E}[|s|^{2p}]/\mathbb{E}[|s|^p]$, and s is a random selection from the PAM symbols alphabet.

In the following, we propose to exploit this algorithm in order to adaptively correct the imperfect channel estimates and hence greatly alleviate the performance degradation due to the contaminated pilots. A blind-tracking algorithm, similar to least mean squares (LMS), and based on the cost function in Eq. (4.28), can be adopted. Extension of the proposed blind equalization technique of Farhang-Boroujeny [8] to massive MIMO application can be straightforwardly derived as

$$\mathbf{w}_{j\ell}(n+1) = \mathbf{w}_{j\ell}(n) - 2\mu\ \text{sign}(\hat{s}_j^{(n)}(\ell))(|\hat{s}_j^{(n)}(\ell)| - R) \cdot \mathbf{x}_j(n) \tag{4.29}$$

where $\hat{s}_j^{(n)}(\ell) = \mathbf{w}_{j\ell}^{\text{H}}(n)\mathbf{x}_j(n)$, the $N \times 1$ vector $\mathbf{x}_j(n)$ is the nth symbol of the received data packet at the BS antenna, the $N \times 1$ vector $\mathbf{w}_{j\ell}(n)$ contains the combiner tap-weights calculated in the nth iteration and μ is the step-size parameter. We initialize the algorithm through the matched filter tap-weight vector

$$\mathbf{w}_{j\ell}(0) = \frac{\hat{\mathbf{h}}_{jj\ell}}{\hat{\mathbf{h}}_{jj\ell}^{\text{H}}\hat{\mathbf{h}}_{jj\ell}} \tag{4.30}$$

where $\hat{\mathbf{h}}_{jj\ell}$ is the estimated channel vector between the user ℓ located in the cell j and the jth BS antenna arrays, in other words the ℓth column of $\hat{\mathbf{H}}_{jj}$ in Eq. (4.27). In the next section, we will show through numerical results that our proposed channel-tracking algorithm is able to effectively converge towards the MMSE linear combination with perfect knowledge of channel responses of all the users in all the considered cells, while starting from matched filter tap-weights with imperfect CSI.

4.4.1 Simulation Results

In this section, we will numerically investigate the performance of our proposed pilot decontamination technique based on Eq. (4.29). This solution extends the blind equalization capability of CMT to massive MIMO networks in order to cope with imperfect channel estimates caused by the pilot contamination effect.

In our simulations, we consider a massive MIMO network comprising seven cells, and where the pilot signals of the cell of interest, cell j, suffer interference from the users of adjacent cells (Figure 4.8). We assume one interferer in each neighbouring cell – 6 interferers in total – whose random cross-gains are less than one. Without loss of generality, in order to expedite our simulations, we consider one user in the jth cell using the same pilot sequence as the users in all neighbouring cells. Thus, the channel estimates at the BS include some residuals from the channel responses of the users of other cells. One transmit antenna is assumed for each user and the number of antennas at the BS $N = 128$. Uncorrelated channel impulse responses between the users and the BS antennas are assumed. The results are for a sample set of channel responses generated based on the COST 207 channel model for a typical urban area with 6 taps. The cross-gain factors $\alpha_{cj\ell}$ are randomly chosen from the range [0,1]. The total bandwidth for this channel is equal to 5~MHz. The number of subcarriers $L = 256$, and their subcarrier spacing is equal to 19.531 kHz. Binary PAM signaling is used in our simulations. This is equivalent to quadrature phase shift keying signaling if OFDM was adopted. The target SINR in our simulations is set equal to 32.

After the BS estimates the channel responses of its users through their pilot sequences, the equalizer tap-weights are initialized using Eq. (4.30). Then filter tap-weight adaptation will be performed. Due to the different channel gains, each subcarrier has a different signal level. Therefore, the step-size μ will be normalized with respect to the instantaneous signal energy

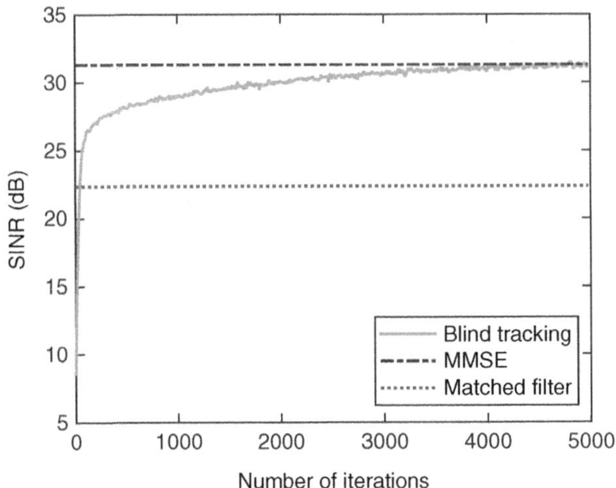

Figure 4.9 SINR comparison of our proposed blind tracking technique with respect to the MF and MMSE detectors having perfect CSI knowledge

in each iteration. Accordingly, for binary PAM signaling, the combiner tap-weights can be updated using

$$\mathbf{w}_{j\ell}(n+1) = \mathbf{w}_{j\ell}(n) - \frac{2\mu}{\mathbf{x}_j^{\mathrm{H}}(n)\mathbf{x}_j(n)+\varepsilon}\,\mathrm{sign}(\hat{s}_j^{(n)}(\ell)) \times(|\hat{s}_j^{(n)}(\ell)| - R)\cdot \mathbf{x}_j(n) \tag{4.31}$$

where ε is a small positive constant which assures numerical stability of the algorithm when the term $\mathbf{x}_j^{\mathrm{H}}(n)\mathbf{x}_j(n)$ has a small value.

Our proposed pilot decontamination technique is evaluated through looking into its SINR performance and comparing it with the SINRs of the MF and MMSE detectors having the perfect CSI knowledge of all the users in all the cells. The SINRs of the MF and MMSE linear combiners are calculated on the basis of Eqs. (4.10) and (4.11), respectively. It is worth mentioning that for MMSE combining, the jth BS needs to know the perfect channel impulse responses of all the users located in its neighbouring cells having the same pilot sequences as its own user.

Figure 4.9, shows the SINR performance of the proposed blind tracking technique in dB with respect to the number of iterations. There is an abrupt SINR improvement during the first 50 iterations, where the output SINR of the blind combiner reaches that of the MF combiner with perfect CSI knowledge. Running larger numbers of iterations has shown that the output SINR of our blind channel tracking technique can suppress the pilot contamination effect and converge towards that of the MMSE combiner. Apart from its high computational complexity, the MMSE detector needs perfect knowledge of the channel impulse responses between the interfering users of the other cells and its array antennas. This clearly is an impossible condition. The methods proposed here, on the other hand, can approach MMSE performance simply by running an LMS-like algorithm.

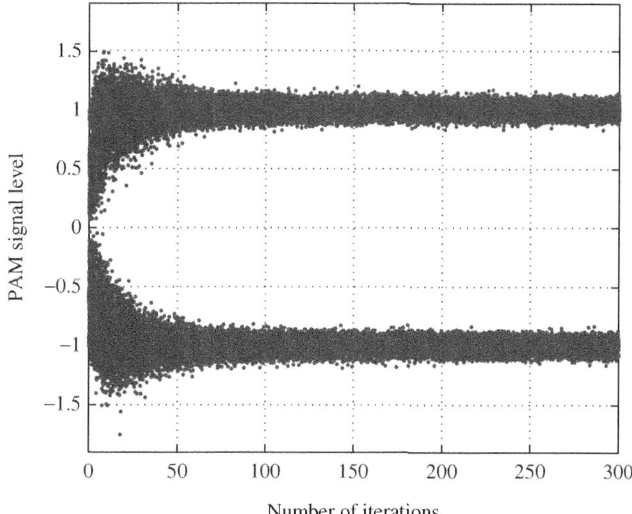

Figure 4.10 Eye pattern of the combined symbols using the proposed blind tracking technique

Figure 4.10 represents the eye pattern of the detected symbols with respect to the number of iterations for our proposed blind combiner. As can be seen, the eye pattern of the detected symbols improves as the number of iterations increases.

Existing techniques that address the pilot contamination problem either need cooperation between the cells or are computationally intensive [20, 21, 22]. The solutions that are applicable to non-cooperative cellular networks are of interest here [20, 22]. The solution proposed by Ngo *et al.* [20] needs eigenvalue decomposition of the covariance of the received signal and the one presented by Müller *et al.* [22] has to calculate the SVD of the received signal matrix. The matrices involved are of size N (128 for the examples given here). On the other hand, our solution simply needs to update the combiner tap-weights using Eq. (4.31) and hence it has a low computational complexity. In addition, it is structurally simple, which makes it attractive from practical implementation point of view.

From Figure 4.9, one may see that a large number of iterations are needed for our algorithm to approach the SINR performance of the MMSE detector. This does not mean that a very long packet of data is needed for this algorithm to converge towards the MMSE detector's performance. MMSE performance can be achieved through multiple runs of the algorithm over a much shorter packet of data. In other words, the LMS algorithm in Eq. (4.31) can be repeated over the same set of data until it converges.

4.5 Conclusion

In this chapter, we introduced FBMC as a viable candidate waveform in the application of massive MIMO. Among various FBMC techniques, CMT was identified as the best choice. It was shown that while FBMC offers the same processing gain as OFDM, it offers the advantages of more flexible carrier aggregation, higher bandwidth efficiency (because of the absence of CP), blind channel equalization and larger subcarrier spacing, and hence less sensitivity to CFO and lower PAPR. The self-equalization property of CMT in massive MIMO channels was also elaborated. The SINR performance of two different linear combination techniques – MF and MMSE – was investigated. We addressed the pilot contamination problem in a TDD multicellular massive MIMO network. The pilot contamination problem can adversely affect the performance of massive MIMO networks and as a result create a great deal of multi-cell interference in both uplink and downlink transmissions. The blind equalization capability of CMT was extended to massive MIMO networks to mitigate the pilot contamination effect. The performance of our proposed solution was analyzed through computer simulations. It was shown that starting from corrupted channel estimates, after running a small number of iterations our algorithm performs as well as or better than the matched filter with perfect CSI. We have shown that the output SINR of our algorithm converges towards that of the MMSE solution, in which the BS should have perfect knowledge of CSI for all users located in its neighboring cells.

References

[1] Marzetta, T. (2010) Noncooperative cellular wireless with unlimited numbers of base station antennas. *IEEE Trans. Wireless Commun.*, **9** (11), 3590–3600.

[2] Chang, R. (1966) High-speed multichannel data transmission with bandlimited orthogonal signals. *Bell Sys. Tech. J.*, **45**, 1775–1796.

[3] Saltzberg, B. (1967) Performance of an efficient parallel data transmission system. *IEEE Transactions on Communication Technology*, **15** (6), 805–811.

[4] Farhang, A., Marchetti, N., Doyle, L., and Farhang-Boroujeny, B. (2014) Filter bank multicarrier for massive MIMO, in *IEEE 80th Vehicular Technology Conference (VTC Fall)*, pp. 1–7.

[5] Aminjavaheri, A., Farhang, A., Marchetti, N., Doyle, L., and Farhang-Boroujeny, B. (2015) Frequency spreading equalization in multicarrier massive MIMO, in *IEEE ICC'15 Workshop on 5G and Beyond*.

[6] Farhang-Boroujeny, B. (2011) OFDM versus filter bank multicarrier. *IEEE Signal Process. Mag.*, **28** (3), 92–112.

[7] Rusek, F., Persson, D., Lau, B.K., Larsson, E., Marzetta, T., Edfors, O., and Tufvesson, F. (2013) Scaling up MIMO: opportunities and challenges with very large arrays. *IEEE Signal Process. Mag.*, **30** (1), 40–60, doi:10.1109/MSP.2011.2178495.

[8] Farhang-Boroujeny, B. (2003) Multicarrier modulation with blind detection capability using cosine modulated filter banks. *IEEE Trans. Commun.*, **51** (12), 2057–2070.

[9] Farhang, A., Aminjavaheri, A., Marchetti, N., Doyle, L., and Farhang-Boroujeny, B. (2014) Pilot decontamination in CMT-based masive MIMO networks, in *11th International Symposium on Wireless Communications Systems (ISWCS)*, pp. 589–593.

[10] Farhang-Boroujeny, B. (2011) *Signal Processing Techniques for Software Radios*, Lulu Publishing.

[11] Farhang-Boroujeny, B. and (George) Yuen, C. (2010) Cosine modulated and offset QAM filter bank multicarrier techniques: a continuous-time prospect. *EURASIP J. Appl. Signal Process. Special issue on filter banks for next generation multicarrier wireless communications*, **2010**, 6.

[12] Madhow, U. and Honig, M. (1994) MMSE interference suppression for direct-sequence spread-spectrum CDMA. *IEEE Trans. Commun.*, **42** (12), 3178–3188.

[13] Farhang-Boroujeny, B. (2013) *Adaptive Filters: Theory and Applications*, John Wiley.

[14] Bellanger, M. (2012) FS-FBMC: An alternative scheme for filter bank based multicarrier transmission, in *5th International Symposium on Communications Control and Signal Processing (ISCCSP)*, pp. 1–4, doi:10.1109/ISCCSP.2012.6217776.

[15] The IEEE 802.16 Broadband Wireless Access Working Group (2003) *Channel models for fixed wireless applications*. http://www.ieee802.org/16/tg3/contrib/802163c-01_29r4.pdf.

[16] Krishnan, N., Yates, R.D., and Mandayam, N.B. (2014) Uplink linear receivers for multi-cell multiuser MIMO with pilot contamination: Large system analysis, in *arXiv:1307.4388*.

[17] Fettweis, G., Krondorf, M., and Bittner, S. (2009) GFDM - generalized frequency division multiplexing, in *IEEE 69th Vehicular Technology Conference, 2009. VTC Spring 2009.*, pp. 1–4.

[18] Jose, J., Ashikhmin, A., Marzetta, T., and Vishwanath, S. (2009) Pilot contamination problem in multi-cell TDD systems, in *IEEE International Symposium on Information Theory, ISIT 2009*, pp. 2184–2188.

[19] Godard, D. (1980) Self-recovering equalization and carrier tracking in two-dimensional data communication systems. *IEEE Trans. Commun.*, **28**, 11.

[20] Ngo, H.Q. and Larsson, E. (2012) EVD-based channel estimation in multicell multiuser MIMO systems with very large antenna arrays, in *IEEE International Conference on Acoustics, Speech and Signal Processing (ICASSP)*, pp. 3249–3252.

[21] Yin, H., Gesbert, D., Filippou, M., and Liu, Y. (2013) A coordinated approach to channel estimation in large-scale multiple-antenna systems. *IEEE J. Select. Areas in Commun.*, **31** (2), 264–273.

[22] Müller, R., Cottatellucci, L., and Vehkaperä, M. (2014) Blind pilot decontamination. *arXiv: 1309.6806*.

5

Bandwidth-compressed Multicarrier Communication: SEFDM

Izzat Darwazeh, Tongyang Xu and Ryan C Grammenos

Signal Processing for 5G: Algorithms and Implementations, First Edition. Edited by Fa-Long Luo and Charlie Zhang.

5.1 Introduction

The physical layers of many of today's communication systems utilize multicarrier transmission techniques, as these offer good spectrum utilization in frequency-selective and multipath channels. The most prominent of the multicarrier systems is orthogonal frequency division multiplexing (OFDM), in which the information signal is carried on parallel orthogonal carriers (termed subcarriers) with frequency separation equal to the symbol rate. OFDM has its origins in the Kineplex system, originally proposed for computer communications in late 1950s [1]. The modern variant of OFDM first appeared in the 1966 and was implemented using digital techniques in 1971 [2, 3]. OFDM is currently the system of choice for many wired and wireless transmission systems [4], currently most notably as the downlink transmission system for the 4G cellular system, long-term evolution (LTE) [5]. In the quest to save spectrum, a new non-orthogonal multicarrier scheme, termed spectrally efficient frequency division multiplexing (SEFDM), was proposed in 2003 [6]. Relative to OFDM systems, SEFDM improves spectral efficiency by reducing the subcarrier spacing whilst maintaining the same transmission rate per subcarrier. This is illustrated in Figure 5.1, where the spectra of OFDM and SEFDM are compared and bandwidth is saved in SEFDM when both systems use the same number of subcarriers and the same subcarrier bandwidth or symbol rate, but in SEFDM the subcarriers are overlapped at spacings below the symbol rate, effectively resulting in compressed subcarrier spacing and leading to improved spectral efficiency.

The history of SEFDM can be traced back to 2002 when a data-rate-doubling OFDM-based technique termed fast OFDM [7] was proposed. The principle of SEFDM was based on fast OFDM and was proposed in 2003 [6]. Since then, a number of researchers have worked on different aspects of SEFDM signals, ranging from novel algorithms and techniques to generate and detect them, through to their practical implementation using state-of-the-art hardware devices. Table 5.1 gives an account of the key milestones achieved during the study of SEFDM signals over the last decade.

SEFDM is a non-orthogonal waveform technique that provides improved spectral efficiency by packing subcarriers less than the symbol rate. The other non-orthogonal technique, termed faster-than-Nyquist (FTN) [36], which violates Nyquist criteria by transmitting data faster than the Nyquist limit in order to achieve its purpose of spectral efficiency. Other spectrally efficient waveform techniques such as filterbank-based multicarrier (FBMC) [37] and generalized

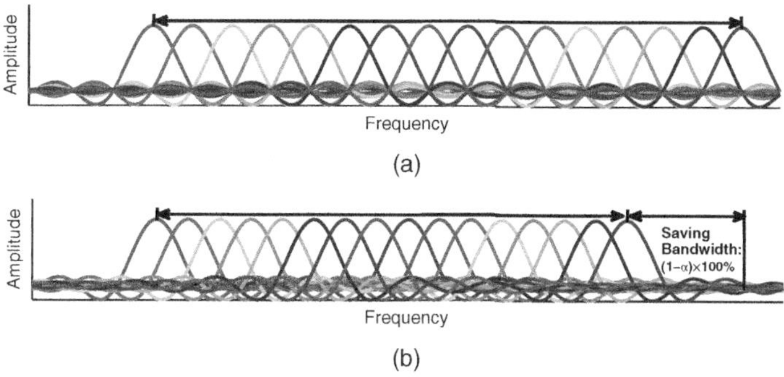

Figure 5.1 Spectra of 16 overlapped subcarriers for OFDM and SEFDM. (a) OFDM spectrum; (b) SEFDM spectrum with bandwidth compression factor $\alpha = 0.8$

Table 5.1 History of SEFDM

2002	[7]	Fast OFDM: A proposal for doubling the data rate of OFDM
2003	[6]	Proposal of the non-orthogonal waveform (SEFDM) concept
Sept. 2008	[8]	A ML SEFDM signal detector
Jun. 2009	[9]	Complexity analysis of SEFDM receivers
Sept. 2009	[10]	Investigation of semidefinite program (SDP) signal detector
Sept. 2009	[11]	An application of SEFDM in physical layer security
Sept. 2009	[12]	A pruned SD signal detector
Jul. 2010	[13]	An IFFT SEFDM signal generator
Sept. 2010	[14]	A joint channel equalization and detection scheme
Sept. 2010	[15]	Precoded SEFDM
Nov. 2010	[16]	The use of a fast constrained SD for signal detection
Mar. 2011	[17]	Proposal of a TSVD detector
May. 2011	[18]	Peak to average power ratio (PAPR) reduction in SEFDM
May. 2011	[19]	Evaluation of FSD detector in SEFDM
May. 2011	[20]	A real-time field programmable gate array (FPGA) based SEFDM signal generator
Sept. 2011	[21]	FPGA implementation of the TSVD detector
Apr. 2012	[22]	A robust partial channel estimation (PCE) scheme for SEFDM
May. 2012	[23]	A reconfigurable hardware based SEFDM transmitter
Jun. 2012	[24]	A hardware verification methodology for SEFDM signal detection
Sept. 2012	[25]	A hybrid DSP-FPGA implementation of the TSVD-FSD detector
May. 2013	[26]	A real-time FPGA based implementation of the TSVD-FSD detector
Jun. 2013	[27]	A pure DSP implementation of the modified TSVD-FSD detector
Oct. 2013	[28]	An enhanced FSD detector with iterative soft mapping scheme
Feb. 2014	[29]	10 Gbit/s optical SEFDM (direct detection)
Jul. 2014	[30]	A higher order modulation scheme with iterative soft mapping scheme
Jul. 2014	[31]	A multi-band SEFDM architecture with simplified signal detector
Nov. 2014	[32]	A soft detector for an LTE-like SEFDM system
Jun. 2015	[33]	Experimental demonstration of CA SEFDM
Jun. 2015	[34]	Experimental demonstration of RoF SEFDM
Oct. 2015	[35]	24 Gbit/s optical SEFDM (coherent detection)

frequency division multiplexing (GFDM) [38] were proposed to reduce out-of-band radiation and avoid harmful interference to adjacent channels by using a pulse-shaping filter on each sub-carrier. Therefore, the protection guard band between two channels can be narrowed leading to improved spectral efficiency.

In addition, the non-orthogonal concept can be applied for multiple access scenarios such as sparse code multiple access (SCMA) [39], non-orthogonal multiple access (NOMA) [40] and multi-user shared access (MUSA) [41]. These techniques can superimpose signals from multiple users in the code domain or the power domain to enhance the system-access performance. All these non-orthogonal multiple access schemes are implemented at the multi-user scheduling level, which aims to superimpose non-orthogonally several orthogonal waveforms (e.g. OFDM signals) from different users. SEFDM may also be used in multiple access applications. Carrier aggregation SEFDM (CA-SEFDM) [33], which is described in Section 5.5.2, is a form of orthogonal multiple access (OMA), since different component carriers (CCs) do not overlap. It is inferred that a non-orthogonal version of CA-SEFDM will lead to further spectral efficiency improvements achieved via overlapping adjacent CCs.

In the context of future wireless systems (5G and beyond) SEFDM will offer significant bandwidth saving advantages if used either as an air-interface technique or for backhauling signals over wireless and/or wired and fibre communication channels. In this chapter, SEFDM is introduced and then the details of SEFDM, from conceptual system modeling to experimental demonstrations, are presented. An SEFDM system with multiple separated blocks is described in Section 5.3. The system architecture can effectively tackle the self-created intercarrier interference (ICI) of the SEFDM system by applying SD to each block, with reduced complexity. Thus it can support systems with large numbers of subcarriers while maintaining an error performance close to that of the OFDM system. A coded SEFDM system based on the turbo principle [42] is described in Section 5.4. The detector is termed a "soft" detector, in that it can iteratively remove interference according to soft information; in other words, probabilities. Finally, the concept of SEFDM is experimentally evaluated in the LTE-Advanced carrier aggregation (CA) scenario on a wireless testbed in Section 5.5.2.

5.2 SEFDM Fundamentals

SEFDM is a non-orthogonal multicarrier technique, which, for a given bit rate, reduces the required bandwidth through using non-orthogonal waveform design. However, the reduction in bandwidth requirement comes at the cost of increased complexity at the transmitting and receiving ends. This section explains the principle of SEFDM and its signals and then presents methods for signal generation and detection.

5.2.1 The Principle of SEFDM

An SEFDM symbol consists of a block of N modulating symbols, generally considered as quadrature amplitude modulated (QAM) complex symbols and denoted by $s = s_{\Re} + js_{\Im}$. Each of these complex symbols is modulated on one of SEFDM's non-orthogonal overlapped subcarriers. Therefore, for a system with N subcarriers, the normalized SEFDM signal is expressed as

$$x(t) = \frac{1}{\sqrt{T}} \sum_{\iota=-\infty}^{\infty} \sum_{n=0}^{N-1} s_{\iota,n} \exp\left(\frac{j2\pi n\alpha(t-\iota T)}{T}\right) \tag{5.1}$$

where $\alpha = \Delta f T$ is the bandwidth compression factor, Δf is the subcarrier spacing and T is the period of one SEFDM symbol. N is the number of subcarriers and $s_{\iota,n}$ is the complex QAM symbol modulated on the nth subcarrier in the ιth SEFDM symbol. α determines bandwidth compression and hence the bandwidth saving equals $(1-\alpha) \times 100\%$, as shown in Figure 5.1. Note that Δf in SEFDM is smaller than that in OFDM; for OFDM signals $\alpha = 1$, and $\alpha < 1$ for SEFDM.

SEFDM can be implemented either in a simple (but impractical) analog form using a bank of modulators – this is termed the "continuous" version – or by using a specially arranged set of inverse fast Fourier transform (IFFT) elements in what is termed the "discrete" version. In this chapter, we consider the more general case, which is the discrete version, where the first SEFDM symbol of $x(t)$ is sampled at T/Q intervals, where $Q = \rho N$ and $\rho \geq 1$ is the oversampling factor. Hence the discrete SEFDM signal is mathematically represented by

$$X[k] = \frac{1}{\sqrt{Q}} \sum_{n=0}^{N-1} s_n \exp\left(\frac{j2\pi nk\alpha}{Q}\right) \tag{5.2}$$

where $X[k]$ is the kth time sample of $x(t)$ where $k = [0, 1, \ldots, Q-1]$, s_n is a QAM symbol modulated on the nth subcarrier and $\frac{1}{\sqrt{Q}}$ is a scaling factor for the purpose of normalization. For presentation, simplification and ease of mathematical manipulation we generally express the signal in matrix form as

$$X = \mathbf{F}S \tag{5.3}$$

In this chapter we use standard notation, in which matrices and vectors are represented by upper case letters in bold font and italic, respectively. Hence, in the equation above X is a Q-dimensional vector of time samples of the transmitted signal, S is an N-dimensional vector of the QAM data symbols and $\mathbf{F}$ is a $Q \times N$ subcarrier matrix with elements equal to the SEFDM complex subcarriers $e^{\frac{j2\pi nk\alpha}{Q}}$.

The simplest transmission case to consider is a hypothetical channel in which only white Gaussian noise is added to the signal; in other words, the AWGN case. The received signal vector Y is then the transmitted vector X contaminated by noise, which in turn is expressed as the vector of sampled AWGN values Z, as in Eq. (5.4).

$$Y = X + Z \tag{5.4}$$

The first step in the reception process is to demodulate the received signal. This is effected through a process of correlation in which the received signal vector Y is correlated with the conjugate subcarrier matrix $\mathbf{F}^*$. The demodulation process is expressed as

$$R = \mathbf{F}^*Y = \mathbf{F}^*X + \mathbf{F}^*Z = \mathbf{F}^*\mathbf{F}S + \mathbf{F}^*Z = \mathbf{C}S + Z_{\mathbf{F}^*} \tag{5.5}$$

where R is an N-dimensional vector of demodulated symbols or, in other words, collected statistics, $\mathbf{C}$ is an $N \times N$ correlation matrix which is defined as $\mathbf{C} = \mathbf{F}^*\mathbf{F}$, where $\mathbf{F}^*$ denotes the $N \times Q$ conjugate subcarrier matrix with elements equal to $e^{\frac{-j2\pi nk\alpha}{Q}}$ for $k = [0, 1, \ldots, Q-1]$ and $Z_{\mathbf{F}^*}$ is a non-white (colored) noise vector resulting from correlating the AWGN with the conjugate subcarriers. Clearly, the demodulated symbols are not only contaminated by noise but also by the interference resulting from the non-orthogonal nature of the subcarriers. Unlike the random noise, this interference is deterministic and dependent on the bandwidth compression factor α. Each subcarrier would have (to a varying degree) added interference from every other subcarrier in the SEFDM signal. Such interference has been extensively studied [43] and can be calculated from the correlation matrix $\mathbf{C}$ whose elements are expressed as

$$\begin{aligned} c_{m,n} &= \frac{1}{Q} \sum_{k=0}^{Q-1} e^{\frac{j2\pi mk\alpha}{Q}} e^{-\frac{j2\pi nk\alpha}{Q}} \\ &= \frac{1}{Q} \times \left\{ \begin{array}{ll} Q & , m = n \\ \frac{1-e^{j2\pi\alpha(m-n)}}{1-e^{\frac{j2\pi\alpha(m-n)}{Q}}} & , m \neq n \end{array} \right\} \end{aligned} \tag{5.6}$$

where m, n are indices of two arbitrary subcarriers and $c_{m,n}$ represents the cross-correlation of these two subcarriers; in other words the relative interference power in subcarrier m resulting from signal on the non-orthogonal subcarrier n. Clearly, when m equals n, $c_{m,n}$ represents the autocorrelation and would have the normalized value of one. The off-diagonal terms in the correlation matrix $\mathbf{C}$ indicate the effect of non-orthogonal overlapping, which results in ICI. This is in contrast to OFDM, where orthogonality dictates all $c_{m,n}$ values to be zero when m and n are different and therefore for OFDM the matrix $\mathbf{C}$ is diagonal.

5.2.2 *Generation of SEFDM Signals*

The key challenge of SEFDM transmitter design is that conventional OFDM transmission techniques cannot directly be applied for SEFDM, due to the non-orthogonal subcarriers used in SEFDM. Some techniques have been developed to generate non-orthogonal multicarrier signals as easily as in OFDM [13, 23]. Three different, but functionally equivalent, SEFDM transmitter types are illustrated in Figure 5.2. All are based on modified inverse discrete Fourier transform (IDFT) designs (or alternatively IFFT) to generate SEFDM signals with subcarrier spacing that can be modified by changing/selecting different IDFT input and output parameters.

In these architectures, the bandwidth compression factor α is expressed as a ratio of two positive integers b and c; $\alpha = \frac{b}{c}$, with $b < c$. The transmitters shown in Figure 5.2 are then designed to offer different trade-offs. Type 1, known as the *proportional inputs* transmitter, is the simplest, using a standard IDFT block of length $M = \frac{Q}{\alpha}$. This means that the size of this block compared to a conventional OFDM system oversampling at the same rate is $\frac{1}{\alpha}$ times larger. The number of input data symbols remains the same, at N, with the remaining $(M - N)$ IDFT inputs padded with zeros. Therefore there will be M inputs to the IDFT with input symbols s'_i given by

$$s'_i = \begin{cases} s_i & 0 \le i < N \\ 0 & N \le i < M \end{cases} \tag{5.7}$$

The output of the IDFT is truncated, hence only Q output samples $(X_0, X_1, \ldots, X_{Q-1})$ are taken forward to construct the SEFDM signal while the remaining $(M - Q)$ outputs are ignored. Then, the SEFDM signal in a new format is expressed as

$$X[k] = \frac{1}{\sqrt{M}} \sum_{n=0}^{M-1} s'_n \exp\left(\frac{j2\pi nk}{M}\right) \tag{5.8}$$

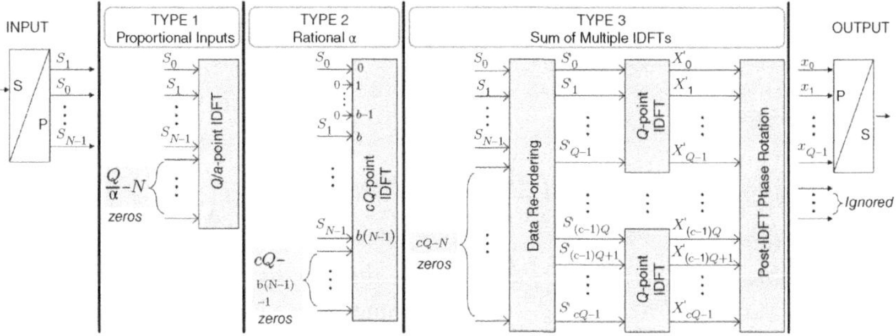

Figure 5.2 IDFT based SEFDM transmitter types

where $k = [0, 1,\ldots,Q - 1]$ and $n = [0, 1,\ldots,M - 1]$. The key drawback of the Type 1 transmitter is that it requires the ratio $\frac{Q}{\alpha}$ to be an integer number. Consequently, this limits flexibility in choosing the value of α.

For this reason, two alternative transmitter schemes – Types 2 and 3 – have been designed, and these are also shown in Figure 5.2. Type 2, known as the *rational* α transmitter, uses a cQ length IDFT block. Only the inputs of the IDFT whose indices are integer multiples of b are linked to the input data symbols; the remaining IDFT inputs are padded with zeros. This is achieved via the following condition

$$s'(i) = \begin{cases} s_{i/b} & i \; mod \; b = 0 \\ 0 & otherwise \end{cases} \tag{5.9}$$

where the notation mod is defined as the modulus of the remainder after division. Complying with the condition of Eq. (5.9), the generation process for the Type 2 transmitter may be expressed as

$$X[k] = \frac{1}{\sqrt{Q}} \sum_{n=0}^{cQ-1} s'(n) \exp\left(\frac{j2\pi nk}{cQ}\right) \tag{5.10}$$

As in the case of Type 1, only Q samples are taken to construct the SEFDM signal. The benefit of this transmitter technique is that b and c can be chosen independently to give any desired value of α for any number of subcarriers N. The IDFT block in this case is c times longer than an equivalent OFDM block and b times longer than the Type 1 SEFDM transmitter.

To avoid using excessively long IDFT blocks while allowing flexibility in the choice of α values, the Type 3 transmitter, known as the *sum of multiple IDFTs*, was designed. This transmitter uses c identical IDFT blocks each of length Q instead of one large IDFT block of length cQ. There are two advantages of this transmitter over the Type 2 architecture:

- It allows multiple IDFT blocks to be configured in parallel, a feature attractive for hardware implementation employing FPGAs.
- It reduces the computation time required for the IDFT processing stage provided that all c IDFT blocks can operate concurrently in parallel.

It is important to note that the Type 3 transmitter requires two additional operations: a reordering stage before the IDFT block and a post-IDFT phase-rotation stage. By substituting with $n = i + lc$, $s'(n)$ in Eq. (5.10) is rearranged as $s'(i + lc)$. The expression for the Type 3 transmitter is therefore given by

$$X[k] = \frac{1}{\sqrt{Q}} \sum_{i=0}^{c-1} \sum_{l=0}^{Q-1} s'(i + lc) \exp\left(\frac{j2\pi k(i + lc)}{cQ}\right) \tag{5.11}$$

which can be rearranged as

$$X[k] = \frac{1}{\sqrt{Q}} \sum_{i=0}^{c-1} \exp\left(\frac{j2\pi ik}{cQ}\right) \sum_{l=0}^{Q-1} s'(i + lc) \exp\left(\frac{j2\pi lk}{Q}\right) \tag{5.12}$$

where the variable $i \in [0,\ldots,c - 1]$ determines the number of IDFT operations (or blocks) and the variable $l \in [0,\ldots,Q - 1]$ determines the size of each IDFT operation. It is apparent that the SEFDM signal can be generated by using c parallel IDFT blocks.

Each one of the IDFT blocks is of length Q (i.e. Q-point IDFT). The first summation term on the right-hand side of Eq. (5.12) determines the number of parallel IDFT operations. The second summation term indicates a Q-point IDFT of the sequence $s'(i + lc)$. This indicates that an SEFDM symbol is generated by effectively combining multiple (of number c) OFDM symbols, with appropriate choice of zero padding at the inputs and phase rotation at the output to effect the required non-orthogonal overlapping.

5.2.3 Detection of SEFDM Signals

The key challenge of SEFDM receiver design is that the signal suffers from ICI resulting from the non-orthogonal structure of the subcarriers. Consequently, the detection of the SEFDM signal requires efficient handling of such ICI. In 2008, work by Kanaras *et al.* showed that the maximum likelihood (ML) detector can produce optimal bit error rate (BER) performance but is prohibitively complex [8]. The history table has shown that simple linear detectors such as the zero-forcing (ZF) [17], minimum mean-squared error (MMSE) [17] and truncated singular value decomposition (TSVD) [17] fail to provide competitive BER performance for moderate bandwidth savings or number of carriers. Non-linear detection techniques such as sphere decoding (SD) [44, 12, 16], iterative detection (ID) [28, 29, 45] and fixed sphere decoding (FSD) [46, 19] are impaired by high realization complexity when a large number of subcarriers and/or high-order modulation are desired. This constitutes a challenge for the hardware implementation of practical SEFDM systems [25, 27, 26]; the pursuit of low-complexity detectors remains a challenge and is the focus of this section.

5.3 Block-SEFDM

In a practical wireless communication system, a large number of subcarriers is desirable to ameliorate multipath effects and fading. SD was examined as a potential solution for the SEFDM detection problem, but its complexity increases rapidly with the enlargement of the system (i.e. the number of subcarriers), rendering it impractical. For similar reasons, the other SEFDM detection algorithms cited in the subsection above, but which will not be discussed in detail here, were found to be only suitable for small systems.

To simplify detection it is therefore sensible to divide the problem into subproblems where the detection of each is more manageable; in other words split the overall SEFDM signal into a collection of spectrally independent ones, each with a small number of subcarriers. A non-orthogonal multicarrier system termed block-spectrally efficient frequency division multiplexing (Block-SEFDM) [31], which divides the whole spectrum into several non-orthogonal blocks, has been devised and will be discussed in this subsection. Symbols in each block can be detected separately by using the SD algorithm. The out-of-block interference is minimized through use of narrow frequency guard bands. This technique makes it practical to detect large non-orthogonal SEFDM signals (say from 128 subcarriers).

5.3.1 Principle of Block-SEFDM

OFDM, standard SEFDM and Block-SEFDM are compared in Figure 5.3. In the top panel (a), shows a typical OFDM spectrum with subcarrier spacing Δf_1 equal to $\frac{1}{T}$, where T is the

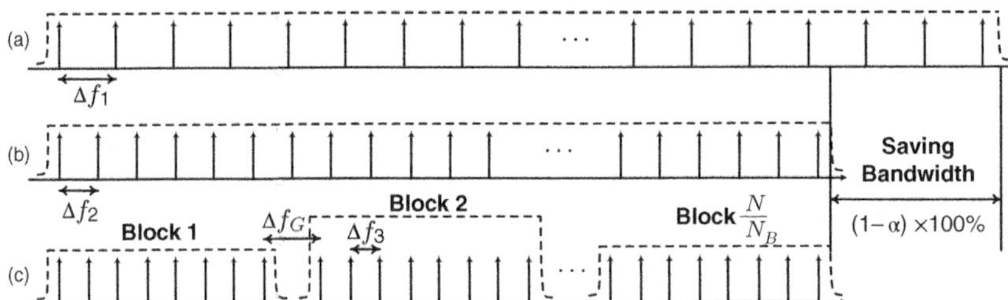

Figure 5.3 Spectral illustrations for (a) OFDM, (b) SEFDM, (c) Block-SEFDM. Each impulse indicates one subcarrier and there are overall N subcarriers for each system, respectively

time period of one OFDM symbol. The middle panel (b) shows a typical SEFDM system with subcarrier spacing $\Delta f_2 = \frac{\alpha}{T}$ where $\alpha < 1$ is the bandwidth compression factor. It is apparent that the SEFDM system can save $(1 - \alpha) \times 100\%$ of bandwidth compared to a typical OFDM system. The typical SEFDM signal is straightforwardly detected using a single detector due to the single band feature. However, this kind of detector is limited by the system size. In order to simplify the detector design and maintain system performance, multiple parallel shorter (small size) detectors are desirable. On the basis of this idea, Block-SEFDM divides the original SEFDM spectrum into several non-orthogonal blocks where a short detector is adopted in each block. The principle of Block-SEFDM is illustrated in the bottom panel (c), where the entire spectrum is evenly partitioned into $\frac{N}{N_B}$ blocks where each subblock comprises N_B subcarriers and every $(N_B + 1)$th subcarrier is removed. Δf_3 is the subcarrier spacing and Δf_G is the guard band between two adjacent blocks and is equal to $2\Delta f = \frac{2\alpha}{T}$. Hence the partitioned blocks are non-orthogonally packed. Due to the introduction of the guard band Δf_G, in order to keep the same occupied bandwidth, subcarriers in each block are more compressed leading to a smaller Δf_3. According to Figure 5.3, the subcarrier spacing relationship is $\Delta f_3 < \Delta f_2 < \Delta f_1$. Considering the guard band, the Block-SEFDM signal is expressed as [31]:

$$X[k] = \frac{1}{\sqrt{Q}} \sum_{l_B=0}^{\frac{N}{N_B}-1} \sum_{i=0}^{N_B-1} s_{i+l_B N_B} \exp\left[\frac{j2\pi k\alpha(i + l_B(N_B + 1))}{Q}\right] \tag{5.13}$$

where $s_{i+l_B N_B}$ is the ith symbol modulated in the l_Bth block. Therefore, N_B determines the size of each block and l_B indicates the index of blocks. The product of N_B and the maximum value of l_B equals N. It should be noted that not all of the subcarriers are evenly overlapped and therefore not all have the same levels of ICI, since there is a deleted subcarrier (additional spacing) between adjacent blocks in order to mitigate the non-orthogonal out-of-block interference.

5.3.2 *Two-stage Signal Detection*

The proposed spectrum segmentation scheme provides a new route to signal detection. As shown in Figure 5.3 (b), the entire signal band is decomposed into several blocks. Although one subcarrier is reserved as a protection gap between two adjacent subblocks, due to the non-orthogonal interblock overlapping, the interference is still challenging with high bandwidth compressions (i.e. small values of α) since the subcarrier spacing is smaller and higher levels of interference exist. Therefore, at the receiver, an iterative detection ID [28] algorithm is executed first to remove the out-of-block interference in each block using the iterative soft mapping scheme. Then, a typical SD [16] is adopted in each interference-free block (e.g. Block 1, Block 2,...) to recover the signals.

It has been proven that ID has a better immunity against interference [28]. Due to the random effects of interference and noise, some symbols are severely degraded by interference, while the other symbols are less affected. The operation of the ID algorithm is based on iterative interference cancellation and symbol detection. The algorithm first determines which symbols are less distorted by interference and then recovers the *more* distorted symbols. The interference is removed gradually after each iteration and is effectively cancelled by the last iteration. Such cancellation requires knowledge of interference powers based on the correlation matrix, as explained in Section 5.2.1. In Block-SEFDM the ID principle devised is based on cancelling interference from adjacent blocks iteratively while ignoring the interference of the subcarriers within each block, which is left for the second stage – the sphere decoders – to deal with.

The ID method for a 4QAM mapping strategy is illustrated in Figure 5.4(a); the gray area is an uncertainty zone that is determined by an uncertainty interval defined by $\Delta d = 1 - \frac{\kappa}{V}$, where κ is the κth iteration and V is the total iteration number. Points outside the grey area are mapped to QAM symbols while other points are unchanged and left to the next iteration. The interval is reduced gradually in the iteration process until it reaches zero, meaning that the effect of ICI has been cancelled. The iterative operation is mathematically expressed as

$$S_n = R - (\mathbf{C} - \mathbf{e})S_{n-1} \tag{5.14}$$

where $\mathbf{e}$ is an $N \times N$ identity matrix, S_n is an N-dimensional vector of recovered symbols after n iterations and S_{n-1} is an N-dimensional vector of estimated symbols after $n-1$ iterations. At the last iteration, the constrained (i.e. hard) estimate can be obtained using the rounding function $\lfloor . \rceil$ as $\bar{S}_{ID} = \lfloor S_n \rceil$. Detailed descriptions of the ID algorithm with respect to the 4QAM and 16QAM modulation schemes can be found elsewhere [28, 30].

The interference cancellation is decomposed into two stages. The first stage, shown in Eq. (5.15), is interference modeling and the second stage, shown in Eq. (5.16), is to cancel out the

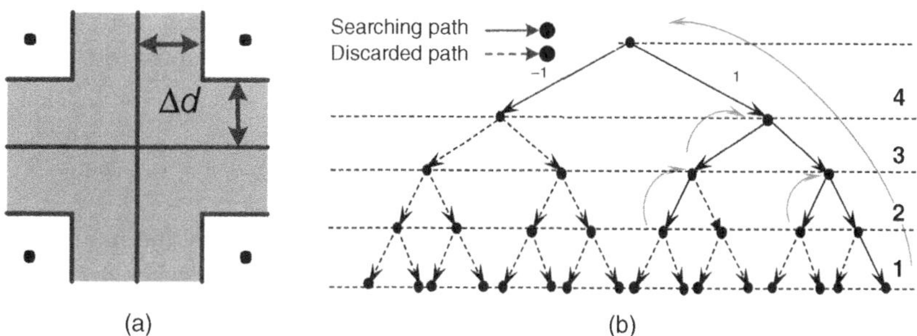

Figure 5.4 The principle of two-stage signal detection. (a) ID detection algorithm that removes interblock interference; (b) SD algorithm used within each block

modeled interference. The entire process is expressed as

$$I[m] = \sum_{n=0}^{N-1} c[m,n] * \bar{S}_{ID}(n) - \sum_{n=U-N_B}^{U-1} c[m,n] * \bar{S}_{ID}(n) \tag{5.15}$$

$$\tilde{R}_{U/N_B-1} = R[m] - I[m] \tag{5.16}$$

where $m \in [U - N_B,\ldots,U-1]$, $U = [N_B, 2N_B,\ldots,N]$. $U/N_B - 1$ represents the block sequence number, which starts from zero and ends with $N/N_B - 1$, $I[m]$ is an N_B-dimensional vector of the out-of-block interference and should be cancelled out in Eq. (5.16). $\tilde{R}_{U/N_B-1}$ is an N_B-dimensional vector of the interference-cancelled symbols and is transferred to the SD algorithm for detection.

After removing interference from each block, the ID detected symbols are fed to the SD algorithm as initial estimates. Henceforth, the recovered SEFDM signals are processed in each block by examining only points that exist within an N_B-dimensional hypersphere of radius g. The recovered (or sphere-decoded) SEFDM symbols S_{SD} are therefore obtained by solving the minimization problem

$$S_{SD} = \arg \min_{S \in O^{N_B}} \left\| \tilde{R} - \tilde{\mathbf{C}} S \right\|^2 \leq g \tag{5.17}$$

where $\tilde{\mathbf{C}} = \sum_{m=U-N_B}^{U-1} \sum_{n=U-N_B}^{U-1} c[m,n]$ is an $N_B \times N_B$ correlation matrix providing interference information within one subblock; O is the constellation cardinality; g is the initial radius, equal to the distance between $\tilde{R}$ and S_{ID}, where S_{ID} is the truncated ID initial estimates as $S_{ID} = \bar{S}_{ID}(n)$ and $n \in [U - N_B,\ldots,U-1]$. Then the initial radius is expressed as

$$g = \left\| \tilde{R} - \tilde{\mathbf{C}} S_{ID} \right\|^2 \tag{5.18}$$

Expanding the norm argument in Eq. (5.17) and substituting by $P = \tilde{\mathbf{C}}^{-1}\tilde{R}$, where P is the unconstrained estimate of S in Eq. (5.17), leads to

$$S_{SD} = \arg \min_{S \in O^{N_B}} \{(P-S)^* \tilde{\mathbf{C}}^* \tilde{\mathbf{C}} (P-S)\} \leq g \tag{5.19}$$

In order to simplify the squared Euclidean norm calculation of Eq. (5.17), Eq. (5.19) can be transformed into an equivalent expression using Cholesky decomposition. The transformation is carried out using $chol\{\tilde{\mathbf{C}}^* \tilde{\mathbf{C}}\} = \tilde{\mathbf{L}}^* \tilde{\mathbf{L}}$ [17], where $\tilde{\mathbf{L}}$ is an $N_B \times N_B$ upper triangular matrix. Hence, Eq. (5.17) can be rewritten as

$$S_{SD} = \arg \min_{S \in O^{N_B}} \left\| \tilde{\mathbf{L}}(P - S) \right\|^2 \leq g \tag{5.20}$$

The SD algorithm proceeds by examining all the nodes that satisfy the radius constraint, starting from level number N_B and moving downwards until reaching level number 1. At each level only the points that satisfy Eq. (5.20) are kept and the radius is updated accordingly. It should be noted that although SD can obtain the ML estimate, the complexity of the algorithm is variable and depends on the noise and the properties of the system.

An alternative detection method is FSD, which fixes the complexity of SD by restricting the search within a limited subspace of the problem in Eq. (5.20). At each level, a fixed number of nodes, termed here the "tree width", are examined. Complexity is reduced at the expense of performance degradation. The FSD algorithm is given by

$$S_{_{FSD}} = \arg \min_{S \in \Phi} \left\| \tilde{\mathbf{L}}(P - S) \right\|^2 \leq g \tag{5.21}$$

where $S \in \Phi$ indicates the solution is within the subspace of $S \in O^{N_B}$. It should be noted that in this section, SD is applied instead of FSD due to its optimal performance.

Figure 5.4(b) illustrates a tree-search diagram showing how the SD detection algorithm works. For the sake of simplicity, a four-subcarrier system with binary phase shift keying (BPSK) symbols is taken as an example. In the Block-SEFDM system used in this section, the SD is configured for an eight-subcarrier system modulated with 4QAM symbols. Each point in the tree is referred to as a node and simply represents a constellation point. The number of branches per node is equal to the constellation size. The number on the right at each level represents the index of subcarriers. The ML algorithm searches for all the nodes, including both retained nodes and discarded nodes, whilst SD only tests nodes within a predefined space. At each level, only points that lie within the space are retained while the rest of the nodes are discarded. A transition from higher to lower levels indicates the decision of one symbol; otherwise, it indicates the discarding of one node and all its children nodes. The initial radius determines the complexity of SD since it determines the size of the search space. However, it should be noted that a small radius would reduce the probability of finding the optimal solution while a large radius will increase the complexity. There are 31 nodes in Figure 5.4(b). It is evident that only 9 nodes are searched, while the rest are discarded along with their children nodes. Therefore, the throughput of SD is much higher than that of ML.

The BER performance of the two-stage signal detector is investigated in Figure 5.5, in which an AWGN channel is assumed. Both typical SEFDM (Figure 5.3 (b)) and Block-SEFDM (Figure 5.3 (c)) systems are demonstrated. The effective bandwidth compression factor is α for both systems. Therefore, the bandwidth saving is calculated as $(1 - \alpha) \times 100\%$. It should be noted that for the typical SEFDM system, the ID-FSD [28] detector is employed. For the Block-SEFDM system, the proposed two-stage detector is described above. Figure 5.5(a)

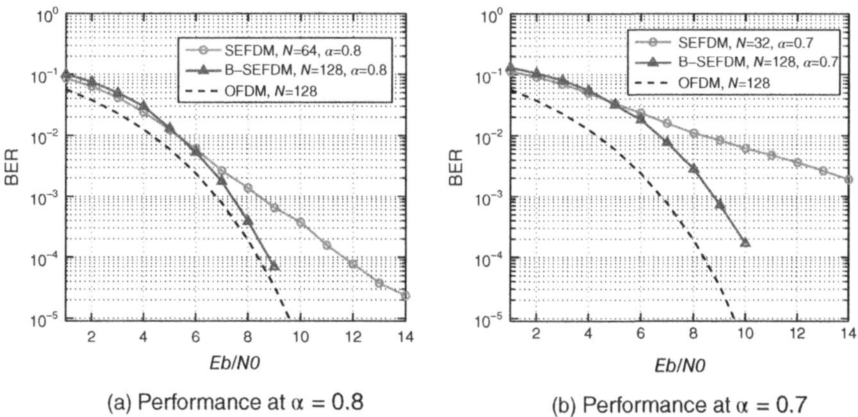

(a) Performance at α = 0.8

(b) Performance at α = 0.7

Figure 5.5 The benefit of Block-SEFDM in terms of BER performance

shows results for 20% bandwidth compression associated with $\alpha = 0.8$. For the typical SEFDM system, 64 subcarriers are considered due to complex signal detection. However, for the Block-SEFDM system, a larger number of subcarriers (say, 128) can be employed. The performance of Block-SEFDM is much better and approaches that of OFDM, while typical SEFDM shows a marked performance degradation for a bit error rate below 10^{-3}.

With a lower bandwidth compression factor of $\alpha = 0.7$, as shown in Figure 5.5(b), the performance gap becomes more obvious for Block-SEFDM due to its higher self-created ICI. But it still outperforms typical SEFDM significantly. The clear performance improvement of Block-SEFDM over SEFDM is attributed to the reduction of interference in the Block-SEFDM system coupled with the use of SD in each block. In summary, these simulation results prove that the Block-SEFDM system architecture and its corresponding detector can support a practical-size wireless system (e.g. 128 subcarriers) with up to 30% bandwidth saving but with only minor performance degradation, making Block-SEFDM a reasonable practical alternative to OFDM.

5.4 Turbo-SEFDM

One challenge of iterative detection methods is that signal detection is dependent on previous iterations and a decision error will affect the subsequent signal decisions, leading to performance degradation. The impact is more serious in SEFDM since self-created ICI is introduced into the system. Error-control coding may therefore be considered to improve data reliability. In this section a turbo principle [42] detector is employed in SEFDM to improve the reliability of symbol decisions iteratively.

The turbo-SEFDM system indicates an SEFDM system employing a turbo equalizer at the receiver to remove iteratively the self-created ICI of SEFDM. The transmitted signal is coded using convolutional coding [32] and the spectral characteristics of the encoded SEFDM transmitted signal follow the typical one as, illustrated in Figure 5.3(b).

The block diagram of the turbo-SEFDM system is shown in Figure 5.6. It consists of the transmitter and the turbo equalization receiver. The turbo-principle receiver includes a feed-forward and feedback loop between an SEFDM detector and a decoder. The soft information, termed "extrinsic information", is updated between the SEFDM detector and the decoder in an iterative process to improve the reliability of the estimation of transmitted symbols. The detailed process within the turbo-SEFDM system is described in the following.

5.4.1 *Principle of Turbo-SEFDM*

At the transmitter, as shown in Figure 5.6, a bit stream $U = [u_1, u_2, \ldots, u_M]$ of length M is encoded in the encoder with a coding rate R_{code} convolutional code. The encoded bits $W = [w_1, w_2, \ldots, w_H]$ of length $H = M/R_{code}$, are interleaved using a random interleaver $\mathbf{\Pi}$ to permute the order of encoded bits W. The interleaved bits $\tilde{S}$ are mapped to the corresponding complex symbols $S = [s_1, s_2, \ldots, s_P]$ with $P = H/log_2 O$, where O is the constellation cardinality. Finally, the SEFDM signal is obtained after the modulator [23].

The turbo-SEFDM receiver consists of a soft detector [32, 47] and a buffer. The proposed soft detector maximizes the a posteriori probability (APP) for a given bit through a process of iteration based on the turbo principle [42]. Soft (i.e. extrinsic) information L^e is exchanged

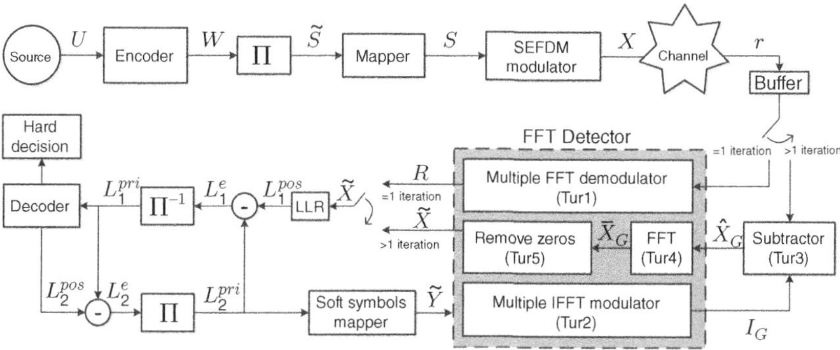

Figure 5.6 Block diagram of turbo-SEFDM. The block labelled **Π** is the interleaver and $\mathbf{\Pi}^{-1}$ represents deinterleaver. Symbols in brackets (.) denote equation indexes. **Tur**1 referrs to Eq. (5.23); **Tur**2 is Eq. (5.25); **Tur**3 is Eq. (5.27); **Tur**4 is Eq. (5.28) and **Tur**5 is Eq. (5.29)

between an FFT detector and a decoder in each iteration. L^e is expressed in the form of LLR where its sign indicates the sign of the bit and its magnitude determines the probability of that sign. In Figure 5.6, a posteriori information L_1^{pos} is generated in the LLR module based on the demodulated information $\tilde{X}$. Then, extrinsic information, L_1^e is obtained by subtracting a priori information L_2^{pri} from a posteriori information L_1^{pos}. After interleaving, the permuted information L_1^{pri} is fed to the decoder as the a priori information. The decoder outputs a posteriori information L_2^{pos}, which then generates extrinsic information L_2^e. After interleaving, the new information is sent back to the soft symbols mapper as the new a priori information L_2^{pri}. Updated symbols $\hat{Y}$ are obtained in the mapper and then fed to the multiple IFFT modulator to retrieve interference I_G. Finally, an interference-free signal $\hat{X}_G$ is generated by subtracting the interference from the original received signal. The above operations are repeated in each iteration until the performance converges to a fixed level. It should be noted that the buffer is introduced to assist the iterative detection. In the first iteration the multiple FFT demodulator is activated while the subtractor is deactivated. After the first iteration, the reverse operations are executed. The buffer stores the received symbols r during one complete symbol detection. After that, the buffer refreshes and new symbols are accepted.

5.4.2 Soft Detection

The standard Bahl-Cocke-Jelinek-Raviv (BCJR) algorithm is employed in the decoder. This section will skip the description of the decoder part and its detailed description can be found elsewhere [48]. The FFT detector is the crucial component since it plays an important role in signal demodulation and ICI cancellation in the received symbols. Therefore, this section focuses on the description of the FFT detector and two other cooperating modules. It is inferred from Section 5.2.2 that the demodulation of an SEFDM signal can be treated as multiple FFT operations, since one SEFDM symbol is composed of multiple overlapped OFDM symbols. It is apparent that one OFDM symbol is an interference signal superimposed on other OFDM symbols. The soft detector therefore aims to remove the superimposed interference from each OFDM symbol.

The demodulation process is an inverse operation of the modulation in Eq. (5.2). In addition, based on Eq. (5.5), the SEFDM signal demodulation can be expressed as

$$R[n] = \frac{1}{\sqrt{Q}} \sum_{k=0}^{Q-1} r(k) \exp\left(\frac{-j2\pi nk\alpha}{Q}\right) \tag{5.22}$$

where $n, k = [0, 1, \ldots, Q-1]$, and n is truncated to N if $N < Q$. Following the same principle shown in Eq. (5.12), Eq. (5.22) can be expressed as the sum of multiple FFTs, represented as

$$R[n] = \frac{1}{\sqrt{Q}} \sum_{i=0}^{c-1} \exp\left(\frac{-j2\pi ni}{cQ}\right) \sum_{l=0}^{Q-1} r^{'}(i+lc) \exp\left(\frac{-j2\pi nl}{Q}\right) \tag{5.23}$$

where r' is a cQ-dimensional vector of symbols as

$$r^{'}(i) = \begin{cases} r_{i/b} & i \bmod b = 0 \\ 0 & otherwise \end{cases} \tag{5.24}$$

The second summation term in Eq. (5.23) is a Q-point FFT of the sequence $r'(i+lc)$. Therefore, the demodulation of the SEFDM signal can be treated as a manipulation of c parallel overlapped OFDM signals. Eq. (5.25) is the interference superimposed to the Gth ($G \in [0, 1, \ldots, c-1]$) OFDM signal with zero padding as in the condition of Eq. (5.26).

$$I_G[k] = \frac{1}{\sqrt{Q}} \sum_{i=0, i \neq G}^{c-1} \exp\left(\frac{j2\pi ik}{cQ}\right) \sum_{l=0}^{Q-1} Y'(i+lc) \exp\left(\frac{j2\pi lk}{Q}\right) \tag{5.25}$$

where

$$Y'(i) = \begin{cases} \hat{Y}_{i/b} & i \bmod b = 0 \\ 0 & otherwise \end{cases} \tag{5.26}$$

The second summation term in Eq. (5.25) is a Q-point IFFT of the sequence $Y'(i+lc)$. After one iteration, the interference $I_G[k]$ is subtracted from the received discrete symbols $r[k]$ to get the more reliable interference-cancelled received symbols $\hat{X}_G[k]$, as shown in Eq. (5.27). It should be noted that accuracy of channel estimation determines the accuracy of the interference generation. Increasing estimation errors would affect interference generation and further degrade the following turbo-principle signal detection, since errors would be passed to the next iteration.

$$\hat{X}_G[k] = r[k] - I_G[k] \tag{5.27}$$

Since the $c-1$ parallel OFDM interference signals have been removed from the Gth OFDM signal, only one Q-point FFT is required to demodulate the signal. Demodulation of the single OFDM signal is shown in Eq. (5.28).

$$\bar{X}_G[k] = \sum_{l=0}^{Q-1} [\hat{X}_G[k] \exp\left(\frac{-j2\pi Gk}{cQ}\right)] \exp\left(\frac{-j2\pi lk}{Q}\right) \tag{5.28}$$

Because the original SEFDM signal is decomposed into c parallel OFDM signals, the same interference cancellation process has to be repeated c times. After that, a $c \times Q$ matrix

$\bar{\mathbf{X}} = [\bar{X}_0, \bar{X}_1, \ldots, \bar{X}_{c-1}]$ interpolated with zeros is obtained. The interpolated zeros are removed in Eq. (5.29) to get a single vector $\breve{X}$ composed of soft symbols:

$$\breve{x}_{i/b} = \bar{x}_i, i \bmod b = 0 \tag{5.29}$$

where useful symbols are extracted every b positions. $\breve{x}_{i/b}$ and $\bar{x}_i$ are the elements of the vector $\breve{X}$ and the matrix $\bar{\mathbf{X}}$, respectively. Then, the obtained vector $\breve{X}$ is delivered to the log-likelihood ratio (LLR) module to generate LLR information.

The LLR information is calculated according to the standard turbo-principle approach [42]. In this section, the LLR of the transmitted bit $\tilde{S}$ conditioned on the demodulation output $\tilde{X}$ is expressed as

$$L(\tilde{s}|\tilde{x}) = ln\frac{P(\tilde{s} = +1|\tilde{x})}{P(\tilde{s} = -1|\tilde{x})} \tag{5.30}$$

Considering a conditioned LLR equation $L(\tilde{s}|\tilde{x}) = L(\tilde{s}) + L(\tilde{x}|\tilde{s})$, the above equation can be further rearranged as Eq. (5.31).

$$\begin{aligned} \underbrace{L(\tilde{s}|\tilde{x})}_{\text{a-posteriori}} &= ln\frac{P(\tilde{s} = +1)}{P(\tilde{s} = -1)} + ln\left\{\frac{\frac{1}{\sqrt{2\pi\sigma^2}}exp\left(\frac{-(\tilde{x}-d)^2}{2\sigma^2}\right)}{\frac{1}{\sqrt{2\pi\sigma^2}}exp\left(\frac{-(\tilde{x}+d)^2}{2\sigma^2}\right)}\right\} \\ &= \underbrace{\ln\frac{P(\tilde{s} = +1)}{P(\tilde{s} = -1)}}_{\text{a-priori}} + \underbrace{\frac{2d\tilde{x}}{\sigma^2}}_{\text{extrinsic}} \end{aligned} \tag{5.31}$$

where d is a parameter associated with fading ($d = 1$ for AWGN channels) and σ^2 is the noise variance. The information that needs to be delivered to the next module is the extrinsic information ${L_1}^e = \frac{2d\tilde{x}}{\sigma^2}$, which can be obtained by subtracting the a priori information from the a posteriori information.

In order to realize the iterative principle, the complex symbols $\hat{Y}$ are required to be regenerated on the basis of the updated L_2^{pri}. The mapping function is realized in the soft symbols mapper. L_2^{pri} is the LLR of the bit $\tilde{s}$. The two possible values of $\tilde{s}$ are $+1$ and -1. Therefore, the LLR of its two possible values are defined as

$$L_2^{pri} = ln\frac{P(\tilde{s} = +1)}{P(\tilde{s} = -1)} \tag{5.32}$$

Due to the relation that $P(\tilde{s} = +1) + P(\tilde{s} = -1) = 1$, the bit probability for $\tilde{s}$ can be calculated as

$$P(\tilde{s} = +1) = \frac{1}{1 + e^{-L_2^{pri}}} \tag{5.33}$$

$$P(\tilde{s} = -1) = \frac{e^{-L_2^{pri}}}{1 + e^{-L_2^{pri}}} \tag{5.34}$$

Therefore, the mapped soft symbols $\hat{Y}$, which are equivalent to the expectation of $\tilde{s}$, are calculated as $\hat{Y} = (+1) \times P(\tilde{s} = +1) + (-1) \times P(\tilde{s} = -1)$.

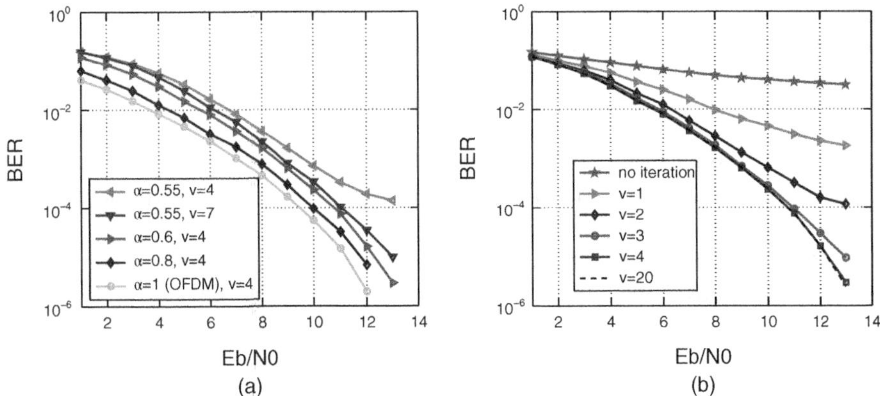

Figure 5.7 Turbo-SEFDM performance. (a) BER performance in the frequency selective channel with $N = 1024$ at various α; (b) Convergence performance at $\alpha = 0.6$ with various iterations

The soft detector is tested in a static frequency-selective channel scenario [32]. An SEFDM signal modulated with 1024 data subcarriers is adopted. Perfect channel state information (CSI) is assumed at the receiver side in order to neglect the channel estimation effect. The channel impulse response is:

$$h(t) = 0.8765\delta(t) - 0.2279\delta(t - T_s) + 0.1315\delta(t - 4T_s)$$
$$-0.4032e^{\frac{j\pi}{2}}\delta(t - 7T_s) \tag{5.35}$$

The results shown in Figure 5.7(a) indicate that transmitting the same amount of data, the SEFDM system can save up to 45% bandwidth compared with OFDM. In other words, in a given bandwidth, the transmission data rate is effectively improved by $82\% \approx (\frac{1}{0.55} - 1) \times 100\%$. The cost of this benefit is a minor performance loss and iterative processing. The convergence performance for the case of $\alpha = 0.6$ is investigated in Figure 5.7(b) which shows that with the increase of iterations, performance is significantly improved. The performance remains stable at four iterations, indicating that this is sufficient to get converged performance.

5.5 Practical Considerations and Experimental Demonstration

The previous sections discussed efficient generation and reception techniques for SEFDM signals. For the purposes of illustration and functional verification, the previous discussion was limited to systems impaired only by AWGN. In realistic systems, performance investigations need to address practical channel issues, particularly the effect of imperfect channels and their estimation in a real wireless communication system. Therefore, this section addresses these issues and presents experimental testing of an SEFDM system using a practical wireless communication platform.

5.5.1 Channel Estimation

Channel estimation is required in a practical system, due to RF effects such as multipath fading channel, imperfect timing synchronization and phase offset. However, estimation of the channel in the case of SEFDM signals is challenged by the non-orthogonal structure of the subcarriers in the system. The standard OFDM channel estimation algorithm is not applicable. Therefore, an SEFDM time-domain channel estimation technique is presented.

Let us assume that the length of CP is N_{CP}. After passing through a multipath fading channel, the received samples Y_{CP} can be expressed in matrix format as

$$Y_{CP} = \mathbf{H}X_{CP} + Z_{CP} \tag{5.36}$$

where X_{CP} is the transmitted signal including both useful data and CP, $\mathbf{H}$ is a $U \times U$ channel matrix, where $U = Q + N_{CP}$, and Z_{CP} is the AWGN of length U. At the receiver, after the removal of CP, the channel matrix $\mathbf{H}$ is transformed to a circulant matrix and, using Eq. (5.3), the signal vector is represented by

$$Y_c = \mathbf{H_c}X + Z = \mathbf{H_c}\mathbf{F}S + Z \tag{5.37}$$

where Y_c, X and Z are the sample vectors after truncating the first N_{CP} samples of Y_{CP}, X_{CP} and Z_{CP}, respectively. $\mathbf{F}$ is the $Q \times N$ subcarrier matrix and $\mathbf{H_c}$ is a $Q \times Q$ circulant matrix, thus its first column gives all the information needed to construct the matrix. After demodulation, the signal vector is expressed as

$$R_c = \mathbf{F}^*\mathbf{H_c}\mathbf{F}S + \mathbf{F}^*Z = \mathbf{G}S + Z_{\mathbf{F}^*} \tag{5.38}$$

According to the characteristic of the circulant matrix $\mathbf{H_c}$, for orthogonal multicarrier signals (e.g. OFDM), $\mathbf{G}$ is a diagonal matrix. Thus, the channel can be estimated through a single tap frequency-domain estimator. However, this is not the case in SEFDM since there are off-diagonal elements in matrix $\mathbf{G}$. This introduces both multiplicative (diagonal elements) and additive (off-diagonal elements) distortions. Therefore, a time-domain channel estimation/equalization algorithm is used to estimate and compensate the channel response. Assuming P is the pilot vector, at the receiver, after CP removal, the faded signal is expressed as

$$Y_{c-pilot} = \mathbf{H_c}\mathbf{F}P + Z \tag{5.39}$$

In order to equalize the faded signal, we need to estimate the channel matrix $\mathbf{H_c}$ and compute its inverse. By rearranging Eq. (5.39), a new expression is produced:

$$Y_{c-pilot} = \mathbf{P}h + Z \tag{5.40}$$

where h is a $Q \times 1$ vector and $\mathbf{P}$ is a $Q \times Q$ circulant matrix whose first column is equal to the vector $X = \mathbf{F}P$. Therefore the matrix $\mathbf{P}$ is expressed as

$$\mathbf{P} = \begin{bmatrix} F_1 & F_Q & \cdots & F_2 \\ F_2 & F_1 & \ddots & \vdots \\ \vdots & \ddots & \ddots & F_Q \\ F_Q & \cdots & F_2 & F_1 \end{bmatrix} \begin{bmatrix} P & 0 & \cdots & 0 \\ 0 & P & \cdots & 0 \\ \vdots & \vdots & \ddots & \vdots \\ 0 & 0 & \cdots & P \end{bmatrix} \tag{5.41}$$

where F_i is the ith row of the subcarrier matrix $\mathbf{F}$. Then the estimate of h is expressed as

$$\hat{h} = \mathbf{P}^*(\mathbf{P}\mathbf{P}^*)^{-1}Y_{c-pilot} \tag{5.42}$$

where $\hat{h}$ is the estimate of the first column of the matrix $\mathbf{H_c}$. Since $\mathbf{H_c}$ is a circulant matrix, the matrix $\mathbf{H_c}$ can be regenerated via copying and shifting $\hat{h}$ repeatedly. The estimate matrix $\hat{\mathbf{H}}_\mathbf{c}$

is then inverted to give $\hat{\mathbf{H}}_{\mathbf{c}}^{-1}$ and used to equalize the distorted SEFDM symbols in Eq. (5.37), as below

$$Y_{eq} = \hat{\mathbf{H}}_{\mathbf{c}}^{-1} Y_c = \hat{\mathbf{H}}_{\mathbf{c}}^{-1} \mathbf{H}_{\mathbf{c}} \mathbf{F} S + \hat{\mathbf{H}}_{\mathbf{c}}^{-1} Z \tag{5.43}$$

The equalized time-domain SEFDM sample vector Y_{eq} can also be expressed in terms of an error factor $\mathbf{\Psi} = \hat{\mathbf{H}}_{\mathbf{c}}^{-1} \mathbf{H}_{\mathbf{c}}$ matrix, attributed to the imperfect channel equalization process as

$$Y_{eq} = \mathbf{\Psi} \mathbf{F} S + \hat{\mathbf{H}}_{\mathbf{c}}^{-1} Z \tag{5.44}$$

It is important to note that the noise itself is now enhanced by the multiplication with $\hat{\mathbf{H}}_{\mathbf{c}}^{-1}$. Y_{eq} is then demodulated as in Eq. (5.45), giving the demodulated (and equalised) symbols vector R_{eq}

$$R_{eq} = \mathbf{F}^* Y_{eq} = \mathbf{F}^* \mathbf{\Psi} \mathbf{F} S + \mathbf{F}^* \hat{\mathbf{H}}_{\mathbf{c}}^{-1} Z = \mathbf{C}_{\mathbf{\Psi}} S + Z_{\mathbf{F}^* \hat{\mathbf{H}}_{\mathbf{c}}^{-1}} \tag{5.45}$$

where $\mathbf{C}_{\mathbf{\Psi}}$ is the correlation matrix contaminated by the error factor matrix $\mathbf{\Psi}$ and $Z_{\mathbf{F}^* \hat{\mathbf{H}}_{\mathbf{c}}^{-1}}$ is the noise vector contaminated by the multiplicative factor $\mathbf{F}^* \hat{\mathbf{H}}_{\mathbf{c}}^{-1}$. Compared with Eq. (5.5), it is apparent that in the condition of multipath fading, the signal is further distorted even if a channel equalization algorithm is adopted. Therefore, the soft detector is essential to ameliorate these degradation factors in the experiment.

The efficacy of the channel estimation and equalization described in this section has been tested in various simulations [22, 33]. As an example of results obtained, Figure 5.8 shows that while frequency-domain channel estimation fails to produce accurate channel estimates (with MSE greater than 7%) the time-domain method gives good channel estimation and therefore would be expected to give good channel equalization. The technique discussed above was tested experimentally using a commercial channel emulator, as will be described in Section 5.5.2.

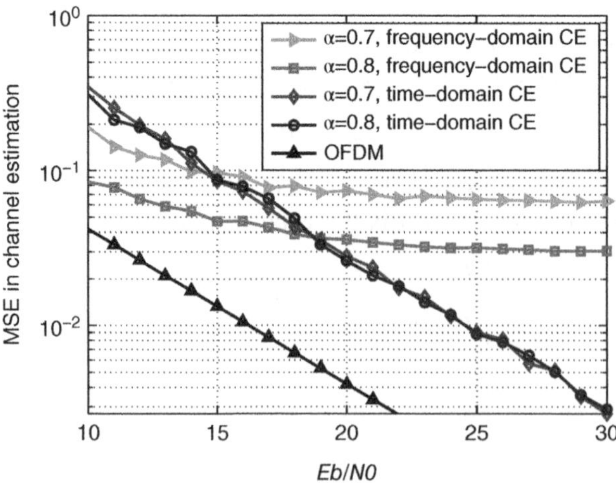

Figure 5.8 MSE of time-domain and frequency-domain channel estimation methods. $N = 72$, $Q = 128$ and CP of length nine samples

5.5.2 Experimental Demonstration

In a real RF environment, a wireless channel is not only modeled by the AWGN but also by the multipath fading, which includes amplitude attenuation, phase distortion and propagation delay. Therefore, in order to practically verify the concept of SEFDM, this section presents the first experimental evaluation of the technique together with carrier aggregation (CA) deployed in a real LTE wireless fading channel scenario. The employment of SEFDM in the CA scenario is discussed conceptually and then evaluated experimentally in a realistic RF scenario.

CA is a bandwidth extension technique proposed in LTE-Advanced. The main idea of CA is to collect legacy fragmented frequency bands (i.e. LTE signal bands) and aggregate them to support a wider transmission bandwidth. Each aggregated frequency band is termed component carriers (CC). The bandwidth of each CC can be 1.4, 3, 5, 10, 15 or 20 MHz, which are all defined in LTE since LTE-Advanced intends to provide backward compatibility to LTE. This section considers the advantages of both CA and SEFDM. Recalling that CA is a bandwidth-extension scheme while SEFDM is a bandwidth-compression technique, the combination of the two results in more aggregated CCs in a given bandwidth.

In a typical LTE-Advanced CA-OFDM scenario, a 10% frequency gap is reserved as a protection band between two CCs. CA-SEFDM aims to compress both the signal band and the protection band. Thus more CCs are aggregated in a given spectral band resulting in a more spectrally efficient transmission. The general CA-SEFDM idea is illustrated in Figure 5.9. A single subcarrier with 15-kHz baseband bandwidth is generated. In the figure, both OFDM and SEFDM subcarrier packing schemes are demonstrated for the purpose of comparison. For OFDM orthogonal multiplexing, multiple subcarriers are orthogonally packed at each frequency with 15-kHz subcarrier spacing. For SEFDM, after non-orthogonal multiplexing, subcarriers are packed more densely, so the spacing between adjacent subcarriers is smaller than 15 kHz (i.e. below the orthogonality limit). It is apparent that by multiplexing the same number of subcarriers, SEFDM will occupy less bandwidth. In the figure, the signal spectra of OFDM and SEFDM CCs are illustrated to show bandwidth compression in SEFDM CCs and the aggregation of a higher number of CCs, with narrower guard bands, in CA-SEFDM whilst maintaining the same data rate per subcarrier. Therefore, for the same bandwidth allocation, CA-SEFDM offers a higher throughput than CA-OFDM.

The experimental setup of CA-SEFDM is illustrated in Figure 5.10. To execute this experiment, both software and hardware are included in this testbed. The software consists of two digital signal processing (DSP) blocks for signal generation and detection at the transmitter and the receiver, respectively. The hardware consists of the PXI-Tx module (i.e. an RF signal generator), the PXI-Rx module (i.e. the RF digitizer) and the channel emulator, emulating an

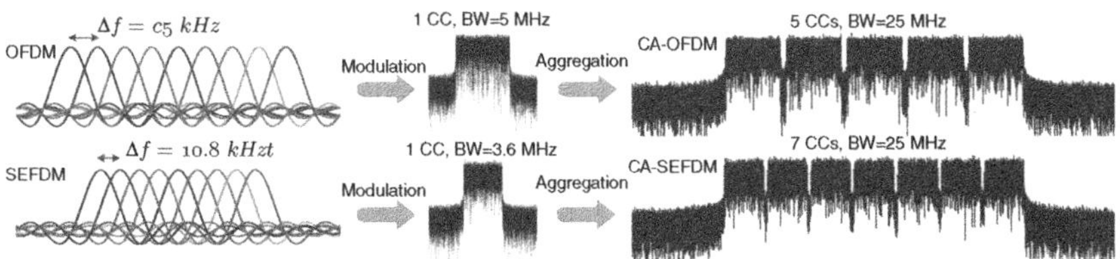

Figure 5.9 Carrier aggregation for both OFDM and SEFDM. BW is the channel bandwidth including data bandwidth and 10% protection bandwidth

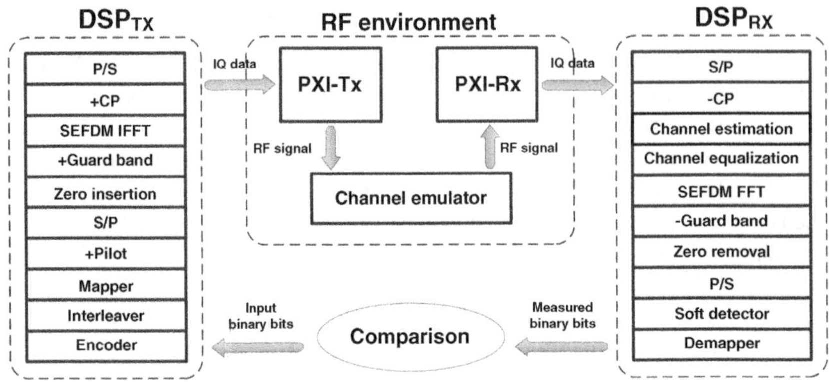

Figure 5.10 Experimental setup for CA-SEFDM transmission in a real multiplath fading channel

LTE wireless fading channel. Signal transmission, wireless channel and signal reception are all implemented in a realistic RF environment.

At the transmitter side, the input binary bits are first encoded in the encoder. Then a random interleaver $\mathbf{\Pi}$ is employed to permute the coded bits. Depending on the specific modulation scheme, the interleaved bits are mapped to the corresponding complex symbols. One uncoded pilot symbol is inserted at the beginning of each subframe (i.e. 1 pilot symbol and 13 complex coded symbols) and is used to estimate CSI, compensate for imperfect timing synchronization and local oscillator phase offset. LTE defines 10% protection subcarriers. Therefore, a gap between adjacent CCs is reserved to combat the Doppler spread encountered in a real world fading channel. This is done by inserting zeros following useful data in each band after the serial to parallel conversion. Then, the guard band is introduced for the purpose of oversampling. The data stream interpolated with pilot symbols is modulated to specific frequencies using SEFDM IFFT. A CP is added to combat multipath delay spread and a serial data stream is obtained at the last stage. It should be noted that all the signal processing within the transmitter side DSP block is operated offline in the Matlab environment. Then, the I and Q data of the SEFDM signal are uploaded to the RF environment. In the RF domain, the PXI-Tx converts the incoming baseband digital signal to an analog one and upconverts the analog signal to a radio frequency. An LTE-defined multipath fading channel model "Extended Pedestrian A" (EPA) [49] is configured in this experiment in order to evaluate the system performance in real wireless conditions.

At the receiver side, the distorted analog signal after experiencing the EPA fading channel is downconverted to baseband and transformed back to digital I and Q signals within the PXI-Rx module. The captured signal is then transferred to the receiver-side DSP block for offline processing. A parallel signal is obtained after the serial-to-parallel block and the CP is stripped away. Due to the multipath fading channel, phase and amplitude distortions are introduced. Therefore, a channel-estimation algorithm is employed to extract the CSI, which is further used to equalize the distorted symbols. The compensated signal is demodulated using either the single FFT or the multiple FFTs method. The raw SEFDM signal is obtained after the removal of the guard band and zeros. Since self-created ICI is introduced in SEFDM, the soft detector described in Section 5.4.2 is required to recover the SEFDM symbols from

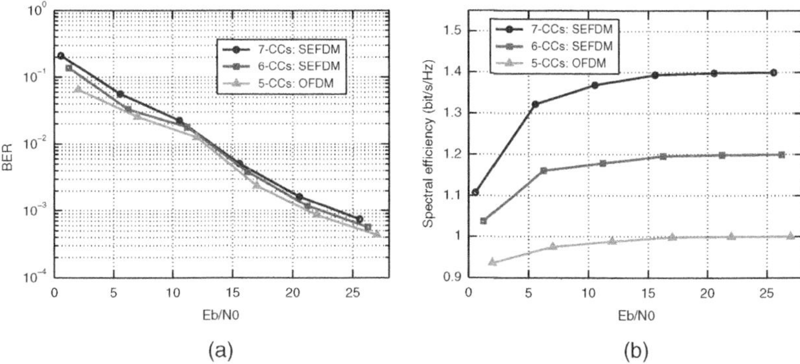

Figure 5.11 Performance of different CA-SEFDM systems in the condition of real RF environment with the LTE EPA fading channel. (a) BER performance; (b) Spectral efficiency

the interference. Then, the recovered complex symbols are demapped to binary bits. Finally, the measured binary bit stream is compared with the original input binary bit stream for the purpose of BER calculation.

This experiment is operated on the basis of the CA scenarios in Figure 5.9 and the testbed setup in Figure 5.10. The CA-OFDM and CA-SEFDM signals are generated offline in the DSP block. In order to maintain compatibility with LTE, for CA-OFDM, the subcarrier spacing is set to be 15 kHz. For CA-SEFDM, the subcarriers are intentionally packed closer, leading to a smaller subcarrier spacing of $\alpha\times$ 15 kHz. Packing flexibility can be easily achieved in the experiment. A total of 25 MHz bandwidth is used in this experiment. Therefore, in a CA-OFDM scenario five CCs are aggregated, with 5 MHz bandwidth for each, while in a CA-SEFDM scenario seven CCs are aggregated, with $\alpha\times$ 5 MHz bandwidth for each.

The performance in terms of BER is shown in Figure 5.11(a). Three systems are considered. The first CA system is based on the typical OFDM concept of the aggregation of five CCs. The other systems are based on SEFDM, where different bandwidth compression factors are employed. In the second CA system, six CCs are aggregated, with each band compressed by $16\% = (1-0.84)\times 100\%$. The third system can aggregate seven CCs, with a higher bandwidth compression corresponding to $28\% = (1-0.72)\times 100\%$ at the expense of more interference introduced by closer packing of subcarriers. However, the results indicate that with proper signal detection (i.e. soft detection), signals can be recovered in the CA-SEFDM system even with higher ICI. It is apparent that the performance of the two CA-SEFDM systems is close to that of CA-OFDM.

Moreover, Figure 5.11(b) shows that CA-SEFDM outperforms CA-OFDM in terms of effective spectral efficiency, which is defined as the non-error bits per second per Hertz that can be achieved. In Figure 5.11(b), spectral efficiencies are plotted for different CA schemes at different E_b/N_o values. The effective spectral efficiency is defined as follows:

$$R_a = (1-BER)\times R_{code}\times B_{(CC,OFDM)}\times N_{CC}\times log_2 O \tag{5.46}$$

$$B = B_{(CC,OFDM)}\times N_{(CC,OFDM)} \tag{5.47}$$

where R_a is transmission data rate, B is occupied bandwidth, BER is the bit error rate at a specific E_b/N_o value, $(1-BER)$ indicates the probability of a non-error received bit stream,

$B_{(CC,OFDM)}$ is the bandwidth of one CC in OFDM, N_{CC} is the number of CCs in either OFDM or SEFDM, $N_{(CC,OFDM)}$ is the number of CCs in OFDM and O is the constellation cardinality. Therefore, the spectral efficiency is computed as $SE = R_a/B$. It is apparent in Figure 5.11(b) that spectral efficiencies of CA-SEFDM with different CCs are higher than that of CA-OFDM. This is because compared with CA-OFDM, in the CA-SEFDM scenario, more CCs are packed in a given bandwidth.

In addition to the wireless demonstration described above, the principle of SEFDM has recently been demonstrated in optical fiber systems at 10 Gbit/s direct detection[29]; 24 Gbit/s coherent detection [35] and in an LTE-like radio-over-fiber environment [34].

5.6 Summary

The orthogonal multicarrier system, OFDM, is a technique that packs overlapping subcarriers orthogonally thus saving half the bandwidth compared to FDM. Hence, its use has been standard in 4G LTE and LTE-Advanced. However, for future 5G or beyond-5G networks, OFDM seems to be out of fashion. The evolution towards the next-generation wireless networks will be based on techniques with significantly higher spectral efficiency characteristics. Non-orthogonal multicarrier techniques have received significant attention from the wireless communication community due to their improved spectral efficiency. This chapter presents a non-orthogonal multicarrier system, termed SEFDM, which packs subcarriers at frequency separation less than the symbol rate while maintaining the same transmission rate per individual subcarrier. This allows it to improve spectral efficiency in comparison with the OFDM system. This chapter shows that transmitting the same amount of data the SEFDM system can save up to 45% bandwidth. In a practical experiment, the SEFDM concept is evaluated in a CA scenario with a realistic fading channel. It is experimentally demonstrated that the data rate is significantly improved with no extra bandwidth requirement. In addition to the benefits achieved, SEFDM also brings challenges, especially in terms of hardware implementation. This is due to the fact that complex signal detection is required at the receiver side, leading to higher computation power requirements and longer processing delays. But with the development of silicon technology, it is anticipated that complicated hardware implementation will become possible. SEFDM is an attractive technology that can tackle spectrum congestion and provide higher network capacity in future networks and much research has been carried out over the past 10 years or so to show its features and efficacy. Notwithstanding this, SEFDM remains an open research topic and many issues relating to system architecture and practical implementation still need further investigation.

References

[1] Mosier, R.R. and Clabaugh, R.G. (1958) Kineplex, a bandwidth-efficient binary transmission system. *Trans. AIEE I: Commun. Electron.*, **76** (6), 723–728.

[2] Chang, R.W. (1966) Synthesis of band-limited orthogonal signals for multichannel data transmission. *Bell System Tech. J.*, **45** (10), 1775–1796.

[3] Weinstein, S. and Ebert, P. (1971) Data transmission by frequency-division multiplexing using the discrete Fourier transform. *IEEE Trans. Commun. Tech.*, **19** (5), 628–634.

[4] Shahriar, C., La Pan, M., Lichtman, M., Clancy, T., McGwier, R., Tandon, R., Sodagari, S., and Reed, J. (2015) PHY-layer resiliency in OFDM communications: a tutorial. *IEEE Commun. Surveys Tut.*, **17** (1), 292–314.

[5] 3GPP (2010) Ts 36.300 version 8.12.0 Release 8. Evolved universal terrestrial radio access (E-UTRA) and Evolved universal terrestrial radio access network (E-UTRAN); Overall description; Stage 2 (Release 8).

[6] Rodrigues, M. and Darwazeh, I. (2003) A spectrally efficient frequency division multiplexing based communications system, in *Proceedings of 8th International OFDM Workshop*, Hamburg, pp. 48–49.

[7] Rodrigues, M. and Darwazeh, I. (2002) Fast OFDM: a proposal for doubling the data rate of OFDM schemes, in *International Conference on Telecommunications*, pp. 484–487.

[8] Kanaras, I., Chorti, A., Rodrigues, M., and Darwazeh, I. (2008) A combined MMSE-ML detection for a spectrally efficient non orthogonal FDM signal, in *5th International Conference on Broadband Communications, Networks and Systems, 2008*, pp. 421–425.

[9] Kanaras, I., Chorti, A., Rodrigues, M., and Darwazeh, I. (2009) Spectrally efficient FDM signals: Bandwidth gain at the expense of receiver complexity, in *IEEE International Conference on Communications, 2009*, pp. 1–6.

[10] Kanaras, I., Chorti, A., Rodrigues, M., and Darwazeh, I. (2009) Investigation of a semidefinite programming detection for a spectrally efficient FDM system, in *IEEE 20th International Symposium on Personal, Indoor and Mobile Radio Communications, 2009*, pp. 2827–2832.

[11] Chorti, A. and Kanaras, I. (2009) Masked M-QAM OFDM: A simple approach for enhancing the security of OFDM systems, in *IEEE 20th International Symposium on Personal, Indoor and Mobile Radio Communications, 2009*, pp. 1682–1686.

[12] Kanaras, I., Chorti, A., Rodrigues, M., and Darwazeh, I. (2009) A new quasi-optimal detection algorithm for a non orthogonal spectrally efficient FDM, in *9th International Symposium on Communications and Information Technology, 2009*, pp. 460–465.

[13] Isam, S. and Darwazeh, I. (2010) Simple DSP-IDFT techniques for generating spectrally efficient FDM signals, in *7th International Symposium on Communication Systems Networks and Digital Signal Processing (CSNDSP), 2010*, pp. 20–24.

[14] Chorti, A., Kanaras, I., Rodrigues, M., and Darwazeh, I. (2010) Joint channel equalization and detection of spectrally efficient FDM signals, in *IEEE 21st International Symposium on Personal Indoor and Mobile Radio Communications (PIMRC), 2010*, pp. 177–182.

[15] Isam, S. and Darwazeh, I. (2010) Precoded spectrally efficient FDM system, in *IEEE 21st International Symposium on Personal Indoor and Mobile Radio Communications (PIMRC), 2010*, pp. 99–104.

[16] Kanaras, I., Chorti, A., Rodrigues, M., and Darwazeh, I. (2010) A fast constrained sphere decoder for ill conditioned communication systems. *IEEE Commun. Lett.*, **14** (11), 999–1001.

[17] Isam, S., Kanaras, I., and Darwazeh, I. (2011) A truncated SVD approach for fixed complexity spectrally efficient FDM receivers, in *IEEE Wireless Communications and Networking Conference (WCNC), 2011*, pp. 1584–1589.

[18] Isam, S. and Darwazeh, I. (2011) Peak to average power ratio reduction in spectrally efficient FDM systems, in *18th International Conference on Telecommunications, 2011*, pp. 363–368.

[19] Isam, S. and Darwazeh, I. (2011) Design and performance assessment of fixed complexity spectrally efficient FDM receivers, in *IEEE 73rd Vehicular Technology Conference (VTC Spring), 2011*, pp. 1–5.

[20] Perrett, M. and Darwazeh, I. (2011) Flexible hardware architecture of SEFDM transmitters with real-time non-orthogonal adjustment, in *Telecommunications (ICT), 2011 18th International Conference on*, pp. 369–374.

[21] Grammenos, R., Isam, S., and Darwazeh, I. (2011) FPGA design of a truncated SVD based receiver for the detection of SEFDM signals, in *IEEE 22nd International Symposium on Personal Indoor and Mobile Radio Communications (PIMRC), 2011*, pp. 2085–2090.

[22] Isam, S. and Darwazeh, I. (2012) Robust channel estimation for spectrally efficient FDM system, in *19th International Conference on Telecommunications (ICT), 2012*, pp. 1–6.

[23] Whatmough, P., Perrett, M., Isam, S., and Darwazeh, I. (2012) VLSI architecture for a reconfigurable spectrally efficient FDM baseband transmitter. *IEEE Trans. Circuits Systems I: Reg. Papers*, **59** (5), 1107–1118.

[24] Perrett, M., Grammenos, R., and Darwazeh, I. (2012) A verification methodology for the detection of spectrally efficient FDM signals generated using reconfigurable hardware, in *IEEE International Conference on Communications (ICC), 2012*, pp. 3686–3691.

[25] Grammenos, R. and Darwazeh, I. (2012) Hardware implementation of a practical complexity spectrally efficient FDM reconfigurable receiver, in *IEEE 23rd International Symposium on Personal Indoor and Mobile Radio Communications (PIMRC), 2012*, pp. 2401–2407.

[26] Xu, T., Grammenos, R.C., and Darwazeh, I. (2013) FPGA implementations of real-time detectors for a spectrally efficient FDM system, in *20th International Conference on Telecommunications (ICT), 2013*, pp. 1–5.

[27] Grammenos, R. and Darwazeh, I. (2013) Performance trade-offs and DSP evaluation of spectrally efficient FDM detection techniques, in *IEEE International Conference on Communications (ICC), 2013*, pp. 4781–4786.

[28] Xu, T., Grammenos, R.C., Marvasti, F., and Darwazeh, I. (2013) An improved fixed sphere decoder employing soft decision for the detection of non-orthogonal signals. *IEEE Commun. Lett.*, **17** (10), 1964–1967.

[29] Darwazeh, I., Xu, T., Gui, T., Bao, Y., and Li, Z. (2014) Optical SEFDM system; bandwidth saving using non-orthogonal sub-carriers. *IEEE Photonics Tech. Lett.*, **26** (4), 352–355.

[30] Xu, T. and Darwazeh, I. (2014) M-QAM signal detection for a non-orthogonal system using an improved fixed sphere decoder, in *9th IEEE/IET International Symposium on Communication Systems, Networks & Digital Signal Processing 2014 (CSNDSP14)*, pp. 623–627.

[31] Xu, T. and Darwazeh, I. (2014) Multi-band reduced complexity spectrally efficient FDM systems, in *9th IEEE/IET International Symposium on Communication Systems, Networks & Digital Signal Processing 2014 (CSNDSP14)*, pp. 904–909.

[32] Xu, T. and Darwazeh, I. (2014) A soft detector for spectrally efficient systems with non-orthogonal overlapped sub-carriers. *IEEE Commun. Lett.*, **18** (10), 1847–1850.

[33] Xu, T. and Darwazeh, I. (2015) Bandwidth compressed carrier aggregation, in *IEEE ICC 2015 - Workshop on 5G & Beyond - Enabling Technologies and Applications (ICC'15 - Workshops 23)*, pp. 1107–1112.

[34] Mikroulis, S., Xu, T., Mitchell, J., and Darwazeh, I. (2015) First demonstration of a spectrally efficient FDM radio over fiber system topology for beyond 4G cellular networking, in *20th European Conference on Networks and Optical Communications - (NOC), 2015*, pp. 1–5.

[35] Nopchinda, D., Xu, T., Maher, R., Thomsen, B., and Darwazeh, I. (2015) Dual polarization coherent optical spectrally efficient frequency division multiplexing. *IEEE Photonics Tech. Lett.*, **28** (1), 83–86.

[36] Mazo, J. (1975) Faster-than-Nyquist signaling. *Bell Syst. Tech. J*, **54** (8), 1451–1462.

[37] Schellmann, M., Zhao, Z., Lin, H., Siohan, P., Rajatheva, N., Luecken, V., and Ishaque, A. (2014) FBMC-based air interface for 5G mobile: challenges and proposed solutions, in *9th International Conference on CROWNCOM, 2014*, pp. 102–107.

[38] Michailow, N., Matthe, M., Gaspar, I., Caldevilla, A., Mendes, L., Festag, A., and Fettweis, G. (2014) Generalized frequency division multiplexing for 5th generation cellular networks. *IEEE Trans. Commun.*, **62** (9), 3045–3061.

[39] Nikopour, H. and Baligh, H. (2013) Sparse code multiple access, in *IEEE 24th International Symposium on Personal Indoor and Mobile Radio Communications (PIMRC), 2013*, pp. 332–336.

[40] Saito, Y., Benjebbour, A., Kishiyama, Y., and Nakamura, T. (2013) System-level performance evaluation of downlink non-orthogonal multiple access (NOMA), in *Personal Indoor and Mobile Radio Communications (PIMRC), 2013 IEEE 24th International Symposium on*, pp. 611–615.

[41] Hara, S. and Prasad, R. (1997) Overview of multicarrier CDMA. *IEEE Commun. Mag.*, **35** (12), 126–133.

[42] Hagenauer, J. (1997) The turbo principle: tutorial introduction and state of the art, in *Proceedings of the International Symposium on Turbo Codes*, pp. 1–11.

[43] Isam, S. and Darwazeh, I. (2012) Characterizing the intercarrier interference of non-orthogonal spectrally efficient FDM system, in *8th International Symposium on Communication Systems, Networks Digital Signal Processing (CSNDSP), 2012*, pp. 1–5.

[44] Hassibi, B. and Vikalo, H. (2005) On the sphere-decoding algorithm I. expected complexity. *IEEE Trans. Signal Process.*, **53** (8), 2806–2818.

[45] Darwazeh, I., Xu, T., Gui, T., Bao, Y., and Li, Z. (2014) Optical spectrally efficient FDM system for electrical and optical bandwidth saving, in *IEEE International Conference on Communications (ICC), 2014*, pp. 3432–3437.

[46] Barbero, L. and Thompson, J. (2008) Fixing the complexity of the sphere decoder for MIMO detection. *IEEE Trans. Wireless Commun.*, **7** (6), 2131–2142.

[47] Xu, T. and Darwazeh, I. (2014) Spectrally efficient FDM: Spectrum saving technique for 5G?, in *1st International Conference on 5G for Ubiquitous Connectivity (5GU), 2014*, pp. 273–278.

[48] Bahl, L., Cocke, J., Jelinek, F., and Raviv, J. (1974) Optimal decoding of linear codes for minimizing symbol error rate (Corresp.). *IEEE Trans. Info. Theory*, **20** (2), 284–287.

[49] 3GPP (2011) TS 36.104 V10.2.0. Evolved Universal Terrestrial Radio Access (E-UTRA). Base Station (BS) Radio Transmission and Reception.

6

Non-orthogonal Multi-User Superposition and Shared Access

Yifei Yuan

6.1 Introduction

Multiple access technologies have been a key way to distinguish different wireless systems from the first generation (1G) to the current fourth generation (4G). For example, frequency division multiple access (FDMA) is used for 1G, time division multiple access (TDMA) mostly for 2G, code division multiple access (CDMA) for 3G, and orthogonal frequency

Signal Processing for 5G: Algorithms and Implementations, First Edition. Edited by Fa-Long Luo and Charlie Zhang.

division multiple access (OFDMA) for 4G. Most of these are orthogonal multiple-access (OMA) schemes, especially for the downlink transmission; in other words different users are allocated with orthogonal resources, either in time, frequency or code domain, in order to alleviate cross-user interference. In this way, multiplexing gain is achieved with reasonable complexity.

The fast growth of mobile Internet has propelled the 1000-fold data-traffic increase that is expected for 2020. Spectral efficiency therefore becomes one of the key challenges to support mobile broadband (MBB) services such as video and virtual reality in the future. Moreover, due to the ever-increased interest in the Internet of things (IoT), 5G needs to support diverse scenarios via machine-type-communication (MTC). MTC can be further divided into two main types: massive machine communication, with low data rates, and MTC, with low latency and high reliability. For massive MTC, the network is expected to accommodate a massive number of connections with sparse short messages [1], which should be of low-cost and energy efficient to enable large-scale deployment.

To satisfy these requirements, enhanced or revolutionary technologies are needed. Among the potential candidates, non-orthogonal multi-user superposition and shared access is a promising technology that can increase the system throughput and simultaneously serve massive connections. Non-orthogonal access allows multiple users to share time and frequency resources in the same spatial layer via simple linear superposition or code-domain multiplexing. Recently, several non-orthogonal access schemes have attracted much interest, not only from academia but also from the wireless industry. They can roughly be divided into two categories:

- without spreading, where modulation symbols are one-to-one mapped to the time/frequency resource elements
- with spreading, where symbols are first spread and then mapped to time/frequency resources.

Their design principles, key features, advantages and disadvantages will be discussed in this chapter. The rest of the chapter is organized into the following four sections. In Section 6.2, we will present basic principles and major features of non-orthogonal transmission and access. Section 6.3 will be devoted to downlink non-orthogonal transmission and Section 6.3 to uplink non-orthogonal access. The final section will give a summary and further discussion on future work about design and implementation of non-orthogonal multi-user transmission and access.

6.2 Basic Principles and Features of Non-orthogonal Multi-user Access

Although the interference-free condition between orthogonally multiplexed users facilitates simple multi-user detection (MUD) at receivers, it is widely known that OMA cannot achieve the sum-rate capacity of a wireless system. OMA also has limited granularity of resource scheduling, so it struggles to handle a large number of active connections.

Non-orthogonal multi-user transmission/access has been recently investigated in a systematic manner to deal with the above problems. Interference is made controllable by non-orthogonal resource allocation, at the cost of slightly increased receiver complexity. Non-orthogonal schemes can be used for the purposes described in the following subsections.

6.2.1 Non-orthogonal Multi-user Superposition for Improved Spectral Efficiency

Figure 6.1 shows an example in which the downlink sum capacity is compared for orthogonal and non-orthogonal schemes. Here, UE1 and UE2 represent users far away and near the serving base station (BS), respectively. Their transmit powers are P_1 and P_2, and channel gains are $|h_1|^2$ and $|h_2|^2$. In the OMA case, UE1 uses α of total time-frequency resources, leaving the rest of the resources $(1 - \alpha)$ to UE2. Therefore, the rate of each user can be written as

$$R_1 < \alpha \cdot \log\left(1 + \frac{P_1 |h_1|^2}{\alpha N_0}\right)$$
$$R_2 < (1 - \alpha) \cdot \log\left(1 + \frac{P_2 |h_2|^2}{(1 - \alpha) N_0}\right) \tag{6.1}$$

In the non-orthogonal case, the resources are commonly shared by UE1 and UE2. The rate of each user becomes:

$$R_1 < \log\left(1 + \frac{P_1 |h_1|^2}{P_2 |h_1|^2 + N_0}\right)$$
$$R_2 < \log\left(1 + \frac{P_2 |h_2|^2}{N_0}\right) \tag{6.2}$$

When the geometries of UE1 and UE2 are 0 dB and 20 dB, respectively, the rate region can be plotted in the right figure in Figure 6.1. It is observed that sum capacity wise, non-orthogonal scheme outperforms OMA.

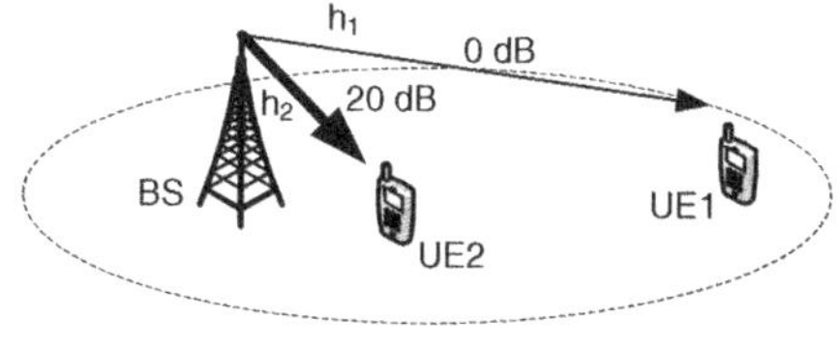

Rate	UE2 (strong)	UE1 (week)
NOMA (A)	3 bps/Hz	0.9 bps/Hz
OMA (B)	3 bps/Hz	0.64 bps/Hz
OMA (C)	1 bps/Hz	0.9 bps/Hz

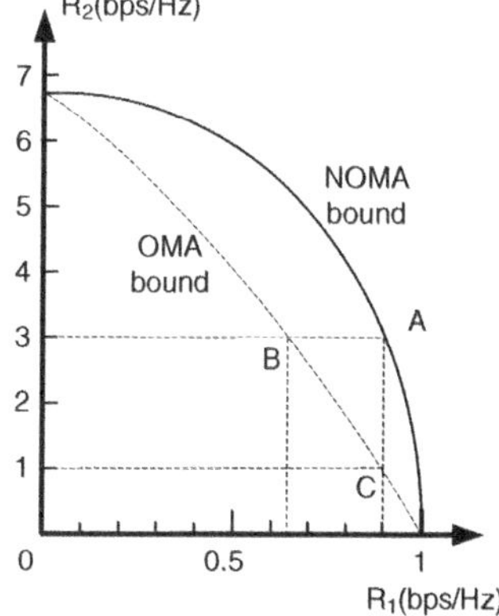

Figure 6.1 An illustrative example of downlink sum capacity comparison between non-orthogonal scheme and OMA

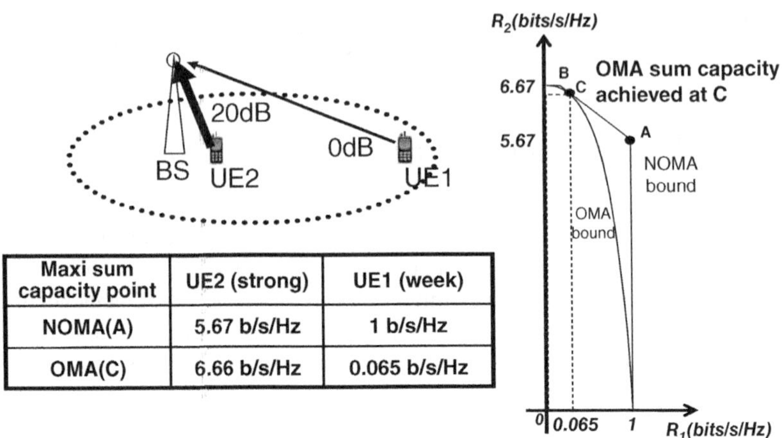

Maxi sum capacity point	UE2 (strong)	UE1 (week)
NOMA(A)	5.67 b/s/Hz	1 b/s/Hz
OMA(C)	6.66 b/s/Hz	0.065 b/s/Hz

Figure 6.2 An example of uplink sum capacity comparison between orthogonal and non-orthogonal access

The uplink capacity can be calculated in the similar fashion as for downlink, although the formula is a little different. Defining $P_{r,1}$ and $P_{r,2}$ as the received power at the BS for UE1 and UE2, respectively, the rate of each user in case of non-orthogonal uplink access can be written as:

$$
\begin{aligned}
R_1 &< \log\left(1 + \frac{P_{r,1}}{N_0}\right) \\
R_2 &< \log\left(1 + \frac{P_{r,2}}{N_0}\right) \\
R_1 + R_2 &< \log\left(1 + \frac{P_{r,1} + P_{r,2}}{N_0}\right)
\end{aligned} \tag{6.3}
$$

When the geometries of UE1 and UE2 are 0 dB and 20 dB, respectively, the uplink sum capacities of non-orthogonal and orthogonal systems can be plotted (see Figure 6.2). Although OMA can achieve the maximum sum rate in Point C, the system would be highly unfair: the rate of UE1 is only 0.065 bps/Hz, compared to 6.66 bps/Hz for UE2. In the case of non-orthogonal access, the maximum sum rate can be achieved in Point A where the fairness is significantly improved.

Generally speaking, there has been less attention to uplink non-orthogonal access for the purpose of capacity improvement, especially in standardization bodies such as 3GPP and IEEE. A possible reason is that most techniques are implementation-specific and can be transparent to air-interface standards. Because of this, we will not spend more time on uplink non-orthogonal schemes that are solely aimed at capacity enhancements of MBB-type services.

It should be emphasized that if the target is to improve the system throughput, proper scheduling and hybrid automatic repeat requests (HARQ) are needed for both downlink and uplink non-orthogonal transmission or access, similar to the case of downlink or uplink OMA.

Dynamic scheduling and HARQ are the key ingredients of link adaptation, so that the transmission format of a radio link matches the instantaneous channel condition even in fast fading.

For spectral efficiency improvement purposes, a macro-coverage scenario is the suitable deployment environment for non-orthogonal multi-user transmission/access. There are three main reasons for this. Firstly, the long intersite distances, ranging from 500–1732 m for macro eNBs (base station of LTE), offer more chance to see drastically different path-loss across different UEs. It is well known that non-orthogonal superposition schemes will provide a significant performance benefit over OMA only when multiplexed UEs experience very different path-loss. Secondly, given the large cell size of a macro cell, the number of active users served by a macro eNB tends to be large, say 20–30. This is a favorable condition for multi-user superposed transmission, since a macro eNB scheduler has more opportunity to find suitable UE pairs for joint scheduling, and therefore can fully unleash the performance potential of non-orthogonal access. Finally, multi-input multi-output (MIMO) antennas and non-orthogonal multi-user superposition are supposed to co-exist well. It is generally believed that only macro eNBs can afford to install sophisticated MIMO, including massive MIMO, which is another key technology of 5G. Therefore, deploying non-orthogonal multi-user schemes in a macro setting can help to get both benefits of both of the technologies that will play an important role in significantly boosting the system capacity.

6.2.2 *Non-orthogonal Multi-user Access for Massive Connectivity*

The design goal of Long-term Evolution (LTE) is to provide high data-rate services for a relatively small number of users. To achieve high spectrum efficiency, LTE adopts strict scheduling and control procedures that require tight control and heavy signaling. For example, the uplink transmission of each terminal is scheduled and granted individually, mostly in orthogonal radio resources. In massive connection scenarios, the payload is very small and the number of connections is huge, so the overhead of LTE becomes significant. Large overhead will increase the energy consumption of devices and the tight control mechanism tends to increase the design complexity and the cost of the terminals. On the other hand, the spectral efficiency requirement is rather relaxed.

The multiple access mechanism in the uplink of IS-95, CDMA2000 and Universal Mobile Terrestrial Services (UMTS) is indeed non-orthogonal. In those systems, the primary service is circuit-switch voice, in which the packet size is small compared to those in LTE, or even high-speed packet access (HSPA). The commonality of those systems is spread-spectrum: modulation symbols are spread before being transmitted. Spreading allows multiple users to share a resource pool, thus eliminating the need for resource indication for each individual user. The idea of spreading is refined in non-orthogonal access for massive connectivity, with more advanced techniques being added on top of it, for example:

Using factor graph to facilitate the design

The spreading operation can be generalized by using a factor graph that contains a number of variable nodes and factor nodes [2], as illustrated in Figure 6.3. The variable nodes normally correspond to the modulation symbols, or coded bits, while the factor nodes are the time-frequency resource units. The connections between variable nodes and factor nodes represent the key characteristic of each non-orthogonal access scheme that ultimately

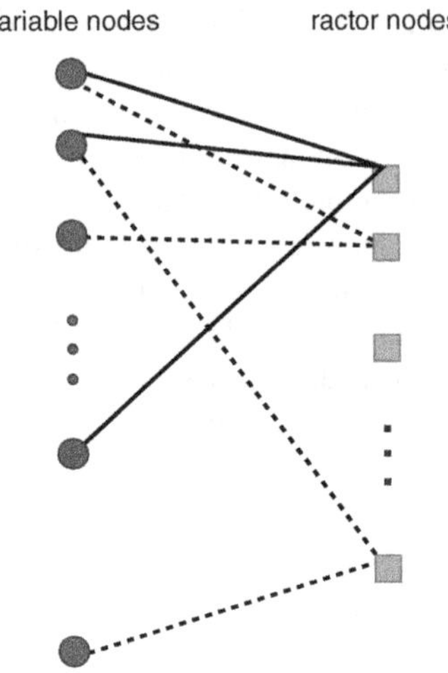

Figure 6.3 Factor graph as the generalized spreading

determines the performance. It is clearly seen that a variable node can connect to multiple factor nodes, which is essentially the spreading process. Also, a factor node can connect to multiple variable nodes, meaning that resources are non-orthogonally shared between users/layers. A factor graph can facilitate at least two design objectives. First, multiple patterns can be designed to map between variable nodes and factor nodes, so that the performance can be effectively optimized. Second, the map provides guidance to the receiver algorithms.

Using non-binary sequences

CDMA systems use binary sequences, while non-orthogonal superposition can rely on a special family of complex spread sequences in order to achieve relatively low cross-correlation, even when they are very short; say 8, or even 4. In one example, the real and imaginary parts of the complex spread sequence are from an M-ary real value set, i.e. $\{-1, 0, 1\}$. The corresponding tri-level constellation is depicted in Figure 6.4.

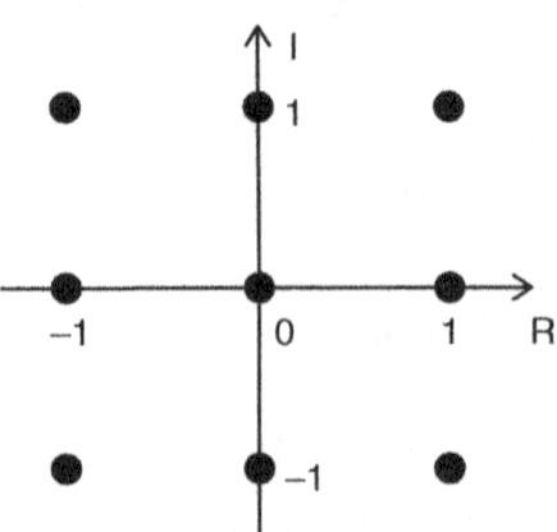

Figure 6.4 Non-binary elements for spreading sequences

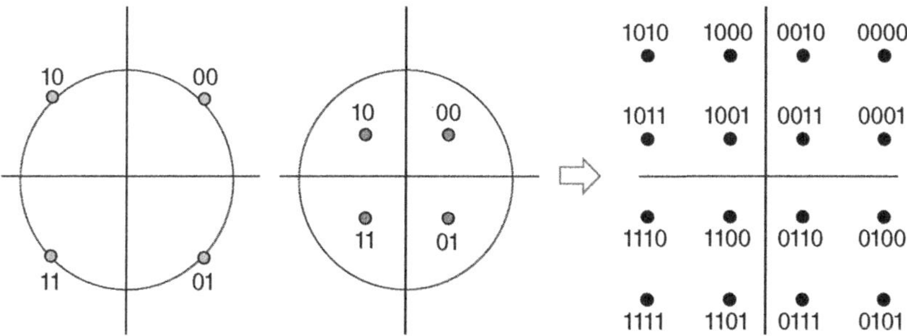

Figure 6.5 Direct superposition of two QPSK constellations, power ratio = 4 : 1

6.3 Downlink Non-orthogonal Multi-user Transmission

One of the key scenarios for downlink non-orthogonal multi-user transmission is MBB, in which the target is to maximize the downlink system capacity. Roughly, the schemes can be divided into two groups: direct superposition without Gray mapping restriction and superposition with Gray mapping.

6.3.1 *Direct Superposition without Gray Mapping*

Figure 6.5 is an example of direct superposition of two quadrature phase-shift keying (QPSKs) with a 4 : 1 power ratio. It is observed that while the constellation of each UE's coded bits follows Gray mapping, the constellation of the directly superimposed signal is not Gray mapping, leading to certain capacity loss. The mapping from coded bit to constellation can vary, depending on the power ratio and the modulation of each UE's signal, and so on. The arbitrary mapping would hamper the further optimization of superposition transmission.

A non-Gray mapped constellation tends to rely heavily on an advanced receiver. Symbol-level interference cancellation (IC) often cannot provide sufficiently good resolution of the blurred constellation due to cross-user interference. More sophisticated receivers – for example the codeword-level IC – are required, so that the performance will be acceptable. Apart from the complexity for implementation, codeword-level IC reduces flexibility in resource allocation. For instance, the resources of paired users should be fully overlapped so that the control-signaling overhead can be minimized. The HARQ mechanism also becomes more complicated with codeword-level IC.

6.3.2 *Superposition with Gray Mapping*

A simple linear superposition scheme does not require significant changes in the standards. The specification impact is control signaling, power control, and so on, leaving receiver type as an implementation issue. Due to the limited optimization at the transmitter, the performance of simple linear superposition without Gray mapping is generally inferior to other non-orthogonal superposition schemes, and requires a more advanced receiver to cancel the cross-layer interference.

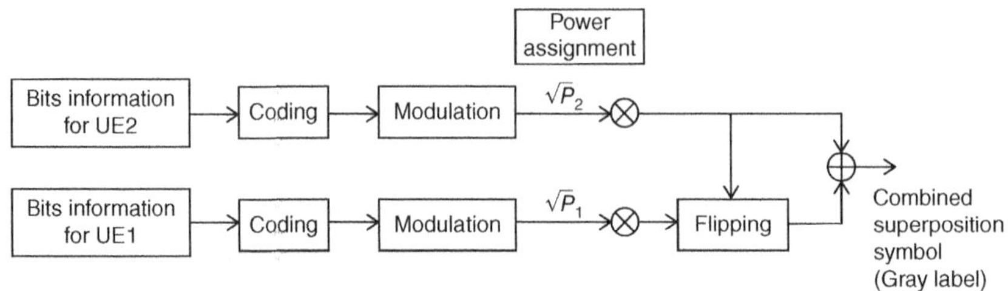

Figure 6.6 Combined superposition symbol, Gray labeled by flipping after modulation

Given the sub-optimality of direct superposition without Gray mapping, the following criteria may be considered as possible enhancements:

- simple shuffling of bits information or modulation symbols before combination, resulting in minor impacts on transmitter-side processing
- being able to support different modulation schemes
- striving to achieve Gray-nature mapping (only one bit flipped between adjacent constellation points) in the superimposed signal
- robust performance of UE receiver; for example, using symbol-level interference cancellation rather than bit-level interference cancellation. This also helps to lower the cost of UE implementation.

A simple scheme is illustrated in Figure 6.6, where a Gray labeled constellation is achieved by flipping the modulated symbols of UE1 [3].

Figure 6.7 shows the superposition QPSKs with Gray mapping. The power ratio is 4 : 1. For the low-powered user, its constellation depends on the constellation point of the high-power user. In the case that the superposition is between the "00" of the high-power user and a constellation point of the low-powered user, the upper-right four points (which are the original Gray mapping of QPSK) should be used for the low-powered user. If the superposition is between the "10" of the high-power user and a constellation point of the low-powered user, the upper-left four points (which are the horizontally flipped version of the original Gray mapping) should be used for the low-powered user. Similarly, for "11", the lower-left four points (a both horizontally-and-vertically flipped version of the original Gray mapping) should be used. For "01", the lower-right four points (vertically flipped) should be used. As a result, the superimposed signal has the characteristics of a Gray constellation.

The Gray-labeled mapping ensures robust performance even when successive interference cancellation (SIC) is implemented at symbol level. In symbol-level SIC, interference cancellation is based on symbol detection without channel decoding. It is significantly less complicated than codeword SIC, in which channel decoding has to be performed during the detection stage in order to generate the interfering signal. Figure 6.8 shows the rate regions of two superposed users via link-level simulations. Several power partitions are simulated. Square-shaped points are for direct superposition without Gray mapping, assuming symbol-level SIC. Diamond-shaped points are with Gray mapping, using symbol-level SIC. Triangular dots are with non-Gray mapping and codeword SIC. It is observed that Gray mapping can approach the performance level of a codeword SIC receiver.

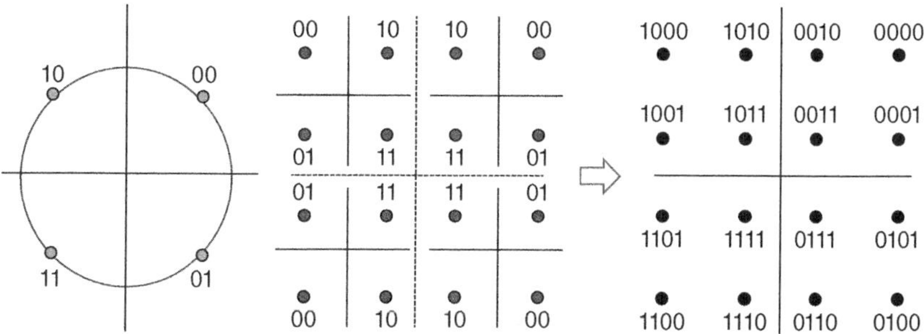

Figure 6.7 An example of bit mapping for the low powered user flipped vertically or/and horizontally, two QPSK signals with power ratio 4 : 1

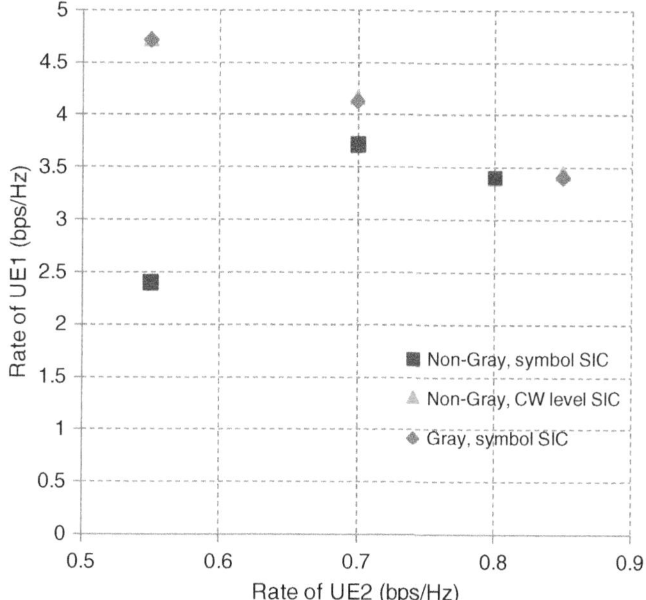

Figure 6.8 Rate pairs achieved by codeword SIC (of direct superposition), with and without Gray mapping through symbol-level SIC. SNR of UE1 is 20 dB, SNR of UE2 is 0 dB

To minimize the impact on the air-interface specification, the superposed constellations can be restricted to the legacy constellations specified in LTE-A. More specifically, the choice of power partition α_0, is limited to those shown in Table 6.1 [4]. For example, when a distant UE is allocated with 80% of power and a nearby UE is allocated with 20% power, the superposition of these two QPSK constellations would form a regular 16-QAM constellation. Since the superposed constellation reuses regular QAM constellations, the resources can also be considered as the bit partition of each combined modulation symbol. For instance, for a 64-QAM symbol that carries 6 coded bits can be partitioned to a QPSK symbol carrying 2 coded bits and a 16-QAM symbol carrying 4 coded bits, if the power partition is 0.762 : 0.238.

Table 6.1 Bit partition and the corresponding power partition

Symbol	Partitioned to Far UE	Near UE	α_0
16QAM	QPSK	QPSK	0.8
64QAM	QPSK	16QAM	0.762
64QAM	16QAM:	QPSK	0.952
256QAM	QPSK	64QAM	0.753
256QAM	16QAM	16QAM	0.941
256QAM	64QAM	QPSK	0.988

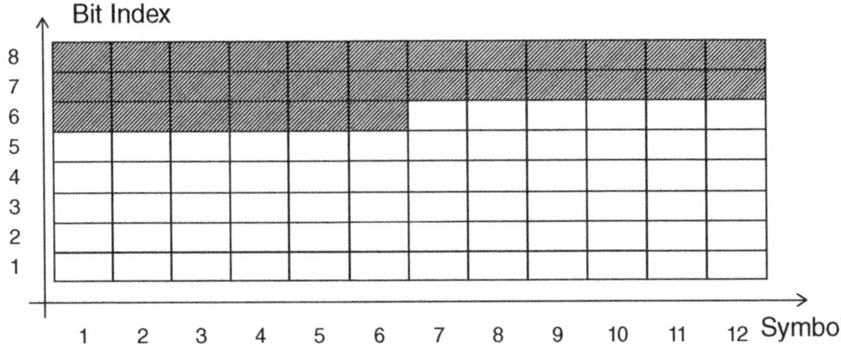

Figure 6.9 An example of superposition based on bit-division of 256 QAM symbols in a code block [5]

In the above example, the bit partition is in the sense of each superposed symbol. In another words, each modulation symbol of superposed signal in a code block uses the same partition pattern of the bits. Such a partition can be generalized to the code-block level. An example is shown in Figure 6.9 [5]. It is seen that for some 256 QAM symbols, the partition is 3 : 5, while for other symbols, the partition is 2 : 6. This will provide more flexibility.

System-level simulation assumptions are provided in Table 6.2. SIC is assumed to be able to completely cancel the interference of the far UE on the near UE. The data rate of the multiplexed UEs of non-orthogonal scheme and the single UE of SU is ideally modelled as Shannon channel capacity.

Preliminary simulation results for NOMA are provided in Figure 6.10. Based on the CDF curves, the gain of cell-edge users' throughput of NOMA over OMA is about 30%. The gain of maximum user throughput of NOMA is about 70%. The performance comparison indicates that the system capacity of NOMA is significantly higher than OMA.

The non-orthogonal superposition transmission concept can readily be applied to downlink broadcast and multicast channels, such as physical multicast channel (PMCH), in LTE. In this case, different rates are transmitted at the same time, frequency and in the same spatial resource/domain. Each rate targets to users whose geometries fall into certain ranges, for example a basic (low) rate is intended for the majority of users, in particular cell-edge users, while an enhanced (higher) rate is for users close to the serving eNBs, as illustrated in Figure 6.11.

Table 6.2 System-level simulation assumptions of DL NOMA (SISO configuration)

Parameter	Value
Cellular layout	Hexagonal grid, 3 sectors per site, 7 macro sites
Intersite distance	500 m
Minimum distance between BS and UE	25 m
Channel model	3GPP Spatial Channel Model UMa
Shadowing correlation	1.0 (intra-site), 0.5 (intersite)
Channel estimation	Ideal
UE speed	30 km/h
BS total transmission power	46 dBm
UE noise figure	7 dB
Carrier frequency	2.0 GHz
Number of antennas	1 TX (BS), 1 RX (UE)
Number of UEs per cell	10
Maximum number of multiplexed UE	1 (OMA), 2 (NOMA)
Scheduling algorithm	Proportional fairness

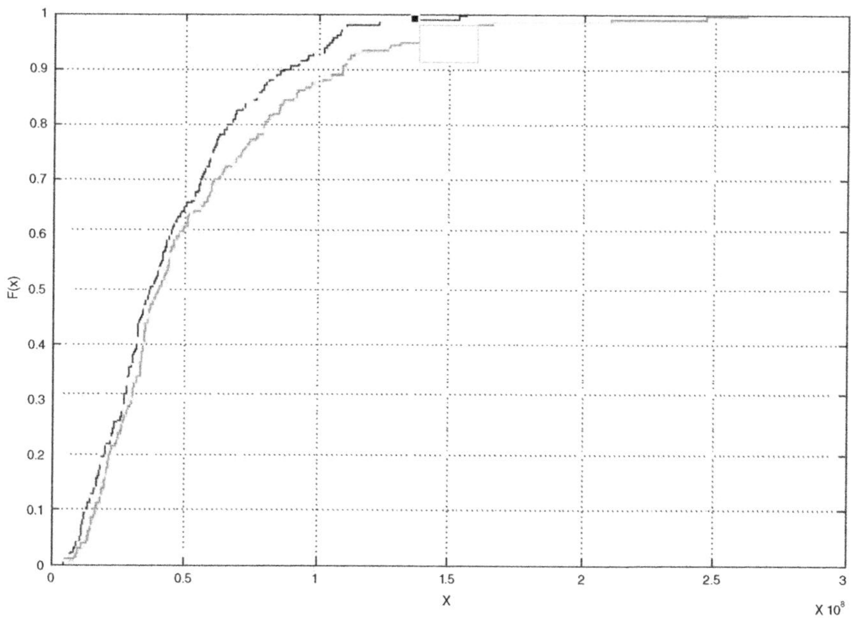

Figure 6.10 User throughput CDFs of DL NOMA and OMA with SISO configuration

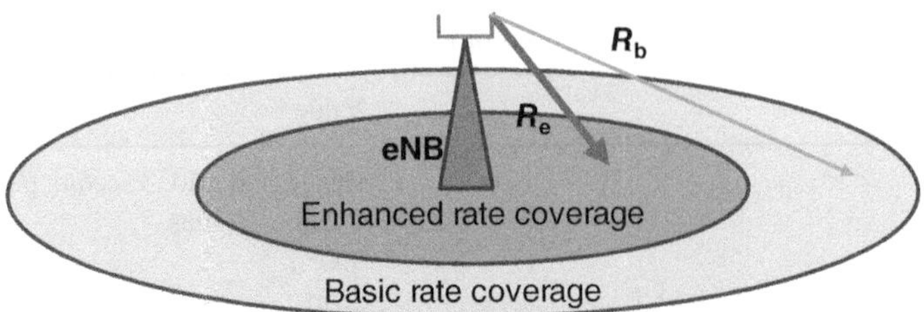

Figure 6.11 Two-layer superposition transmission for PMCH

PMCH does not have link adaptation or HARQ. Its performance metric at system level is the coverage percentage for a certain data rate, instead of system capacity as for unicast services/channels. At the link level, long-term performance – block error rate (BLER) vs. the average SNR – is often used, for the same reason. This is in contrast to unicast channels such as physical downlink shared channel (PDSCH) where elaborate link adaptation and HARQ allow the channel to operate close to the additive white Gaussian noise (AWGN) channel, even in fast fading.

A preliminary simulation has been carried out to study the performance potential of superposition transmission for PMCH [6]. Key parameters for system simulations are as follows. The single-frequency network (SFN) cluster contains 19 macro sites of 57 sectors, with wrap-around. Ten UEs are randomly dropped per sector, and there are a total of 570 UEs in the system. The number of transmit antennas at eNB is two, and the number of receive antennas at UE is two. Due to the SFN operation – all eNBs are synchronized and transmitting the same signal – the only interference seen at the UEs is thermal noise, as long as the cyclic prefix is long enough to absorb the propagation delay from different eNBs. Two scenarios are simulated: single PMCH vs (PMCH + PMCH). In the (PMCH + PMCH) scenario, the transmit power partition between the enhanced layer and basic layer can have five alternatives: [10% : 90%], [20% : 80%], [30% : 70%], [40% : 60%] and [50% : 50%]. Figure 6.12 shows the signal to interference and noise ratio (SINR) CDF for single-layer PMCH. Since the only interference is thermal noise, most UEs enjoy >15 dB SINR.

In Figure 6.13, SINR CDFs of enhanced-layer PMCH are compared, under different power ratios. As expected, when less power is allocated to the enhanced layer, its SINR becomes worse; in other words, shifted to the left.

SINR CDFs of the basic-layer PMCH are plotted in Figure 6.14. It is observed that as the power ratio of the basic layer goes down from 90% to 50%, its SINR quickly deteriorates.

Figure 6.15 shows the spectral efficiency vs coverage performance of the basic and enhanced layers of PMCH, under different power partitions. Here we mainly address spectral efficiency, which is considered more appropriate and universal for broadcast/multicast services. There are three sets of curves: single-layer PMCH, basic-layer PMCH, and enhanced-layer PMCH. The single-layer PMCH performance curve is flat, corresponding to the supported spectral efficiency with 95% coverage. It is seen that with single-layer PMCH, or the traditional PMCH, 4 bps/Hz can be supported. The basic-layer curve is the supported spectral efficiency when 95% coverage is guaranteed. When half of power is allocated to the basic layer, the spectral efficiency is about 0.8 bps/Hz. It increases to around 2.8 bps/Hz when 90% of the power is allocated. The enhanced layer curves are the supported spectral efficiencies under different

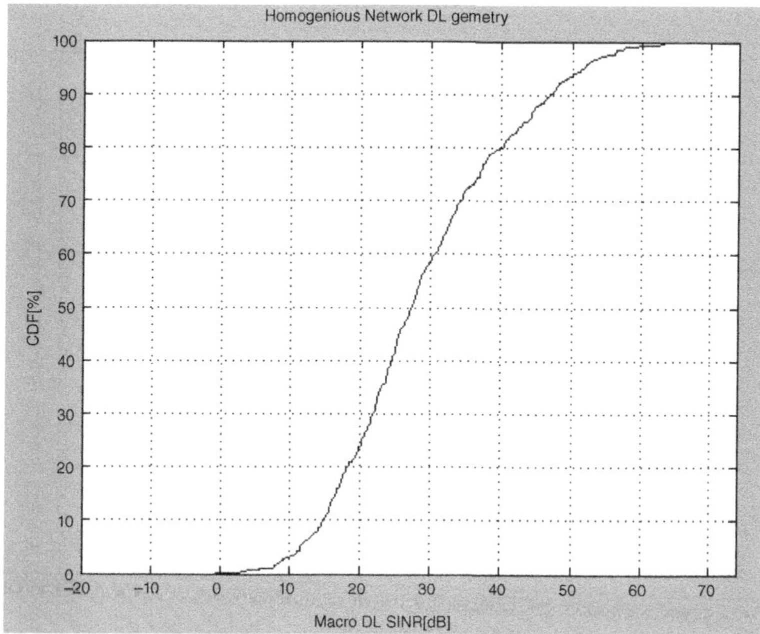

Figure 6.12 DL SINR CDF of single-layer PMCH

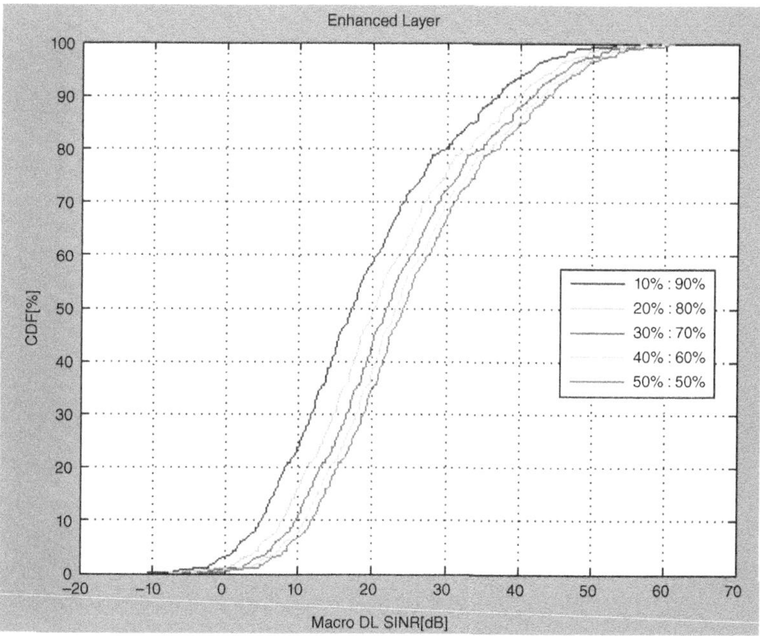

Figure 6.13 DL SINR CDF of enhanced-layer PMCH

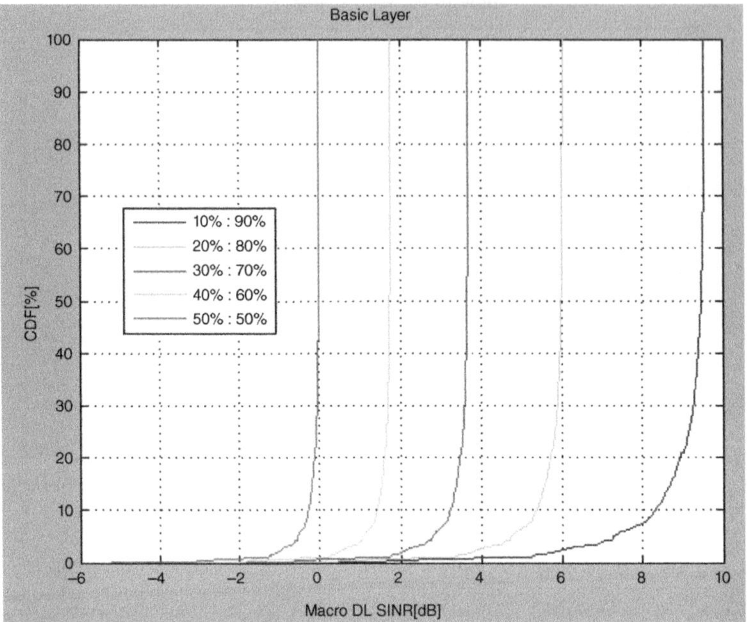

Figure 6.14 DL SINR CDF of the basic layer of PMCH

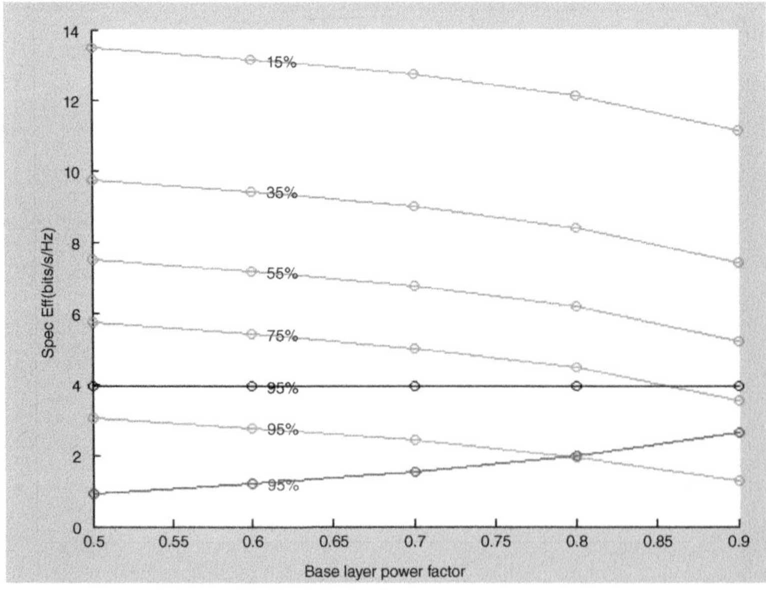

Figure 6.15 Coverage of supported spectral efficiencies under different power partitions

coverages. For example, with 35% coverage, the enhanced layer can support nearly 8 bps/Hz while its share of transmit power decreases to 10%.

Figure 6.15 provides operators with choices as to how to balance coverage and services, which can be either tuned towards a basic or to an enhanced service. Note that the simulation here is rather preliminary, while the link to system mapping uses ideal Shannon formula, without the modeling of fast-fading channels. The cancellation at the enhanced layer is also assumed ideal; in other words that the interference coming from the basic-layer transmission is completely eliminated. Nevertheless, the results reflect the basic trends of PMCH performance with superposition transmission.

6.4 Uplink Non-orthogonal Multi-user Access

Non-orthogonal resource allocation allows a large number of users/devices to be simultaneously served. It facilitates grant-free transmission so that the system will not be strictly limited by the amount of available resources and their scheduling granularity. To mitigate the potential impact of resource collision in non-orthogonal transmission, spreading can be used. Examples of uplink non-orthogonal schemes based on spreading are considered in this section:

- low-density spreading (LDS) with CDMA or OFDM
- sparse-code multiple access (SCMA)
- multi-user shared access (MUSA)
- pattern-defined multiple access (PDMA).

Note that the detectors of LDS, SCMA and PDMA do not have strong error correction capabilities, and thus a channel decoder is needed after the sequence detection.

In many of these non-orthogonal schemes, especially when used for grant-free uplink transmission, an important issue is that users' activity or instantaneous system loading is not readily known to the receiver. This would have negative impact on the performance. Compressive sensing (CS) is a promising technique to estimate the resource occupancy. Some work on CS-based random access has been done recently; for example asynchronous random access protocol [7] and compressive random access [8].

6.4.1 LDS-CDMA/OFDM

Consider a classical synchronous CDMA system in the downlink with K users and N chips (N equals the number of observations at the receiver). The transmitted symbol of the kth user is first generated by mapping a sequence of independent information bits to a constellation alphabet. Then the modulation symbol is mapped to a unique spreading sequence $\mathbf{S}_k$ and then all users' symbols are combined for transmission. In conventional CDMA systems, the spreading sequences contain many non-zero elements; in other words, they are not sparse, which means that each user would see interference from many other users in each chip. While orthogonal spreading sequences would significantly reduce the interuser interference, orthogonal sequences are generally not designed for overloading of users. The basic idea behind LDS-CDMA is to use sparse spreading sequences instead of conventional dense spreading

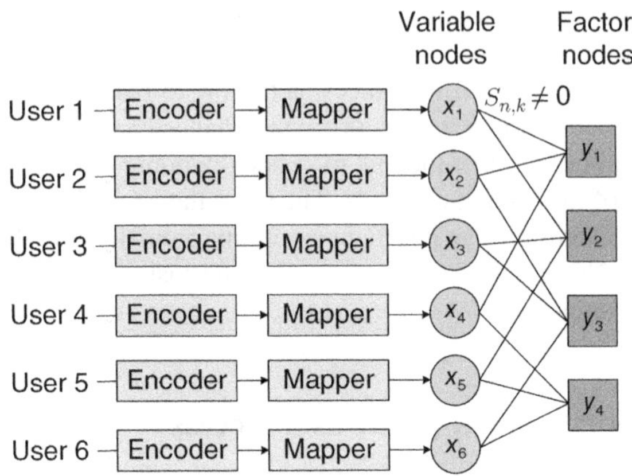

Figure 6.16 An example of LDS-CDMA with 6 users and 4 chips per symbol, 150% overloading

sequences [9], where the number of non-zero elements in the spreading sequence is much less than N, resulting in reduced cross-sequence interference at each chip. LDS-CDMA can improve the robustness of the receiver by exploiting LDS structure, which is the key feature that distinguishes conventional CDMA and LDS-CDMA. In the example shown in Figure 6.16, the number of users $K = 6$, and the spreading factor $N = 4$.

The factor graph corresponding to Figure 6.16 can be written in a matrix as:

$$F = \begin{bmatrix} 1 & 1 & 1 & 0 & 0 & 0 \\ 1 & 0 & 0 & 1 & 1 & 0 \\ 0 & 1 & 0 & 1 & 0 & 1 \\ 0 & 0 & 1 & 0 & 1 & 1 \end{bmatrix} \tag{6.4}$$

It is observed that the maximum number of users per chip is three.

At the receiver, a message-passing algorithm (MPA) can be used for MUD; this is essentially a simplified sequence detector. MPA assumes the factor graph. In the LDS-CDMA system, a variable node represents the transmitted symbol, and a factor node corresponds to the received signal at each chip. Messages, representing the reliability of the symbols, are passed between variable nodes and factor nodes through the edges. Assuming that the maximum number of users superposed at the same chip is w, due to the LDS structure, the receiver complexity is $O(Q^w)$ instead of $O(Q^K)$ where $K > w$ for conventional CDMA, and Q denotes the constellation order. Note that MPA is often followed by a channel decoder such as Turbo since LDS itself has limited capability for error correction. As both MPA and Turbo decoding are iterative processes, they can be conducted independently – in other words the iteration loop is constrained to itself – or jointly, looping over two. The latter has better performance, but with increased complexity.

It should be pointed out while low-density codes make advanced symbol-level detectors such as MPA more affordable for real implementations, the necessity or the urgency to use advanced symbol-level detectors is somewhat less when codeword-level SIC is used.

LDS-CDMA can directly be converted to LDS-OFDM, in which the chips are replaced by subcarriers in OFDM. The transmitted symbols are first mapped to certain LDS sequences

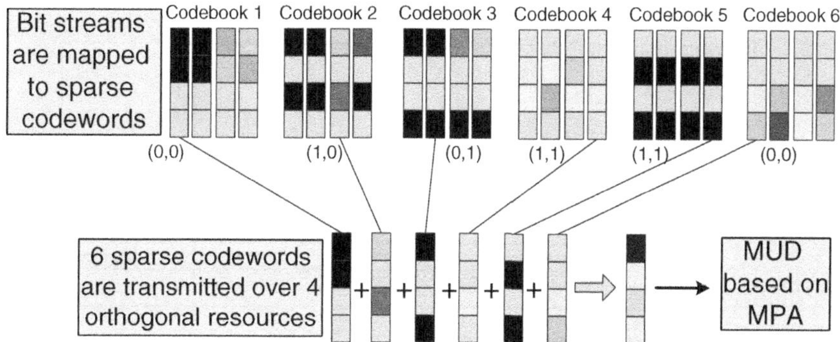

Figure 6.17 An example of SDMA with six users and 150% loading

and then transmitted on different OFDM subcarriers. The number of symbols can be larger than the number of subcarriers; in other words overloading is allowed to improve the spectral efficiency. MPA in LDS-CDMA can also be used in an LDS-OFDM receiver.

6.4.2 SCMA

SCMA, which was proposed recently [10], is an enhanced LDS. The fundamental of LDS and SCMA is the same: to use a low-density or sparse non-zero element sequence to reduce the complexity of MPA processing at the receiver. However, in SCMA, bit streams are directly mapped to different sparse codewords. This is illustrated in Figure 6.17, where each user has a codebook and there are six users. All codewords in the same codebook contain zeros in the same two dimensions, and the positions of the zeros in the different codebooks are distinct so as to facilitate the collision avoidance of any two users [10]. For each user, two bits are mapped to a complex codeword. Codewords for all users are multiplexed over four shared orthogonal resources, for example OFDM subcarriers.

The key difference between LDS and SCMA is that a multi-dimensional constellation for SCMA is designed to generate codebooks, which brings the "shaping" gain that is not possible for LDS. In order to simplify the design of the multi-dimensional constellation, a mother constellation can be generated by minimizing the average alphabet energy for a given minimum Euclidian distance between constellation points, and also taking into account the codebook-specific operations such as phase rotation, complex conjugate and dimensional permutation.

Figure 6.18 shows the link performance of SCMA under the different loading factors. The MCS level is QPSK with code rate of 1/2. Turbo codes are used. There are 12 physical resource blocks (PRBs; 2.16 MHz, 1 ms) to be used. Each PRB has 144 resource elements for traffic use. So the total number of resource elements in 12 PRBs is $144 \times 12 = 1728$. The spectral efficiency of each user is 0.5 bps/Hz since 864 information bits map to 1728 resource elements. The carrier frequency is 2 GHz. The ITU Urban Macro fast-fading model is assumed. The average SNR of different users is the same, although each undergoes fast fading independent from the others. Ideal channel estimation is assumed. In the LTE baseline setting, there are two users, each occupying six resource blocks (1.08 MHz, 1 ms). In SCMA, 100%, 150%, 200%

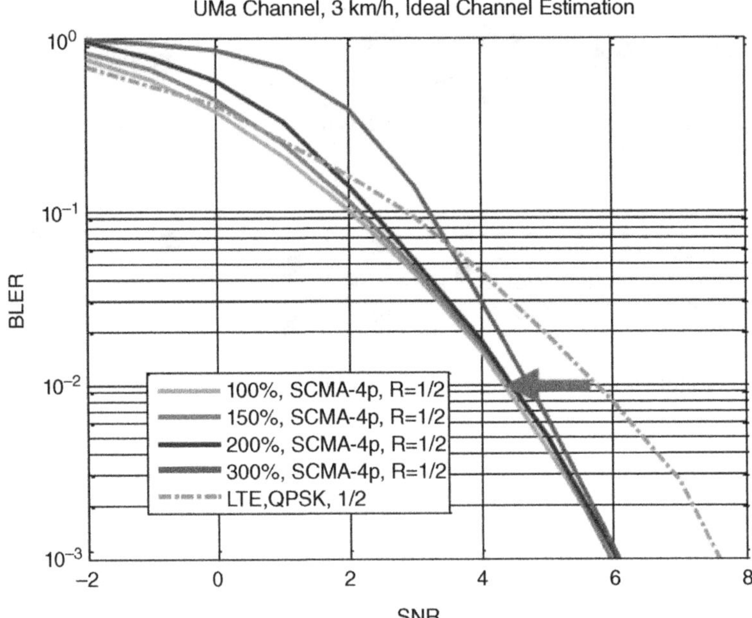

Figure 6.18 Average BLER of users with various loading factors; spectral efficiency of each user is 0.5 bps/Hz

and 300% loading factors correspond to 2, 3, 4 and 6 users, respectively, sharing 12 physical resource blocks.

It can be seen that SCMA outperforms LTE's orthogonal transmission when BLER is 1%. Note that 1% BLER is a reasonable operating point, considering the absence of HARQ. Even with 300% loading, the required SNR is 6 dB at BLER of 1%. As the overloading is reduced, BLER performance is slightly improved, but quickly saturating to the case of 100% loading,

Table 6.3 compares the uplink performance of SCMA and LTE baseline. Small packets of 160 bits are generated according to the Poisson arrival process. The average arrival interval is 300 ms. Four PRBs are used, their resources spanning 720 kHz in frequency and 1 ms in time. In the LTE baseline, each packet occupies one PRB that has 144 resource elements to accommodate modulation symbols. With QPSK modulation, the code rate is roughly 0.56. Higher-layer retransmission is assumed, the latency of which is of the order of 40 ms. While ARQ can mitigate the packet loss, it comes with the cost of latency. The latency is defined as the time between the first transmission and the time when the packet is successfully decoded. It is observed that at a 1% average packet loss, the number of supported devices can be increased from 68 to 270 by using SCMA. The latency is also significantly reduced.

For grant-free small packets, four PRBs

6.4.3 *MUSA*

Multi-user shared access (MUSA) [11, 12] is a non-OMA scheme operating in the code domain. Conceptually, each user's modulated data symbols are spread by a specially designed

Table 6.3 Uplink system performance comparison between SCMA and LTE baseline

	Number of devices supported at 1% packet loss	Latency
LTE baseline	68	~100 ms
SCMA	270	~1 ms

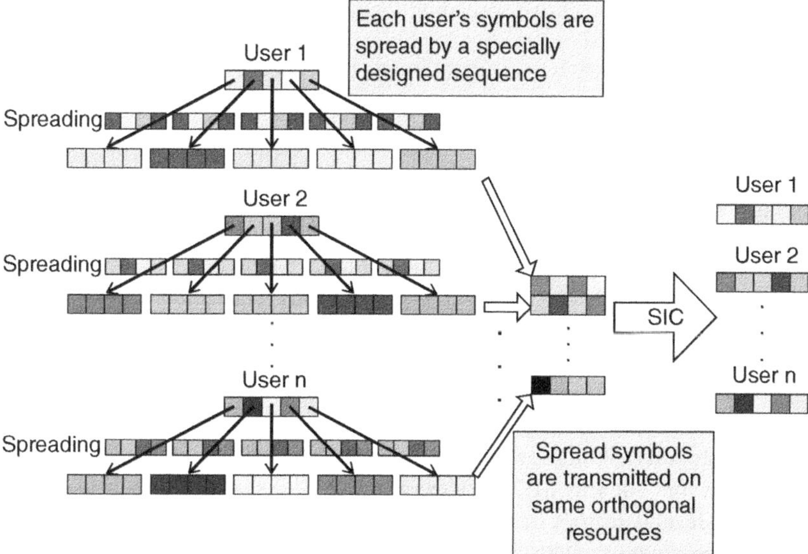

Figure 6.19 An example of MUSA with four resources shared by multiple users

sequence that can facilitate robust SIC implementation compared to the sequences employed by traditional direct-sequence CDMA (DS-CDMA). Then each user's spread symbols are transmitted concurrently on same radio resource by means of "shared access", which is essentially a superposition process. Finally, decoding of each user's data from superimposed signal can be performed at the BS side using SIC technology.

The major processing blocks of MUSA transmitter and receiver are illustrated in Figure 6.19. Symbols of each user are spread by a spreading sequence. Multiple spreading sequences constitute a pool from which each user can randomly pick one. Note that for the same user, different spreading sequences may also be used for different symbols. This may further improve the performance via interference averaging. Then, all spreading symbols are transmitted over the same time–frequency resources. The spreading sequences should have low cross-correlation and can be non-binary. At the receiver, codeword-level SIC is used to separate data from different users. The complexity of codeword-level SIC is less of an issue in the uplink as the receiver needs to decode the data for all users anyway. The only noticeable impact on receiver implementation would be that the pipeline of processing may be changed in order to perform the SIC operation.

The design of the spreading sequence is crucial to MUSA since it determines the interference between different users and the system performance. Moreover, the impact on the complexity of the SIC implementation also needs to be considered when designing the spreading sequence. The long pseudorandom spreading sequences used for the traditional DS-CDMA standard seems a good choice, as it exhibits relatively low cross-correlation even if the number of sequences is greater than the length of sequences. This property is desirable since those sequences can offer a soft capacity limit on the system rather than a hard capacity limit. Note that the spreading sequences in CDMA are long, which is less efficient when being used in conjunction with SIC. Since the system is expected to be heavily overloaded with a large number of MTC links, the excessive spreading factor introduced by long sequences may not be suitable.

Therefore, a short spread sequence with relatively low cross-correlation would be very helpful to MUSA. Since grant-free transmission can minimize the overhead of control signaling, users should generate the spreading sequence locally, without the coordination by the BS. For MUSA, a family of complex spreading sequences can be studied that would achieve relatively low cross-correlation at very short length. Complex sequences exhibit lower cross-correlation than traditional pseudorandom noise (PN) since they utilize the additional freedom of the imaginary part. The real and imaginary parts of the complex element in the spreading sequence are drawn from a multilevel real-value set with a uniform distribution. For example, for a three-value set $\{-1, 0, 1\}$, every bit of the complex sequence is drawn from the constellation depicted in Figure 6.4 with equal probability.

It should be noted that the spreading sequences of MUSA are different from the spreading codes outlined by Nikopour and Baligh [9], in the sense that MUSA spreading does not have the low-density property. While low-density codes can reduce the complexity of advanced symbol-level detectors such as MPA, codeword-level SIC somewhat reduces the need for advanced symbol-level detectors.

The typical deployment scenario is that a large number of users are distributed across the entire cell. To keep the signaling overhead low, no closed-loop power control is implemented to compensate for fast fading. Because of this, the received SNR of different users tends to be widely distributed, the so called near–far effect. This is actually a favor to MUSA since users with high SNR can be demodulated and decoded first, and then subtracted from the received signals.

For MUSA, the user load is essentially the ratio between the number of simultaneously transmitting users and the length of spread code. For example, when the length of spread code is 4 and the number of users is 12, the user load is 300%. Turbo coding with a code rate of 1/2 and QPSK modulation are used by each user.

Figure 6.20 shows the user overloading performance of MUSA in AGWN channels under different sequence lengths. More details of the link-level simulation parameters are listed in Table 6.4. Single transmitting antenna and single receiving antenna are used. The block size of information bits of each user is 256. QPSK modulation is used. Simulation results using binary PN sequence in traditional CDMA system are also plotted for comparisons. SNR of the accessing users are uniformly distributed in the range of 4~20dB, which characterizes the typical situation when tight power control procedure is absent and the receiver see certain disparity in received SNR among different users. Here the SNR is defined as the signal to noise ratio after de-spreading at the receiver. It can be observed that MUSA with tri-level complex spreading sequence of length 4 and length 8 can achieve 225% and 300% user overloading,

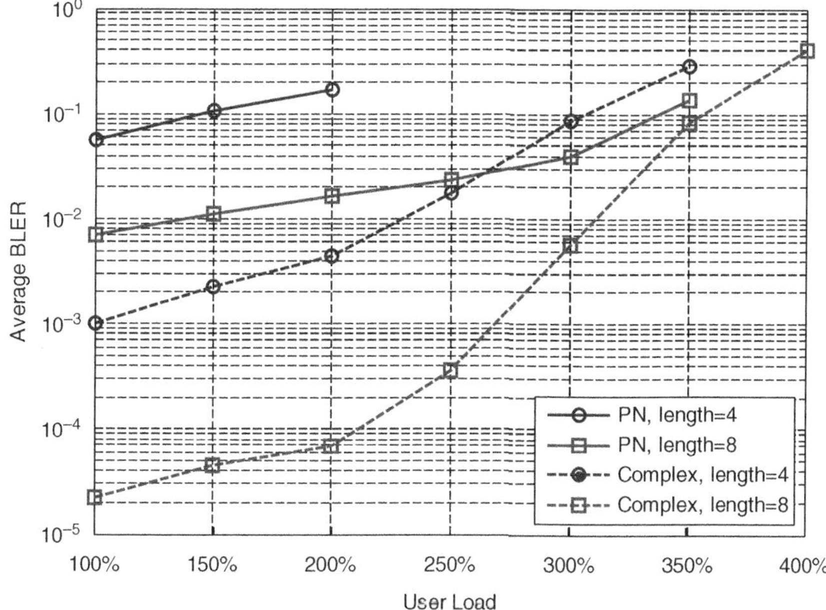

Figure 6.20 User overloading performance of MUSA in the AGWN channel with different sequence lengths

Table 6.4 Link simulation parameters for MUSA in the AWGN channel

Parameters	Assumptions
Channel model	AWGN
Modulation and coding scheme	QPSK, LTE Turbo rate 1/2 and interleaver length 256
Structure of the spreading sequence	Real and imaginary part taken from a 3-value set, $\{-1, 0, 1\}$. Randomly generated.
Spreading sequence length	4, 8, 16
User load	100%, 150%, 200%, 250%, 300%, 350%, 400%
User SNR distribution	Uniformly distributed within [4, 20] dB
Channel estimation	Ideal
Receiver algorithm	SIC

respectively at the average BLER = 1%. The tri-level sequence outperforms PN sequence with same length. This performance gain can be explained by the fact that the additional degree of freedom reduces the cross-correlation between different users.

Figure 6.21 shows the user overloading performance of MUSA under ITU UMa channels with different SNR distributions. One transmitting antenna and two receiving antennas are used. Again, there are six PRBs, equivalent to 1.08 MHz and 1 ms, are used by each user. The block size of each user's information bits is 216. Given the $144 \times 6 = 832$ resource elements, the spectral efficiency of each user is 0.25 bps/Hz. The length of the complex spreading sequence

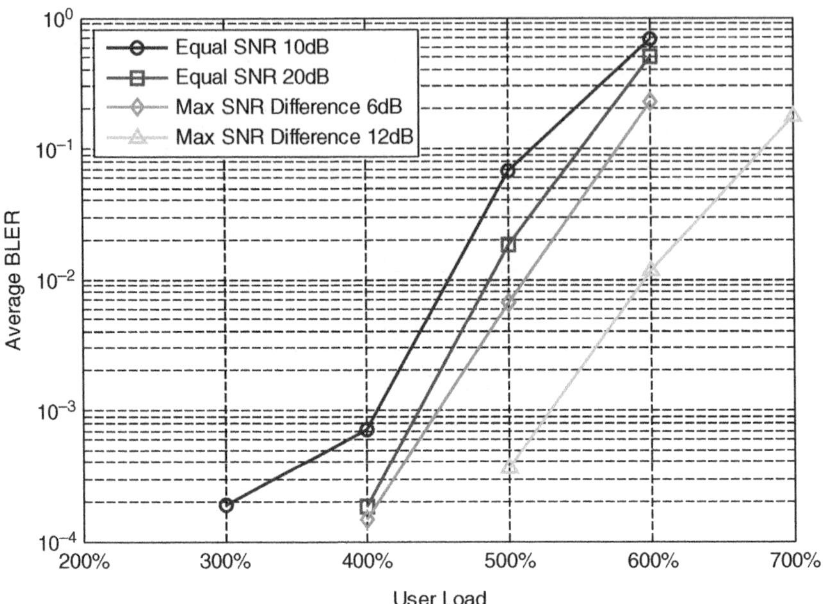

Figure 6.21 User overloading performance of MUSA in ITU UMa channels with different SNR distributions; spectral efficiency of each user is 0.25 bps/Hz

is 4. Therefore, 300% user load means that 12 users share the resources spanning 1.08 MHz in frequency and 1 ms in time. Four cases of SNR distribution are evaluated:

- same average SNR of 10 dB
- same average SNR of 20 dB
- maximum SNR difference between users is 6 dB
- maximum SNR difference between users is 12 dB.

Other simulation settings and parameters are similar to those in Table 6.4. From the simulation results, it can be observed that MUSA works better when long-term average SNRs are different between users. It implies that MUSA can take advantage of near–far effect. It is expected that near–far SNR differences should be a typical situation for MUSA since tight power control would be disabled to reduce signaling overhead. From the comparison between Figure 6.20 and Figure 6.21, it is also seen that MUSA performs better with two receiving antennas. Since massive MIMO is a potential key technology in 5G, it is reasonable to expect that UL MUSA would work well with massive MIMO.

The average BLER vs SNR is shown in Figure 6.22 at different loadings. Comparing Figs 6.22 and 6.18, in the MUSA simulation, there are 12 users sharing 12 PRBs, each user with spectral efficiency of 0.25 bps/Hz. In the SCMA simulation, there are 6 users sharing 12 PRBs, each of spectral efficiency of 0.5 bps/Hz. Hence, the sum spectral efficiency at 300% overload is the same between MUSA and SCMA: $0.5\times6 = 0.25\times12 = 3$ bps/Hz. The required SNRs are also similar at about 4.7 dB.

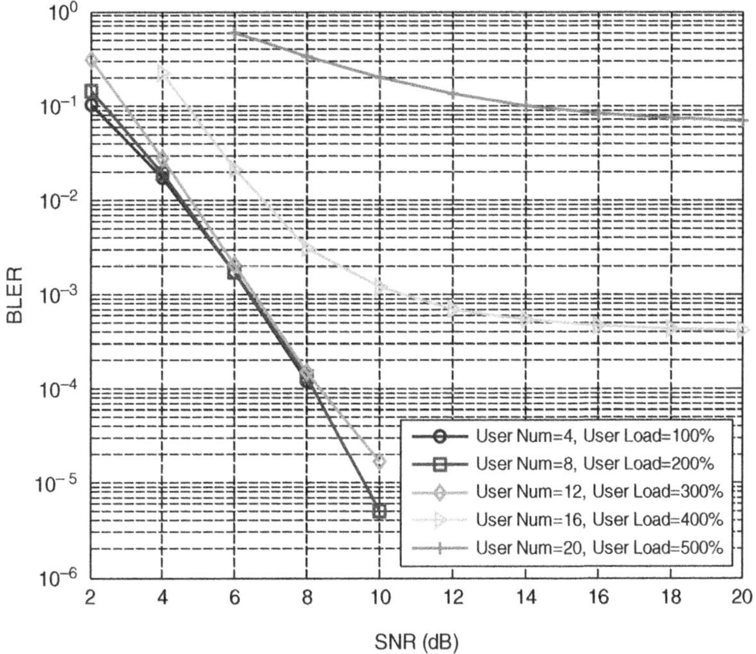

Figure 6.22 User overloading performance of MUSA in ITU UMa channels with different SNR distributions; spectral efficiency of each user is 0.25 bps/Hz

Uplink MUSA system simulation is carried out with the parameters shown in Table 6.5. Note that only open-loop power control is implemented, without closed-loop control. This means that the instantaneous SNR at the eNB receiver would fluctuate, since channels of different users fade independently. In each cell, four PRBs are used for the resource pool of MUSA, and different cells use the same set of PRBs, resulting in certain intercell interference. The average packet arrival rate is the ratio of the number of active users per cell vs the Poisson arrival interval.

For comparison, an OFDMA-based scheme is also simulated, also with four PRBs. Each user occupies one PRB, randomly selected from the four PRBs. Since MMSE receive is assumed for the OFDMA scheme, when the resources of two users collide. It is likely that either one or both links are dropped.

Figure 6.23 shows the packet-loss rate vs traffic load for UL MUSA and OFDMA. The average packet arrival period is 80 ms. It is seen that in order to satisfy a 1% packet-loss rate, MUSA can support 1 packet/ms traffic load, which is about 10 times the level of OFDMA. If the packet-loss rate is relaxed to 2%, MUSA can support 1.5 packets/ms, while OFDMA can support 0.22 packets/ms of traffic load. As the packet-loss rate is further relaxed, the difference in supported traffic load between MUSA and OFDMA becomes smaller.

Since the average packet arrival interval is 80 ms, the number of active users under the traffic load of 1 packet/ms is 80. Roughly speaking, if the packet arrival interval were 300 ms, the number of users that could be supported per cell for MUSA would be 300, slightly higher than that for SCMA (270) as seen in Table 6.3.

Table 6.5 System simulation parameters for UL MUSA

Parameters	Assumptions
Cell layout	Hexagonal, 19 sites, 3 cells/site
Scenario	ITU UMa, ISD = 500 m
Carrier frequency	2 GHz
Number of active users per cell	Depending on packet arrival rate
UE mobility	3 km/h
UE max TX power	23 dBm
UL power control	Open loop, alpha = 1, P0 = −95 dBm
Antenna configuration	1TX, 2RX
Channel estimation	Ideal
Traffic model	Small packet, 160 bits, Poisson arrival
MCS	Fixed, QPSK
HARQ/ARQ	No
Number of PRBs	4

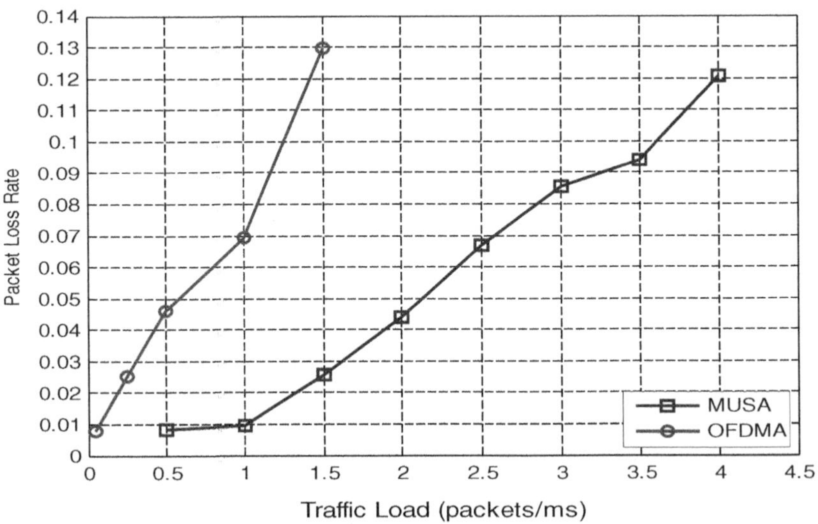

Figure 6.23 Packet loss rate vs traffic load for UL MUSA and OFDMA

6.4.4 PDMA

The original motivation of pattern division multiple access (PDMA) is to reap the benefit of diversity antenna [13], yet without losing spectral efficiency. Its concept is generalized to any diversity, not only in spatial domain, but also in time or frequency domain, the very reason the "pattern" stands for; in other words, generic enough to cover many kinds of radio resources. The key technique can be well represented by factor graph. For example, in the case of 5 users

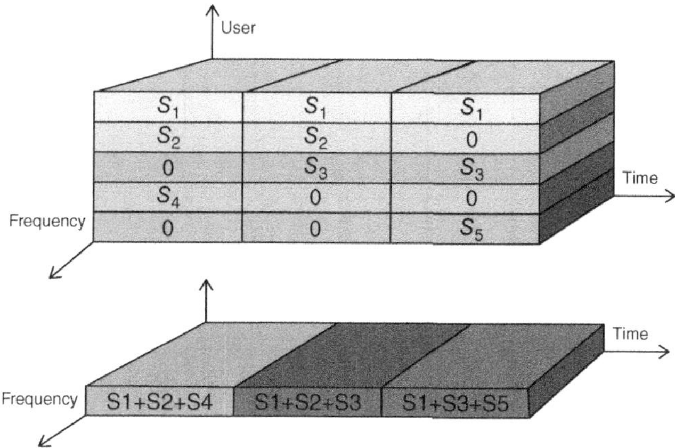

Figure 6.24 An example of PDMA with three resources shared by five users. The factor graph matrix of PDMA is constructed by combinatory of non-zero elements, as shown below

sharing 3 resources, the factor graph for PDMA can be written as the following matrix:

$$B = \begin{bmatrix} 1 & 1 & 0 & 1 & 0 \\ 1 & 1 & 1 & 0 & 0 \\ 1 & 0 & 1 & 0 & 1 \end{bmatrix} \tag{6.5}$$

Figure 6.24 illustrates the superposition of users in three resources. Specifically:

- User 1, User 2 and User 4 share Resource 1
- User 1, User 2 and User 3 share Resource 2
- User 1, User 3 and User 5 share Resource 3.

$$B_{\text{PDMA}}^{(4\times15)} = \left[\underbrace{\begin{matrix} 1 \\ 1 \\ 1 \\ 1 \end{matrix}}_{\binom{4}{4}=1} \; \underbrace{\begin{matrix} 1 & 0 & 1 & 1 \\ 1 & 1 & 0 & 1 \\ 1 & 1 & 1 & 0 \\ 0 & 1 & 1 & 1 \end{matrix}}_{\binom{4}{3}=4} \; \underbrace{\begin{matrix} 1 & 1 & 1 & 0 & 0 & 0 \\ 1 & 0 & 0 & 1 & 1 & 0 \\ 0 & 1 & 0 & 1 & 0 & 1 \\ 0 & 0 & 1 & 0 & 1 & 1 \end{matrix}}_{\binom{4}{2}=6} \; \underbrace{\begin{matrix} 1 & 0 & 0 & 0 \\ 0 & 1 & 0 & 0 \\ 0 & 0 & 1 & 0 \\ 0 & 0 & 0 & 1 \end{matrix}}_{\binom{4}{1}=4} \right] \quad \boxed{K = \binom{N}{1} + \binom{N}{2} +, \ldots, + \binom{N}{N} = 2^N - 1} \tag{6.6}$$

The detection also uses a MPA to compute the marginal functions of the global code constraint by iterative computation of a local code constraint. Although the codewords of PDMA do not have the low-density property in general, appropriate diversity order disparity can be observed in Eq. (6.6) where the first column has diversity order 4, Columns 2–5 have diversity order 3, Columns 6–11 have diversity order 2, and Columns 12–16 have diversity order 1. Diversity order disparity leads to faster convergence of MPA.

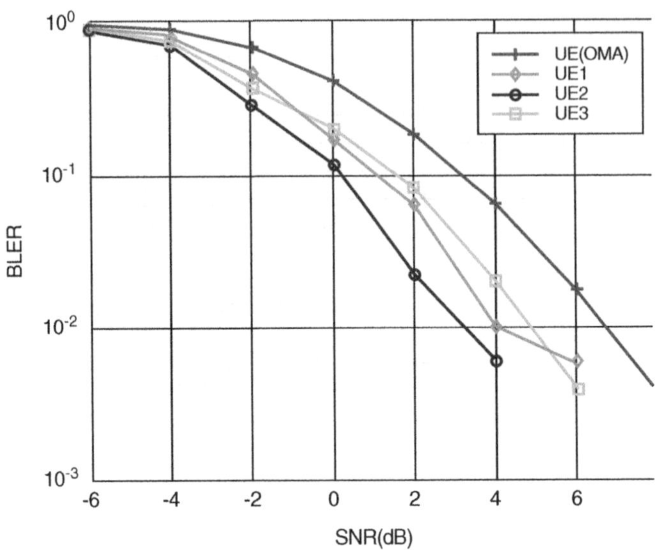

Figure 6.25 BLER performance of PDMA when three users share six PRBs; 150% overloading, spectral efficiency of each user is 0.5 bps/Hz

Essentially, PDMA uses non-orthogonal patterns, which are designed by maximizing the diversity and minimizing the overlaps among multiple users. Then the actual multiplexing can be carried out in the code domain, spatial domain or their combinations.

Figure 6.25 is the link simulation result for PDMA. The simulation parameters are similar to those for MUSA and SCMA for fading channel. Note that three users share six PRBs (1.08 MHz and 1 ms). The packet size of each user is 432 bits. Each PRB contains 144 resource elements for traffic. The corresponding spectral efficiency of each user is 432/144/6 = 0.5 bps/Hz. The baseline OMA corresponds to the two-user case in which each user occupies three PRBs without overlapping. Each user in the OMA case has 216 bits in each packet, with corresponding spectral efficiency of 216/144/6 = 0.5 bps/Hz. The improved performance is reflected in two aspects:

- reduced BLER for each user, due to the diversity provided by the PDMA factor graph
- more users can be supported in the same resource pool.

For PDMA, the required SNR for 1% BLER ranges from 3.2 to 4.8 dB, which is ~1 dB lower than in the MUSA and SCMA simulations in Figure 6.22 and Figure 6.18. Note that there are only three PDMA users sharing six PRBs in Figure 6.25, resulting in 1.5 bps/Hz sum spectral efficiency, which is only half of those in Figs 6.22 and 6.18. In other words, the traffic loading in Figure 6.25 is 150%, only half of the 300% overloading in Figure 6.22 and Figure 6.18.

6.5 Summary and Future Work

This chapter has addressed major technology aspects of non-orthogonal multi-user superposition transmission and shared access. The discussion covered basic principles and individual schemes for both downlink and uplink cellular systems, for both mobile broadband and Internet-of-things scenarios. Several key techniques were described: LDS, SCMA, MUSA and PDMA, together with simulation results. Future work on design and implementation can be listed as follows:

Theoretical Work

In NOMA schemes, theoretical analyses are needed in order to get more insights for system design. Capacity bounds of multiple access is a key metric to system performance. The capacity bounds and the rate region of code-domain NOMA with LDS need to be studied in a more comprehensive manner, so that LDS parameters such as scarcity of the code and the overloading factor can be taken into account. On the other hand, interference cancellation will play an important role in the overall performance, for example the maximum overloading that the system can support. It will be desirable that the capacity bound analysis can take into account the interference-cancellation capability at the receiver.

Design of Spreading Sequences or Codebooks

In LDS systems, due to the non-orthogonal resource allocation, interference exists among multiple users. The factor graph in MPA should be optimized to get a good tradeoff between overloading factor and receiver complexity. In addition, it has been proved that MPA can obtain an exact marginal distribution with a cycle-free factor graph and an accurate solution with a "locally tree-like" factor graph. Graph theory can be used to design a cycle-free or "local tree-like" factor graph in NOMA without compromising the spectral efficiency. In addition, the matrix design principle and methods in low-density-parity-check (LDPC) can be considered when designing the factor graph for NOMA. Apart from the factor graph design, the non-zero values in each sequence should also be optimized. One promising method is to take different values from a complex-valued constellation for these non-zero elements to maintain the maximum Euclidean distance.

Receiver Design

For an MPA-based receiver, the complexity will still be high for the massive connectivity in 5G. Therefore, some sub-optimal solutions of MPA can be used to reduce receiver complexity. A good example of such approximation solutions is the Gaussian approximation of the interference (GAI), which models the interference-plus-noise as Gaussian distributed and tends to be more accurate as the amount of connectivity becomes larger in 5G. In addition, MPA can be used to jointly detect and decode the received symbols, in which the constructed graph consists of variable nodes, observation nodes and check nodes corresponding to the check equations of the LDPC code. In this way, intrinsic information between the decoder and the demodulator can be used more efficiently so as to improve the detector's performance. For a SIC-based receiver, error propagation may degrade the performance of some users. Therefore, at each stage of SIC, some non-linear detection algorithms with higher detection accuracy can be considered to suppress error propagation.

Grant-free Transmission

As discussed earlier, uplink scheduling requests and downlink resource assignments cause the large transmission latency and signaling overhead in the granted transmission. NOMA shared access can achieve grant-free transmission with small transmission latency and low signaling overhead, and can support massive connectivity, especially for small packet transmission in 5G. A contention-based NOMA scheme is a promising solution, in which preconfigured or semistatically configured resource pools can be assigned to competing users. Integrated protocols, including random back-off schemes, can be considered to minimize resource collisions and reduce packet drop rates. Additionally, compressive sensing recovery algorithms can be used to detect the instantaneous system loading or user activity in the absence of a grant procedure.

Other Challenges

There are also some other engineering aspects to non-orthogonal multi-user superposition and shared access:

- reference signal design, channel estimation and channel state information mechanisms that can deliver robust performance when cross-user interference is severe
- resource allocation signaling that can support different transmission modes for NOMA
- extension to MIMO that can reap the performance benefits of both non-orthogonal transmission and multi-user MIMO
- power-amplifier-friendly transmission-side schemes to limit peak-to-average-power ratios in multicarrier non-orthogonal schemes such as LDS-OFDM, so that non-orthogonal transmission can work properly when the error vector magnitude of radio-frequency components is poor
- system scalability that can support different loading and radio environments.

References

[1] Yuan, Y. and Zhu, L. (2014) Application scenarios and enabling technologies of 5G. *China Commun.*, **13** (11), 69–79.

[2] Kschischang, F.R., Frey, B.J., and Loeliger, H.-A. (2001) Factor graphs and the sum-product algorithm. *IEEE Trans. Info. Theory*, **47** (2), 498–519.

[3] GPP (2015) R1-152974, Potential transmission schemes for MUST, ZTE.

[4] 3GPP (2015) R1-152806, Multi-user superposition schemes, Qualcomm Inc.

[5] Huang, J., Peng, K., Pan, C., Yang, F., and Jin, H. (2014) Scalable video broadcasting using bit division multiplexing. *IEEE Trans. Broadcast.*, **60** (4), 701–706.

[6] 3GPP (2015) RP-150979, Multi-rate superposition transmission of PMCH, ZTE.

[7] Shah-Mansouri, V., Duan, S., Chang, L.-H. Wong, V.W., and Wu, J.-Y. (2013) Compressive sensing based asynchronous random access for wireless networks. in *Proc. IEEE WCNC 2013*, Apr. 2013, pp. 884–888.

[8] Wunder, G., Jung, P., and Wang, C. (2014) Compressive random access for post-LTE systems, in *Proc. IEEE ICC 2014*, June 2014, pp. 539–544.

[9] Nikopour, H. and Baligh, H. (2013) Sparse code multiple access, in *Proc. IEEE PIMRC 2013*, Sep. 2013, pp. 332–336.

[10] Taherzadeh, M., Nikopour, H., Bayesteh, A., and Baligh, H. (2014) SCMA codebook design, in *Proc. IEEE VTC Fall 2014*, Sep. 2014, pp. 1–5.

[11] Yuan, Z., Yu, G., and Li, W. (2015) Multi-user shared access for 5G, *Telecommun. Network Tech.*, **5** (5), 28–30.

[12] Dai, L., Wang, B., Yuan, Y., S., Han, C-L., and Wang, Z. (2015). Non-orthogonal multiple access for 5G: solutions, challenges, opportunities, and future research, *IEEE Commun. Mag.*, **53** (9), 74–81.

[13] Dai, X., Chen, S., Sun, S., Kang, S., Wang, Y., Shen, Z., and Xu, J. (2014) Successive interference cancelation amenable multiple access (SAMA) for future wireless communications, in *Proc. IEEE ICCS 2014*, Nov. 2014, pp. 1–5.

7

Non-Orthogonal Multiple Access (NOMA): Concept and Design

Anass Benjebbour, Keisuke Saito, Anxin Li, Yoshihisa Kishiyama and Takehiro Nakamura

7.1 Introduction

The design of multiple access schemes is one important aspect of cellular system design. It aims to provide the means for multiple users to share the radio resources in a spectrum-efficient

Signal Processing for 5G: Algorithms and Implementations, First Edition. Edited by Fa-Long Luo and Charlie Zhang.

and cost-effective manner. In 1G, 2G, and 3G, frequency division multiple access (FDMA), time division multiple access and code division multiple access were introduced, respectively. In Long-Term Evolution (LTE) and LTE-Advanced, orthogonal frequency division multiple access (OFDMA) and single-carrier (SC)-FDMA are adopted as an orthogonal multiple access (OMA) approach [1]. Such an orthogonal design has the benefit that there is no mutual interference among users, and therefore good system-level performance can be achieved even with simplified receivers.

In recent years, non-orthogonal multiple access (NOMA) has been attracting a lot of attention as a novel and promising multiple-access scheme for LTE enhancements and 5G systems [2–11]. NOMA introduces power-domain user multiplexing, exploits channel difference among users to improve spectrum efficiency, and relies on more advanced receivers for multi-user signal separation at the receiver side. In fact, NOMA, as indicated by its name, is a non-orthogonal multiple access approach in which mutual interference is intentionally introduced among users. With NOMA, multiple users are paired and share the same radio resources, either in time, frequency or in code. From an information-theoretic perspective, it is well-known that non-orthogonal user multiplexing using superposition coding at the transmitter and successive interference cancellation (SIC) at the receiver not only outperforms orthogonal multiplexing, but also is optimal, in the sense of achieving the capacity region of [2, 3]:

- downlink or broadcast channel
- uplink or multiple access channel.

On the other hand, NOMA captures the evolution of device-processing capabilities, which generally follows Moore's law, by relying on more advanced receiver processing. For the purpose of intercell interference (ICI) mitigation, network-assisted interference cancellation and suppression (NAICS), including SIC, was discussed and specified in LTE Release 12 [12]. Thus NOMA is indeed a promising direction to extend the Third Generation Partnership Project (3GPP) work on NAICS, as it should be much easier, from the point of view both of synchronization and signaling, to apply more advanced receivers to deal with intracell than intercell interference. For downlink NOMA, for example, the signals of NOMA multiplexed users depart from the same transmitter and so there is no issue related to synchronization and the overhead owing to signaling can be minimized since the information related to the demodulation and decoding of other users multiplexed with a particular user can be jointly transmitted with the information of that user. Downlink NOMA is currently being discussed at 3GPP as a study item of LTE Release 13, under the name of "Downlink multiuser superposition transmission" [13].

In this chapter, we introduce the concept of NOMA for both downlink and uplink channels, where it enables power-domain user multiplexing and the exploitation of the channel-gain difference among users in a cellular system, two features that were not exploited in current and past cellular systems. We also explain the interface-design aspects of NOMA, including multi-user scheduling and multi-user power control, and its combination with MIMO (multiple-input, multiple-output). The performance evaluation of downlink and uplink NOMA is assessed. For downlink NOMA, we show the link-level performance and system-level performance of NOMA combined with MIMO for OFDMA. For uplink NOMA, we present a system-level performance comparison of NOMA with SC-FDMA and show the impact of fractional frequency reuse (FFR). Moreover, our experimental testbed for downlink NOMA is introduced. The simulation results and the measurements obtained from the testbed show that under multiple configurations the cell throughput achieved by NOMA is more than 30% higher than OFDMA.

The rest of this chapter is organized as follows. Sections 7.2 and 7.3 describe the concept and the benefits of NOMA for both downlink and uplink, respectively. Sections 7.4 and 7.5 discuss the interface design of NOMA and its combination with MIMO, respectively. Section 7.6 provides the performance evaluation results based on computer simulations and the measurements obtained from the NOMA testbed. Finally, Section 7.7 presents the conclusions of this chapter.

7.2 Concept

7.2.1 Downlink NOMA

Figure 7.1 illustrates downlink NOMA for the case of one BS (base station) and two UE (user equipment).

For simplicity, in this section we assume the case of single transmit and receive antennas. The overall system transmission bandwidth is assumed to be 1 Hz. The BS transmits a signal for UE-i ($i = 1, 2$), x_i, where $E[|x_i|^2] = 1$, with transmit power P_i and the sum of P_i is equal to P. In downlink NOMA, the signals of the two UEs, x_1 and x_2, are superposed as follows:

$$x = \sqrt{P_1}x_1 + \sqrt{P_2}x_2 \tag{7.1}$$

The received signal at UE-i is represented as

$$y_i = h_i x + w_i \tag{7.2}$$

where h_i is the complex channel coefficient between UE-i and the BS. Term w_i denotes additive white Gaussian noise including ICI. The power-spectral density of w_i is $N_{0,i}$. In downlink

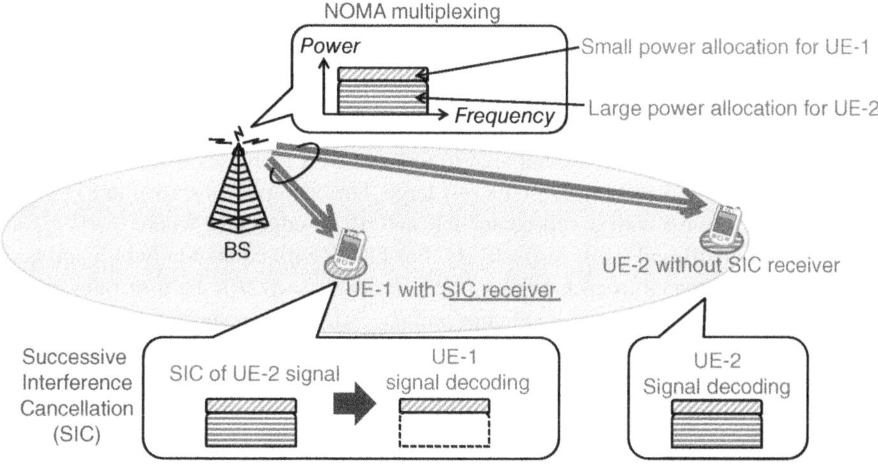

Figure 7.1 Illustration of downlink NOMA for the case of SIC applied at UE (two-UE case)

NOMA, the transmit signal from the BS and the received signal at both UE receivers is composed of a superposition of the transmit signals of both UEs (see Figure 7.1). Thus multi-user signal separation needs to be implemented at the UE side so that each UE can retrieve its signal and decode its own data. This can be achieved by non-linear receivers such as maximum likelihood detection [14] or SIC [15]. For the case of SIC, the optimal order for decoding is in the order of the decreasing channel gain normalized by noise and ICI power, $|h_i|^2/N_{0,i}$, simply referred to as "channel gain" in the rest of this chapter. Based on this order, we can actually assume that any user can correctly decode the signals of other users whose decoding order comes before the corresponding user. Thus, UE-i can remove the interuser interference from the j-th user whose $|h_j|^2/N_{0,j}$ is lower than $|h_i|^2/N_{0,i}$. In a two-UE case, assuming that $|h_1|^2/N_{0,1} > |h_2|^2/N_{0,2}$, UE-2 does not perform interference cancellation since it comes first in the decoding order. UE-1 first decodes x_2 and subtracts its component from received signal y_1; then it decodes x_1 without interference from x_2. Assuming successful decoding and no error propagation, the throughput of UE-i, R_i, is represented as

$$R_1 = \log_2\left(1 + \frac{P_1|h_1|^2}{N_{0,1}}\right), R_2 = \log_2\left(1 + \frac{P_2|h_2|^2}{P_1|h_2|^2 + N_{0,2}}\right) \tag{7.3}$$

From Eq. (7.3), it can be seen that power allocation for each UE greatly affects the user throughput performance and thus the modulation and coding scheme (MCS) used for data transmission of each UE. By adjusting the power allocation ratio, P_1/P_2, the BS can flexibly control the throughput of each UE such that the signal designated to each UE is decodable at its corresponding receiver. Also, since the channel gain of the cell-center UE is higher than cell-edge UE, as long as the cell-edge UE signal is decodable at cell-edge UE receiver, its decoding at the cell-center UE receiver can be successful with high probability.

7.2.1.1 Comparison with OMA

For OMA with orthogonal user multiplexing, the bandwidth of α $(0 < \alpha < 1)$ Hz is assigned to UE-1 and the remaining bandwidth, $1-\alpha$ Hz, is assigned to UE-2. The throughput of UE-i, R_i, is represented as

$$R_1 = \alpha\log_2\left(1 + \frac{P_1|h_1|^2}{\alpha N_{0,1}}\right), R_2 = (1-\alpha)\log_2\left(1 + \frac{P_2|h_2|^2}{(1-\alpha)N_{0,2}}\right) \tag{7.4}$$

The performance gain of NOMA compared to that of OMA increases when the difference in channel gain – the path loss between UEs – is large. For example, as shown in Figure 7.2, we assume a two-UE case with a cell-center UE and a cell-edge UE, where $|h_1|^2/N_{0,1}$ and $|h_2|^2/N_{0,2}$ are set to 20 dB and 0 dB, respectively. For OMA with equal bandwidth and equal transmission power allocated to each UE ($\alpha = 0.5$, $P_1 = P_2 = 1/2P$), the user rates are calculated according to Eq. (7.4) as $R_1 = 3.33$ bps and $R_2 = 0.50$ bps, respectively. On the other hand, when the power allocation is conducted for NOMA ($P_1 = 1/5P$ and $P_2 = 4/5P$) the user rates are calculated according to Eq. (7.3) as $R_1 = 4.39$ bps and $R_2 = 0.74$ bps, respectively. The corresponding gains of NOMA over OMA are 32% and 48% for UE-1 and UE-2, respectively. According to this simple two-UE case, NOMA provides a higher sum rate than OMA. In fact, the cell-center UE gains in terms of rate since this UE is bandwidth-limited and thus benefits more from being able to use double bandwidth, even if this comes at the price

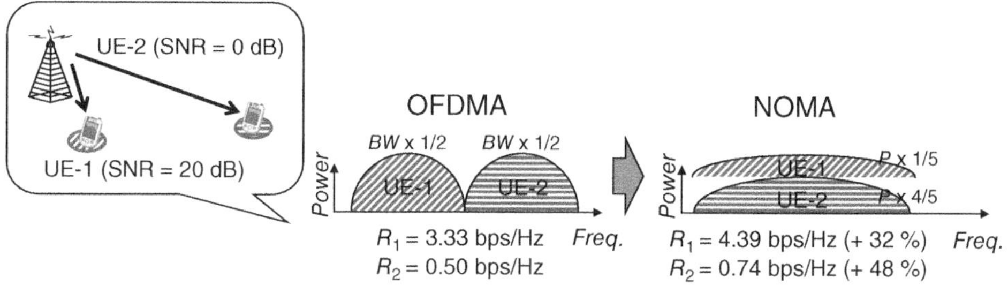

Figure 7.2 Simple comparison example of NOMA and OMA (OFDMA) for downlink

of much lower transmit power. Meanwhile, the cell-edge UE also gains in terms of rate since it is power-limited; its transmit power is only slightly reduced under NOMA but its transmit bandwidth can be doubled. These results and observations can be generalized to the case of multiple UEs and multi-user scheduling, as shown later in the system-level simulation results (see Section 7.6.1.2).

7.2.2 Uplink NOMA

Figure 7.3 illustrates uplink NOMA with two UEs transmitting signals to the BS on the same frequency resource and at the same time, and SIC conducted at the BS for UE multi-user signal separation.

Similar to downlink, we assume the case of single transmit and receive antennas, and the overall system transmission bandwidth is 1 Hz. The signal transmitted by UE-i ($i = 1, 2$) is denoted x_i, where $E[|x_i|^2] = 1$, with transmit power P_i. In uplink NOMA, the received signal at the BS is a superposed signal of x_1 and x_2 as follows:

$$y = h_1\sqrt{P_1}x_1 + h_2\sqrt{P_2}x_2 + w \tag{7.5}$$

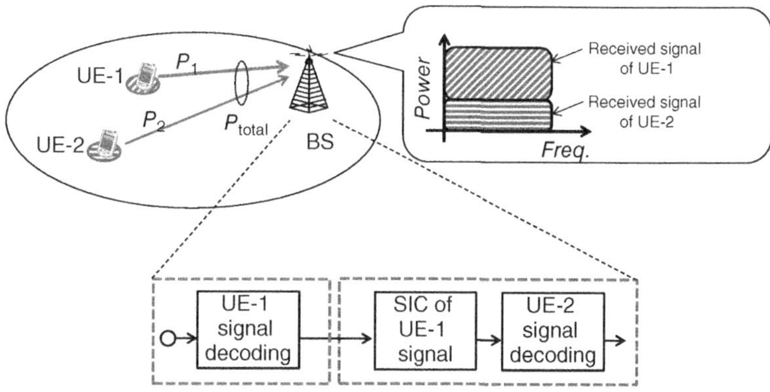

Figure 7.3 Uplink NOMA with SIC applied at BS receiver

where h_i denotes the complex channel coefficient between UE-i and the BS. Term w denotes ICI and noise observed at the BS with a power-spectral density of N_0. We assume UE-1 is the cell-center user and UE-2 is the cell-edge user – $|h_1|^2/N_0 > |h_2|^2/N_0$ – and the BS conducts SIC according to the descending order of channel gains. The throughput of UE-i, is denoted R_i, and assuming no error propagation can be calculated as

$$R_1 = \log_2\left(1 + \frac{P_1|h_1|^2}{P_2|h_2|^2 + N_0}\right), \quad R_2 = \log_2\left(1 + \frac{P_2|h_2|^2}{N_0}\right) \tag{7.6}$$

If BS conducts SIC according to ascending order of channel gains, the throughput of UE-i can be calculated as

$$R_1 = \log_2\left(1 + \frac{P_1|h_1|^2}{N_0}\right), \quad R_2 = \log_2\left(1 + \frac{P_2|h_2|^2}{P_1|h_1|^2 + N_0}\right) \tag{7.7}$$

Interestingly, the total UE throughput is the same regardless whether the SIC order is descending or ascending order of channel gain:

$$R_1 + R_2 = \log_2\left(1 + \frac{P_1|h_1|^2 + P_2|h_2|^2}{N_0}\right) \tag{7.8}$$

However, it needs to be pointed out that the conclusion that total UE throughput of different SIC orders is the same only holds under the assumption of no error propagation. In practical systems where we have error propagation, the optimal SIC order is decreasing order of channel gains.

7.2.2.1 Comparison with OMA

For OMA, we assume the bandwidth of α $(0 < \alpha < 1)$ Hz is assigned to UE-1 and the remaining bandwidth, $1-\alpha$ Hz, is assigned to UE-2. The throughput of UE-i can be calculated as

$$R_1 = \alpha \log_2\left(1 + \frac{P_1|h_1|^2}{\alpha N_0}\right), \quad R_2 = (1-\alpha)\log_2\left(1 + \frac{P_2|h_2|^2}{(1-\alpha)N_0}\right) \tag{7.9}$$

One comparison example of OMA and NOMA is shown in Figure 7.4 by assuming a two-UE case with a cell-center UE and a cell-edge UE, and where $|h_1|^2/N_0$ and $|h_2|^2/N_0$ are set to 20 dB and 0 dB, respectively. For OMA, we assume equal bandwidth is allocated to each UE ($\alpha = 0.5$), and that the user rates are calculated according to Eq. (7.9) as $R_1 = 3.33$ bps and $R_2 = 0.50$ bps, respectively. On the other hand, in NOMA, assuming the total transmission power of each UE is the same as that in OMA, the user rates are calculated according to Eq. (7.6) as $R_1 = 5.10$ bps and $R_2 = 0.59$ bps, respectively. The total UE throughput gain of NOMA over OMA is 46%. Therefore, for uplink NOMA, we can obtain similar performance gains as for downlink NOMA.

7.3 Benefits and Motivations

NOMA is a novel multiplexing scheme that improves spectrum efficiency by utilizing an additional new domain – the power domain – which is not sufficiently utilized in previous systems.

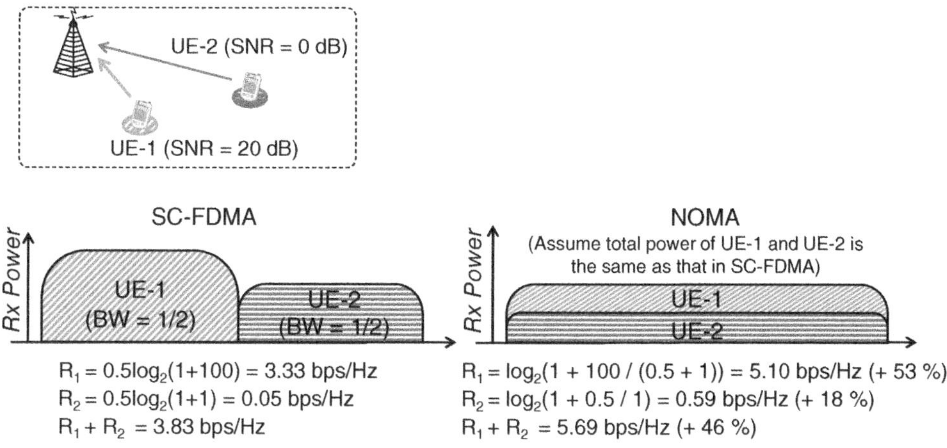

Figure 7.4 Simple comparison example of NOMA and OMA (SC-FDMA) for uplink

Figure 7.5 Example of non-orthogonal multiplexing in frequency/power domains

We show an example of non-orthogonal multiplexing in the frequency/power domains using NOMA in Figure 7.5.

Unlike OMA (OFDMA), where the channel gain difference is translated into multi-user diversity gains via frequency-domain scheduling, in NOMA the channel-gain difference is translated into multiplexing gains by superposing in the power-domain the transmit signals of multiple users of different channel gains. By exploiting the channel gain difference in NOMA, both UEs of high and low channel gains are in a win-win situation. In addition, the large power difference allocated to cell-center and cell-edge users also facilitates the successful decoding of the signals designated to both UEs, by enabling the use of relatively low-complexity receivers at the receiver side.

On the other hand, NOMA user multiplexing has the potential to further increase system capacity in a fashion similar to multi-user MIMO (MU-MIMO), even for the case in which the number of transmit and receive antennas is the same, a feature which is very attractive for deployment scenarios with limited space for antenna installation. Moreover, NOMA does not rely heavily on the knowledge of the transmitter of the instantaneous channel state information (CSI) of frequency-selective fading. As a matter of fact, with NOMA, CSI is used at the receiver for user demultiplexing and at the transmitter mainly for deciding user pairing and multi-user power allocation. Thus a robust performance gain in practical wide-area deployments can be expected, irrespective of UE mobility or CSI feedback latency.

7.4 Interface Design

7.4.1 Downlink NOMA

7.4.1.1 Multi-user Scheduling, Pairing and Power Allocation

In downlink NOMA, the scheduler at the BS searches and pairs multiple users for simultaneous transmission at each subband. To decide on the set of paired users and the allocated power set at each subband, an approximated version of the multiuser scheduling version of the Proportional Fairness (PF) scheduler is used [16]. Among all candidate user sets U and power sets, Ps, the PF scheduling metric maximizing candidate user set U_{max}, and power set $Ps_{max,}$ is selected over each subband s as follows:

$$Q(U, Ps) = \sum_{k \in U, Ps} \left(\frac{R_s(k, U, Ps, t)}{L(k, t)} \right)$$

$$\{U_{\max}, Ps_{\max}\} = \max_{U, Ps} Q(U, Ps) \tag{7.10}$$

Term $Q(U, Ps)$ denotes the PF scheduling metric for candidate user set U being allocated power set Ps and is given by the summation of the PF scheduling metric of all users in user set U. Term $R_s(k|U,t)$ is the instantaneous throughput of user k in subband s at time instance t (the time index of a subframe), while $L(k,t)$ is the average throughput of user k. For the NOMA case, by assuming dynamic switching between NOMA and OMA, we can use NOMA only when it provides gains over OMA. Also, the number m of users to be multiplexed over each subband is also decided by searching all possible candidate user sets of different sizes up to m. The number of candidate user sets to be searched is given by:

$$C = \binom{K}{1} + \binom{K}{2} + \cdots + \binom{K}{m}$$

7.4.1.2 User Scheduling and MCS Selection: Wideband vs Subband

In LTE and LTE-Advanced, the same MCS is selected over all subbands allocated to a single user. Therefore the averaged signal-to-interference and noise power ratio (SINR) over all allocated subbands is used for MCS selection. However, when NOMA is applied over each subband, user pairing and power allocation are conducted over each subband as well. With such a mismatch between MCS selection granularity (i.e. wideband) and power allocation granularity (i.e. subband), the full exploitation of NOMA gains would not be possible. Also, the higher the power allocation granularity the more signaling overhead we have. Therefore, the following three different granularities can be candidates for user scheduling/pairing and MCS selection (see Figure 7.6):

- Case 1: Subband scheduling and subband MCS selection
- Case 2: Subband scheduling and wideband MCS selection
- Case 3: Wideband scheduling and wideband MCS selection.

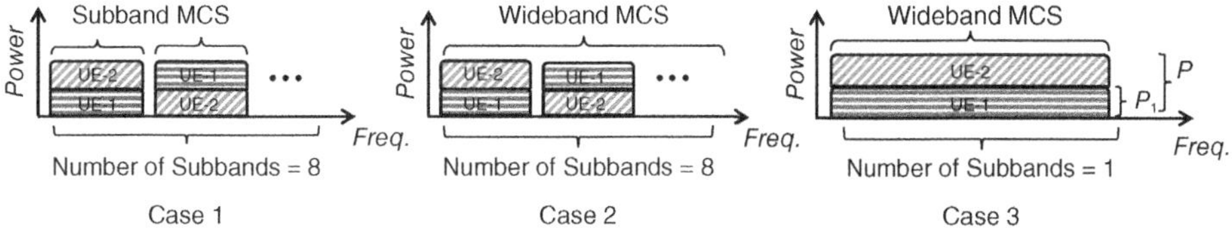

Figure 7.6 Granularity of user scheduling and MCS selection

7.4.2 Uplink NOMA

The interface design of uplink NOMA has may commonalities with that of downlink except for two major differences. One is scheduling design and the other is power control. These aspects are elaborated in detail in the following.

7.4.2.1 Scheduling Design for Uplink NOMA

For uplink, SC-FDMA is adopted in LTE/LTE-Advanced specifications owing to its low peak to average power ratio (PAPR) characteristic, which better relieves the power consumption issue at the UE compared to OFDMA. In order to achieve low PAPR, consecutive resource allocation for each UE is required in SC-FDMA. When NOMA is applied, such a constraint of SC-FDMA requires new designs for scheduling algorithms. For example, since multiple UEs are non-orthogonally multiplexed and share the same radio resource, the transmission power limitation of one UE may limit the resource allocation of the whole NOMA UE group, which is a problem that does not exist in either downlink NOMA scheduling with OFDMA or uplink SC-FDMA scheduling without NOMA. One example is shown in Figure 7.7, where widening is conducted in order to achieve consecutive resource allocation for a NOMA UE group {UE-1, UE-2, UE-3} in which UE-2 reaches maximum transmission power during the widening procedure. For SC-FDMA, if a UE reaches its maximum transmission power, the BS can simply stop allocating subbands to it. For NOMA, the situation becomes much more complicated since the BS allocates subbands to a group of UEs rather than to one UE and the transmission power of UEs can be adjusted among the paired UEs keeping the same total transmission power. From this point of view, the optimal scheduling scheme should have a joint design of transmission power control (TPC), UE selection algorithm and subband allocation algorithm, within an acceptable computational complexity. Three suboptimal solutions are potentially applicable to deal with this problem. One is stopping the widening procedure of the UE group; in other words, the NOMA UE group and its subsets do not join in competition for subband allocation anymore. The second is stopping the widening procedure of the UE reaching maximum transmission power while continuing the widening procedure of other UEs; in other words the NOMA UE group does not join in competition for subband allocation anymore but its subsets *can*. The third is continuing the widening procedure of the UE group with all UEs until subband allocation is finished; in other words both the NOMA UE group and its subsets can join in competition for subband allocation. For the UE that reaches maximum transmission power, the $P_{\max}$ is equally assigned to all of its allocated subbands.

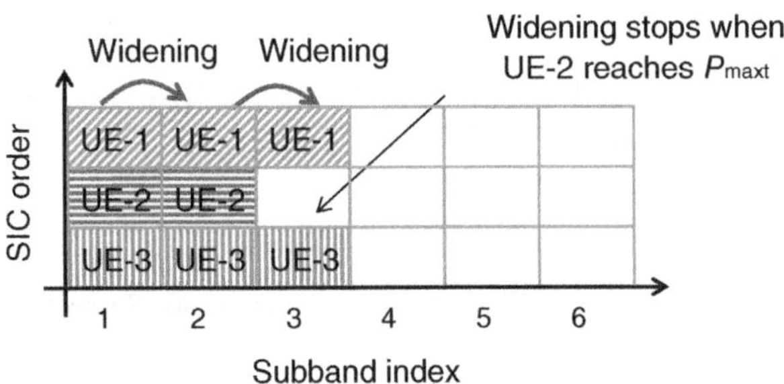

Figure 7.7 Illustration of the UE power limitation problem in uplink NOMA scheduling

7.4.2.2 Power Control for Uplink NOMA

Power control of uplink NOMA is different from that of downlink in two aspects. Firstly the transmission power constraint for optimization is different. In downlink, the transmission power only has one constraint: the maximum transmission power of the BS; in uplink transmission power optimization is constrained by the maximum transmission power of each individual UE. Secondly, the design methodology of the TPC is different. In downlink, the superposed signal received at a UE experiences the same channel: the signals of different UEs have the same channel gain at each UE receiver. Therefore, the design of downlink power control aims to artificially create enough difference among the signals of different UEs in the power domain so as to enable signal separation, for example SIC, at the receiver. For uplink, because the signals transmitted by the different UEs that are received at the BS experience different channels, the received signal powers of different UEs already have differences in the power domain. On the other hand, when NOMA is applied in uplink, the ICI greatly increases because multiple UEs are allowed for transmission simultaneously, while in downlink ICI does not increase when NOMA is applied because generally the BS has fixed transmission power regardless of the number of multiplexed UEs. Therefore, the design of uplink power control consists of two aspects:

- to create additional differences among the received signals of different UEs when the channel-gain difference is not enough
- to control the transmission power of UEs in order not to cause too large ICI to neighboring cells [5].

An example of an uplink NOMA power-control algorithm based on maximizing the sum PF scheduling metric is now presented in the three steps. In the first step, the fractional TPC (FTPC) method of LTE [22] is utilized to determine the basic transmission power of user k (P_k):

$$P_k = \min\{P_{\max}, 10\log_{10}(M_k + P_0 + \alpha \cdot PL_k)\} \tag{7.11}$$

where $P_{\max}$ denotes the maximum transmission power of a user, M_k denotes the number of allocated frequency blocks, PL_k denotes the path-loss including distance-dependent path-loss and log-normal shadowing, P_0 denotes the target received signal level when the path-loss from a cell site is zero and α denotes the path-loss attenuation factor of FTPC. In the second step, the total transmission power is determined for each candidate NOMA user set. For each candidate NOMA user set U, the total transmission power ($P_{\text{total, U}}$) is determined by

$$P_{total,U} = \beta \times N \times P_{avg} \tag{7.12}$$

$$P_{avg} = \frac{1}{K}\sum_{k=1}^{K} P_k \tag{7.13}$$

where $\beta \varepsilon$ (0, 1) denotes a parameter to be optimized considering ICI to other cells, N denotes the number of users in the candidate NOMA user set, and P_{avg} denotes the average basic transmission power of users in the cell. In the third step, after the total transmission power of each NOMA user set is determined, exhaustive search is utilized to select the best power set (Ps_{max}) and user set (U_{max}) using the same method as that in the downlink; in other words according to Eq. (7.10). Note that the point of parameter β is to balance the signal power to the serving BS and the ICI to neighboring cells. The optimal value of β should be cell-specific and determined by joint optimization with multiple cells. If the cells have similar traffic and the users are uniformly distributed in the cell, a fixed value of β can be adopted for all cells for simplicity.

It is important to point out here that downlink NOMA and uplink NOMA have different tolerances to error propagation, which should be considered in scheduler design for practical systems. To be more specific, for downlink NOMA, the MCS of a cell-edge UE is selected according to the channel gain of cell-edge UE for a targeted block error rate (BLER), say 10%. Because a cell-center UE has a higher channel gain than a cell-edge UE, when the cell-center UE detects the signal of a cell-edge UE, the BLER will be much lower than 10%. However, for uplink NOMA, if the MCS of the cell-center UE is selected with a targeted BLER of 10%, when the BS detects the signal of the cell-center UE, the BLER will be similar to 10%. Therefore, uplink NOMA will suffer more from error propagation than downlink NOMA especially when the multiplexing order m is large.

7.5 MIMO Support

7.5.1 Downlink NOMA

7.5.1.1 NOMA Extension from SIMO to MIMO

MIMO is one of the key technologies identified to improve spectrum efficiency in LTE/LTE-Advanced. In general, MIMO techniques can be categorized under two headings:

- single-user MIMO (SU-MIMO), where only one UE is served in data transmission
- multi-user MIMO (MU-MIMO), where more than one UE is served in data transmission.

Because MIMO technology exploits the spatial domain and NOMA exploits the power domain, the two technologies can be combined to further boost the system spectrum efficiency. In single-input, single-output (SISO) and single-input, multiple-output (SIMO) downlink the broadcast channel is degraded and superposition coding with SIC and dirty paper coding (DPC) are equivalent and both optimal from the viewpoint of the achievable capacity

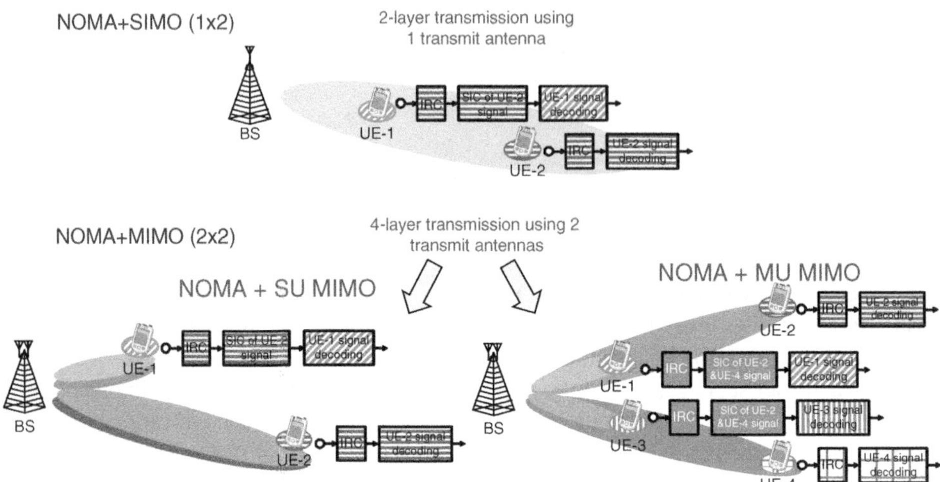

Figure 7.8 NOMA extension from SIMO to MIMO (SU-MIMO and MU-MIMO)

region. However, for the downlink MIMO case, the broadcast channel is non-degraded and the superposition coding with a SIC receiver becomes non-optimal; although DPC remains optimal [2, 3, 11].

As illustrated in Figure 7.8, there are two major approaches to combine NOMA and MIMO technologies.

One approach is to use the NOMA technique to create multiple power levels and apply the SU-MIMO and/or the MU-MIMO technique inside each power level. For example, for NOMA with SU-MIMO (2×2), with up to two-user multiplexing in the power domain, non-orthogonal beam multiplexing enables up to four-beam multiplexing using only two transmit antennas. In addition, the combination of NOMA with SU-MIMO can involve both open-loop MIMO – for example space frequency block coding or large delay cyclic delay diversity – and closed-loop MIMO based on CSI – such as the precoder indicator, channel quality indicator (CQI) or rank indicator feedback by users. Open-loop MIMO schemes, when combined with NOMA, are expected to provide robust performance in high mobility scenarios.

The other approach is to convert the non-degraded 2×2 MIMO channel to two degraded 1×2 SIMO channels, where NOMA is applied over each equivalent 1×2 SIMO channel separately, as shown in Figure 7.9 [8]. For this scheme, multiple transmit beams are created by opportunistic beamforming, and superposition coding of signals designated to multiple users is applied within each transmit beam (i.e. intrabeam superposition coding). At the user terminal, the interbeam interference is first suppressed by spatial filtering only by using multiple receive antennas, then multisignal separation such as SIC) is applied within each beam. This scheme can be seen as a combination of NOMA with MU-MIMO where fixed rank 1 transmission is applied to each user. Thus SU-MIMO with rank adaptation is not supported, and a large number of users would be required to obtain sufficient gains [10].

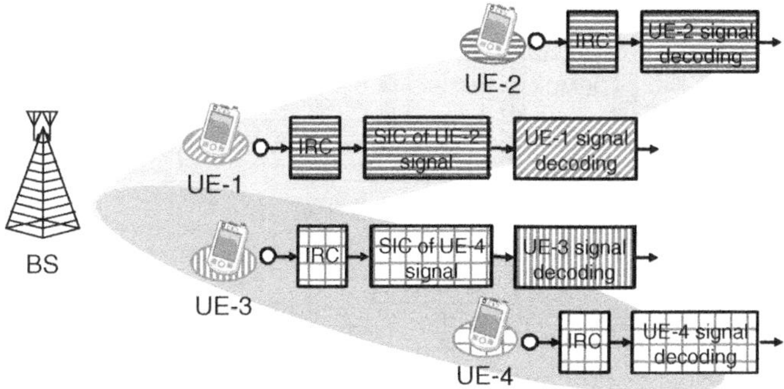

Figure 7.9 NOMA combined with 2×2 MIMO using random beamforming and applying IRC-SIC receivers

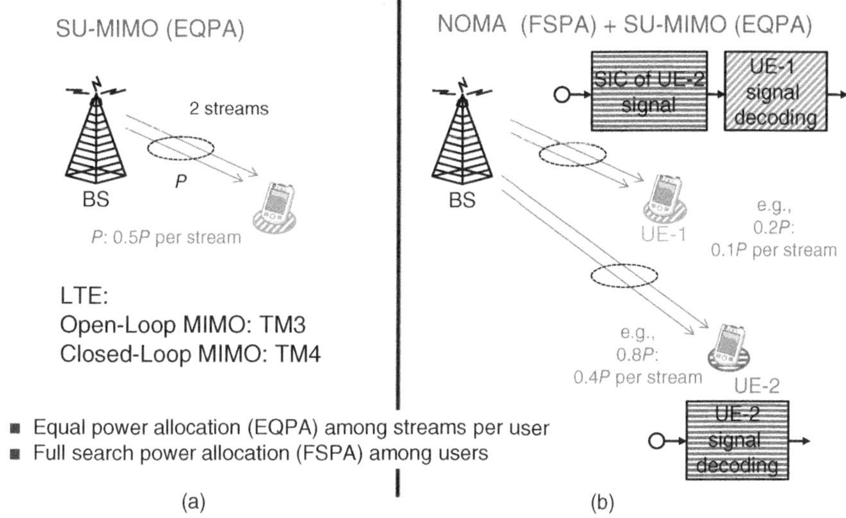

Figure 7.10 Downlink NOMA with SIC combined with SU-MIMO (2×2 MIMO, 2 UE)

7.5.1.2 Combination of NOMA with SU-MIMO

Here, we describe our proposal for NOMA combined with SU-MIMO [17]. Figure 7.10(a) illustrates the case of 2×2 SU-MIMO, while Figure 7.10(b) shows the proposed combination of NOMA with 2×2 SU-MIMO ($N_t = N_r = 2$), where $m = 2$. UE-1 and UE-2 are NOMA paired cell-center and cell-edge users, respectively.

By combining NOMA with SU-MIMO, up to four-layer (four-beam) transmission is enabled using 2×2 MIMO. One example of multi-user/stream separation receiver is based on codeword-level SIC (CWIC) and is depicted in Figure 7.11.

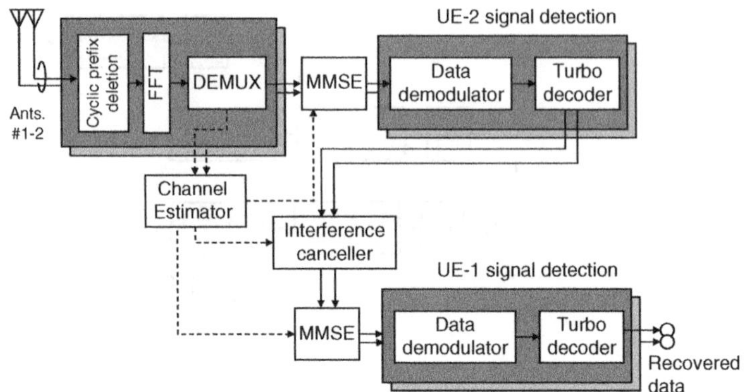

Figure 7.11 Structure of CWIC receiver for cell-center UE

7.5.2 Uplink NOMA

NOMA can be readily combined with MIMO in the uplink. In the following, we discuss the combination of NOMA with MIMO for two cases.

7.5.2.1 SIMO

For a SIMO system in downlink, only diversity gain can be achieved because the degree of freedom is 1. However for uplink, not only diversity gain but also multiplexing gain can be achieved using so-called "virtual MIMO" or collaborative MIMO technology [18]. NOMA can be applied regardless of whether multiple antennas are used for diversity gain or multiplexing gain. Examples are shown in Figure 7.12 assuming a 1×2 antenna configuration. In Figure 7.12(a), UE-1 and UE-2 are separated in the power domain and multiple receive antennas are used for diversity gain by maximum ratio combining. In Figure 7.12(b), UE groups {UE-1, UE-3} and {UE-2, UE-4} are separated in the power domain and in each power layer UEs are separated in spatial domain by minimum mean-squared error (MMSE); in other words UE-1 and UE-3 in the first power layer and UE-2 and UE-4 in the second power layer. For a SC-FDMA and SIMO configuration, only one or two data streams are supported, while

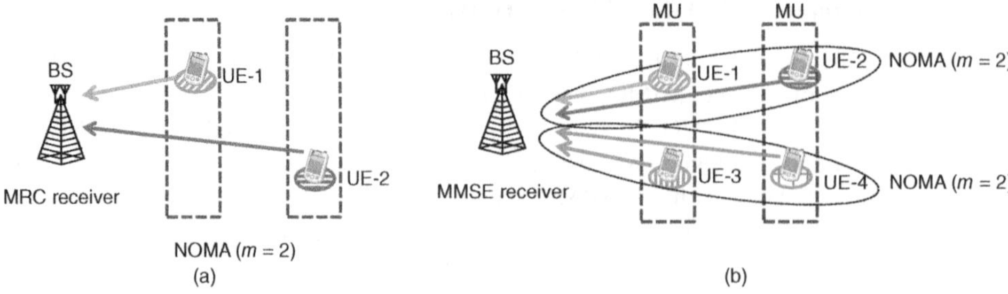

Figure 7.12 Uplink NOMA with SC-FDMA and SIMO. (a) UE separation in power domain; (b) UE separation in power and spatial domain

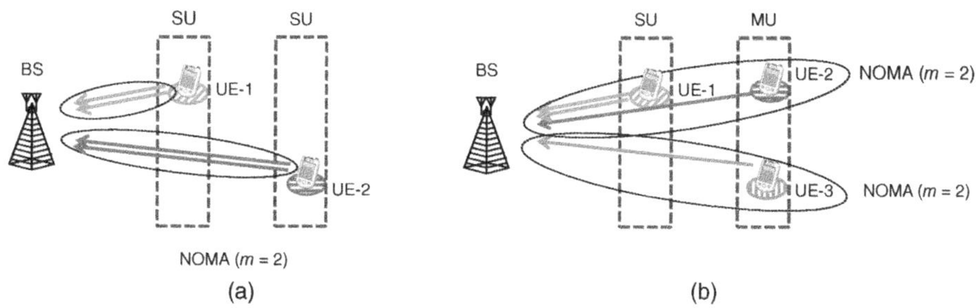

Figure 7.13 Uplink NOMA with SC-FDMA and MIMO. (a) UE separation in power domain; (b) UE separation in power and spatial domain

up to two and four data streams are supported by the approaches in Figure 7.12(a) and (b), respectively. It can be seen that for the SIMO antenna configuration, more data streams can be supported in uplink than downlink. For example, for one receive antenna at the UE and two transmit antennas at the BS (1×2 uplink MIMO), up to two data streams are supported in downlink NOMA, while up to four data streams are supported in uplink. This is shown in Figure 7.12(b).

7.5.2.2 MIMO

When multiple antennas are also adopted at the UE, the degree of freedom of each point-to-point link becomes more than 1. Thus, multiple data streams can be transmitted by a UE to the BS. Examples of NOMA with MIMO assuming a 2×2 antenna configuration are shown in Figure 7.13. In Figure 7.13(a), UEs are separated in the power domain, and the spatial domain is used to multiplex multiple data streams of a single UE. In Figure 7.13(b), UEs are separated in both power and spatial domain: {UE-1} and {UE-2, UE-3} are separated in the power domain, and UE-2 and UE-3 are further separated in the spatial domain. For UE-1, SU-MIMO transmission is applied and for {UE-2, UE-3}, MU-MIMO transmission is applied. It can be seen that for the MIMO antenna configuration, the same number of data streams are supported in uplink and downlink. For example, for a 2×2 system, up to four data streams are supported in downlink and the same number for the uplink as shown by either Figure 7.13(a) or Figure 7.13(b).

7.6 Performance Evaluations

7.6.1 Downlink NOMA

7.6.1.1 Link-level Simulations

First, we verify the effectiveness of CWIC for NOMA and the BLER performance of cell-center UE by evaluating link-level simulations for the case of two UEs [20]. Table 7.1 shows the evaluation assumptions. The number of transmit and receive antennas are both set to two and TM3 (transmission mode) is assumed for the SU-MIMO transmission. The number of iterations for turbo decoding is set to six. The channel model is an instantaneous multipath

Table 7.1 Link-level simulation parameters

System bandwidth	10 MHz
Number of subcarriers	1200
Subcarrier separation	15 kHz
Subframe length	1.0 ms
Symbol duration	66.67 μs + cyclic prefix: 4.69 μs
Antenna configuration	2-by-2 (uncorrelated)
Channel estimation	CRS-based channel estimation
Channel coding/decoding	Turbo coding (constraint length: 4 bits)/max-log-MAP decoding (6 iterations)
Receiver	MMSE + SIC

Rayleigh fading model with a six-path exponentially decayed power delay profile model with a root mean squared (r.m.s.) delay spread value of 0.29 μs, where the relative path power is decayed by 2 dB and each path is independently Rayleigh-faded with maximum Doppler frequency of 10 Hz. The impact of channel estimation is also taken into account assuming cell-specific reference signal (CRS)-based channel estimation.

Figure 7.14(a) shows the BLER performances of the cell-center UE for ideal SIC and CWIC as a function of signal-to-noise power ratio (SNR) where the signal power after transmission power allocation is considered. Note that ideal SIC here refers to the ideal generation of the replica of the interfering signal. However, interference cancellation may still remain imperfect due to channel estimation error.

The BLER performance of the decoding of the cell-edge UE signal in cell-center UE is also plotted and named as "Before SIC". Transmit power ratios of $\beta_1 = 0.32$ for cell-center UE and $\beta_2 = 0.68$ for cell-edge UE are considered for the rank combinations of 1-1, 2-1, and 2-2. We assume 64QAM modulation and a coding rate of $R = 0.51$ for cell-center UE and QPSK modulation and coding rate of $R = 0.49$ for cell-edge UE. For this setup, we can see that almost the same BLER performance is obtained for NOMA with ideal SIC and CWIC. This means that error propagation can be very limited for β_1 smaller than 0.32. The performance gap between 1-1 and 2-2 is due to the increase in the umber of transmit streams and thus interstream interference. This can be confirmed from the BLER of cell-edge user signal decoding, which corresponds to "Before SIC". Nevertheless, SIC error propagation will increase for power sets with β_1 larger than 0.35 for 2-2 and 0.4 for 1-1 and 0.45 for 2-1 as shown in Figure 7.14(b), which provides the required SNR for achieving BLER of 10^{-1} as a function of the power ratio of a cell-center UE. Also note that the degradation of performance for small values of β_1 is due to the increase of channel estimation errors for the cell-center UE.

Therefore, it is be important to limit the power sets to be used by the scheduler in order to

- limit error propagation
- reduce the impact of channel estimation errors
- ensure that all MCS combinations are decodable.

However, this would limit the NOMA gains. Alternatively adjustments can be made to the combination of power sets that can be selected according to the MCS combinations selected by the scheduler.

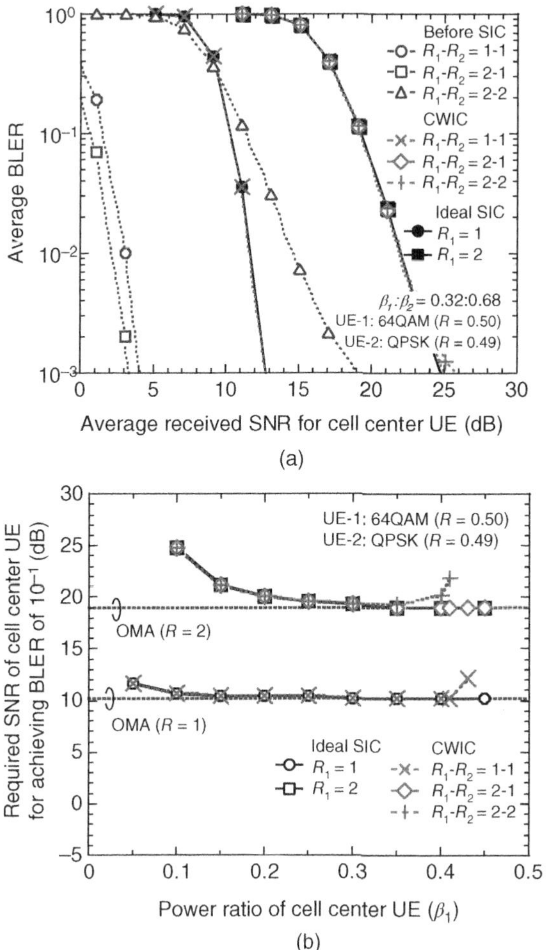

Figure 7.14 Link-level performance of CWIC. (a) Average BLER vs received SNR for cell-center UE; (b) Required SNR for achieving BLER$=10^{-1}$ vs power ratio for cell-center UE

7.6.1.2 System-level Simulations

In the following, we present system-level simulation results of our investigations on the performance gains of NOMA over OMA. The simulator used consists of a system-level model utilizing exponential SINR link-to-system level mapping [15]. At the UE receiver, the effective SINR is calculated for each user using the exponential effective SNR mapping (EESM) model, where the weighting factor beta is optimized for each MCS [21]. Based on the effective SINR, MCS decoding is attempted using the BLER-vs-SINR link-level mapping table. Note here that OMA follows the same procedure as NOMA but with $m = 1$. To evaluate the performance gain of NOMA, a multicell system-level simulation is conducted. The simulation parameters are basically compliant with existing LTE specifications [1], as summarized in Table 7.2. We employed a 19-hexagonal macro cell model with three cells per cell site. The

Table 7.2 System-level simulation parameters

Parameter	Value
Carrier frequency	2.0 GHz
System bandwidth	10 MHz
Resource block bandwidth	180 kHz
Number of resource blocks	48
Cell layout	Hexagonal grid, 19-cell sites, 3 cells/site, wrap around
Number of users per cell	$K = 10$
Inter-site distance	500 m
Minimum distance between UE and cell site	35 m
Maximum transmission power	46 dBm
Distance dependent path loss	$128.1 + 37.6\log_{10}(r)$, r. kilometers (dB)
Shadowing standard deviation	8 dB
Shadowing correlation	0.5 (intersite), 1.0 (intrasite)
Channel model	3GPP Spatial Channel Model (SCM), Urban macro
Receiver noise density	−174 dBm/Hz
Antenna configuration	Cross-polarized antenna (CPA) eNB: 2Tx: X (+45, −45) UE: 2Rx: X (+90, 0)
Antenna gain	14 dBi @ eNB, 0 dBi @ UE
User speed (max. Doppler freq.)	3 km/h (5.55 Hz)
Maximum number of multiplexed UEs	$m = 1$ (OMA) and $m = 2$ (NOMA)
Number of subbands	$S = 1$ (wideband scheduling), or 8 (subband scheduling)
MCS selection	Targeted BLER: below 10%
Number of power sets for FSPA	10 sets: (0.88, 0.12); (0.85, 0.15); (0.83, 0.17); (0.81, 0.19); (0.78, 0.22); (0.75, 0.25); (0.73, 0.27); (0.71, 0.29); (0.68, 0.32); (0.65, 0.35)
Traffic model	Full buffer
HARQ	Yes
CQI/PMI feedback interval	10 ms
Rank report interval	100 ms

cell radius of the macro cells is set to 289 m, with an intersite distance of 500 m. Ten UEs are dropped randomly, following a uniform distribution. In the propagation model, we take into account distance-dependent path loss with a decay factor of 3.76, log-normal shadowing with the standard deviation of 8 dB, and instantaneous multipath fading. The shadowing correlation between the sites (cells) is set to 0.5 (1.0).

The 3GPP spatial channel model (SCM) for urban macro with a low angle spread is assumed. The system bandwidth is 10 MHz and the total transmission power of the BS in each cell is 46 dBm. A 2×2 MIMO configuration is assumed. The antenna gain at the BS and UE is 14 dBi and 0 dBi, respectively. For NOMA and OMA we assume ideal channel and ICI estimation. The CQI feedback is quantized and has a control delay of 6 ms. The system-level performance of NOMA is investigated without SIC error propagation [22].

A full-buffer traffic model is assumed and hybrid automatic repeat request (HARQ) is applied to retransmit packets that are not successfully decoded. For NOMA, the maximum number of simultaneously multiplexed users is set to $m = 2$ and the number of power sets for FSPA is set to ten. In order to assess the performance gains of NOMA, cell throughput (Mbps) and cell-edge user throughput (Mbps) are evaluated on the basis of the following definitions. The cell throughput is defined as the average aggregated throughput for users scheduled per a single cell, while the cell-edge user throughput is defined as the 5% value of the cumulative distribution function (CDF) of the user throughput.

In Figure 7.15, the CDF of the user throughput of NOMA with SU-MIMO is compared to that of OMA with SU-MIMO for open-loop (TM3) and closed-loop (TM4) MIMO assuming ten UEs per cell, ten power sets and genie-aided CQI estimation and Case 1 (see also Figure 7.6). We can see that NOMA with SU-MIMO provides gains over OMA with SU-MIMO over the entire user throughput region for both TM3 and TM4. In addition, it can be seen that due to larger precoding gains owing to closed-loop transmission, OMA with TM4 and NOMA with TM4 have better performance than both OMA with TM3 and NOMA with TM3, respectively. Table 7.3 shows the performance gains for different granularities of scheduling and MCS selection (Case 1, Case 2, and Case 3) for both TM3 and TM4.

7.6.1.3 Experimental Trials

We developed a NOMA testbed in order to confirm NOMA performance with a real SIC receiver taking into account hardware impairments such as error vector magnitude and the number of quantization bits of analog/digital (A/D) converter, etc. We assume we have two UEs. The testbed parameters are as in Table 7.1 with carrier frequency of 3.9 GHz and

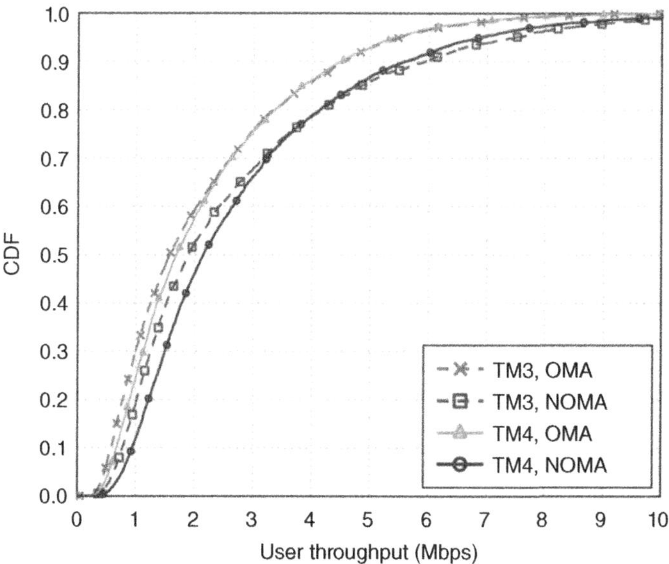

Figure 7.15 CDF of user throughput for OMA ($m = 1$) and NOMA ($m = 2$) for Case 1

Table 7.3 Comparison of cell throughput and cell-edge user throughput (Mbps) between NOMA and OMA with 2×2 MIMO TM3 and TM4

	2×2 MIMO, TM3			2×2 MIMO, TM4		
	OMA ($m = 1$)	NOMA ($m = 2$)	Gain	OMA ($m = 1$)	NOMA ($m = 2$)	Gain
Case 1: Subband scheduling and subband MCS selection						
Cell throughput	21.375	27.053	26.56%	21.97	27.866	26.84%
Cell-edge user throughput	0.472	0.633	34.11%	0.544	0.777	42.83%
Case 2: Subband scheduling and wideband MCS selection						
Cell throughput	21.59	26.29	21.77%	22.291	27.499	23.36%
Cell-edge user throughput	0.476	0.62	30.25%	0.552	0.769	39.31%
Case 3: Wideband scheduling and wideband MCS selection						
Cell throughput	19.068	24.894	30.55%	19.577	25.515	30.33%
Cell-edge user throughput	0.401	0.538	34.16%	0.451	0.649	43.90%

bandwidth per user is 5.4 MHz for NOMA and 2.7 MHz for OFDMA. The LTE Release 8 frame structure is adopted and channel estimation is based on CRS. For MIMO transmission LTE TM3 is utilized for open-loop 2×2 SU-MIMO transmission. At the transmitter side, for each UE data, Turbo encoding, data modulation and multiplication by precoding vector are applied, then the precoded signal of the two UEs is superposed according to a predefined power ratio (UE-1:UE-2 = P_1:P_2, $P_1 + P_2 = 1.0$, $P_1 < P_2$) and goes through a digital/analog (D/A) converter before upconversion to the carrier frequency of 3.9 GHz and transmission from two antennas. At the receiver side, two receive antennas are used to receive the RF signal, which is first downconverted and then goes through a 16-bit A/D converter. At the cell-center UE (UE-1), CWIC is applied. After channel estimation based on CRS, UE-1 decodes the signal of UE-2 using the max-log-MAP algorithm for turbo decoding (six iterations), then re-encodes using the Turbo encoder and modulates in order to generate the UE-2 signal replica, which is subtracted from the UE-1 received signal. At the cell-edge UE (UE-2), no SIC is applied and the decoding is applied directly, since the power ratio of UE-2 is higher than that of UE-1. Both UEs apply MMSE-based stream separation.

Using the fading emulator, for simplicity we set each link of the 2×2 MIMO channel to a one-path channel with maximum Doppler frequency of 0.15 Hz. In Figure 7.16, we compared the user throughput of UE-1 with NOMA and SIC applied (29 Mbps) and with OFDMA applied (18 Mbps).

For UE-1, NOMA gains compared to OFDMA are about 61%. These gains are obtained when the SNR of UE-1 is set to 33 dB and that of UE-2 to 0 dB, while we set MCS of UE-1 to 64QAM (coding rate of 0.51) and UE-2 to QPSK (coding rate of 0.49). The transmit ranks of UE-1 and UE-2 are set to 2 and 1, respectively. For UE-2, the cell-edge user, the same rate is set for both OFDMA and NOMA. NOMA gains are the result of enabling three-layer transmission using a 2×2 MIMO channel and still being able to use twice the bandwidth compared to OFDMA.

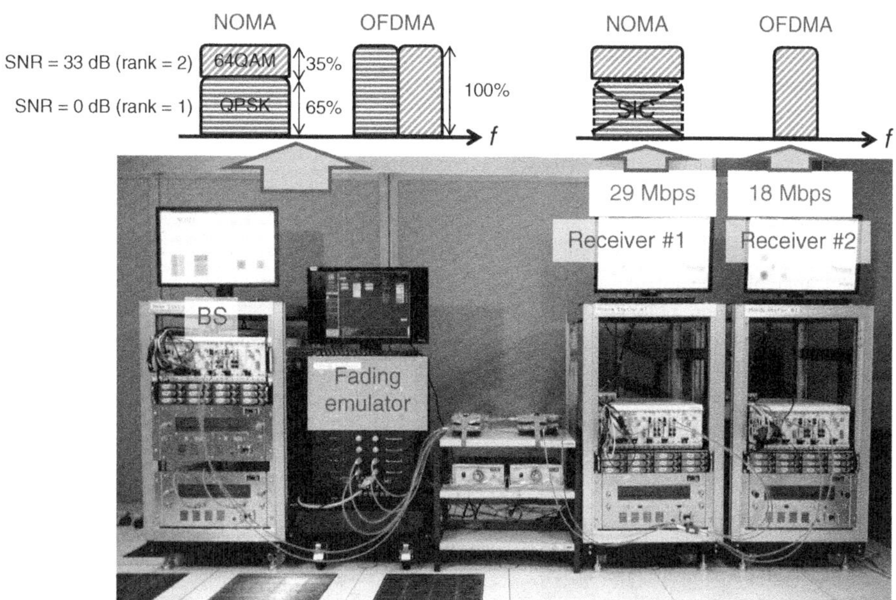

Figure 7.16 NOMA testbed

7.6.2 Uplink NOMA

In this section, the system-level performance of uplink NOMA is presented. The major simulation parameters are shown in Table 7.4. These are well aligned with existing LTE specifications [1]. The locations of the UEs are randomly generated with a uniform distribution within each cell. The same MCS sets are used for both SC-FDMA and NOMA in the simulations. In frame error rate evaluations, sixteen resource blocks are defined as edge bands for each cell; these are non-overlapped among the three neighboring cells as shown in Figure 7.17. Within each cell, one third of the UEs are categorized as cell-edge based on their reference-signal receiving power from the serving evolved Node B (eNB). Both the average UE throughput and cell-edge UE throughput are evaluated, where the cell-edge UE throughput is defined as the 5% value of the CDF of the UE throughput.

Figure 7.18 presents the overall cell throughput of SC-FDMA and NOMA. The subband number is set to eight and the maximum multiplexing order, m, for NOMA is set to two. It can be seen that the cell throughput of SC-FDMA is almost saturated when the number of UEs is larger than ten. However, the cell throughput of NOMA still increases as the number of UEs per cell becomes larger. When number of UEs per cell is larger than 40, the cell throughput of NOMA reaches saturation and NOMA achieves about 28% cell throughput gain compared to SC-FDMA. It is important to emphasize here that such a large performance gain for NOMA is achieved with very practical assumptions and can be further increased by applying a larger number of multiplexed users m and/or enhanced schemes such as advanced TPC. The large gain of NOMA mainly comes from the non-orthogonal multiplexing of users with large channel-gain differences, which improves the resource utilization efficiency compared to SC-FDMA where one UE occupies the radio resource exclusively.

Table 7.4 Major simulation parameters

Parameter	Value range
Cell layout	Hexagonal 19-cell sites, 3 cells per site, wrap-around
Intersite distance	500 m
Carrier frequency	2.0 GHz
Overall transmission bandwidth	10 MHz
Resource block bandwidth	180 kHz
Number of resource blocks	48
Subband bandwidth	1080 kHz without FFR 1440 kHz with FFR
Number of UEs per cell	10, 20, 30, 40, 50
eNB receive antenna	
Number of antennas	2
Antenna gain	14 dBi
UE transmit antenna	
Number of antennas	1
Antenna gain	0 dBi
Maximum transmission power	23 dBm
Distance-dependent path loss	$128.1 + 37.6\log_{10}(r)$, r. kilometers (dB)
Shadowing standard deviation	8 dB
Channel model	6-ray typical urban
Channel estimation	Ideal
Receiver noise density	−174 dBm/Hz
Noise figure of cell site	5 dB
UE speed(Doppler frequency)	3 km/h (5.55 Hz)
Scheduling interval	1 msec
Averaging interval of throughput	200 ms
Traffic model	Full buffer

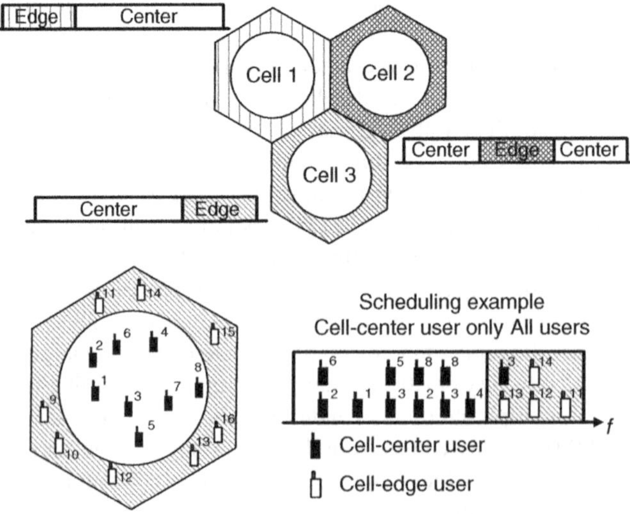

Figure 7.17 FFR scheme and its application to NOMA

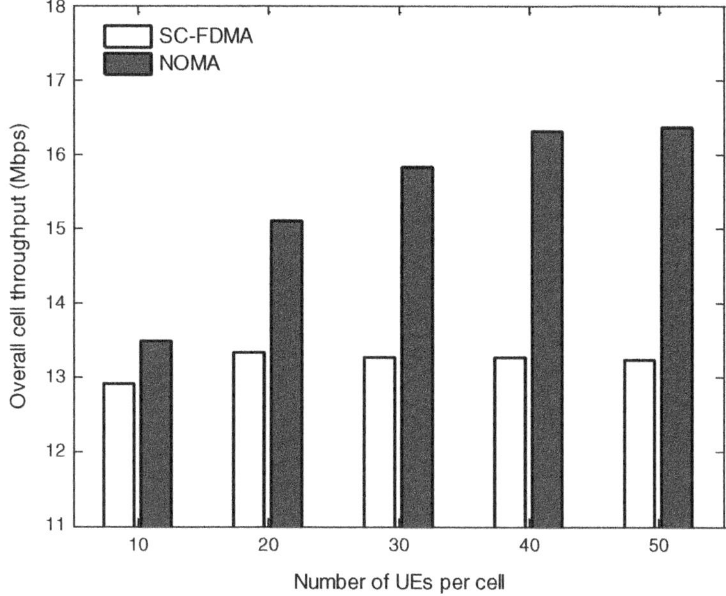

Figure 7.18 Comparison of cell throughput of SC-FDMA and NOMA

Figure 7.19 compares the UE throughput of SC-FDMA and NOMA with $m = 2$ and 3. It can be observed that:

- NOMA can achieve higher UE throughput than SC-FDMA for most regions of the CDF curve.
- For the cell-edge throughput, i.e. 5% UE throughput, NOMA performance is worse than that of SC-FDMA.

This is mainly for two reasons. One is the increase of ICI in NOMA compared with SC-FDMA because more than one UE can be scheduled for simultaneous uplink transmission. The other is the TPC algorithm used [5]. This is not fully optimized and the total transmission power is controlled by a predefined parameter and the UEs in non-orthogonal transmission get less transmission power than they get in SC-FDMA. Furthermore the transmission power of the UEs is determined from large-scale fading, without considering instantaneous channel condition or scheduling metrics. Therefore, sophisticated TPC algorithms must be designed for NOMA or the combination of NOMA with other cell-edge performance-enhancing technologies such as FFR.

Figures 7.20(a) and (b) show the performance of NOMA when FFR is applied. The maximum multiplexing order m is set to two and the subband number is set to six in the evaluations. It can be seen that by applying FFR, the cell throughput and cell-edge throughput of NOMA improve due to the reduction in ICI for both the cell-center and cell-edge UEs. Figure 7.20(c)

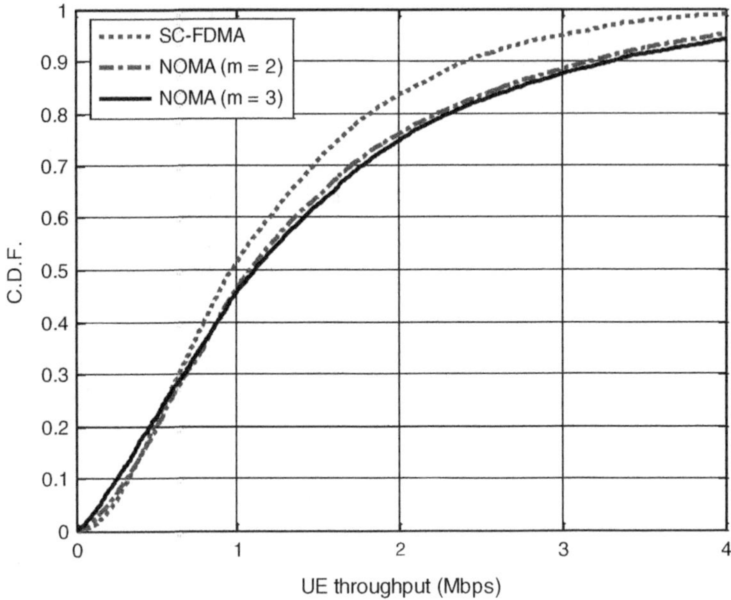

Figure 7.19 Comparison of UE throughput of SC-FDMA and NOMA

compares the UE throughput of SC-FDMA and NOMA with and without FFR when m is set to 3. It can be seen that:

- the cell-edge throughput of NOMA becomes better than that of SC-FDMA when FFR is applied
- NOMA with FFR improves not only the cell-edge throughput gain but also the overall cell throughput gain.

NOMA with FFR performs better than SC-FDMA in the whole UE throughput region. Therefore, FFR is a good candidate technology to combine with NOMA in the uplink.

7.7 Conclusion

This chapter presented an overview, performance evaluation and the ongoing experimental trials of our NOMA concept. In contrast to OFDMA, NOMA superposes multiple users in the power domain, exploiting the channel-gain difference between multiple UEs. We showed, using our NOMA testbed, the potential gains for both system-level and link-level simulations when NOMA is combined with SU-MIMO.

NOMA is currently under study in 3GPP RAN1. NOMA involves several aspects that need careful design, including the granularity in time and frequency of multi-user power allocation, signaling overhead, feedback enhancements and receiver design. NOMA can also be applied to the uplink. For uplink, there are design issues related to balancing intracell and intercell

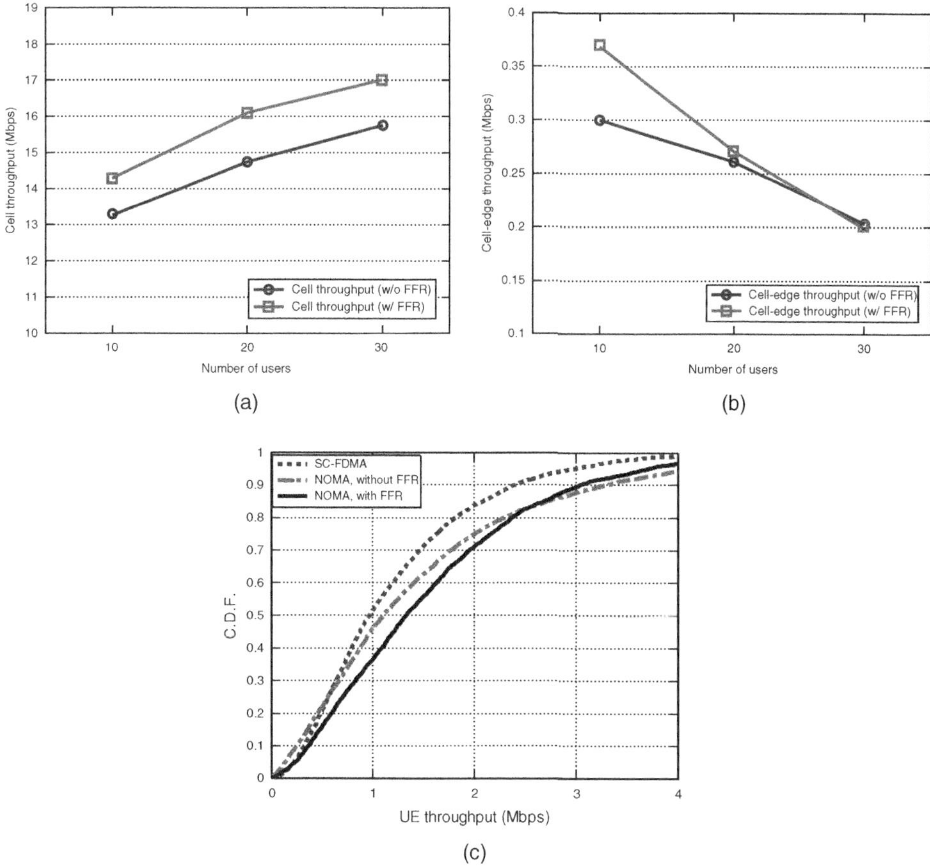

Figure 7.20 Comparison of NOMA and SC-FDMA when FFR is applied. (a) Overall cell throughput ($m = 2$); (b) Cell-edge throughput ($m = 2$); (c) UE throughput CDF ($m = 3$)

interference and the design of the scheduling algorithm where consecutive resource allocation of non-orthogonally multiplexed UEs is taken into account. Based on our performance evaluation results, we showed that NOMA has promising gains for both downlink and uplink. These gains are expected to increase with more users, which is the case for mobile broadband (MBB) transmission for an ultra-dense urban area, the case of massive sensors, or the Internet of things, with devices with small packets being simultaneously transmitted on the cellular network.

References

[1] 3GPP (2006), Physical layer aspects for Evolved UTRA, *TR* **25**.814,V7.1.0.

[2] Tse, D. and Viswanath, P. (2005) *Fundamentals of Wireless Communication*. Cambridge University Press.

[3] Caire, G. and Shamai, S. (2003) On the achievable throughput of a multi-antenna Gaussian broadcast channel. *IEEE Trans. Info. Theory*, **49** (7), 1692–1706.

[4] Higuchi, K. and Kishiyama, Y. (2012) Non-orthogonal access with successive interference cancellation for future radio access. *APWCS (2012)*.

[5] Endo, Y., Kishiyama, Y., and Higuchi, K. (2012) Uplink non-orthogonal access with MMSE-SIC in the presence of intercell interference. *ISWCS2012*, pp. 261–265.
[6] Benjebbour, A., Saito, Y., Kishiyama, Y., Li, A., Harada, A., and Nakamura, T. (2013) Concept and practical considerations of non-orthogonal multiple access (NOMA) for future radio access. *Proc. IEEE ISPACS*, pp. 770–774.
[7] Saito, Y., Benjebbour, A., Kishiyama, Y., and Nakamura, T. (2013) System-level performance evaluation of downlink non-orthogonal multiple access (NOMA). *Proc. IEEE PIMRC 2013*, pp. 611–615.
[8] Benjebbour, A., Li, A., Saito, Y., Kishiyama, Y., Harada, A., and Nakamura, T. (2013) System-level performance of downlink NOMA for future LTE enhancements, *IEEE Globecom*, pp. 66–70.
[9] Higuchi, K. and Kishiyama, Y. (2013) Non-orthogonal access with random beamforming and intra-beam SIC for cellular MIMO downlink. *Proc. IEEE VTC2013-Fall*, pp. 1–5.
[10] Li, A., Benjebbour, A., and Harada, A. (2014) Performance evaluation of non-orthogonal multiple access combined with opportunistic beamforming. *Proc. IEEE VTC Spring 2014*,.
[11] Higuchi, K. and Benjebbour, A. (2015) Non-orthogonal multiple access (NOMA) with successive interference cancellation for future radio access. IEICE *Trans. Commun.* **E98-B** (3), pp. 403–414.
[12] 3GPP (2014) TR 36.866 V.12.0.1, Study on network-assisted interference cancellation and suppression (NAICS) for LTE.
[13] 3GPP (2015) RP-150496, MediaTek, New SI Proposal: Study on downlink multiuser superposition transmission for LTE.
[14] Zelst, A.V. Nee, R.V., and Awater, G.A. (2000) Space division multiplexing (SDM) for OFDM systems. *Proc. IEEE VTC2000-Spring*, pp. 1070–1074.
[15] Wolniansky, P.W., Foschini, G.J., Golden, G.D., and Valenzuela, R.A. (1998) V-BLAST: An architecture for realizing very high data rates over the rich-scattering wireless channel. *Proc. ISSUE 1998*, pp. 295–300.
[16] Weingarten, H., Steinberg, Y., and Shamai (Shitz), S. (2004) The capacity region of Gaussian MIMO broadcast channel, *Proc. IEEE ISIT 2004*, p. 174.
[17] Benjebbour, A., Li, A., Kishiyama, Y., Jiang, H., and Nakamura, T. , (2014) System-level performance of downlink NOMA combined with SU-MIMO for future LTE enhancements, *IEEE Globecom*, pp. 706–710.
[18] 3GPP (2005) TSG-RAN1 42, R1-0501162, UL virtual MIMO transmission for E-UTRA.
[19] Saito, K., Benjebbour, A., Kishiyama, Y., Okumura, Y., and Nakamura, T. (2015) Performance and design of SIC receiver for downlink NOMA with open-loop SU-MIMO, *IEEE ICC 2015*, pp. 1161–1165.
[20] Brueninghaus, K., Astely, D., Salzer, T., Visuri, S., Alexiou, A., Karger, S., and Seraji, G.A. (2005) Link performance models for system level simulations of broadband radio access systems, *Proc. IEEE PIMRC, pp.* 2306–2311.
[21] Lan, Y., Benjebbour, A., Chen, X., Li, A., and Jiang, H. (2014) Considerations on downlink non-orthogonal multiple access (NOMA) combined with closed-loop SU-MIMO, *Proc. ICSPCS*, pp. 1–5.
[22] 3GPP (2015) TS 36.213 V12.5.0, Evolved universal terrestrial radio access (E-UTRA), physical layer procedures.

8

Major 5G Waveform Candidates: Overview and Comparison

Hao Lin and Pierre Siohan

Signal Processing for 5G: Algorithms and Implementations, First Edition. Edited by Fa-Long Luo and Charlie Zhang.

8.1 Why We Need New Waveforms

In each decade a new mobile communication system is invented to meet, thanks to use of novel technological features, exponentially growing market demand. For 5G[1] the objectives targeted by the European Union METIS project[2] are to provide, at the 2020 horizon, 1000 times more mobile data volume per area, 10–100 times more connected devices, 10–100 times higher user data rates, 10 times longer battery life for low-power massive machine communication, and 5 times reduced end-to-end latency [1].

Naturally, all these huge increases will only be made possible by the combination of several complementary factors: better use of the already available spectrum, authorization to use new spectrum above 6 GHz, generalization of small cells, introduction of massive multiple-input multiple-output (MIMO), and so on. Our feeling is also that the 4G LTE modulation scheme is not well suited to meeting some essential 5G requirements and that consequently a new air interface needs to be defined. Indeed, LTE and LTE-Advanced have been conceived for the mobile broadband (MBB) application and are based on cyclic prefix OFDM (CP-OFDM) modulation. In this context, the multicarrier modulation alternatives to CP-OFDM can only bring improvements by removing the CP interval in time and reducing the guard bands in frequency, which may be considered marginal with respect to the 5G expectations.

However, beyond MBB, the two main drawbacks of CP-OFDM – its bad spectral confinement and its lack of waveform flexibility – may be serious ones from the perspective of future 5G multiservice communication systems. Firstly, to get optimal usage of the available frequency bandwidths below 6 GHz, dynamic spectrum aggregation will be a key issue [2]. In this respect an LTE-OFDM based approach will suffer from its high out-of-band (OOB) emission and also from its resource block granularity which, for example, prevents allocation of a single subcarrier to low data-rate machine type communication services. The "Internet of Things" expansion will also feature a huge amount of unsynchronized machine-to-machine communication, thus destroying orthogonality. OOB emissions will then severely pollute the services and users operating in the frequency vicinity if, as is the case for OFDM, the spectrum is not properly contained. Similar problems may also occur with frequency shifts and/or spreading due to the Doppler effects that will characterize 5G high-mobility applications.

Let us end with a last example in relation with the extremely low latency which is demanded for mission-critical communication (MCC) in 5G. As mentioned in the paper by Wunder *et al.* [3], to get a significant latency reduction may be problematic with CP-OFDM. Indeed, when reducing the CP-OFDM duration, in order to avoid interference problems, one has to keep the CP length unchanged. This naturally leads to a strong reduction of the spectral efficiency. Moreover, if one thinks a step further, to simultaneously support MBB and MCC with different transmission time interval durations, strong inter-service interference is created if the CP-OFDM system is used.

For all these reasons, it becomes urgent to find a system based on a flexible waveform that can solve the different challenges 5G poses to the physical layer. The next section will describe the OFDM alternatives that are today's best candidates.

---

[1] All acronyms are defined either the first time they are used and/or at the end of this chapter.

[2] Mobile and Wireless Communications Enablers for the Twenty-twenty Information Society

The rest of this chapter is organized into three sections. In Section 8.2, we will mainly present the major multicarrier modulation (MCM) candidates for 5G and their categorization. Section 8.3 is devoted to high-level comparisons of these candidate algorithms in terms of performance, efficiency, complexity, compatibility and integration. The final section of this chapter will give a summary and further discussion of future work on the design and implementation of MCM schemes for 5G standardization and deployment.

8.2 Major Multicarrier Modulation Candidates

Since the mid-nineties CP-OFDM has become the indisputable reference in terms of MCM. However, at the same time, two of its variants, though presenting a far lower degree of maturity, were also available. These two other schemes are known by several names, FBMC/OQAM and FMT being nowadays the most common. More recently, cyclic convolution based implementations of these three well-known MCM schemes have led to new proposals, namely GFDM, FBMC/COQAM, FB-OFDM and CB-FMT. Finally, two new schemes named UFMC, or more recently UF-OFDM, and f-OFDM are also now entering the 5G debate. In contrast to all the previously mentioned schemes, instead of being subcarrier-wise filtered schemes these newcomers correspond to subband filtered schemes. In this chapter, we briefly go through this list of waveform proposals by separating them into three groups:

- subcarrier filtered MCM using linear convolution
- subcarrier filtered MCM using circular convolution
- subband windowed MCM.

In addition, to be complete, we also present the baseline waveform: CP-OFDM.

To get a unified formulation, in the following, M denotes the maximum number of subcarriers; in other words it also corresponds to the size of the (inverse) fast Fourier transform ((I)FFT)) that can be used for their implementation. K denotes the number of subsymbols in the case of block transmission. To simplify the notation, equations are expressed in discrete-time, assuming a unitary sampling period. However, in order to provide a link with the time and frequency physical parameters, we denote by T the duration of each single OFDM symbol. Note also that the symbols transmitted over each subcarrier, denoted by symbols c (complex constellations) or a (real-valued constellations), can also take zero values, for example for guard bands. Furthermore, without loss of generality, the baseband discrete-time expression of the different modulated signals is expressed in a non-causal form.

8.2.1 CP-OFDM Modulation

OFDM is the simplest MCM system and it is widely adopted in many applications. In contrast to single-carrier modulation, for which the transmitted data is spread over the entire available bandwidth, in OFDM the data are modulated on a set of narrow subcarriers, the bandwidth of which is smaller than the channel coherent bandwidth, leading to quasi-flat fading at each subcarrier. Moreover, a CP is inserted in front of each OFDM symbol, which is the copy of the tail samples. Let L_{cp} denote its length. This further ensures a true flat fading at each subcarrier, therefore enhancing the robustness against frequency selective fading. OFDM is a symbol-by-symbol transmission, in other words $K = 1$, leading to a baseband CP-OFDM symbol that can be expressed, for $k \in [-L_{\text{cp}}, M-1]$, as

$$s_{\text{CP-OFDM}}[k] = \sum_{m=0}^{M-1} c_m e^{\frac{j2\pi mk}{M}} \tag{8.1}$$

with c_m the complex-valued symbols transmitted at subcarrier m, for example QAM constellations. The overall CP-OFDM system can be efficiently realized by FFT and IFFT. Another important advantage of the CP-OFDM is to maintain a full orthogonality for transmissions over channels that are only dispersive in time. Consequently, simple channel estimation and equalization methods can be used to recover orthogonality at the receiver side. However, CP-OFDM employs a rectangular pulse which has several disadvantages that will be analyzed in Section 8.3. It is worth noting that since a rectangular pulse is used in the OFDM system, the pulse shaping is implicitly realized by the Fourier transform.

8.2.2 *Subcarrier Filtered MCM using Linear Convolution*

CP-OFDM does not have good spectrum localization. In a perfect transmission condition, for example perfect time-frequency synchronization or sufficient length of CP, each subcarrier is orthogonal to or independent of the others. Nevertheless, if the transmission condition becomes imperfect, severe performance degradation is caused because intercarrier interference spreads over a wide subcarrier range. To improve the spectrum localization, a subcarrier filtering is used on the top of OFDM basis, leading to some advanced MCM schemes, such as FBMC/OQAM and FMT.

8.2.2.1 FBMC/OQAM Modulation

The remarkable contribution of the FBMC/OQAM concept is the introduction of a staggered transmission structure, which allows it to escape from requirements of the Balian–Low theorem [4]. Hence the FBMC/OQAM scheme can simultaneously employ an improved pulse shape, keep full orthogonality, and transmit at the Nyquist rate. In contrast to the OFDM scheme, which transmits complex valued symbols at subcarriers, in the staggered structure the real and imaginary parts of the complex-valued symbols are transmitted separately with a delay of half the OFDM symbol duration, in other words $T/2$. More details of the FBMC/OQAM concept can be found in Le Floch *et al.* and the references therein [5]. The baseband FBMC/OQAM-modulated signal can be written for any integer k as [6]:

$$s_{\text{OQAM}}[k] = \sum_{n\in\mathcal{Z}} \sum_{m=0}^{M-1} a_{m,n} \underbrace{g[k-nN_1]e^{j\frac{2\pi}{M}mk}e^{j\phi_{m,n}}}_{g_{m,n}[k]} \tag{8.2}$$

where g is the prototype filter, $N_1 = M/2$ is the discrete-time offset, $\phi_{m,n}$ is an additional phase term at subcarrier m and symbol index n, which can be expressed as $\frac{\pi}{2}(n+m)$. The transmitted symbols $a_{m,n}$ are real-valued, and can be obtained from a QAM constellation by taking the real and imaginary parts. To address a perfect reconstruction of real symbols, the prototype filter must satisfy the orthogonality condition:

$$\Re\left\{\sum_{k\in Z} g_{m,n}[k]g^*_{p,q}[k]\right\} = \delta_{m,p}\delta_{n,q} \tag{8.3}$$

where * denotes the complex conjugation, $\delta_{m,p} = 1$ if $m = p$ and $\delta_{m,p} = 0$ if $m \neq p$.

As plain OFDM, FBMC/OQAM, being a Gabor-based scheme,[3] also uses a complex exponential kernel and therefore can also take advantage of fast IFFT/FFT algorithms. However, it carries an extra cost compared to OFDM. This cost comes first from the necessity to operate with real symbols of half duration and perform IFFT/FFT at a rate twice as fast and, second, from the addition of filtering blocks. This added complexity for the modem depends more precisely on the selected implementation scheme. Indeed, as noted by Siohan *et al.* [6], it may be based either on an M-point IFFT/FFT plus a polyphase network (PPN) for filtering or, as in Bellanger's work [7], on a bM-point IFFT, with b being the overlapping factor of the prototype filter plus overlap and sum operations leading to so-called frequency-spreading FBMC (FS-FBMC) [8]. The added complexity in using FS-FBMC is higher than for PPN-FBMC by a factor of around b. More details of the implementation costs are provided in Section 8.3.

Regarding the receiver design, channel estimation methods based on scattered pilots or preamble sequence have been proposed in various publications [9–13]. For the scattered pilots design, as FBMC implements OQAM signaling, real-valued pilots, such as BPSK, are usually used. Moreover, channel estimation performance can be improved by inserting auxiliary pilots, which provide interference pre-cancelation. For the preamble design, special care can be taken over the preamble pattern in order to boost the receiver pilot power, leading to enhanced noise immunity. With respect to channel equalization, a 1-tap equalizer is generally used in practice due to its low complexity. Nevertheless, various multi-tap/step-equalization approaches have also been proposed [14–16] in more difficult channel conditions, resulting naturally in more complex receivers.

8.2.2.2 FMT Modulation

From filter bank (FB) theory [17], it is known that in a critically sampled FB system only a rectangular prototype filter can give a PR. This is an alternative interpretation of the Balian–Low theorem. Therefore the FMT shifts to a non-critical sampled system by introducing an oversampling at each subcarrier so that the additional degree of freedom can be traded for some better designed prototype filters [18]. The baseband FMT signal can be written, for any integer k, as

$$s_{\mathrm{FMT}}[k] = \sum_{n\in Z}\sum_{m=0}^{M-1} c_{m,n}\underbrace{h[k-nN_2]e^{j\frac{2\pi km}{M}}}_{h_{m,n}[k]} \tag{8.4}$$

[3] In discrete-time, a Gabor family of functions can be expressed as $f_{m,n} = f[k-nN]\exp(j\frac{2\pi mk}{M})$, with $(m,n) \in Z^2$. f denotes the prototype filter and N, M are two positive integer parameters defining the corresponding time-frequency lattice.

where $c_{m,n}$ denotes the complex-valued data symbols, and N_2 is the expansion factor, $N_2 > M$. To hold a PR, the prototype filter h must meet[4]

$$\sum_{k \in Z} h_{m,n}[k] h^*_{p,q}[k] = \delta_{m,p}\delta_{n,q} \tag{8.5}$$

While PR systems exactly satisfy the above set of equations [22–24], nearly-PR systems only approximately fulfill Eq. (8.5) [18]. If the employed pulse has a spectrum that is restrictively limited within the increased subcarrier spacing, no spectra-crossing happens between adjacent subcarriers, which confirms an intercarrier-interference-free transmission. Thus the orthogonality condition given in Eq. (8.5) is changed to one-dimensional only; in other words, any traditional Nyquist pulse can be used. This can be interpreted as an alternative way to relax the orthogonality condition, which is made easier for increasing values of the oversampling ratio, in other words increasing the subcarrier spacing $\frac{N_2}{M}\frac{1}{T}$. Moreover, like FBMC/OQAM, the FMT only holds a PR in the distortion-free case as well.

FFT algorithms can also be used to build the FMT modem, while various implementation schemes are available for the filtering subsection [25] and again the extra complexity with respect to CP-OFDM is due to the filtering operations. Concerning the equalization cost, it may naturally depend greatly upon the channel and system characteristics, leading to proposals either with relatively high computational complexity [18], or, as in some more recent contributions, to simple 1-tap equalizers [26–28].

8.2.3 *Subcarrier Filtered MCM using Circular Convolution*

Linear-convolution-based MCM effectively improves the signal spectrum localization feature. However, due to the additional filtering operation, which is operated in an oversampled symbol basis, consecutive time-domain symbols are overlapped. This makes each MCM symbol no longer independent but rather correlated with its neighbors. This fact shifts the block processing, as in OFDM, to continuous processing. As a consequence, some processing "tricks", for instance CP or space-time block coding (STBC), cannot be easily used for linear-convolution-based MCM. To retake advantage of block processing properties, circular-convolution-based MCM is introduced. This technique groups several MCM symbols together to form a "block" using circular convolution. Thus inside each block linear convolution remains, while amongst the blocks independence is maintained, so that CP or STBC can be applied at the block level. In contrast to the CP-OFDM case, for these CC-based systems, the CP can be shared by K OFDM subsymbols, thus limiting the loss of spectral efficiency. The CC concept has been introduced for OFDM, FMT and FBMC/OQAM, leading to their new versions: GFDM, CB-FMT and FBMC/COQAM, which we will now describe, but without CP insertion.

8.2.3.1 GFDM Modulation

The idea of GFDM is to group a set of complex-valued symbols from a time-frequency lattice into one block. Then, for each block, subcarrier-wise processing is carried out. This includes

[4] Using, as proposed for 5G random access by Kasparick *et al.* [19], a pair of biorthogonal prototype filters is another means to get PR with biorthogonal FDM [20, 21].

up-sampling, pulse shaping and tail biting, and finally is followed by a modulation operation to a set of subcarrier frequencies (see Fig. 1 in Fettweis *et al.* [29]). The baseband GFDM-modulated signal of one block, in other words for $k \in [0, MK - 1]$, is expressed as

$$s_{\text{GFDM}}[k] = \sum_{m=0}^{M-1} \sum_{n=0}^{K-1} c_{m,n} \tilde{f}[k - nM] e^{\frac{j2\pi km}{N}} \tag{8.6}$$

The pulse shape $\tilde{f}[k]$ indicates a periodic repetition of the prototype filter $f[k]$ with a period of MK:

$$\tilde{f}[k] = f[\text{mod}(k, MK)]$$

The periodic filter is used to realize the circular convolution at the transmitter, which is equivalent to the tail-biting process [29]. From the implementation point of view, as the filtering may be implemented as a direct product in the frequency domain [30], for each block computation the transmitter realization requires M FFTs of size K, filtering and an MK-point IFFT to finally get the baseband modulated signal. Note that, as no orthogonality or biorthogonality constraint is imposed for the filter design, the GFDM is a non-orthogonal system. Naturally the self-interference of GFDM needs to be compensated for. As illustrated in Datta *et al.* [31], the GFDM promoters propose a serial interference cancellation (SIC) method. This method is iterative and its complexity is directly proportional to the number of subcarriers. Therefore it results in rather high computational costs to totally (with double-sided SIC) or partially (basic SIC) restore orthogonality.

8.2.3.2 CB-FMT Modulation

The CB-FMT scheme adopts circular filtering on the top of traditional FMT modulation. Through this change, the FMT can turn into a block transform process such that a CP can be easily appended in front of each block, which therefore enhances robustness against multi-path fading. The baseband expression of CB-FMT is similar to Eq. (8.4), only with a circular filtering instead [32]:

$$s_{\text{CB-FMT}}[k] = \sum_{m=0}^{M-1} \sum_{n=0}^{K-1} c_{m,n} \tilde{h}[k - nN_2] e^{j\frac{2\pi km}{M}} \tag{8.7}$$

with

$$\tilde{h}[k] = h[\text{mod}(k, N_2 K)]$$

Note that if we set $N_2 = M$, then the CB-FMT turns into GFDM. Thus, the only filter that can guarantee orthogonality is the rectangular one. Otherwise, if $N_2 > M$, the PR conditions [33, 34] can be satisfied by non-rectangular prototype filters. Note also that these orthogonality constraints are different from the ones that have to be checked to get linear-convolution-based FMT systems.

On the implementation side, CB-FMT, which involves, for each block, running a K-point FFT M-times, filtering in the frequency domain, and a KN_2-point IFFT to get the modulated signal, is also very similar to GFDM as regards the modulation and demodulation block. However, due to its orthogonality property, and in contrast to GFDM, there is no need to operate a SIC processing at the receiver side. For equalization, different schemes can be envisioned depending on whether the channel is time-invariant or not [35].

8.2.3.3 FBMC/COQAM Modulation

The suggestion of the replacement of linear convolution with circular convolution for FBMC/OQAM appeared recently [36–39]. The interest of this scheme is to combine the advantages of OQAM signaling and the block processing feature from the circular convolution. To be more specific, a direct usage of circular convolution, such as for GFDM, will cause non-orthogonality, which is complicated to deal with at receiver side, while trading the Nyquist rate to improve the orthogonality, such as for CB-FMT, will also result in a non-negligible throughput loss. On the other hand, a combination of circular convolution and OQAM signaling seems a good solution: it does not require a symbol extension in the time domain to maintain an orthornomal system. Partly derived from Eq. (8.2), the discrete-time COQAM signal $s[k]$ defined in a block interval such that $k \in [0, MK' - 1]$ is expressed as

$$s_{\text{COQAM}}[k] = \sum_{m=0}^{M-1} \sum_{n=0}^{K'-1} a_m[n]\tilde{g}[k - nN_1]e^{j\frac{2\pi}{M}mk}e^{j\phi_{m,n}} \tag{8.8}$$

with K' the number of real symbol slots per block. Note that the real symbols are obtained from taking the real and imaginary parts of QAM constellations. Thus the relation to the symbol slot K introduced in the GFDM system is $K' = 2K$. This also implies that FBMC/ COQAM and GFDM have a same block length $K'N_1 = KM$. The rest of the parameters are identical to those presented in the FBMC/OQAM scheme. To implement circular convolution with a prototype filter g of length $L = KM$, we introduce a pulse-shaping filter denoted $\tilde{g}$, obtained by the periodic repetition of duration KM of the prototype filter g; that is $\tilde{g}[k] = g[\text{mod}(k, MK)]$.

Based on the implementation schemes of Lin and Siohan [39], it can be easily seen that the linear and circular convolution implementations of FBMC/OQAM can both benefit from the Hermitian symmetric property [40]. Furthermore, for a prototype filter of identical length, both schemes lead to the same overall complexity for the modulator. The complexity at the receiver side naturally depends on the precise equalizer being used. It appears that in most cases the FBMC/COQAM receiver allows the implementation to be achieved at reduced cost. Finally, note that, as shown recently [41], the PR conditions are the same for FBMC/OQAM and FBMC/COQAM.

8.2.3.4 FB-OFDM Modulation

One of the issues with circular-convolution-based MCM, such as in GFDM, CB-FMT and FBMC/COQAM, is the scattered pilot insertion. This is due to the fact that in the frequency domain the data symbols and the pilots are mixed up by performing an horizontal FFT [30, 33, 39]. Thus the insertion of the pilots becomes non-trivial. Moreover, for the same reason, the compatibility with OFDM-based MIMO solutions, notably space frequency block coding and codebook-based precoding, cannot be simply reused. To address this compatibility issue, which is also related to the complexity issue, an enhanced version of FBMC/COQAM called filter-bank-based OFDM (FB-OFDM) was proposed recently [42]. In the proposal, the author showed that by utilizing a symbol extension we can form an extended symbol basis that follows a certain symmetric property. With such a property, the symbols can be filtered in the extended frequency domain, resulting in a set of combinational symbols. This is indeed an alternative way to do the frequency-domain filtering without involving a mixture of data symbols and pilots, in contrast to to the alternatives above. Moreover, the data symbols can easily be recovered from the combinational symbol sequence by exploiting the symmetric property. Furthermore, a two-layer processing has been proposed [42], in which symbol recovery and fading channel compensation (e.g. channel estimation, equalization and MIMO decoding) are done separately at different layers, namely the original symbol and extended symbol layers. In this way, it can be shown that the FB-OFDM system is becoming completely compatible with OFDM systems.

The implementation of the FB-OFDM system is rather simple. The FB-OFDM signal is generated by an extended IFFT in which the inputs are the extended symbols filtered in the frequency domain. CP and windowing can be applied in the same manner as FBMC/COQAM. The receiver structure is dual to the transmitter, which does not involve much complexity except for the extended FFT operation. The added complexity in terms of the arithmetical computation is only 30% higher than of OFDM. Meanwhile, receiver signal processing, such as channel estimation and equalization, is the same as for OFDM systems.

8.2.4 Subband Filtered MCM

Another MCM scheme trend was introduced recently for 5G waveforms. The concept is to use time-domain filtering after OFDM modulation. As a convolution operation in the time domain is equal to a multiplication operation in the frequency domain, this system can be seen as a frequency-domain windowed MCM, in which the window width is indeed the filter bandwidth, which is intentionally designed to cover a certain subband. Thus we name this type of scheme 'subband windowed MCM'. Two proposals have been put forward:

- universal filtered multicarrier (UFMC) [43] or, synonymously, universal filtered OFDM
- filtered OFDM [44].

8.2.4.1 UF-OFDM Modulation

In the UF-OFDM case the total available bandwidth is partitioned into B subbands; each subband is separately modulated using classical OFDM modulation [43]. Then a FIR filtering of length L is applied to each subband-modulated signal. Finally, the UF-OFDM signal is a summation of B filtered subband-modulated signals. For each resulting block of length $L + M - 1$, the baseband UF-OFDM signal can be written, for $k \in [0, M + L - 1]$, as follows

$$s_{\text{UF-OFDM}}[k] = \sum_{i=1}^{B} \sum_{l=0}^{L-1} \sum_{m=0}^{M-1} c_m^i e^{j\frac{2\pi(k-l)m}{M}} f_i[l] \tag{8.9}$$

with c_m^i the complex-valued symbols for subcarrier m and subband i. Note that in the UF-OFDM proposal, the definition of subband is imposed as one physical resource block (PRB).

As the successive blocks do not overlap, for a back-to-back system, orthogonality in time is ensured, while orthogonality in frequency depends on the precise features of the filter f being used. Note also that the transition interval between consecutive blocks, resulting from filtering, plays the role of a guard interval and protects the transmitted symbols as long as L is greater or equal to the maximum delay spread introduced by a multipath channel.

More details on this scheme can be found in Vakilian *et al.* [43]. However, concerning the implementation aspects, one has to refer to a more recent publication [45], which proposes a realization scheme that outperforms the one resulting from a direct implementation based on Eq. (8.9). Nevertheless, it appears that even with this later implementation scheme, the extra cost compared to a LTE-OFDM reference goes up at least to a factor two in the uplink case, assuming the usage of a minimum number of resource blocks, and up to eight or ten for the downlink.

8.2.4.2 f-OFDM Modulation

Filtered OFDM (f-OFDM) is another newly proposed MCM scheme [44, 46]. It has a similar flavor to UF-OFDM in the sense that it also introduces a filtering in the time domain. But in contrast to UF-OFDM, the filter bandwidth, designed for a certain subband, may not necessarily be equal to 1 PRB. Moreover, f-OFDM does not impose that the filter length must be equal to the CP length, which naturally gives more degrees of freedom for the filter design, leading to a narrower filter transition bandwidth. The f-OFDM signal constitutes blocks of K CP-OFDM subsymbols of length $M + L_{\text{cp}}$. Then, applying a time-filtering f_i of length L for each subband of index i produces a f-OFDM signal that can be written for $k \in [-L_{\text{cp}}, KM + (K-1)L_{\text{cp}} + L - 2]$ as

$$s_{\text{f-OFDM}}[k] = \sum_{i=1}^{B} \sum_{n=0}^{K-1} \sum_{l=0}^{L-1} \sum_{m=0}^{M-1} c_{m,n}^{i} e^{j\frac{2\pi(k-l-n L_{\text{cp}})m}{M}} f_i[l] \tag{8.10}$$

where $c_{m,n}^{i}$ are complex valued symbols for the subcarrier m, the subsymbol n and subband i. As mentioned in the introduction of this section, depending on the different modulation schemes, only a fraction of the M subcarriers may need to be activated. As f-OFDM mainly targets the up-link with subbands as narrow as a few tens of subcarriers [46], this can be taken into account at the realization step. In order to improve the spectrum in the case of asynchronous uplink multiple access, one can introduce a pulse shaping, resulting from a soft window truncation of the sinc-pulse, with a length of $T/2$, centered at the desired frequency and sufficiently concentrated in time to limit the resulting interference.

8.3 High-level Comparison

In this section, we provide a high-level comparison of the different waveform candidates. The comparison includes several aspects:

- spectral efficiency
- spectrum confinement
- mobility

- tail issue
- latency
- modem complexity
- compatibility with LTE.

With this comparison, we intend to provide a global view of the advantages and drawbacks of each scheme.

8.3.1 Spectral Efficiency

The spectral efficiency (SE) analysis for part of the waveform candidates has been reported by Lin and Siohan [39]. For a multicarrier system, denoting F as the spacing between subcarriers and T as the symbol duration, the modulated signal can be written as a linear combination of a Gabor family, reflecting a lattice-form time-frequency representation. The maximum SE for an orthogonal system is reached when the symbol duration and the subcarrier spacing satisfy

$$T \cdot F = 1 \tag{8.11}$$

Indeed, the measure of SE is inversely proportional to the product of Eq. (8.11). If the product value gets greater, it means that there exists an SE loss in either the time domain – taking a longer time to transmit one symbol – or in frequency domain – using more frequency band for the transmission or the combination of both causes. Denoting the spectral efficiency indicator as SEI, we have $0 \leq \text{SEI} \leq 1$ with SEI$= 1$ being the optimum. CP-OFDM cannot achieve this maximum value due to the addition of a CP of length T_{CP} [47]. This leads to an overall efficiency reduction:

$$\text{SEI}_{\text{CP-OFDM}} = \frac{1}{(T + T_{\text{CP}}) \cdot F} = \frac{T}{T + T_{\text{CP}}} < 1$$

In contrast, the FBMC scheme respects the Nyquist rate and no CP is used. Hence it is possible to achieve the maximum efficiency [5]:

$$\text{SEI}_{\text{OQAM}} = \frac{1}{T \cdot F} = 1$$

Although the FMT also does not use any CP, it cannot achieve maximum spectral efficiency due to its increased subcarrier spacing. In an FMT realization, this is generally embodied in the addition of a frequency-domain roll-off factor $\alpha = (N_2 - M)/M$, yielding the time-frequency efficiency [18]:

$$\text{SEI}_{\text{FMT}} = \frac{1}{T \cdot F_0'} = \frac{T}{(1 + \alpha) \cdot T} < 1$$

Since the circularly convolved modulation uses the same CP length as in the CP-OFDM modulation, the two approaches normally address similar time-frequency efficiency. Nevertheless, it is possible for GFDM to group K-modulated symbols to one block with one CP appending ahead [29, 39]. One may understand that the inserted CP is shared by K-modulated symbols, which is one of the main differences between CP-OFDM and the variants of circularly convolved modulation, leading the time-frequency efficiency to

$$\mathrm{SEI}_{\text{GFDM}} = \mathrm{SEI}_{\text{COQAM}} = \mathrm{SEI}_{\text{FB-OFDM}} = \frac{1}{(T + T_{\text{CP}}/K) \cdot F} = \frac{T}{T + \frac{T_{\text{CP}}}{K}} < 1$$

In a special case where $K = 1$, this SEI is identical to that of CP-OFDM.

Likewise, the CB-FMT scheme also employs a CP with the same length as the CP-OFDM. Moreover, as it keeps the FMT kernel, the subcarrier spacing is also extended due to the over-sampled nature. This gives CB-FMT a two-dimensional SE loss [32]. The time-frequency efficiency of CB-FMT system yields

$$\mathrm{SEI}_{\text{CB-FMT}} = \frac{1}{(T + T_{\text{CP}}/K) \cdot F_0'} = \frac{T}{(T + \frac{T_{\text{CP}}}{K})(1+\alpha)} < 1$$

The UF-OFDM scheme does not imply any CP. But nevertheless, a zero-padding is employed after OFDM modulation, which ensures an isolation between consecutive UF-OFDM symbols after time-domain FIR filtering. The number of padded zeros is equal to the FIR filter length minus 1, making the overall spectral efficiency the same as CP-OFDM:

$$\mathrm{SEI}_{\text{UF-OFDM}} = \frac{1}{(T + T_{\text{ZP}}) \cdot F} = \frac{T}{T + T_{\text{CP}}} < 1$$

where T_{ZP} is the zero-padding duration, which is equal to T_{CP} of the CP-OFDM case [43]. Finally, f-OFDM completely retains the CP-OFDM process and employs a FIR filtering on top. The filtering does not reduce spectral efficiency due to the fact that the f-OFDM symbols are overlapped in the time domain, with the overlap length being the filter length minus 1. Therefore the spectral efficiency of f-OFDM is also equal to CP-OFDM. It is, however, worth noting that the above analysis based on the Gabor concept assumes a continuous transmission. In the case of burst transmission, additional spectral efficiency losses will be introduced due to the symbols overlapping in the time domain; this is called the “tail issue”.

8.3.2 Tail Issue

The tail issue indicates a phenomenon that the MCM symbols are not completely isolated in the time domain but instead part of the symbols are overlapped. In a burst transmission, the tail issue mainly reflects the potential overlap between two bursts; that is to say between the tail of the first burst and head of the next one. This issue has been identified for the FBMC scheme as, in theory, FBMC is able to achieve full time/frequency efficiency through the use of OQAM. However, this holds only in case of symbol sequences with infinite length. In realistic cellular scenarios, data transmission is divided into smaller time-direction chunks. For example, in LTE a single TTI spans 1 ms. Ramp-up and ramp-down times at the edges of these intervals caused by filtering reduce the actual efficiency. Hence this effect may become a major issue in applications such as machine-type communications, where the packets to be transmitted are expected to be rather short.

Some solutions have been proposed to overcome this problem, such as burst truncation [48] and the border reconstruction method [49] (the latter only valid for an overlapping factor equal to 1). It is obvious that burst truncation will introduce some performance degradation, which further depends on the prototype filter length. A filter with higher order – say overlapping factor 4 – is more suitable for burst truncation; filters with smaller overlapping factor are more sensitive to the truncation. Contradictorily, an FBMC system using higher overlapping factor prototype filter leads to higher latency, which impacts delay-constrained applications. For circularly convolved MCM schemes the tail issue is completely solved, as the tail is folded back to the block head, an approach also named "tail biting" [29], so that full independence between blocks is maintained. The UF-OFDM controls the filter length and utilizes zero padding to absorb the filtering tails. UF-OFDM has similar properties to CP-OFDM in this respect: each UF-OFDM symbol is completely isolated in the time domain so that it does not have any tail issue. In contrast to UF-OFDM, f-OFDM uses CP instead of ZP. Moreover, the filter length is longer than the CP length [46]. Therefore, the additional filtering will naturally widen the filtered symbols causing the tail issue.

8.3.3 Spectrum Confinement

Radio signal spectrum confinement may become even more important in future radio systems. Because of spectrum scarcity, physical layer modulation must provide improved radio signal spectrum confinement to enable multi-numerology supported air interface; one example is the multi-service scenario envisioned for 5G networks [50]. The waveform candidates discussed above have all claimed to produce better spectrum confinement than CP-OFDM systems can deliver. There are two main problems with CP-OFDM that lead to bad spectrum confinement:

- Spectral leakage due to the waveform discontinuity [51], which happens at the edge of each CP-OFDM symbol.
- The rectangular pulse shape for CP-OFDM, which leads to a sinc-pulse property in the frequency domain with a very low second lobe attenuation of -13 dB.

The first problem can be solved when the envelope of the symbol edges are smoothly decreasing to zero. To solve the second problem one can employ a filtering that has better frequency localization. For linearly convolved subcarrier filtered (SCF)-MCM – i.e. FBMC and FMT – the rectangular filter is replaced with a prototype filter that has good frequency localization and a length longer than (or at least equal to) the FFT size. This is equivalent to employing a windowing on the over-sampled OFDM symbol. Since the prototype filter has better frequency localization and, moreover, in the time domain the amplitude of its two edges is smoothly decreasing to zero, SCF-MCM can simultaneously solve both problems.

On the other hand, for circularly convolved SCF-MCM approaches, such as GFDM, CB-FMT, COQAM and FB-OFDM, the waveform discontinuity still exists between successive blocks, albeit it happens less often than for CP-OFDM. To solve this problem, a windowing process is needed for the circularly convolved schemes to relieve the waveform discontinuity, leading for example to WCP-COQAM [39] and windowed GFDM [52]. For UF-OFDM and f-OFDM, the waveform discontinuity issue can be overcome by time-domain FIR filtering. However, the improvement in the signal spectrum confinement is in general less than for FBMC and FMT due to the limited filter length involved. Relatively speaking, f-OFDM has better spectrum confinement than UF-OFDM since it utilizes a filter longer than that of UF-OFDM [46].

8.3.4 Mobility

Mobility robustness is also a very important criterion for 5G, particularly regarding prospective future V2X services. The linear-convolution-based subcarrier filtered schemes have, in general, better robustness against the Doppler effect due to the subcarrier filtering process. However, for the circular-convolved schemes, the block processing, especially with a lengthened block, will make the system quite vulnerable when there is high mobility. For this reason, a hybrid receiver structure is proposed for WCP-COQAM [39], which will provide a compromise between the Doppler effect and delay spread. On the other hand, the subband-filtered schemes in general handle the Doppler effect in the same manner as CP-OFDM, because the signal within a subband has the same character as the CP-OFDM signal. One claimed advantage is that the subband filtered schemes can have more degrees of freedom to realize different subband numerologies in a frame [53], which means that the subcarrier spacing can be made wider for a particular subband in order to serve high-mobility users.

8.3.5 Latency

Latency is another important consideration for 5G networks, in particular for applications such as remote driving and surgery, in which there are stringent latency requirementsd. Thus it is worth analyzing the latency of the waveform candidates.

In a multicarrier system, assuming the same number of subcarriers for all the above candidates, CP-OFDM is advantageous because of its short transceiver latency, which is mainly due to the FFT transform and CP, in other words $T + T_{\text{CP}}$. Any additional filtering will naturally increase the latency. Moreover, as described above, latency and spectrum confinement are two competing factors: better spectrum confinement comes at the cost of using a longer filter length, which, however, leads to increased latency. The lowest latency for FBMC is when it uses the shortest filter: the one with optimized time–frequency localization features for a length equal to the FFT size [6]. In this case the resulted latency is $1.5T$, which is due to the offset QAM signaling with a delay of half of the symbol duration, $T/2$.

The latency of FMT is also higher than that of OFDM due to its oversampled property. The increased latency not only depends on the filter length but also on the oversampling factor. On the other hand, the latency of the circularly convolved SCF-MCM is increased by the block processing. Usually one block may contain several MCM symbols. Therefore, the increased latency is proportional to the number of the composed symbols within one block, $KT + T_{\text{CP}}$. In this sense, this increased latency seems a major issue for this type of scheme.

Regarding the subband filtered MCM, UF-OFDM trades CP with the filtering transition period, which makes UF-OFDM have the same latency as CP-OFDM. Unlike UF-OFDM, f-OFDM does not use ZP, so that it needs additional buffers to absorb the filter transition period, which naturally increases the latency. This increased latency can be quantified as the filter length minus 1.

8.3.6 Modem Complexity

Modem complexity has always been a debate in new waveform investigations. One important advantage of the CP-OFDM is its low modem complexity. It is obvious that when working on some new advanced modulation simplicity might need to be sacrificed. Nevertheless, the ideal is to add enhanced properties while still being able to keep the additional complexity at an acceptable level. So although all the new waveform candidates yield higher complexity than CP-OFDM, it is important to compare them. In the FBMC literature, there are two implementation fashions: PPN-FBMC and FS-FBMC [7]. The modem complexity of PPN-FBMC is more than double that of CP-OFDM, mainly due to the offset QAM signaling that is required to separately treat the real and the imaginary parts of the complex-valued QAM symbols. Although another algorithm can be used on the modulation side [40], resulting in a similar modulation complexity to CP-OFDM, it cannot relieve the pressure on the demodulation. On the other hand, the modem complexity of FS-FBMC is more than $2b$ times higher than CP-OFDM [7], where b is the prototype filter overlapping factor, with a typical value for an FS-FBMC system of 4. This is because the FS-FBMC modulation uses an extended FFT (with an FFT size of bM instead of M) to produce FBMC signals.

Circularly convolved SCF-MCM, for instance GFDM, requires N FFTs with size K (a small size compared to M), where N stands for the number of modulated subcarriers. It needs a large FFT (size KM) to generate one block of GFDM signal. Note that this large FFT is only operated once per block. Therefore its overall modem complexity is in general similar to that of PPN-FBMC but less than that of FS-FBMC. Nevertheless, note that GFDM is not an orthogonal system, and therefore requires iterative interference cancellation to obtain the same match filtered performance as FBMC in the AWGN case [54]. Hence, the complexity issue for GFDM mainly arises from data detection.

CB-FMT and WCP-COQAM have similar modem complexity to GFDM. However, FB-OFDM has much less modem complexity than the others as it completely avoids the small FFTs, only one FFT of size KM being needed for each block. Therefore, its complexity is close to that of CP-OFDM [42].

Regarding the UF-OFDM, a direct implementation such as that outlined by Vakilian *et al.* [43] will lead to significant complexity increases, mainly due to the fact that the modulation of each resource block will consume one FFT transform. Thus, in an LTE setting where 100 resource blocks are to be modulated, the additional complexity will easily reach a factor of hundreds.

According to a recent report [45], a frequency-domain realization can remarkably reduce modulation complexity, by a factor of 30. The reported modulation complexity increase in comparison with the CP-OFDM modulation is a factor of 8–10 for downlink on the base-station side, and the demodulation complexity is slightly more than doubled in comparison to CP-OFDM demodulation.

For the f-OFDM modulation, a FIR filtering is employed on the top of CP-OFDM modulation, which can be realized in the frequency domain using the overlap–save method [46]. Although the authors have not reported concrete complexity values, we could infer that its modulation complexity should be between of those of PPN-FBMC and UF-OFDM. On the demodulation side, since f-OFDM uses a matched FIR filtering to the modulation side, its demodulation complexity might be higher than that for PPN-FBMC and UF-OFDM, and a similar order of demodulation complexity to FS-FBMC.

8.3.7 Compatibility with LTE

Compatibility, another consideration that is of much concern, does not mean that a receiver based on new waveform is able to decode LTE signals; it rather means that the new system should preferably be able to reuse existing LTE techniques, for example reference signal (RS) design and MIMO pre-/decoding, in a *straightforward* manner; so that less redesign effort is needed to allow adoption of the new technique. However, for FBMC systems, only real-valued symbols are transmitted due to offset QAM signaling. It thus cannot directly reuse the LTE RS, leading to some new designs [9, 10].

Similar to MIMO design, the LTE space frequency block coding (SFBC) is another example of a technique that cannot directly be adopted for the FBMC systems [55]. For GFDM, CB-FMT and WCP-COQAM systems, the channel estimation and equalization is performed at the block level and, due to the small FFTs, which mix the data and pilots up, the LTE RS cannot be directly reused. In addition, to realize the diversity transmission, space time block coding is preferred over the SFBC.

In contrast, FB-OFDM was invented to solve the compatibility issue. It has been shown [42] that FB-OFDM is compatible with LTE solutions. Similarly, for the UF-OFDM and f-OFDM schemes, it is claimed that the signal within the subband has the same character as the OFDM signal. Moreover, the additionally introduced filtering effect will be treated, together with the physical channel impulse response, at the receiver by the channel equalizer. Therefore, regardless of the potential performance degradation, the systems can maintain a good compatibility with LTE, ensuring a straightforward reuse of the LTE-based solutions like LTE RS as well as the MIMO pre-/decoder.

8.4 Conclusion

The 5G mobile communication system is now being intensively prepared. Ideally, 5G must not only bring improvements with respect to 4G concerning the data rate for the MBB service, but 5G must also provide an appropriate technical answer for forthcoming new services, such as M2M, MTC and MCC. To this end, new waveforms need to be envisioned for transmission environments, bringing new characteristics compared to 4G: absence of synchronization, very high mobility, very low latency.

Based on several studies, mainly in Europe [1, 3] and Asia, it is now widely recognized that, due to its lack of flexibility, the existing 4G air interface cannot be directly reused for 5G. Therefore, several flexible waveforms that meet 5G requirements are now being studied and promoted. These waveforms can be classified into three categories:

- subcarrier filtered MCM using linear convolution (FBMC/OQAM, FMT)
- subcarrier filtered MCM using circular convolution (GFDM, CB-FMT, WCP-COQAM, FB-OFDM)
- subband windowed MCM (UF-OFDM, f-OFDM).

Each scheme has its pros and cons. All of them, together with CP-OFDM, have been presented in a unified framework. We have also identified a list of essential criteria for 5G (spectral efficiency, tail issue, spectrum confinement, mobility, latency, modem complexity, compatibility with LTE) and we have proceeded to a qualitative comparison of the different waveform candidates with respect to these criteria. It is still too early to say if a single air interface will be able to cope with requirements going into different, and sometimes contradictory, directions, or if different waveforms may be used for different services. Anyway, as there is a real need for a new generation of mobile communication, an answer to this type of question is expected in a near future.

List of acronyms

5G	Fifth generation
BPSK	Binary phase shift keying
CB-FMT	Cyclic block FMT
CC	Circular convolution
COQAM	Circular OQAM
CP	Cyclic prefix
CR	Cognitive radio
DSA	Dynamic spectrum access
FB-OFDM	Filter-bank OFDM
FBMC	Filter bank multicarrier
FFT	Fast Fourier transform
FIR	Finite impulse response
FMT	Filtered multitone
f-OFDM	Filtered OFDM
FS-FBMC	Frequency spreading FBMC
GFDM	Generalized frequency division multiplex
ICI	Inter-carrier interference
IFFT	Inverse FFT
IoT	Internet of things
LTE	Long-term evolution
M2M	Machine to machine
MBB	Mobile broadband
MCC	Mission critical communication
MCM	Multicarrier modulation
METIS	Mobile and wireless communications enablers for the twenty-twenty information society
MIMO	Multiple-input, multiple-output
MTC	Machine type communication
OFDM	Orthogonal frequency division multiplex
OOB	Out of band
OQAM	Offset QAM
PPN	Polyphase network
PR	Perfect reconstruction
PRB	Physical resource block
QAM	Quadrature amplitude modulation
RB	Resource block
RS	Reference signal
SCF	Subcarrier filtered
SE	Spectral efficiency
SEI	SE indicator

SFBC	Space frequency block coding
SIC	Serial interference cancellation
STBC	Space time block coding
TTI	Transmission time interval
UFMC	Universal filtered multicarrier
UF-OFDM	Universal filtered OFDM
V2X	Vehicular to anything
WCP-COQAM	Windowed CP-Cyclic OQAM
ZP	Zero padding

References

[1] Metis Project (2013), Metis deliverable D1.1: scenarios, requirements and KPIs for 5G mobile and wireless systems.

[2] Bogucka, H., Kryszkiewicz, P., Jiang, T., and Kliks, A. (2015) Dynamic spectrum aggregation for future 5G communications. *IEEE Commun. Mag.*, **53** (5), 35–43.

[3] Wunder, G., Jung, P., Kasparick, M., Wild, T., Schaich, F., Chen, Y., Brink, S.T., Gaspar, I., Michailow, N., Festag, A., Mendes, L., Cassiau, N., Ktenas, D., Dryjanski, M., Pietrzyk, S., Eged, B., Vago, P., and Wiedmann, F. (2014) 5GNOW: Non-orthogonal, asynchronous waveforms for future mobile applications. *IEEE Commun. Mag.*, **52**, 97–105.

[4] Feichtinger, H. and Strohmer, T. (1998) *Gabor Analysis and Algorithm – Theory and Applications*, Birkhäuser.

[5] Le Floch, B., Alard, M., and Berrou, C. (1995) Coded Orthogonal Frequency Division Multiplex. *Proc. IEEE*, **83**, 982–996.

[6] Siohan, P., Siclet, C., and Lacaille, N. (2002) Analysis and design of OFDM/OQAM systems based on filterbank theory. *IEEE Trans. Signal Process.*, **50** (5), 1170–1183.

[7] Bellanger, M., FBMC physical layer: a primer, http://www.ict-phydyas.org/teamspace/internal-folder/FBMC-Primer_06-2010.pdf.

[8] Bellanger, M. (2012) FS-FBMC: an alternative scheme for filter bank based multicarrier transmission, in *Proceedings of ISCCSP 2012*.

[9] Javaudin, J.P., Lacroix, D., and Rouxel, A. (2003) Pilot-aided channel estimation for OFDM/OQAM, in *Vehicular Technology Conference (VTC) 2003 – Spring*.

[10] Zhao, Z., Vucic, N., and Schellmann, M. A simplified scattered pilot for FBMC/OQAM in highly frequency selective channels, in *International Symposium on Wireless Communications Systems (ISWCS)*.

[11] Lélé, C., Javaudin, J.P., Legouable, R., A. Skrzypczak, and Siohan, P. (2007) Channel estimation methods for preamble-based OFDM/OQAM modulations, in *Proceedings European Wireless*, Paris, France.

[12] Lélé, C., Siohan, P., and Legouable, R. (2008) Channel estimation with scattered pilots in OFDM/OQAM, in *Proceedings SPAWC'08*, Recife, Brazil.

[13] Kofidis, E., Katselis, D., Rontogiannis, A.A., and Theodoridis, S. (2013) Preamble-based channel estimation in OFDM/OQAM systems: A review. *Signal Process.*, **93** (7), 2038–2054.

[14] Ihalainen, T., Stitz, T., Rinne, M., and Renfors, M. (2007) Channel equalization in filter bank based multicarrier modulation for wireless communications. *EURASIP J. Adv. Signal Process.*, **2007**, pp. 1–18 (2007(ID 49389)), doi:10.1155/2007/49389.

[15] Ihalainen, T., Ikhlef, A., Louveaux, J., and Renfors, M. (2011) Channel equalization for multi-antenna FBMC/OQAM receivers. *IEEE Trans. Vehic. Tech.*, **60** (5), 2070–2085.

[16] Ndo, G., Lin, H., and Siohan, P. (2012) FBMC/OQAM equalization: Exploiting the imaginary interference, in *PIMRC'12*, Sydney, Australia.

[17] Vaidyanathan, P.P. (1993) *Multirate Systems and Filter Banks*, Prentice Hall.

[18] Cherubini, G., Eleftheriou, E., and Ölçer, S. (2002) Filtered multitone modulation for very high-speed subscriber lines. *IEEE J. Select. Areas Commun.*, **20** (5), 1016–1028.

[19] Kasparick, M., Wunder, G., Jung, P., and Maryopi, D. Bi-orthogonal waveforms for 5G random access with short message support, in *Proceedings of European Wireless* 2014.

[20] Kozek, W. and Molish, A.F. (1998) Nonorthogonal pulseshapes for multicarrier communication over doubly dispersive channels. *IEEE J. Select. Areas Commun.*, **16** (8), 1579–1589.

[21] Siclet, C. and Siohan, P. (2000) Design of BFDM/OQAM systems based on biorthogonal modulated filter banks, in *Proceedings GLOBECOM'00*, San Francisco, USA.
[22] Hleiss, R., Duhamel, P., and Charbit, M. (1997) Oversampled OFDM systems, in *International Conference on Digital Signal Processing*, Santorini, Greece.
[23] Siclet, C., Siohan, P., and Pinchon, D. (2006) Perfect reconstruction conditions and design of oversampled DFT modulated transmultiplexers. *EURASIP J. Appl. Signal Process.*, **2006**, 1–14. doi:10.1155/ASP/2006/15756.
[24] Rahimi, S. and Champagne, B. (2011) Perfect reconstruction DFT modulated oversampled filter bank transceivers, in *European Signal Processing Conference (EUSIPCO'12)*, Barcelona, Spain.
[25] Tonello, A. (2006) Time domain and frequency domain implementations of FMT modulation architectures, in *Proceedings ICASSP 2006*, Toulouse, France.
[26] Beaulieu, F.D. and Champagne, B. One-tap equalizer for perfect reconstruction DFT filter bank transceivers, in *ISSSE'07*, Montreal, Canada.
[27] Moret, N. and Tonello, A. (2010) Design of orthogonal filtered multitone modulation systems and comparison among efficient realizations. *EURASIP J. Adv. Signal Process.*, **2010**, 1–18. special Issue on filter banks for next generations multicarrier wireless communications, doi:0.1155/2010/141865.
[28] Roque, D., Siclet, C., and Siohan, P. (2012) A performance comparison of FBMC modulation schemes with short perfect reconstruction filters, in *International Conference on Telecommunications (ICT)*, Jounieh, Lebanon.
[29] Fettweis, G., Krondorf, M., and Bittner, S. (2009) GFDM – Generalized frequency division multiplexing, in *IEEE Vehicular Technology Conference (VTC Spring'09)*, Barcelona, Spain.
[30] Michailow, N., Gaspar, I., Krone, S., Lentmaier, M., and Fettweis, G. Generalized frequency division multiplexing: analysis of an alternative multi-carrier technique for next generation cellular systems, in *International Symposium on Wireless Communiation System (ISWCS'12)*, Paris, France.
[31] Datta, R., Michailow, N., Lentmaier, M., and Fettweis, G. (2012) GFDM interference cancellation for flexible cognitive radio PHY design, in *IEEE Vehicular Technology Conference (VTC Fall'12)*, Quebec city, Canada.
[32] Tonello, A. (2013) A novel multi-carrier scheme: Cyclic block filtered multitone modulation, in *ICC 2013*, Budapest, Hungary.
[33] Girotto, M. and Tonello, A. (2014) Orthogonal design of cyclic-block filtered multitone modulation, in *Proceedings European Wireless 2014*, Barcelona, Spain.
[34] Pinchon, D. and Siohan, P. (2015) A general analysis of cyclic block transmultiplexers with cyclic convolution, in *Proceedings Signal Processing Advances in Wireless Communications (SPAWC)*, Stockholm, Sweden.
[35] Tonello, A. and Girotto, M. (2014) Cyclic-block filtered multitone modulation. *EURASIP J. Adv. Signal Process.*, doi: 10.1186/1687-6180-2014-109.
[36] Gao, X., Wang, W., Xia, X., Au, E., and You, X. (2011) Cyclic prefixed OQAM-OFDM and its application to single-carrier FDMA. *IEEE Trans. Commun.*, **59**, 1467–1480.
[37] Abdoli, M., Jia, M., and Ma, J. (2013) Weighted circularly convolved filtering in OFDM/OQAM, in *PIMRC'13*, London, UK.
[38] Lin, H. and Siohan, P. (2014) An advanced multi-carrier modulation for future radio systems, in *Proceedings of ICASSP*, Firenze, Italy.
[39] Lin, H. and Siohan, P. (2014) Multi-carrier modulation analysis and WCP-COQAM proposal. *EURASIP J. Adv. Signal Process.* Special issue in flexible multicarrier waveforms for future wireless communications, **2014**, 1–19. doi:10.1186/1687-6180-2014-79.
[40] Dandach, Y. and Siohan, P. (2011) FBMC/OQAM modulators with half complexity, in *Proceedings Globecom'11*, 1077–1082, Houston, USA.
[41] Chen, D., Xia, X.G., Jiang, T., and Gao, X. Properties and power spectral densities of CP-based OQAM-OFDM systems. *IEEE Trans. Signal Process.*, **63** (4), 3561–3575.
[42] Lin, H. (2015) Filter bank OFDM: a new way of looking at FBMC, in *ICC'2015 Workshop on 5G Enablers and Applications*, London, UK.
[43] Vakilian, V., Wild, T., Schaich, F., ten Brink, S., and Frigon, J.F. (2013) Universal-filtered multi-carrier technique for wireless systems beyond LTE, in *IEEE Globecom 2013 Workshop – Broadband Wireless Access*.
[44] Zhu, P. (2014) 5G enabling technologies: An unified adaptive software defined air interface, in *Keynote presentation proceedings PIMRC*, Washington DC, USA.
[45] Wild, T. and Schaich, F. (2015) A reduced complexity transmitter for UF-OFDM, in *Proceedings VTC Spring 2015*, Glasgow, UK.
[46] Abdoli, M., Jia, M., and Ma, J. (2015) Filtered OFDM: a new waveform for future wireless, in *Proceedings Signal Processing Advances in Wireless Communications (SPAWC)*, Stockholm, Sweden.

[47] Zou, W. and Wu, Y. (1995) COFDM: an overview. *IEEE Trans. Broadcast.*, **41** (1), 1–8.

[48] Bellanger, M. (2010) Efficiency of filter bank multicarrier techniques in burst radio transmission, in *IEEE Globecom 2010.*

[49] Dandach, Y. and Siohan, P. Packet transmission for overlapped offset QAM, in *Proceedings IEEE International Conference on Wireless Communications and Signal Processing*, Suzhou, China.

[50] Schaich, F., Sayrac, B., Schubert, M., Lin, H., Pedersen, K., Shaat, M., Wunder, G., and Georgakopoulo, A. (2015) FANTASTIC-5G - 5GPPP project on the air interface below 6 GHz, in *EuCnC 2015 Conference Proceedings*, Paris, France.

[51] Harris, F. (1978) On the use of windows for harmonic analysis with the discrete Fourier transform. *Proc. IEEE*, **66** (1), 51–83.

[52] Michailow, N., Matthe, M., Gaspar, I., Caldevilla, A., Mendes, L., Festag, A., and Fettweis, G. (2014) Generalized frequency division multiplexing for 5th generation cellular networks. *IEEE Trans. Commun.*, **62**, 3045–3061.

[53] Schaich, F. and Wild, T. (2015) Subcarrier spacing – a neglected degree of freedom, in *Proceedings Signal Processing Advances in Wireless Communications (SPAWC)*, Stockholm, Sweden.

[54] Michailow, N., Krone, S., Lentmaier, M., and Fettweis, G. (2012) Bit error rate performance of generalized frequency division multiplexing, in *IEEE Vehicular Technology Conference (VTC Fall'12)*, Quebec city, Canada.

[55] Zakaria, R. and Ruyet, D.L. (2013) On interference cancellation in Alamouti coding scheme for filter bank based multicarrier systems, in *ISWCS*.

Part Two

New Spatial Signal Processing for 5G

9

Massive MIMO for 5G: Theory, Implementation and Prototyping

Ove Edfors, Liang Liu, Fredrik Tufvesson, Nikhil Kundargi and Karl Nieman

Signal Processing for 5G: Algorithms and Implementations, First Edition. Edited by Fa-Long Luo and Charlie Zhang.

9.1 Introduction

Modern communication systems rely upon multiple antennas at the transmitter and/or receiver to enhance link performance. This class of techniques, known as multiple input, multiple output (MIMO), exploits the spatial dimension by employing spatial encoding and/or decoding.

Since their emergence in the mid-1990s and perhaps earlier, MIMO technologies have been successful in practice, leading to deployments in WiFi and cellular systems. More specifically, MIMO methods introduced in 802.11ac support up to 8 transmit and 8 receive antennas. Cellular systems based on LTE Release 10 support up to 8 antennas in the base station per sector and 8 at the mobile station. Multiple antennas in these systems can be used to increase link robustness using space-time block codes or data rate by applying spatial-multiplexing.

MIMO techniques can be extended beyond point-to-point to multi-user applications with multi-user MIMO (MU-MIMO). MU-MIMO can be used to separate users by their spatial position, allowing for further network densification and increased capacity. MU-MIMO provides higher guarantees for spatial multiplexing than a point-to-point system but inherits challenges in near-far power control and time/frequency synchronization. MU-MIMO modes have been provisioned as part of the LTE and LTE-A standards.

Radically departing from existing MIMO is a new generation of large antenna array techniques, commonly referred to as "Massive MIMO", where the number of antennas at the base station is increased drastically (by an order of magnitude or more over current MIMO systems) to harvest further gains. Massive MIMO theory has promised vast gains in spectral efficiency, increase in energy efficiency, and reduction in network interference, all of which are key to address the demands of a data-centric world where spectrum and energy are increasingly precious.

The basic concept of Massive MIMO is shown in Fig. 9.1, where a base station is using M antennas to spatially multiplex $K \ll M$ single-antenna terminals. The success of such a spatial multiplex, in both uplink and downlink, relies on several important concepts. One of the most important requirements is that the base station should have sufficiently good knowledge of the propagation channel in both directions, on which efficient downlink precoders and uplink detectors can be based.

Since acquisition of channel-state information (CSI) is generally infeasible in the downlink [1], Massive MIMO systems typically rely on channel reciprocity, uplink channel estimation, and time-division duplex (TDD). With the massive number of channels to estimate between base station and mobile stations, a long-enough channel coherence time is needed to allow for

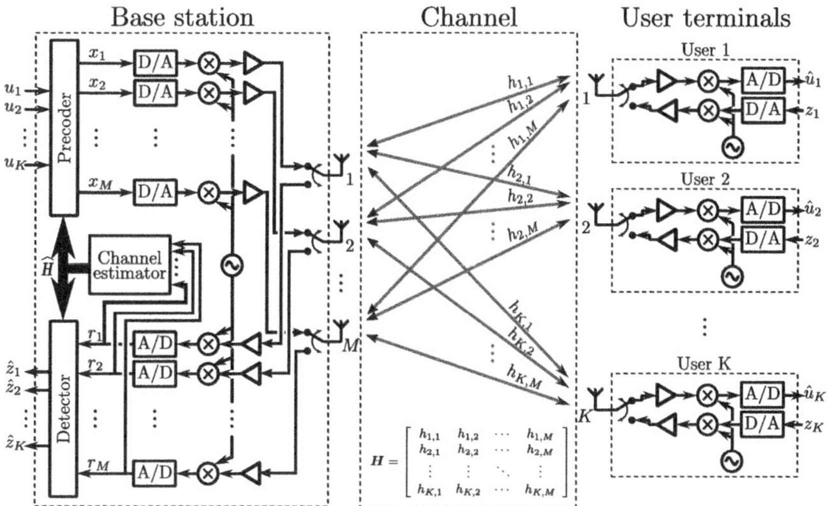

Figure 9.1 Simplified multi-user Massive MIMO system model, assuming time-division duplex and channel reciprocity

efficient operation. The accuracy at which we can estimate the channel and the time interval over which it can be assumed constant bring fundamental limitations to Massive MIMO [1].

Many of the algorithms required for Massive MIMO are also found in other wireless communication systems, such as traditional MIMO systems, with the essential difference that a much larger number of transceiver chains have to be processed in parallel. While this expands the processing complexity in one dimension, properties of Massive MIMO also allow many of the processing algorithms to be linear rather than non-linear, which helps to balance the massive increase of transceiver chains.

When implementing any communication system, it is essential to select the correct hardware platforms. Depending on the requirements on flexibility and energy efficiency, different choices come into play. For the prototype development and proof-of-concept work in the EU FP7 project Massive MIMO for Efficient Transmission (MAMMOET) [2], it is quite natural to use as flexible platforms as possible. A typical choice would include a combination of software defined radios (SDRs) complemented by additional computational resources. Due to the large number of transceiver chains and high requirements on synchronized real-time processing, often with an exchange of large amounts of data between processing nodes, it is important that the chosen platform has a high enough data shuffling capacity.

An important part of selecting the appropriate hardware platform deals with how Massive MIMO algorithms can and will be mapped onto computational hardware resources. In some cases it is quite sufficient that low-performing generic processors execute an algorithm, while in other cases much more advanced combinations of accelerators and/or specific computational structures are required. An important part of the work in MAMMOET is to find algorithms that ensure high communication performance, which can be efficiently mapped onto appropriate hardware and thereby make Massive MIMO a proven alternative for future communications standards.

Although Massive MIMO theory holds great promise, further development requires prototype systems to test the theory under real-world conditions. Unfortunately, real-time testbeds capable of Massive MIMO are not publicly available, not real-time, or both. In this chapter, we present our work on a flexible platform that supports prototyping up to 20 MHz bandwidth 128-antenna MIMO.

This chapter is devoted to theory, design, implementation, and prototyping of Massive MIMO for 5G communication systems. The rest of this chapter is organized into the following six sections. Section 9.2 presents an overview of principles and theory of Massive MIMO. Section 9.3 presents channel modeling system with a focus on specific phenomena of large scale 3D channels. Section 9.4 discusses the practical implementation aspects of Massive MIMO systems including analog circuit imperfections and digital baseband processing. Section 9.5 describes the design and architecture of building an 128 antenna Massive MIMO testbed. Section 9.6 presents the time synchronization and phase coherency aspects of a Massive MIMO base station. Section 9.7 is the final section which discusses future challenges and concludes with a summary of this chapter.

9.2 Massive MIMO Theory

A simplified model of a Massive MIMO base station, with M antennas serving K single antenna users, is shown in Fig. 9.1. Time-division duplex (TDD) operation and narrow-band channel conditions are assumed, which is sufficient for a basic description of the properties of Massive MIMO. The wide-band case, with a frequency selective propagation channel, can be analyzed either as a collection of narrow-band systems, i.e. as an Orthogonal Frequency-Division Multiplexing (OFDM) system where each subcarrier is modeled as in Fig. 9.1, or by a more proper extension where a discrete-time or continuous-time dispersion is introduced in the channel. For the sake of brevity, we only treat the single-cell case in detail. Some cellular aspects of Massive MIMO are covered, such as pilot contamination, but the reader is referred, e.g., [3–5] for a more comprehensive treatment.

The symbols transmitted to the K users are denoted $\boldsymbol{u} \triangleq (u_1, \ldots, u_K)^\mathsf{T}$, where u_k is directed towards user k. The resulting transmit signals, after precoding, are denoted $\boldsymbol{x} \triangleq (x_1, \ldots, x_M)^\mathsf{T}$, where x_m is the transmitted signal on antenna m. The analog-to-digital (A/D) and digital-to-analog (D/A) conversions are assumed to represent all the required transmit and receive filtering and their influences are, for now, considered a part of the propagation channel[1].

9.2.1 Downlink

The signals received by the users are collectively denoted by $\hat{\boldsymbol{u}} \triangleq (\hat{u}_1, \ldots, \hat{u}_k)^\mathsf{T}$, where $\hat{u}_k$ is the signal received at user k.

The channel is described by the $K \times M$ matrix $\boldsymbol{H}$, whose elements are defined in Fig. 9.1. Given this, the received signal is given by

$$\hat{\boldsymbol{u}} = \boldsymbol{H}\boldsymbol{x} + \boldsymbol{w}, \tag{9.1}$$

[1] Naturally, transmit and receive chains are not identical and the up- and downlink channels would therefore be different, even if the true propagation channel is. This particular problem is treated under the implementation section, where reciprocity calibration is addressed.

where $\boldsymbol{w} \sim \mathcal{CN}(0, \mathbf{I}_K)$ is an i.i.d. complex Gaussian receive noise vector with covariance matrix $\mathbf{I}_K$. The transmitted signals $\boldsymbol{x}$ on the antennas are a result of some precoding function

$$\boldsymbol{x} = f_{\text{pre}}(\widehat{\boldsymbol{H}}, \boldsymbol{u}) \tag{9.2}$$

depending on the channel estimate $\widehat{\boldsymbol{H}}$, the symbols $\boldsymbol{u} \triangleq (u_1, \ldots, u_K)^{\mathsf{T}}$ transmitted to the users, and the available transmit power P. We primarily address linear precoding schemes, where

$$\boldsymbol{x} = \boldsymbol{F}(\widehat{\boldsymbol{H}})\boldsymbol{u}, \tag{9.3}$$

and $\boldsymbol{F}(\widehat{\boldsymbol{H}})$ is a linear precoding matrix calculated from the channel estimate and other system parameters, such as available transmit power.

9.2.2 Linear Precoding Schemes

In general, the transmit signals should be chosen such that users receive their own symbols, with suppressed interference caused by the symbols intended for other users. There are many different strategies for this, including the optimal dirty-paper coding [6]. Given the Massive MIMO assumption of an excessive number of base station antennas, $M \gg K$, we can assume that linear precoding methods will work well in scenarios where we have favorable propagation conditions [7].

In this section we will address three linear precoders and discuss their properties. Computational issues are detailed in the section on Massive MIMO Implementation.

Channel estimation is included in the system model but, for simplicity, we assume the base station knows the channel perfectly, i.e. $\widehat{\boldsymbol{H}} = \boldsymbol{H}$. Detailed analysis of performance under different channel estimation conditions can be found in, e.g., [8, 9].

Maximum-Ratio Transmission Precoder

In maximum-ratio transmission (MRT) [1], the precoding matrix is given by

$$\boldsymbol{F}_{\text{MRT}} = \alpha_{\text{MRT}} \boldsymbol{H}^{\mathsf{H}}, \tag{9.4}$$

where α_{MRT} is a normalizing scalar controlling, e.g., transmit power and receive SNR.

MRT maximizes the array gain of the transmission, but the interference to other users will still be present in the received signal since there is no active interference mitigation. In typical scenarios (e.g., line-of-sight propagation and non-line-of-sight Rayleigh fading), MRT achieves interference suppression passively as the number of base station antennas increase and user channels become orthogonal in the limit [1].

Zero-Forcing Precoder

A linear precoding scheme that nulls all the interference, both inter-symbol and inter-user interference, is called zero-forcing (ZF). The precoding matrix of ZF is given by the pseudo-inverse

$$\boldsymbol{F}_{\text{ZF}} = \alpha_{\text{ZF}} \boldsymbol{H}^{\mathsf{H}}(\boldsymbol{H}\boldsymbol{H}^{\mathsf{H}})^{-1} \tag{9.5}$$

of the channel, where α_{ZF} is a normalizing scalar.

The main difference between ZF and MRT is the matrix inversion, which provides the desired interference suppression. This inverse computation can result in a major complexity increase but, as we will see in the section on implementation, the properties of Massive MIMO channels allow us to significantly reduce the computational complexity, compared to performing general matrix inverses.

Regularized Zero-Forcing Precoder

In addition to the MRT and ZF precoders, it is possible to use a regularized form of ZF precoding (RZF). This is a linear precoder situated "between" MRT and ZF, sharing properties with both. The RZF precoding matrix can be written as

$$\boldsymbol{F}_{\text{RZF}} = \alpha_{\text{RZF}} \boldsymbol{H}^{\mathsf{H}} (\boldsymbol{H}\boldsymbol{H}^{\mathsf{H}} + \beta_{\text{reg}} \boldsymbol{I}_{KN})^{-1} \tag{9.6}$$

where the regularization contant β_{reg} can be used to trade between array gain and interference suppression. If β_{reg} is selected to minimize mean-squared error (MSE) $E\|\boldsymbol{u} - \frac{1}{\sqrt{\rho}}\hat{\boldsymbol{u}}\|^2$, where ρ is a scaling constant, we obtain the minimum MSE (MMSE) precoder [10].

9.2.3 *Uplink*

The uplink signal model is easily derived by following a similar argumentation as for the downlink, given that we assume the propagation channel to be reciprocal. A notable difference is that we do not perform any precoding on the user side, since users are assumed not to cooperate to reduce interference. The only thing the users can control is their own transmitted power level. Using the notation from Fig. 9.1, letting the power levels of the K users build the $K \times K$ diagonal matrix $\boldsymbol{P}_{\text{ul}}$, and collecting the transmitted user symbols in $\boldsymbol{z} \triangleq (z_1, \ldots, z_K)^{\mathsf{T}}$, the received signals $\boldsymbol{r} \triangleq (r_1, \ldots, r_M)^{\mathsf{T}}$ on the M antennas become

$$\boldsymbol{r} = \boldsymbol{H}\sqrt{\boldsymbol{P}_{\text{ul}}}\boldsymbol{z} + \boldsymbol{w}, \tag{9.7}$$

where $\boldsymbol{w} \sim \mathcal{CN}(0, \mathbf{I}_M)$ is i.i.d. zero-mean complex Gaussian noise and the received user symbols at the base station are determined by some detector function

$$\hat{\boldsymbol{z}} = g_{\text{det}}(\boldsymbol{r}, \widehat{\boldsymbol{H}}) \tag{9.8}$$

primarily depending on the channel estimate $\widehat{\boldsymbol{H}}$ and the received signal $\boldsymbol{r}$, but also on implicit parameters like receive signal-to-noise ratio (SNR). Again, there is a class of detectors

$$\hat{\boldsymbol{z}} = \boldsymbol{G}(\widehat{\boldsymbol{H}})\boldsymbol{r} \tag{9.9}$$

using only linear processing (before applying, e.g., a simple slicer to extract symbols), where $\boldsymbol{G}(\widehat{\boldsymbol{H}})$ is a $K \times M$ matrix, depending on the channel estimate and other parameters (such as SNR), combining received signals from all antennas.

9.2.4 *Linear Detection Schemes*

Proper combining of signals from the M antennas can amplify desired signals and reject interfering ones. Since the down- and uplink transmissions in TDD systems take place over the same reciprocal channels, the same rates are typically achievable in both directions—known as uplink-downlink duality [11–13]. The main precoding schemes (e.g., MRT, ZF, and RZF) therefore have direct counterparts in the uplink detection. Again we assume, for simplicity, that channel estimation provides true channels $\widehat{\boldsymbol{H}} = \boldsymbol{H}$. We only give expressions up to proportionality, since scaling constants do not change the processed signal quality when using linear processing[2].

Maximum-ratio Combiner

The uplink counterpart to MRT is called maximum-ratio combining (MRC) and the combining matrix is

$$\boldsymbol{G}_{\text{MRC}}(\boldsymbol{H}) \propto \boldsymbol{H}^{\mathsf{H}}. \tag{9.10}$$

Similar to MRT, MRC maximizes the array gain, but the interference between user signals is still present since, like in MRT precoding, there is only passive interference mitigation.

Zero-forcing Combiner

The counterpart of the ZF precoder is the ZF combiner

$$\boldsymbol{G}_{\text{ZF}}(\boldsymbol{H}) \propto \boldsymbol{H}^{\mathsf{H}}(\boldsymbol{H}\boldsymbol{H}^{\mathsf{H}})^{-1}, \tag{9.11}$$

which removes all interference between user signals, given that the channel has full rank. To be able to remove intererence, array gain is sacrificed.

Regularized Zero-forcing Combiner

As for precoding, there is also a regularized version

$$\boldsymbol{G}_{\text{RZF}} \propto \boldsymbol{H}^{\mathsf{H}}(\boldsymbol{H}\boldsymbol{H}^{\mathsf{H}} + \beta_{\text{reg}}\mathbf{I}_K)^{-1} \tag{9.12}$$

of the zero-forcing combiner, where the regularization constant β_{reg} can be used to trade between array gain and interference mitigation. If β_{reg} is selected to minimize the mean-squared error (MSE) between the transmitted signals $\boldsymbol{z}$ and the processed received signal $\hat{\boldsymbol{z}}$, $E\|\boldsymbol{z} - \frac{1}{\sqrt{\rho}}\hat{\boldsymbol{z}}\|^2$, where ρ is a scaling constant, we obtain the minimum MSE (MMSE) combiner.

In addition to being computationally efficient when precoding and detecting, the strong similarities between linear precoding matrices and their receive-combining counterparts also makes the calculation of precoders/combiners more efficient. The strong similarities are seen by comparing Eq. (9.4) $\sim$ Eq. (9.10), Eq. (9.5) $\sim$ Eq. (9.11), and Eq. (9.6) $\sim$ Eq. (9.12). More details on computational aspects are found in Sec. 9.4.

9.2.5 Channel Estimation

The above description of precoding and detection techniques assumed that the channel estimator delivered the true channel. This is of course not the case and we need to estimate the channel before we can precode or detect data [14].

[2] Scaling will, however, influence numerical precision and efficiency of hardware implementation.

There are two major concerns when it comes to channel estimation in Massive MIMO systems; the large number of channel coefficients to estimate and the amount of pilots required. The first of these is a complexity issue and the second a radio-resource issue. Transmission of pilots dilute the fraction of resources used for data and, together with the rate of change of the channel, this constitutes one of the fundamental limits of Massive MIMO [1]. By using TDD, assuming reciprocity, and only performing channel estimation in the uplink, this problem is greatly simplified. There are approaches investigated for the FDD case, where we need channel estimates for both up- and downlink, but they are often relying on very specific properties of the channel.

Number of Channel Coefficients to Estimate

The number of channel coefficients, even in the simplified narrowband system of Fig. 9.1, can be quite massive. Assuming that we have N parallel narrowband systems, using e.g. N carrier OFDM, the total number of coefficients is NMK. As an example, a 100-antenna base station, operating under 3GPP LTE-like conditions with 1200 OFDM subcarriers, serving 30 users, needs to estimate 3.6 million channel coefficients. Considering that aging of the channel estimate takes place due to movements in the environment [15], it has to be updated on a regular basis. Even at walking speed, at 3-4 GHz carrier, the update frequency needs to be in the range of 100 times per second[3]. Under this assumption, 360 million channel coefficients per second need to be calculated in our example. Efficient channel estimation procedures are essential for Massive MIMO.

Number of Pilots to Transmit

Let us again start with the simplified narrowband system in Fig. 9.1, where we have MK channel coefficients that need to be estimated. Assuming ideal independent coefficients in the narrowband channel matrix, there is no room for reducing dimensionality of the estimation problem there. However, having an OFDM system with N such narrowband carriers, using an L sample cyclic prefix, allows us to reduce the dimensionality from N to roughly L per transmit/receive antenna pair, in the worst case, across all subcarriers [16]. This results in LMK parameters to estimate, which can then be interpolated to the NMK channel coefficients. Given this dimensionality of the channel, we need to excite the channel with a minimum of L pilots from each of the K terminals. These LK pilots are received by all M base station antennas, giving the minimal number of LMK samples needed for channel estimation.

Pilot Contamination

We have not addressed any specific cellular aspects above, but there is one case that we need to discuss in the context of Massive MIMO channel estimation. Since channel properties are used in the extreme to precode and combine signals to and from antennas, ensuring that signal energy to and from each user terminal is as concentrated as possible, we need to avoid any systematic estimation errors leading to distortion of these processing functions. One such systematic type of error is the contamination of pilot signals transmitted by terminals in the own cell by other terminals outside the cell non-orthogonal pilots. Signals from the contaminating terminals bear all the characteristics of a real pilot, making its contribution to the channel

[3] Calculation done assuming a need to update the channel estimate when time-correlation has dropped to 90% in a Jakes' fading environment, with 1 m/s walking speed.

estimate almost impossible to distinguish from the pilot transmitted by the wanted user[4]. The effect of pilot contamination in the downlink is that the contaminated channel estimate result in a precoder that will focus a certain portion of its transmit energy towards the contaminating terminal, causing increased interference. In the uplink, the detector will combine the received signals so that not only the signal from the wanted terminal is amplified, but to a certain extent also the signal from the contaminating terminal. The strength of these effects depend primarily on how strong the contaminating signal is compared to the wanted pilot signal [1, 18]. Means to mitigate pilot contamination, including e.g. pilot coordination between neighboring cells, have been proposed in, e.g., [19, 20].

9.3 Massive MIMO Channels

When modeling the behavior of the radio channel for Massive MIMO systems, some system specific phenomena needs to be considered that does not play such a large role in conventional MIMO systems. The radio channel is of course the same, independent of system and antenna configuration used, but some propagation effects become more pronounced or important when using physically large arrays employing many antenna elements at the base station, and additionally, have many closely located users. These effects are important and we need to capture the detailed behavior that can explain, for example, user separability, temporal behavior, as well as the possibility for significant increase in energy efficiency and spectral efficiency. Among the important specific propagation effects for Massive MIMO spherical wavefronts, variations of statistics over physically large arrays, and the limited lifetime of individual multipath components (MPCs) for moving users may be mentioned noticeable.

It is important to model the spherical wavefronts when physically large arrays are used at the BS as not only directions to users and scatterers are important, but also the distances to scatterers and users. The inherent beamforming capability of Massive MIMO systems makes it possible to focus the signal energy to a specific point in the environment rather than just in a certain direction. For example, if two line-of-sight (LOS) users are located in the same direction but at different distances from the BS, the spherical wavefronts make it possible to focus the signal individually and separate those users. This is typically not the case for conventional MIMO using smaller antenna arrays since there is no difference in the phase characteristics from those users over the array.

The variation in statistics of the received signal from a specific user over the array also contributes to the ability of the system to separate users. The variations include, e.g., received signal power, angular power spectra, and the power delay profile between different antenna elements [21], and also apply in the cases where the array elements have identical antenna patterns aimed in the same direction. Furthermore, it has been found that the power contribution of the individual clusters varies across the array, and clusters may not be "seen" over the whole array [22].

The limited lifetime of individual MPCs, or rather a limited area where a specific MPC can be observed, is another important effect to be considered when analyzing user separability of closely located users. In conventional MIMO models, as discussed in next section, all the scatterers in a cluster are visible from all positions in the visibility region of the cluster. In practice,

[4] Non-linear estimation techniques can circumvent this problem under certain conditions [17].

however, this is not the case for Massive MIMO. Each of the MPCs typically has a limited area inside the visibility region where they can be seen. Clusters provide a very effective way of modeling antenna correlation for a single user, but our observations show that conventional MIMO models tend to overestimate correlation between users in Massive MIMO systems. Measurement campaigns have been performed to analyze the case of closely-located users for both outdoor and indoor scenarios [23–25], so called crowd scenarios. This is seen as one of the more challenging scenarios for conventional MIMO systems as the user channels are typically highly correlated, but it has been shown that Massive MIMO offers the possibility to spatially separate those users.

9.3.1 Existing Conventional MIMO Models

For the Massive MIMO channel modeling we start with models for conventional MIMO and consider the group of channel models called Geometric Stochastic Channel Models (GSCMs) as they can provide a straightforward way to capture and describe correlation effects between users and between antenna elements through the concept of clusters and visibility regions. It is not straightforward to capture the specific effects important for Massive MIMO mentioned above by any of the other conventional modeling approaches such as the stochastic wideband models based on the so called tap delay line models or by spatio-temporal MIMO models such as the Kronecker model. Ray tracing simulations on the other hand, work well in general for Massive MIMO, given that the environment description is detailed enough. Ray-tracing based investigations and models can provide useful insights for system design and performance assessment. It should also be mentioned that there are theoretical geometrical model proposals in the literature, see e.g. [26]. As we aim for a model connected to a physical reality those models are out of scope of our treatment here, though these theoretical models can provide useful insights into, e.g., correlation characteristics.

Within the group of GSCMs there are two basic approaches, the COST 2100 approach [27] and the WINNER approach [28], both having the same origin in the COST 259 model. In the COST 2100 approach the scatterers have fixed physical positions in the simulated environment, whereas in the WINNER approach the channel simulation is based on angles to the scatterers as seen from BS and UE. From a Massive MIMO perspective the latter has the drawback that the angles vary as long as we are not in the far field of the array and this far field distance can be very large for physically large arrays. For the COST 2100 approach, as the scatterers and clusters are placed at specific locations in the simulation environment, the effect of spherical-wave fronts are inherently included in the model. Due to this reason and since we aim for a consistent model showing realistic correlations between users in a Massive MIMO context we recommend to base the model on the COST 2100 modeling approach where the clusters and scatterers are described by their physical locations rather than directions in the simulation area. The extensions we describe below are not specific to Massive MIMO only but can be useful to realistically represent physical propagation mechanisms when taking wireless communication beyond the conventional cellular scenario with one or several base stations, e.g. for modeling of wireless peer-to-peer channels. However, in the remainder we focus on the Massive MIMO scenario with one base station equipped with many antennas and several users having mobile stations with one or a few antennas.

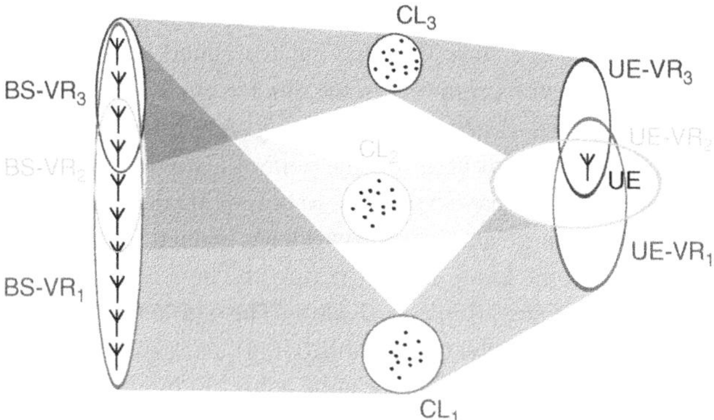

Figure 9.2 An illustration of the cluster visibility regions at the base station side (BS-VR) and at the user side (UE-VR). A cluster (CL) is active when both the BS antenna and the UE antenna are inside the respective visibility regions [24]

9.3.2 Necessary Model Extensions

For the Massive MIMO channel model we apply a cluster centric approach, where the visibility regions belong to the clusters and both visibility regions and clusters are randomly placed in the simulation area according to specific distributions. We use visibility regions (VRs) at both sides of the link, both for the BS and for the UE, see Fig. 9.2. As the visibility regions belong to the clusters and not to the base station we get a more symmetric approach which has many advantages in scenarios with multiple base stations, with or without Massive MIMO, and also for peer-to-peer communication such as vehicle-to-vehicle or device-to-device communication. For the extension it is important to be backwards compatible with the conventional COST 2100 model, which can be achieved by setting the size of the base station visibility regions to infinity, so that all BS antennas see all the clusters that are active from the mobile side, i.e. clusters where the users are inside the corresponding UE visibility regions.

Another aspect of Massive MIMO systems is the inherent higher angular resolution, which requires the channel model to provide detailed description of the channel in the sub-cluster or even multipath component levels. Therefore, despite the fact that the cluster based channel models describe a cluster as a collection of physical scattering points, more details are required to describe the behavior of individual scattering points when a user terminal is moving within the corresponding clusters visibility regions. From a Massive MIMO perspective it is, e.g., important to capture the effect that not all multipath components are visible in the whole VR. Therefore the concept of gain functions of the MPCs within the cluster is introduced [24]. The gain function is a smooth function allowing power variations of single MPCs within the cluster. As a first approach, gain functions that have a Gaussian shape are used. The width of the gain function (determined by the standard deviation of the Gaussian function) controls the variability of the MPC gain within the visibility region. Each MPC has its own gain function and the specific gain function of an MPC has a peak location (in 2D or 3D as discussed below) within the VR. For a specific user location the Euclidean distance to the peak of the gain function is calculated and the gain of the MPC is given as a function of this distance.

The gain function has also the advantage that it can be used to create sub-clusters out of clusters if this is desired. This is achieved by having correlated locations of the peaks of the gain functions with closely located scatterers in the cluster. Backwards compatibility with the conventional COST 2100 model is achieved by setting the variance of the gain function to infinity so that there are no variations at all for the gain of a single MPC inside the VR.

A third feature necessary for Massive MIMO is to have a 3D description of the propagation paths and the environment. As large antenna arrays are envisioned for the base station it is quite likely that those arrays have some kind of 2D structure and hence have good resolvability and beamforming capability both in elevation and azimuth. The whole modeling framework should be in 3D so that locations of users, clusters, visibility regions, base station(s) are described in 3D. Backward compatibility with 2D models can be achieved by projection of the 3D parameters on the ground floor.

Besides the requirement of capturing the behavior of the wireless channels in the different spatio-temporal, angular, and delay domains, the channel model should also be consistent in both the spatial domain and the frequency domain. In the spatial domain, the channel model has to be able to capture the propagation behavior over small distances in the range of a few wavelengths to very large distances (hundreds or thousands of meters), for both the terminal side and the base station side. The model should also cover the cases where the user terminals are closely spaced and cases where user terminals are widely separated. Similarly, it has to realistically cover cases where the base station array is physically small and cases where it is physically large. The consistency in the spatial domain makes it possible to compare Massive MIMO with conventional MIMO. In the frequency domain, the channel model should support lower frequencies below 6 GHz as well as higher frequencies above 6 GHz. The consistency in the frequency domain has become ever more important due to the trend of 5G communications in using the higher frequency bands above 6 GHz. Cluster-based stochastic channel models such as the COST 2100 model can meet the requirement of the consistency in both the spatial and frequency domains. We, however, limit the efforts here to the case below 6 GHz.

9.3.3 A Massive MIMO Extension of the COST 2100 Channel Model

In the following a measurement based effort for extending the COST 2100 model for Massive MIMO is described. The description is somewhat brief; full details of the model, the implementation and its parameters can be found in [24].

It is important to model the effect of the polarization on the performance of Massive MIMO systems in serving, e.g., closely-spaced users. In Massive MIMO, as in any other wireless system, both vertically and horizontally polarized antennas at the base station are useful for separating the users. The polarization at the user side is usually dual but random, especially when having a portable device as UE. Therefore, the model includes the polarization states of the MPCs. The polarization states of the individual MPCs are combined with the polarization of the BS/UE antenna pattern to generate channel transfer functions. To describe the polarization states of the MPCs we adopt the cross polarization ratio (XPR) parameters used for the WINNER II channel models. The XPR for each MPC is modeled by a log-normal distribution $\kappa = 10^{\frac{X}{10}}$, where $X \sim N(\mu, \sigma)$. For indoor office scenario, the mean $\mu = 11$ dB for LOS and $\mu = 10$ dB for NLOS, and the standard deviation $\sigma = 4$ dB. For large indoor hall hotspot, $\mu = 9$ dB for LOS and $\mu = 6$ dB for NLOS, $\sigma = 4$ dB and $\sigma = 3$ dB for LOS and NLOS, respectively. For urban micro-cell, $\mu = 9$ dB for LOS and $\mu = 8$ dB for NLOS, and $\sigma = 3$ dB. Furthermore, the model is extended to 3D, in a similar way as for 3D MIMO [29], by including parameters such as intra-cluster angular spread in elevation for both the BS and the UE side.

In order to capture the effect of having a physically-large array at the BS and be compatible with the convectional COST 2100 model, the simulation area at the base station side is extended to several meters, and cluster visibility regions at the BS side are used. The cluster contributes in the channel between a UE and a BS antenna only when the UE is within its UE visibility region and the BS antenna is within its BS visibility region.

The extension for physically-large arrays is consistent with physically-compact arrays where a whole array is within the same BS-VR and does not experience cluster power variations. This allows direct comparison of physically-large and compact arrays in the simulation.

When a user is moving, the effects of changing position, orientation and tilt, and the effect of other users around the user can be captured as follows:

- The effect of changing the orientation and the tilt of the antenna is captured by the user antenna pattern through the complex antenna gain associated with the direction of arrival of each MPC.
- The effect of nearby users, that cause local scattering or absorption, can be captured by utilizing the concept of local clusters. It could be needed to modify the concept of the local clusters compared to the conventional COST 2100 model, e.g., by allowing local clusters to have different parameters compared to the far clusters, or adding shadowing objects around the users. This issue needs, however, further investigation.
- The variation due to the specific position within the cluster visibility region is modeled both by the fact that the different MPCs add up differently at various positions but also by using the MPC gain functions, which have a symmetric Gaussian shape.

The gain functions are characterized by their coordinates in the simulation area and their width. The gain of an MPC is determined by the distance between the location of the peak of the gain function in the visibility region and the location of the antenna, i.e.

$$g_{\mathrm{MPC}}(d) = \exp\left(-\frac{d^2}{2\sigma_g^2}\right), \tag{9.13}$$

where $d = \sqrt{(x - x_i)^2 + (y - y_i)^2 + (z - z_i)^2}$ and (x_i, y_i, z_i) is the coordinate for the peak of the gain function for MPC i.

Each MPC has its own gain function and the peak of the MPC gain function is randomly distributed in the cluster visibility region, see Fig. 9.3 for an illustration. From our preliminary observations, the lifetime of the MPCs, measured as the movement over which a specific MPC can be constantly observed, is about 2 m. Therefore, we set $\sigma_g = 2.37$, which corresponds to 3 dB power decay when $d = 2$ m. When introducing the MPC gain functions, backward compatibility with the conventional model can be achieved by setting a large standard deviation for the gain functions, i.e., according to the size of the cluster visibility region. In accordance with the conventional MIMO models COST 2100 and WINNER II, we aim for an average number of effective MPCs per cluster to be 16. Hence, with $R_c = 10$ m and $r_g = 2$ m, the total number of MPCs per cluster has to be increased to around 400 in

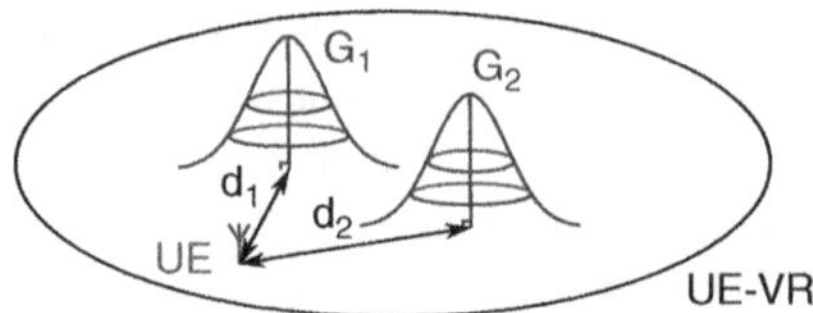

Figure 9.3 An illustration of gain functions G_1 and G_2 inside the user visibility region (UE-VR)

our case. Only a few of those MPCs are, however, strong and constitute dominant specular reflections; the MPCs with weak power constitute diffuse or dense multipath components (DMC).

9.4 Massive MIMO Implementation

Due to its great potential for upcoming 5G systems, Massive MIMO has attracted a lot of research activities in both academia and industry and substantial theoretical progress has been achieved. On the other hand, the research activity from the practical implementation perspective is still in its infancy. To efficiently realize and deploy this promising class of systems and facilitate the corresponding standardization, many critical practical issues have to be studied and addressed. Motivated by this objective, this section will build on the theory of Massive MIMO introduced in the last section and will discuss the practical implementation aspects of Massive MIMO systems. First, the impact of non-ideal phenomenon in analog front-end circuitries on the system performance will be discussed. Then, the digital baseband processing of Massive MIMO systems (mainly focusing on the base station side) will be introduced, including methods and techniques for cost and power efficient hardware implementation. The section will continue with the challenges and requirements for building up Massive MIMO prototyping systems, which is crucial to bring this promising technology from theory to practice. Finally, the potential deployment scenarios of Massive MIMO will be analyzed to envision the role of this technology in future 5G network.

9.4.1 Antennas and Analog Front-ends

A real world implementation of Massive MIMO, must address issues such as RF impairments, phase noise, antenna calibrations, reciprocity mismatches, and mutual coupling among antenna elements, etc. It will also focus on the unique advantages of Massive MIMO in practical deployments, such as use of low cost power amplifiers with significant output power reductions and relaxed linearity constraints etc.

Analog Imperfections

The large number of antenna elements in Massive MIMO systems require a large number of analog components. To enable commercial deployments, the implementation of Massive MIMO systems should be economically feasible. Thereby, the cost of these analog elements per antenna element must be significantly less expensive than that in existing systems, to keep the overall cost reasonable. Fortunately, one of the potential benefits of Massive MIMO is in enabling the extensive use of many lost-cost analog components, shifting from the use of a few expensive ones in conventional systems. For instance, hundreds of power amplifiers with output power in the milli-Watt range can be used [4]. The consequence of this is the introduction of analog front-end imperfections (or impairments), including frequency offset, I/Q imbalance, quantization noise in analog-to-digital/digital-to-analog converters, non-linearity, etc. Analyzing the system performance in the presence of these imperfections is an important research topic for Massive MIMO.

In the presence of analog imperfections, the signals are distorted and interfered with during the transmit and receive processing, leading to deviations between the desired (ideal) signal and that which is actually transmitted and received. To evaluate the corresponding system performance, it is crucial to have proper models of the imperfections. Existing published studies on the topic are based on two different impairment models: stochastic (both additive and multiplicative) and deterministic [30].

STOCHASTIC MODELS: The more abstract and analytical method is to model the residual analog imperfections using a stochastic approach. In this model, the distortion from the analog front-end and thc transmitted signal waveform are assumed (and simplified to some extent) to be uncorrelated. In the additive stochastic model, the residual analog front-end impairments are treated as additive Gaussian noise. The corresponding system model for the downlink signal relationship can be written as [14]

$$\hat{\boldsymbol{u}} = \boldsymbol{H}_{dl}(\boldsymbol{x} + \boldsymbol{\beta}_t^{BS}) + \boldsymbol{\beta}_r^{UE} + \boldsymbol{w}, \tag{9.14}$$

where $\boldsymbol{\beta}^{BS} \in \mathbb{C}^{M\times 1}$ and $\boldsymbol{\beta}^{UE} \in \mathbb{C}$ are the additive distortion noise terms modelling the residual analog component impairments of the transmitter at the base station and the receiver at the single-antenna user equipment, respectively. The corresponding uplink system model can be described in a similar way, as

$$\boldsymbol{r} = \boldsymbol{h}_{ul}(z + \boldsymbol{\beta}_t^{UE}) + \boldsymbol{\beta}_r^{BS} + \boldsymbol{w}. \tag{9.15}$$

In practical systems, some of the analog imperfections cannot be treated as completely independent of the transmitted signals. For instance, the nonlinearity and distortion in power amplifier are related to the transmitted signals. Quantization of the data converter can be modelled as additive noise, while the saturation due to the limited number of bits in converters is amplitude dependent. Another approach to model such signal-depended imperfections is using multiplicative (instead of the additive) error terms, e.g.,

$$\hat{\boldsymbol{u}} = \boldsymbol{H}_{dl}\boldsymbol{\Theta}\boldsymbol{x} + \boldsymbol{w}, \tag{9.16}$$

where

$$\boldsymbol{\Theta} = diag((1 + \alpha_m)e^{(-j\varphi_m)_{m=1}^{M}}) \tag{9.17}$$

is the multiplicative imperfection error matrix. In Eq. (9.17), α_m and φ_m are the stochastic errors modelling the impairments effects on amplitude and phase, respectively.

It is worthwhile to mention that the impairment errors in Eq. (9.14) and Eq. (9.16) may change the transmitting power. To model the imperfections more accurately and for more reasonable performance analysis, a constant κ is added to ensure that no transmitting energy is added by the imperfections [30], e.g. Eq. (9.14) is modified as:

$$\hat{\boldsymbol{u}} = \kappa \boldsymbol{H}_{dl}(\boldsymbol{x} + \boldsymbol{\beta}_t^{BS}) + \boldsymbol{\beta}_r^{UE} + \boldsymbol{w}. \tag{9.18}$$

Similar transmitting power normalization can also be applied to the multiplicative stochastic model in Eq. (9.16).

Deterministic Model: The previous two imperfection models are simplified to a certain extent based on statistics. They are helpful for system performance analysis but may fail to capture all the important impacts from analog components. Thereby, for components having high impact on the system performance, researchers have created a deterministic behavior model to enable more accurate performance analysis. For instance, authors in [30] propose a power amplifier model that captures both nonlinearity and mutual coupling, and thus is more accurate in the context of Massive MIMO system. In the model, the output signal from the mth power amplifier has been modeled

$$o_{PA}^{m} = \mathcal{F}(x^{m}, x^{m,n}). \tag{9.19}$$

Here, the function $\mathcal{F}()$ models the non-linearity in the PA and the coupling from other $M - 1$ branches (mainly contributed by neighboring PAs) has been summarized in $x^{m,n}$.

Based on aforementioned three imperfection models, system performance (in terms of system capacity, average data rate, EVM, etc.) has been studied analytically or with simulation. Based on the stochastic model, the studies in [14] show that the analog hardware quality has fundamental impacts on the achievable spectral efficiency of a Massive MIMO system. More specifically, due to the analog imperfections the minor spectral efficiency improvement is observed with the growth of base station antennas after $M \geq 100$. One important conclusion from [14] is that the imperfections at the UE side limits the capacity growing with M, while the analog components quality can be decreased with the growth of M. This is mainly due to the average-out at the base station side, implying that huge degrees-of-freedom in signal paths brings robustness to such imperfections, leaving room for the extensive use of low-cost, low-power antenna branches.

Non-reciprocity and Calibration

As aforementioned, in TDD Massive MIMO systems we rely on channel reciprocity to perform downlink precoding based on uplink channel estimation. Because the propagation channel is typically reciprocal, the uplink channel estimation can be directly used in this ideal case. However, responses of analog components need to be taken into account in practical systems. Practical impairments such as circuit variations, manufacturing process, voltage supply, environment temperature, etc. may destroy the nice reciprocal property. In this practical case, differences in analog chains between uplink and downlink have to be estimated and compensated. This process is commonly referred to as TDD channel reciprocity calibration.

Fig 9.4 illustrates how a typical equivalent duplex channel is experienced by a signal in a wireless transmission.

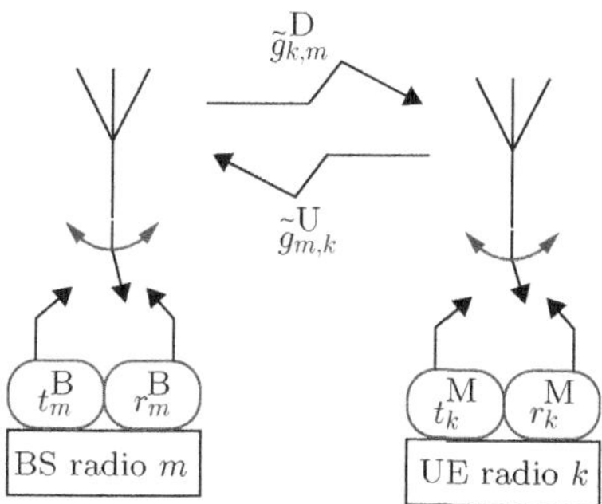

Figure 9.4 Illustration of uplink and downlink equivalent channels

We denote the uplink and downlink narrow-band radio channels between the base station and mobile station (UE) as

$$\begin{aligned} g^{\mathrm{U}}_{m,k} &= r^{\mathrm{B}}_{m}\,\tilde{g}^{\mathrm{U}}_{m,k}\,t^{\mathrm{M}}_{k} \\ g^{\mathrm{D}}_{k,m} &= r^{\mathrm{M}}_{k}\,\tilde{g}^{\mathrm{D}}_{k,m}\,t^{\mathrm{B}}_{m}, \end{aligned} \tag{9.20}$$

where $m \in [1, \ldots, M]$ is the BS antenna index, $k \in [1, \ldots, K]$ is the mobile station antenna index, r^{B} and r^{M} represent the response of BS and UE receiver RF chains, t^{B} and t^{M} represent the corresponding transmitter RF chains, and $\tilde{g}^{\mathrm{U}}$ and $\tilde{g}^{\mathrm{D}}$ are the uplink and the downlink propagation channels, respectively.

Assuming that the propagation channel is reciprocity, i.e. $\tilde{g}^{\mathrm{U}}_{m,k} = \tilde{g}^{\mathrm{D}}_{k,m}$, the relation between the uplink and downlink equivalent channels can be written as

$$b_{m,k} = \frac{r^{\mathrm{M}}_{k}\,\tilde{g}^{\mathrm{D}}_{k,m}\,t^{\mathrm{B}}_{m}}{r^{\mathrm{B}}_{m}\,\tilde{g}^{\mathrm{U}}_{m,k}\,t^{\mathrm{M}}_{k}} = \frac{r^{\mathrm{M}}_{k}\,t^{\mathrm{B}}_{m}}{r^{\mathrm{B}}_{m}\,t^{\mathrm{M}}_{k}}. \tag{9.21}$$

In Eq. (9.21), $b_{m,k}$ denotes the calibration coefficient between link m and k. With the knowledge of $b_{m,k}$, one can overcome the channel non-reciprocity and compute the equivalent downlink channel based on the uplink channel estimates.

Calculating Eq. (9.21) directly requires training signals between BS and UEs, which can be expensive in terms of bandwidth and will also incur processing latency. Another way is to obtain the coefficient indirectly. Let us introduce the channel between two BS antenna links as

$$h_{\ell,m} = r^{\mathrm{B}}_{\ell}\,\tilde{h}_{\ell,m}\,t^{\mathrm{B}}_{m}, \tag{9.22}$$

where $\ell \neq m$, $\ell \in [1, \ldots, M]$, and $\tilde{h}_{\ell,m}$ is the propagation channel between the BS antennas ℓ and m. The calibration coefficient between BS antenna links can be introduced as

$$h_{\ell,m} = b_{m\to\ell}\,h_{m,\ell}, \tag{9.23}$$

which by assuming perfect reciprocity yields:

$$b_{m\to\ell} = \frac{h_{\ell,m}}{h_{m,\ell}} = \frac{r^{\mathrm{B}}_{\ell}t^{\mathrm{B}}_{m}}{r^{\mathrm{B}}_{m}t^{\mathrm{B}}_{\ell}} = \frac{1}{b_{\ell\to m}}. \tag{9.24}$$

Here we denote the calibration coefficients between two BS antenna links using "→" to distinguish from the calibration coefficient between a BS-UE link which uses ",".

Authors of [31] proposed an internal reciprocity calibration method for Massive MIMO base stations. The method has two main points as basis: 1) Calibration between the radio link m and k can also be achieved if their forward and reverse channels to another BS radio n are jointly processed, i.e.,

$$b_{m,k} = \frac{t_m^{\rm B}}{r_m^{\rm B}} \frac{r_k^{\rm M}}{t_k^{\rm M}} = \frac{r_n^{\rm B}\, t_m^{\rm B}}{r_m^{\rm B}\, t_n^{\rm B}} \frac{r_k^{\rm M}\, t_n^{\rm B}}{r_n^{\rm B}\, t_k^{\rm M}} = b_{m\to n} b_{n,k}, \tag{9.25}$$

where n can be any index in the base station antennas. Without loss of generality, we set $n = 1$ for convenience and denote this radio as the reference radio. 2) As long as the equivalent downlink channel from all BS antennas deviates from the real one by a same complex factor, the resulting downlink beam pattern shape does not change. As results, the transceiver response of any UE shows up as a constant factor to all BS antennas, which can be compensated with downlink pilots.

Combining Eq. (9.21) with the aforementioned two points yields

$$g_{k,m}^{\rm D} = b_{m,k}\, g_{m,k}^{\rm U} \tag{9.26}$$

$$\overset{1)}{=} b_{m\to 1}\, b_{1,k}\, g_{m,k}^{\rm U} \tag{9.27}$$

$$\overset{2)}{\Longleftrightarrow} g_{k,m}^{\prime\rm D} = b_{m\to 1} g_{m,k}^{\rm U}, \tag{9.28}$$

where $g_{k,m}^{\prime\rm D}$ is a relative downlink channel that takes $b_{1,k}$ into consideration. The relative downlink channels can be obtained by multiplying the uplink channel estimates with their corresponding calibration coefficients to the reference antenna.

With this concept in mind, the authors in [32] applied the relative method to calibrate access points of a distributed MIMO network and the proposed approach is

$$g_{k,m}^{\prime\rm D} = b_{m\to 1}\, g_{m,k}^{\rm U} \tag{9.29}$$

$$\Longleftrightarrow g_{k,m}^{\prime\prime\rm D} = b_m\, g_{m,k}^{\rm U}, \tag{9.30}$$

where $b_m = \frac{r_m^B}{t_m^B} = \frac{1}{b_{m\to 1}} \frac{t_1^B}{r_1^B}$, and $g_{k,m}^{\prime\prime\rm D}$ is another relative downlink channel. This relative equivalence not only relaxes the double-indexing overhead, but allows different calibration coefficients to be treated as mutually independent.

Worthwhile to be mentioned again that the absolute reference to the terminals was lost in the derivation step 2), which makes $b_{m\to 1}$ or b_m valid calibration coefficients up to a complex factor. Thus, pre-coded downlink pilots still need to be broadcast through the beam to compensate for this uncertainty, as well as for the analogy chain responses of the UEs. The overhead of these supplementary pilots is reported as very small, given that the frequency selectivity of in-band analogy chain responses can be much less comparing to multi-path prorogations. The calibration coefficients can be valid over long periods of time if BS radios share the same synchronization references.

In [33], the authors proposed to estimate the calibration coefficients b_m by sounding and straining the M antenna links one-by-one, i.e., by transmitting pilot sequences from each one and receiving on the other $M - 1$ silent antennas.

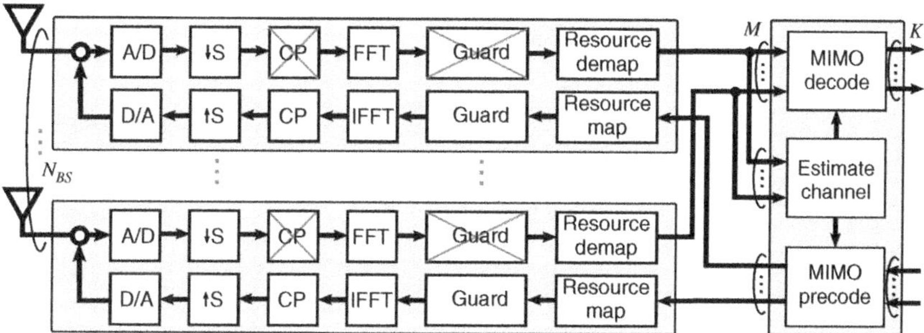

Figure 9.5 The processing performed by an OFDM Massive MIMO base station. M antennas of synchronized uplink baseband samples are acquired by an ADC and processed using an OFDM receiver (CP removal, FFT, guard subcarrier removal, resource demap) and then passed to a MIMO detector and channel estimator. Channel estimates are used to precode downlink data. Precoded symbols are then distributed to the M OFDM transmitters (resource map, guard subcarrier insertion, IFFT, add CP) and transmitted out the antenna ports

9.4.2 *Baseband Processing*

The processing for Massive MIMO base station that uses OFDM signaling is shown in Figure 9.5 [34]. In this diagram, the base station processes M channels that consist of M antenna ports connected to M OFDM transceivers. The entire processing flow is as follows. First, uplink signals are demodulated from RF and digitized using an analog-to-digital (ADC) converter and associated RF hardware. Second, the digital samples are then downsampled to the desired sampling rate that is some fraction $\frac{1}{S}$ times the A/D sample rate. The typical OFDM signal processing chain is then applied, i.e., the signal is synchronized, some of the analog impairments are compensated, the cyclic prefix (CP) is removed, the FFT is taken after serial-to-parallel conversion, guard subcarriers are removed, and uplink data and pilots are deallocated according to the resource map, or schedule, observed by the link.

Pilot and data symbols for all M receive chains are then passed to the channel estimator. Channel estimates are then used to decode the K uplink data streams and precode the K downlink data streams (with calibration is necessary). The K downlink data streams are precoded using the ZF precoder shown in Eq. (9.5) or the MRT result shown in Eq. (9.4). Other more advanced precoding may be adopted in certain scenarios. Precoded data for the M antennas is then input to a resource mapper, guard symbols are inserted, serial-to-parallel conversion takes place, the IFFT is taken, the CP is added, samples are digitally upconverted by S times the baseband rate, and digital samples are converted to analog using a digital-to-analog converter (DAC) and conditioned for transmission out of the base station antenna ports.

Matrix Multiplication and Inversion

As mentioned before, the concept of zero-forcing can be used in both precoding and detection in the context of Massive MIMO. In this section, we demonstrate how ZF processing can be efficiently implemented, with the constraint in hardware resources, processing speed, and power consumption. ZF contributes to the the bulk of the processing complexity, due to the inversion of large channel matrices. In terms of latency, precoding is more critical compared to detection (especially TDD processing for fast-changing channels), since detection may be

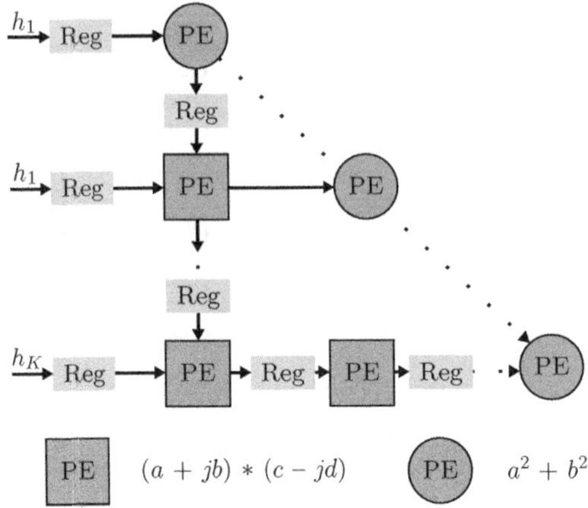

Figure 9.6 Systolic array to perform hermitian symmetric matrix multiplication

performed by buffering the symbols. The ZF precoding/detection requires a pseudo-inverse of the channel matrix, which requires two matrix multiplications, one matrix inversion, and one matrix-vector multiplication.

A traditional matrix multiplication has complexity of $\mathcal{O}(N^3)$. Other divide-and-conquer algorithms have lower complexity, e.g., Strassen algorithm with complexity of $\mathcal{O}(N^{2.8})$. However, these algorithms have been shown to be efficient only for very large matrices. Moreover they have a very high (routing) overhead in hardware, leading to potential long latency. In case of pseudo-inverse, the matrix multiplication is to compute a Gram matrix (Hermitian symmetric matrix, i.e., $\boldsymbol{H}^H\boldsymbol{H}$), hence the complexity can be reduced almost by half by exploring the symmetric property. This can be implemented with the traditional multiply accumulate (MAC) units and a controller to handle the data-flow. An alternative high-throughput solution to exploit this is to use systolic arrays as shown in Fig. 9.6. The corresponding hardware cost (in terms of arithmetic operations) for the multiplication unit is detailed in Table 9.1, with a 100×10 channel matrix as an example. Systolic array architecture is also flexible (e.g., by folding) to support the matrix multiplication of different dimensions.

Implementing matrix inversion in hardware is much more complicated and expensive (mainly due to the involved higher-complexity arithmetical operations), although in theory it has same order of complexity as matrix multiplication. The matrix inversion operation can be divided into three approaches: explicit computation, implicit computation, and polynomial expansion. In the first approach the matrix inversion is performed explicitly, whereas in the

Table 9.1 Hardware Details for Matrix Multiplication

	Systolic Array	MAC-unit Based
Matrix Size	10×100	10×100
# of multipliers	200	40
# of adders	200	40
# Internal Accumulators	110	20

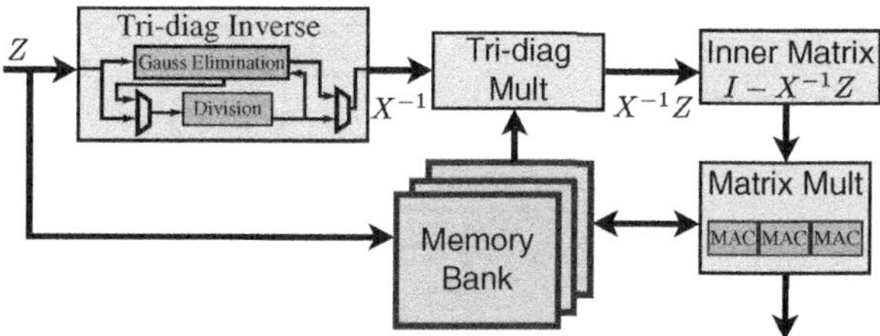

Figure 9.7 Neumann series based matrix inversion with tri-diagonal matrix as initial condition

implicit approach the inversion can be treated as the solution to linear equations. From the computational complexity point of view, the complexity has a crossover point (in terms of matrix size), after which performing explicit inversion would have lower complexity. On the other side, the implicit inversion would have a lower processing latency, since a full matrix inversion is avoided. The third approach is based on rewriting the inversion as a matrix polynomial, which leads to reduced computational complexity if the order of the polynomial is truncated with some loss in the inversion accuracy. Brief details of these three approaches are discussed in the following:

EXPLICIT INVERSION: Explicit inversions can be performed using various methods like QR-decomposition, LU-factorization, Cholesky-decomposition, Gauss-elimination, etc. These methods generally own higher complexity. Another approach for explicit inversion is the Neumann series [35]. This is a strong candidate for performing inversions since it requires only a series of matrix multiplications, which can be highly parallelized and is efficient in hardware.

One possible top-level architecture for implementing the Neumann series based ZF precoder is shown in Fig. 9.7 [36]. The special channel properties arising in Massive MIMO, e.g. the Gram matrix is diagonally dominated, can be leveraged to reduce the computational complexity without significant loss in accuracy. Computational complexity can be further optimized by balancing between the accuracy of pre-condition matrix and the convergence iterations. In the case shown in Fig. 9.7, the Neumann series convergence speed is improved by using tri-diagonal matrix as initial condition. The inversion of tri-diagonal matrix is performed directly using Gauss Elimination. The tri-diagonal multiplication is implemented in an FIR filter like structure as shown in Fig. 9.8.

IMPLICIT INVERSION: Implicit inversion can be performed by using standard linear-solvers like conjugate-gradient and coordinate-descent [37], etc. In addition to lower computational complexity, implicit inversion approaches also have a lower memory requirements. The processing latency may also be reduced, depending on the initial state and the convergence speed of the algorithms.

POLYNOMIAL EXPANSION: The inverse of any invertible $M \times M$ matrix $\boldsymbol{G}$ can be expressed as a matrix polynomial expansion of order M:

$$\boldsymbol{G}^{-1} = \boldsymbol{G}^{M} + c_{M-1}\boldsymbol{G}^{M-1} + \cdots + c_1\boldsymbol{G} + c_0\boldsymbol{I} \tag{9.31}$$

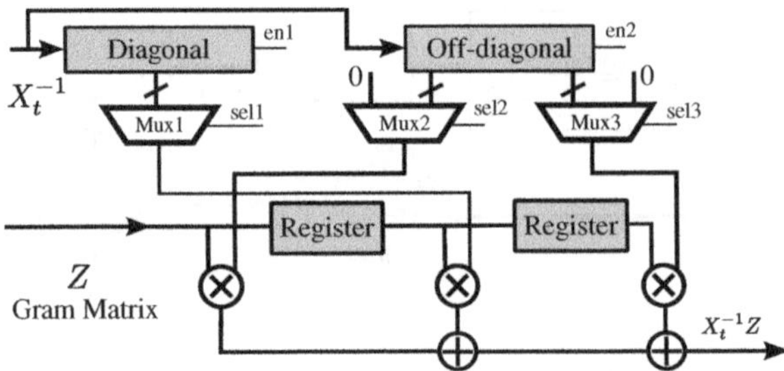

Figure 9.8 Circuit description of tri-diagonal matrix multiplication

where $c_0, \ldots, c_{M-1}$ are the coefficients of the characteristic polynomial of the matrix. This is a consequence of the Cayley-Hamilton theorem, which also provides the exact expressions for the coefficients as sums and products of the eigenvalues of $\boldsymbol{G}$. Computing all the coefficients precisely could be more complex than an explicit inversion of the matrix. Instead, various polynomial approximations are available in the literature. In these methods, only a handful of the terms in the polynomial matrix expansions are needed to obtain a good approximation of the matrix inversion. Furthermore, the number of terms can be tuned to provide a tradeoff between inversion accuracy and complexity. In the context of Massive MIMO, this approximation can be explored to large extent without sacrificing too much on the system performance.

9.4.3 Prototyping

To prove a new wireless technique, it is crucial to build testbeds to conduct verification with over-the-air transmission. For Massive MIMO it is even more important, since it heavily relies on properties of the propagation environment. While channel modeling is not mature yet for this new system, tests need to be performed in reality to give correct results. Although the aforementioned signal processing flow is standard in reciprocity-based MIMO, challenges arise in scaling M beyond conventional systems. Practical Massive MIMO requires $M \geq 64$ for the asymptotic results to begin to apply, imposing a high cost to prototype such systems. Impressive theoretical results have prompted interest in experimentally validating Massive MIMO. Ideally, the processing shown in Figure 9.5 could be specified and readily deployed in hardware. Unfortunately, scaling antennas and thereby data rate requirements an order of magnitude or more beyond conventional systems has led to challenges in availability of prototyping hardware.

To address limited availability of Massive MIMO platforms, we have developed an extensible platform to realize up to 20 MHz bandwidth 128-antenna MIMO. This platform consists exclusively of commercial off-the-shelf hardware, making it accessible and modifiable (The details of the platform will be introduced in Section 9.5). Due to the highly specific processing requirements on latency and throughput of a typical Massive MIMO architectures, the prototype has a number of unique challenges that have to be addressed:

- Phase alignment and sample level time synchronization for all the M antennas
- Triggering mechanisms which are crucial for timing synchronized transmission
- Aggregation and processing of the massive data with high precision

- Providing low-latency data transfer to support channel reciprocity
- Effective architectures for high dimension matrix multiplication, inversion, MMSE receivers etc.
- Channel reciprocity calibration, preferably at runtime.

In Section 9.5, with the help of a reference prototype, we will introduce and provide pathways to an implementation to tackle these challenges.

9.4.4 Deployment Scenarios

The potential benefits of Massive MIMO can only be fully understood when the corresponding use cases and deployment scenarios have been studied and identified. This session discusses the possible deployment scenarios for Massive MIMO in the coming 5G networks, especially those that cannot be efficiently supported by the systems existing today. In this aspect, EU FP7 Mobile and wireless communications Enablers for Twenty-twenty Information Society (METIS) project [38] has been working on definition of 5G mobile broadband scenarios. EU FP7 MAMMOET project [39] has been focusing specifically on the Massive MIMO technology with the aim to advance the development of Massive MIMO and bring initial promising concepts to a very attractive technology for deployment in future broadband mobile networks. Here, we summarize the potential scenarios for Massive MIMO, from the studies of these two EU projects.

High-performance Connectivity with Crowded UEs

One of the main benefits from Massive MIMO over conventional wireless systems is the (large-scale) multi user support, with the ability to focus transmission energy into smaller spaces. A nature deployment scenario for Massive MIMO is high UE density cases, which can be further divided into out-door and in-door scenarios. One out-door deployment example can be ***open exhibition***. In this case, outdoor-deployed base stations serve outdoor-located UEs, with high density and low mobility (e.g., up to pedestrian speeds). UEs are in principle distributed randomly, although some correlation may exist depending on UE positions. In such propagation environments, the channels can have both line-of-sight (LOS) and non-line-of-sight (NLOS) components. This scenario has high potential to demonstrate the capacity advantages of Massive MIMO, due to the low path loss and close-to-favorable propagation.

The in-door counterpart of this high-density UE scenario can be ***concert halls***, where both base stations and UEs are located indoors. UEs are also randomly distributed with high density. In this case, UEs are almost static in most cases but the possibility of having correlated UEs is higher. Both LOS and NLOS propagation conditions are expected, depending on the deployment of the base station (can be on the roof, seating in a corner, or using wall antenna array). Other examples of such scenario are office, class rooms, and indoor arenas/conference centers. The corresponding cell geometry and size are decided by the scenario boundaries and can vary significantly among cases.

Wider Connectivity with Mobility

As aforementioned, Massive MIMO is able to provide better UE separation in the spatial domain, due to its capability of focusing transmitting energy. This feature, on the other side, may also be leveraged to expand the coverage of future networks. In this scenario, cell sizes are substantially larger, such as in suburban and rural areas. The channel delay spread is longer in these scenarios. On the other side, higher UE mobility conditions are expected and should be considered. These conditions lead to relatively shorter time-domain coherence intervals, which is one of the main challenges for Massive MIMO technology, especially for TDD mode.

Massive Connectivity with Heterogeneous Buildings
Another feature provided by Massive MIMO system (base station with antenna array of many antenna elements) is the potential to cover both the horizontal and the vertical dimensions during transmission. This can be explored to compensation for the deployment scenarios that existing wireless networks are not well supported. One example can be in-door UEs are served by out-door base stations. According to [39], around 90% of today's mobile traffic comes from indoors and this trend is highly likely to increase with the explosively increase in the number of connected devices. Some of the devices are requiring ever more mobile traffic because of the evolution in multimedia technologies, for instance, ultra-high resolution videos. Massive MIMO represents a promising method to this challenges, with out-door macro base stations serving in-door UEs at different layers. In recent year, many in-door solution have been studied and proposed, including distributed antenna systems, advanced WiFi, and (ultra) small scale cells. These technologies, together with Massive MIMO, may form future heterogeneous network, providing elegant solutions to this challenging scenario.

Ubiquitous Connectivity with Dense Urban Environments
This scenario can be typical modern cities with massive and diverse mobile connections are demanded everywhere and at any time. Manhattan has been a representative example of such a scenario in many studies, while other typical big cities in Europe and Asia also capture many aspects. According to the analysis in MAMMOET project, this scenario represents the most ambitious case of providing, with an outdoor deployed base stations, uniform user experience in realistic ultra-dense urban environments. This can also be one of the most interesting cases in terms of business opportunities for operators. UEs are distributed randomly, which can be located both indoors and outdoors (e.g., residential areas, parks, shopping malls, urban public transportation, or office buildings). The UE speed can also vary in the large range with speed limitation up to approximately 30 km/h. The main challenge of this scenario is to provide same quality of experience in diverse environments and use cases. Massive MIMO provides extra degree-of-freedom for multi-user scheduling and thus will be one of the key technical solutions to tackle this challenge.

9.5 Testbed Design

Prototyping is a fundamental aspect of validating a complex system design and it is at the forefront of establishing the feasibility of the potential 5G technologies, including Massive MIMO, in a realistic deployment scenario. With the goal of demonstrating that the advantages of scaling up the number of base station antennas as postulated in the previous sections carry over to the real world, the authors have created one of the world's first fully real time Massive MIMO testbed with up to 128 antennas at the base station and up to 12 UEs over a 20 MHz channel.

We will present the design choices and architectural tradeoffs of building such a testbed with commercially available off-the-shelf radio prototyping components. Such a description will allow the next generation of implementations to be based on a sound and reusable Software Defined Radio foundation. More details of the hardware components are available here [40].

For an ideal Massive MIMO system, a potential base station (BS) architecture designed to yield low processing latency, transport latency, and high transport reliability would

- use a powerful central controller Baseband Unit (BBU) aggregating and processing data from/to 128 antennas;
- be architected in a star-like fashion, yielding hundreds of input/output ports to the Remote Radio Heads (RRH);
- shuffle large amounts of baseband data between the BBU and RRH front ends through high-bandwidth/low-latency interconnects;
- operate with hundreds of synchronized RF chains with measurable and correctable RF impairments.

While the second point imposes a tight hardware constraint, potentially preventing the flexibility and scalability of the system, the first is the toughest to meet with today's off-the-shelf solutions since 128 antennas of baseband data far exceeds the input/output (IO) capabilities of most practical hardware. Flexible implementations of Massive MIMO base stations with real-time processing requirements are thus non-trivial designs. The detailed specifications of the testbed are described in Table 9.2.

The testbed uses a fully reconfigurable frame structure. On a superficial level, the numerology is similar to LTE systems. This allows us to reuse the existing infrastructure for sampling rates, sub-carrier spacing, FFT implementations, etc. We define slots and subframes in the standard manner, but on a per OFDM symbol basis within each 10 ms frame, we can allocate each symbol to be either an uplink or downlink symbol. This is an important design consideration for a Massive MIMO system as it is sensitive to the coherence time of the channel

Table 9.2 Specifications for Massive MIMO 128-antenna testbed

Parameter	Variable	Value
# of BS antennas	M	2-128
# of UEs	K	1-12
Baseband sample rate (MS/s)	f_s	30.72
FFT size	N_{FFT}	2048
Data subcarriers (excluding DC)	N_{SC}^D	1200
Guard subcarriers	N_{SC}^G	848
Occupied channel bandwidth (MHz)	B	18.015
OFDM symbols per radio slot	N_S	7
Modulation	—	OFDMA (downlink) OFDMA (uplink)
Length of CP	L_i	160 (symbol 0), 144 (symbols 1-6)
Uplink pilot receive to downlink transmit turnaround time	T_t	$< 500\ \mu s$

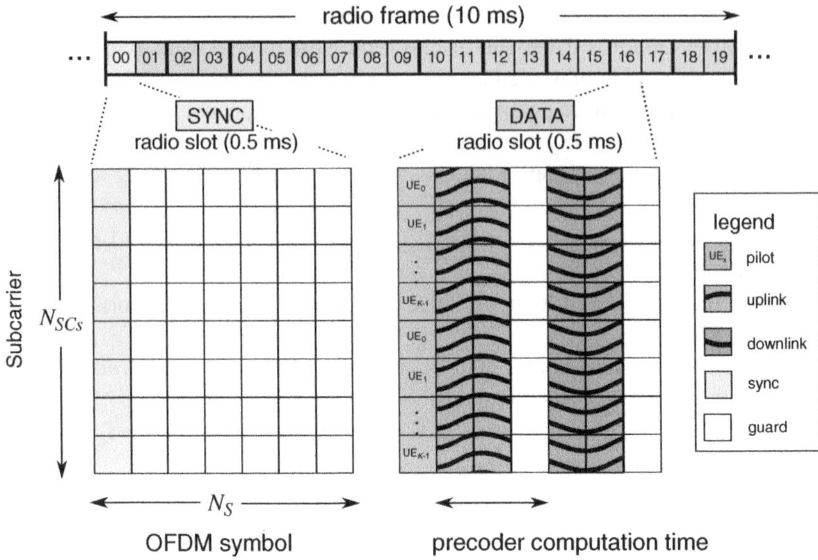

Figure 9.9 TDD radio subframe structure that supports channel reciprocity for even highly mobile channels. In a SYNC radio slot, one OFDM symbol is used for a downlink synchronization signal. The following six OFDM symbols are blank, allowing for sample shifting to align the clocks. In a DATA slot, the first OFDM symbol is uplink pilot (different subcarriers for each UE) with uplink data following shortly thereafter. Computation for the downlink precoder takes place over OFDM symbols 1-3 (indexing by zero)

for using reciprocity based precoding. Future implementations should allow exploration of flexible UL/DL periods that are adaptive to the channel coherence time. The detailed frame structure implemented is described in Figure 9.9.

9.5.1 Hierarchical Overview

Fig. 9.10 shows the hierarchical overview of our system, whose main blocks are detailed as follows:

9.5.1.1 Central Controller (CC)

An industrial form factor PXIe chassis (NI PXIe-1085) embeds a x64 controller (NI PXIe-8135), which runs LabVIEW software environment on a Windows 7 64-bit OS and serves three primary functions: (*i*) provision of user interface for radio configuration, deployment of FPGA bitfiles, system control, and the visualization of the system, (*ii*) acts as source and sink for the user data—e.g., HD video streams—sent across the links, and (*iii*) measures link quality with metrics such as bit error rate (BER), Error Vector Magnitude (EVM), and packet error rate (PER). It connects to three switches through cabled Gen 2x8 PCI Express in a star fashion.

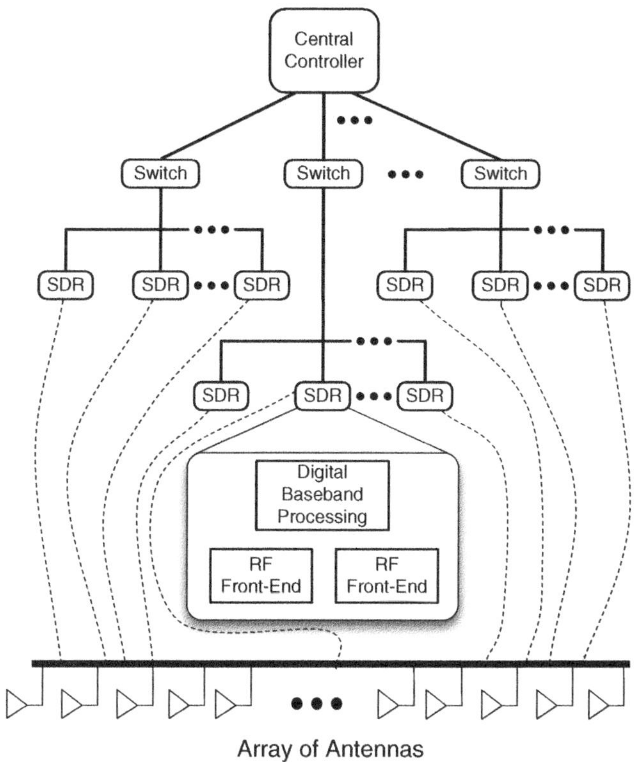

Figure 9.10 Hierarchical overview of the base station

9.5.1.2 Switches

The switches in the switching fabric consist of a PXI Express chassis backplane in three 18-slot chassis. The chassis is capable of 3.2 GB/s bidirectional per-slot bandwidth. Figure 9.11(b) shows the dual-switch backplane architecture using a very high throughput low latency link setup between the slots in the chassis. Also, multiple chassis can be deployed in a star configuration to aggregate more and more antenna data when building higher channel-count systems.

Switches yield no processing but allow data to be transferred between RRH SDRs using peer-to-peer direct memory access (DMA) streaming and between RRH SDRs and the BBU using DMA transfers.

9.5.1.3 Software Defined Radios

The RRH is implemented using a generic Software Defined Radio (NI 2943R USRP-RIO). Each RRH that we use contains a reconfigurable Xilinx Kintex-7 FPGA and two independent 40 MHz RF bandwidth transceivers that can be configured for center frequencies of 1.2-6 GHz, and can transmit with up to 15 dBm transmit power. The RF transceivers connect to the antenna

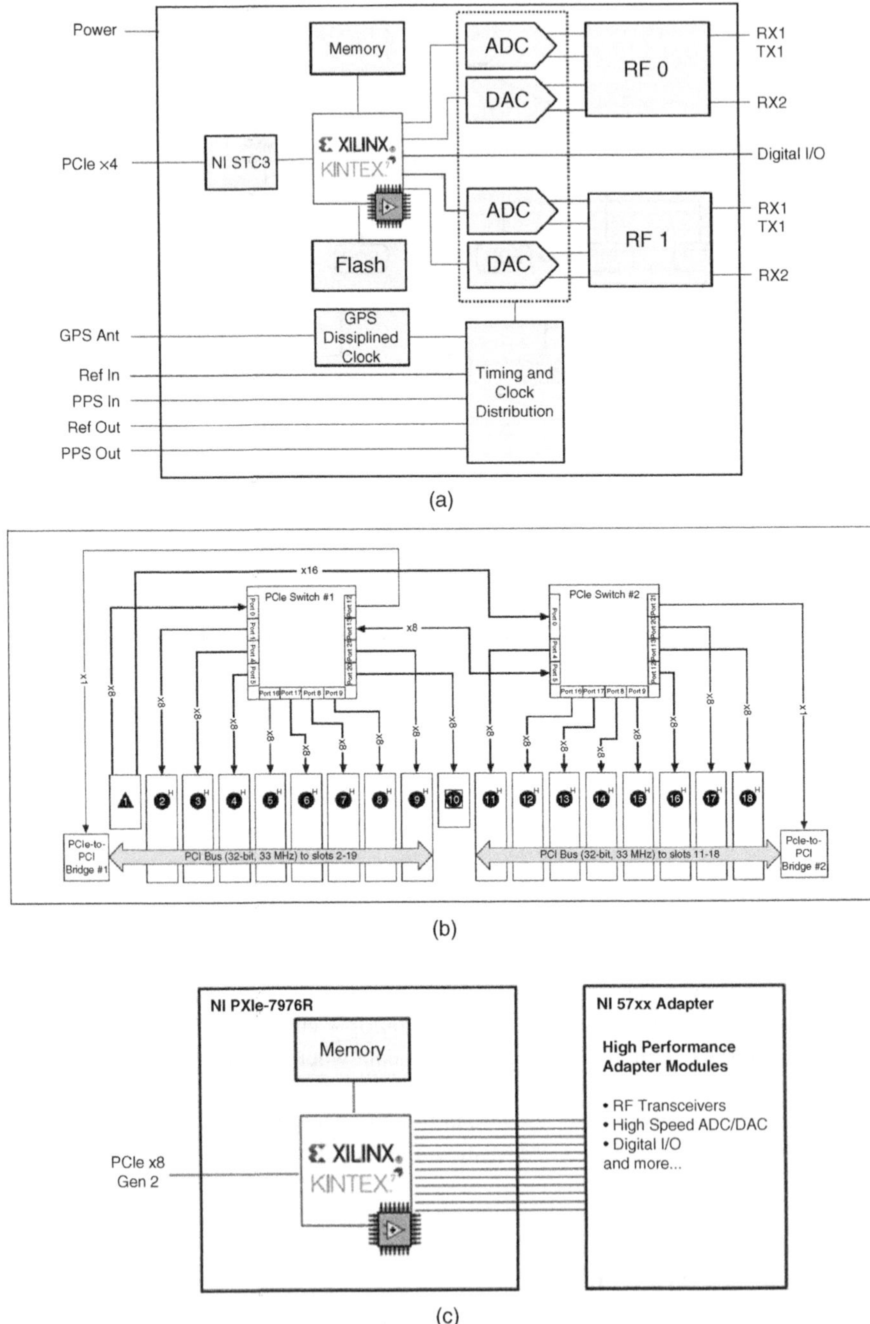

Figure 9.11 Hardware Components of the Massive MIMO testbed. (a) USRP-RIO SDR block diagram (b) 18-Slot NI PXIe-1085 chassis block diagram (c) NI PXIe-7976R FlexRIO module block diagram

Table 9.3 Detailed specifications for USRP-2943R software defined radio

System Parameter	Value
Center frequency	1.2–6.0 GHz
RF bandwidth	40 MHz
Number of RF channels per device	2
ADC sampling rate	120 MS/s
ADC resolution	14 bit
DAC sampling rate	400 MS/s
DAC resolution	16 bit
Onboard FPGA	Xilinx Kintex-7 XC7K410T
Digital backend interface	PCI-Express Gen 1 x4
Max bidirectional peer-to-peer bandwidth per device	830 MB/s
Max unique DMA endpoints per device	16

array described in Section 9.5.4. Figure 9.11(a) provides a block diagram overview of the USRP-RIO hardware. The interface between the chassis backplane and the RRHs is via cabled PCIe Gen 1x4. More hardware specifications of the RRH are described in Table 9.3.

9.5.1.4 FPGA Coprocessing Module

The aggregated data is processed in the BBU which is implemented on dedicated FPGA coprocessor modules (NI FlexRIO). Each BBU module (see Figure 9.11(c)) provides a large Xilinx Kintex-7 410T FPGA with PCIe Gen 2x8 connectivity to aggregate data.

9.5.2 *Streaming IO Rates*

For a proper baseband processing partition, the throughput constraints of the hardware components implementing the system in Fig. 9.10 include the following:

- Each Gen 2x8 PCIe interface linking the three chassis handles up to 3.2 GB/s bidirectional traffic.
- Two PCIe Gen 2x8 switches link the interface cards through the backplane of the chassis. Their streaming rate is bounded to 3.2 GB/s of bidirectional traffic between the switches.
- Each RRH has 16 available DMA channels (including one used for the radio configuration channel) that share the total IO rate for Gen 1x4 PCIe of 830 MB/s bidirectional traffic.

These limitations will inform the distribution of processing functionality across the different components of the Massive MIMO system.

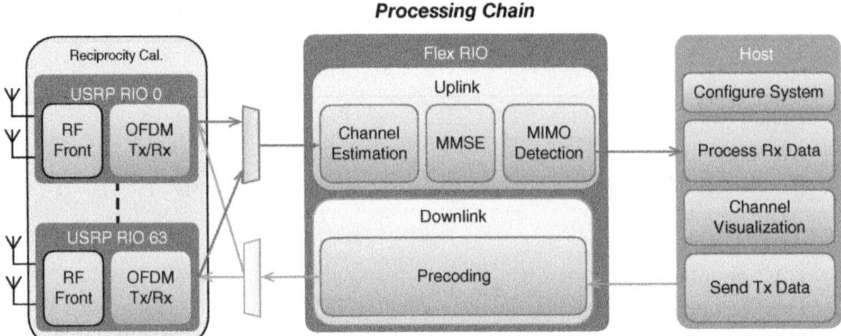

Figure 9.12 Testbed architecture with functional representation of an OFDM based Massive MIMO system

9.5.3 *Architecture and Functional Partitioning*

The testbed architecture with functional representation of an OFDM based Massive MIMO system is shown in Figure 9.12. The functional blocks are implemented on either the RRH units, the BBU unit, or the floating point host controller.

9.5.3.1 Remote Radio Head and Subsystem

On each RRH, the on-board FPGA implements the OFDM modulator and demodulator. This is on a per-antenna basis and there is a one-to-one mapping between each transceiver unit and each antenna. Thus on the boundary between the RRH and the BBU, the data is transferred in the subcarrier domain. As previously shown in Figure 9.5, for each receiver chain, the received RF signals are digitized, followed by analog front-end calibration and time/frequency synchronization. From the synchronized data, the CP is removed, followed by FFT OFDM demodulation and guard-band removal. Note that the OFDM symbols contain the superposition of the transmitted signals by all users. Reciprocity calibration and compensation is also applied at this stage, on a per-antenna basis.

To achieve a modular design, and scale the number of antennas in a parameterizable manner, the system is composed of multiple subsystems, and each subsystem consists of a maximum of 8 RRH units (16 antennas). Figure 9.13 describes a single subsystem. Since the testbed has 128 antennas, it is divided into 8 subsystems.

The wide-band subcarriers from all the antennas in a single subsystem are aggregated on one designated RRH in each subsystem (USRP RIO 0 in Figure 9.13). The aggregated subcarrier data is reshaped into four streams where each stream contains a contiguous number of subcarriers that correspond to one fourth of the wide-band occupied subcarrier bandwidth. Thus, we parallelize the MIMO processor into multiple subbands. Similar processing split occurs on the downlink for the precoded data.

9.5.3.2 MIMO Processor

Unfortunately, large MIMO detectors and many channels of processing can easily exceed the size of many current state-of-the-art FPGAs. For reference, a reference design produced

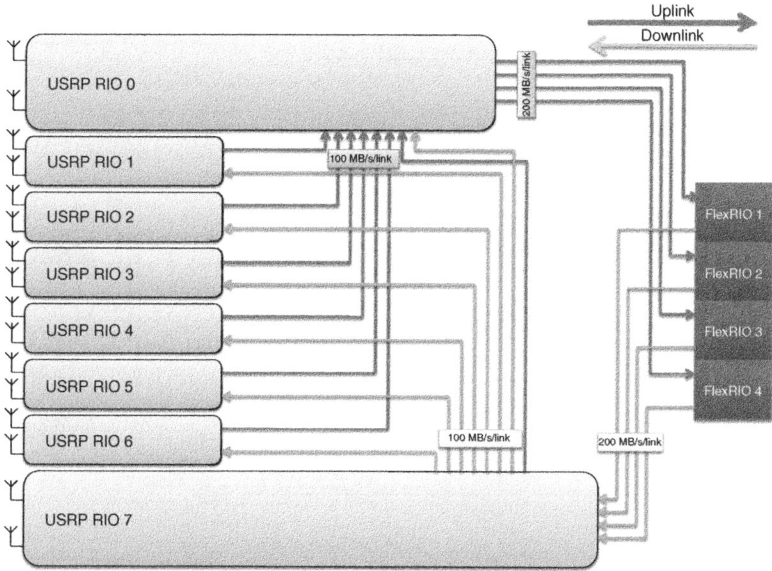

Figure 9.13 An 8-FPGA (16-antenna) subsystem used to build up to a 128-antenna system. Each arc represents data paths inter- or intra-FPGA. Inter-FPGA paths use a DMA interface implemented over PXI Express

by Xilinx achieves 2×2 MIMO over 20 MHz with the full LTE physical layer for uplink and downlink. Even with only 2×2 MIMO, the resource usage is 544 DSP blocks, 151,800 Lookup Tables (LUTs), and 923 18 kbit block RAMs. This is an appreciable percentage of even large FPGA targets such as the Kintex-7 XC7K410T, where this translates to 544/1540 = 35.3% of available DSPs, 151,800/254,200 = 59.7% of available LUTs, and 923/1590 = 58.1% of available 18 kbit block RAMs [41]. Massive MIMO systems must process much larger matrix sizes (e.g. 128x12 in the testbed), due to the extremely high channel count, with stricter latency constraints. To meet these system rate requirements and finite resource constraints, the MIMO processor signal processing is performed on each of the 4 subbands in parallel since after OFDM processing, the signals are separable and mutually orthogonal in frequency. Each subband corresponds to a "bandwidth chunk" of the total system bandwidth.

The BBU is comprised of four coprocessor FPGAs (FlexRIOs) that we have previously described. Each BBU module processes one fourth of the total system bandwidth. The overall implementation can be seen in Fig. 9.15.

The MIMO Detector block collects the data of a given subband (bandwidth chunk) from all the subsystems. Using the channel matrix estimated from uplink pilots, the MIMO Detector cancels interference and detects the frequency-domain symbols from each user equipment. The testbed software has highly optimized MMSE, ZF and MRC implementations which can efficiently do the MIMO decoding matrix operations. The detected symbols are then sent to the host controller for further processing, such as link quality evaluation.

At the downlink, the channel estimates and estimated reciprocity calibration weights are passed to the MIMO Precoder block, and reciprocal processing is performed, e.g., modulation instead of demodulation. The MIMO processor has to aggregate data from all the antennas

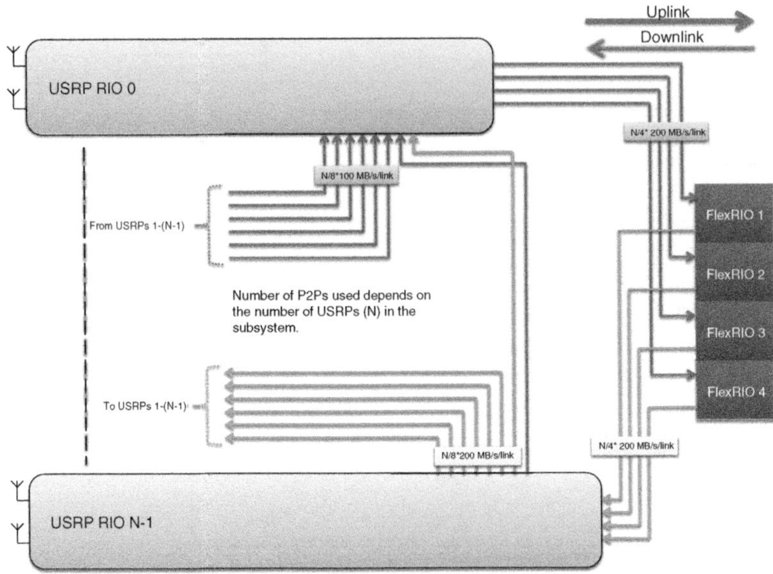

Figure 9.14 Inter subsystem data communication using FIFOs for upto 8 subsystems

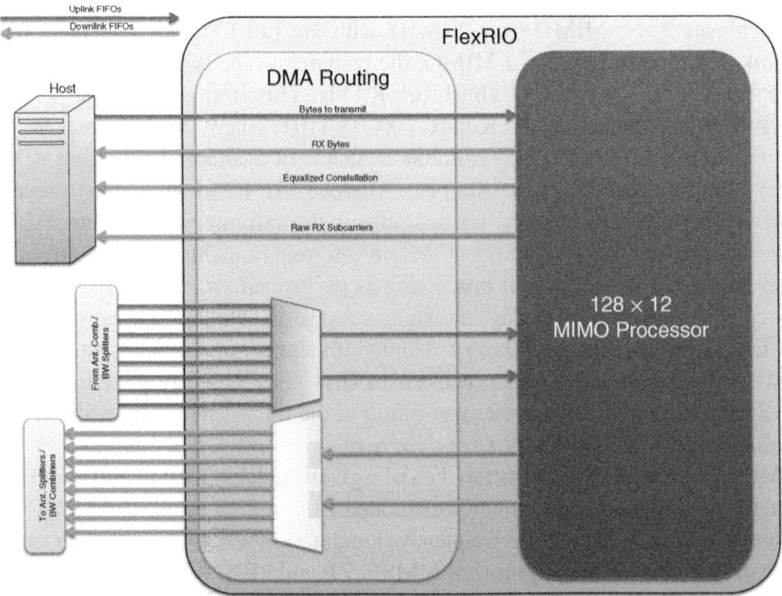

Figure 9.15 MIMO Processor implementation

in the system. Each subcarrier data sample is quantized with 12 bits for each in-phase and quadrature component.

The subsystems, each of which implement the architecture described above, are inter connected as shown in Figure 9.14.

9.5.4 Antenna Array

In principal, the testbed can be used with any antenna array. Here, we describe the antenna array developed for the Lund Massive MIMO (LuMaMi) testbed [42]. In the following, the three different stages of the array building process are described.

9.5.4.1 Material and Characterization

We chose Diclad 880 with a thickness of 3.2 mm as the printed circuit board substrate. The dielectric constant and dissipation factor were confirmed using a trapped waveguide characterization method [43]. To verify the substrate characterization, a 6-element patch array with slightly different element sizes was built, measured, and compared with the simulated data. To fit the final results, a final re-characterization of the substrate was performed, and the simulated and measured bandwidth matched within 1 MHz.

9.5.4.2 Design

A planar T-shaped antenna array was built with 160 dual polarized $\lambda/2$ shorted patch elements. The T upper horizontal rectangle has 4×25 elements and the central square has 10×10 elements (see Fig. 9.16 right). This yields 320 possible antenna ports that can be used to explore different antenna array arrangements. All antenna elements are center shorted, which improves isolation and bandwidth and reduces the risk of static shock traveling into the active components if the elements encounter a static electric discharge. The feed placement shifts by 0.52 mm from the center of the array elements to the outer edge elements in order to maintain a match with changing array effects that impact individual elements differently. The size of the element changes by 0.28 mm from the center of the array to the outer elements. This maintains a constant center frequency of 3.7 GHz throughout the entire array.

9.5.4.3 Measurements

The final 160-element array was simulated at 3.7 GHz. Results showed an average match of -51 dB and an average 10dB-bandwidth of 185 MHz. Similar tests were done to the manufactured array, which yielded an average 10dB-bandwidth of 183 MHz centered at 3.696 GHz and the average antenna match was found to be −28 dB.

9.5.5 Mechanical Structure and Electrical Characteristics of LuMaMi Testbed

Two rack mounts contain all BS components with the combined measures of $0.8 \times 1.2 \times 1$ m shown in Fig. 9.16. They are attached on top of a 4-wheel trolley to avoid compromising

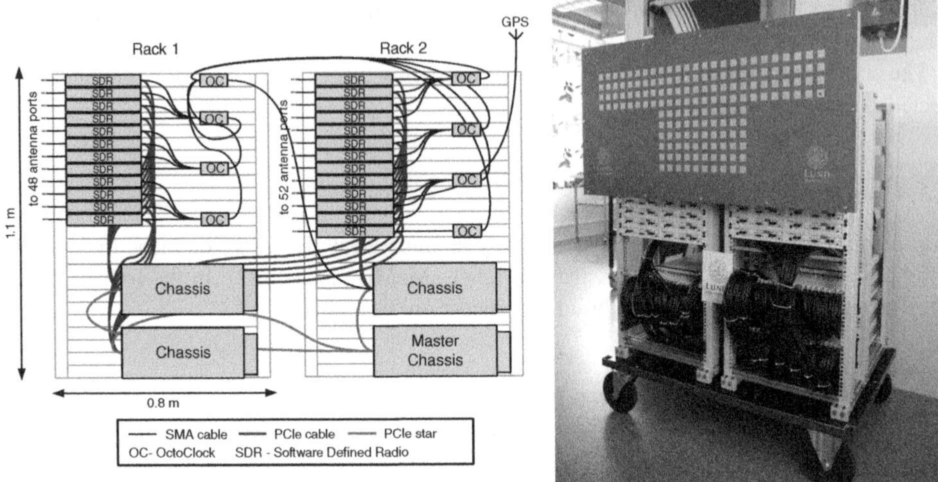

Figure 9.16 Left: Side view of the mechanical assembly of the BS. The two racks sit side by side (not as shown) with the SDRs facing the same direction (towards the antenna array). Two columns of URSP SDRs are mounted in each rack, totaling 50 of them. Right: The assembled LuMaMi testbed at Lund University, Sweden

its mobility when testing different scenarios. The approximate combined weight and average power consumption are 300 kg and 2.5 kW, respectively.

9.6 Synchronization

A MIMO base station requires time synchronization and phase coherence between each RF chain. This is achieved using a reference clock and trigger distribution network. The reference clocks are used as the source of each radio local oscillator. This provides phase coherence among devices. The trigger signal is used to provide a time reference to all the radios in the system. All FPGAs receive this time-aligned trigger so processing starts simultaneously. State synchronization is ensured by invoking a state reset on all devices prior to the trigger event.

9.6.1 Types of Synchronization

Due to the TDD nature of the system, time across all radios must be synchronized to within the accuracy of 1 sample of the A/D and D/A converters. There are two common methods for radio synchronization that include time based synchronization using a periodic pulse per second, and signal synchronization based on a shared trigger architecture. Each presents certain advantages but also challenges on the system architecture.

9.6.1.1 Time-based Synchronization

Time-based synchronization requires two shared signals, a 10 MHz reference and pulse per second (PPS) signal. The 10 MHz reference provides a periodic 100 ns pulse with 50% duty cycle that disciplines and increments an internal time-keeping register within the FPGA. The periodic PPS signal is then used for establishing the beginning of a second. A command is typically issued to each radio to set a common time on the next rising edge of PPS, establishing a common time base among the radios. The shared time can be set to an arbitrary time or Coordinated Universal Time (UTC) queried from a GPS. Commands can then be issued to the radios for future events based on the time register. The advantages of this approach include the ability to synchronize one or many devices using industry standard, periodic signals provided by many commercial off-the-shelf GPS-disciplined oscillators such as the Trimble Thunderbolt or Ettus Research OctoClock [40]. The first challenge of time-based synchronization is the start-up calibration procedure coordinated by a host controller to establish a common time among devices. This calibration would be required each time the system is started up. The second is the latency introduced by the clock cycles that are necessary to load a future event into the system, which prevents the immediate triggering of events.

9.6.1.2 Trigger-based Synchronization

Trigger-based synchronization utilizes a shared event trigger for event execution with no prior knowledge of the system time in each radio. A shared 10 MHz reference then keeps time among devices through the deterministic execution of the code across FPGAs rather than storing the time in a register. A master provides an output digital trigger that is amplified and divided among all the radios. Upon receipt of the rising edge of the event trigger, each radio executes the specified event. The advantages of this approach include the ability to immediately trigger as well as simplified FPGA design in each radio. The challenges include the requirement for a master radio to coordinate the start trigger and the fact that there is no reference to real-world GPS or UTC time.

9.6.1.3 Frequency and Phase Synchronization

In addition to providing a reference for system time synchronization, a shared 10 MHz oven-controlled crystal oscillator (OCXO) reference provides frequency disciplining for the internal voltage-controlled oscillator (VCO) used to generate the system local oscillator and A/D and D/A channels, both of which must be synchronized across the entire array. This ensures subsample synchronization among radios for both transmit and receive operations. However, because each radio has an independent PLL - VCO disciplining circuit for both transmit and receive, the system can be considered phase coherent but not phase aligned. Some clock-deriving chips do enable a sync pin, but generally channels must be considered coherent and not aligned. However, through periodic calibration, alignment can be achieved by digitally adjusting the real and imaginary signal component (I and Q) phase in software.

9.6.1.4 Testbed Synchronization Implementation

The hardware used to build the clock and trigger distribution network is available commercially off the shelf. This consists of nine OctoClock modules in a tree structure with a top OctoClock feeding 8 lower OctoClocks (see Figure 9.17). Figure 9.18 shows a system overview of the

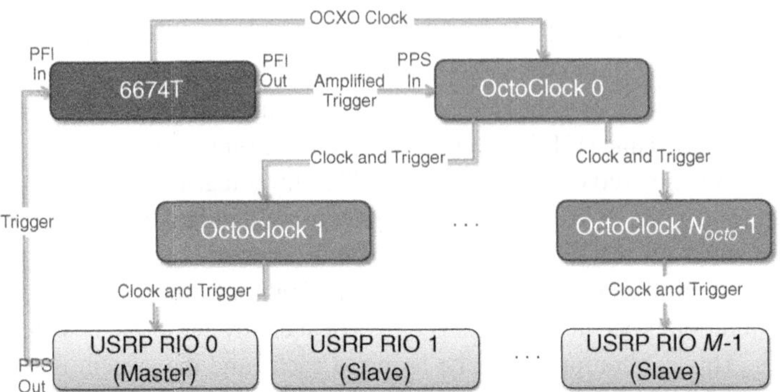

Figure 9.17 Clocks and reference signals are routed to each NI USRP-2943R device using this timing and synchronization tree. The master *USRP RIO 0* sends a trigger signal that is fed into the NI PXIe-6674T module. The NI PXIe-6674T provides a conditioned and amplified trigger along with an OCXO reference clock. These signals are fed to the top OctoClock, which fans out these signals to the lower branch of OctoClocks

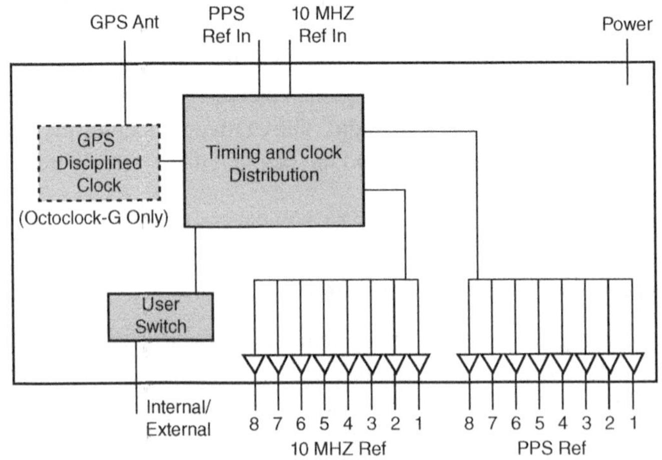

Figure 9.18 OctoClock-G Module block diagram

OctoClock. Low-skew buffering circuits and matched-length transmission cables ensure that there is low skew between the reference clock input at each USRP RIO. The source clock for the system is an OCXO within an NI PXIe-6674T timing module [40]. Triggering is achieved by instigating a start pulse within the master USRP RIO via a software trigger. This trigger is then outputted from an output port on the master and inputted to the NI PXIe-6674T, which conditions and amplifies the trigger. The trigger is then propagated to the top OctoClock and distributed down the tree to each USRP device in the system (including the master itself). This signal sets the reference clock edge to use for the start of acquisition for the transmit

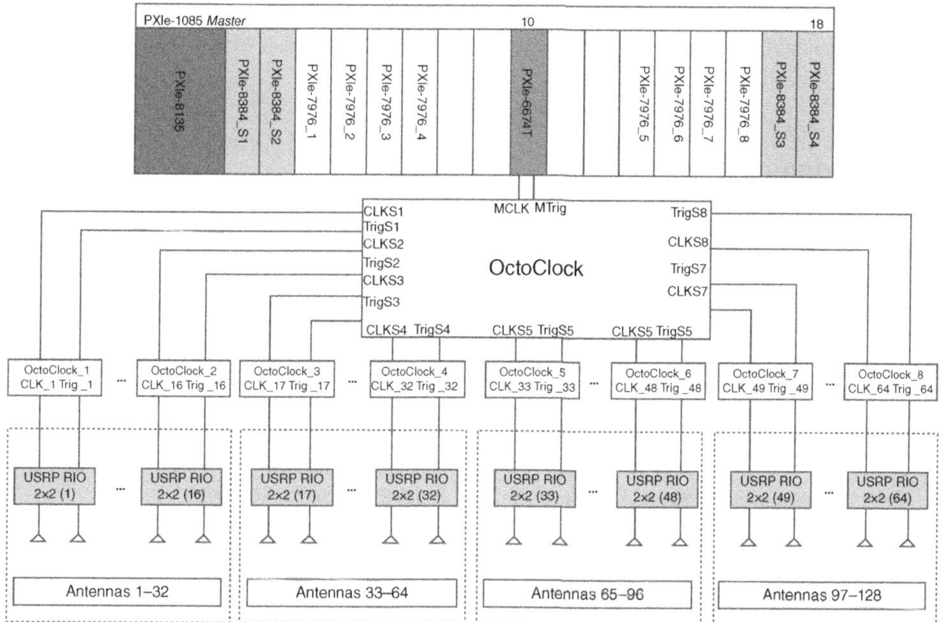

Figure 9.19 Massive MIMO clock distribution diagram

and receive within each channel. Initial results show that the reference clock skew is within 100 ps and the trigger skew is within 1.5 ns. Additional signal alignment calibration is done to digitally compensate for front-end delay. This ensures the TDD signals are properly aligned over the air. This further allows the design to be modularly used on top of any radio hardware. The complete system architecture for the distribution of clocks and triggers in the Massive MIMO testbed is shown in Figure 9.19.

9.7 Future Challenges and Conclusion

Although successful Massive MIMO prototype systems have been demonstrated, many challenges confront their path to deployment. Increased interconnection, processing, and overall system complexity and cost are some of the few challenges that must be addressed. Fortunately, new classes of highly integrated Radio Frequency Integrated Circuits (RFICs) could be adopted to reduce size and complexity. New lower cost interconnects and cabling for synchronization and data buses could be adopted to build lower cost and more compact systems. In addition, highly scalable baseband processing strategies must be developed. Addressing these challenges is essential to realizing deployed Massive MIMO systems. The most critical, however, is support from the wireless standardization bodies to adopt and specify these systems. A combination of innovative technical solutions and support from standardization bodies will allow Massive MIMO to be realized in practice.

Massive MIMO will play a crucial role in 5G systems as spectral and radiated power efficiency become increasingly critical. Initial theoretical and implementation work has led to promising results. Additional challenges remain on the path to commercial viability. Testbeds will play a crucial role in addressing these challenges and help push Massive MIMO systems from theory to reality.

Acknowledgments

The authors would like the acknowledge the contributions of co-authors Steffen Malkowsky and Joao Vieira from Lund University and Ian Wong from National Instruments for their work preparing the chapter. The authors would like to additionally thank Hemanth Prabhu for his contributions to Section 9.4.2, Zachary Miers for contributions to Sections 9.5.4.1–9.5.4.3, as well as Ghassan Dahman, Jose Flordelis and Xiang Gao for their contributions to Section 9.3.

References

[1] Marzetta, T. (2010) Noncooperative cellular wireless with unlimited numbers of base station antennas. *IEEE Trans, Wireless Commun.*, **9** (11), 3590–3600, doi:10.1109/TWC.2010.092810.091092.

[2] MAMMOET Project (2014–2016), Massive MIMO for efficient transmission, EU FP7 FP7-ICT-2013-11 programme, project ref. 619086. URL http://mammoet-project.eu/.

[3] Hoydis, J., Hosseini, K., Brink, S.t., and Debbah, M. (2013) Making smart use of excess antennas: massive MIMO, small cells, and TDD. *Bell Labs Tech. J.*, **18** (2), 5–21.

[4] Larsson, E.G., Edfors, O., Tufvesson, F., and Marzetta, T.L. (2014) Massive MIMO for next generation wireless systems. *IEEE Commun. Mag.*, **52** (2), 186–195. URL http://arxiv.org/abs/1304.6690.

[5] Nam, Y.H., Ng, B.L., Sayana, K., Li, Y., Zhang, J., Kim, Y., and Lee, J. (2013) Full-dimension MIMO (FD-MIMO) for next generation cellular technology. *IEEE Commun. Mag.*, **51** (6), 172–179.

[6] Costa, M. (1983) Writing on dirty paper. *IEEE Trans. Intell. Transp. Syst.*, **IT-29** (3), 439–441.

[7] Ngo, H.Q., Larsson, E.G., and Marzetta, T.L. (2014) Aspects of favorable propagation in massive MIMO, in *European Signal Processing Conference (EUSIPCO), Proceedings of the 22nd*, pp. 76–80. URL http://ieeexplore.ieee.org/stamp/stamp.jsp?arnumber=6951994.

[8] Hoydis, J., ten Brink, S., and Debbah, M. (2013) Massive MIMO in the UL/DL of cellular networks: how many antennas do we need? **31** (2), 160–171, doi:10.1109/JSAC.2013.130205. URL http://ieeexplore.ieee.org/stamp/stamp.jsp?arnumber=6415388.

[9] Bjornson, E., Hoydis, J., Kountouris, M., and Debbah, M. (2014) Massive MIMO systems with non-ideal hardware: energy efficiency, estimation, and capacity limits. *IEEE J. Info. Theory*, **60** (11), 7112–7139, doi:10.1109/TIT.2014.2354403. URL http://ieeexplore.ieee.org/stamp/stamp.jsp?arnumber=6891254.

[10] Björnson, E., Bengtsson, M., and Ottersten, B. (2014) Optimal multiuser transmit beamforming: a difficult problem with a simple solution structure. *IEEE Signal Process. Mag.*, **31** (4), 142–148, doi:10.1109/MSP.2014.2312183.

[11] Boche, H. and Schubert, M. (2002) A general duality theory for uplink and downlink beamforming, in *Proc. IEEE VTC-Fall*, pp. 87–91.

[12] Viswanath, P. and Tse, D. (2003) Sum capacity of the vector Gaussian broadcast channel and uplink-downlink duality. *IEEE J. Info. Theory*, **49** (8), 1912–1921.

[13] Björnson, E. and Jorswieck, E. (2013) Optimal resource allocation in coordinated multi-cell systems. *Found. Trends Commun. Inf. Theory*, **9** (2-3), 113–381.

[14] Bjornson, E., Hoydis, J., Kountouris, M., and Debbah, M. (2014) Massive MIMO systems with non-ideal hardware: energy efficiency, estimation, and capacity limits. *IEEE Trans. Info. Theory*, **60** (11), 7112–7139.

[15] Truong, K. and Heath, R. (2013) Effects of channel aging in massive MIMO systems. *J. Commun. Networks*, doi:10.1109/JCN.2013.000065.

[16] Edfors, O., Sandell, M., van de Beek, J.J., Wilson, S.K., and Borjesson, P.O. (1998) OFDM channel estimation by singular value decomposition. **46** (7), 931–939, doi:10.1109/26.701321. URL http://ieeexplore.ieee.org/stamp/stamp.jsp?arnumber=701321.

[17] Müller, R., Vehkaperä, M., and Cottatellucci, L. (2013) Blind pilot decontamination, in *Proc. ITG Workshop on Smart Antennas (WSA)*.

[18] Ngo, H.Q., Marzetta, T., and Larsson, E. (2011) Analysis of the pilot contamination effect in very large multicell multiuser MIMO systems for physical channel models, in *IEEE International Conference on Acoustics, Speech and Signal Processing (ICASSP)*, pp. 3464–3467, doi:10.1109/ICASSP.2011.5947131.

[19] Ashikhmin, A. and Marzetta, T. (2012) Pilot contamination precoding in multi-cell large scale antenna systems, in *Information Theory Proceedings (ISIT), 2012 IEEE International Symposium on*, pp. 1137–1141.

[20] Yin, H., Gesbert, D., Filippou, M., and Liu, Y. (2013) A coordinated approach to channel estimation in large-scale multiple-antenna systems. *IEEE J. Select. Areas Commun.*, **31** (2), 264–273.

[21] Payami, S. and Tufvesson, F. (2012) Channel measurements and analysis for very large array systems at 2.6 GHz, in *Antennas and Propagation (EUCAP), 2012 6th European Conference on*, pp. 433–437, doi:10.1109/EuCAP.2012.6206345.

[22] Gao, X., Edfors, O., Tufvesson, F., and Larsson, E.G. (2015) Massive MIMO in real propagation environments: do all antennas contribute equally? *IEEE Wireless Commun.* In press.

[23] Flordelis, J., Gao, X., Dahman, G., Rusek, F., Edfors, O., and Tufvesson, F. (2015) Spatial separation of closely-spaced users in measured massive multi-user MIMO channels, in *IEEE International Conference on Communications (ICC)*, IEEE.

[24] MAMMOET Project (2015) Deliverable D1.2: MaMi channel characteristics: measurement results, *Tech. Rep.*, MAMMOET.

[25] Martinez, A.O., De Carvalho, E., and Nielsen, J.O. (2014) Towards very large aperture massive MIMO: a measurement based study, in *Globecom Workshops (GC Wkshps), 2014*, IEEE, pp. 281–286.

[26] Wu, S., Wang, C.X., Aggoune, E.H., Alwakeel, M., and He, Y. (2014) A non-stationary 3-D wideband twin-cluster model for 5G Massive MIMO channels. *IEEE J. Select Areas Commun.*, **32** (6), 1207–1218, doi:10.1109/JSAC.2014.2328131.

[27] Liu, L., Oestges, C., Poutanen, J., Haneda, K., Vainikainen, P., Quitin, F., Tufvesson, F., and Doncker, P. (2012) The COST 2100 MIMO channel model. *IEEE Wireless Commun.*, **19** (6), 92–99, doi:10.1109/MWC.2012.6393523.

[28] Kyösti, P., Meinilä, J., Hentilä, L., Zhao, X., Jämsä, T., Schneider, C., Narandzić, M., Milojević, M., Hong, A., Ylitalo, J. *et al.* (2008) Winner II channel models, in *Radio Technologies and Concepts for IMT-Advanced* (eds M. Döttling, W. Mohr, and A. Osseiran), John Wiley.

[29] 3GPP Project (2015) Study on 3D channel model for LTE, *Tech. Rep. TR 36.873 V12.1.0*, 3GPP.

[30] Gustavsson, U., Sanchéz-Perez, C., Eriksson, T., Athley, F., Durisi, G., Landin, P., Hausmair, K., Fager, C., and Svensson, L. (2014) On the impact of hardware impairments on massive MIMO, in *Globecom Workshops (GC Wkshps), 2014 IEEE*, pp. 294–300.

[31] Shepard, C., Yu, H., Anand, N., Li, E., Marzetta, T., Yang, R., and Zhong, L. (2012) Argos: practical many-antenna base stations, in *Proceedings of the 18th Annual International Conference on Mobile Computing and Networking*, ACM, pp. 53–64.

[32] Rogalin, R., Bursalioglu, O.Y., Papadopoulos, H., Caire, G., Molisch, A.F., Michaloliakos, A., Balan, V., and Psounis, K. (2014) Scalable synchronization and reciprocity calibration for distributed multiuser MIMO. *IEEE Trans. Wireless Commun.*, **13** (4), 1815–1831.

[33] Vieira, J., Rusek, F., and Tufvesson, F. (2014) Reciprocity calibration methods for massive MIMO based on antenna coupling, in *Global Communications Conference (GLOBECOM), 2014 IEEE*, pp. 3708–3712, doi:10.1109/GLOCOM.2014.7037384. URL http://ieeexplore.ieee.org/stamp/stamp.jsp?arnumber=7037384.

[34] Nieman, K.F. (2014) Space-time-frequency methods for interference-limited communication systems, Ph.D. thesis, University of Texas at Austin.

[35] Prabhu, H., Rodrigues, J., Edfors, O., and Rusek, F. (2013) Approximative matrix inverse computations for very-large MIMO and applications to linear pre-coding systems, in *Wireless Communications and Networking Conference (WCNC), 2013 IEEE*, pp. 2710–2715, doi:10.1109/WCNC.2013.6554990. URL http://ieeexplore.ieee.org/stamp/stamp.jsp?arnumber=6554990.

[36] Prabhu, H., Edfors, O., Rodrigues, J., Liu, L., and Rusek, F. (2014) Hardware efficient approximative matrix inversion for linear pre-coding in massive MIMO, in *Circuits and Systems (ISCAS), 2014 IEEE International Symposium on*, pp. 1700–1703, doi:10.1109/ISCAS.2014.6865481.

[37] Prabhu, H., Edfors, O., Rodrigues, J., Liu, L., and Rusek, F. (2014) A low-complex peak-to-average power reduction scheme for OFDM based massive MIMO systems, in *Communications, Control and Signal Processing (ISCCSP), 2014 6th International Symposium on*, pp. 114–117, doi:10.1109/ISCCSP.2014.6877829.

[38] METIS Project (2013) Deliverable D1.1: Scenarios, requirements and KPIs for 5G mobile and wireless systems, *Tech. Rep.*, METIS.

[39] MAMMOET Project (2014) Deliverable D1.1: System scenarios and requirements specifications, *Tech. Rep.*, MAMMOET.

[40] Luther, E. (2014), 5G massive MIMO testbed: From theory to reality. URL http://www.ni.com/white-paper/52382/en/.

[41] Xilinx (2013), LTE Small Cell Baseband Solutions, http://www.xilinx.com/publications/prod_mktg/lte_small_cell_baseband.pdf.

[42] Vieira, J., Malkowsky, S., Nieman, K., Miers, Z., Kundargi, N., Liu, L., Wong, I., Owall, V., Edfors, O., and Tufvesson, F. (2014) A flexible 100-antenna testbed for massive MIMO, in *Globecom Workshops (GC Wkshps), 2014*, pp. 287–293, doi:10.1109/GLOCOMW.2014.7063446. URL http://ieeexplore.ieee.org/stamp/stamp.jsp?arnumber=7063446.

[43] Namba, A., Wada, O., Toyota, Y., Fukumoto, Y., Wang, Z.L., Koga, R., Miyashita, T., and Watanabe, T. (2001) A simple method for measuring the relative permittivity of printed circuit board materials. *IEEE Trans. Electromag. Compat.*, **43** (4), 515–519, doi:10.1109/15.974630.

10

Millimeter-Wave MIMO Transceivers: Theory, Design and Implementation

Akbar M. Sayeed and John H. Brady

Signal Processing for 5G: Algorithms and Implementations, First Edition. Edited by Fa-Long Luo and Charlie Zhang.

10.1 Introduction

Wireless communication technology is approaching a spectrum crunch with the proliferation of data-hungry applications enabled by broadband mobile wireless devices, such as smartphones, laptops and tablets [1–4]. On the one hand, the data-hungry applications will soon outgrow the megabits/sec speeds offered by current networks and require aggregate data transport capability of tens or hundreds of gigabits/sec (Gbps) [5–8]. On the other hand, the wireless electromagnetic spectrum is essentially limited, due to technological, physical and regulatory constraints [5, 9]. This has led to new approaches for the efficient use of the available spectrum, such as advanced interference management techniques [10, 11], multi-antenna technology [12–16], cognitive radio [17–19] and the current industry approach of *small-cell* technology to maximize the spatial reuse of the limited spectrum [4, 20]. However, despite these advances there is a growing realization that current wireless networks, operating at frequencies below 5 GHz, will not be able to meet the growing bandwidth requirements – there simply is not enough physical spectrum [5–9, 21].

Two technological trends offer new synergistic opportunities for meeting the exploding bandwidth requirements in 5G wireless. First, millimeter-wave (mmW) operating in the 30–300 GHz band offer orders-of-magnitude larger chunks of physical spectrum [5, 6, 9, 21–24]. The second trend involves multi-antenna MIMO (multiple input, multiple output) transceivers that exploit the spatial dimension to significantly enhance capacity and reliability over traditional single-antenna systems [12–16]. MIMO systems represent a particularly promising opportunity at mmW frequencies: high-dimensional MIMO operation [25–27] is possible with physically compact antennas due to small wavelengths. The large number of MIMO degrees of freedom can be exploited for a number of critical capabilities, including [5, 6, 25, 26, 28–32]:

- higher antenna or beamforming gain for enhanced power efficiency
- higher spatial multiplexing gain for enhanced spectral efficiency
- highly directional communication with narrow beams for enhanced spatial reuse.

However, despite intensive MIMO research in the last fifteen years [12–16, 33–35], conventional state-of-the-art approaches [36–42] fall significantly short of harnessing these opportunities because of their failure to address the fundamental performance–complexity tradeoffs inherent to high-dimensional MIMO systems. In particular, the hardware complexity of the mmW beamforming front-end and the software complexity of the back-end digital signal processing challenge the current paradigm and result in significant technology gaps that require a fresh look at the design of the high-dimensional spatial analog–digital interface.

In this chapter, we present a framework for the design, analysis, and practical implementation of a new MIMO transceiver architecture: continuous aperture phased MIMO (CAP-MIMO). This combines the directivity gain of traditional antennas, the beam-steering capability of phased arrays, and the spatial multiplexing gain of MIMO systems to realize the multi-Gbps capacity potential of mmW technology as well as the unprecedented operational functionality of dynamic multi-beam steering and data multiplexing [25, 26, 29]. It is based on the concept of beamspace MIMO communication – multiplexing data into multiple orthogonal spatial beams – which it uses to optimally exploit the spatial antenna dimension

[25, 26, 35, 43]. In order to realize the full potential of mmW technology, the CAP-MIMO transceiver leverages beamspace MIMO theory through two key elements

- a lens-based front-end antenna for analog beamforming
- a multi-beam selection architecture that enables joint hardware–software optimization of transceiver complexity.

As a result, CAP-MIMO promises significant advantages over competing technologies, including:

- significant improvements in capacity, power efficiency and functionality.
- optimum performance with the lowest transceiver complexity – the number of transmit/receive (T/R) modules, and digital signal processing (DSP) complexity.
- electronic multi-beam steering and data multiplexing (MBDM) capability.

CAP-MIMO also offers a broad application footprint spanning point-to-point (P2P) and point-to-multipoint (P2MP) network operation in line of sight (LoS) and/or multipath (MP) propagation conditions [25, 26, 30, 44, 45].

10.1.1 Millimeter-Wave MIMO Technology: Background and Promise

Advances in device, integrated circuit and antenna technology are enabling mmW wireless communication [9, 21, 24, 46–48]. Recently, mmW wireless backhaul systems[1] (20–90 GHz) have emerged as a promising alternative to traditional fiber-based solutions for connecting a local enterprise network to the wired backbone. Emerging mmW systems (60 GHz) are also being envisioned for delivering multi-Gbps speeds in indoor applications, such as HDTV, and at smaller scales [20, 47, 49–53], as well as for mobile broadband networks (30–90 GHz) [5, 6, 9, 21, 24, 28] as part of the emerging vision for 5G.

The current trend for increasing the spectral efficiency of cellular wireless (below 5 GHz) is to increase spatial reuse of spectrum through smaller cell sizes [4, 20, 53]. However, this approach is inherently limited by the available spectrum and introduces new challenges, such as making the inter-cell interference more pronounced. Figure 10.1 illustrates the potential of beamspace CAP-MIMO transceivers to provide a strong complement to small-cell technology by exploiting the large number of spatial degrees of freedom at mmW frequencies. For a given antenna size, the large antenna directivity gains shown in Figure 10.1(a) more than offset the increased path loss and atmospheric absorption at mmW frequencies compared to lower frequencies. The extremely narrow beamwidths at mmW frequencies, illustrated in Figure 10.1(b), offer dramatically enhanced spatial reuse through dense spatial multiplexing: the spectrum resources can be reused across distinct beams. The resulting spatial multiplexing gain promises unprecedented gains in spectral efficiency. This is illustrated in Figure 10.1(c), which shows idealized spectral efficiency

[1] For example, Bridgewave Communications (http://www.bridgewave.com) and Siklu Communications (http://www.siklu.com).

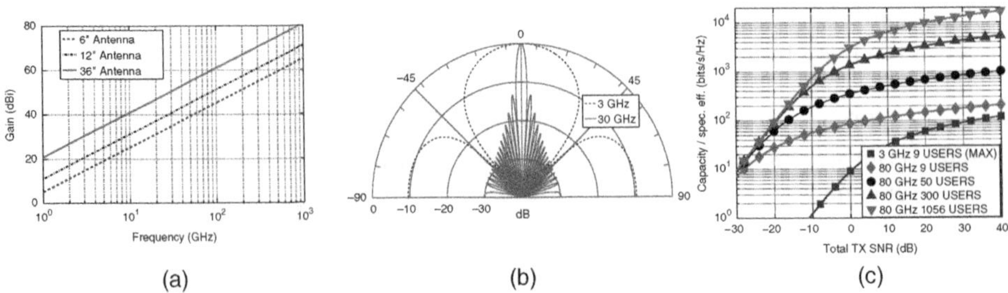

Figure 10.1 The potential of beamspace CAP-MIMO. (a) Antenna gain vs frequency, (b) antenna beam patterns for a 6″ antenna at 3 GHz vs 30 GHz, (c) potential spectral efficiency gains due to spatial multiplexing at 80 GHz vs 3 GHz with a 6″ antenna

(bits/s/Hz) upper bounds for downlink communication from an access point (AP) with a 6″ × 6″ antenna [30]: while at 3 GHz a maximum of 9 users can be spatially multiplexed, at 80 GHz orders-of-magnitude improvements are possible in signal-to-noise ratio (SNR) due to antenna gain, and in spectral efficiency due to spatial multiplexing gain. These gains in power and spectral efficiency, coupled with larger available bandwidths, promise very large gains in overall network throughput. Indeed, 100–10,000 Gbps aggregate rates, over 100–300 users, seem attainable with 1–10 GHz of available mmW bandwidth. In contrast, much lower aggregates rates of 30–50 Gbps, over a maximum of 9 users, are possible with 300-500 MHz of bandwidth at 3 GHz. Furthermore, the required total transmit SNR at 3 GHz is about 30 dB higher than that at 80 GHz.[2]

Unleashing this potential of mmW technology for 5G wireless requires the critical functionality of electronic MBDM and associated MIMO communication techniques. This, in turn, presents significant challenges in terms of transceiver complexity for conventional MIMO designs due to the high dimension of the spatial signal space. Systems with fewer but widely spaced antennas have been proposed to reduce complexity in P2P links [36–39, 42], but they suffer from significantly lower array gain and grating beams that also increase interference and compromise security [25, 29]. Antenna selection is another sub-optimal mechanism for reducing complexity [54–57].

The CAP-MIMO transceiver architecture optimally reduces complexity through beam selection, can deliver significantly superior performance, and is squarely aimed at addressing the challenges in realizing the potential of mmW technology. The underlying beamspace MIMO theory also provides a unifying framework for comparing and analyzing other state-of-the-art architectures for mmW MIMO, as overviewed in Section 10.2. It builds on cutting-edge research on MIMO theory, transceiver design and wireless channel modeling performed in the Wireless Communication and Sensing Laboratory[3] over the last 15 or more years [25, 26, 35, 43, 58–65]. This work shows the optimality of beamspace MIMO communication and suggests powerful strategies for practical system design.

[2] The TX SNR values in Figure 10.1(c) ignore propagation/absorption losses, which would be higher at 80 GHz. However, for a given antenna size, the significantly higher antenna gains at 80 GHz versus 3 GHz (Figure 10.1(a)) more than compensate for such losses. While the actual SNR values will depend on the link characteristics (e.g. the length), the *relative* differences are valid.

[3] See http://dune.ece.wisc.edu.

10.1.2 Organization

The next section provides an overview of three main transceiver architectures for realizing mmW MIMO communication. All three architectures can be designed, analyzed and compared in the common framework of beamspace MIMO communication that is developed in subsequent sections. Section 10.3 develops the key ideas in the context of single-user P2P links. Section 10.4 extends the framework to P2MP network settings. The focus is on the spatial dimension and on 1D antenna arrays. Extensions to 2D arrays and time- and frequency-selective channels are briefly discussed in Section 10.5. Overall, the beamspace framework developed in this chapter enables the optimization of fundamental performance–complexity tradeoffs inherent to high-dimensional mmW MIMO systems. The CAP-MIMO transceiver architecture suggests a novel practical realization of beamspace MIMO concepts. The basic theory developed in this chapter applies to systems equipped with uniform linear arrays (ULAs), or uniform planar arrays (UPAs), as well as continuous aperture antennas, such as the lens antenna used in CAP-MIMO, via the concept of critical spatial sampling.

10.2 Overview of Millimeter-Wave MIMO Transceiver Architectures

MIMO techniques have been the focus of intense research since the mid-1990s, aimed at enhancing the capacity of wireless communication systems without additional bandwidth or power [12, 13, 33]. The key MIMO operation for capacity enhancement is spatial multiplexing: transmission of simultaneous data streams to a single receiver in a P2P link or multiple receivers in a P2MP configuration. In contrast to the assumption of rich multipath in traditional MIMO [12, 13, 33, 43], the highly directional nature of propagation at mmW frequencies makes LoS (and sparse multipath) propagation the dominant mode of communication [5, 44]. Thus, the beamspace approach to optimal MIMO communication, first proposed for multipath channels [43], is particularly relevant at mmW frequencies [25, 26]. The following observations and insights underlie the CAP-MIMO transceiver architecture for approaching optimal performance and MBDM functionality with the lowest transceiver complexity [25, 26, 29, 43].

Optimality of Beamspace Communication
Orthogonal spatial beams serve as approximate channel eigen-modes for optimal spatial communication.

High-dimensional Signal Space
For a given antenna size, the dimension of the signal space, n, increases quadratically with frequency; for a $6'' \times 6''$ antenna, $n \approx 9$ at 3 GHz and $n \approx 6000$ at 80 GHz. The antenna gain and the maximum multiplexing gain are proportional to n; see Figure 10.1. This leads to an extremely high, $\mathcal{O}(n)$, transceiver complexity in conventional MIMO; see Figure 10.2(a).

Low-dimensional Communication Subspace
The multiplexing gain, p, or the number of spatial data streams, is typically much smaller than n. In P2P links, this is due to the expected channel sparsity in beamspace [5, 28, 44]. In P2MP links, $p \sim K$, the number of mobile stations (MSs) [5, 30].

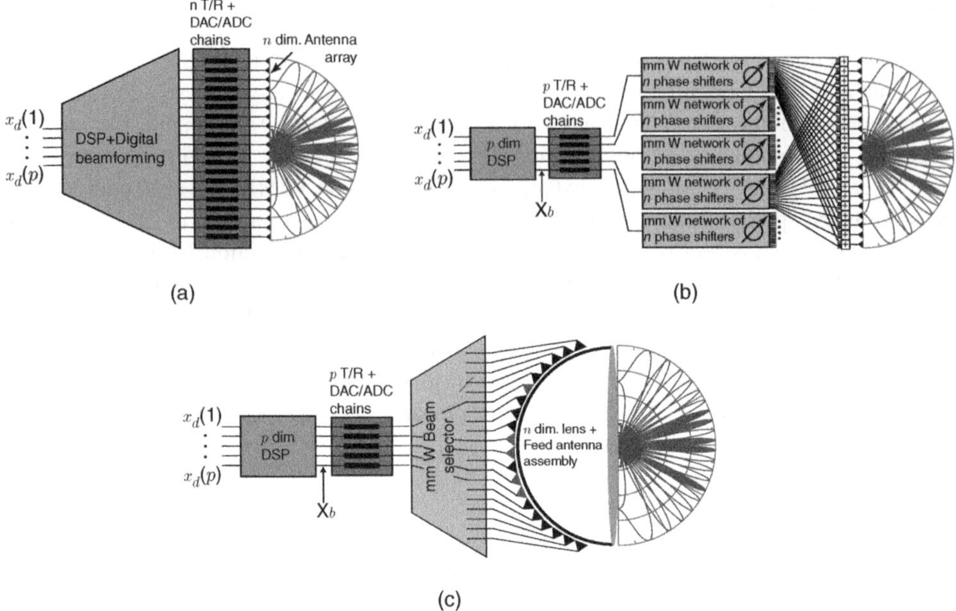

Figure 10.2 Three main mmW MIMO transceiver architectures. (a) A conventional MIMO transceiver that uses an n-element antenna array and baseband digital beamforming; (b) A phased array-based transceiver that uses a network of n mmW phase shifters, one for each of the p data streams, to drive the n-element antenna array for generating p beams; (c) The CAP-MIMO transceiver that uses a lens antenna for analog beamforming to directly map the p data streams to $O(p)$ beams via the mmW beam selector

Analog Beamforming and Beam Selection
This enables near-optimal performance with the lowest, $\mathcal{O}(p)$, transceiver complexity by providing direct beamspace access to the p channel modes through the beam selector network; see Figure 10.2(c).

Figure 10.2 shows the three main mmW MIMO transceiver architectures. Given the narrow beams and highly directional nature of propagation at mmW frequencies, all architectures are based on beamspace MIMO communication. The differences are in the hardware implementation. Each transceiver has four components:

- DSP
- beam selector
- transceiver hardware consisting of multiple analog-to-digital convertor (ADC) and digital-to-analog convertor (DAC) modules and T/R chains
- beamforming mechanism.

Each T/R chain includes mixers, filters, and amplifiers.

A conventional MIMO transceiver, shown in Figure 10.2(a), uses an n-element array with half-wavelength spaced antennas and employs baseband digital beamforming to map the p

data streams onto the n antenna signals. Each antenna element is driven by a dedicated ADC/DAC module and a T/R chain. The conventional MIMO transceiver has full flexibility but suffers from a prohibitively high $\mathcal{O}(n)$ complexity of the front-end hardware (T/R chains and ADCs/DACs), regardless of the number of data streams p, as well as a corresponding $\mathcal{O}(n)$ computational DSP complexity.

A phased-array-based transceiver, illustrated in Figure 10.2(b), has been proposed for reducing the complexity of conventional architecture [31, 48, 66]. In this architecture, each data stream is associated with a network of n mmW phase shifters to map it into a particular beam direction. Thus the mapping of the p data streams is accomplished in passband via the overall network of np phase-shifting elements. Compared to the conventional architecture, the number of ADC/DAC modules and T/R chains can be reduced from n to p via beam selection. But this approach still suffers from the high complexity of the np-element mmW phase-shifting network, especially as p gets larger. The design of this phase-shifting network becomes even more challenging when fully utilizing the wider bandwidths (1–10 GHz) available in the mmW band.

A CAP-MIMO transceiver, shown in Figure 10.2(c), uses a continuous aperture lens-based front-end for mmW analog beamforming. Unlike the other two architectures, CAP-MIMO samples the spatial dimension in beamspace via an array of feed antennas arranged on the focal surface of the lens antenna. With a properly designed front-end, different feed antennas excite orthogonal spatial beams that span the coverage area. The number of ADC/DAC modules and T/R chains tracks the number of data streams p, as in the phased array-based transceiver. However, the mapping of the p data streams into corresponding beams is accomplished via the mmW beam selector that maps the mmW signal for a particular data stream into feed antenna(s) representing the corresponding beam. A wideband lens antenna can be designed in a number of efficient ways, including a discrete lens array (DLA) for lower frequencies or a dielectric lens at higher frequencies [26].

10.3 Point-to-Point Single-User Systems

This section develops the beamspace MIMO system model, including channel models for LoS and multipath propagation environments, in a single-user P2P setting. The design of optimum and low-complexity beamspace MIMO transceivers is also discussed, along with numerical results on their performance. For simplicity, we consider systems with 1D ULAs.

10.3.1 Sampled MIMO System Representation

Consider a linear antenna of length L. If the aperture is sampled with critical spacing $d = \frac{\lambda}{2}$, where λ is the wavelength, there is no loss of information and the sampled points on the aperture are equivalent to an n-dimensional ULA, where $n = \lfloor \frac{2L}{\lambda} \rfloor$ is the maximum number of spatial modes supported by the ULA [25, 67]. A MIMO system with ULAs at the transmitter and the receiver can be modeled as

$$\boldsymbol{r} = \boldsymbol{H}\boldsymbol{x} + \boldsymbol{w} \tag{10.1}$$

where $\boldsymbol{H}$ is the $n_R \times n_T$ aperture domain channel matrix representing coupling between the transmitter and receiver ULA elements, $\boldsymbol{x}$ is the n_T-dimensional transmitted signal vector, $\boldsymbol{r}$ is the n_R-dimensional received signal vector, and $\boldsymbol{w} \sim \mathcal{N}(\boldsymbol{0}, \boldsymbol{I})$ represents the Gaussian noise vector.

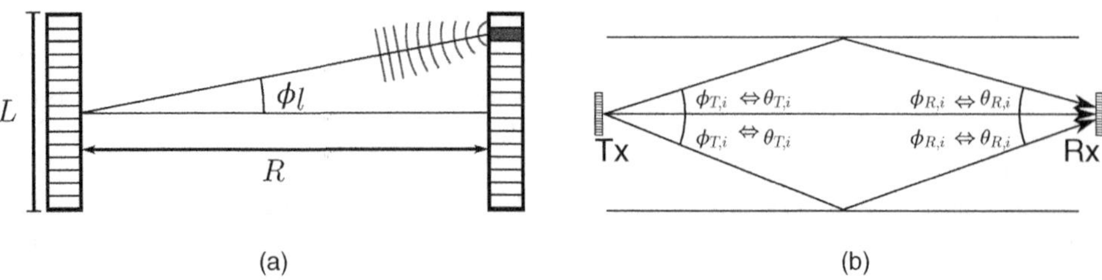

Figure 10.3 (a) Illustration of a LoS channel; (b) Illustration of a single-bounce sparse multipath channel

10.3.2 Beamspace MIMO System Representation

Beamspace MIMO system representation is obtained from Eq. (10.1) via fixed beamforming at the transmitter and the receiver. Each column of the beamforming matrix, $\boldsymbol{U}_n$, is a steering/response vector for a specified angle [25, 35, 43]. For a critically spaced ULA, a plane wave in the direction of angle $\phi \in [-\pi/2, \pi/2]$ (see Figure 10.3) corresponds to a spatial frequency, $\theta \in [-1/2, 1/2]$, given by

$$\theta = \frac{d}{\lambda}\sin(\phi) = 0.5\sin(\phi) \tag{10.2}$$

and the corresponding array steering/response (column) vector is given by [25, 43]

$$\boldsymbol{a}_n(\theta) = \left[e^{-j2\pi\theta i}\right]_{i\in\mathcal{I}(n)} \quad ; \quad \mathcal{I}(n) = \left\{i - \frac{(n-1)}{2} : i = 0, \cdots, n-1\right\} \tag{10.3}$$

where $\mathcal{I}(n)$ is a symmetric set of indices, centered around 0, for a given n. The columns of beamforming matrix $\boldsymbol{U}_n$ correspond to n fixed spatial frequencies/angles with uniform spacing $\Delta\theta_o = \frac{1}{n}$:

$$\boldsymbol{U}_n = \frac{1}{\sqrt{n}}\left[\boldsymbol{a}_n\left(\Delta\theta_o i\right)\right]_{i\in\mathcal{I}(n)}, \ \Delta\theta_o = \frac{1}{n} = \frac{\lambda}{2L} \tag{10.4}$$

which represent n orthogonal beams that cover the entire spatial horizon ($-\pi/2 \le \phi \le \pi/2$) and form a basis for the n-dimensional spatial signal space [25, 35, 43]. In fact, $\boldsymbol{U}_n$ is a unitary discrete Fourier transform (DFT) matrix: $\boldsymbol{U}_n^H\boldsymbol{U}_n = \boldsymbol{U}_n\boldsymbol{U}_n^H = \boldsymbol{I}$. The beamspace system representation is obtained from Eq. (10.1) as

$$\boldsymbol{r}_b = \boldsymbol{H}_b\boldsymbol{x}_b + \boldsymbol{w}_b\ , \ \boldsymbol{H}_b = \boldsymbol{U}_{n_R}^H\boldsymbol{H}\boldsymbol{U}_{n_T} \tag{10.5}$$

where $\boldsymbol{x}_b = \boldsymbol{U}_{n_T}^H\boldsymbol{x}$, $\boldsymbol{r}_b = \boldsymbol{U}_{n_R}^H\boldsymbol{r}$, and $\boldsymbol{w}_b = \boldsymbol{U}_{n_R}^H\boldsymbol{w}$ are the transmitted, received and noise signal vectors, respectively, in beamspace. Since $\boldsymbol{U}_{n_T}$ and $\boldsymbol{U}_{n_R}$ are unitary DFT matrices, $\boldsymbol{H}_b$ is a 2D DFT of $\boldsymbol{H}$ and thus a completely equivalent representation of $\boldsymbol{H}$ [25, 35, 43].

10.3.3 Channel Modeling

Due to the highly directional nature of propagation at mmW frequencies, LoS propagation is expected to be the dominant mode, with some additional sparse (single-bounce) multipath components possible in urban environments [5, 32]. For LoS channels, $\boldsymbol{H}$ can be represented in terms of the array response vectors [25, 26]. As illustrated in Figure 10.3(a), the n_T columns of $\boldsymbol{H}$ can be constructed via the receiver array response vectors corresponding to the spatial frequencies, $\theta_{R,\ell} = 0.5\sin(\phi_{R,\ell})$, induced by the different transmitter ULA elements [25, 26]

$$\boldsymbol{H} = [\boldsymbol{a}_{n_R}(\theta_{R,\ell})]_{\ell\in\mathcal{I}(n_T)}, \quad \theta_{R,\ell} = \Delta\theta_{ch}\ell \approx \frac{\lambda}{4R}\ell \tag{10.6}$$

The rank p_{los} of the LoS channel matrix is typically much smaller than $\min(n_R, n_T)$ if the link length R is large compared to the antenna lengths L_T and L_R. Given the above construction of $\boldsymbol{H}$, the rank can be accurately estimated by the number of orthogonal transmit beams that span the receiver aperture [25, 26]:

$$p_{los} \approx \frac{\Delta\theta_{max,R}}{\Delta\theta_{o,R}} + 1 = \frac{\Delta\theta_{max,T}}{\Delta\theta_{o,T}} + 1 \approx \frac{L_T L_R}{R\lambda} + 1 \tag{10.7}$$

where $\Delta\theta_{max,R} = \max_\ell \theta_{R,\ell} - \min_\ell \theta_{R,\ell} \approx \frac{n_T\lambda}{4R}$ is the range of spatial frequencies induced by the transmitter elements at the receiver, and similarly $\Delta\theta_{max,T} \approx \frac{n_R\lambda}{4R}$ is the maximum range of spatial frequencies induced by the receiver elements at the transmitter. The orthogonal steering vectors also serve as approximate eigenfunctions of $\boldsymbol{H}$: only an approximately $p_{los} \times p_{los}$ sub-matrix of $\boldsymbol{H}_b$ is non-zero and approximately diagonal (see Figure 10.4(a)).

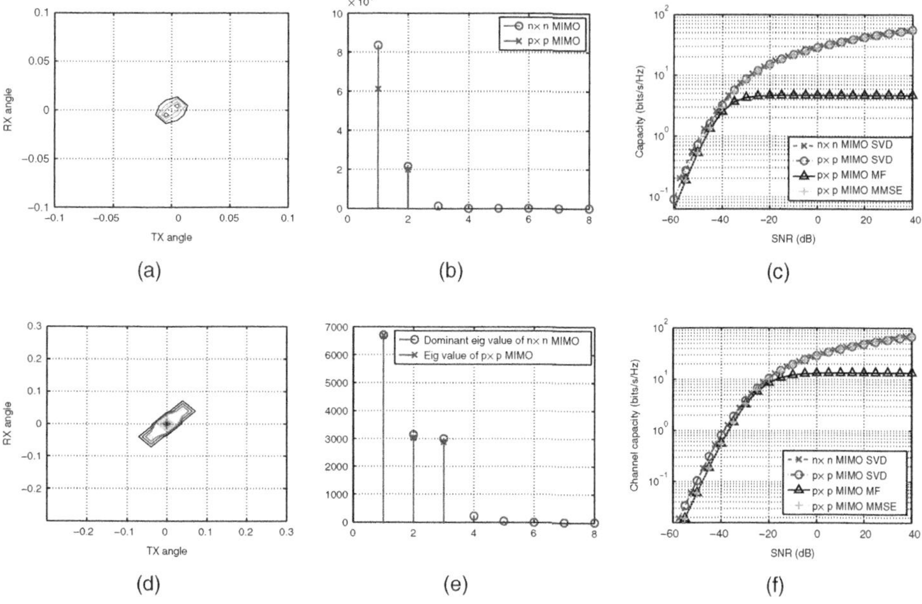

Figure 10.4 Upper row: LoS channel. (a) Contour plot of $|\boldsymbol{H}_b|^2$ with $n = 326$ and $p_{los} = 2$; (b) Eigenvalues of $\tilde{\boldsymbol{H}}_b^H \tilde{\boldsymbol{H}}_b$ and $\boldsymbol{H}_b^H \boldsymbol{H}_b$; (c) Capacity of different transceivers. Lower row: Multipath channel. (d) Contour plot of $E[|\boldsymbol{H}_b|^2]$ with $n = 81$ and $p_{mp} = 3$; (e) Eigenvalues of $E[\tilde{\boldsymbol{H}}_b^H \tilde{\boldsymbol{H}}_b]$ and $E[\boldsymbol{H}_b^H \boldsymbol{H}_b]$; (f) Capacity of different transceivers

A multipath channel can be modeled as [35, 43]:

$$\boldsymbol{H} = \sum_{i=0}^{N_p} \beta_i \boldsymbol{a}_{n_R}(\theta_{R,i}) \boldsymbol{a}_{n_T}^H(\theta_{T,i}) \tag{10.8}$$

where N_p denotes the number of paths and β_i, $\theta_{R,i}$ and $\theta_{T,i}$ represent the complex amplitude, angle of arrival and angle of departure of the i-th path, respectively. The $i = 0$ path is the LoS path with $\beta_0 = 1$ and $\theta_{T,0} = \theta_{R,0} = 0$. For the other paths, β_i can be modeled as $\beta_i = |\beta_i| \exp(-j\psi_i)$, where $|\beta_i|^2$ represents path loss (between -5 and -10 dB [32, 68]), and ψ_i is uniformly distributed in $[0, 2\pi]$.

Figure 10.3(b) shows a simple physical model of a sparse multipath channel: the buildings alongside the road create single-bounce multipath propagation paths. We assume that the link length R is large enough that $|\theta_{R,i}| \in [\Delta\theta_{o,R} - \frac{\Delta\theta_{o,R}}{4}, \Delta\theta_{o,R} + \frac{\Delta\theta_{o,R}}{4}]$ and $|\theta_{T,i}| \in [\Delta\theta_{o,T} - \frac{\Delta\theta_{o,T}}{4}, \Delta\theta_{o,T} + \frac{\Delta\theta_{o,T}}{4}]$, where $\Delta\theta_{o,T} = 1/n_T$ and $\Delta\theta_{o,R} = 1/n_R$. This leads to approximately $p_{mp} = 3$ orthogonal beams that couple the the transmitter and the receiver, resulting in an $\boldsymbol{H}$ with approximate rank $p_{mp} = 3$; $\boldsymbol{H}_b$ only has a $p_{mp} \times p_{mp}$ non-zero sub-matrix that is approximately diagonal (see Figure 10.4(d)).

10.3.4 Beam Selection: Low-dimensional Beamspace MIMO Channel

A key feature of mmW MIMO is that while the system dimension n is high, the dimension of the communication subspace p is typically much smaller, $p \ll n$. Efficient access to the communication subspace is critical from a practical viewpoint. We now outline a framework for beamspace MIMO transceiver design for achieving near-optimal performance with complexity that tracks the lower dimension of the communication subspace.

Let $\sigma_c^2 = \mathrm{tr}(\boldsymbol{H}_b^H \boldsymbol{H}_b) = \mathrm{tr}(\boldsymbol{H}^H \boldsymbol{H})$ denote the channel power. For a given channel realization, the low-dimensional communication subspace is captured by the singular value decomposition (SVD) of $\boldsymbol{H}$: $\boldsymbol{H} = \boldsymbol{U}\boldsymbol{\Lambda}\boldsymbol{V}^H$, where $\boldsymbol{\Lambda}$ is a diagonal matrix of (ordered) singular values: $\lambda_1 \geq \lambda_2 \cdots \lambda_{\min(n_T,n_R)}$. We define the effective channel rank, p_{eff}, as the number of singular values that capture most of channel power: $\sum_{i=1}^{p_{eff}} \lambda_i^2 \geq \eta\sigma_c^2$, for some η close to 1 (say 0.8 or 0.9). Optimal communication over the p_{eff}-dimensional communication subspace is achieved through the corresponding right and left singular vectors in $\boldsymbol{V}$ and $\boldsymbol{U}$ [25].

Beamspace MIMO naturally enables access to the low-dimensional communication subspace through the Fourier basis vectors that serve as approximate singular vectors for the spatial channel [25, 43]. The channel power is concentrated in a low-dimensional sub-matrix, $\tilde{\boldsymbol{H}}_b$, of $\boldsymbol{H}_b$ whose entries capture most of the channel power. Let $\boldsymbol{\Sigma}_{T,b} = \boldsymbol{H}_b^H \boldsymbol{H}_b$ and $\boldsymbol{\Sigma}_{R,b} = \boldsymbol{H}_b \boldsymbol{H}_b^H$ denote the transmit and receive beamspace correlation matrices, respectively [35, 43]. We define $\tilde{\boldsymbol{H}}_b$ as

$$\tilde{\boldsymbol{H}}_b = [H_b(i,j)]_{i \in \mathcal{M}_R, j \in \mathcal{M}_T} \tag{10.9}$$

$$\mathcal{M}_T = \{i : \boldsymbol{\Sigma}_{T,b}(i,i) > \gamma_T \max_i \boldsymbol{\Sigma}_{T,b}(i,i)\} \tag{10.10}$$

$$\mathcal{M}_R = \{i : \boldsymbol{\Sigma}_{R,b}(i,i) > \gamma_R \max_i \boldsymbol{\Sigma}_{R,b}(i,i)\} \tag{10.11}$$

where the thresholds $\gamma_T, \gamma_R \in (0, 1)$ are chosen so that

$$\sum_{i \in \mathcal{M}_T} \boldsymbol{\Sigma}_{T,b}(i,i) \approx \sum_{i \in \mathcal{M}_R} \boldsymbol{\Sigma}_{R,b}(i,i) \geq \eta_b \sigma_c^2 \tag{10.12}$$

for some η_b close to 1. $\mathcal{M}_T$ and $\mathcal{M}_R$ denote the transmit and receive sparsity masks representing the beams selected for communication. The overall channel sparsity mask $\mathcal{M}$ is given by

$$\mathcal{M} = \{(i,j) : i \in \mathcal{M}_R \ , \ j \in \mathcal{M}_T\} \tag{10.13}$$

Finally, define $p_{eff,T} = |\mathcal{M}_T|$, $p_{eff,R} = |\mathcal{M}_R|$ and $p_{eff,b} = \min(p_{eff,T}, p_{eff,R})$. The low-complexity beamspace MIMO transceivers access the low-dimensional communication subspace by selecting the $p_{eff,T}$ transmit beams in $\mathcal{M}_T$ and $p_{eff,R}$ receive beams in $\mathcal{M}_R$. For example, in the CAP-MIMO architecture, shown in Figure 10.2(c), this corresponds to activating the corresponding feed antennas via the beam selector. By choosing the thresholds appropriately, $p_{eff,b} \approx p_{eff}$ and because of Eq. (10.12) the resulting performance is near optimum. The above discussion applies to deterministic LoS channels. For random multipath channels, $\mathcal{M}_T$ and $\mathcal{M}_R$ are determined using statistical channel covariance matrices, $\boldsymbol{\Sigma}_{T,b} = E[\boldsymbol{H}_b^H \boldsymbol{H}_b]$ and $\boldsymbol{\Sigma}_{R,b} = E[\boldsymbol{H}_b \boldsymbol{H}_b^H]$ with $\sigma_c^2 = \text{tr}(\boldsymbol{\Sigma}_{T,b}) = \text{tr}(\boldsymbol{\Sigma}_{R,b})$.

10.3.5 Optimal Transceiver

The performance benchmark is provided by the SVD transceiver in which independent data streams are communicated over channel singular vectors to eliminate interference. The transmitted signal in Eq. (10.1) is precoded as $\boldsymbol{x} = \boldsymbol{V}\boldsymbol{x}_e$ and the received signal is transformed as $\boldsymbol{r}_e = \boldsymbol{U}^H \boldsymbol{r}$ to result in non-interfering eigen channels: $\boldsymbol{r}_e = \boldsymbol{\Lambda}\boldsymbol{x}_e + \boldsymbol{w}_e$, where $\boldsymbol{w}_e \sim \mathcal{N}(\boldsymbol{0}, \boldsymbol{I})$. The capacity-achieving transmitted signal $\boldsymbol{x}_e$ consists of independent Gaussian signals: $\boldsymbol{x}_e \sim \mathcal{N}(\boldsymbol{0}, \boldsymbol{\Lambda}_s)$ with $\boldsymbol{\Lambda}_s = \text{diag}\ (\rho_1, \cdots, \rho_n)$ representing the allocation of total transmit power ρ over different channels, $\rho = E[\|\boldsymbol{x}\|^2] = \sum_i \rho_i$. For a given channel realization, the conditional link capacity is

$$C(\rho|\boldsymbol{H}) = \max_{\rho_i : \sum \rho_i = \rho} \sum_{i=1}^{n_T} \log(1 + \text{SNR}_i(\boldsymbol{H})) \tag{10.14}$$

which represents optimal power allocation via water-filling [69], with $\text{SNR}_i(\boldsymbol{H}) = \rho_i \lambda_i^2$. The power would be mainly allocated to the p_{eff} dominant channels. For stochastic multipath channels, the ergodic capacity is obtained by averaging over the channel statistics: $C(\rho) = E[C(\rho|\boldsymbol{H})]$.

10.3.6 Beamspace MIMO Transceivers

The low-dimensional beamspace MIMO (B-MIMO) transceivers operate on the $p_{eff,R} \times p_{eff,T}$ subsystem induced by $\tilde{\boldsymbol{H}}_b$: $\tilde{\boldsymbol{r}}_b = \tilde{\boldsymbol{H}}_b \tilde{\boldsymbol{x}}_b + \tilde{\boldsymbol{w}}_b$. We focus on the class of linear transceivers that use a precoding matrix $\boldsymbol{G}_b$ at the transmitter ($\tilde{\boldsymbol{x}}_b = \boldsymbol{G}_b \boldsymbol{s}_b$) and a filter matrix $\boldsymbol{F}_b$ at the receiver ($\boldsymbol{y}_b = \boldsymbol{F}_b^H \tilde{\boldsymbol{r}}_b$) to yield

$$\boldsymbol{y}_b = \boldsymbol{F}_b^H \tilde{\boldsymbol{H}}_b \boldsymbol{G}_b \boldsymbol{s}_b + \boldsymbol{z}_b; \quad \text{where} \quad \boldsymbol{z}_b = \boldsymbol{F}_b^H \tilde{\boldsymbol{w}}_b \sim \mathcal{CN}(\boldsymbol{0}, \boldsymbol{F}_b^H \boldsymbol{F}_b) \tag{10.15}$$

The optimal low-dimensional transceiver is now determined by the SVD of $\tilde{\boldsymbol{H}}_b$, $\tilde{\boldsymbol{H}}_b = \tilde{\boldsymbol{U}}_b \tilde{\boldsymbol{\Lambda}}_b \tilde{\boldsymbol{V}}_b^H$, and choosing $\boldsymbol{G}_b = \tilde{\boldsymbol{V}}_b$ and $\boldsymbol{F}_b = \tilde{\boldsymbol{U}}_b$ to create $p_{eff,b} \approx p_{eff}$ non-interfering channels. The optimal transmitted signal $\boldsymbol{s}_b$ is again an independent Gaussian vector and, for a given $\tilde{\boldsymbol{H}}_b$, the system capacity is

$$C(\rho|\tilde{\boldsymbol{H}}_b) = \max_{\rho_i : \sum \rho_i = \rho} \sum_{i=1}^{p_{eff,b}} \log(1 + \mathrm{SNR}_i(\tilde{\boldsymbol{H}}_b)) \tag{10.16}$$

with $\mathrm{SNR}_i(\tilde{\boldsymbol{H}}_b) = \rho_i \tilde{\lambda}_{b,i}^2$, and the ergodic capacity $C(\rho) = E[C(\rho|\tilde{\boldsymbol{H}}_b)]$. By increasing $p_{eff,R}$ and $p_{eff,T}$ – that is, by including a progressively larger number of dominant beams by choosing γ_R and γ_T appropriately – the performance of an SVD-based B-MIMO transceiver can be made arbitrarily close to the optimal transceiver.

The optimal SVD transceiver (antenna or beam domain) requires channel-state information (CSI) at both the transmitter and receiver, which is impractical in many situations. Thus we present two simple suboptimal B-MIMO transceivers for $\tilde{\boldsymbol{H}}_b$ that only require CSI at the receiver. For the suboptimal receivers we assume that $p_{eff,b} = \min(p_{eff,T}, p_{eff,R}) = p_{eff,T} \leq p_{eff,R}$. In both transceivers $\boldsymbol{G}_b = \boldsymbol{I}$ and the transmitted signal $\boldsymbol{s}_b$ in Eq. (10.15) is an independent Gaussian vector with equal power allocation over the $p_{eff,b}$ beams: $\boldsymbol{s}_b \sim \mathcal{N}(\boldsymbol{0}, \rho\boldsymbol{I}/p_{eff,b})$. The transceivers differ in their choice of $\boldsymbol{F}_b$ for suppressing interference at the receiver. In the matched filter (MF) transceiver, $\boldsymbol{F}_{b,\mathrm{MF}} = \tilde{\boldsymbol{H}}_b$ and Eq. (10.15) becomes $\boldsymbol{y}_b = \tilde{\boldsymbol{H}}_b^H \tilde{\boldsymbol{H}}_b \boldsymbol{s}_b + \boldsymbol{z}_b$. Since $\tilde{\boldsymbol{H}}_b^H \tilde{\boldsymbol{H}}_b$ is diagonally dominant, the interference is limited but still present. In the minimum mean-squared error (MMSE) receiver, $\boldsymbol{F}_b$ is chosen to minimize the MSE at the receiver to further suppress interference

$$\boldsymbol{F}_{b,\mathrm{MMSE}} = \arg \min_{\boldsymbol{F}_b} E\left[\left\|\boldsymbol{F}_b^H \tilde{\boldsymbol{r}}_b - \boldsymbol{s}_b\right\|^2\right] \tag{10.17}$$

$$\boldsymbol{F}_{b,\mathrm{MMSE}}^H = \boldsymbol{Q}_b \tilde{\boldsymbol{H}}_b^H \left(\tilde{\boldsymbol{H}}_b \boldsymbol{Q}_b \tilde{\boldsymbol{H}}_b^H + \boldsymbol{I}\right)^{-1} \tag{10.18}$$

where $\boldsymbol{Q}_b = E[\boldsymbol{s}_b \boldsymbol{s}_b^H]$ is the covariance matrix of $\boldsymbol{s}_b$ which equals $\boldsymbol{Q}_b = \frac{\rho}{p_{eff,b}} \boldsymbol{I}$ under independent and equal-power signaling. For both suboptimal transceivers, the conditional capacity is

$$C(\rho|\tilde{\boldsymbol{H}}_b) = \sum_i \log\left(1 + \mathrm{SINR}_i(\tilde{\boldsymbol{H}}_b)\right) \tag{10.19}$$

where the interference is treated as noise and the signal-to-interference-and-noise-ratio (SINR) for the ith data stream is

$$\mathrm{SINR}_i(\tilde{\boldsymbol{H}}_b) = \frac{\left|\boldsymbol{f}_{b,i}^H \tilde{\boldsymbol{H}}_b \boldsymbol{g}_{b,i}\right|^2 \rho_i}{\sum_{j \neq i} \left|\boldsymbol{f}_{b,i}^H \tilde{\boldsymbol{H}}_b \boldsymbol{g}_{b,j}\right|^2 \rho_i + \left\|\boldsymbol{f}_{b,i}\right\|^2} \tag{10.20}$$

where $\rho_i = \rho/p_{eff,b}$, $\boldsymbol{f}_{b,i}$ is the ith column of $\boldsymbol{F}_b$, $\boldsymbol{g}_{b,i}$ is the ith column of $\boldsymbol{G}_b$. For stochastic channels, the ergodic capacity is $C(\rho) = E[C(\rho|\tilde{\boldsymbol{H}}_b)]$.

10.3.7 Numerical Results

We present numerical results for beampace MIMO transceivers at $f_c = 80$ GHz for both LoS and multipath channels. Figure 10.4(a)–(c) correspond to a LoS channel with antennas of length $L = 0.6$ m and link length $R = 100$ m, which yields $n = 326$ and $p_{los} = 2$. Figure 10.4(a) shows a contour plot of $|\boldsymbol{H}_b|^2$. Using $p_{eff} = 2$ channel singular values yields $\eta = 0.98$. Using $\gamma_T = \gamma_R = 0.1$ results in a 2×2 $\tilde{\boldsymbol{H}}_b$ with $\eta_b = 0.8$. As shown in Figure 10.4(b), $\tilde{\boldsymbol{H}}_b$ captures the two dominant channel singular values. Figure 10.4(c) shows the capacity of the different transceivers for the LoS channel. The capacity of the low-dimensional $p \times p$ SVD system based on $\tilde{\boldsymbol{H}}_b$ is nearly identical to that of the $n \times n$ SVD system based on $\boldsymbol{H}_b$. Furthermore, the low-dimensional beamspace MMSE transceiver closely approximates the performance of SVD transceiver by effectively suppressing interference. The interference is not negligible as evident from the performance degradation in the MF transceiver at higher SNRs.

Figure 10.4(d)–(f) correspond to a multipath channel with $L = 0.15$ m and $R = 200$ m, yielding $n = 81$. In addition to the LoS path, there are additional $N_p = 10$ single-bounce multipath components, as illustrated in Figure 10.3(b), which result in a total of $p_{mp} = 3$, as discussed in Section 10.3.3. Figure 10.4(d) shows a contour plot of $E[|\boldsymbol{H}_b|^2]$, from which three dominant diagonal entries are evident. Three dominant channel singular values result in $\eta = 0.98$. Using $\gamma_T = \gamma_R = 0.1$ yields a 3×3 $\tilde{\boldsymbol{H}}_b$. As shown in Figure 10.4(e), the eigenvalues of $E[\tilde{\boldsymbol{H}}_b^H \tilde{\boldsymbol{H}}_b]$ are nearly identical to the eigenvalues of $E[\boldsymbol{H}_b^H \boldsymbol{H}_b]$ and capture 95% of the channel power ($\eta_b = 0.95$). Again, despite the dimensionality reduction from $n = 81$ to $p = 3$, the capacities of the low-dimensional and full-dimensional SVD systems are nearly identical. Furthermore, the low-complexity MMSE transceiver closely approximates the SVD transceiver, while the MF transceiver degrades at high SNR due to residual interference. The results for the multipath channel were obtained by averaging over 1000 channel realizations.

10.4 Point-to-Multipoint Multiuser Systems

We now consider a P2MP multiuser MIMO (MU-MIMO) link in which an AP equipped with an n-element ULA communicates with K single-antenna MSs. We focus on the more challenging scenario of downlink communication; the uplink problem is well-studied [34] and can be formulated easily along the lines discussed here.

The received signal at the ith MS is given by $r_i = \boldsymbol{h}_i^H \boldsymbol{x} + w_i$, where $\boldsymbol{x}$ is the n-dimensional transmitted signal, $\boldsymbol{h}_i$ is the n-dimensional channel vector, and $w_i \sim \mathcal{N}(0, \sigma^2)$ is additive white Gaussian noise. Stacking the signals for all MSs in a K-dimensional vector $\boldsymbol{r} = [r_1, \cdots, r_K]^T$ we get the antenna domain system equation

$$\boldsymbol{r} = \boldsymbol{H}^H \boldsymbol{x} + \boldsymbol{w} \ , \ \boldsymbol{H} = [\boldsymbol{h}_1, \cdots, \boldsymbol{h}_K] \tag{10.21}$$

where $\boldsymbol{H}$ is the $n \times K$ channel matrix that characterizes the system. Our focus is on the design of the linear precoding matrix $\boldsymbol{G} = [\boldsymbol{g}_1, \boldsymbol{g}_2, \cdots, \boldsymbol{g}_K]$ for the transmitted signal, $\boldsymbol{x} = \boldsymbol{G}\boldsymbol{s} = \sum_{i=1}^{K} \boldsymbol{g}_i s_i$, where $\boldsymbol{s}$ is the K-dimensional vector of independent symbols for different MSs.

The overall system equation becomes

$$r = H^H G s + w \ , \ E[\|x\|^2] = \text{tr}(G \Lambda_s G^H) \le \rho \tag{10.22}$$

where the second equality represents the constraint on total transmit power, ρ, and $\Lambda_s = E[ss^H]$ denotes the diagonal correlation matrix of s.

10.4.1 Channel Model

The channel matrix H governs the performance of the MU-MIMO link. Due to the highly directional and quasi-optical nature of propagation at mmW frequencies, LoS propagation is the predominant mode of propagation, with possibly a sparse set of single-bounce multipath components [32, 68]. We assume that LoS paths exist for all MSs. Let $\theta_{k,0}$, $k = 1, \cdots, K$, denote the LoS directions (spatial frequencies) for the K MSs. Then the LoS channel for the kth MS is $h_k = \beta_{k,0} a_n(\theta_{k,0})$, where $\beta_{k,0}$ is the complex path loss. In general, for sparse multipath channels

$$h_k = \beta_{k,0} a_n(\theta_{k,0}) + \sum_{i=1}^{N_p} \beta_{k,i} a_n(\theta_{k,i}) \tag{10.23}$$

where $\{\theta_{k,i}\}$ denote the path angles and $\{\beta_{k,i}\}$ represent the complex path losses associated with the different paths for the kth MS. The amplitudes $|\beta_{k,i}|$ for multipath components are typically 5–10 dB weaker than the LoS component $|\beta_{k,0}|$ [32]. For numerical results, we focus on purely LoS channels with $\theta_{k,0} = \theta_k$, $|\beta_{k,0}| = 1$ and $\beta_{k,i} = 0$ for $i \neq 0$ for all MSs.

10.4.2 Beamspace System Model

The beamspace MIMO system representation is obtained from Eq. (10.21) via fixed beamforming at the transmitter using the beamforming matrix U_n defined in Eq. (10.4), and by using the beamspace representation of the precoding matrix $G = U_n G_b$ in Eq. (10.22)

$$r = H_b^H G_b s_b + w \ , \ H_b = U_n^H H = [h_{b,1}, \cdots, h_{b,K}] \tag{10.24}$$

where $s_b = s$ represents the beamspace symbol vector, and G_b is the beamspace precoder. $x_b = G_b s_b$ represents the precoded beamspace transmit signal vector. Since U_n is a unitary matrix, the beamspace channel matrix H_b is a completely equivalent representation of H.

10.4.3 Beam Selection: Low-dimensional Channel

The most important property of H_b is that it has a sparse structure representing the directions of the different MSs, as illustrated in Figure 10.5(a) for LoS links. The kth column $h_{b,k} = U_n^H h_k$ (the rows in Figure 10.5(a)) is the beamspace representation of the kth MS channel and has a few dominant entries near the true LoS direction θ_k of the MS. This sparse nature of the beamspace channel is exploited for designing reduced-complexity beamspace precoders that deliver near-optimal performance through the concept of beam selection.

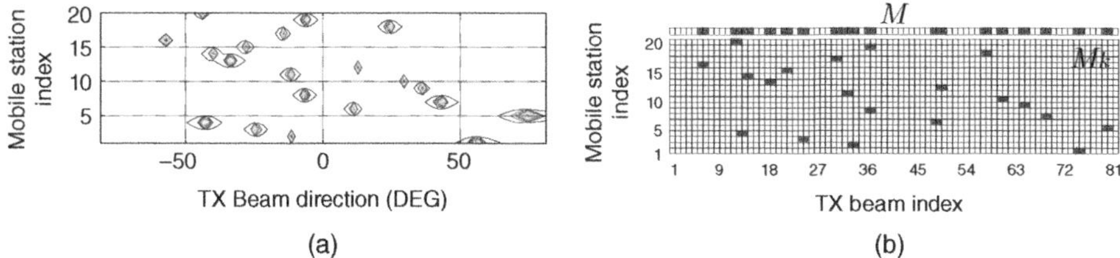

Figure 10.5 (a) Contour plot of $|\boldsymbol{H}_b^H|^2$ for a ULA with $n = 81$, representing the beamspace channel vectors (rows) for 20 MSs randomly distributed between $\pm 90°$. (b) Illustration of beamspace channel sparsity masks $\mathcal{M}_k$ and $\mathcal{M}$ for the $\boldsymbol{H}_b$ in (a)

We define the following sets of beam indices – channel sparsity masks – that represent the dominant beams selected for transmission at the AP (see Figure 10.5(b)):

$$\mathcal{M}_k = \left\{ i : |h_{b,k}(i)|^2 \geq \gamma_k \max_i |h_{b,k}(i)|^2 \right\}, \mathcal{M} = \bigcup_{k=1,\cdots,K} \mathcal{M}_k \tag{10.25}$$

where $\mathcal{M}_k$ is the sparsity mask for the kth MS, determined by the threshold $\gamma_k \in (0, 1)$, and $\mathcal{M}$ is the overall beamspace sparsity mask representing the beams activated by the AP. This beam selection is equivalent to selecting a subset of $p = |\mathcal{M}|$ rows of $\boldsymbol{H}_b$ resulting in the following low-dimensional system equation

$$\boldsymbol{r} = \tilde{\boldsymbol{H}}_b^H \tilde{\boldsymbol{G}}_b \boldsymbol{s}_b + \boldsymbol{w} \ , \ \tilde{\boldsymbol{H}}_b = [\boldsymbol{H}_b(\ell, :)]_{\ell \in \mathcal{M}} \tag{10.26}$$

where $\tilde{\boldsymbol{H}}_b$ is the $p \times K$ beamspace channel matrix corresponding to the selected beams, and $\tilde{\boldsymbol{G}}_b$ is the corresponding $p \times K$ precoder matrix, where $p \leq n$.

For a given $\boldsymbol{H}$, the total multiuser channel power is defined as $\sigma_c^2 = \mathrm{tr}(\boldsymbol{H}\boldsymbol{H}^H) = \mathrm{tr}(\boldsymbol{H}_b \boldsymbol{H}_b^H)$, which under the simple LoS model is $\sigma_c^2 = n \sum_{k=1}^{K} |\beta_k|^2 = nK$. The beam selection thresholds $\{\gamma_k\}$ can be chosen so that $\mathcal{M}_k$ captures a significant fraction η_k of the power of $\boldsymbol{h}_{b,k}$ (e.g. $\eta_k \geq 0.9$). This in turn implies that the fraction η of the total channel power captured by $\tilde{\boldsymbol{H}}_b$ is at least $\min_{k=1,\ldots,K} \eta_k$.

Conversely, the sparsity masks $\mathcal{M}_k$ can be chosen to select the m dominant (strongest) beams for each MS. This choice implicitly defines γ_k as the ratio between the power of the mth strongest beam to the power of the strongest beam for the kth user. For the simple LoS channel model this corresponds to selecting the m orthogonal beams closest to the true LoS direction of the MS θ_k. For numerical results, we use a two-beam mask for complexity reduction (see Figure 10.5(b)). The expected value of η for the two-beam mask can be lower bounded as [30]:

$$E[\eta] \geq \frac{2}{n} \int_0^{\frac{\Delta\theta_o}{2}} f_n^2(\delta) + f_n^2(\delta - \Delta\theta_o) d\delta \tag{10.27}$$

where $f_n(\theta) = \sin(n\pi\theta)/\sin(\pi\theta)$ is the Dirichlet sinc function.

10.4.4 Multiuser Beamspace MIMO Precoders

Generally, achieving true sum capacity requires dirty paper coding, which suffers from high complexity [34]. We thus focus on simple linear precoders. There are three main types of linear

MU-MIMO precoders: the matched filter (MF), zero-forcing (ZF), and Wiener filter (WF). For the full-dimensional antenna domain system Eq. (10.22), the three precoders are given by [35, 70, 71]

$$G = \alpha F = \alpha[f_1, f_2, \cdots, f_K] \ , \ \alpha = \sqrt{\frac{\rho}{\mathrm{tr}(F \Lambda_s F^H)}} \tag{10.28}$$

$$F_{MF} = H \ , \ F_{ZF} = H(H^H H)^{-1} \tag{10.29}$$

$$F_{WF} = (HH^H + \zeta I)^{-1} H \ , \ \zeta = \frac{\mathrm{tr}(\Sigma_w)}{\rho} = \frac{\sigma^2 K}{\rho} \tag{10.30}$$

where $\Sigma_w = E[ww^H]$ and the precoder matrix G is an $n \times K$ matrix. In beamspace, the equivalent full-dimensional precoder G_b can be obtained via the above equations by replacing H with H_b. Similarly, the reduced-complexity B-MIMO precoder matrix $\tilde{G}_b$ ($p \times K$) is obtained via Eqs. (10.28)–(10.30) by replacing H with $\tilde{H}_b$. As we demonstrate in the numerical results section, the reduced-complexity B-MIMO precoder can deliver the performance of the full-dimensional precoder with a complexity that tracks the number of MSs K. The computational complexity of the full dimensional precoders is driven by the $n \times K$ matrix H for the determination of $n \times K$ G as evident from Eqs. (10.28)–(10.30). However, the computational complexity of the low-dimensional B-MIMO system is driven by the $p \times K$ matrix $\tilde{H}_b$, and is thus significantly lower.

10.4.5 Numerical Results

We present numerical results to assess the sum capacity of the multiuser B-MIMO precoders. Let ρ denote the total transmit power, which equals the total transmit SNR for $\sigma^2 = 1$. We use the following idealistic upper bound for the sum capacity

$$C_{ub}(\rho, K, n) = K \log_2 \left(1 + \rho \frac{n}{K}\right) \ \text{bits/s/Hz} \tag{10.31}$$

which corresponds to K MSs with orthogonal channels (MS directions coincident with a subset of the fixed beams in U_n). The received SNR associated with each MS is given by $\rho n / K$, reflecting the n-fold array/beamforming gain of the AP antenna. For the general full-dimensional precoder in Eq. (10.28) we assess the conditional sum capacity for a given channel realization (random MS directions $\{\theta_k\}$) as

$$C(\rho, G|H) = \sum_{i=1}^{K} \log_2(1 + \mathrm{SINR}_i(\rho, G|H)) \ \text{bits/s/Hz} \tag{10.32}$$

where the interference is treated as noise and the SINR for the ith user is

$$\mathrm{SINR}_i(\rho, G|H) = \frac{\rho \frac{|\alpha|^2}{K} |h_i^H f_i|^2}{\rho \frac{|\alpha|^2}{K} \sum_{m \neq i} |h_m^H f_i|^2 + \sigma^2} \tag{10.33}$$

We can use the same relations for assessing the sum capacity of B-MIMO precoders as well by replacing H with H_b (full-dimensional) or $\tilde{H}_b$ (low-dimensional), and G with G_b or $\tilde{G}_b$. The ergodic sum capacity for a given precoder (determined by G) is given by $C(\rho, G) =$

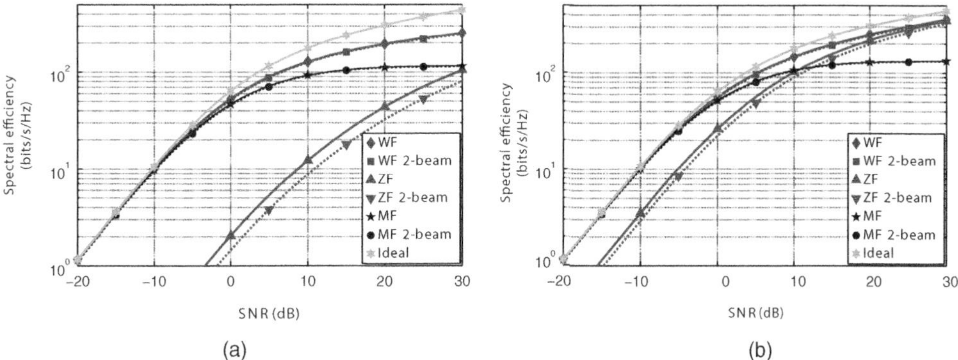

Figure 10.6 (a) Capacity of three different B-MIMO precoders for downlink communication from an AP ($n = 81$) to $K = 40$ randomly distributed single-antenna MSs, and minimum MS separation of $\Delta\theta_{min} = 0$. (b) With minimum separation $\Delta\theta_{min} = \frac{\Delta\theta_o}{4}$

$E[C(\rho, \boldsymbol{G}|\boldsymbol{H})]$, where the averaging is done over the random MS directions. Note that the precoder $\boldsymbol{G}_b$ changes with the channel realization.

Figs 10.6 and 10.7 show numerical ergodic sum capacity results for the B-MIMO precoders generated by averaging over 2000 channel realizations for an AP equipped with a ULA of dimension $n = 81$ (linear 6″ antenna at 80 GHz) communicating with $K = 40$ (Figure 10.6) or $K = 60$ (Figure 10.7) single-antenna MSs over LoS links. The idealized upper bound is also included for comparison. A two-beam mask is used for complexity reduction, which from Eq. (10.27) captures at least 90% of the channel power on average ($E[\eta] \geq 0.9$). The MSs are randomly located over the entire spatial horizon ($-0.5 \leq \theta \leq 0.5$). The curves in Figure 10.6(a) and Figure 10.7(a) were generated with no restrictions on the MS LoS directions $\{\theta_k\}$, while the curves in Figure 10.6(b) and Figure 10.7(b) were generated with a minimum MS separation of $\Delta\theta_{min} = \frac{\Delta\theta_o}{4}$.

These plots show that the simplest MF precoder performs well at lower SNRs due to the approximate orthogonality of high-dimensional user channels, as noted elsewhere [27, 34, 35, 71]. However, there is always interference for finite n, resulting in performance loss at higher SNR. The ZF precoder completely eliminates interference, but significantly reduces the received signal power when the interference is high, resulting in performance degradation. The WF precoder achieves the best performance in all cases by adapting to the operating SNR (see Eq. (10.30)). Most importantly, the reduced-complexity B-MIMO precoders ($\tilde{\boldsymbol{G}}_b$) are able to closely approximate their full-dimensional counterparts ($\boldsymbol{G}_b$).

Table 10.1 summarizes the performance of the reduced-complexity WF precoder when operating at an SNR of 20 dB with 5-GHz bandwidth for $K = 20$, 40 or 60 MSs and $\Delta\theta_{min} = 0$ or $\frac{\Delta\theta_o}{4}$. In the best case, the reduced-complexity WF precoder achieves an average per-user rate of 39.8 Gbps ($K = 20$, $\Delta\theta_{min} = \Delta\theta_o/4$). Even in the worst interference case ($K = 60$, $\Delta\theta_{min} = 0$), the reduced-complexity WF precoder supports an average per-user data rate of 18.8 Gbps. The aggregate sum rates range between 670–1415 Gbps with corresponding spectral efficiencies ranging between 134 and 283 bps/Hz.

The table shows that enforcing a minimum user separation increases the data rate. However this comes at the cost of increased complexity. Figure 10.8(a) plots $E[p]$ (average number

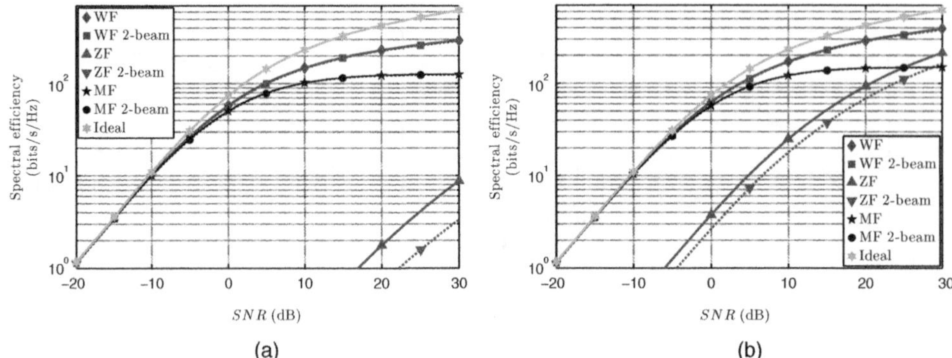

Figure 10.7 (a) Capacity of three different B-MIMO precoders for downlink communication from an AP ($n = 81$) to $K = 60$ randomly distributed single-antenna MSs, and minimum MS separation of $\Delta\theta_{min} = 0$. (b) With minimum separation $\Delta\theta_{min} = \frac{\Delta\theta_o}{4}$

Table 10.1 Performance of the reduced-complexity B-MIMO WF precoders at an SNR of 20 dB with 5 GHz of system bandwidth

	Spectral efficiency (bits/s/Hz)	Aggregate rate (Gbps)	Average per-user rate (Gbps)
$K = 20$, $\Delta\theta_{min} = 0$	134	670	33.5
$K = 20$, $\Delta\theta_{min} = \Delta\theta_o/4$	159	795	39.8
$K = 40$, $\Delta\theta_{min} = 0$	192	960	24
$K = 40$, $\Delta\theta_{min} = \Delta\theta_o/4$	243	1215	30.4
$K = 60$, $\Delta\theta_{min} = 0$	226	1130	18.8
$K = 60$, $\Delta\theta_{min} = \Delta\theta_o/4$	283	1415	23.6

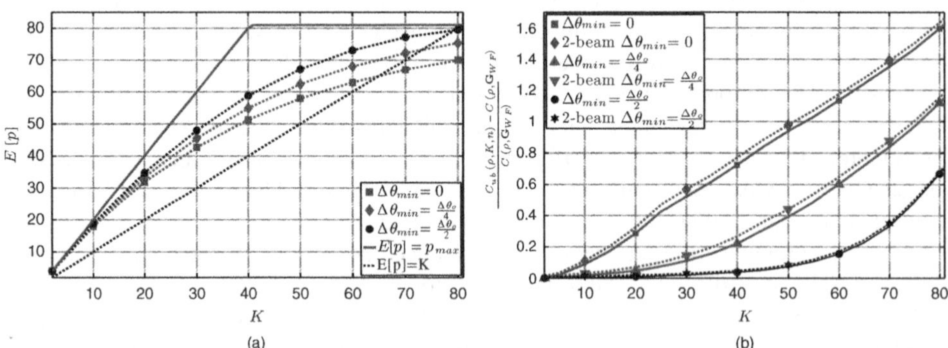

Figure 10.8 (a) $E[p]$ (average number of selected beams) using the two-beam sparsity mask when $n = 81$. (b) Normalized capacity gap between the idealized upper bound and the WF precoder at an SNR of 30 dB

of selected beams) for the two-beam mask as a function of K. Enforcing a minimum user separation requires the reduced-complexity precoders to select more beams on average. The maximum number of beams that the two-beam mask can select (corresponding to the minimum complexity reduction) is $p_{max} = \min(2K, n)$. Figure 10.8(a) shows that for small K, $E[p] \approx p_{max}$. However, for larger K, $E[p]$ is generally much smaller than p_{max} with the largest gap when $K \approx \frac{n}{2}$.

While the reduced-complexity WF precoder has the best performance, there is still residual interference between closely spaced MSs. This results in the idealized upper bound C_{ub} overestimating the sum capacity achieved by the system. However, as shown in Figure 10.8(b), when the interference is limited ($K \ll n$ and/or there is a minimum separation between the MSs) the difference between C_{ub} and the sum capacity for the WF precoders is minimal.

The design and analysis of CAP-MIMO APs equipped with UPAs for small-cell applications is discussed in a paper by Brady and Sayeed [45]. In particular, it is shown that using a UPA of dimension $2.3'' \times 11.5''$, the spectral efficiency of a CAP-MIMO transceiver with a four-beam mask and servicing $K = 100$ MSs is 1067 bps/Hz at a transmit power of 20 dBm. For a system with 5 GHz of bandwidth, this corresponds to an aggregate rate of 5335 Gbps or an average per user rate of about 53 Gbps. Currently, LTE Advanced using 8×8 MIMO spatial multiplexing can provide a peak downlink rate of 3.3 Gbps over 100-MHz bandwidth under ideal conditions [72]. Thus the combination of the fifty-fold increase in bandwidth, the increased array gain, and the dense spatial multiplexing of $K = 100$ MSs results in a more than thousand-fold increase in the aggregate downlink rate. On the other hand, since the four-beam mask is used for complexity reduction, when using a system with an analog beamforming front-end, such as CAP-MIMO, there is only a fifty-fold increase in transceiver hardware complexity (from 8 to 400 transceiver chains).

10.5 Extensions

For simplicity, we have focussed on MIMO systems equipped with 1D ULAs in a frequency non-selective setting. In this section, we briefly discuss extensions to 2D UPAs and frequency-selective channels.

Consider a critically sampled n-dimensional UPA with $n = n_{\rm az} \times n_{\rm el}$, where $n_{\rm az}$ and $n_{\rm el}$ represent the number of critical samples in the azimuth and elevation planes. For a UPA, the 2D steering vector can be represented as an $n \times 1$ vector given by the Kronecker product of the steering vectors for the azimuth and elevation angles [45]:

$$\boldsymbol{a}_n(\theta^{\rm az}, \theta^{\rm el}) = \boldsymbol{a}_{n_{\rm az}}(\theta^{\rm az}) \otimes \boldsymbol{a}_{n_{\rm el}}(\theta^{\rm el}) \tag{10.34}$$

where $\theta^{\rm az} \in [-1/2, 1/2]$ and $\theta^{\rm el} \in [-1/2, 1/2]$ are the azimuth and elevation spatial frequencies, and $\boldsymbol{a}_n(\theta)$ is the steering vector for a 1D ULA defined in Eq. (10.3). A multipath channel can then be developed using these steering vectors along the lines of Eq. (10.8) for P2P links and along the lines of Eq. (10.23) and Eq. (10.21) for P2MP links. The beamspace system representation is obtained considering n orthogonal spatial directions, $n_{\rm az}$ in azimuth and $n_{\rm el}$ in elevation, with orthogonal spacings $\Delta\theta_o^{\rm az} = \frac{1}{n_{\rm az}}$ and $\Delta\theta_o^{\rm el} = \frac{1}{n_{\rm el}}$. The columns of the beamforming matrix $\boldsymbol{U}_n$ are steering vectors corresponding to n fixed spatial frequencies in azimuth and elevation [45]:

$$\boldsymbol{U}_n = \frac{1}{\sqrt{n}}\left[\boldsymbol{a}_n\left(i\Delta\theta_o^{\mathrm{az}}, \ell\Delta\theta_o^{\mathrm{el}}\right)\right]_{i\in\mathcal{I}(n_{\mathrm{az}}),\ell\in\mathcal{I}(n_{\mathrm{el}})} \tag{10.35}$$

that represent n orthogonal beams covering the spatial horizon $(-\frac{\pi}{2} \le \phi^{\mathrm{az}} \le \frac{\pi}{2}, -\frac{\pi}{2} \le \phi^{\mathrm{el}} \le \frac{\pi}{2})$ and form a basis for the n-dimensional spatial signal space. In fact, $\boldsymbol{U}_n$ can also be represented as a Kronecker product of the beamforming matrices in azimuth and elevation: $\boldsymbol{U}_n = \boldsymbol{U}_{n_{\mathrm{az}}} \otimes \boldsymbol{U}_{n_{\mathrm{el}}}$; $\boldsymbol{U}_n^H\boldsymbol{U}_n = \boldsymbol{U}_n\boldsymbol{U}_n^H = \boldsymbol{I}$.

A time- and frequency-selective channel model can be obtained by including path delays and Doppler shifts [35, 73] in the models in Eq. (10.8) and Eq. (10.23). A number of signaling strategies can be developed for the resulting time-varying wideband MIMO channel as discussed elsewhere [35]. However, new constraints relevant to mmW channels (e.g. sparsity) need to be incorporated into the models and signaling schemes.

10.6 Conclusion

In this chapter we have presented a framework for beamspace MIMO communication for optimizing the performance–complexity tradeoffs inherent to high-dimensional MIMO systems encountered at mmW frequencies. A key insight that drives complexity reduction is that the MIMO channel is expected to exhibit a sparse structure in beamspace due to the predominantly LoS and single-bounce modes of multipath propagation and the high dimension of the spatial signal space. MIMO system representation in the beamspace naturally reveals the channel sparsity. The concept of beamspace sparsity masks – which capture the dominant beams through power thresholding – is introduced to characterize the low-dimensional sparse channel subspace.

We have considered the design and analysis of both P2P single-user links and P2MP multiuser links. In particular, we have focussed on the development of low-complexity transceiver architectures that leverage the channel sparsity masks to design low-dimensional precoding schemes at the transmitter and the corresponding processing strategies at the receiver. By choosing the thresholding parameters of the sparsity masks appropriately, the performance of the low-complexity beamspace MIMO transceivers can be made to approach the optimal performance arbitrarily closely. This performance–complexity optimization afforded by the beamspace MIMO transceivers is illustrated through representative numerical results.

The beampace MIMO framework outlined in this chapter applies to the design and analysis of all leading architectures for high-dimensional MIMO systems overviewed in Section 10.2, including phased array-based systems and lens-based systems. In particular, the lens-based CAP-MIMO architecture is a natural candidate for realizing the performance–complexity tradeoffs afforded by beamspace MIMO theory. The CAP-MIMO architecture achieves the key operational functionality of electronic MBDM through the combination of front-end lens antenna, focal surface feed antennas, and the mmW beam selector network. The resulting transceiver architectures enable performance–complexity optimization from both hardware and computational perspectives.

An initial proof-of-concept demonstration of CAP-MIMO has been achieved via a 10-GHz prototype link that can support four spatial channels at a 1-Gbps rate [26, 74]. These results provide a compelling demonstration in a fixed P2P link and indicate that the hybrid analog-digital CAP-MIMO transceiver can potentially deliver near-optimal performance and unprecedented operational functionality with dramatically lower complexity than competing mmW MIMO designs.

While the recent developments are promising, much work needs to be done to realize the potential of mmW MIMO in emerging 5G systems. The ideas outlined in this chapter provide a solid foundation for addressing the technical challenges and harnessing the tremendous opportunities offered by emerging mmW MIMO systems.

References

[1] Plumb, M. (2012) Fantastic 4G. *IEEE Spectrum (Top Tech 2012)*, **49** (1), 51–53.

[2] Hilbert, M. and Lopez, P. (2011) The world's technological capacity to store, communicate, and compute information. *Science*, **332** (6025), 60–65.

[3] Baraniuk, R. (2011) More is less: signal processing and data deluge. *Science*, **331** (6018), 717–719.

[4] Bleicher, A. (2013) A surge in small cells [2013 Tech To Watch]. *IEEE Spectrum*, **50** (1), 38–39, doi:10.1109/MSPEC.2013.6395306.

[5] Pi, Z. and Khan, F. (2011) An introduction to millmeter-wave mobile broadband systems. *IEEE Commun. Mag.*, pp. 101–107.

[6] Adhikari, P. (2008) *White Paper: Understanding Millimeter Wireless Communication*, Loea Corporation.

[7] BuNGEE Project (2010) *Beyond Next Generation Mobile Broadband, IR 1.1*. URL http://www.ict-bungee.eu/).

[8] Nokia Solutions and Networks (2013), Looking ahead to 5G, URL http://nsn.com/file/28771/nsn-5g-white-paper.

[9] Wells, J. (2010) *Multigigabit Microwave and Millimeter-Wave Wireless Communications*, Artech House.

[10] Kadame, V. and Jafar, S. (2008) Interference alignment and the degrees of freedom for the K user interference channel. *IEEE Trans. Info. Theory*, pp. 3425–3441.

[11] Andrews, J., Jindal, N., Haenggi, M., Berry, R., Jafar, S., Guo, D., Shakkottai, S., Heath, R., Neely, M., Weber, S., and Yener, A. (2008) Rethinking information theory for mobile ad hoc networks. *IEEE Commun. Mag.*, pp. 94–101.

[12] Foschini, G.J. (1996) Layered space-time architecture for wireless communication in a fading environment when using multi-element antennas. *Bell Labs Tech. J.*, **1** (2), 41–59.

[13] Telatar, I.E. (1999) Capacity of multi-antenna Gaussian channels. *Eur. Trans. Telecommun.*, **10**, 585–595.

[14] Goldsmith, A. (2006) *Wireless Communications*, Cambridge University Press.

[15] Tse, D. and Viswanath, P. (2005) *Fundamentals of Wireless Communication*, Cambridge University Press.

[16] Jensen, M. and Wallace, J. (2004) A review of antennas and propagation for MIMO wireless communications. *IEEE Trans. Antennas Propag.*, **52** (11), 2810–2824, doi:10.1109/TAP.2004.835272.

[17] Haykin, S. (2005) Cognitive radio: brain-power wireless communications. *IEEE J. Select. Areas Commun.*, pp. 201–220.

[18] Fette, B. (ed.) (2006) *Cognitive Radio Technology*, Elsevier.

[19] Ma, J., Li, G.Y., and Juang, B.W. (2009) Signal processing in cognitive radio. *Proc. IEEE*, pp. 805–823.

[20] Chandrasekhar, V., Andrews, J., and Gatherer, A. (2007) Femtocell networks: a survey. *IEEE Commun. Mag.*, **331** (6018), 717–719.

[21] Wells, J. (2009) Faster than fiber – the future of multi-Gb/s wireless communication. *IEEE Microwave Mag.*, pp. 104–112.

[22] FCC (2004) Public notice. *DA 04-1493*.

[23] Bleicher, A., Millimeter waves may be the future of 5G phones, *IEEE Spectrum*, 13 June 2013. URL http://spectrum.ieee.org/telecom/wireless/millimeter-waves-may-be-the-future-of-5g-phones.

[24] Huang, K.C. and Wang, Z. (2011) *Millimeter Wave Communication Systems*, IEEE-Wiley.

[25] Sayeed, A. and Behdad, N. (2010) Continuous aperture phased MIMO: basic theory and applicatons. *Proc. 2010 Annual Allerton Conference on Communications, Control and Computers*, pp. 1196–1203.

[26] Brady, J., Behdad, N., and Sayeed, A. (2013) Beamspace MIMO for millimeter-wave communications: system architecture, modeling, analysis and measurements. *IEEE Trans. Antennas Propag.*, pp. 3814–3827.

[27] Marzetta, T. (2010) Noncooperative cellular wireless with unlimited numbers of base station antennas. *IEEE Trans. Wireless Commun.*, pp. 3590–3600.

[28] Pi, Z. and Khan, F. (2011) Millimeter-wave mobile broadband: the new frontier for mobile communication, in *Proc. 2011 Texas Wireless Summit (presentation).*

[29] Sayeed, A. and Behdad, N. (2011) Continuous aperture phased MIMO: a new architecture for optimum line-of-sight links. *Proc. 2011 IEEE Antennas and Propagation Symposium*, pp. 293–296.

[30] Sayeed, A. and Brady, J. (2013) Beamspace MIMO for high-dimensional multiuser communication at millimeter-wave frequencies. *2013 IEEE Global Communications Conference*, pp. 3785–3789.

[31] Ayach, O., Heath, R., Abu-Surra, S., Rajagopal, S., and Pi, Z. (2012) Low complexity precoding for large millimeter wave MIMO systems, in *Communications (ICC), 2012 IEEE International Conference on*, pp. 3724–3729, doi:10.1109/ICC.2012.6363634.

[32] Rappaport, T., Sun, S., Mayzus, R., Zhao, H., Azar, Y., Wang, K., Wong, G., Schulz, J., Samimi, M., and Gutierrez, F. (2013) Millimeter wave mobile communications for 5G cellular: It will work! *Access, IEEE*, **1**, 335–349, doi:10.1109/ACCESS.2013.2260813.

[33] Goldsmith, A., Jafar, S., Jindal, N., and Vishwanath, S. (2003) Capacity limits of MIMO channels. *IEEE J. Selec. Areas Commun. (special issue on MIMO systems)*, **21** (5), 684–702.

[34] Gesbert, D., Kountouris, M., Heath, R., Chae, C.B., and Salzer, T. (2007) From single user to multiuser communications: Shifting the MIMO paradigm. *IEEE Signal Process. Mag.*, **24** (5), 36–46.

[35] Sayeed, A. and Sivanadyan, T. (2010) Wireless communication and sensing in multipath environments using multi-antenna transceivers, in *Handbook on Array Processing and Sensor Networks* (eds J. Liu and S. Haykin), IEEE-Wiley.

[36] Torkildson, E., Ananthasubramaniam, B., Madhow, U., and Rodwell, M. (2006) Millimeter-wave MIMO: wireless links at optical speeds, in *Proc. 44th Allerton Conference on Communication, Control and Computing*, pp. 36–45.

[37] Sheldon, C., Seo, M., Torkildson, E., Rodwell, M., and Madhow, U. (2009) Four-channel spatial multiplexing over a millimeter-wave line-of-sight link, in *Proc. IEEE MTT-S Int. Microwave Symp.*, pp. 389–393.

[38] Bohagen, F., Orten, P., and Oien, G. (2007) Design of optimal high-rank line-of-sight MIMO channels. *IEEE Trans. Wireless Commun.*, **6** (4), 1420–1425.

[39] Mecklenbrauker, C.F., Matthaiou, M., and Viberg, M. (2011) Eigenbeam transmission over line-of-sight MIMO channels for fixed microwave links, in *Proc. 2011 International ITG Workshop on Smart Antennas (WSA)*, pp. 1–5.

[40] Papadogiannis, A. and Burr, A. (2011) Multi-beam assisted MIMO; a novel approach to fixed beamforming, in *Future Network Mobile Summit (FutureNetw), 2011*, pp. 1–8.

[41] Hara, Y., Taira, A., and Sekiguchi, T. (2003) Weight control scheme for MIMO system with multiple transmit and receive beamforming. *IEEE Vehicular Technology Confeference (Spring 2003)*, pp. 823–827.

[42] Sarris, I. and Nix, A.R. (2005) Maximum MIMO capacity in line-of-sight. *2005 International Conference on Information, Communications, and Signal Processing*, pp. 1236–1240.

[43] Sayeed, A.M. (2002) Deconstructing multi-antenna fading channels. *IEEE Trans. Signal Processing*, **50** (10), 2563–2579.

[44] Song, G.H., Brady, J., and Sayeed, A. (2013) Beamspace MIMO transceivers for low-complexity and near-optimal communication at mm-wave frequencies. *2013 IEEE International Conference on Acoustics, Speech and Signal Processing*, pp. 4394–4398.

[45] Brady, J. and Sayeed, A. (2014) Beamspace MU-MIMO for high-density gigabit small cell access at millimeter-wave frequencies, in *2014 IEEE International Workshop on Signal Processing Advances for Wireless Communications*, pp. 80–84.

[46] Niknejad, A. (2010) Siliconization at 60 GHz. *IEEE Microwave Mag.*, pp. 78–85

[47] Daniels, R., Murdock, J., Rappaport, T., and Heath, R. (2010 (supplement)) 60 GHz wireless: Up close and personal. *IEEE Microwave Mag.*, pp. 44–50

[48] Shin, D. and Rebeiz, G.M. (2010) Low-power low-noise 0.13 μm CMOS X-band phased array receivers, in *Proceedings of the IEEE International Microwave Symposium*, vol. 1, Anaheim, CA, vol. 1, pp. 956–959.

[49] Wireless Gigabit (WiGig) Alliance (2010) White paper: Defining the future of multi-gigabit wireless communications.

[50] Babakhani, A., Rutledge, D., and Hajimiri, A. (2009) Near-field direct antenna modulation. *IEEE Microwave Magazine*, pp. 36–46.

[51] Broadcom Corp. Press Release (2014), Millimeter Wave SOC offers 10 Gbps capacity. URL http://news.thomasnet.com/fullstory/Millimeter-Wave-SoC-offers-10-Gbps-capacity-20022362.

[52] Talbot, D. (2014), Intel touts new ultra-high-speed wireless data technology. URL http://www.technologyreview.com/news/524961/intel-touts-new-ultra-high-speed-wireless-data-technology/.

[53] J. Andrews (moderator) (2011) The rapid evolution of cellular networks: femto, pico and all that, in *Proc. 2011 Texas Wireless Summit (panel discussion)*.

[54] Sanayei, S. and Nosratinia, A. (2004) Antenna selection in MIMO systems. *IEEE Commun. Mag.*, pp. 68–73.

[55] Renzo, M., Haas, H., and Grant, P. (2011) Spatial modulation for multi-antenna wireless systems: A survey. *IEEE Commun. Mag.*, pp. 182–191.

[56] Mohammadi, A. and Ghannouchi, F. (2011) Single RF front-end MIMO transceivers. *IEEE Commun. Mag.*, pp. 104–109.

[57] Chen, R., Shen, Z., Heath, R.W., and Andrews, J.G. (2007) Transmit selection diversity for unitary precoded multiuser spatial multiplexing systems. *IEEE Trans. Signal Processing*, pp. 1159–1171.

[58] Deckert, T. and Sayeed, A.M. (2003) A continuous representation of multi-antenna fading channels and implications for capacity scaling and optimum array design. *Proc. IEEE Global Telecommunications Conference*, **4**, 1786–1790.

[59] Hong, Z., Liu, K., Heath, R., and Sayeed, A. (2003) Spatial multiplexing in correlated fading via the virtual channel representation. *IEEE J. Select. Areas Commun. (special issue on MIMO systems)*, **21** (5), 856–866.

[60] Zhou, Y., Rondineau, S., Popovic, D., Sayeed, A., and Popovic, Z. (2005) Virtual channel space-time processing with dual-polarization discrete lens arrays. *IEEE Trans. Antennas Propag.*, pp. 2444–2455.

[61] Sayeed, A.M. and Raghavan, V. (2007) Maximizing MIMO capacity in sparse multipath with reconfigurable antenna arrays. *IEEE J. Select. Topics Signal Process (special issue on adaptive waveform design for agile sensing and communication)*, pp. 156–166.

[62] Sayeed, A. (2006) Sparse multipath channels: modeling and implications, in *Proc. Adaptive Sensor Array Processing Workshop*, pp. 1–41.

[63] Liu, K., Raghavan, V., and Sayeed, A.M. (2003) Capacity scaling and spectral efficiency in wideband correlated MIMO channels. *IEEE Trans. Info. Theory (special issue on MIMO systems)*, pp. 2504–2526.

[64] Veeravalli, V., Liang, Y., and Sayeed, A. (2005) Correlated MIMO wireless channels: capacity, optimal signaling and asymptotics. *IEEE Trans. Info. Theory*, pp. 2058–2072.

[65] Raghavan, V. and Sayeed, A. (2011) Sublinear capacity scaling laws for sparse MIMO channels. *IEEE Trans. Info. Theory*, pp. 345–364

[66] Chen, P., Floyd, B., Lai, J., Natarajan, A., Nicolson, S., Reynolds, S., Tsai, M., Valdes-Garcia, A., and Zhan, J. (2013), Phased-array transceiver for millimeter-wave frequencies. URL https://www.google.com/patents/US8618983, US Patent 8,618,983 (IBM).

[67] Mailloux, R.J. (2005) *Phased Array Antenna Handbook*, Artech House, 2nd edn.

[68] Xu, H., Rappaport, T., Boyle, R., and Schaffner, J. (2000) Measurements and models for 38-Ghz point-to-multipoint radiowave propagation. *IEEE J. Select. Areas Commun.*, pp. 310–321.

[69] Cover, T.M. and Thomas, J.A. (1991) *Elements of Information Theory*, Wiley.

[70] Joham, M., Utschick, W., and Nossek, J. (2005) Linear transmit processing in MIMO communication systems. *IEEE Trans. Signal Processing*, pp. 2700–2712.

[71] Sivanadyan, T. and Sayeed, A. (2008) Space-time reversal techniques for wideband MIMO communication. *2008 Asilomar Conference on Signals, Systems and Computers*, pp. 2038–2042

[72] Rumney (Ed.), M. (2013) *LTE and the Evolution to 4G Wireless: Design and Measurement Challenges*, Wiley.

[73] Sayeed, A.M. (2003) A virtual representation for time- and frequency-selective correlated MIMO channels. *Proc. 2003 International Conference on Acoustics, Speech and Signal Processing*, **4**, 648–651.

[74] Brady, J., Thomas, P., Virgilio, D., and Sayeed, A. (2014) Beamspace MIMO prototype for low-complexity Gigabit/s wireless communication, in *2014 IEEE International Workshop on Signal Processing Advances for Wireless Communications*, pp. 135–139.

11

3D Propagation Channels: Modeling and Measurements

Andreas F. Molisch

Signal Processing for 5G: Algorithms and Implementations, First Edition. Edited by Fa-Long Luo and Charlie Zhang.

11.1 Introduction and Motivation

Multiple-input multiple-output (MIMO) systems can give tremendous performance improvements over single antenna systems because of their beamforming, spatial multiplexing and multi-user capabilities (see for example Chapter 20 of [1]). Since the original papers suggesting MIMO for wireless communications [2, 3], thousands of papers have been written on this topic, and a large variety of transmission schemes, receiver structures and signal processing methods have been developed. MIMO has been adopted as a key component of both wireless LAN and 4G cellular systems, and will similarly play a key role in 5G systems.

Recently, massive MIMO has gained a lot of attention because of its ability to improve the spectrum efficiency (SE) and energy efficiency (EE) as well as to simplify signal processing, and has been regarded as a promising technique for next generation wireless communication networks [4, 5]. The base station (BS), which may be equipped with hundreds of antenna elements, can be used to serve several tens of users simultaneously. Additionally, signal processing may be simplified in massive MIMO, and energy consumption can be reduced since the transmit energy can be focused very precisely towards the intended receivers.

11.1.1 Full-dimensional MIMO

The large number of antenna elements in the BS array poses its own unique challenges. One possibility for realizing such large arrays is use of linear arrays, but these suffer from both practical problems, such as wind load, and finding a sufficiently long structure to support them. Three-dimensional (3D) arrays provide an attractive way of implementation.[1] Such structures can serve users distributed in the elevation and azimuth domains. Both beamforming and spatial multiplexing can be applied [6]. Under the name of full-dimensional MIMO (FD-MIMO), this approach has been explored in the scientific literature and in an international standardization project: 3GPP [7]. For more details on FD-MIMO, please refer to the other chapters in this book.

The performance gains of any MIMO communication system depend on the propagation channel in which it is deployed. It is now well established that correlation between signals at different antenna elements decreases the capacity of a single-user MIMO system, and this correlation is usually higher between antenna elements that are spaced in the vertical, rather than the horizontal, dimension. Thus, for a long time, the main interest of the communications community has been in the analysis of systems with antennas strung out along the horizontal plane. For multiuser systems, antenna arrays in the horizontal plane are also the logical choice, since users located in different parts of the cell are "seen" at different azimuth angles at the BS, so that separation of the users through beamforming from a horizontal array is relatively straightforward.

Since FD-MIMO makes use of the elevation domain, in order to assess the possible system performance and allow efficient design of FD-MIMO systems, a first requirement is an understanding and quantitative characterization of the 3D propagation channel. This chapter aims to provide an overview of the fundamental issues of measuring, characterizing and modeling such channels. The emphasis is on fundamental insights; extensive references to the current

[1] Strictly speaking, a planar array is a 2D structure. But since it allows exploitation of multipath components (MPCs) coming from all directions, it has become common to also subsume it in the 3D array structures

literature are also given. However, we caution that due to the rapid evolution of the area, new papers are likely to appear by the time this book goes to print.

The rest of this chapter is organized as follows. In Subsection 11.1.2, fundamental channel descriptions are presented. Section 11.2 is devoted to advanced measurement techniques for 3D propagation channels. Fundamental propagation effects that influence 3D channel behavior are elucidated in Section 11.3. Key measurement results from the literature are summarized in Section 11.4, and channel models are described in Section 11.5. Conclusions and further discussions on future work are made in Section 11.6.

11.1.2 Fundamental Channel Descriptions

Wireless propagation channels are characterized first and foremost by the channel gain: the ratio of the received power to the transmitted power. Without going into details (which can be found elsewhere [1, 8]), channel gain averaged over small-scale fading can be written as the sum (on a dB scale) of a (distance-dependent) path gain, and a (stochastic) shadowing term that describes the variations of the strengths of the multipath components (MPCs).

For modern cellular systems, both the delay dispersion and angular dispersion play an important role. In a deterministic description – in other words, for a particular location of transmitter, receiver and scattering objects – those quantities can be described by the double-directional impulse response [9], which consists of a sum of contributions from the MPCs:

$$\begin{aligned} h(t, \mathbf{r}_{\mathrm{TX}}, \mathbf{r}_{\mathrm{RX}}, \tau, \Omega, \Psi) &= \sum_{l=1}^{L} h_l(t, \mathbf{r}_{\mathrm{TX}}, \mathbf{r}_{\mathrm{RX}}, \tau, \Omega, \Psi) \\ &= \sum_{l=1}^{L} a_l \delta(\tau - \tau_\ell)\delta(\Omega - \Omega_\ell)\delta(\Psi - \Psi_\ell), \end{aligned} \tag{11.1}$$

where: $\mathbf{r}_{\mathrm{TX}}$ and $\mathbf{r}_{\mathrm{RX}}$ are the locations of the transmitter (TX) and receiver (RX), respectively; Ω and Ψ are the direction-of-departure (DoD) and direction-of-arrival (DoA), respectively, each of which consists of an azimuth and an elevation component; τ is the delay; L is the number of MPCs. The phases of the a_l change quickly, while all other parameters – absolute amplitude $| a_l |$, delay, DoA and DoD – vary slowly with the transmit and receive locations (over many wavelengths). It is noteworthy that in this representation, the MPC amplitudes represent the complex gain of the propagation channel only, without any consideration of the antennas.

For multiple-antenna systems, we are also often interested in the impulse response or channel transfer function of the radio channel (i.e. including the antenna characteristics) from each of the N_{TX} transmit antenna elements to each of the N_{RX} receive antenna elements. This is given by the impulse response matrix. We denote the transmit and receive element coordinates $\mathbf{r}_{\mathrm{TX}}^{(1)}, \mathbf{r}_{\mathrm{TX}}^{(2)}, \ldots, \mathbf{r}_{\mathrm{TX}}^{(N_{\mathrm{TX}})}$, and $\mathbf{r}_{\mathrm{RX}}^{(1)}, \mathbf{r}_{\mathrm{RX}}^{(2)}, \ldots, \mathbf{r}_{\mathrm{RX}}^{(N_{\mathrm{RX}})}$, respectively, so that the impulse response from the ith transmit to the mth receive element becomes

$$\begin{aligned} & h_{i,m} = h\left(\mathbf{r}_{\mathrm{TX}}^{(i)}, \mathbf{r}_{\mathrm{RX}}^{(m)}\right) = \\ & \sum_{\ell} h_\ell\left(\mathbf{r}_{\mathrm{TX}}^{(1)}, \mathbf{r}_{\mathrm{RX}}^{(1)}, \tau_l, \Omega_\ell, \Psi_\ell\right) \widetilde{G}_{\mathrm{TX}}(\Omega_\ell)\widetilde{G}_{\mathrm{RX}}(\Psi_\ell) \\ & \exp\left(j\langle \mathbf{k}(\Omega_\ell), (\mathbf{r}_{\mathrm{TX}}^{(i)} - \mathbf{r}_{\mathrm{TX}}^{(1)})\rangle\right) \exp\left(j\langle \mathbf{k}(\Psi_\ell), (\mathbf{r}_{\mathrm{RX}}^{(m)} - \mathbf{r}_{\mathrm{RX}}^{(1)})\rangle\right) \end{aligned} \tag{11.2}$$

where $\mathbf{k}$ is the wave vector and $\langle\cdot\rangle$ denotes the inner product; $\widetilde{G}$ is the complex antenna pattern. Note that Eq. (11.2) implicitly assumes that the DoD (or DoA) at each antenna element is the same, which is well-fulfilled for concentrated antenna arrays, but might not be valid in distributed or physically large arrays. We also note that the above description is simplified in that it ignores wavefront curvature, diffuse components, as well as the effect of polarization.

11.2 Measurement Techniques

Measurements are the essential basis for our understanding of wireless propagation channels, and are required to parameterize realistic channel models. This section provides an overview of measurement techniques relevant for massive MIMO. We first discuss basic methods to obtain impulse responses in single-antenna systems, and then show how those can be modified to measure multiple-antenna channels. This is followed by a discussion of techniques for extracting channel parameters from measurement results. A brief discussion of ray tracing concludes this section.

11.2.1 Basic Channel Measurement Techniques

All measurement devices ("channel sounders") for obtaining elevation information are based on simpler devices that measure $h(t, \tau)$ or its equivalents, and use one of the following principles.

1. Pulse generator: A direct measurement of the impulse response with short, intense excitation pulses is most straightforward conceptually but is difficult to implement as the required peak-to-average power is very high.
2. Correlation sounders: This is the most widespread sounding technique. The sounder transmits a pseudorandom sequence, and the receiver correlates with the same sequence. Alternatively, chirp signals or multitone signals can be used.
3. Swept time delay cross-correlator: This method transmits the sounding signal multiple times, and samples the received signal at a lower rate, at slightly delayed sampling times at each transmit repetition. This reduces the requirements for the analog-to-digital converter but at the same time reduces the maximum admissible Doppler frequency for the channel to remain identifiable.
4. Network analyzer: This measurement device performs essentially a slow frequency sweep of an exciting sine wave, thus directly measuring the transfer function. Network analyzers are readily available in most laboratories and thus very popular.

11.2.2 MIMO Sounders

In order to get directional information, we can combine the basic channel sounder described above with one of the following two approaches: measurements with directional antennas and array measurements.

11.2.2.1 Measurements with directional antennas

A highly directive antenna is installed at the receiver. This antenna is then connected to the receiver of a "regular" channel sounder. The output of the receiver is thus the impulse response of the combination of channel and the antenna pointed in a specific direction. By stepping through different values of Ψ, we can obtain an approximation to the directionally resolved impulse response. One requirement for this measurement is that the channel stays constant during the *total* measurement duration, which encompasses the measurements into all the different Ψ. As the antenna has to be rotated mechanically in order to point in a new direction, the total measurement duration can be several seconds or even minutes. The better the directional resolution, the longer the measurement duration. This measurement method is mostly used for millimeter-wave channels (both for 2D and 3D measurements), since at microwave ranges, strongly directional antennas are too unwieldy. In the 3D case, the need to step through both azimuth and elevation can lead to extremely long measurement times, and often shortcuts, such as limiting the scanned elevation range, are used.

11.2.2.2 Measurement with an antenna array

The impulse response is measured at all antenna elements of the array (quasi-) simultaneously. The resulting vector of impulse responses is either useful by itself, or the directional impulse response can be extracted from it by appropriate signal-processing techniques, as described below. The measurement of the impulse response at the different antenna elements can be done with three different approaches (described here for the receiver; the same principles apply at the transmitter).

Real arrays
In this case, one demodulator chain exists for each receive antenna element. The measurement of the impulse response thus truly occurs at all antenna elements simultaneously. The drawbacks include high costs, as well as the necessity to calibrate multiple demodulator chains. In the case of 3D measurements, some 100 RF chains are required, something that has only recently been implemented for information transmission [10], and not yet for channel sounding.

Multiplexed arrays
In this technique, multiple antenna elements, but only one demodulator chain, exist. The different antenna elements are connected to a demodulator chain (conventional channel sounder) via a fast RF switch [11]. The receiver thus first measures the impulse response at the first antenna element $h_{1,1}$, then it connects the switch to the second element, measure $h_{1,2}$, and so on. This technique has become the most popular for "standard" MIMO measurements, due to its attractive tradeoff between cost and speed. However, for 3D channel measurements, the main challenge is one of cost: large switches are expensive, in particular at the transmitter when they should be able to deal with high power.

Virtual arrays
In this technique, there is only a single antenna element, which is moved mechanically from one position to the next, measuring the impulse responses at the different positions. The placement of the antennas should occur with high precision (better than a small fraction of a wavelength), typically with a mechanical position condition. For a 3D measurement, obviously a 2D or 3D positioning is required. Note that for obtaining azimuth and elevation information at the RX alone, some 100 positions need to be taken by the RX antenna element.

For MIMO measurements that want to get azimuth and elevation information at both link ends, this implies some 10,000 position combinations. Since moving from one position to the next requires several seconds, the time to obtain a transfer function matrix at just one measurement position can take several hours. For this reason, virtual arrays are not attractive for large-scale 3D MIMO campaigns.

Hybrid array

This approach combines the multiplexed array with a virtual array. A small multiplexed array (using as large a switch as possible within the given financial and other constraints) is moved by a positioner to different locations. The number of these mechanically scanned locations is much smaller than that in a pure virtual-array approach (obviously, the total "effective" number of antenna elements $N_{\text{tot}} = N_{\text{switch}} \times N_{\text{position}}$).

A basic assumption for evaluation with virtual and hybrid arrays is that the environment does not change during the measurement procedure. This precludes scenarios in which cars or moving persons are significant interacting objects; in the latter case, only real or switched arrays can be used.

A further challenge encountered in virtual and hybrid approaches is the impact of frequency drift and loss of synchronization between TX/RX. The longer a measurement lasts, the higher the impact of these impairments. This can be resolved either by having a cable connection between the clocks at TX and RX, or, if that is not possible, using high-precision clocks, such as Rubidium clocks, at both link ends. In the latter case, the residual frequency drift can still be problematic although an approach to circumvent this by using reference antennas has been described [12].

11.2.3 Parameter Extraction Techniques

The measurements with the above-mentioned sounders yield the transfer function matrix (or impulse response matrix). From these, the parameters of the MPCs can be extracted through further signal processing. A variety of processing methods are available, which represent a tradeoff between accuracy and complexity.

The simplest method is Fourier processing. It is well known that the location vector and the directional cosine (i.e. the cosine of the DoA or DoD) constitute a Fourier pair. For a uniform linear array, it is thus straightforward to form beams in one dimension (usually azimuth) by means of a fast Fourier transform (FFT). Similarly, a planar rectangular array allows the formation of beams in azimuth and elevation through a 2D FFT. The resolution of this type of direction determination is limited by the antenna aperture: as a rough rule of thumb, the resolution is $4\pi/N_{\text{tot}}$.

More accurate results can be achieved through the use of "high-resolution algorithms". There are two main categories: subspace-based algorithms and parametric methods. The former encompasses MUSIC and ESPRIT, which were widely used in the 1990s, but we will not discuss these any further here, since they nowadays have mostly been superseded. The latter category starts with the assumption that the received signal can be described through a set of parameters, in particular, the delays, DoAs and DoDs of the MPCs. The algorithms then aim to estimate those parameters.

The simplest of the parametric methods is the CLEAN algorithm. It is an iterative deconvolution technique first introduced for the enhancement of radio-astronomical maps of the sky [13], and widely used in microwave and ultrawideband communities as an effective post-processing method for time-domain channel measurements. However, the principle can also be used to extract the delay and direction information from the channel transfer functions. It is a grid-search-based algorithm and hence the resolution is limited by the grid size. A 3D version of the algorithm is described in [12].

Even higher accuracy can be achieved through a maximum-likelihood estimation of the multipath parameters. Due to the large number of parameters – in particular in the case of 3D, but even for 2D – a simultaneous maximum-likelihood estimation of all parameters is not possible; instead iterative methods have to be used.

An efficient, iterative implementation is the space-alternating generalized expectation-maximization (SAGE) algorithm [14]. Since 2000, this has become the most popular method for channel sounding evaluations. The drawbacks are that the iterations may converge to a local optimum, not the global one, and the relatively slow convergence speed, which again becomes most noticeable for 3D measurements. The algorithm has been used for a majority of the measurement campaigns mentioned in Section 11.4.

An alternative approach is the RiMax algorithm [15]. It uses a somewhat different type of iteration, based on gradient methods such as Levenberg–Marquardt [16]. This algorithm has been used in the evaluation of recent measurement campaigns [17].

11.2.4 Ray Tracing

An attractive complement to measurement campaigns is ray tracing, which solves (using a high-frequency approximation) Maxwell's equations for given boundary conditions (i.e. given the locations and electromagnetic properties of objects in the environment). Ray tracing inherently provides DoAs and DoDs, as well as the delay and amplitude, of the MPCs. Its main drawbacks lie in the errors introduced by inaccurate databases: often buildings are described as smooth polygons, neglecting the impact of "fine structure" and, furthermore, the dielectric properties of the buildings are often not known. On the positive side, ray tracing allows analysis of a much larger number of locations than an actual measurement campaign. For lack of space, we will not discuss details of ray tracing but just refer to the extensive literature on this topic.

11.3 Propagation Effects

One of the outputs of measurement campaigns and ray tracing, which we will discuss below, is an understanding of the fundamental propagation processes determining the joint azimuth and elevation characteristics. As always in propagation research, the peculiarities of the environment have to be taken into account.

Consider first the situation in which most of the scatterers surround the mobile station (MS, also called "user equipment" or UE in 3GPP). The mean elevation observed at the BS is therefore given by the quasi-LOS (line-of-sight) direction under which the BS "sees" the MS. The angular spectrum at the MS is determined by the distribution of the scatterers along the z-axis. The model was widely used in the early days of FD-MIMO research. It led investigators to conclude that users at different distances from the BS can be well separated, as they are seen under different elevation angles. However, few measurements have shown agreement with this idealized scenario, related to the fact that the simple "circle of scatterers" model is mostly applicable for rural environments, for which massive MIMO is not of paramount interest.

11.3.1 Urban Macrocells

Propagation conditions are more complex in urban environments. Here, there are three main effects dominating the propagation from TX to RX.

MPC diffraction over rooftops

MPCs propagating over the rooftops, and then being diffracted in the vertical plane towards the MS are characterized by a small angular spread, which is determined mostly by the angles under which the BS "sees" the relevant rooftop edges. More importantly, the mean co-elevation[2] shows a relatively weak dependence on the distance between BS and MS [18]. Finally, the elevation at the MS is determined by the angle under which the MS "sees" the roof edge in the street canyon it is located in. These angles can be quite large (on the order of 60°, typically).

MPC waveguides in street canyons

A street canyon essentially constitutes a "waveguide", with a missing top, and rough and "hole-affected" (due to cross streets) sidewalls. Such a structure can be quite effective in guiding waves with relatively small losses; in particular for microcells, where the TX is below the rooftop, and peer-to-peer communications such as car-to-car, this propagation process can be dominant. Such waves propagate essentially in the horizontal plane, leading to small elevation angles at BS and MS. Again, the elevation angle shows little dependence on the actual distance between TX and RX.

MPCs reflected at "far scatterers"

In a number of environments, high-rise structures can constitute an efficient "relay point" for MPCs. Such a structure would have LOS to both the BS and the MS, and a specularly reflecting surface. Thus, the MPCs suffer relatively little attenuation, proportional to $(d_{\text{BS-scatt}} + d_{\text{scatt-MS}})^2$. The elevation angle under which such scatterers are seen is usually positive, but with a relatively small value.

These propagation mechanisms (or a subset of them) have been found in a number of measurements [19, 20, 21, 22, 17]. An example is shown in Fig. 11.1.

In many cases, a clear clustering of the MPCs can be observed, and the overall angular power spectrum in the elevation domain can thus be modeled as a sum of those contributions [21, 23].

11.3.2 Outdoor-to-indoor

The elevation spectrum can also be changed when the MS is located indoors; a situation that occurs nowadays for a majority of cellular links. Since MPCs can be reflected off different points on the walls surrounding the MS, very high elevation angles (and, overall, very high angular spread) are possible, see Fig. 11.2. The impact on the elevation spectrum at the BS, on the other hand, is small: all the above-mentioned propagation effects (over the rooftop, wave guiding, far scatterers) result in angles that are essentially independent of whether the antenna

[2] Co-elevation is the offset from the horizontal plane, while the elevation angle is defined as being zero when looking straight up.

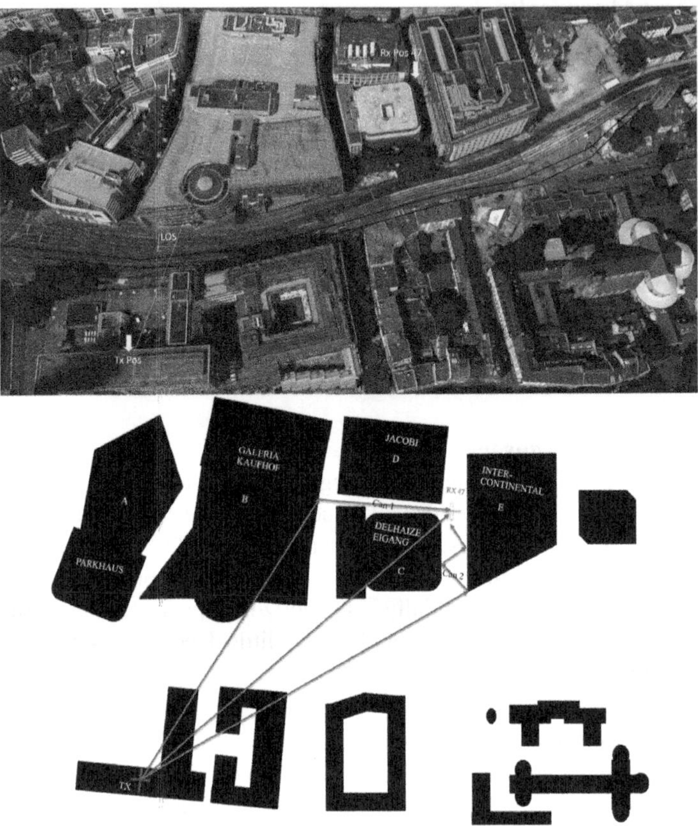

Figure 11.1 Photo and propagation processes in an urban macrocell in Cologne. *Source:* Sangodoyin 2015. Reproduced with permission from IEEE [17]

Figure 11.2 Figure showing the direction of arrival of MPCs for a macrocell measurement

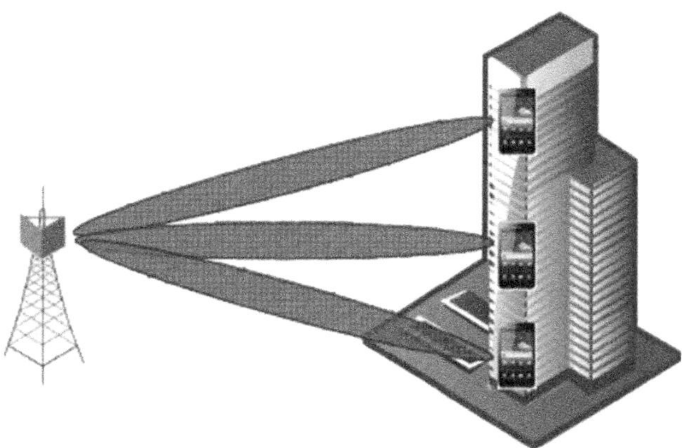

Figure 11.3 Beamforming from BS to MSs on different floors of high-rise building. *Source:* Zhang 2015. Reproduced with permission of IEEE [24]

is inside or outside, and even are mostly independent of how high the MS is located above ground (i.e. which floor it is on). We note, however, that the *efficiencies* of the different processes, and thus their relative weights, *do* depend on the MS height. This is most noticeable for over-the-rooftop propagation, since the diffraction angle with respect to the rooftops changes as the MS height increases. In extreme cases, an MS on an upper floor can have LOS to the BS, even though the MS at street level does not.

A special case occurs when a BS is located outside a high-rise building, and has LOS to the different floors (see Fig. 11.3). In this case, we can expect that the mean elevation at the BS is approximately the LOS direction (plus some variations), a fact confirmed in the measurements of [24]. Another paper describes a measurement campaign in such a scenario, but provides only single- and multi-user capacities, and not angular spreads [25].

11.3.2.1 Indoor

A number of papers have investigated elevation spreads in indoor scenarios [26], showing larger elevation spreads in particular at the BS (access point) than in outdoor environments. This is intuitive, since reflections from the ceiling can greatly increase the elevation spread. It has also been observed that diffuse multipath (DMC) is very important in indoor environments [26]; again, this is intuitive, since the objects within a room typically have rougher surfaces, and a number of scatterers might be close to the TX and RX, giving rise to spherical wavefronts (deviations from the "sum of plane waves" model are usually incorporated into the DMC).

11.4 Measurement Results

This section will review the key measurement results for angular spreads, in particular elevation spreads. We first concentrate on spreads at the MS, followed by the spreads at the BS, which are most important for FD-MIMO systems.

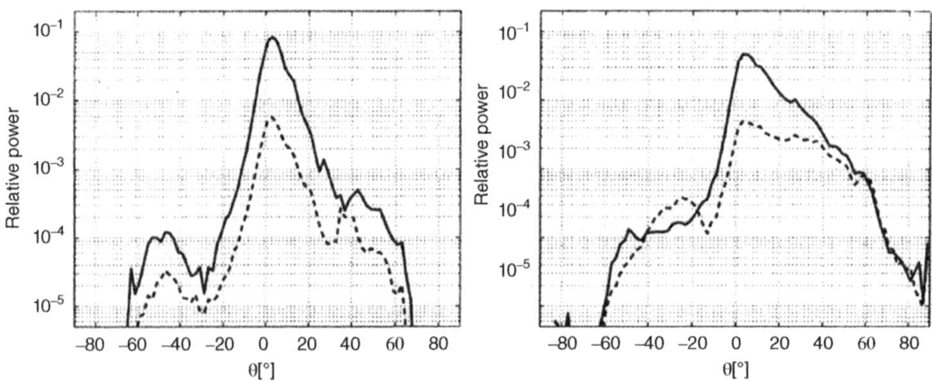

Figure 11.4 Elevation spectra for urban microcell (left) and macrocell (right) at the MS. *Source:* Kalliola 2002. Reproduced with permission of IEEE [28]

11.4.1 Angular Spreads at the Mobile Station

Measurements of the elevation characteristics of propagation channels fall into two categories: elevation spectrum at the BS and at the MS; depending on whether we consider uplink or downlink, direction at the BS might be DoA or DoD, respectively. Historically, elevation spectra at the MS were measured first, as they showed greater spreads and provided valuable insights into propagation mechanisms. MPCs arriving at the MS via over-the-rooftop propagation in urban environments tend to have co-elevation angles of up to 60°, while wave-guided waves have co-elevation on the order of 10–20° [27, 19]. Extensive measurements have shown that an (asymmetrial) double-exponential function provides a good fit for the elevation power spectrum measured at the MSs (see Fig. 11.4 and [28]). Rms elevation spreads measured in that reference were typically less than 10–15°; similar values have been obtained for an indoor hotspot [29]. Others have measured somewhat larger values in both outdoor [30] and indoor [31] environments; the differences are most likely due to the different building structures in the measured cities.[3] It has also been found that the MS elevation spread increases as the BS height increases [28]. The cross-polarization discrimination has been measured as a function of the elevation [33] .

11.4.2 Angular Spreads at the Base Station

There is a considerable number of measurements in urban environments. An evaluation of a number of measurements with rectangular antenna arrays at the BS [20, 21] found that over-the-rooftop propagation, wave guiding in street canyons, and reflections off high-rise buildings and dominant scatterers provide different contributions to the elevation spectrum, as

[3] A measurement setup for SIMO measurements has also been described [32], but only a few sample measurements were shown.

described in Section 11.3. The measured co-elevations of the MPCs are typically around −10 to 10° in both macro- and microcellular environments. Other groups have measured elevation spreads:

- on the order of 2–4° in LOS and 5–15° in NLOS (non-line-of-sight) in urban environments [34]
- 2 and 7° for LOS and NLOS in macrocells, and 3 and 6° in microcells [35]
- between 1 and 4° in macrocell NLOS [23]
- around 10° for LOS and 5° in NLOS in an urban macrocell [18].

Measurements and simulations have also shown that the elevation spread has a distance dependence: generally a smaller spread at larger distances [36]. Elevation spreads have been found to decrease (in LOS) from approx. 3–5° to 1–2° [37]. Furthermore, a significant difference in elevation spreads between the LOS and NLOS propagation environments exists. Other interesting sample measurements are described by Jungnickel *et al.* [38]

Further ray tracing studies [39] investigated the impact of the BS height on the elevation spreads at the BS and MS. The BS spread showed only minor variations (only in the far tails of the elevation spread). A slight decrease of elevation spread with increasing BS height has been found. Following [45], Fig. 11.5 shows that the elevation spread in outdoor-to-indoor scenarios tends to be lower when the MS is on a higher floor, as obtained from ray tracing.

For the microcell, where the BS has LOS to different floors in a high-rise building, elevation spreads have been measured at between 5 and 20° for BS–MS distances between 120 and 150 m [24]. Ref. [40] investigated whether a cluster model is better suited to represent the channel than a continuous distribution of the dispersion characteristics. They find, not unexpectedly,

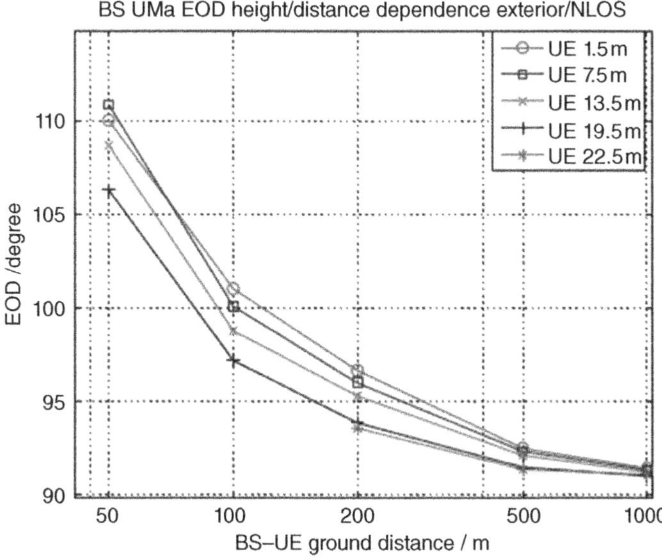

Figure 11.5 Distance dependence of the elevation spread for different MS(UE) heights. *Source:* Wang 2014. Reproduced with permission of IEEE [45]

that including the additional domain of elevation results in a better separability of clusters, and thus a larger number of clusters. They also find that the larger the height difference between the TX and RX, the more clusters can be found. The mean value of the *cluster* elevation spread at the BS is 5° , and 10° at the MS.

The impact of vegetation in an urban microcell was investigated in [41], and a mean elevation spread of 5–7° (depending on BS height and polarization), and a variance of 1–2° was measured.

Ref. [42] analyzed the elevation spread (as well as other parameters) for an ultrawideband channel in several apartments and houses, and observed that the elevation spread decreases with frequency, from 28 to 23°. Narrowband measurements investigating the distance dependence of the elevation spread in an indoor office environment were done in Ref. [64]. It was found that the angular spread decreased from 10 to 2° (for LOS) and from 18 to 12° (for NLOS). However, in an atrium, elevation spread *increased* with distance, at least for NLOS situations (from 15 to 25°), although it (slightly) decreased in LOS situations. For a hotspot, values of 3–15° were measured in [29].

Thomas *et al.* provide a model for elevation spreads at mm-waves in urban microcells where the elevation spread decreases with distance, up to a saturation value [43]. Sun *et al.* provide results for indoor mm-wave channels [44].

11.5 Channel Models

System simulations require channel models that are realistic yet simple enough to allow efficient large-scale simulations. This section describes how the addition of a new dimension – elevation – is handled in both geometry-based channel models and in 3GPP-type (tapped delay line) models. These discussions are preceded by a brief description of fundamental modeling methods.

11.5.1 Fundamental Modeling Methods

The characteristics of 3D channels are captured by various channel models. In order to reflect the variety of channel realizations in nature, stochastic channel descriptions that provide probability density functions (PDFs) of the impulse responses, are required. One common form is the "tapped delay line" model, which, in its generalization to the double-directional case, fixes the delay, DoA and DoD of the MPCs, while allowing the phase and amplitudes to be chosen stochastically according to their PDFs with the common assumption of wide sense stationarity–uncorrelated scattering (WSS-US); in other words that the statistics of the fading do not change with time, and fading of each MPC is uncorrelated. Such an approach underlies the 3GPP spatial channel model (SCM) as well as the ITU/Winner models. An alternative approach is the geometry-based stochastic channel model (GSCM) [46], in which locations of scatterers or interacting objects are defined according to a given probability density function, and the characteristics of the MPCs are finally obtained through a simple ray-tracing procedure that allows only single-interaction processes or, as in the COST 273 [47] and COST 2100 [48] models, double interactions with so-called "twin clusters" [49]. In many cases, tapped delay line and geometry-based approaches are combined, such that the "locations" (i.e. delays, DoAs and DoDs) of the taps are obtained from geometric considerations.

Another important concept in modern channel modeling is clustering. Measurement results show that in many environments, MPCs arrive in clusters i.e., groups with similar characteristics. Intra-cluster parameters, such as the decay time constant of the cluster power decay, do not change, even when an MS moves over a larger area, though the inter-cluster parameters, such as the delay of one cluster with respect to another, might change. Furthermore, a cluster shape function – the squared magnitude of the double-directional impulse response of the cluster, averaged over the small-scale fading –can often be decomposed [50]:

$$P(\tau,\theta,\varphi,\theta',\varphi') = P_{\tau}(\tau)P_{\theta}(\theta)P_{\varphi}(\varphi)P_{\theta'}(\theta',\tau)P_{\varphi'}(\varphi',\tau) \tag{11.3}$$

where θ, φ are the elevation and azimuth at the BS (and analogously for θ', φ' at the MS). although in other work the angular spectra at the BS are also assumed to depend on the delay. It is common to assume that the PDP is a single-exponential function, while the angular power spectra – both elevation and azimuth – are Laplacian functions. The second central moments of these functions – the delay spread, azimuth spreads and elevation spreads – are then commonly used for the characterization of the environments. The different propagation processes described in Section 11.3 lead to MPCs that are typically in different clusters (see Fig. 11.6).

11.5.2 Regular and Irregular GSCMs

A number of reference models have been suggested for simulating the performance of MIMO systems including the elevation. Ref. [51] develops a model based on the assumption that scatterers are uniformly distributed inside an ellipsoidal volume; some comparisons with ray tracing results showed good agreement. Scatterers on spherical surfaces around BS and MS are assumed by [52], which also takes polarization into account, and computes in closed form

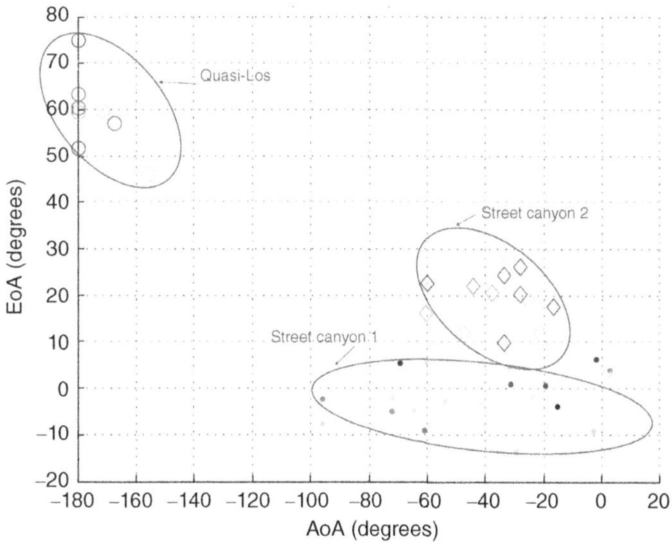

Figure 11.6 Clusters observed at a location in Cologne (compare Fig. 11.1)

the correlation between the signals at the antenna elements. Scatterers on cylindrical surfaces are discussed in [53, 54]. Ref. [55] developed a theoretical model including dependence on BS height, and compared it to the measurements of [56].

11.5.3 3GPP Channel Models

For system simulations, a number of extensions of the ITU/Winner model (which itself is an extension of the 3GPP-SCM model) have been proposed. A number of studies in the open literature, which were based on measurements and ray tracing [36, 57, 58], suggested elevation spreads and mean elevations for indoor, hotspot, outdoor-to-indoor, urban microcell and urban macrocell; elevation spreads at the BS are typically less than 10°. Several papers discuss system considerations for extending the 3GPP-SCM to elevation [59, 60, 61].

From 2013 to 2015, a subgroup within 3GPP has developed the "official" generalization of the SCM model to include 3D propagation. The parameterization was done on the basis of three types of results:

- ray tracing results in "virtual" cities with a regular street grid and buildings chosen from a prescribed probability density function
- ray tracing based on 3D maps of actual cities
- measurement campaigns (several of which are among the above-cited papers).

Since the ITU/Winner model is based on clusters, we have to distinguish between cluster spreads and composite spreads. Also noteworthy is the nomenclature of these models: the ITU/Winner models implicitly assume a downlink, so that "departure" really means "BS" and "arrival" means "MS"; naturally the spread at a particular station should be independent of whether it operates in transmit or receive mode. The elevation spread is modeled as a log-normally distributed variable whose mean and variance are parameterized for the different environments, consistent with the modeling of the azimuthal spread.

The final 3GPP report [62] specifies the simulation methodology – which is very similar to the "standard" 3GPP model [63], and not repeated here for the sake of brevity – as well as the parameterization for three scenarios:

- urban micro cell with high MS density
- urban macro cell with high MS density
- urban macro cell with one high-rise per sector and 300-m inter-site distance.

In each of the scenarios, 80% of users are indoors, and the remainder are outdoors.

11.6 Summary and Open Issues

Despite the fact that measurement and modeling of 3D channel characteristics has been going on for a long time, there are still many open issues. Firstly, measurement techniques need to be refined to allow measurements with additional dimensions within a reasonable time and at reasonable cost. Some progress has been made in the construction of hybrid channel sounders, but further improvements are required. More importantly, a great many additional measurements need to be performed, in particular in outdoor-to-indoor scenarios, which account for the majority of traffic nowadays. A number of important scenarios with high densities of users – shopping malls, sports stadia and so on – have not been measured at all up to now. Also, the relationship between polarization and elevation (and other parameters) needs to be explored further.

On the systems side, the interplay between the elevation characteristics and massive MIMO requires more investigation. A proper distinction has to be made between the possibility of spatial multiplexing for one user (associated with the elevation spread at one location) and the multi-user capacity (associated with the separability of users at different locations in the elevation domain). While FD-MIMO looks promising for a number of applications, it is essential that performance limits be explored before widespread deployment, to avoid any "trough of disappointment" that might be created by deploying in the wrong scenarios.

Acknowledgements

Discussions with Dr. Charlie Zhang, Prof. Reiner Thomae, Prof. Fredrik Tufvesson, Dr. Martin Toeltsch, Vinod Kristem and Seun Sangodoyin are gratefully acknowledged. Parts of this work were supported by a gift from Samsung and grants from the National Science Foundation.

Disclaimer

No warranty is given for the factual correctness or usability for any purpose of the material included in this chapter and neither the authors or the publisher shall be liable for any claims, losses, damages or expenses, whether direct or indirect, arising from actions taken by the reader that result in physical injury or any other damage.

References

[1] Molisch, A.F. (2010) *Wireless Communications*, 2nd. ed., IEEE Press-Wiley.

[2] Winters, J. (1987) On the capacity of radio communication systems with diversity in a Rayleigh fading environment. *IEEE J. Select. Areas Commun.*, **5** (5), 871–878.

[3] Foschini, G.J. and Gans, M.J. (1998) On limits of wireless communications in a fading environment when using multiple antennas. *Wireless Pers. Commun.*, **6** (3), 311–335.

[4] Marzetta, T. (2010) Noncooperative cellular wireless with unlimited numbers of base station antennas. *IEEE Trans. Wireless Commun.*, **9** (11), 3590–3600.

[5] Rusek, F., Persson, D., Lau, B.K., Larsson, E., Marzetta, T., Edfors, O., and Tufvesson, F. (2013) Scaling up MIMO: opportunities and challenges with very large arrays. *IEEE Signal Process. Mag.*, **30** (1), 40–60.

[6] Cheng, X., Yu, B., Yang, L., Zhang, J., Liu, G., Wu, Y., and Wan, L. (2014) Communicating in the real world: 3D MIMO. *IEEE Wireless Commun.*, **21** (4), 136–144.

[7] Kim, Y., Ji, H., Lee, J., Nam, Y.H., Ng, B.L., Tzanidis, I., Li, Y., and Zhang, J. (2014) Full dimension MIMO (FD-MIMO): the next evolution of MIMO in LTE systems. *IEEE Wireless Commun.*, **21** (2), 26–33.

[8] Molisch, A.F., Greenstein, L.J., and Shafi, M. (2009) Propagation issues for cognitive radio. *Proc. IEEE*, **97** (5), 787–804.

[9] Steinbauer, M., Molisch, A.F., and Bonek, E. (2001) The double-directional radio channel. *IEEE Antennas Propag. Mag.*, **43** (4), 51–63.

[10] Vieira, J., Malkowsky, S., Nieman, K., Miers, Z., Kundargi, N., Liu, L., Wong, I., Owall, V., Edfors, O., and Tufvesson, F. (2014) A flexible 100-antenna testbed for massive MIMO, in *IEEE GLOBECOM 2014 Workshop on Massive MIMO: from theory to practice*, pp. 287–293.
[11] Thoma, R.S., Hampicke, D., Richter, A., Sommerkorn, G., Schneider, A., Trautwein, U., and Wirnitzer, W. (2000) Identification of time-variant directional mobile radio channels. *IEEE Trans. Instrument. Measure.*, **49** (2), 357–364.
[12] Kristem, V., Sangodoyin, S., and Molisch, A. (2015) Channel measurements and modeling for 3D MIMO outdoor-to-indoor propagation, in *Proc. of 2015 IEEE Global Communications Conference (GLOBECOM)*, pp. 1–6.
[13] Hogbom, J.A. (1974) Aperture synthesis with a non-regular distribution of interferometer baselines. *Astron. Astrophys.*, **15**, 417–426.
[14] Fleury, B., Tschudin, M., Heddergott, R., Dahlhaus, D., and Ingeman Pedersen, K. (1999) Channel parameter estimation in mobile radio environments using the sage algorithm. *IEEE J. Select. Areas Commun.*, **17** (3), 434–450.
[15] Thomä, R., Landmann, M., and Richter, A. (2004) Rimax? A maximum likelihood framework for parameter estimation in multidimensional channel sounding, in *Proceedings of the International Symposium on Antennas and Propagation (ISAP'04)*, pp. 53–56.
[16] Richter, A., Salmi, J., and Koivunen, V. (2006) Distributed scattering in radio channels and its contribution to MIMO channel capacity, in *Antennas and Propagation, 2006. EuCAP 2006. First European Conference on*, IEEE, pp. 1–7.
[17] Sangodoyin, S., He, R., Molisch, A., and Kristem, V. (2015) Cluster-based analysis of 3D MIMO channel measurement in an urban environment, in *Proc. of 2015 IEEE International Conference on Communications,*, pp. 2277–2282.
[18] Pei, F., Zhang, J., and Pan, C. (2013) Elevation angle characteristics of urban wireless propagation environment at 3.5 GHz, in *IEEE 78th Vehicular Technology Conference (VTC Fall), 2013*, pp. 1–5.
[19] Kuchar, A., Rossi, J.P., and Bonek, E. (2000) Directional macro-cell channel characterization from urban measurements. *IEEE J. Antennas Propag.*, **48** (2), 137–146.
[20] Laurila, J., Kalliola, K., Toeltsch, M., Hugl, K., Vainikainen, P., and Bonek, E. (2002) Wideband 3D characterization of mobile radio channels in urban environment. *IEEE J. Antennas Propag.*, **50** (2), 233–243, doi: 10.1109/8.998000.
[21] Toeltsch, M., Laurila, J., Kalliola, K., Molisch, A., Vainikainen, P., and Bonek, E. (2002) Statistical characterization of urban spatial radio channels. *IEEE J. Select. Areas Commun.*, **20** (3), 539–549.
[22] Ghoraishi, M., Ching, G., Lertsirisopon, N., Takada, J., Nishihara, A., Imai, T., and Kitao, K. (2009) Polar directional characteristics of the urban mobile propagation channel at 2.2 GHz, in *Antennas and Propagation, 2009. EuCAP 2009. 3rd European Conference on*, pp. 892–896.
[23] Medbo, J., Asplund, H., Berg, J.E., and Jalden, N. (2012) Directional channel characteristics in elevation and azimuth at an urban macrocell base station, in *Antennas and Propagation (EUCAP), 2012 6th European Conference on*, pp. 428–432.
[24] Zhang, R., Cai, L., Lu, X., Yang, P., and Zhou, J. (2015) Elevation domain measurement and modeling of UMa uplink channel with UE on different floors, in *Computing, Networking and Communications (ICNC), 2015 International Conference on*, pp. 679–684, doi: 10.1109/ICCNC.2015.7069427.
[25] Zhang, C., Tian, L., Zhang, J., Sun, T., and Pan, C. (2014) Measurement-based performance evaluation of 3D MIMO in high rise scenario, in *Wireless Personal Multimedia Communications (WPMC), 2014 International Symposium on*, pp. 718–723, doi: 10.1109/WPMC.2014.7014909.
[26] Mani, F., Quitin, F., and Oestges, C. (2012) Directional spreads of dense multipath components in indoor environments: Experimental validation of a ray-tracing approach. *IEEE Trans. Antennas Propag.*, **60** (7), 3389–3396, doi: 10.1109/TAP.2012.2196942.
[27] Fuhl, J., Rossi, J.P., and Bonek, E. (1997) High-resolution 3-D direction-of-arrival determination for urban mobile radio. *IEEE J. Antennas Propag.*, **45** (4), 672–682.
[28] Kalliola, K., Sulonen, K., Laitinen, H., Kivekas, O., Krogerus, J., and Vainikainen, P. (2002) Angular power distribution and mean effective gain of mobile antenna in different propagation environments. *IEEE Trans. Vehic. Tech.*, **51** (5), 823–838.
[29] Huang, C., Zhang, J., Nie, X., and Zhang, Y. (2009) Cluster characteristics of wideband MIMO channel in indoor hotspot scenario at 2.35GHz, in *Vehicular Technology Conference Fall (VTC 2009-Fall), 2009 IEEE 70th*, pp. 1–5.

[30] Schneider, C., Narandzic, M., Kaske, M., Sommerkorn, G., and Thoma, R. (2010) Large scale parameter for the WINNER II channel model at 2.53 GHz in urban macro cell, in *Vehicular Technology Conference (VTC 2010-Spring), 2010 IEEE 71st*, pp. 1–5.
[31] Quitin, F., Oestges, C., Horlin, F., and De Doncker, P. (2010) A polarized clustered channel model for indoor multiantenna systems at 3.6 GHz. *Vehicular Technology, IEEE Transactions on*, **59** (8), 3685–3693.
[32] de Jong, Y. (2010) Results of mobile channel sounding measurements in the 4.9 GHz public safety band, in *IEEE Radio and Wireless Symposium (RWS), 2010*, pp. 585–588.
[33] Dunand, A. and Conrat, J.M. (2007) Dual-polarized spatio-temporal characterization in urban macrocells at 2 GHz, in *Vehicular Technology Conference, 2007. VTC-2007 Fall. 2007 IEEE 66th*, pp. 854–858.
[34] Wang, J., Zhang, R., Duan, W., Lu, S., and Cai, L. (2014) Angular spread measurement and modeling for 3D MIMO in urban macrocellular radio channels, in *IEEE International Conference on Communications Workshops (ICC), 2014*, pp. 20–25.
[35] Zhong, Z., Yin, X., Li, X., and Li, X. (2013) Extension of ITU IMT-advanced channel models for elevation domains and line-of-sight scenarios, in *Vehicular Technology Conference (VTC Fall), 2013 IEEE 78th*, pp. 1–5.
[36] Thomas, T.A., Vook, F.W., Mellios, E., Hilton, G.S., Nix, A.R., and Visotsky, E. (2013) 3D extension of the 3GPP/ITU channel model, in *IEEE 77th Vehicular Technology Conference (VTC Spring), 2013*, pp. 1–5, doi: 10.1109/VTCSpring.2013.6691813.
[37] Roivainen, A., Hovinen, V., Meinila, J., Tervo, N., Sonkki, M., and Dias, C. (2014) Elevation analysis for urban microcell outdoor measurements at 2.3 GHz, in *5G for Ubiquitous Connectivity (5GU), 2014 1st International Conference on*, pp. 176–180.
[38] Jungnickel, V., Brylka, A., Krueger, U., Jaeckel, S., Narandzic, M., Kaeske, M., Landmann, M., and Thomae, R. (2013) Spatial degrees of freedom in small cells: Measurements with large antenna arrays, in *IEEE 77th Vehicular Technology Conference (VTC Spring), 2013*, pp. 1–6, doi: 10.1109/VTCSpring.2013.6692761.
[39] Kitao, K., Imai, T., Saito, K., and Okumura, Y. (2013) Elevation directional channel properties at BS to evaluate 3D beamforming, in *International Symposium on Intelligent Signal Processing and Communications Systems (ISPACS), 2013*, pp. 640–644.
[40] Du, D., Zhang, J., Pan, C., and Zhang, C. (2014) Cluster characteristics of wideband 3D MIMO channels in outdoor-to-indoor scenario at 3.5 GHz, in *Proc. IEEE ICC 2014*, pp. 1–6.
[41] Le, H.V., Takada, J., Ghoraishi, M., Phakasoum, C., Kitao, K., and Imai, T. (2012) Angular spread of the radio wave propagation in foliage environment, in *6th European Conference on Antennas and Propagation (EUCAP), 2012*, pp. 3356–3360.
[42] Pajusco, P., Malhouroux-Gaffet, N., and El Zein, G. (2015) Comprehensive characterization of the double directional UWB residential indoor channel. *IEEE Trans Antennas Propag.*, **63** (3), 1129–1139.
[43] Thomas, T., Nguyen, H.C., MacCartney, G., and Rappaport, T. (2014) 3D mmWave channel model proposal, in *IEEE 80th Vehicular Technology Conference (VTC Fall), 2014*, pp. 1–6.
[44] Sun, S., Rappaport, T., Thomas, T., and Ghosh, A. (2015) A preliminary 3D mm wave indoor office channel model, in *International Conference on Computing, Networking and Communications (ICNC), 2015*, pp. 26–31, doi: 10.1109/ICCNC.2015.7069289.
[45] Wang, R., Sangodoyin, S., Molisch, A., Zhang, J., Nam, Y.H., and Lee, J. (2014) Elevation characteristics of outdoor-to-indoor macrocellular propagation channels, in *IEEE 79th Vehicular Technology Conference (VTC Spring), 2014*, pp. 1–5.
[46] Molisch, A.F., Kuchar, A., Laurila, J., Hugl, K., and Schmalenberger, R. (2003) Geometry-based directional model for mobile radio channels – principles and implementation. *Europ. Trans. Telecommun.*, **14** (4), 351–359.
[47] Molisch, A., Hofstetter, H. (2006) *The COST273 channel model*, in Correia, L. (ed.) Mobile Broadband Multimedia Networks (COST 273 Final Report) Springer.
[48] Liu, L., Oestges, C., Poutanen, J., Haneda, K., Vainikainen, P., Quitin, F., Tufvesson, F., and Doncker, P. (2012) The COST 2100 MIMO channel model. *IEEE Wireless Commun.*, **19** (6), 92–99.
[49] Hofstetter, H., Molisch, A.F., and Czink, N. (2006) A twin-cluster MIMO channel model, in *First European Conference on Antennas and Propagation, 2006. EuCAP 2006.*, IEEE, pp. 1–8.
[50] Asplund, H., Glazunov, A.A., Molisch, A.F., Pedersen, K.I., and Steinbauer, M. (2006) The COST259 directional channel model part II: Macrocells. *IEEE Trans. Wireless Commun.*, **5**, 3434–3450.
[51] Alsehaili, M., Noghanian, S., Buchanan, D., and Sebak, A. (2010) Angle-of-arrival statistics of a three-dimensional geometrical scattering channel model for indoor and outdoor propagation environments. *IEEE Antennas Wireless Propag. Lett.*, **9**, 946–949.

[52] Dao, M.T., Nguyen, V.A., Im, Y.T., Park, S.O., and Yoon, G. (2011) 3D polarized channel modeling and performance comparison of MIMO antenna configurations with different polarizations. *IEEE Trans. Antennas Propag.*, **59** (7), 2672–2682.

[53] Cheng, X., Wang, C.X., Laurenson, D.I., Salous, S., and Vasilakos, A.V. (2009) An adaptive geometry-based stochastic model for non-isotropic MIMO mobile-to-mobile channels. *IEEE Trans. Wireless Commun.*, **8** (9), 4824–4835.

[54] Zajic, A.G. and Stuber, G. (2008) Three-dimensional modeling, simulation, and capacity analysis of space–time correlated mobile-to-mobile channels. *IEEE Trans Vehic. Tech.*, **57** (4), 2042–2054.

[55] Blaunstein, N., Toeltsch, M., Laurila, J., Bonek, E., Katz, D., Vainikainen, P., Tsouri, N., Kalliola, K., and Laitinen, H. (2006) Signal power distribution in the azimuth, elevation and time delay domains in urban environments for various elevations of base station antenna. *IEEE J. Antennas Propag.*, **54** (10), 2902–2916.

[56] Toeltsch, M., Laurila, J., Kalliola, K., Molisch, A.F., Vainikainen, P., and Bonek, E. (2002) Statistical characterization of urban spatial radio channels. *IEEE J. Select. Areas Commun.*, **20** (3), 539–549.

[57] Kaya, A. and Calin, D. (2013) Modeling three dimensional channel characteristics in outdoor-to-indoor LTE small cell environments, in *IEEE Military Communications Conference, MILCOM 2013*, pp. 933–938.

[58] Almesaeed, R., Ameen, A., Doufexi, A., Dahnoun, N., and Nix, A. (2013) A comparison study of 2D and 3D ITU channel model, in *IFIP Wireless Days (WD), 2013*, pp. 1–7.

[59] Jiang, M., Hosseinian, M., Lee, M.i., and Stern-Berkowitz, J. (2013) 3D channel model extensions and characteristics study for future wireless systems, in *IEEE 24th International Symposium on Personal Indoor and Mobile Radio Communications (PIMRC), 2013*, pp. 41–46, doi: 10.1109/PIMRC.2013.6666101.

[60] Hentila, L., Kyosti, P., and Meinila, J. (2011) Elevation extension for a geometry-based radio channel model and its influence on MIMO antenna correlation and gain imbalance, in *Proceedings of the 5th European Conference on Antennas and Propagation (EUCAP)*, pp. 2175–2179.

[61] Kammoun, A., Khanfir, H., Altman, Z., Debbah, M., and Kamoun, M. (2013) Survey on 3D channel modeling: From theory to standardization. *arXiv preprint arXiv:1312.0288.*

[62] 3rd Generation Partnership Project (2015), Study on 3D channel model for LTE (release 12).

[63] ITU (2008), ITU-R Report M.2135: Guidelines for evaluation of radio interface technologies for IMT-Advanced. http://www.itu.int/publ/R-REP-M.2135-2008/en.

[64] Yalong, Zhang, Zhang, R., Lu, S. X., Weiming, Duan, and Lin, Cai (2014) Measurement and modeling of indoor channels in elevation domain for 3D MIMO applications, in *2014 IEEE International Conference on Communications Workshops (ICC)*, Sydney, NSW, pp. 659–664.

12

3D-MIMO with Massive Antennas: Theory, Implementation and Testing

Guangyi Liu, Xueying Hou, Fei Wang, Jing Jin and Hui Tong

Signal Processing for 5G: Algorithms and Implementations, First Edition. Edited by Fa-Long Luo and Charlie Zhang.

12.1 Introduction

The explosive increase in mobile internet traffic has put stringent requirements on the improvement of spectrum efficiency of wireless communication systems. The 5G mobile communications system is envisioned to meet new and unprecedented demands that are beyond the capabilities of previous systems [1].

MIMO technologies can significantly improve the capacity and reliability of wireless systems [2], and have been widely applied in current cellular mobile communication systems, such as long-term evolution (LTE) systems [3]. In order to meet the demand for higher spectrum efficiency in 5G systems, more antennas can be deployed at base stations (BS) to increase capacity, which is referred to as massive MIMO in academia [4]. Specifically, the current commercialized antenna products have a limited number of active antennas: usually 2, 4 or a maximum of 8. To exploit the benefit of more antennas in future 5G systems, the number of active antennas can be increased to 64, 128 or more.

Considering that the antenna panel size is proportional to the number of antennas, if the massive antennas are deployed in a horizontal line, the integrated antenna panel will be quite large [5]. Placing the antenna in a two-dimensional grid is an effective way to reduce the antenna panel size, an approach that is termed 3D-MIMO in this chapter, and which has been studied in the Third Generation Partnership Project (3GPP) Release-13 for possible future applications [6]. 3D-MIMO not only can exploit the degree of freedom of massive transmit antennas, as studied by Nam *et al.* [4], but can also adjust the direction of the transmit beam in both the horizontal and vertical dimensions, which helps in enhancing the spatial resolution in the third dimension, so improving the signal power and reducing intercell interference, as shown in Figure 12.1.

In consideration of the potential benefits of 3D-MIMO with massive antennas and its crucial role in future 5G systems, information-theoretic analyses and implementation issues related to channel estimation, detection and precoding schemes have been the subject of several academic studies [7, 8, 9, 10, 11, 12]. In [7], asymptotic arguments based on random matrix theory demonstrate that as the number of antennas tends to infinity, simple linear signal processing approaches, such as matched filter precoding/detection, can be used to achieve the benefits of massive antennas. The required number of antennas for such massive antenna systems has been shown to be proportional to the number of users (UEs) [8]. Mohammed and Larson designed a precoding for massive MIMO by considering the per-antenna power constant [9]. For time division duplexing (TDD) systems, the pilot contamination is seen to be a vital problem for channel state information (CSI) acquisition in massive MIMO systems, and various solutions have been designed to solve this problem [10, 11]. A survey on the related studies can be found in the paper written by Lu *et al.* [12].

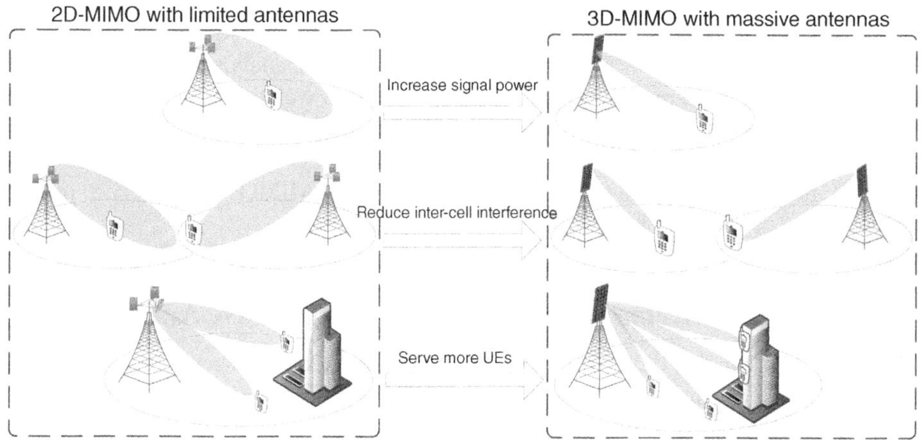

Figure 12.1 Benefits of 3D-MIMO with massive antennas

In this chapter, rather than focusing on specific algorithm designs, we concentrate on assessing the performance of 3D-MIMO in large-scale network simulations, and use field trials to demonstrate their performance in practical networks. The contents of this chapter include the following three focuses.

Performance in System-level Simulations

The performance of 3D-MIMO is evaluated and analyzed by exploiting a system-level simulation platform and the well-established 3D channel model. The simulation results show that 3D-MIMO with 64 antennas can enhance the cell-average and cell-edge throughput by 117% and 228%, respectively, which verifies the promised gains to be brought by 3D-MIMO for future 5G systems. Moreover, the evaluation results show that 3D-MIMO performs better in urban-micro (UMi) scenarios than in urban-macro (UMa) scenarios. The performance of 3D-MIMO with different antenna shapes is compared, and it is shown that placing more antennas in the horizontal domain provides more performance gain in the UMa and UMi scenarios.

Performance in Field Trials

The performance of 3D-MIMO is tested in a field trial, based on the prevalent LTE specification and commercial terminals. To verify the performance gain seen in the system-level simulations, 3D-MIMO performance in a practical UMa scenario is tested. The test results in field trials show even larger performance gains than the simulation results. In addition, a test in a typical high-rise scenario is also performed, and the results verify the benefits of 3D-MIMO for providing satisfying performance, especially for UEs located in high buildings.

Deployment Issues

The way to take 3D-MIMO from theory to practice in terms of transmit hardware design and commercialization considerations is analyzed. To balance the tradeoff between cost and performance, it is suggested that active antenna systems (AAS) are used, which integrate the active transceivers and the passive antenna array into one unit. Three promising AAS structures are provided and can be considered for further study.

The rest of the chapter is organized as follows. Section 12.2 analyzes the application scenarios of 3D-MIMO with massive antennas. In Section 12.3 we will describe how to exploit 3D-MIMO gain based on techniques in current standards. Section 12.4 provides the performance evaluation by system-level simulations. Section 12.5 is devoted to the field trials of 3D-MIMO. In Section 12.6, we will address how to take 3D-MIMO from theory to practice. Conclusions are provided in Section 12.7.

12.2 Application Scenarios of 3D-MIMO with Massive Antennas

By exploiting the benefits of 3D-MIMO shown in Figure 12.1, the scenarios of 3D-MIMO with massive antennas that should see performance enhancement are:

- macro and micro coverage scenario
- high-rise scenario
- indoor scenario.

These are illustrated in Figure 12.2.

12.2.1 *Macro- and Micro-coverage Scenario*

Considering the urgent requirement for cell throughput enhancement in current cellular systems, one promising scenario of 3D-MIMO with massive antennas is the macro-cell and

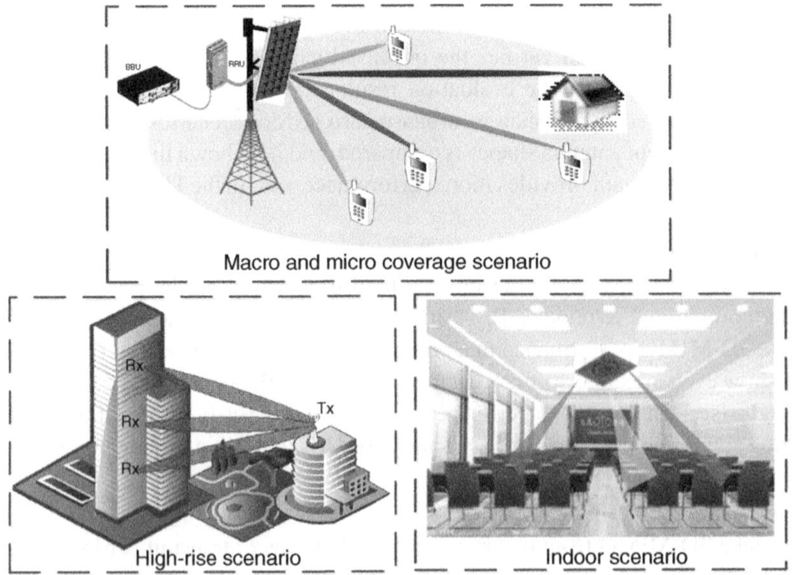

Figure 12.2 Deployment scenarios for 3D-MIMO with massive antennas

micro-cell coverage scenario. By exploiting horizontal and vertical domain beamforming to simultaneously serve more UEs by multi-user MIMO (MU-MIMO) transmission, cell throughput can be greatly improved. By forming narrower and more directional beams, the signal power can be increased and intercell interference is more likely to be avoided, which helps in promoting spectral efficiency.

12.2.2 High-rise Scenario

The height of buildings can be vastly different around the world. One typical urban environment sees some higher buildings (around thirty floors) surrounded by many lower buildings (4–8 floors). Nonetheless, the UEs in the high-rise scenario (more than 8 floors) cannot be covered properly by typical macro BSs, because these normally use downtilt antennas. In such scenarios, 3D-MIMO can solve the coverage problem of UEs in higher floors by flexibly adjusting the beam direction in the elevation domain. In addition, it is possible to apply elevation-domain MU-MIMO between UEs at lower and higher floors due to the large elevation angular separation, which should greatly improve spectral efficiency.

12.2.3 Indoor Scenario

According to published statistics [13], 70% of current cellular traffic is indoors: at home or in an office. In the near future, this figure is envisioned to increase to 90% or more. In indoor scenarios, the scatters are rich, which makes it suitable for exploiting the multiplexing gain of MIMO transmission. As shown in Figure 12.2, by placing the 3D-MIMO array under the roof, the forming of beams in different directions, together with the rich scatters, allows more UEs to be served, so as to meet the explosive growth of traffic in indoor.

12.3 Exploiting 3D-MIMO Gain Based on Techniques in Current Standards

In current LTE standards, MIMO transmission can be achieved in two different manners: codebook-based and non-codebook-based transmission [14]. Codebook-based solutions aim at solving the feedback problems in frequency division duplexing (FDD) systems. Although 3D-MIMO with massive antennas can inherently use such codebook-based transmission, the codebooks in current standards do not exploit the characteristics of 3D-MIMO well. This is mainly because the channel characteristics of 2D-MIMO differs from that of 3D-MIMO [5]. Furthermore, with the increase in the number of transmit antennas, the downlink-training and channel-feedback overhead for CSI acquisition at the BS side will increase too [8], which uses both downlink and uplink resources and reduces the throughput. Consequently, the application of 3D-MIMO in FDD transmission requires substantially new designs.

For TDD systems, channel reciprocity can be exploited for the BS to acquire downlink CSI for precoding, which eliminates the overhead for training and feedback. Thanks to the fact that the cost of downlink channel acquisition does not scale with the number of transmit antennas, TDD transmission is more suitable and scalable for 3D-MIMO with massive antennas [12]. In this chapter, we focus on TDD systems with non-codebook transmission. We discuss the important components that will be promising in TDD systems with massive

3D-MIMO antennas. These include uplink feedback of channel quality information (CQI), uplink channel estimation for downlink precoding to exploit channel reciprocity, downlink precoding schemes and modulation and coding selection (MCS) calculations at the BS that exploit the CQI and channel reciprocity.

12.3.1 System Model

Consider a cellular MIMO system with N BSs, where each BS is equipped with a massive MIMO antenna array placed in a plane. Denote the number of antenna elements in the array as N_a and the number of transceivers as N_t, where one transceiver includes a radio frequency unit and the corresponding digital processing degree of freedom in baseband. One transceiver corresponds to one antenna port that can be identified in baseband. The number of transceivers is less than that of antenna elements: $N_t \leq N_a$. If $N_a = N_t$, the one-to-one mapping from transceiver to antenna elements is considered. If $N_a > N_t$, each transceiver is mapped to several antenna elements.

The comparison of traditional 2D-MIMO with 8 antenna ports and 3D-MIMO with 16, 32 and 64 antenna ports is shown in Figure 12.3. The number of antenna elements in the four structures shown is the same, namely 64. The difference lies in the number of antenna ports N_t. Regarding traditional 2D-MIMO with 8 antenna ports, one port is mapped to 8 vertical antenna elements in a column. 3D-MIMO with 16, 32 and 64 antenna ports is constructed by mapping one antenna port to 4, 2 and 1 vertical antenna elements, respectively.

Denote the number of subcarriers as K and the number of resolvable paths of channel impulse response (CIR) as L. Assuming that on each subcarrier, a maximum of M UEs can be served simultaneously, where each UE is equipped with N_r antennas, and that a total of S data streams are transmitted to each UE. In the following, we will omit the index of subcarrier for the sake of brevity. The received signal of UE_m served by the nth BS is

$$\begin{aligned} \boldsymbol{y}_{m,n} =& \boldsymbol{H}_{m,n}\boldsymbol{W}_{m,n}\boldsymbol{P}^{1/2}_{m,n}\boldsymbol{s}_{m,n} + \sum_{l=1,l\neq m}^{M} \boldsymbol{H}_{m,n}\boldsymbol{W}_{l,n}\boldsymbol{P}^{1/2}_{l,n}\boldsymbol{s}_{l,n} \\ & + \mathbf{Inf}_{m,n} + \boldsymbol{z}_{m,n} \end{aligned} \tag{12.1}$$

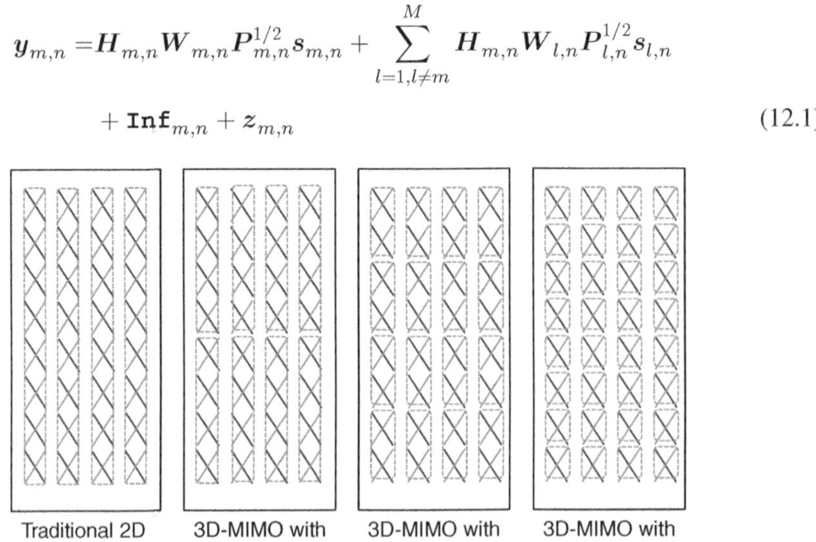

Figure 12.3 Comparison of traditional 2D-MIMO with 8 antenna ports and 3D-MIMO with 16, 32 and 64 antenna ports

where $\boldsymbol{H}_{m,n} \in \mathbb{C}^{N_r \times N_t}$ is the equivalent channel between the UE_m and the transceivers of the nth BS. It worthwhile to note that when $N_a > N_t$, the channel matrix includes the mapping from transceivers to antenna elements. $\boldsymbol{W}_{m,n} \in \mathbb{C}^{N_t \times S}$ is the precoding matrix for the data transmission to UE_m with the norm of each column as 1, $\boldsymbol{P}_{m,n}$ is the power allocation matrix, and $\boldsymbol{s}_{m,n}$ is data intended for UE_m. The part of $\sum_{l=1,l\neq m}^{M} \boldsymbol{H}_{m,n}\boldsymbol{W}_{l,n}\boldsymbol{s}_{m,n}$ represents the intra-cell interference, while the term $\mathbf{Inf}_{m,n}$ represents the inter-cell interference received by UE_m. $\boldsymbol{z}_{m,n}$ is additive white Gaussian noise (AWGN) with zero mean and covariance σ_z^2.

In this chapter, the 3D-MIMO channel model with elevation domain parameters that has been finalized in 3GPP [15] is used to model the channel between the BS and UE.

12.3.2 Uplink Feedback for TDD Systems with 3D-MIMO

Current TDD operation is based on channel reciprocity, in which the downlink channel can be estimated on the basis of uplink training sequences. An example is the sounding reference signal (SRS) in LTE sytems. Nonetheless, it should be noted that downlink interference power experienced at the UE (mainly intercell interference) cannot be estimated via uplink training sequences, which is important information for MCS selection at the BS side. To feedback downlink interference power information, it is necessary to calculate the CQI at the UE and then feed back to the BS. In this section, one possible calculation method is provided.

Considering that the UE can not predict the precoding matrix that is applied for downlink data transmission, the UE and BS need to predefine certain SU-MIMO precoding and power allocation calculation principles to allow the CQI calculation at the UE, and MCS refining at the BS according to the feedback of CQI.

Denote $\boldsymbol{W}_{m,n}^{\text{RxCQI}}$, $\boldsymbol{P}_{m,n}^{\text{RxCQI}}$ as a predefined single-user MIMO (SU-MIMO) precoding matrix and power allocation for the UE, which are known by both the BS and the UE. One possible solution is to let the $\boldsymbol{W}_{m,n}^{\text{RxCQI}}$ be the eigen-beamforming with SU-MIMO transmission, i.e., $\boldsymbol{W}_{m,n}^{\text{RxCQI}}$ is the eigenvector corresponding to the largest eigenvalue of the downlink channel matrix $\boldsymbol{H}_{m,n}$. The $\boldsymbol{P}_{m,n}^{\text{RxCQI}}$ can be predefined with maximum transmit power of the BS: $\boldsymbol{P}_{m,n}^{\text{RxCQI}} = P_n$, where P_n is the maximum transmit power of BS_n. Then the estimation of the receive signal power for CQI can be obtained as

$$S_{m,n}^{\text{RxCQI}} = P_n \|\boldsymbol{W}_{m,n}^{\text{RxCQI}}\|^2 \tag{12.2}$$

The interference power can be calculated according to the long-term average interference power as

$$I_{m,n}^{\text{RxCQI}} = \frac{1}{T}\text{tr}\left\{\sum_{t=1}^{T} \mathbf{Inf}_{m,n}(t)\mathbf{Inf}_{m,n}(t)^H\right\} \tag{12.3}$$

where $\mathbf{Inf}_{m,n}(t)$ represents the interference vector experienced at the tth observation time instance, and T denotes the averaging window size of the average interference power.

Then the receive signal-to-interference-plus-noise ratio (SINR) of the UE can be calculated as

$$\text{SINR}_{m,n}^{\text{RxCQI}} = \frac{S_{m,n}^{\text{RxCQI}}}{I_{m,n}^{\text{RxCQI}} + \sigma_z^2} \tag{12.4}$$

UE can feedback the quantization of $R_{m,n}^{\text{RxCQI}} \triangleq \log_2(1 + \text{SINR}_{m,n}^{\text{RxCQI}})$, which is denoted as $\hat{R}_{m,n}^{\text{RxCQI}}$, as its CQI.

12.3.3 Uplink Channel Estimation for Downlink Precoding

During the uplink training period, each BS needs to estimate the CSI between itself and all UEs. We consider that all UEs send training sequences, and assume that the training sequences of all UEs and the multiple transmit antennas of each UE are orthogonal. Orthogonality can be achieved in the time, frequency or code domain. To simplify the notation, time-domain orthogonal training sequences are considered here.

Assume that the transmit power is the same for all the UEs, and it is denoted as p_u. Denote the frequency-domain training sequence of $\text{UE}_{m,n}$ as $\mathbf{S}_{m,n} \in \mathbb{C}^{K\times N_r}$, where the K represents the length of the training sequences. Then, the received signal matrix at the nth BS during the uplink training phase can be expressed as

$$\mathbf{Y}_n = \sqrt{p_u}\mathbf{H}_n\mathbf{S}^T + \mathbf{N} \tag{12.5}$$

where $\mathbf{Y}_n = [\mathbf{y}_b^1, \dots, \mathbf{y}_b^K] \in \mathbb{C}^{N_t\times K}$, $\mathbf{y}_b^k \in \mathbb{C}^{N_t\times 1}$, $\mathbf{H}_n = [\mathbf{H}_{1,n}, \dots, \mathbf{H}_{M,n}] \in \mathbb{C}^{N_t\times MNr}$ is the channel matrix between BS_b and all UEs when channels are estimated, $\mathbf{S} = [\mathbf{S}_{1,n}, \dots, \mathbf{S}_{M,n}] \in \mathbb{C}^{K\times MNr}$ is the training matrix formed by the training sequences of all UEs, $\mathbf{N} \in \mathbb{C}^{N_t\times K}$ is the AWGN matrix, whose elements are random variables with zero mean and covariance σ_n^2.

By vectorizing the received signal in Eq. (12.5) and applying $\text{vec}(\mathbf{ABC}) = (\mathbf{C}^T \otimes \mathbf{A})\text{vec}(\mathbf{B})$, we can obtain the vectorization of $\mathbf{Y}_n$ as $\text{vec}(\mathbf{Y}_n) = \sqrt{p_u}\tilde{\mathbf{S}}\text{vec}(\mathbf{H}_n) + \text{vec}(\mathbf{N})$, where $\tilde{\mathbf{S}} = \mathbf{S} \otimes \mathbf{I}_{N_t}$. Then $\mathbf{H}_n$ can be estimated as

$$\text{vec}(\hat{\mathbf{H}}_n) = \left(\tilde{\mathbf{S}}^H\tilde{\mathbf{S}} + \mu \cdot \frac{\sigma_n^2}{p_u}\mathbf{R}_n^{-1}\right)^{-1} \tilde{\mathbf{S}}^H \text{vec}(\mathbf{Y}_n) \tag{12.6}$$

where $\mathbf{R}_n = \mathbb{E}\{\text{vec}(\mathbf{H}_n)\text{vec}(\mathbf{H}_n)^H\}$ is the correlation matrix of the vectorized channel matrix $\mathbf{H}_n$. The estimator in Eq. (12.6) is the minimum mean-square error (MMSE) estimator when $\mu = 1$ and least square (LS) estimator when $\mu = 0$.

Since $\mathbf{H}_n$ is the concatenation of channel matrices from all the M UEs to the nth BS, with the estimation of $\mathbf{H}_n$ as shown in Eq. (12.6), it is easy to obtain the estimation of the channel matrix from each UE to the BS, in other words $\hat{\mathbf{H}}_{m,n}$.

12.3.4 Downlink Precoding for TDD Systems with 3D-MIMO

In the following, we discuss the SU-MIMO and MU-MIMO precoding solutions that can be promising for application in future 5G systems with massive 3D-MIMO antennas.

12.3.4.1 SU-MIMO Transmission

SU-MIMO transmission transmits a signal to a single UE on each subcarrier, so $M = 1$. Then the intracell interference part in Eq. (12.1) does not exist, and intercell interference $\texttt{Inf}_{m,n}$ and receive AWGN noise $z_{m,n}$ dominate the performance.

The precoding on each carrier of the BS can be designed to achieve the channel capacity by eigen-beamforming and water-filling power allocation [16]. Considering the complexity of water-filling power allocation in practical systems, we assume equal power allocation between the S streams of UE_m, so $\boldsymbol{P}_{n,m} = \frac{P_0}{S}\mathbf{I}_S$, where $\mathbf{I}_S \in \mathbb{C}^{S\times S}$ is the identity matrix, and P_0 is the total transmit power of the BS.

Denote the estimation of $\boldsymbol{H}_{m,n}$ at the BS as $\hat{\boldsymbol{H}}_{m,n}$. The SVD decomposition of $\hat{\boldsymbol{H}}_{m,n}$ can be expressed as

$$\hat{\boldsymbol{H}}_{m,n} = \boldsymbol{U}_{m,n}\boldsymbol{\Lambda}_{m,n}\boldsymbol{V}^H_{m,n} \tag{12.7}$$

where $\boldsymbol{U}_{m,n} \in \mathbb{C}^{N_r\times N_r}$, $\boldsymbol{\Lambda}_{m,n} \in \mathbb{C}^{N_r\times N_t}$ and $\boldsymbol{V}_{m,n} \in \mathbb{C}^{N_t\times N_t}$. Then the eigen-beamforming matrix $\boldsymbol{W}_{m,n}$ in Eq. (12.1) is the first S columns of $\boldsymbol{V}_{m,n}$, so $\boldsymbol{W} = \boldsymbol{V}_{m,n}(:,1:S)$ [17].

12.3.4.2 MU-MIMO Transmission

MU-MIMO transmission transmits signals to multiple UEs simultaneously on each subcarrier, which means $M > 1$. By exploiting massive antennas for MU-MIMO transmission, the performance can be improved by low-complexity linear precoding [4].

The optimal MU-MIMO precoding is based on dirty paper coding, which is of high complexity [2]. In this chapter, we consider block diagonalized zero-forming (ZFBD) [18], which is sub-optimal but of low complexity. The basic idea is to transmit multiple data streams to UE_m at the null-space of the channel of all the other co-schedule UEs.

Denote the estimated composed channel matrices of all UEs except UE_m as

$$\overline{\boldsymbol{H}}_{m,n} = [\hat{\boldsymbol{H}}^T_{1,n}, \ldots, \hat{\boldsymbol{H}}^T_{m-1,n}, \hat{\boldsymbol{H}}^T_{m+1,n}, \ldots, \hat{\boldsymbol{H}}^T_{M,n}]^T \tag{12.8}$$

where $\overline{\boldsymbol{H}}_{m,n} \in \mathbb{C}^{(M-1)N_r\times N_t}$.

Let the $R_{m,n} = \text{rank}(\overline{\boldsymbol{H}}_{m,n})$. The SVD decomposition of $\hat{\overline{\boldsymbol{H}}}_{m,n}$ is

$$\overline{\boldsymbol{H}}_{m,n} = \overline{\boldsymbol{U}}_{m,n}\overline{\boldsymbol{\Lambda}}_{m,n}[\overline{\boldsymbol{V}}^{(1)}_{m,n}\overline{\boldsymbol{V}}^{(0)}_{m,n}]^H \tag{12.9}$$

where $\overline{\boldsymbol{V}}^{(1)}_{m,n}$ holds the first $R_{m,n}$ right singular vectors, and $\overline{\boldsymbol{V}}^{(0)}_{m,n}$ holds the last $(N_t - R_{m,n})$ right singular vectors. Thus, $\overline{\boldsymbol{V}}^{(0)}_{m,n}$ forms an orthogonal basis for the null space of $\overline{\boldsymbol{H}}_{m,n}$.

The projection of channel of UE_m on the null space of $\overline{\boldsymbol{H}}_{m,n}$ can be denoted as $\hat{\boldsymbol{H}}_{1,n}\overline{\boldsymbol{V}}^{(0)}_{m,n}$, and its rank is $\tilde{R}_{m,n} = \text{rank}(\hat{\boldsymbol{H}}_{1,n}\overline{\boldsymbol{V}}^{(0)}_{m,n})$. The SVD decomposition of the projected channel matrix can be expressed as

$$\hat{\boldsymbol{H}}_{1,n}\overline{\boldsymbol{V}}^{(0)}_{m,n} = \tilde{\boldsymbol{U}}_{m,n}\tilde{\boldsymbol{\Lambda}}_{m,n}[\tilde{\boldsymbol{V}}^{(1)}_{m,n}\tilde{\boldsymbol{V}}^{(0)}_{m,n}]^H \tag{12.10}$$

where $\tilde{\boldsymbol{V}}^{(1)}_{m,n}$ represents the first $\tilde{R}_{m,n}$ right singular vectors, and $\tilde{\boldsymbol{V}}^{(0)}_{m,n}$ represents the last $(N_t - \tilde{R}_{m,n})$ right singular vectors.

The product of $\overline{\boldsymbol{V}}^{(0)}_{m,n}$ and $\tilde{\boldsymbol{V}}^{(1)}_{m,n}$ produces an orthogonal basis of dimension $\tilde{R}_{m,n}$ and represents the transmission vectors that maximize the information rate for UE subject to producing zero interference.

Then the beamforming matrix $\boldsymbol{W}_{m,n}$ in Eq. (12.1) can be denoted as

$$\boldsymbol{W}_{m,n} = \overline{\boldsymbol{V}}^{(0)}_{m,n}\tilde{\boldsymbol{V}}^{(1)}_{m,n} \tag{12.11}$$

12.3.5 MCS Calculation at BS by Exploiting the CQI and Channel Reciprocity

Considering that the CQI is based on the pre-defined SU-MIMO precoding scheme, while the actual downlink can be either SU-MIMO or MU-MIMO based on the scheduling and precoding schemes applied at the BS, it is necessary for the BS to calculate the MCS for each UE to match the downlink transmission of data. One way to achieve such a goal is to let the BS predict the demodulation algorithm and derive the intercell interference based on the CQI feedback from the UE. With this, the downlink transmission SINR of each UE can be estimated by the BS.

Specifically, the BS can first derive the intercell interference based on the CQI feedback from the UE. With the estimated channel matrix for $\text{UE}_{m,n}$, the predefined SU-MIMO precoding matrix and power allocation for the UE that applies for CQI estimation can be obtained. For example, if the precoding is predefined as the eigen-beamforming with SU-MIMO transmission, then the predefined SU-MIMO precoding can be calculated at the BS as the eigenvector corresponding to the largest eigenvalue of the estimated downlink channel matrix $\hat{\boldsymbol{H}}_{m,n}$. Then the receive signal power estimation for calculating the CQI can be obtained as

$$S_{m,n}^{\text{TxCQI}} = P_n \|\hat{\boldsymbol{W}}_{m,n}^{\text{RxCQI}}\|^2 \tag{12.12}$$

The intercell interference of UE_m can be estimated at the BS as

$$\hat{I}_{m,n}^{\text{RxCQI}} = \frac{S_{m,n}^{\text{TxCQI}}}{2^{\hat{R}_{m,n}^{\text{RxCQI}}} - 1} - \sigma_z^2 \tag{12.13}$$

Since the UE is equipped with multiple antennas, the BS needs to imitate the receive equalization algorithms to obtain the downlink transmission SINR of each data stream for the UE. One possible way of doing this is to assume that UE applies the MMSE criterion for equalization. Then the equalization can be estimated as

$$\hat{\boldsymbol{G}}_{m,n} = (\hat{\boldsymbol{H}}_{m,n}\boldsymbol{W}_{m,n})^H \left[\sum_{l=1}^{M} \hat{\boldsymbol{H}}_{l,n}\boldsymbol{W}_{l,n}(\hat{\boldsymbol{H}}_{l,n}\boldsymbol{W}_{l,n})^H + \left(\frac{\hat{I}_{m,n}^{\text{RxCQI}}}{N_r} + \sigma_z^2\right)\boldsymbol{I}_{N_r}\right]^{-1} \tag{12.14}$$

where $\boldsymbol{I}_{N_r}$ is the identity matrix with dimension $N_r \times N_r$.

With $\hat{\boldsymbol{G}}_{m,n}$, the signal power of the kth data stream of $\text{UE}_{m,n}$ can be estimated at the BS as

$$\widehat{\text{S}}_{m,n}(k) = |[\hat{\boldsymbol{G}}_{m,n}\hat{\boldsymbol{H}}_{m,n}\boldsymbol{W}_{m,n}]_{k,k}|^2 \tag{12.15}$$

where $[\boldsymbol{A}]_{k,k}$ denotes the (k,k)th elements of matrix $\boldsymbol{A}$.

The interference plus noise power can be estimated as

$$\widehat{\text{I}}_{m,n}(k) = \left[\hat{\boldsymbol{G}}_{m,n}\left(\sum_{l=1}^{M}\hat{\boldsymbol{H}}_{l,n}\boldsymbol{W}_{l,n}\boldsymbol{W}_{l,n}^H\hat{\boldsymbol{H}}_{l,n}^H\right)\hat{\boldsymbol{G}}_{m,n}^H\right]_{k,k} - \hat{\text{S}}_{m,n}(k) + \left(\frac{\hat{I}_{m,n}^{\text{RxCQI}}}{N_r} + \sigma_z^2\right)\left[\hat{\boldsymbol{G}}_{m,n}\hat{\boldsymbol{G}}_{m,n}^H\right]_{k,k} \tag{12.16}$$

Then the SINR of the kth data stream of $\text{UE}_{m,n}$ can be estimated at the BS as

$$\widehat{\text{SINR}}_{m,n}(k) = \frac{\widehat{\text{S}}_{m,n}(k)}{\widehat{\text{I}}_{m,n}(k)}. \tag{12.17}$$

With the estimated SINR, the BS can estimate the data rate of each data stream for the UE and select one MCS that matches the data rate best.

12.4 Evaluation by System-level Simulations

In this section, the performance evaluation of 3D-MIMO systems is provided by large-scale system-level simulations. The deployment scenarios and the evaluation methodology that were agreed in 3GPP are adopted [19], so as to reflect the deployment of practical networks. Specifically, we evaluate 3D-MIMO in UMa and UMi scenarios with 500-m and 200-m intersite distances (ISD). We utilize the 3D channel model that was discussed and approved in 3GPP [19], so as to incorporate the channel statistics in the 3D case.

12.4.1 Simulation Assumptions

We consider a 19-cell site deployment, with 3 sectors per site, so that 57 sectors are simulated. The number of UEs in each sector is 10. Most of the simulation assumptions are summarized in Table 12.1.

Table 12.1 Summary of evaluation assumptions

Parameters	Values
Homogeneous scenarios	3D-UMa, ISD 500 m 3D-UMi, ISD 200 m
Polarized antenna	Model-2 from TR 36.873
Wrapping method	Geographical distance based
Handover margin	3 dB
System bandwidth	10 MHz (50 PRBs)
UE attachment	Based on RSRP from CRS port 0
Carrier frequency	2 GHz
Downtilt	100°
UE speed	3 km/h
UE antenna pattern	Isotropic
UE RX configuration	2 RX x-polar (+90/0)
Feedback	Reciprocity based operation CQI reporting per 5 ms Feedback delay is 5 ms
SRS	Periodicity: 10 ms
Overhead	3 symbols for DL CCHs, 2 CRS ports and DM-RS with 12 REs per PRB
Scheduler	Frequency selective scheduling (multiple UEs per TTI allowed)

In contrast to the simulation results in the literature, where full-buffer transmission is assumed, we consider non-full-buffer transmission to reflect the traffic in practical real networks. Specifically, the traffic model applied in the simulation is the FTP Model 1 [3], with a packet size of 0.5 Mbytes. Medium resource utilization (RU) of 50% is considered.

Other than assuming ideal uplink channel estimation, as usually considered in system-level simulations, in this chapter we model the uplink SRS channel estimation error [20]. Specifically, the estimation of small-scale fading channel at the BS can be modeled as

$$\hat{h} = \alpha(h + e) \tag{12.18}$$

where α is the scaling factor to maintain normalization of the estimated channel, h is the small-scale fading channel and e is a white complex Gaussian variable, with zero mean and variance σ_e^2, which incorporates the channel estimation error. The variance σ_e^2 is given by $\sigma_e^2 = 1/(\mathtt{SINR} + \Delta)$, where SINR denotes the received SINR of SRS at the BS, and Δ is the gain obtained from time-domain filtering during channel estimation. The scaling factor is given as $\alpha = \sqrt{\frac{1}{1+\sigma_e^2}} = \sqrt{\frac{\Delta * \mathtt{SINR}}{\Delta * \mathtt{SINR} + 1}}$. According to a 3GPP analysis of SRS transmission in LTE systems and intercell interference modeling [20], we set Δ to 9 dB.

The receiver in the simulation for UE is MMSE with interference rejection combination (MMSE-IRC). Denote $\tilde{\boldsymbol{H}}_{l,n} \triangleq \boldsymbol{H}_{m,n}\boldsymbol{W}_{l,n}$, $l = 1, \ldots, M$, as the equivalent channel matrix of all the M scheduled UEs in the cell. Assume that UE_m can estimate $\tilde{\boldsymbol{H}}_{l,n}$, $l = 1, \ldots, M$, while only knowing the interference power from other interfering sectors. Then the receiver can be denoted as

$$\boldsymbol{U}_{m,n}^{\mathrm{IRC}} = \hat{\tilde{\boldsymbol{H}}}_{m,n}^{H}\left(\sum_{l=1}^{M} \hat{\tilde{\boldsymbol{H}}}_{l,n}\hat{\tilde{\boldsymbol{H}}}_{l,n}^{H} + \sigma_{\mathrm{I}_{m,n}}^2\right)^{-1} \tag{12.19}$$

where $\hat{\tilde{\boldsymbol{H}}}_{l,n}$ is the estimation of $\tilde{\boldsymbol{H}}_{l,n}$ at UE_m, and $\sigma_{\mathrm{I}_{m,n}}^2$ is the estimation of interference power from other interfering sectors, for example the estimation of $(\mathrm{tr}\{\mathbf{Inf}_{m,n}\mathbf{Inf}_{m,n}^{H}\} + \sigma_z^2)$.

12.4.2 Performance of 3D-MIMO with Massive Antennas

The four antenna structures illustrated in Figure 12.3 are considered in simulation, with an antenna panel equipped with 64 antenna elements. The first one is the traditional 2D-MIMO with 8 antenna ports. The other three are 3D-MIMO with 16, 32 and 64 antenna ports, where one transceiver is mapped to 4, 2 and 1 vertical antenna elements, respectively.

Figures 12.4 and 12.5 show the simulation results of the four antenna structures under UMa and UMi scenarios. Both SU-MIMO and MU-MIMO are evaluated by system-level simulation, where up to two-UE pairing and up to dual-layer transmission per UE are considered for MU-MIMO. The performance metric is the UE perceived throughput (UPT) [3], which is defined as

$$\mathrm{UPT} = \frac{\text{amount of data (file size)}}{\text{time needed to download data}} \tag{12.20}$$

where the time needed to download data starts when the packet is received in the transmit buffer, and ends when the last bit of the packet is correctly delivered to the receiver.

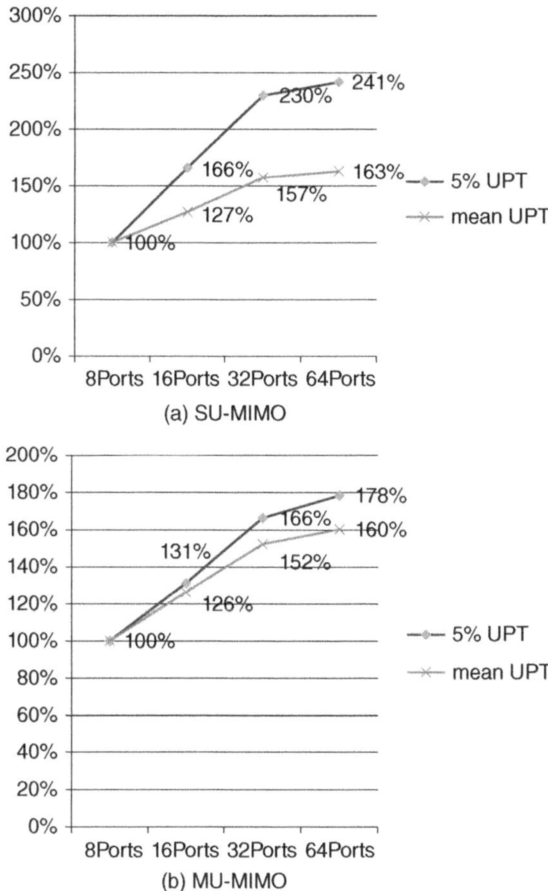

Figure 12.4 Performance comparison of different numbers of ports under UMa scenarios, where one port corresponds to one transceiver. Distance = 500 m

From the simulation results we can observe that 3D-MIMO significantly improves system performance. In the UMa scenario, the performance gains of 3D-MIMO with 64 antenna ports compared with 2D-MIMO with 8 antenna ports are

- 63% and 141% for mean UPT and 5% UPT when SU-MIMO is considered
- 60% and 78% for mean UPT and 5% UPT when 2UE-MU-MIMO is considered.

By contrast, the performance gain in the UMi scenario is larger:

- 125% and 299% for mean UPT and 5% UPT when applying SU-MIMO
- 117% and 228% for mean UPT and 5% UPT when employing MU-MIMO.

By analyzing the simulation results, we can observe that the capability of 3D-MIMO can be best exploited by MU-MIMO transmission, since more UEs can be spatially multiplexed

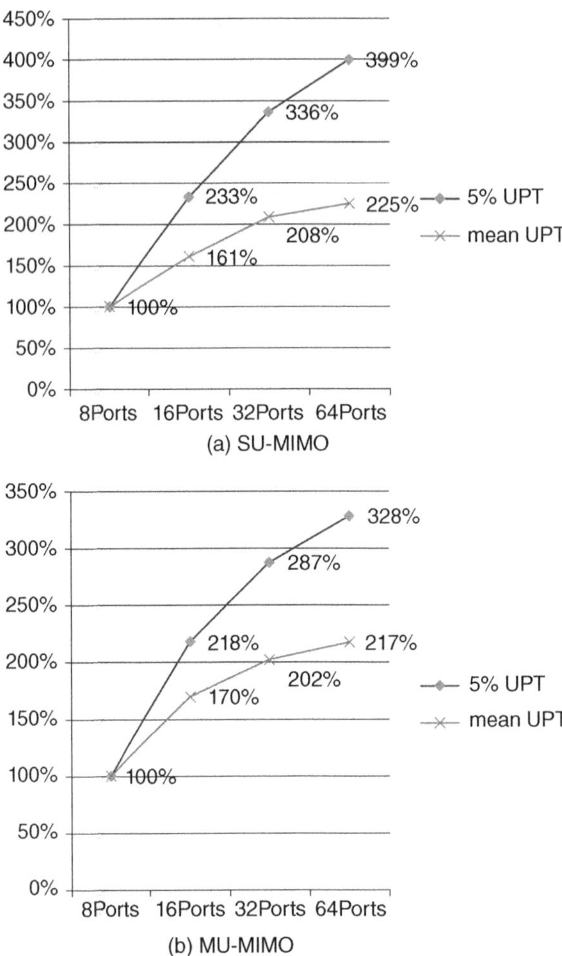

Figure 12.5 Performance comparison of different numbers of ports under UMi scenarios, where one port corresponds to one transceiver. Distance = 200 m

with more antennas at the BS side. Moreover, the performance gain of 3D-MIMO in UMi is larger than that in UMa. The reason lies in the fact that the BS heights in UMa and UMi are 25 m and 10 m, respectively, which indicates that the performance of UEs located in high-rise buildings cannot be ensured by 2D-MIMO in UMi. When 3D-MIMO is considered, since the beamforming direction can be adjusted in both the horizontal and vertical domains, using 3D-MIMO can greatly enhance performance for UEs located in high-rise buildings in UMi.

12.4.3 Performance Comparison of Different Antenna Structures

Different numbers of antenna elements in the horizontal and vertical dimensions determine the size of the antenna directly, which not only affects the feasibility and difficulty of network construction, but also impacts the performance. Figure 12.6 illustrates three different antenna array structures, including 8×8, 16×4 and 32×2, where the first number represents the element number in the horizontal domain and the second is the element number in the vertical domain. If we consider the one-to-one mapping between antenna elements and transceivers, then the structure of the 8×8 structure in Figure 12.6 is the same as the fourth sub-figure in Figure 12.3.

The performance of these three different antenna element structures under UMa and UMi are depicted in Figs 12.7 and 12.8. Compared to 8×8, a 16×4 structure increases mean UPT and 5% UPT by 18% and 22% in the UMa scenario, and enhances the mean UPT and 5% UPT by 3% and 20% in the UMi scenario. Compared to 8×8, a 32×2 structure increases mean UPT and 5% UPT by 33% and 65% in UMa, and improves mean UPT and 5% UPT by 11% and 35% in the UMi scenario.

In summary, 32×2 structure provides the best performance, which indicates that placing more antennas in the horizontal domain provides more performance gains. This is owing to the fact that in both UMa and UMi scenarios, the angle spread and UE distribution range in the horizontal domain are larger than in the vertical domain. The antenna array with more horizontal-domain antennas can exploit the degree of freedom in the horizontal, which is suitable for the channel statistic in the UMa and UMi scenarios. In practical deployment, the

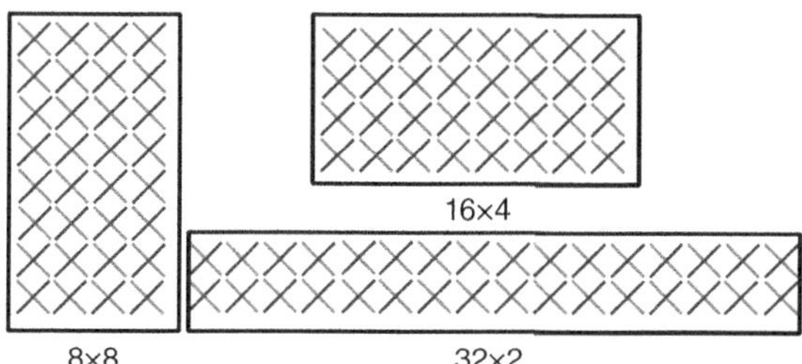

Figure 12.6 Three different antenna array element structures including 8×8, 16×4 and 32×2

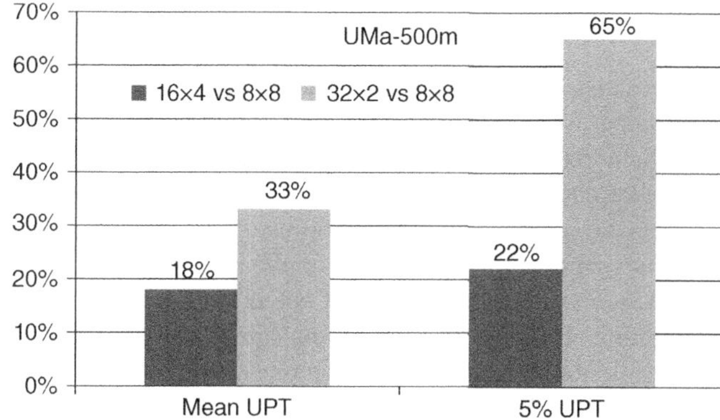

Figure 12.7 Performance comparison of these three different antenna element structures under UMa-500 m

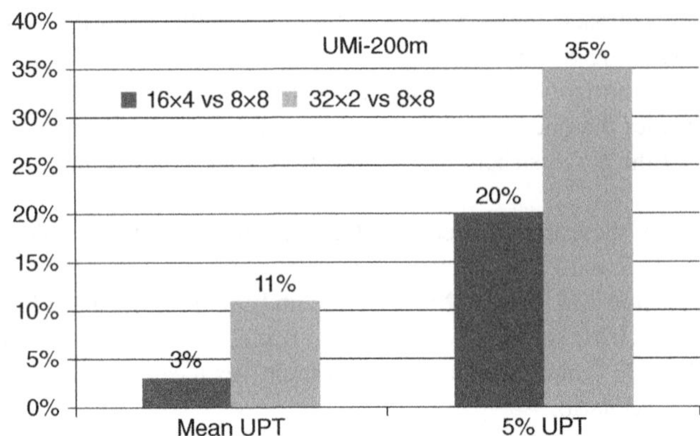

Figure 12.8 Performance comparison of these three different antenna element structures under UMi-200 m

tradeoff between antenna size in the horizontal domain and the performance gain should be balanced to ensure good performance while keeping the antenna size small.

12.5 Field Trials of 3D-MIMO with Massive Antennas

In this section, the test results of 3D-MIMO with massive antennas in field trials are analyzed. The performance is tested both in an anechoic chamber and in two typical scenarios: the urban micro scenario and the high-rise scenario.

The test is based on a demo with 64 antenna elements, each connected with a transceiver, in other words there is a total of 64 transceivers in the system. The 64 antenna elements are placed in a rectangle with 8 in the horizontal and 8 in the vertical domain. The antenna self-calibration module is integrated in the antenna.

12.5.1 Test Performance in Anechoic Chamber

In the test performed in an anechoic chamber, the antenna is located in the center of the chamber, and the test instrument is located in front of the antennas. The antennas can be rotated from -90° to 90° in the horizontal domain and -30° to 30° in the vertical domain, so as to simulate the different locations of the receiver. Since the horizontal beam pattern of 3D-MIMO with massive antennas is the same as that of the traditional 8 antennas, herein we only provide the vertical-domain beam patterns for the sake of simplicity.

The vertical beam patterns are tested by way of broadcast weighting and discrete Fourier transform (DFT) weighting. The broadcast weighting can be considered as the beam pattern used for downlink transmission of broadcast information: downlink control information, common reference signal and so on. DFT weighting can be regarded as the beam pattern used for downlink data transmission.

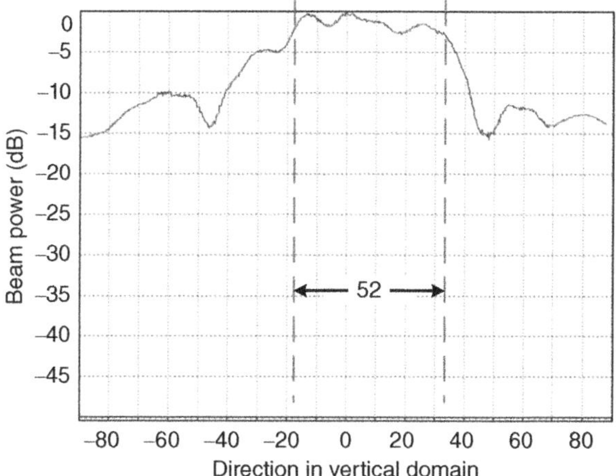

Figure 12.9 Vertical beam pattern with broadcast weighting

12.5.1.1 Vertical Beam Pattern with Broadcast Weighting

Figure 12.9 shows the beam pattern with broadcast weighting. The amplitude and phase of the weighting vector is carefully designed to cover a wide range in the vertical domain. From the results we can observe that the 3-dB bandwidth of the vertical broadcast beam pattern is 52°, which can cover most of the vertical-domain range for typical building heights.

12.5.1.2 Vertical Beam Pattern with DFT Weighting

Figure 12.10 shows the beam pattern with DFT weighting. The kth element of the vector can be expressed as $w_k = \sqrt{1/N_v} \exp\{-2\pi jk\frac{d_v}{\lambda}\cos(\theta\frac{2\pi}{360})\}$, where $N_v = 8$ is the number of antennas in the vertical domain, λ is the wave length, $d_v = 0.5\lambda$ is the antenna element distance and θ is the desired beam direction in degrees. The results show that the direction of the beams can be adjusted as expected. Meanwhile, the sidelobe can also be suppressed: the suppression of sidelobe for beam directions of 0°, 15° and 30° are 13 dB, 9–10 dB and 6 dB, respectively.

12.5.2 Field Trial in Typical Urban Micro Scenario and High-rise Scenario

12.5.2.1 Measurement Setup

The measurement scenario is showed in Figure 12.11, where the traditional antenna and 3D-MIMO antennas with 6° downtilt are deployed at the top of an eight-storey building with a height of 23 m, and commercial UE is scattered over the floors of a high-rise office building on the opposite side. The horizontal distance between the BS and UE is about 110 m, as

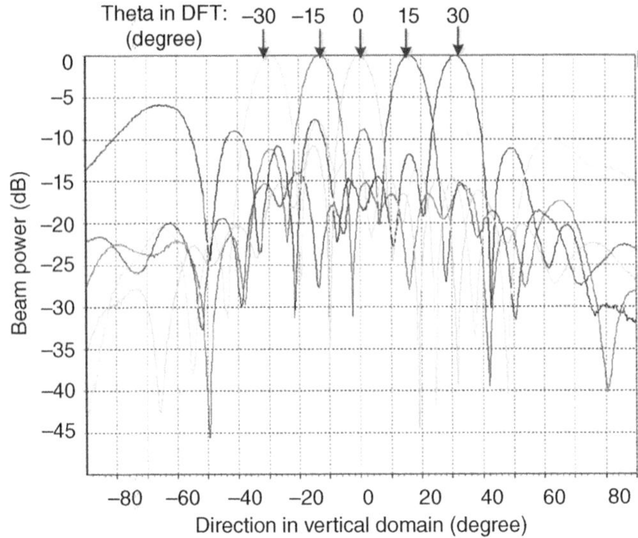

Figure 12.10 Vertical beam pattern with DFT weighting

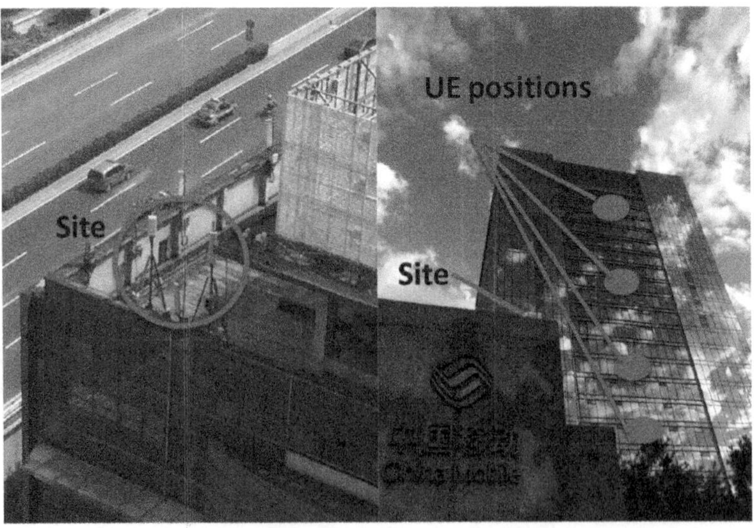

Figure 12.11 Test environment in an typical urban scenario located at the China Mobile offices, Beijing, China

illustrated in Figure 12.12. Both the coverage and throughput of the traditional 2D antenna and 3D-MIMO are compared through this trial [21].

In order to guarantee fairness, the transmit power of both BS are set at the same value. Since 3D-MIMO has 64 transceivers and the BS with traditional MIMO only has 8, the transmit power of each transceiver in 3D-MIMO is one eighth of that of traditional 2D-MIMO. The carrier frequency used for the test is 2615–2635 MHz. The trial is based on

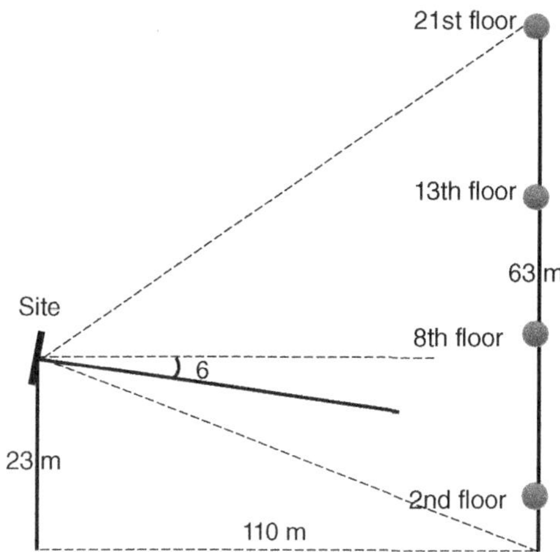

Figure 12.12 The relative location of the BS and UE

commercial TD-LTE systems and is configured with uplink and downlink configuration 2 and special subframe configuration of 3 : 9 : 2 [21]. The MIMO transmission mode 8 in the LTE specification is configured for the test UE.

12.5.2.2 Measurement in Typical Urban Micro Scenario

The MU-MIMO transmission is tested by serving multiple terminals in the right-hand building illustrated in Figure 12.12. For 2D-MIMO transmission, four terminals are simultaneously served; two terminals are located on the second floor and two terminals are on the fourth floor. When 3D-MIMO is applied, eight terminals are simultaneously served; four terminals are located on the fifth floor, two are on the second floor and the other two are on the first floor. The performance of traditional 2D- and 3D-MIMO are measured for 3 minutes to obtain an average. The cell-average throughput is obtained by the summation of data rates of the simultaneously served terminals.

The cell-average throughputs are shown in Figure 12.13, where the performance obtained by system-level simulation is provided for comparison. From the test results we can observe that 3D-MIMO can enhance the cell-average throughput. Comparing the results obtained by simulation and by field trial, we can observe that the performance obtained by field trial is better than that in system-level simulation. This is because in the field test, the location and beam direction of interfering BSs are properly adjusted so that the intercell interference in the field trial is smaller than in the system-level simulation.

12.5.2.3 Measurement in Typical High-rise Scenario

For performance testing in the high-rise scenario, SU-MIMO transmission is considered. The performance of terminals located on the 2nd, 8th, 13th and 21st floors is tested. In each location, the performance of both traditional 2D- and 3D-MIMO are measured for 3 minutes. The final results for each floor are averaged.

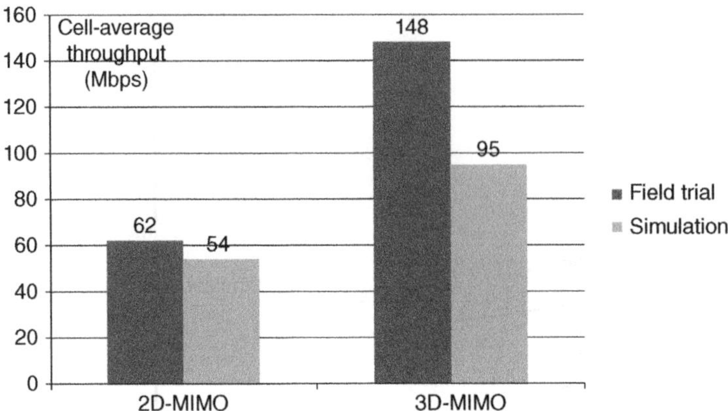

Figure 12.13 Field trial test results for 2D-MIMO with 8 transceivers and 3D-MIMO with 64 transceivers in a typical urban micro scenario. For comparison, the performance obtained by system-level simulation is provided

The test results in terms of per-terminal data rate for both traditional 2D-MIMO and the 3D-MIMO are shown in Figure 12.14. From the measurement results we can observe that 3D-MIMO can significantly improve the throughput on higher floors (13th and 21st floor): it can achieve 23 Mbps and 21 Mbps when the UE is located at the 13th and 21st floors, respectively, while 2D-MIMO only can provide 8 Mbps and 5 Mbps, respectively. The performance gain of 3D-MIMO comes from its capability of elevation beamforming. On the lower floors, both antennas have similar performance: 3D-MIMO provides 17 Mbps and 37 Mbps, while 2D-MIMO offers 22 Mbps and 33 Mbps, when the UE is located on the 2nd and 8th floors, respectively.

12.6 Achieving 3D-MIMO with Massive Antennas from Theory to Practice

In this section, we discuss how to bring 3D-MIMO from theory to practice by analyzing transmit hardware design and commercialization considerations.

12.6.1 AAS: a Key for Commercialization of 3D-MIMO with Massive Antennas

The traditional site deployed in commercial networks consists of distributed baseband unit (BBU), remote radio unit (RRU) and passive antennas, which are connected by cables. To enable complete freedom in the spatial domain, 3D-MIMO with massive antennas requires more transceivers to be integrated in the BS. If the traditional architecture with distributed BBU, RRU and antennas is kept, more cables are required to connect the components, which not only increases the cost but also involves more trouble in arranging so many wires on the same tower.

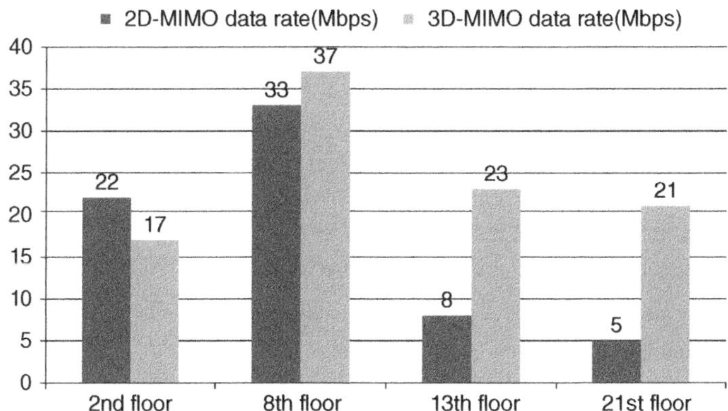

Figure 12.14 Performance comparison of traditional 2D-MIMO with 8 transceivers and 3D-MIMO with 64 transceivers in a typical high-rise scenario

Table 12.2 Benefits and challenges of the three BS architectures with different levels of integration

Architeture	Benefits	Challenges
BBU+AAS	Reduce cost and attenuation of cables	High bandwidth requirements of CPRI
BBU+AAS+Part of BBU in AAS	Low bandwidth requirement of CPRI	New CPRI interface to be defined
Single unit with BBU and AAS	No CPRI, cleaner sites and lower sit cost	Stringent requirements on the size, weight and heat dissipation

Integrating the active transceivers and the passive antenna array into one unit helps avoid the cabling between antennas and RRU. Such an integrated unit is termed an "active antenna system" (AAS), and BBU+AAS can be a promising structure for future application of 5G systems. To further relieve the burden of interaction of RF signals between AAS and BBU, some functionality of the BBU can be moved into the AAS, with the newly defined common public radio interface (CPRI). To eliminate the CPRI for cleaner sites and reduce site costs, the BBU, RRU and antennas can all be integrated in the same unit. The benefits and challenges of the three architectures with different level of integration are summarized in Table 12.2.

12.6.2 Mapping from Transceivers to Antenna Elements in AAS

In current commercialized TD-LTE networks, a total of 8 transceivers are supported. Considering the cost of transceivers is much higher than that of the antenna elements, each transceiver is connected to a column of passive antenna elements in the vertical domain by well designed mapping, so as to form a downtilt beam with a narrow beam bandwidth. For 3D-MIMO with massive antennas, in order to exploit the complete freedom in the spatial domain, not only the number of antenna elements but also the number of transceivers need to be increased.

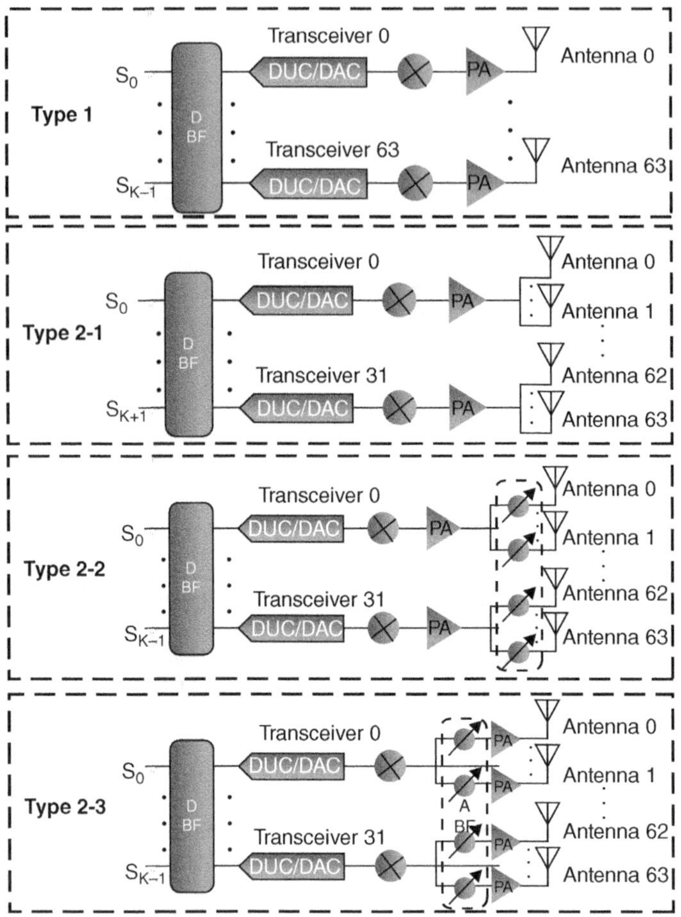

Figure 12.15 Possible types of transmitter design for 3D-MIMO with massive antennas

By balancing the tradeoff between cost and performance, two types of transceiver structure can be considered for commercialization of 5G systems (see Figure 12.15). Type 1 involves one-to-one mapping from transceivers to antenna elements with a full spatial degree of freedom, by digital beamforming but with more expense in transceivers. By contrast, Type 2 provides a tradeoff by keeping a large number of antenna elements while reducing the number of transceivers, with each transceiver mapped to several antenna elements. Analog beamforming can be achieved by some phase shifters in the RF module, as illustrated in the Type 2 structure of Figure 12.15.

12.7 Conclusions

In this chapter, we evaluated the performance of 3D-MIMO with massive antennas by system-level simulation using practical assumption and channel model, and by testing the

3D-MIMO in field trial with commercial terminal and networks. In addition, we compared system-level simulation results and field trial test measurements. The system-level simulation results showed that by equipping 3D-MIMO with 64 active antennas, the performance can be significantly improved. 3D-MIMO shows better performance in the UMi scenario than than in the UMa scenario. The testing results from the field trial in a typical urban micro scenario verify the performance gain of 3D-MIMO obtained by simulation. Furthermore, measurements in typical high-rise scenarios show that 3D-MIMO can significantly improve the data rate of UEs located on higher floors due to its capability of elevation beamforming. Finally, we analyzed how to take 3D-MIMO from theory to practice in terms of transmit hardware design and commercialization considerations. By balancing the tradeoff between cost and performance, the AAS antenna is considered a key for commercialization of 3D-MIMO with massive antennas in future 5G systems.

References

[1] IMT-2020 (2014) 5G vision and requirements.

[2] Tse, D. and Viswanath, P. (2005) *Fundamentals of Wireless Communication*, Cambridge University Press.

[3] 3GPP Long Term Evolution (LTE) (2010) Further advancements for E-UTRA physical layer aspects. *TSG RAN TR 36.814 v9.0.0.*

[4] Larsson, E.G., Tufvesson, F., Edfors, O., and Marzetta, T.L. (2014) Massive MIMO for next generation wireless systems. *IEEE Commun. Mag.*, **52** (2), 186–195.

[5] Nam, Y.H., Ng, B.L., Sayana, K., Li, Y., Zhang, J.C., Kim, Y., and Lee, J. (2013) Full-dimension MIMO (FD-MIMO) for next generation cellular technology. *IEEE Commun. Mag.*, **51** (6), 172–179.

[6] Samsung and Nokia Networks (2014) RP-141644 New SID Proposal: Study on elevation beamforming full-dimension (FD) MIMO for LTE. *3GPP TSG RAN Meeting 65.*

[7] Marzetta, T.L. (2010) Noncooperative cellular wireless with unlimited numbers of base station antennas. *IEEE J. Sel. Areas Commun.*, **9** (11), 3590–3600.

[8] Hoydis, J., ten Brink, S., and Debbah, M. (2013) Massive MIMO in the UL/DL of cellular networks: How many antennas do we need? *IEEE J. Sel. Areas Commun.*, **31** (2), 160–171.

[9] Mohammed, S.K. and Larsson, E.G. (2013) Per-antenna constant envelope precoding for large multi-user MIMO systems. *IEEE Trans. Commun.*, **61** (3), 1059–1071.

[10] Jose, J., Ashikhmin, A., Marzetta, T.L., and Vishwanath, S. (2011) Pilot contamination and precoding in multi-cell TDD systems. *IEEE Trans. Wireless Commun.*, **10** (8), 2640–2651.

[11] Zhang, J., Zhang, B., Chen, S., Mu, X., El-Hajjar, M., and Hanzo, L. (2014) Pilot contamination elimination for large-scale multiple-antenna aided OFDM systems. *IEEE J. Sel. Topics Signal Process.*, **8** (5), 759–772.

[12] Lu, L., Li, G.Y., Swindlehurst, A.L., Ashikhmin, A., and Zhang, R. (2014) An overview of massive MIMO: Benefits and challenges. *IEEE J. Select. Topics Signal Process.*, **8** (5), 742–758.

[13] Cisco Visual Networking (2014) Global mobile data traffic forecast update, 2013–2018.

[14] 3GPP Long Term Evolution (LTE) (2012) Evolved universal terrestrial radio access (E-UTRA): Physical layer procedures. *TSG RAN TR 36.213 v10.6.0.*

[15] 3GPP Long Term Evolution (LTE) (2014) Study on 3D channel model for LTE. *TSG RAN TR 36.873 v1.3.0.*

[16] Paulraj, A., Gore, D.A., Nabar, R.U., and Bolcskei, H. (2004) An overview of MIMO communications–a key to gigabit wireless. *Proc. IEEE*, **92** (2), 198–218.

[17] Liu, G., Liu, X., and Zhang, P. (2005) QoS oriented dynamical resource allocation for eigen beamforming MIMO OFDM. *IEEE VTC-Fall*, pp. 1450–1454.

[18] Spencer, Q.H., Swindlehurst, A.L., and Haardt, M. (2004) Zero-forcing methods for downlink spatial multiplexing in multiuser MIMO channels. *IEEE Trans. Signal Process.*, **52** (2), 461–471.

[19] 3GPP Long Term Evolution (LTE) (2014) 3GPP RAN1_78bis chairman notes. *TSG RAN.*

[20] 3GPP Long Term Evolution (LTE) (2014) SRS and antenna calibration error modeling. *R1-144943, 3GPP TSG RAN1_79.*

[21] 3GPP Long Term Evolution (LTE) (2013) Physical channels and modulation. *TSG RAN TS 36.211 v11.4.0.*

13

Orbital Angular Momentum-based Wireless Communications: Designs and Implementations

Alan. E. Willner, Yan Yan, Yongxiong Ren, Nisar Ahmed and Guodong Xie

Signal Processing for 5G: Algorithms and Implementations, First Edition. Edited by Fa-Long Luo and Charlie Zhang.

One property of electromagnetic (EM) waves that has recently been explored is the ability to multiplex multiple beams carrying orbital angular momentum (OAM) such that each beam has a unique helical phase front. Such OAM-based multiplexing can potentially increase the system capacity and spectral efficiency of wireless communication links by transmitting multiple coaxial data streams. This chapter will introduce the basic concept of OAM and the principle of using OAM for spatial multiplexing in the wireless communications radio frequency (RF) range. In this chapter, we will briefly review the research and techniques related to OAM, including the generation, multiplexing and detection of OAM channels. We will also review the demonstration of wireless communication links using OAM and the propagation effects of OAM channels.

13.1 EM Waves Carrying OAM

Although EM waves have been studied for well over a century, one property of EM waves, OAM, was only identified in the 1990s. In 1992, Les Allen and his colleagues discovered that the OAM of EM waves was associated with the helical transverse phase structure $\exp(i\ell\phi)$, in which ϕ is the transverse azimuthal angle and ℓ is an unbounded integer [1]. The amount of phase front "twisting" indicates the OAM number, and beams with different OAM are spatially orthogonal. Note that OAM relates to the spatial phase profile rather than to the state of polarization of the beam, which is associated with the spin angular momentum (SAM). An EM beam carries SAM if the electrical field rotates along the beam axis, in other words circularly polarized EM waves. It carries OAM if the wave vector spirals around the beam axis, leading to a helical phase front, as shown in Figure 13.1 and (b) [2]. In its analytic expression, this helical phase front is usually related to a phase term of $\exp(i\ell\theta)$ in the transverse plane, where θ refers to the angular coordinate and ℓ is an integer indicating the number of intertwined helices; in other words, the number of 2π phase shifts along the circle around the beam axis. ℓ is an integer that can take a positive, negative, or even a zero value, corresponding to clockwise phase helices, counterclockwise phase helices, or no helix, respectively [3]. Although the SAM and OAM of EM waves can be coupled to each other under certain scenarios [4], they can be clearly distinguished for a paraxial EM beam. Therefore, in the paraxial limit, OAM and polarization can be considered as two independent properties of EM beams. Figure 13.10(c) shows the intensity and phase of the wavefront of OAM beams with different ℓ. When $\ell \neq 0$, the beam wavefront has a spiral phase distribution and intensity null at the center due to the phase singularity. In general, the OAM beams diverge faster as $|\ell|$ increases [5].

In general, an OAM-carrying beam could refer to any helically phased light beam irrespective of its radial distribution. Laguerre–Gaussian (LG) beams are a special subset among all OAM-carrying beams; their radical distribution is characterized by the fact that they are paraxial eigensolutions of the wave equation in cylindrical coordinates and in homogeneous media, such as free space. For an LG beam, both azimuthal and radial wavefront distributions are well defined and indicated by two indices, ℓ and p, in which ℓ has the same meaning as that of a general OAM beam – azimuthal phase dependence – and p refers to the radial nodes in the intensity distribution. LG beams form an orthogonal and complete mode in the spatial domain. In contrast, a general OAM beam may be expanded into a group of LG beams, each with the same ℓ but a different p index, due to the absence of radial definition. Henceforth, the term "OAM beam" refers to all helically phased beams and is to be distinguished from LG beams.

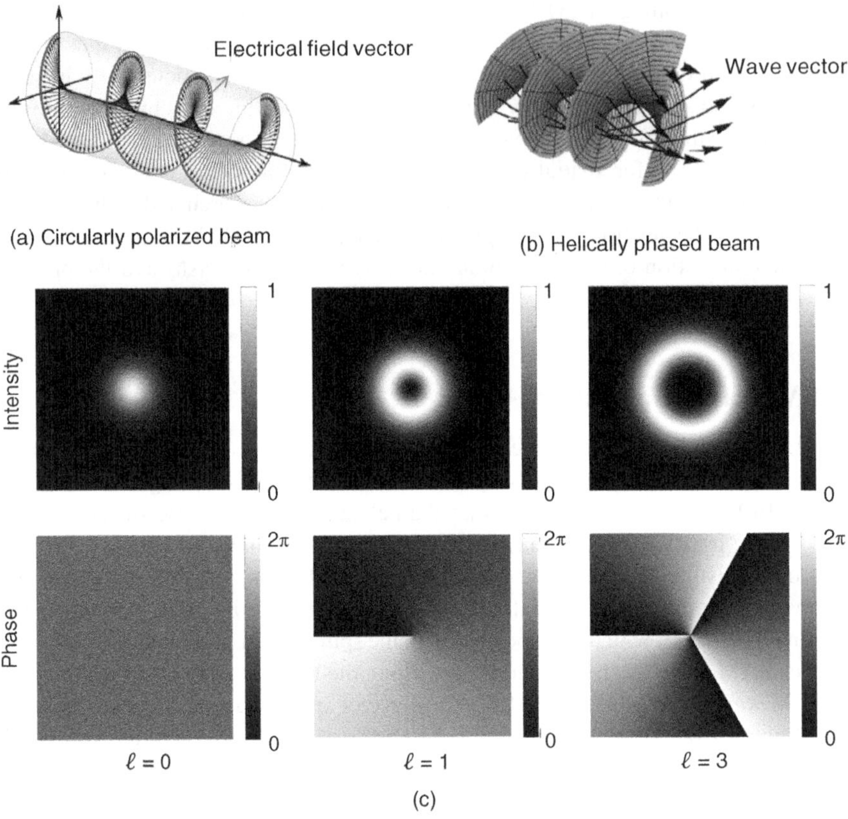

Figure 13.1(a) Circularly polarized beam carrying SAM. (b) Phase structure of a light beam carrying OAM. (c) Intensity and phase of the wavefront of OAM beams with different values of ℓ

The rest of this chapter is organized into the following four sections. Section 13.2 serves as a brief introduction to the application of OAM to wireless communications. In Section 13.3, we present the generation, detection of multiplexing and demultiplexing of OAM beams. Section 13.4 is devoted to the use of OAM-based multiplexing in a wireless communication system. Section 13.5 presents the summary of this chapter and discusses future work and perspectives on the implementation and integration of an OAM-based system.

13.2 Application of OAM to RF Communications

The demand for capacity and spectral efficiency continues to grow exponentially due to the multitude of wireless applications, including indoor communications, data centers, and front-haul and back-haul connections. Such a multitude of applications produces a great interest in devising advanced multiplexing approaches in RF, one of which is OAM multiplexing. OAM beams with different ℓ values are mutually orthogonal in space, allowing them to be multiplexed together along the same beam axis and demultiplexed with low crosstalk [6–9]. Utilization of OAM for communications is based on the fact that coaxially propagating EM beams with different OAM states can be efficiently separated. For instance, consider two OAM beams U_1 and U_2 having an azimuthal index of ℓ_1 and ℓ_2, respectively. Relying only on the azimuthal phase, the two OAM beams can be expressed as:

$$U_1(r, \theta, z) = A_1(r, z)\ \exp(i\ell_1\theta) \tag{13.1}$$

$$U_2(r, \theta, z) = A_2(r, z)\ \exp(i\ell_2\theta) \tag{13.2}$$

where r and z refer to the radial position and propagation distance, respectively. From the above expressions, one can conclude that these two beams are spatially orthogonal in the following sense:

$$\int_0^{2\pi} U_1 U_2^* d\,\theta = \begin{cases} 0 & \text{if } \ell_1 \neq \ell_2 \\ A_1 A_2^* & \text{if } \ell_1 = \ell_2 \end{cases} \tag{13.3}$$

Consequently, one can establish a well-defined line-of-sight (LoS) link, for which each OAM beam at the same carrier frequency can carry an independent data stream, thereby increasing the capacity and spectral efficiency by a factor that is equal to the number of OAM states. Figure 13.2 illustrates a prospective application scenario, using OAM multiplexing as well as polarization multiplexing for short-range, high-speed wireless information exchange between a transmitter and a receiver. As mentioned above, in the paraxial limit, OAM and polarization can be considered as two independent properties of an EM wave. Therefore, polarization multiplexing is compatible with OAM multiplexing and can be used to produce another twofold increase in the capacity and spectral efficiency of a transmission link.

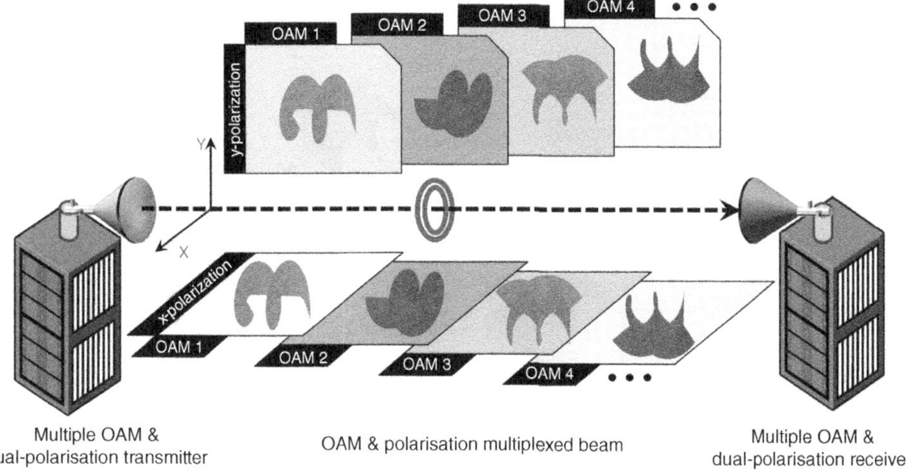

Figure 13.2 Utilizing OAM and polarization multiplexing in a free space mm-wave communication link

13.3 OAM Beam Generation, Multiplexing and Detection

13.3.1 OAM Beam Generation and Detection

Many approaches to creating OAM beams have been proposed and demonstrated. For example, one could convert a fundamental Gaussian beam into an OAM beam using an OAM converter. The OAM converter could be a spiral phase plate [10, 11], helicoidal parabolic antenna [12, 13] or a metamaterial structure [14, 15]. Also, the antenna arrays can be used to generate and detect OAM beams [16–18].

13.3.1.1 Spiral Phase Plate

To generate an OAM beam with a specific ℓ, a spiral phase plate (SPP) can be designed by azimuthally varying the thickness of plate, following the relation $\boldsymbol{h}(\phi) = \frac{(\frac{\phi}{2\pi})\ell\lambda}{n-1}$. Such an SPP acquires a maximum thickness difference of $\boldsymbol{\Delta h} = \frac{\ell\lambda}{n-1}$, where ϕ is the azimuthal angle varying from 0 to 2π, *n* is the refractive index of the plate material, and λ is the wavelength of the mm wave. Figure 13.3(a) illustrates such an SPP. In general, one surface of an SPP is flat, so the required thickness can be obtained simply by controlling the height of the other surface.

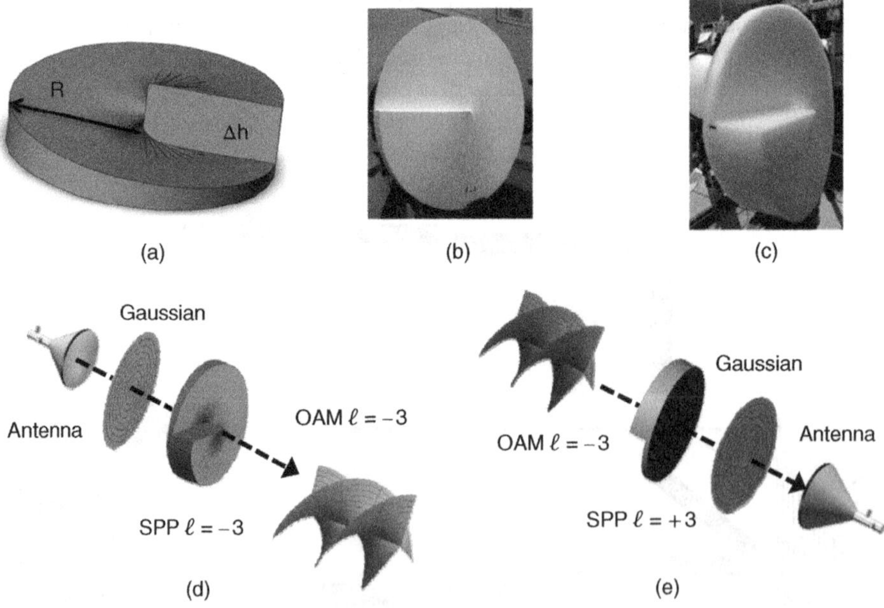

Figure 13.3 Spiral phase plates: (a) illustration of an SPP to generate and detect OAM beams; SPPs made of HDPE materials with (b) $\ell = \mathbf{1}$ and (c) $\ell = \mathbf{3}$ for operation at 28 GHz; (d) the principle of converting a Gaussian beam into an OAM beam; (e) the conversion of an OAM beam back to a Gaussian beam

Any material that is transparent at the RF frequency of interest can be used to fabricate the element. For instance, high-density polyethylene (HDPE) can be used to generate mm wave OAM beams at 28 GHz. Figures 13.3(b) and (c) show the SPPs made of HDPE for $\ell = +1$ and $\ell = +\mathbf{3}$. An example of the generation and detection of an OAM beam of $\ell = -\mathbf{3}$ is shown in Figs 13.3(d) and (e). A regular Gaussian beam passing through the SPP will be transformed into an OAM beam with a helical wavefront. Conversely, an SPP of the opposite OAM number will convert an incoming OAM beam back into a Gaussian beam, which can then be detected by a regular antenna.

13.3.1.2 Helicoidal Parabolic Antenna

A regular parabolic antenna can be mechanically modified to generate and receive RF OAM beams. The azimuthal elevation of the parabolic surface is used to obtain the shape of the helicoidal parabolic antenna from the initial parabolic shape [12, 13]. Due to the reflection of the input beam onto the helicoidal parabolic surface, an elevation value of $\frac{\ell\lambda}{2}$ results in an OAM beam of ℓ. In one published example [12], the elevation value for an OAM of $\ell = 1$ was 6.25 cm to accommodate the OAM beam at a center frequency of 2.4 GHz.

13.3.1.3 Antenna Arrays

Circular array antennas can also be used to generate OAM beams [16–18]. Each antenna element is placed on a circle with equidistant spacing. All the antenna elements are fed by the same signal, but with an incremental phase shift of $2\pi\ell/\mathrm{N}$, where N is the number of the antenna array elements and $\ell$is the OAM number. Hence a radio OAM beam of ℓ can be generated. Figure 13.4 shows an example of a circular antenna array to generate an OAM beam of $\ell = +1$ at a frequency of 10 GHz. The circular array consists of eight identical rectangular patches, which are excited by an asymmetrical microstrip corporate feeding network and back-fed with a 50-Ω SMA connector. The phase difference between each adjacent antenna

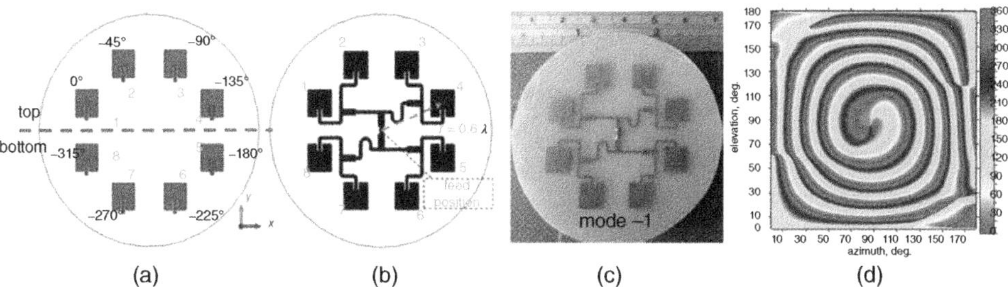

Figure 13.4 Circular antenna array to generate the OAM beam of $\ell = +1$ at 10 GHz frequency: (a) phase distribution of each antenna element; (b) feeding network with phase delay lines of each antenna element; (c) the fabricated circular antenna array; (d) the measured wavefront phase of the generated OAM beam of $\ell = +1$. *Source:* Bai *et al.* [17]

element is 45°. Microstrip power splitters and delay lines are designed to achieve the desired phase delay of each antenna element. Such an antenna array can be fabricated by the standard printed circuit board process, which is considered a more compact and cost-effective approach for the generation of OAM beams.

13.3.2 Multiplexing and Demultiplexing of OAM Beams

13.3.2.1 Beam Splitter and Inverse Phase Hologram

A straightforward way of multiplexing is to use cascaded 3-dB beam splitters (BSs). Each BS can coaxially multiplex two beams that are properly aligned, and *N* such BSs, when cascaded, can multiplex $N+1$ independent OAM beams at most. Figure 13.5(a) illustrates such a configuration. Similarly, at the receiver end, the multiplexed beams are divided into four copies. To demultiplex the data channel on one of the beams (e.g. with $\ell = \ell_i$), a phase plate with a spiral charge of $-\ell_i$ is applied to all the multiplexed beams. As a result, the target beam is transformed into a fundamental Gaussian beam, as shown in Figure 13.5(b). The downconverted beam can be separated from the other OAM beams, which still have helical phase fronts, by using a spatial mode filter. Accordingly, each of the multiplexed beams can be demultiplexed by changing the spiral phase plate. Although this method is power-inefficient because of the power loss incurred by the BSs, it was used in the initial lab demonstrations of OAM multiplexing/demultiplexing.

13.3.2.2 OAM Multiplexing /Demultiplexing Using Antenna Arrays

Another approach to multiplexing OAM channels is to use circular antenna arrays of different sizes having same center. Two antenna arrays of different sizes are designed to generate OAM beams of $\ell = +1$ and $\ell = +2$ at 8.3 GHz [18]. Based on the two antennas, a stacked antenna is used to combine two OAM beams. The antenna includes two boards with a separation of 10 mm. The top board is for an OAM beam of $\ell = +1$, while the bottom one is for the OAM beam of $\ell = +2$. In order for the boards to have the same central axis, the two boards are fixed with screws, and a hole is drilled in the bottom board, through which the feeding cable from

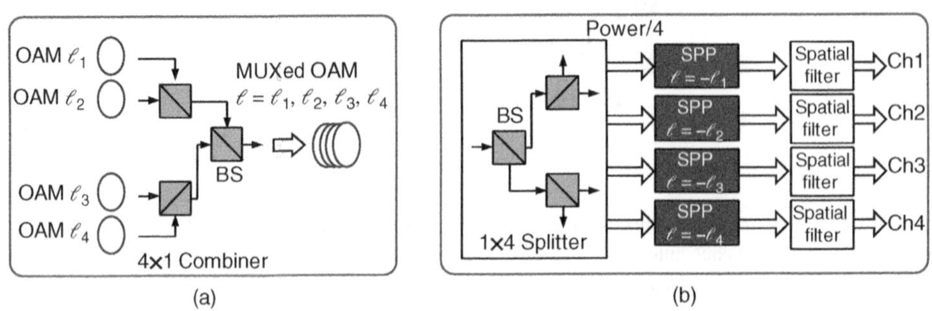

Figure 13.5 (a) Spatial multiplexing using cascaded BS. (b) Demultiplexing using cascaded BS and conjugated SPPs

the top board can be inserted. The mutual coupling between the two modes is measured to be less than −40 dB.

13.4 Wireless Communications Using OAM Multiplexing

13.4.1 Wireless Communications Using Gaussian and OAM Beams

An outdoor radio experiment in the 2.4-GHz WiFi band has been demonstrated over a distance ~442 m by using one Gaussian beam of $\ell = 0$ and one OAM beam of $\ell = +1$ [12]. Two identical WiFi frequency-modulation transmitters were tuned to the carrier frequency of 2.414 GHz to feed two antennas. The Gaussian mode was radiated with linear polarization by a commercial Yagi–Uda antenna, and the OAM mode of $\ell = +1$ was radiated by the helical parabolic antenna. The transmitted frequency modulated signal bandwidths of both channels were 27 MHz. At the receiver, a phase interferometer constructed from two identical Yagi–Uda antennas was deployed along a baseline perpendicular to the direction of the transmitters to recover the signal from each channel.

13.4.2 32-Gbit/s mm-wave Communications using OAM and Polarization Multiplexing

OAM multiplexing and other multiplexing techniques, such as polarization multiplexing, can be used to further increase the capacity and spectral efficiency of the RF communications link. In an experimental demonstration, a high-capacity mm wave communications link was realized by transmitting eight multiplexed OAM beams (four OAM beams on each of the two orthogonal polarizations), each carrying a 1-Gbaud 16-quadrature amplitude modulation (QAM) signal, thereby achieving a capacity of 32 Gbit/s^{-1} and a spectral efficiency of ~16 bit/s^{-1}Hz^{-1} at a single carrier frequency of 28 GHz [9]. Four different OAM beams of state numbers −3, −1, +1, and +3 on each of two polarizations are generated and multiplexed using SPPs made of HDPE and specially designed BSs at the transmitter such that all of the beams are co-propagating from a single transmitter aperture. After propagating through 2.5 m, the OAM channels are demultiplexed at the receiver using SPPs. The on-axis propagation of all OAM channels and the orthogonality among them allow their demultiplexing to be accomplished using physical components with low crosstalk, reducing the need for further signal processing to cancel channel interference.

Figure 13.6(a) depicts the ring-shaped intensity profile and interferogram of the OAM beams generated using horn antennas and SPPs. The OAM number ℓ of the beams can be deduced from the number of rotating arms in their interferograms, shown in Figure 13.6(b), which are generated by interfering the different OAM beams with a Gaussian beam ($\ell = \mathbf{0}$) through a coherent superposition using a beamsplitter. Figure 13.6 (b) shows the normalized measured intensity distributions of the coherent superposition of different OAM beams, which are in good agreement with the corresponding simulation results.

It is expected that the power from other channels would leak into the channel under detection due to the imperfections of OAM generation, multiplexing and setup misalignment, which would essentially result in channel crosstalk when a specific channel is recovered at the receiver. The crosstalk for a specific OAM channel ℓ_1 can be measured by $P_{\ell \neq \ell_1}/P_{\ell = \ell_1}$, where

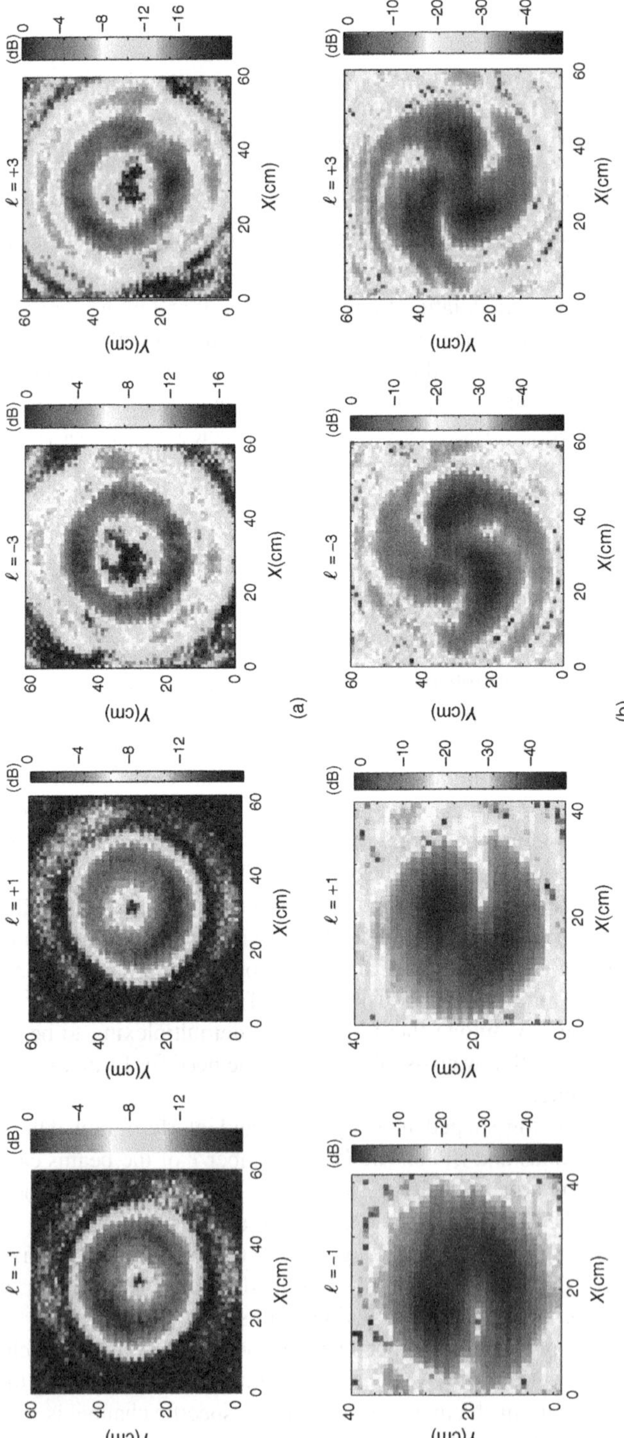

Figure 13.6 (a) Normalized measured intensity of four mm wave OAM beams of charge $\ell = \pm 1$ and $\ell = \pm 3$. (b) Interferogram images taken by combining a Gaussian beam and OAM beams by using a beam splitter

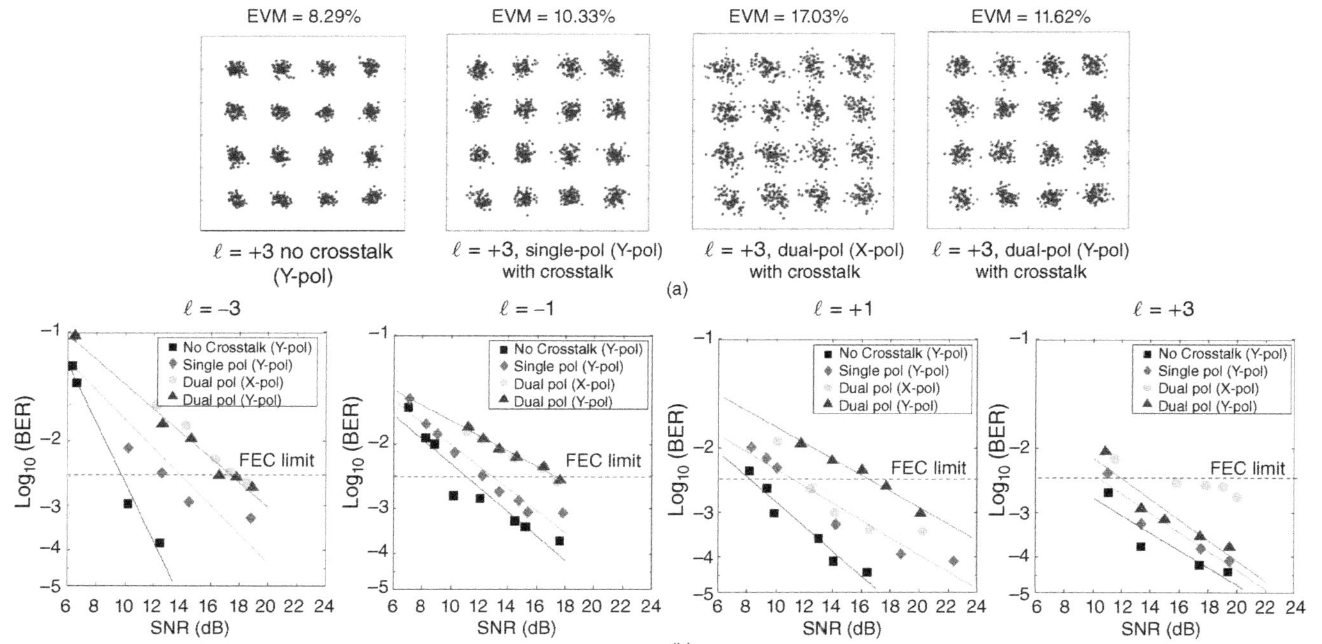

Figure 13.7 Results of 32-Gbit/s^{-1} data transmission using eight pol-mixed mm-wave OAM channels; (a) constellations of the received 1-Gbaud 16QAM signals for OAM channel $\ell = +3$ under single Y-pol and dual-pol conditions, respectively; (b) measured BER curves of 1-Gbaud 16-QAM signals of OAM channels

$P_{\ell \neq \ell_1}$ is the received power of channel ℓ_1 when all channels except channel ℓ_1 are transmitted and $P_{\ell=\ell_1}$ is the received power of channel ℓ_1 when only channel ℓ_1 is transmitted. We find that the crosstalk deteriorates significantly when two polarizations are included, mainly due to the birefringence of the SPPs. Figure 13.7(a) shows constellations and error vector magnitudes (EVMs) of the received 1-Gbaud 16-QAM signals with a signal-to-noise ratio (SNR) of 19 dB for channel $\ell = +3$ in both single-polarization (single-pol) and dual-polarization (dual-pol) cases. $\ell_1 P_{\ell \neq \ell_1}/P_{\ell=\ell_1}$ $P_{\ell \neq \ell_1} \ell_1 \ell_1$ $P_{\ell=\ell_1} \ell_1 \ell_1 \ell = +3$The constellations for channels $\ell = +3$of both X-pol and Y-pol become much worse in the dual-pol case, as they experience higher crosstalk. The measured bit error rates (BERs) for each channel as functions of SNR under both single Y-pol and dual-pol cases are shown in Figure 13.7(b). It is also observed that the OAM channel with higher crosstalk will consequently have worse BER performance. It is clear that each channel is able to achieve a raw BER below 3.8×10^{-3}, which is a level that allows for extremely low block error rates through the application of efficient forward error correction (FEC) codes [19]. The spectral efficiency in the experiment is ~16 bit/s^{-1}Hz^{-1}.

13.4.3 16-Gbit/s mm-wave Communications by Combining Traditional Spatial Multiplexing and OAM Multiplexing with MIMO Processing

OAM multiplexing through a single aperture pair employs the orthogonality of OAM beams to minimize interchannel crosstalk and enable recovery of different data streams, thereby avoiding the use of MIMO processing. This is different from conventional spatial multiplexing, for which each data-carrying beam is received by multiple spatially separated receivers. MIMO-based signal processing is critical for reducing the crosstalk among channels and thus allows data recovery [20–22]. However, MIMO-based signal processing becomes more onerous for conventional spatial multiplexing systems, as the number of antenna elements increases, especially at high data rates of the order of Gbit/s^{-1} [23]. On the other hand, for OAM multiplexing systems, the detection of high-order OAM modes presents a challenge for the receiver because OAM beams with larger ℓ values diverge more during propagation. Therefore, the achievable number of data channels for each type of multiplexing technique might be limited, and achieving a larger number of channels by using any one approach might be significantly more difficult. There might be implementation benefits by combining two, such that they complement each other and enhance system performance.

If each antenna aperture in a conventional spatial multiplexing system can transmit multiple independent information-carrying OAM beams, the total number of channels accommodated could be further increased, thereby increasing system transmission capacity. Furthermore, the complexity of implementing MIMO processing in such a system could be potentially reduced by exploiting the orthogonality of OAM beams.

The concept of a high-capacity LoS wireless link using OAM multiplexing combined with a MIMO-based spatial multiplexing architecture is depicted in Figure 13.8. The system consists of N transmitter/receiver aperture pairs that are arranged in a uniform linear structure. Each of the transmitter apertures $T_i(i = 1, 2, \ldots, N)$transmit M multiplexed OAM beams, resulting in a total number of NM OAM data channels. The N receiver apertures $R_i(i = 1, 2, \ldots, N)$ are used to capture the fields that are transmitted from the N transmitter apertures. Because of divergence along the propagation distance, the OAM beams from each transmitter aperture may spatially overlap at the receiver.

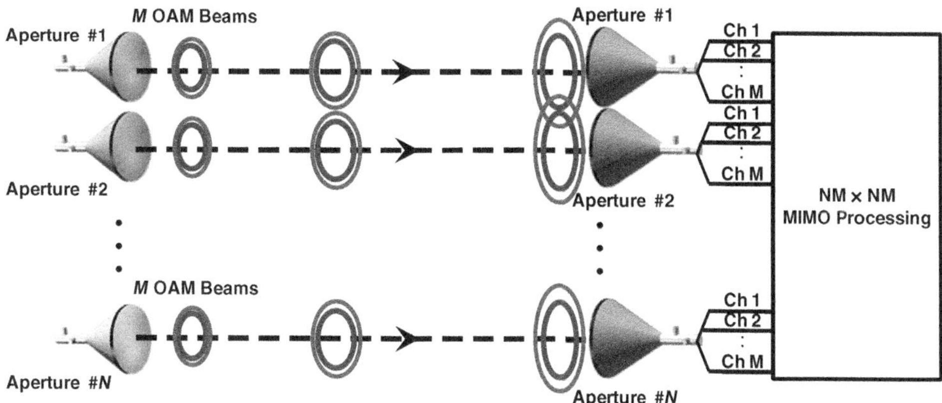

Figure 13.8 Concept of an LoS mm-wave communications link employing OAM multiplexing over conventional spatial multiplexing. The system consists of N transmitter/receiver apertures, with each transmitter aperture containing M OAM modes

A 16-Gbit/s mm-wave communication link using OAM multiplexing combined with conventional spatial multiplexing has been demonstrated [24]. A spatial multiplexing system with a 2×2 antenna aperture architecture, each transmitter aperture containing multiplexed OAM beams with $\ell = +1$ and $+3$, is implemented. Each of the four OAM channels carries a 1-Gbaud 16-QAM signal at a carrier frequency of 28 GHz, thereby achieving a capacity of 16 Gbit/s. After propagating through 1.8 m, the OAM beams from one transmitter aperture spatially overlap those from the other apertures in the receiver aperture planes, resulting in crosstalk among non-coaxial OAM channels. A 4×4 MIMO signal processing is used to mitigate the channel interference. This indicates that OAM multiplexing and conventional spatial multiplexing, combined with MIMO processing, can be compatible with and complement each other, thereby providing the potential to enhance system performance.

Figure 13.9 shows the measured normalized intensity of the generated OAM beams and their normalized spatial coherent superposition. The ring-shaped intensity profile of the generated mm-wave OAM beams, clearly depicted in Figure 13.9 (a), confirms that an OAM beam with $\ell = +3$ diverges faster than one with $\ell = +1$. The state number of the OAM beams ($\ell = +1$ and $\ell = +3$) can be deduced from the number of rotating arms in Figure 13.9 (b), which shows the measured intensity distribution of coherent superposition of OAM beams $\ell = +1$ and $\ell = +3$ at the T_1 or T_2 planes. Also, the coherent superposition of all four OAM beams measured at a 0.5-m or 1-m plane from the transmitter, as shown in Figure 13.9 (c), illustrates that the overlaps between OAM beams from apertures T_1 and T_2 increase with the transmission distance. The measured OAM beam superposition is confirmed by the corresponding simulation results.

Due to the divergence of the transmitted OAM beams and a small separation of 32 cm between the two receiver apertures, the OAM beams from the two transmitter apertures spatially overlap at the receiver aperture planes. Therefore, part of the non-axial OAM beams from the neighboring transmitter aperture would also be coupled into the receiver antenna, resulting in channel crosstalk. Table 13.1 shows the power transfer matrix of the 4×4 MIMO system and the total crosstalk of each channel.

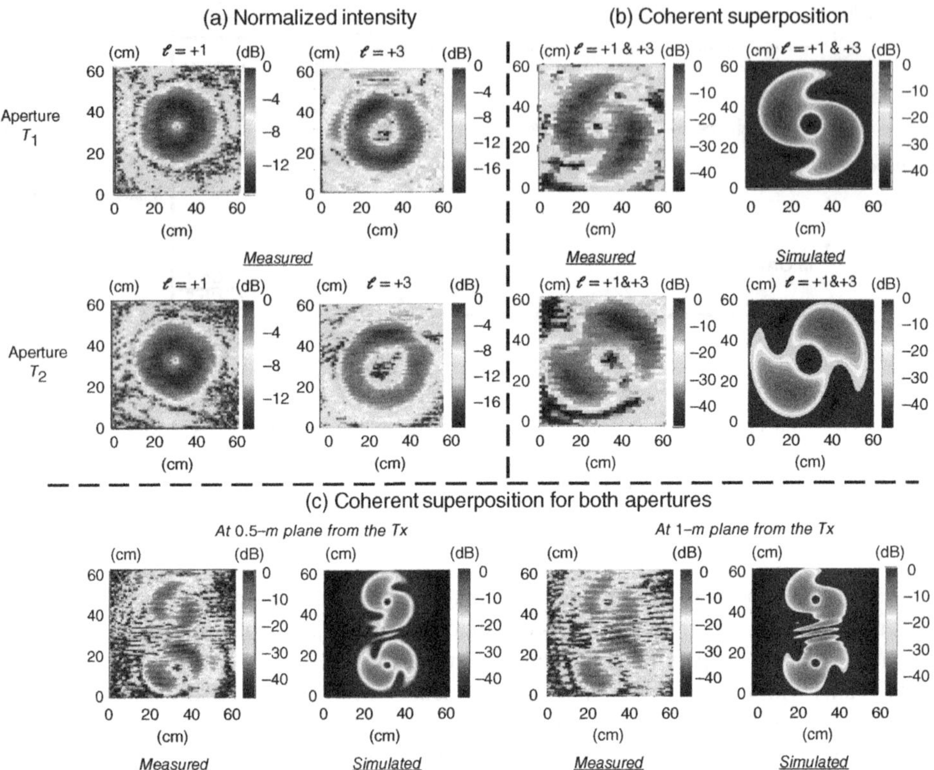

Figure 13.9 Normalized intensity and superposition of mm wave OAM beams; (a) normalized measured intensity of mm wave OAM beams $\ell = +1$ and $\ell = +3$ in transmitter apertures T_1 or T_2; (b) normalized measured and simulated superposition of OAM beams $\ell = +1$ and $\ell = +3$ at T_1 or T_2 plane; (c) normalized measured and simulated superposition of all four OAM beams at a 0.5-m or 1-m plane from the transmitter. Tx: transmitter

Table 13.1 The power transfer and crosstalk of each OAM channel in R_1 and R_2. The crosstalk of each OAM channel is measured at $f = 28$ GHz

		Aperture T_1		Aperture T_2		Total crosstalk
		$\ell = +1$ (dB)	$\ell = +3$ (dB)	$\ell = +1$ (dB)	$\ell = +3$ (dB)	(dB)
Aperture R_1	$\ell = +1$	−31.2	−49.2	−52.8	−55.3	***−15.7***
	$\ell = +3$	−51.1	−31.8	−42.3	−46.7	***−8.8***
Aperture R_2	$\ell = +1$	−54.3	−45.2	−28.8	−50.3	***−14.8***
	$\ell = +3$	−43.6	−44.3	−49.2	−29.2	***−11.1***

$R_1 R_2$

At the receiver, multi-modulus algorithm (MMA)-based equalization [25, 26] is used for MIMO processing and 4 × 4 MMA–MIMO digital signal processing (DSP) [27] is implemented to recover the four data streams. The signal received from each receiver antenna is amplified and downconverted to a 4-GHz carrier frequency. Figure13.10 (a) shows the BER measurements of 1-Gbaud 16-QAM signals for channels $\ell = +1$ and $\ell = +3$ when only one transmitter/receiver aperture pair is on. In this case, there is no mutual crosstalk from channels of the other transmitter/receiver pair. It is clear that without interference from other apertures, each channel can achieve a raw BER of 3.8×10^{-3}, which is a level that allows extremely low block error rates through the application of efficient FEC codes. When all four OAM channels are turned on, they experience crosstalk from both their coaxial and non-coaxial channels. Figure 13.10 (b) shows the measured BERs of all four OAM channels without MIMO processing. The BER curves for all four channels exhibit the "error floor" phenomenon due to the strong crosstalk from other channels. Figure 13.10 (c) depicts the BER curves of all

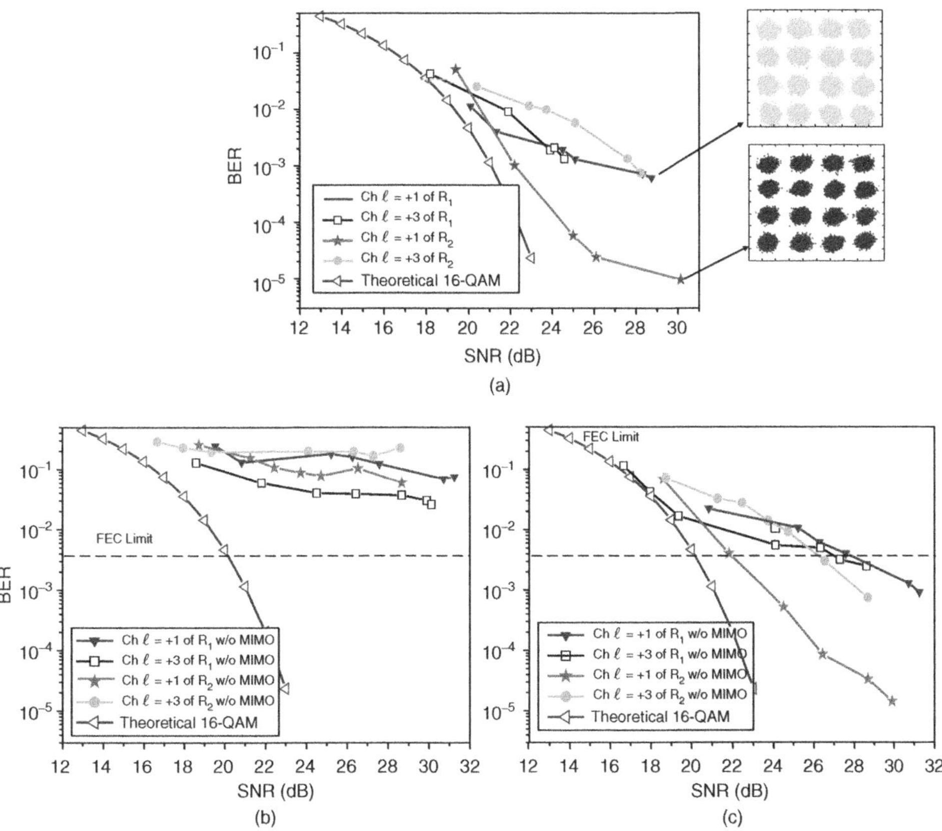

Figure 13.10 (a) BER measurements of 1-Gbaud 16-QAM signal for channels $\ell = +1$ and $\ell = +3$ for R_1 (when T_2 is off) and R_2 (when T_1 is off). Measured BERs as a function of received SNR for all four OAM channels ($\ell = +1$ and $\ell = +3$ in both $R_1$and R_2); (b) without MIMO processing; (c) with MIMO processing

four channels under the same conditions using MIMO equalization processing. The BERs decrease dramatically and can reach below 3.8×10^{-3} for all four channels after MIMO processing.

13.4.4 Multipath Effects of OAM Channels

As for all wireless communication systems, multipath effects [28–31] are likely to have significant effects on OAM multiplexing systems, particularly when the divergence of OAM beams with $\ell \neq 0$ is greater than that of conventional beams with $\ell = 0$, such as. Gaussian beams. Several unique factors associated with an OAM-multiplexed link present the following interesting technical challenges.

Intra- and interchannel crosstalk The reflected energy can be coupled not only into the same data channel of the same OAM value – as in the case of a conventional single beam link – but also into another data channel with a different OAM value. Therefore, both intra- and interchannel crosstalk can occur.

Beam intensity and phase An OAM beam has a doughnut-shaped intensity profile showing low power in the center and a peak showing maximum power around a ring; furthermore, the beam exhibits an azimuthal phase change with a value of $2\pi\ell$. In addition, the detection of a specific OAM beam requires a spatial filter to filter out energy of the other beams, which may reduce the received power from the reflected beam [32].

Figure 13.11 shows the concept of the multipath effects of an OAM beam caused by the specular reflection from a reflector parallel to the link. An OAM channel with an OAM number of ℓ_1 is transmitted along the link. At the receiver end, an SPP with an OAM number of ℓ_2 and an antenna is used to receive the OAM channel with an OAM number of ℓ_2. Ideally, power can only be recovered when $\ell_1 = \ell_2$, owing to the orthogonality of OAM beams in a LoS link. However, the orthogonality no longer holds when the receiver receives the reflected beam.

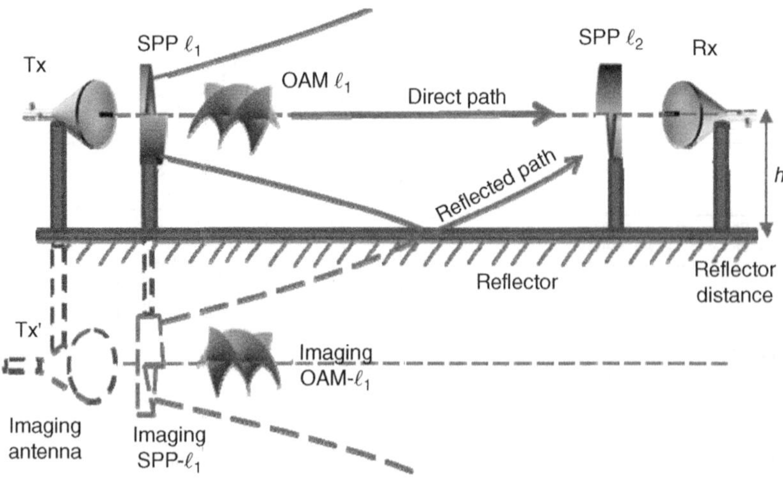

Figure 13.11 Concept of the multipath effects of an OAM beam caused by the specular reflection from a reflector parallel to the link

Assuming a reflector is placed a distance of h away from the beam center, the reflected beam can be observed as an OAM beam from an imaging antenna Tx′ and an imaging SPP with an OAM number of $-\ell_l$ [33] (reflection changes the sign of the OAM value). Therefore, the receiver will receive an OAM beam with an OAM number of ℓ_l from the original link as well as a reflected beam with an OAM number of $-\ell_l$ from an offset link, which is placed at a distance of $2h$.

Reflection causes the distortion of the OAM beam wavefront, giving rise to both intra- and interchannel crosstalk. To illustrate this phenomenon, we use an OAM beam with $\ell_1 = +3$ as an example. In Figure 13.12, the left-hand column shows the intensity, phase and OAM spectrum of the OAM beam in the direct path. The entire power is in the OAM state of $\ell_1 = +3$. The middle column shows the reflected OAM beam. The reflected OAM beam exhibits an OAM number of $\ell_1{}' = -3$, and it is offset to the direct link. As a result, when the reflected OAM beam is decomposed with respect to the OAM basis along the direct path axis, power diverges onto a wide range of OAM states, leading to intrachannel crosstalk with an OAM channel with $\ell_1 = +3$ and interchannel crosstalk with the other OAM channels with $\ell_1 \neq +3$

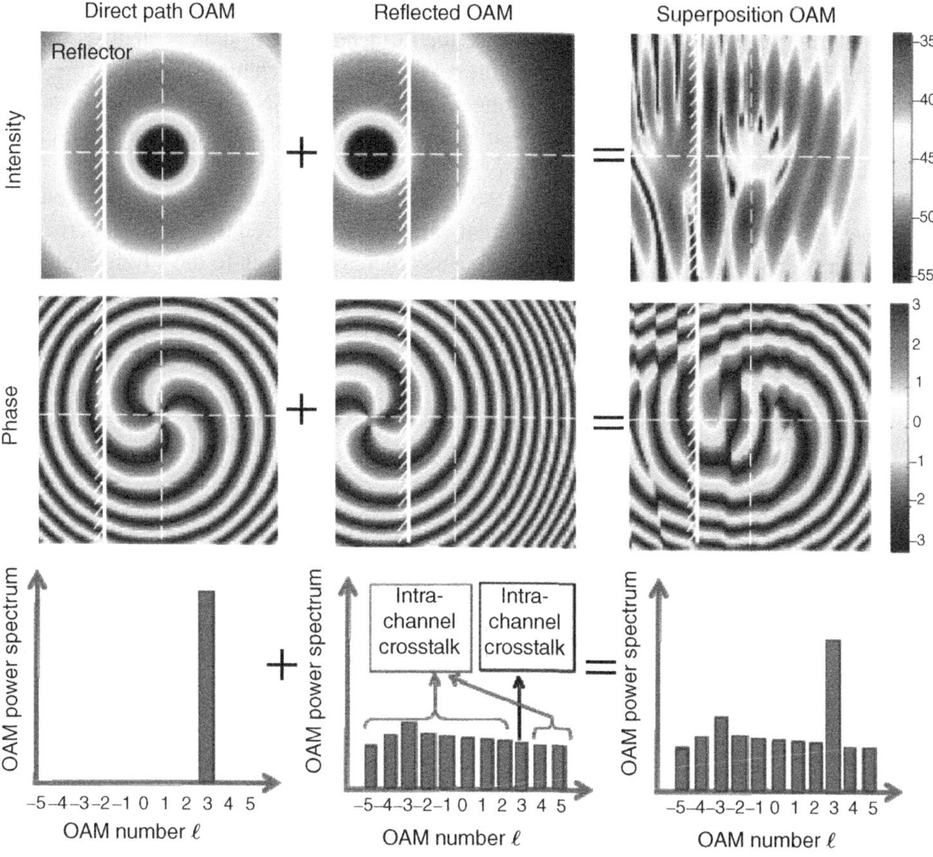

Figure 13.12 Concept of OAM multipath-induced intra- and interchannel crosstalk

[24]. The right-hand column in Figure 13.12 shows the actual beam at the receiver, which is the superposition of the direct and the reflected beams. The intensity exhibits a fringing pattern owing to the interference between the direct and the reflected beams. The wavefront phase is also distorted owing to the multipath effect. The power of the actual received OAM with $\ell_1 = +3$ differs from that of the directed path because of the intrachannel crosstalk from the reflected beam, and the received power of the other OAM beams with values of $\ell_1 \neq +3$ is nothing but the interchannel crosstalk from the reflected beam.

In the experiment, a planar aluminum sheet (2.5 m long × 1.5 m high) is mounted on a cart and placed parallel to the propagation path to investigate the multipath effects. The distance between the path and the reflector can be changed by moving the cart. As the distance between the path and reflector decreases, the multipath effect becomes stronger. The intrachannel crosstalk is measured when only the OAM beam of $\ell = \ell_1$ is transmitted, and at the Rx, only $\ell = \ell_1$ is received. Two effects are observed: First, the received power for OAM beams with a higher ℓ is lower, which is explained by the divergence property of OAM beams [34]. Second, when the reflector distance is small, the received power for the OAM channels varies as the reflector distance changes, which is explained by the interference between the signals from the direct and reflected paths. Figure 13.13 shows the BER and SNR as functions of the reflector distance for OAM channels $\ell = +1$ and $\ell = +3$ when a 1-Gbaud 16-QAM signal is transmitted on each channel. For $\ell = +1$, since the intrachannel crosstalk is smaller, weaker SNR and BER variations are observed when the reflector distance changes. For OAM $\ell = +3$, stronger channel interference and much stronger fluctuations of SNR and BER are observed when the reflector distance changes. The results show that an OAM channel with a larger ℓ is more affected by intrachannel crosstalk.

Interchannel crosstalk is measured when the transmitter SPP OAM number ℓ_1 is fixed and the receiver SPP OAM number ℓ_2 takes different values. It is observed that with an increase in h, the power received from the other OAM channels increases owing to the strong multipath effects. For example, when $\ell_1 = +1$, the received power difference between $\ell_2 = +1$ and the

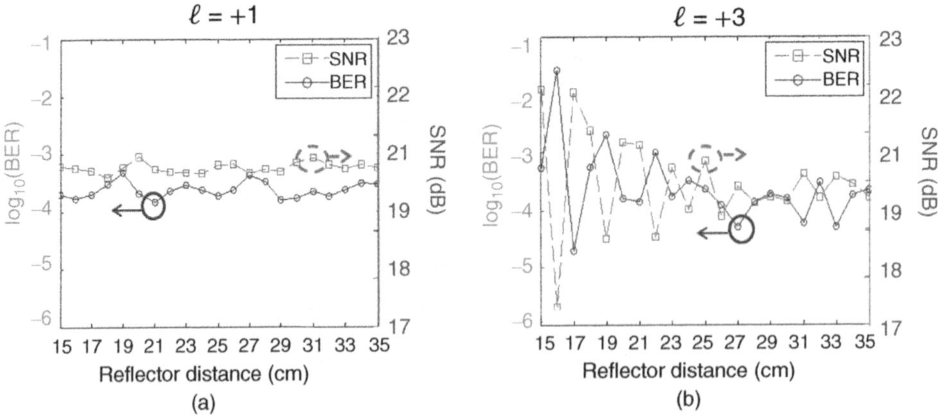

Figure 13.13 Measured BER and SNR as functions of the reflector distance for (a) $\ell = +1$ and (b) $\ell = +3$. Stronger fluctuations of BER and SNR are observed for OAM $\ell = +3$ because of the stronger intrachannel crosstalk induced by the multipath effect

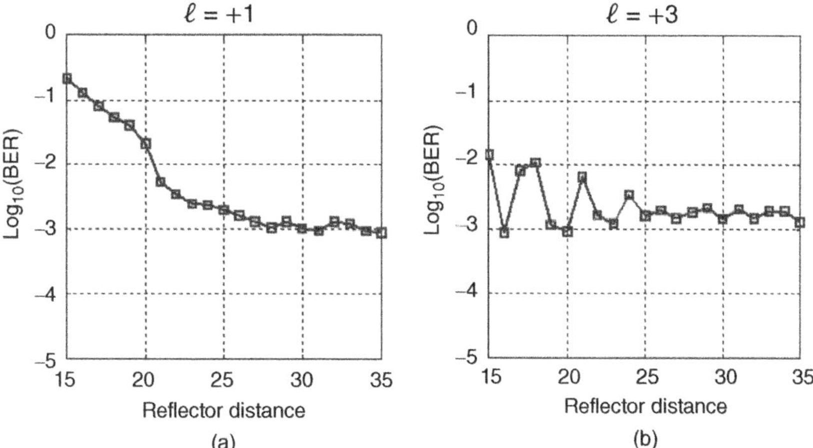

Figure 13.14 Measurement of BER as a function of the reflector distance when (a) $\ell = +1$ and (b) $\ell = +3$

other channels is more than 20 dB, indicating that the interchannel crosstalk from the channel with $\ell = +1$ to the other channels is less than 20 dB. In the case of the OAM channel with $\ell_1 = +3$, the interchannel crosstalk from the channel with $\ell = +3$ to the other channels is 2–9 dB. Next we characterize the channel performance when two OAM channels with $\ell = +1$ and $\ell = +3$ are multiplexed, and each channel experiences both intrachannel and interchannel crosstalk. The signal on each channel is a 1-Gbaud 16-QAM data stream. Figures 13.14 (a) and (b) show the BER as a function of the reflector distance for $\ell = +1$ and $\ell = +3$; the SNR for both channels is ~22 dB when there is no reflector. The BER of OAM channel $\ell = +1$ is observed to significantly increase when the reflector is close to the path, due to the interchannel crosstalk from the OAM channel $\ell = +3$. For the OAM channel $\ell = +3$, Figure 13.14(b) shows a similar BER variation, as observed in Figure 13.13, because the degradation of OAM channel $\ell = +3$ is mainly caused by intrachannel crosstalk and is not much affected by the interchannel crosstalk from OAM channel $\ell = +1$.

13.4.5 OAM Communications based on Bessel Beams

In LoS links, an obstacle moving across the beam path causes loss of power due to blockage of the beam. The link may experience outages, and retransmission of data might be required. In links that utilize spatial multiplexing of multiple modes to carry distinct data channels, such as OAM multiplexing, the effects of the obstructed beam path are twofold. First, similar to single-mode links, part of the energy in the multiplexed modes scatters and does not reach the receiver. Secondly, since the information contained in the transverse field is crucial for discriminating between different received modes, an obstruction inadvertently blocking the beam path will change the spatial profiles of the multiplexed beams due to diffraction effects. Thus a power leakage among neighboring modes occurs, which in turn gives rise to interchannel crosstalk.

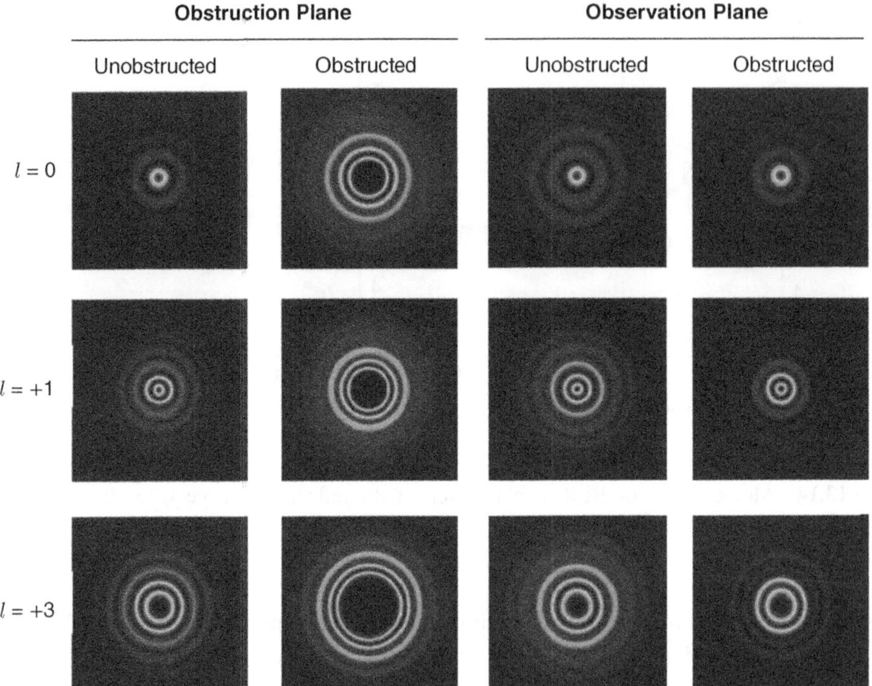

Figure 13.15 Simulated evolution of different obstructed OAM Bessel beams at 28 GHz along the propagation direction. In all cases, the innermost rings of the Bessel beams are obstructed by an opaque circular obstruction

One potential approach to resolve such an issue could be using Bessel beams, which have the "self-healing" feature that their phase can be "reconstituted" after passing through a partial obstruction [34]. The simulation results in Figure 13.15 show the self-healing property of obstructed approximate Bessel beams. In each case, the central spot (for $\ell = 0$) or the inner-most ring (for $\ell > 0$) are blocked by opaque circular obstructions. A comparison between the images of obstructed and unobstructed beams in the observation plane reveals the self-healing of the inner-most rings of the obstructed Bessel beams.

One method to generate approximate Bessel beams is to pass Gaussian or OAM beams through a conical lens (axicon) [35, 36]. As shown in Figure 13.16, an input beam is passed through a conical lens with a cone angle and is transformed into a Bessel beam over a propagation distance for which Bessel beam remains propagation invariant (i.e. the Bessel region). Figure 13.16(b) shows transverse profiles of obstructed and unobstructed Bessel beam $\ell = +1$ at a distance of 1 m. The obstruction radius in this case is 5 cm. The intensity images of unobstructed beams are also shown below for comparison.

Figure 13.17 shows the BER curves for each of the two transmitted channels with and without obstruction. In all of the cases, the obstructions are placed at the beam center, the point where maximum power loss occurs for the desired channel. Each channel can achieve a raw BER of 3.8×10^{-3} (the FEC limit). We observe that the power penalties for OAM channels $\ell = +1$ and $\ell = +3$ at the FEC limit are 2.3 dB and 1.7 dB for 5-cm obstruction, respectively.

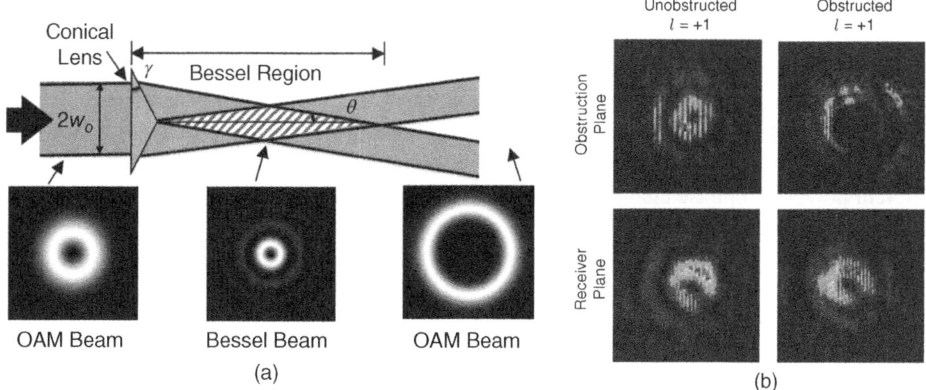

Figure 13.16 (a) Generation of approximate Bessel beams by passing an OAM beam through a conical lens. (b) Intensity images of obstructed and unobstructed Bessel beam $\ell = +1$

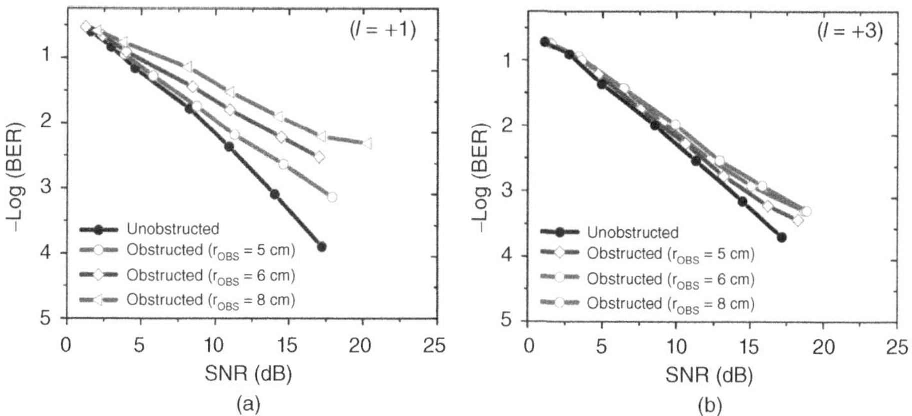

Figure 13.17 BER as a function of received SNR for the multiplexed OAM channels $\ell = +1$ and $\ell = +3$ with and without obstruction. In all cases, the obstructions are placed at the beam center. The radii of obstructions are 5 cm, 6 cm, and 8 cm

Also, the power penalties for both the channels increase with increasing obstruction sizes, but they remain less than 10 dB and less than 3 dB for Bessel beams with $\ell = +1$ and $\ell = +3$, respectively. Due to the smaller spot size of the $\ell = +1$ channel, it experiences a larger power penalty than the channel on $\ell = +3$ for various obstruction diameter cases.

13.5 Summary and Perspective

In this chapter, we introduced the basic concept of the OAM of EM waves and its applications in wireless communications. Different approaches can be used to generate and multiplex OAM beams, such as SPPs, patch antenna arrays, and metamaterial structures. Wireless communication links using OAM multiplexing are reviewed. OAM multiplexing can be also combined with polarization multiplexing and traditional spatial multiplexing to further increase the capacity and spectral efficiency. Other propagation effects of OAM beams, such as multipath effects and the self-healing of Bessel OAM beams are also discussed.

Most RF OAM transmission experiments are demonstrated over short distances. However, OAM multiplexing techniques could potentially be expanded to longer distances as long as sufficient power and phase change of each OAM beam is captured. In general, the number of OAM beams that can be accommodated in a system is limited by different factors, including receiver aperture size and intermodal crosstalk. Given a fixed receiver aperture size, a larger OAM ℓ value may result in a larger beam size at the receiver such that the recovered power decreases and the BER increases. In addition, higher intermodal crosstalk leads to an increase in power penalty and a limited number of available OAM states. Intermodal crosstalk can arise due to many factors, including imperfection of components, system misalignment and effects from the transmission medium. The OAM-based communication system faces several challenges as to its ability to scale to higher numbers of OAM beams. Issues include reducing intermodal crosstalk, improving the demultiplexer, and developing integrated devices.

From a theoretical perspective, OAM multiplexing can be considered as another form of spatial multiplexing, and the implementation of OAM mode-multiplexing is different from traditional RF spatial multiplexing [37–39]. The latter employs multiple spatially separated transmitter and receiver aperture pairs for the transmission of multiple data streams. Since each of the antenna elements receives a different superposition of the different transmitted signals, each of the original channels can be demultiplexed through the use of electronic DSP. In the implementation of OAM multiplexing, the multiplexed beams are completely co-axial throughout the transmission medium and use only one transmitter and receiver aperture (though with certain minimum aperture sizes), employing OAM beam orthogonality to achieve efficient demultiplexing without the need for further digital signal post-processing to cancel channel interference; thus there are significant implementation differences between these two approaches. However, it has been shown that OAM multiplexing and conventional spatial multiplexing can be combined, and tradeoffs (within the constraints of a given aperture size) can be made between the spatial degrees of freedom used for OAM and spatial multiplexing in order to achieve the most advantageous implementation.

References

[1] Allen, L., Beijersbergen, M.W., Spreeuw, R.J.C., and Woerdman, J.P. (1992) Orbital angular-momentum of light and the transformation of Laguerre-Gaussian laser modes. *Phys. Rev. A* **45**, 8185–8189

[2] Barnett, S. and Allen, L. (1994) Orbital angular momentum and non paraxial light beams. *Opt. Commun.*, **110**, 670–678.

[3] Allen, L., Padgett, M., and Babiker, M. (1999) The orbital angular momentum of light. *Prog. Opt.* **39**, 291–372.

[4] Zhao, Y., Edgar, J.S., Jeffries, G.D.M., McGloin, D., and Chiu, D.T. (2007) Spin-to-orbital angular momentum conversion in a strongly focused optical beam. *Phys. Rev. Lett.*, **99**, 073901.

[5] Padgett, M.J., Miatto, F.M., M. P.J. Lavery, Zeilinger, A., and Boyd, R.W. (2015) Divergence of an orbital-angular-momentum-carrying beam upon propagation. *New J. Phys.*, **17**, 023011.

[6] Gibson, G., Courtial, J., Padgett, M., Vasnetsov, M., Pas'ko, V., Barnett, S., and Franke-Arnold, S. (2004) Free-space information transfer using light beams carrying orbital angular momentum. *Opt. Express*, **12**, 5448–5456.

[7] Yao, A.M., and Padgett, M.J. (2011) Orbital angular momentum: origins, behavior and applications. *Adv. Opt. Photonics*, **3**,161–204.

[8] Molina-Terriza, G., Torres, J.P., and Torner, L. (2007) Twisted photons. *Nat. Phys.*, **3**, 305–310.
[9] Yan, Y., Xie, G., Lavery, M.P.J., Huang, H., Ahmed, N., Bao, C., Ren, Y., Cao, Y., Li, L., Zhao, Z., Molisch, A.F., Tur, M., Padgett, M.J. and Willner, A.E. (2014) High-capacity millimetre-wave communications with orbital angular momentum multiplexing. *Nat. Commun.*, **5**, 4876.
[10] Turnbull, G.A., Robertson, D.A., Smith, G.M., Allen, L., and Padgett, M.J. (1996) The generation of free-space Laguerre-Gaussian modes at millimeter-wave frequencies by use of a spiral phase plate. *Opt. Commun.*, **127**, 183–188.
[11] Mahmouli, F.E. and Walker, D. (2013) 4-Gbps uncompressed video transmission over a 60-GHz orbital angular momentum wireless channel, *IEEE Wireless Commun. Lett.*, **2**, 223–226.
[12] Tamburini, F., Mari, E., Sponselli, A., Thidé, B., Bianchini, A., and Romanato, F. (2012) Encoding many channels on the same frequency through radio vorticity: first experimental test. *New J. Phys.***14**, 033001.
[13] Tamburini, F., B. Thidé, Mari, E., Parisi, G., Spinello, F., Oldoni, M., Ravanelli, R.A., Coassini, P., Someda, C.G., Romanato, F. (2013) N-tupling the capacity of each polarization state in radio links by using electromagnetic vorticity. *arXiv*:1307.5569v2
[14] Zhao, Z., Ren, Y., Xie, G., Yan, Y., Li, L., Huang, H., Bao, C., Ahmed, N., Lavery, M.P., Zhang, C., Ashrafi, N., Ashrafi, S., Talwar, S., Sajuyigbe, S., Tur, M., Molisch, A.F. and Willner, A.E. (2015) Experimental demonstration of 16-Gbit/s millimeter-wave communications link using thin metamaterial plates to generate data-carrying orbital-angular-momentum beams, in *Proceedings of IEEE International Communication Conference*, London, UK.
[15] Cheng, L., Hong, W., and Hao, Z. (2014) Generation of electromagnetic waves with arbitrary orbital angular momentum modes. *Scientific Reports*, **4,** 4814.
[16] Thidé, B., Then, H., Sjöholm, J., Palmer, K., Bergman, J., Carozzi, T.D., Istomin, Y.N. Ibragimov, N.H., and Khamitova, R. (2007) Utilization of photon orbital angular momentum in the low-frequency radio domain. *Phys. Rev. Lett.*, **99**, 087701.
[17] Bai, Q., Tennant, A. and Allen, B. (2004) Experimental circular phased array for generating OAM radio beams. *Electron. Lett.*, **50**, 1414–1415.
[18] Li, Z., Ohashi, Y., and Kasai, K. (2014) A dual-channel wireless communication system by multiplexing twisted radio wave. *Proceedings of the 44th European Microwave Conference*, pp. 235-238.
[19] Richter, T., Palushani, E., Schmidt-Langhorst, C., Ludwig, R., Molle, L., Nölle, M., and Schubert, C. (2012) Transmission of single-channel 16-QAM data signals at terabaud symbol rates. *J. Lightwave Technol.* **30**, 504–511.
[20] Andrews, J.G., Wan, C., and Heath, R.W. (2007) Overcoming interference in spatial multiplexing MIMO cellular networks. *IEEE Trans. Wireless Commun.*, **14**, 95–104.
[21] Caire, G. and Shamai, S. (2003) On the achievable throughput of a multiantenna Gaussian broadcast channel. *IEEE Trans. Info. Theory*, **49**, 1691–1706.
[22] Molisch, A.F. (2011) *Wireless Communications*, 2nd edn. Wiley.
[23] Sibille, A., Oestges, C.. and Zanella, A. (2009) *MIMO: From Theory to Implementation*. Academic Press.
[24] Ren, Y., Li, L., Xie, G., Yan, Y., Cao, Y., Huang, H., Ahmed, N., Zhao, Z., Liao, P., Zhang, C., Lavery, M.P.J., Tur, M., Caire, G., Molisch, A.F., and Willner, A.E. (2014) Experimental demonstration of 16 Gb/s millimeter-wave communications using MIMO processing of 2 OAM modes on each of two transmitter/receiver antenna apertures, in *Proc. of IEEE Global Telecommunications Conference*, paper 1569944271.
[25] Yang, J., Werner, J.-J. and Dumont, G.A. (2002) The multimodulus blind equalization and its generalized algorithms. *IEEE J. Select. Areas Commun.*, **20**, 997–1015.
[26] Yuan, J.-T. and Tsai, K.-D. (2005) Analysis of the multimodulus blind equalization algorithm in QAM communication systems. *IEEE Trans. Commun.* **53**, 1427–1431.
[27] Ready, M., and Gooch, R (1990) Blind equalization on radius directed adaptation. *Proc. of International Conference on Acoustics, Speech, and Signal Processing (ICASSP-90)* **3**, 1699.
[28] Jakes, W.C. (1994) *Microwave Mobile Communications*. Wiley-IEEE Press.
[29] Andersen, J.J.B., Rappaport, T.S., and Yoshida, S. (1995) Propagation measurements and models for wireless communications channels. *IEEE Commun. Mag.* **33**, 42–49.
[30] Iskander, M.F. and Yun, Z. (2002) Propagation prediction models for wireless communication systems. *IEEE Trans. Microwave Theory Tech.*, **50**, 662–673.
[31] Saleh, A.A.M., and Valenzuela, R.A. (1987) A statistical model for indoor multipath propagation. *IEEE J. Select. Areas Commun.*, **5**, 128–137.

[32] Yan, Y., Li, L., Xie, G., Bao, C., Liao, P., Huang, H., Ren, Y., Ahmed, N., Zhao, Z., Lavery, M.P., Ashrafi, N. Ashrafi, S., Talwar, S., Sajuyigbe, S., Tur, M., Molisch, A.F., and Willner, A.E. (2015) Experimental measurements of multipath-induced intra- and inter-channel crosstalk effects in a millimeter-wave communications link using orbital-angular-momentum multiplexing, in *Proceedings of IEEE International Communication Conference*, London, UK.

[33] Byun, S.H., Hajj, G.A., and Young, L.E. (2002) Development and application of GPS signal multipath simulator. *Radio Sci.*, **37**, 1098.

[34] Ahmed, N., Zhao, Z., Yan, Y., Xie, G., Wang, Z., Lavery, M.P., Ren, Y., Almaiman, A., Huang, H., Li, L., Bao, C., Liao, P., Ashrafi, N., Ashrafi, S., Tur, M., Molisch, A.F., and Willner, A.E. (2015) Demonstration of an obstruction-tolerant millimeter-wave free-space communications link of two 1-Gbaud 16-QAM channels using Bessel beams containing orbital angular momentum, *Third International Conference on OAM (ICOAM)*, 4–7 August, 2015, New York, USA.

[35] Arlt, J. and Dholakia, K. (2000), Generation of high–order Bessel beams by use of an axicon. *Opt. Commun.*, **177**, 297–301.

[36] Monk, S., Arlt, J., Robertson, D.A., Courtial, J., and Padgett, M.J. (1999) The generation of Bessel beams at millimetre-wave frequencies by use of an axicon. *Opt. Commun.*, **170**, 213–215.

[37] Winters, J.H. (1987) On the capacity of radio communication systems with diversity in a Rayleigh fading environment. *IEEE. J. Select. Areas Commun.*, **5**, 871–878.

[38] Foschini, G.J. and Gans, M.J. (1998) On limits of wireless communications in a fading environment when using multiple antennas. *Wireless Pers. Commun.*, **6**, 311–335.

[39] Gesbert, D., Shafi, M., Shiu, D.S., Smith, P.J. and Naguib, A. (2003) From theory to practice: an overview of MIMO space-time coded wireless systems. *IEEE J. Select. Areas Commun.*, **21**, 281–302.

Part Three

New Spectrum Opportunities for 5G

14

Millimeter Waves for 5G: From Theory To Practice

Malik Gul, Eckhard Ohlmer, Ahsan Aziz, Wes McCoy and Yong Rao

14.1 Introduction

Looking back a generation or two in the wireless industry and fast-forwarding to our world today, one can clearly understand the impact that wireless data has made on our everyday

Signal Processing for 5G: Algorithms and Implementations, First Edition. Edited by Fa-Long Luo and Charlie Zhang.

lives. The uses of wireless data have surpassed our imaginations. Some experts argue that the generational leap in wireless from 3G to 4G came earlier than anticipated, primarily driven by the Internet, social media, and smart devices, in addition to the ever-increasing demand for high-quality video. It is clear that the wireless industry needs to make some significant advances in the next decade to meet the ever-increasing demand for data; a thousand times by 2020 and ten thousand times by 2025 [1–3]. There is consensus in the wireless industry that meeting this demand calls for significant technical changes, hence the push for 5G.

It is important to emphasize that 5G is not only about thousand- or even ten-thousandfold increases in data or thousandfold increase in system capacity, but also about the reliability of the network needed to guarantee the quality of service (QoS) with ultra-low latency. There will be hundreds and thousands of connected devices, with different throughput, latency and reliability requirements. The reliability of the 5G network should satisfy the needs of every user. While a significant amount of research is ongoing to improve spectral reuse and efficiency below 6 GHz, this alone will not be sufficient to meet the requirements for 5G. This is due to the fact that spectrum below 6 GHz is very limited. Most of the spectrum below 6 GHz has already been allocated and the available bands are very narrow. Therefore, the natural progression is to look for larger chunks of spectrum above 6 GHz. This is where mmWave provides the biggest benefits. The largest chunks of spectrum are available in various bands above 28 GHz, where the wavelength is in the millimeter range [4]. Spectrum between 6 and 28 GHz (cmWaves) is also being investigated for use, but the available bandwidth is smaller than that in the mmWave bands.

This chapter is devoted to mmWave technology for 5G, with emphasis on the key requirements for building a 5G mmWave proof-of-concept (PoC) system. The rest of this chapter is organized into the following four sections. Section 14.2 is a brief introduction to mmWave PoC system building. In Section 14.3, desirable features of a mmWave PoC system are outlined. Section 14.4 presents a case study of the design of a mmWave PoC system with detailed discussion on its hardware and software architecture. Section 14.5 is the final section, which presents the summary of this chapter.

14.2 Building a mmWave PoC System

In recent years, researchers have started to address the technical challenges of developing a mmWave cellular system. Research in mmWaves includes designing mmWave integrated circuits (ICs) with integrated chip-scale antennas, minimizing power consumption and designing beam-steering systems. Similarly, some of the system design research areas include measuring, characterizing, and modeling wireless channels at different mmWave bands to enable accurate system design and deployment, designing new air interfaces, and reducing system latency. Network layer research includes designing new network topologies to realize the high–throughput, low-latency 5G systems with guaranteed QoS.

While theoretical research provides the foundation for the enabling technology for 5G, it is the PoC systems that will truly enable researchers to understand the challenges of the 5G mmWave systems and devise a suitable engineering solution to bring a mmWave cellular system to reality. PoC systems and field trials have always been an integral part of all wireless system development. However, until recently, these systems were primarily custom-developed. Developing and maintaining PoC prototyping platforms requires a substantial amount of engineering effort and expense, in addition to the application development time. It is only recently that a segment of the wireless industry has focused on developing the new tools and technologies needed to enable researchers to quickly prototype their ideas. Significant research and investment from academia and industry is being made to create platform-based solutions [5] that can produce highly reliable and repeatable experimental results with systems that are maintainable, extensible and programmable. Advances in software-defined radio (SDR) platforms, new hardware/software interfacing tools and hardware abstraction languages have enabled high-level abstractions of complex hardware architectures. Additionally, programming of devices such as field programmable gate arrays (FPGA) and digital signal processors (DSP) has been made significantly simpler for wireless researchers. These new tools and technologies are redefining the design flow for building PoC systems and have now become an integral component of 5G research.

Prototyping platform design has a very different set of requirements compared to the actual 5G devices, as these platforms need to accommodate many different variants of 5G concepts with "flexibility and ease". The last statement "flexibility and ease" has very profound implications for platform design as it impacts:

- RF front ends
- data converters
- data management and distribution
- DSP algorithms and architectures
- software architecture for real-time operation
- platform connectivity to network
- hardware abstraction language
- high-level language for hardware implementation.

In the following, we will attempt to cover some of these topics as they relate to building a mmWave PoC system for 5G.

14.3 Desirable Features of a mmWave Prototyping System

As outlined in the previous Section 14.2, prototyping is essential for understanding the practical challenges and design tradeoffs of mmWave communications if they are to serve as a key enabler of 5G. As mmWave communication has several design frontiers, it is important that the prototyping platform gives enough flexibility to the end user so that the challenges of mmWave systems can be addressed across different domains, and the design/performance of mmWave systems can be explored ranging from the front end, through the physical layer to higher layers of the protocol stack. In this section, we identify the key requirements for a platform suited for mmWave system prototyping.

A system diagram for a mmWave prototyping system, showing its key components, is given in Figure 14.1. A software-defined prototyping platform gives researchers the ability to test their algorithms and protocols, analyze the performance, quickly re-configure the platform and iterate their algorithms. In the context of 5G, as the throughputs soar to tens of Gbps and the latency reduces to the sub–millisecond level, high-bandwidth analog front-ends and real-time signal-processing elements are critical requirements. The RF bandwidth is

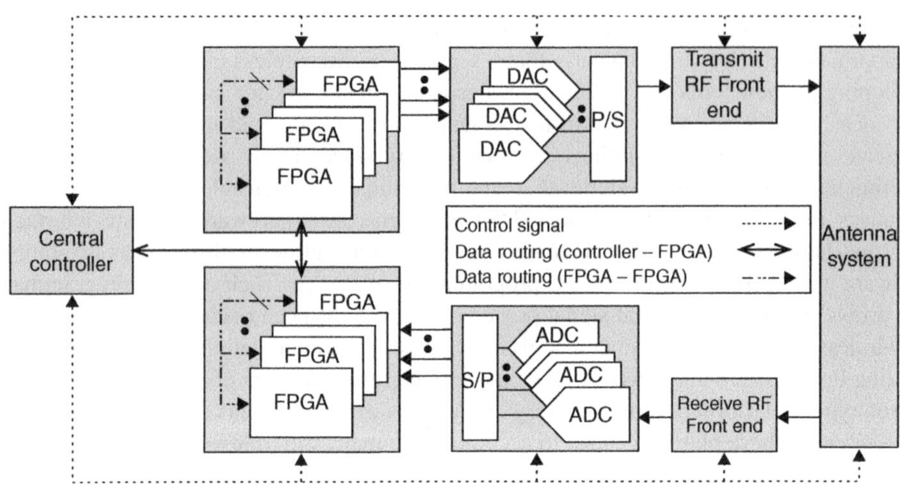

Figure 14.1 mmWave prototype system diagram

extraordinary compared to current standards, requiring up to 4 GHz or more; current standards utilize bandwidths of the order of 160 MHz and below. The mmWave standards are expected to introduce advanced hardware features and requirements that have not existed in earlier cellular standards. It is expected that advanced multi-element antenna architectures such as phased arrays will be used because these support selectable or steerable beams switched at a very high rate. Thus an SDR prototyping platform should have a flexible radio front-end capable of transmitting/receiving very high bandwidth signals with a software-controlled RF component switching times in the sub-microsecond range.

14.3.1 RF Front End Requirements

Accommodating different mmWave bands such as the 28 GHz, 38 GHz, 45 GHz, 60 GHz, 70 GHz and 80 GHz bands is essential, as active research is ongoing in all these bands for mmWave cellular and WiFi applications. It is therefore desirable for the mmWave prototyping system to be able to attach to new RF heads without substantial incremental costs or engineering effort. The power levels of a prototyping system should be sufficient to provide similar coverage as the final deployed system. The equivalent isotropic radiated power (EIRP) of the system is given as

$$EIRP|_{log} = P_T - L_c + G_a, \tag{14.1}$$

where *EIRP* and P_T (output power of transmitter) are in dBm, cable losses (L_c) are in dB, and antenna gain (G_a) is expressed in dBi relative to a (theoretical) isotropic reference antenna. The EIRP will determine the achievable link budget of the system. It should be possible to configure the RF front end of the system for multiple input multiple output (MIMO) operation. This can be accomplished through the use of multiple polarizations. MIMO support requires a common clock reference and possibly even a local oscillator daisy-chaining option. An appropriate antenna interface at the mmWave head should be available so that a variety of antennas may be supported.

14.3.2 *Real-time Control of the RF Front End*

Real-time RF front-end control has three subsystems: beam control, automatic gain control (AGC) and automatic output control (AOC). These three subsystems are tied inextricably together. As beams switch in a mmWave system, receive power levels can shift dramatically as line of sight (LoS) and multipath components form and disappear. In one configuration, a beam may be directed exactly at a receiver; in another configuration that may not be the case. This is a natural part of the beam-discovery process built into the standards protocol, where it is expected that beams will switch on regular time boundaries designed for latencies that are ten times lower than current technologies. The antenna beam must be settled or switched on a per-slot basis in a fraction of that time. This implies that the beam will have to be settled or switched on a sub-microsecond basis. A desirable PoC prototyping platform should be able to control a variety of antennas through fast analog and digital controls from the baseband processor.

Similarly, at the same time as the beams switch, the AGC settings must be reconfigured on the same time-slot basis. This is because as receive power levels shift, so must the AGC settings. The AGC must be settled in order for the over-the-air payload signal part to be decoded correctly. Saturation or insufficient signal level will cause increased noise levels or saturation in baseband, leading to inefficient use of the wireless medium and hard-to-predict signal quality properties. Thus, the AGC must settle at roughly the same rate as the beam switching.

Finally, it may be desirable to have power control in the mmWave protocol. As a mobile device gets close to a base station, power levels may begin to saturate at the receiver. In mmWave systems, where radio heads may be mounted on top of a pole, it is easy to imagine that the user device could be within a few meters of the base station; without some mechanism of transmit power control, saturation is inevitable. Power control is enabled by an AOC subsystem in the mmWave transmit radio head with some digital control mechanism. Lastly, to achieve sub-microsecond switching times with low timing jitter, there is a need for dedicated front-end control logic. This task may be difficult to achieve with shared processors, for example, but may require dedicated FPGA resources instead, as shown in Figure 14.1.

14.3.3 *Converter Requirements*

Analog-to-digital converters (ADC) and digital-to-analog converters (DAC) should support greater than 1 GHz RF bandwidth with 8–14 bits of resolution, depending on the application. This will enable the platform to support 802.11ad as well as most emerging cellular technologies. The system should ideally be able to support multiples of 802.11ad and LTE rates. To support sampling rates of the order of Giga samples per sec (GS/s), an architecture using interleaved ADCs and DACs is often employed, with each ADC/DAC operating at a fraction of the actual sampling rate but with a time offset with respect to the other ADC/DAC as shown in Figure 14.1. In that way, each ADC/DAC operates at a different time instant of the signal within one clock cycle to achieve effective sampling rates of the order of GS/s. This architecture implies that the baseband processing subsystem needs to operate on multiple samples in each clock cycle. Even if a single ADC/DAC with higher sampling rate is used, the interface of the ADC/DAC to the baseband processing unit (usually an FPGA) is in the form of multiple input/output (I/O) pins instead of a serial interface. This is because the FPGA technology cannot support clock rates of the order of GHz to enable serial data transfer between baseband processing and ADC/DACs. Thus, the baseband processing architecture must be designed to operate on array of samples in each clock cycle.

14.3.4 Distributed Multi-processor Control and Baseband Signal-processing Requirements

FPGAs provide dedicated processing resources to meet hard time deadlines and achieve absolute time guarantees for implementation of various air-interface protocols. They also bring the luxury of the ability to reconfigure them through software. Hence FPGAs have become a desirable component of software-defined radios, and will prove to be critical in 5G prototyping platforms.

"RF-to-bits" conversion, or the demodulation, is a complex process including synchronization, RF-impairment corrections, channel equalization, and error correction. It is usually not possible to realize the whole demodulation process within one FPGA for such wide-band systems. Hence multiple FPGAs are deployed to realize a transceiver as shown in Figure 14.1. Splitting the transceiver into different functional blocks implemented on separate FPGAs also helps in efficient system management, parallel code development and easier iterations. For example, one can split a MIMO receiver into synchronization, demodulation, MIMO detection and data decoding, with well defined interfaces between these four functional blocks. However, a distributed FPGA architecture entails a high-speed data-routing mechanism to share data and control information across multiple FPGAs, as shown in Figure 14.1.

Not all processes in a transceiver require strict time guarantees. Also, code development for FPGA is a sophisticated process, which takes time and requires expertise in fixed-point architecture. Hence FPGAs alone do not provide enough flexibility for a baseband processing unit, and a general-purpose processor (GPP) is also an integral part of an SDR prototyping platform, as shown in Figure 14.1 as the central controller. A GPP can be used for distribution of control information, which usually requires lower throughputs than data signals. Also, the processing for layers above the PHY layer of the protocol stack is usually implemented on GPPs.

To summarize, an SDR prototyping platform needs to be a heterogeneous processing platform supporting different RF front ends, FPGAs and GPPs, along with a data and control-signal transport mechanism. The system parameters of different air interfaces defined for mmWave communications have been listed in Table 14.1 for reference. This table can be used to define the key requirements of a mmWave prototyping platform in terms of the bandwidth support, sampling rates, throughput and latencies.

14.4 Case Study: a mmWave Cellular PoC

In this section, a case study is presented of a 1-GHz mmWave cellular PoC system developed using commercial off-the-shelf (COTS) components. The air interface of the PoC is derived from the specifications detailed by Cudak *et al.* [8]. A brief description of the physical layer parameters is provided in Section 14.4.1 followed by the hardware and software architecture of the PoC system in Section 14.4.2. The signal-processing architecture and data flow

Table 14.1 System parameters of different air interfaces defined in the mmWave band

	E-band cellular PoC [6]	28 GHz mmWave PoC [7]	802.11ad
Carrier frequency	71–76 GHz E-band, 28 GHz,	27.925 GHz	60 GHz V-band
Channel bandwidth	2 GHz	520 MHz	2.16 GHz
MIMO spatial streams	Yes (up to 2 × 2)	Yes (up to 2 × 2)	N/A
Typical distance	100–200 m	200–300 m	1–10 m
Modulation	Single carrier	OFDM	Single carrier, OFDM
Symbol rate	1.536 Gsym/s	1 Gsym/s	1.76 Gsym/s (SC) 2.64 Gsym/s (OFDM)
Sampling rate	3.072 GS/s (2× over sampling)	1 Gsym/s	Depends on over sampling (SC) 2.64 GS/s (OFDM) 1.386 GS/s (After FFT; only data subcarriers)
OFDM parameters	N/A	FFT size: 4096 Subcarrier spacing: 244.14 kHz CP length: 0.18 × OFDM symbol length	FFT Size: 512 Subcarrier spacing: 5.15625 MHz Data subcarriers: 336 Null subcarriers: 157 Pilot subcarriers: 16 DC nulls: 3 CP length: 128
Highest modulation order	16QAM	16QAM	16QAM (SC) 64QAM (OFDM)
Highest coding rate	7/8	1/2	3/4 (SC) 13/16 (OFDM)
Peak data rate	10.1 Gbps	1.056 Gbps	4.62 Gbps (SC; 1 stream) 1.76 Gsym/s × 448/512 (512 length block with 64 length guard interval) × 4 bits × 3/4 (max. coding rate) 6.75675 Gbps (OFDM; 1 stream) 5.15625 MHz (subcarrier spacing) × 336 (Data subcarriers) × 1/1.25 (1/4 of the symbol is CP) × 6 bits × 13/16 (max. coding rate)

in the modulation and demodulation process is described, followed by two specific in-depth examples on high-speed parallel signal processing and high-throughput coding architectures in Sections 14.4.3 and 14.4.4, respectively. Section 14.4.5 details the PoC extension to support 2 GHz bandwidth with multiple data streams through MIMO transmissions and the required architectural changes. The section concludes with the experimental results obtained from the PoC system using over-the-air transmissions.

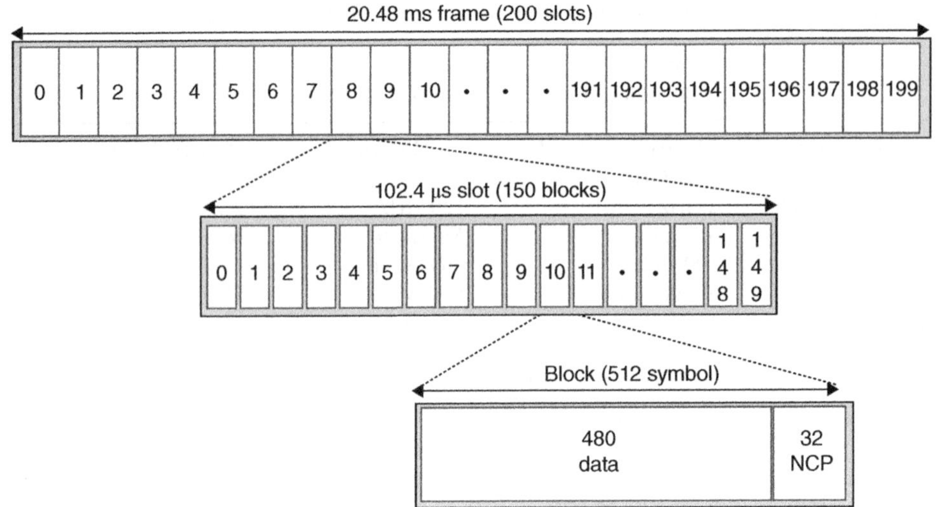

Figure 14.2 Frame structure of the mmWave cellular PoC system

14.4.1 Description of the Air Interface

14.4.1.1 Frame Structure

The air interface supports a point-to-point link between an access-point (AP) and a user-device (UD) supporting bi-directional TDD-based communications with 1 GHz of bandwidth over the E-band (76 GHz). The transmission is carried out in frames, with the frame structure as shown in Figure 14.2. The duration of the frame is 20.48 ms, where each frame is composed of 200 slots, each of 102.4 μs duration. TDD is supported by designating each slot to be either a downlink (DL → AP-to-UD) or uplink (UL → UD-to-AP) slot.

Single-carrier modulation is employed with a symbol rate of 750 mega symbols per second (Msym/s). To enable block-based transmissions, modulation symbols are grouped into blocks, where each block is composed of 512 symbols comprising 480 data symbols with a zero-padding of 32 null symbols, termed as the null-cyclic-prefix (NCP). The purpose of NCP insertion is to combat intersymbol interference (ISI), inter-block-interference, and enable low-complexity frequency-domain channel equalization [9]. Thus the modulation scheme is referred to as "single-carrier with NCP" (SC-NCP). Each slot is composed of 150 SC-NCP blocks or $150 \times 512 = 76{,}800$ symbols. The system parameters are summarized in Table 14.2 along with the raw data-rate supported over-the-air (OTA) with different modulation schemes. As shown, the maximum uncoded data rate is 2.812 Gbps over a single data stream using 16QAM. The actual data throughput is reduced by the presence of channel coding, NCP and pilot blocks, as explained below.

14.4.1.2 Slot Structure

The slot configuration used in the PoC system is shown in Figure 14.3. Out of 150 SC-NCP blocks, the first 10 blocks are reserved; in other words no data are transmitted in these blocks.

Table 14.2 System parameters for SC-NCP modulation in the mmWave PoC

Parameter	Value
Symbol rate	0.75×10^9
Slot duration	102.4 μs
Number of SC-NCP blocks per slots	150
Symbols per SC-NCP block	512
Max. uncoded SISO data rate with BPSK	0.703 Gbps
Max. uncoded SISO data rate with 4QAM	1.406 Gbps
Max. uncoded SISO data rate with 16QAM	2.812 Gbps

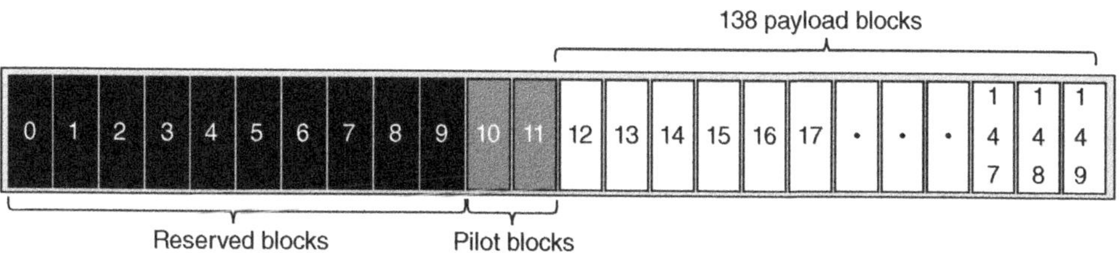

Figure 14.3 Slot structure

The reserved blocks are followed by 2 identical pilot blocks to facilitate timing and frequency synchronization as well as channel equalization. The following 138 blocks are used for payload transmission.

14.4.1.3 Modulation and Coding Scheme

Four modulation and coding schemes (MCS) are supported in the PoC system using BPSK, QPSK, and 16QAM modulation with the different coding rates outlined in Table 14.3. Each codeword (CW) comprises 1440 data symbols. Thus, one CW maps to three SC-NCP blocks, and each slot consists of 46 CWs (as each slot has 138 payload blocks and each block has 480 data symbols). As shown in the table, the PoC system is capable of achieving 2.3179 GBps peak throughput using 7/8 16QAM. The CW generation and turbo encoding/decoding architecture is explained in Section 14.4.4.

The peak data throughput can be increased by using the reserved blocks for payload transmission. If nine out of ten reserved blocks shown in Figure 14.3 are also used for payload, it will add three more CWs per slot, increasing the peak throughput to 2.4691 Gbps. This is the maximum throughput that can be supported by the PoC system, with three blocks per slot still available for transmitting pilots or control information.

The link establishment starts with the transmission of DL slots from the AP to the UD. Once the UD detects the DL signal and establishes synchronization, it starts transmission of the UL slots. The link is established when the AP detects the UD's signal and starts receiving it.

Table 14.3 Modulation and coding schemes in the mmWave PoC

MCS	1/5 BPSK	½ QPSK	½ 16QAM	7/8 16QAM
TBS (bits)	328	1480	3112	5160
CW size (bits)	1440	2880	5760	5760
CWs per slot	46	46	46	46
Payload bits per slot	$328 \times 46 = 15088$	680580	143152	237360
Throughput (Gbps)	0.1473	0.6648	1.43	2.3179

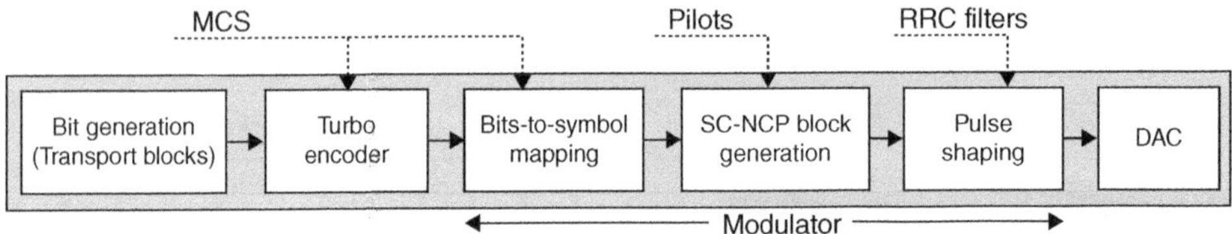

Figure 14.4 Transmitter block diagram

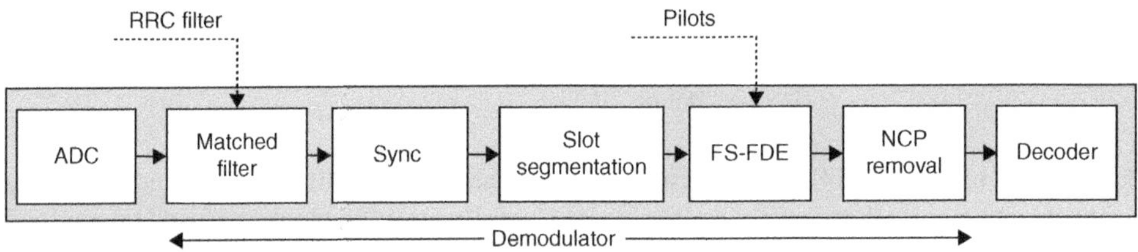

Figure 14.5 Receiver block diagram

Baseband processing is identical in the AP and the UD except for the different designations of slot types; in other words, the AP is required to transmit the DL slot and the UD is supposed to receive it, and vice-versa. The block diagrams of transmit and receive chains of the transceiver are shown in Figs 14.4 and 14.5, respectively.

14.4.1.4 Transmitter Subsystem

The information bits coming from a source are grouped into transport blocks (TBs) (see Figure 14.4). The TB sizes (TBS) for different MCS are listed in Table 14.3. For error recovery, forward error correction (FEC) is applied to map the TBs into CWs according to the MCS. Blocks of SC-NCP are then assembled by appending 480 data symbols with zero padding of 32 zeros symbols or nulls. If the specified block is a pilot block, predefined pilot symbols are mapped instead of payload symbols. The blocks of SC-NCP symbols are pulse shaped and interpolated from symbol rate (0.75 Gsym/s) to the DAC input rate (1.25 GS/s) using a root raised cosine filter with a roll-off factor of 0.3.

14.4.1.5 Receiver Subsystem

On the receiver end, the received signal is sampled and then matched filtered with the same RRC filter specified above, followed by synchronization (see Figure 14.5). Timing synchronization estimates the slot boundaries while frequency synchronization estimates and compensates the carrier frequency offset (CFO) between the transmitter and the receiver. The received slot is segmented into blocks according to the slot boundary estimates and is passed on to the channel equalization process on a block-by-block basis. Fractionally-spaced frequency-domain equalization (FS-FDE) is employed for channel equalization, followed by the NCP removal. The equalized payload symbols are finally passed to the turbo decoder, as explained in Section 14.4.4. The subprocesses of synchronization and FS-FDE are briefly explained below.

Synchronization

Timing and frequency synchronization is carried out using the two identical pilot blocks in the beginning of the slot. Timing synchronization is carried out in two steps: coarse and fine timing synchronization. For coarse timing synchronization and CFO estimation, auto-correlation of the received signal is used, based on published algorithms [10, 11]. Specifically, if y[n] denotes the *n*th sample of the received signal, a timing metric $r[n]$ is calculated as:

$$r[n] = \frac{|p[n]|^2}{q[n]q[n-N]} \tag{14.2}$$

where $N = 512$ is the length of the SC-NCP block,

$$p[n] = \sum_{l=0}^{M-1} y[n+l]y^*[n+l-N] \tag{14.3}$$

$$q[n] = \sum_{l=0}^{M-1} |y[n+l]|^2 \tag{14.4}$$

and $M = 480$ is the number of data symbols in the SC-NCP block. Thus the timing metric $r[n]$ is composed of the M-point auto-correlation of the received signal N samples apart, normalized by the received signal energy. As identical pilots are employed in the beginning of the slot, the timing metric will exhibit a peak at the start of the pilot blocks. Thus, peak detection of the timing metric is carried out to estimate the coarse timing estimate. At the correct starting-point estimate, the phase of $r[n]$ also provides an estimate of the CFO. Specifically, the CFO estimate $\hat{f}$ is calculated as:

$$\hat{f} = \frac{\arg r[n]}{2\pi 512T} \tag{14.5}$$

where $T = \frac{1}{750 \times 10^6}$ is the symbol duration. CFO compensation of the received signal samples is performed as

$$y_c[n] = y[n]e^{-j\frac{2\pi}{NT}\hat{f}n} \tag{14.6}$$

The compensated pilot blocks are then used for fine timing synchronization. If the pilot blocks have ideal autocorrelation properties, cross-correlation of the received pilot blocks with a local copy of the pilot block provides an estimate of the channel impulse response (CIR). The index of the strongest tap of the CIR provides the estimate of the starting index of the slot [12].

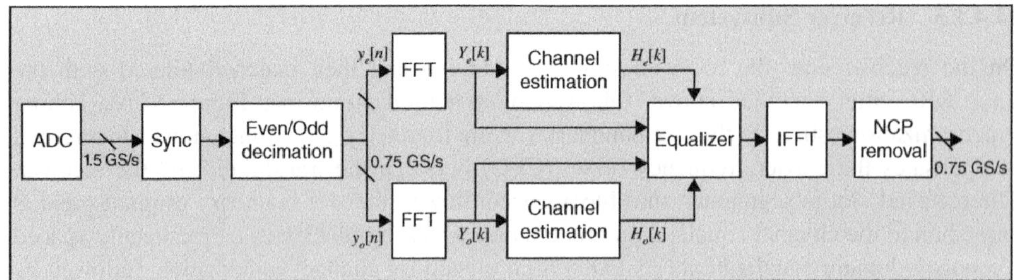

Figure 14.6 Block diagram of fractionally-spaced frequency-domain equalization

(a) *Fractionally-spaced Frequency-Domain Equalization*
As the single-carrier modulation is more sensitive to sampling time instants and ISI, fractionally-spaced equalizers are used to achieve improved performance compared to symbol-spaced equalizers [13]. Also, the addition of NCP enables frequency-domain equalization by eliminating interblock interference. Thus, fractionally-spaced frequency-domain equalization is used for the PoC system. The block diagram is shown in Figure 14.6. The received signal is oversampled by a factor of two: the ADC sampling rate is 1.5 GS/s. The synchronized received signal samples are split into two parallel data streams, each at the symbol rate of 0.75 GS/s. In each parallel path, 512-point fast Fourier transform (FFT) of the received SC-NCP block is calculated. MMSE channel estimation of the pilot blocks is carried out to estimate the channel frequency response [14]. Channel estimates of the two streams are then used to equalize the following payload blocks in the received slot. Specifically, denoting the two parallel streams by subscript $[\cdot]_e$ and $[\cdot]_o$, the MMSE equalizer output is calculated as

$$\widehat{X}(k) = q(w_o(k)Y_o(k) + w_e(k)Y_e(k)) \tag{14.7}$$

where

$$w_o(k) = \frac{H_o^*(k)}{|H_e(k)|^2 + |H_o(k)|^2 + \sigma_n^2} \tag{14.8}$$

$$w_e(k) = \frac{H_e^*(k)}{|H_e(k)|^2 + |H_o(k)|^2 + \sigma_n^2} \tag{14.9}$$

$$q = \frac{1}{\sum_{k=0}^{511} w_e[k]H_e[k] + w_o[k]H_o[k]}, \tag{14.10}$$

$Y[k]$ denotes the output of the FFT for the kth bin and $H[k]$ denotes the corresponding channel estimate. Finally, the inverse fast Fourier transform (IFFT) of the equalized block $\widehat{X}[k],\ 0 \leq k \leq 511$ is computed to transform the signal back into the time domain. The equalization is followed by NCP removal, after which the equalized payload symbols are passed to the turbo decoder.

14.4.2 PoC System Architecture

In this subsection, the hardware and software architecture of the mmWave PoC system is presented. The mmWave PoC system is implemented on a COTS platform developed by National Instruments. Each transceiver (AP or UD) is contained in a PXIe chassis. The PXIe chassis houses all modules and provides a PCI backplane with special circuitry to provide accurate timing and triggering across all the modules, and routing data and control signals between them. The modules residing within the PXIe chassis include a central controller, FPGAs, DAC modules for signal generation on the transmit side, ADC modules for signal analysis on the receive side, digital control modules to control RF front ends, and an antenna system similar to the block diagram in Figure 14.1. All modules, including the FPGAs, are user-programmable using the graphical programming language LabVIEW. LabVIEW provides both a simulation and hardware programming environment. It provides a higher layer of abstraction for efficiently programming and controlling hardware. It also provides an environment to integrate different programming languages, such as C, textual math (.m), VHDL and so on.

14.4.2.1 Hardware Architecture

The hardware architecture of the PoC system is shown in Figures 14.7 and 14.8. Figure 14.7 is the actual snapshot of the baseband transceiver, while an equivalent slot-accurate block diagram is shown in Figure 14.8. A brief description of each module is given below.

Central Controller

The central controller (NI PXIe 8135) is an ×64 bit Intel i7 quad core processor that is capable of running a real-time operating system (OS). It provides an interface for the user to control and configure different modules in the system. It downloads the bitfiles to the different FPGAs in the transceiver, configures the personality of the transceiver to be an AP or UD, and acts as the source/sink for the data. Upper layer (MAC and above) stack and some statistical analysis for the physical layer (PHY) operation are also performed on the controller.

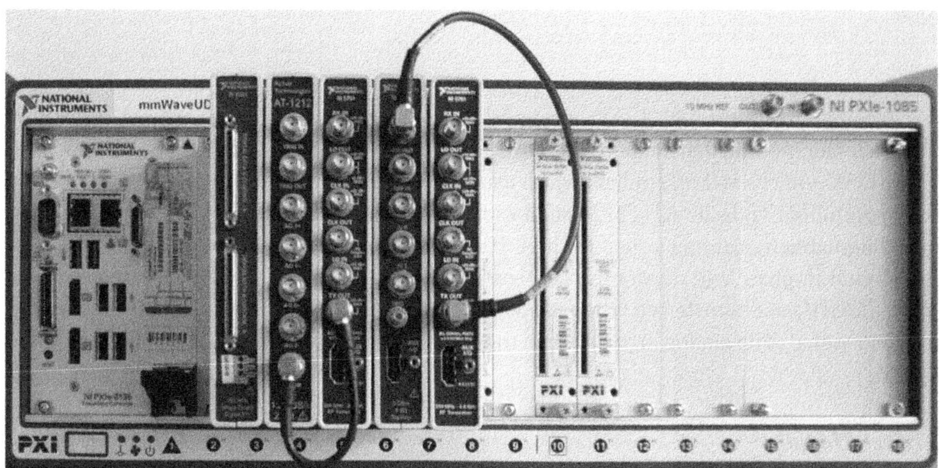

Figure 14.7 Single baseband transceiver node of the mmWave PoC hardware setup

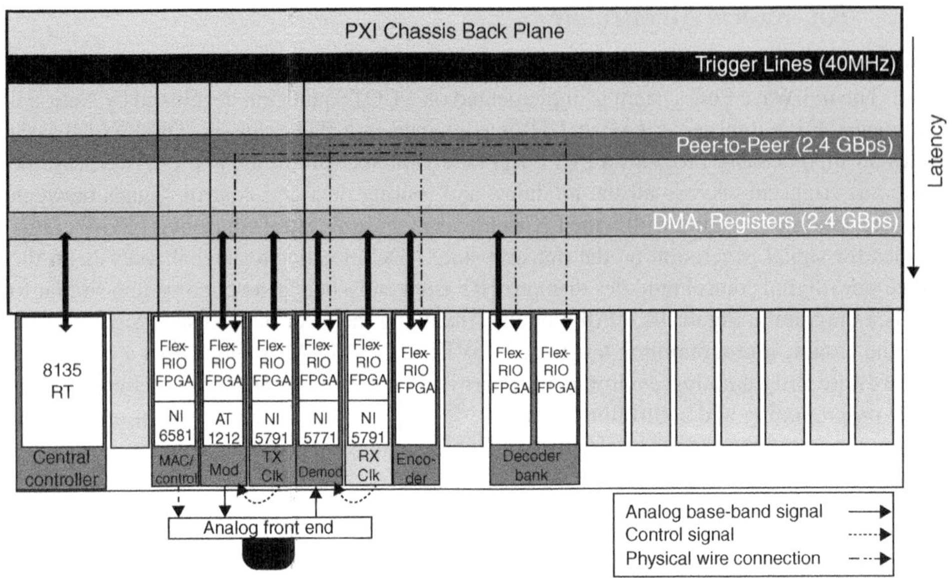

Figure 14.8 mmWave PoC hardware architecture

FPGA Module

The FPGA module (NI PXIe 7976) is a Xilinx® Kintex-7 FPGA, programmable with LabVIEW [15]. It has 2 GB of onboard DDR3 memory for high-speed data storage and retrieval, at a peak throughput of 10.2 Giga Bytes per second (GBps). Kintex-7 FPGA contains 1540 DSP slices that can be used for implementing high-speed signal-processing operations such as multiply-and-accumulate, digital filter and so on. The FPGA module can either act as a standalone processing module for signal-processing applications or it can be mated with adapter modules, which plug into the FPGA module, to form a reconfigurable instrument as explained below.

Transmitter Module

The transmitter module (AT-1212), made by Active Technologies™, is a signal-generation FPGA adapter module that contains the DAC. It supports over 800 MHz (∼1 GHz compensating for DAC *sin(x)/x* roll off) of bandwidth using a DAC with a sampling rate of 1.25 GS/s and 12-bit resolution. Thus the AT-1212 mated with the FPGA module (NI FlexRIO 7976) acts like a programmable transmitter suited for high-bandwidth applications of the mmWave PoC system. Digital in-phase and quadrature (I/Q) samples of the SC-NCP modulation are generated on the FlexRIO and transferred to the adapter module (AT-1212), which generates the analog baseband signal. The analog signal is then transferred to the RF front end for upconversion to 76 GHz.

Receiver Module

Similar to the transmitter module, the receiver module (NI 5771) is a signal analysis FPGA adapter module. It supports 900 MHz of bandwidth with an ADC of 1.5 GS/s sampling rate and

8-bit resolution. The analog received signal is converted to baseband by the front end of the receiver. The baseband signal is then provided to the NI 5771 adapter module and I/Q samples generated by its ADC are transferred to the attached FPGA module for SC-NCP demodulation.

Digital Control Module
The digital control module (NI 6581) is also an FPGA adapter module. It provides 54 digital (I/O) channels at a clock rate of up to 100 MHz. These I/O channels are used to generate control signals for the RF front end and the antenna system of the mmWave PoC.

Clock Generation Module
Two FPGA adapter modules (NI 5791) are used to generate the external clock signals for the transmitter and receiver modules.

System Chassis
The hardware modules mentioned above are installed in a PXIe chassis (NI PXIe 1085) and the actual system is shown in Figure 14.7. An equivalent block diagram for the system in Figure 14.7 is shown in Figure 14.8. The figures show an 18-slot chassis that features PCI express Gen2 $\times 8$ technologies in every slot for high-throughput and low-latency applications. In addition to housing the modules in a single platform, the PCIe bus in the backplane of the PXI chassis routes data between FPGAs or FPGA to the central controller with a maximum one-way throughput of 3.2 GBps per slot and two-way throughput of 2.4 GBps per slot. The backplane also features the dedicated custom hardware for timing, synchronization and hardware triggers that can be received and generated from any slot in the chassis with latency in the order of pico seconds. The modules within a PXI chassis are driven by a common 10-MHz reference clock. The chassis can also receive and export the 10-MHz reference clock. Finally, multiple PXI chassis can be daisy-chained for high-slot-count applications.

Front End and Antenna
The RF frontend is connected to the baseband system with differential or single ended IQ signals. The antenna system is connected to the RF separately. Both the RF and the antenna systems are controlled with low-voltage differential signaling lines directly from the digital control module (NI 6581). The RF front end and the antenna system can be controlled from the FPGA mated with NI 6581 at the slot boundaries, or at any interval as needed.

The various modules described above act as the building blocks of the mmWave PoC system. These modules can be mixed and matched to build a reconfigurable transceiver capable of generating and analyzing the very high bandwidth signals pertinent to mmWave prototyping. The central controller and multiple FPGA modules along with the data-routing mechanism provide a framework for processing signals with high fidelity while meeting the real-time constraints.

14.4.2.2 Software Architecture

In this subsection, the software architecture of the mmWave PoC is described. As discussed in the previous subsection, the mmWave PoC system is a heterogeneous processing platform. Layer 1 (L1), signal-processing functionality and upper-layer protocol stack such as the MAC layer and above can be partitioned between the controller (with Intel i7 processor) running a real time OS and multiple FPGAs.

Controller vs FPGA

FPGAs and controllers provide different benefits in developing a real-time PoC system. FPGAs provide relatively faster execution with very low jitter on execution time, which makes FPGAs suitable for performing large numbers of time-critical computations that can only tolerate relatively low jitter. However, knowledge of the available hardware resources on a FPGA and reliance on lower-level programming are required to make efficient use of the hardware resources. Therefore, efficient FPGA programming requires a steep learning curve. On the other hand, the controller is suited to performing upper-layer functionality and sophisticated signal processing with less stringent timing requirements. Typically, design iterations are easier on the controller, as FPGA code changes require re-compilation. This can be time consuming because the place and route operations need to take effect. In addition, FPGAs computations are typically fixed-point. Thus, porting an algorithm from a floating-point simulation to an FPGA requires a float-to-fixed-point conversion, during which it is important to ensure that the performance is not significantly compromised with fixed-point operations. New software tools are available to make the transition from floating point to fixed point easier [16].

Tasks with sophisticated signal processing, strict throughput and latency requirements are generally suited for FPGAs as these provide dedicated memory and computational hardware resources enabling the designer to meet certain time guarantees. On the other hand, resources on a processor are shared between different tasks (as scheduled by the OS) and therefore, it is hard to achieve strict timelines for a subprocess. Nonetheless, the real-time OS on the central controller provides more determinism than non-real-time OS such as Windows. The designer has the option to set priorities for different tasks, gather information on the execution time, jitter, delay and so on for certain processes. Moreover, in a multicore processor, a particular core can be assigned to a particular task to achieve higher level of determinism.

Data and Control Signaling Transport Mechanism

As mentioned in the previous subsection, the PXI chassis backplane can be used to route data and control signals across the modules in different slots. Various data-transfer media exist, as shown in Figure 14.8. Bidirectional data transfer between the controller and an FPGA takes place through direct memory access (DMA) channels. Each PXIe7976 FPGA module supports 32 DMA channels that can be used for high-speed data transfer at a two-way throughput of 2.4 GBps. In addition to DMA FIFOs, the controller can get read/write access to FPGA registers as well, and these can be used to distribute control information/data at a lower rate.

Inter-FPGA communications happen through the PCIe backplane with dedicated peer-to-peer (P2P) streaming technology. This allows direct data transfer between FPGA modules with very low latency without the involvement of the controller. Thus, P2P streaming provides a mechanism to distribute the transceiver implementation to multiple FPGAs while allowing them to share two-way data at a rate of 2.4 GBps and one-way data at a rate of 3.2 GBps from each FPGA device.

Multi-FPGA synchronization is supported using 8 PXI backplane trigger lines. Digital triggers can be passed from one module/FPGA to any number of modules (including itself) within the PXI chassis in a master–slave configuration.

The software architecture for the mmWave PoC system is shown in Figure 14.9. The central controller oversees the operation of the mmWave transceiver. It distributes the control and configuration information to different modules, downloads the bitfiles to the FPGAs, and acts as the source/sink for data. It is also used to configure the personality of the transceiver to act as an AP or UD, as explained below. Using the central controller, the frame configuration is exposed to the end-user in the form of a frame table where the user can define the slot types and MCS for each slot in the frame. The frame table is shared between the AP and UD. Based on the frame table, the central controller configures transmit and receive chains. In order to meet the throughput requirements and maintain data integrity, most of the baseband signal processing shown in Figs 14.4 and 14.5 is performed on the FPGA modules. The architectures of the transmitter and the receiver are detailed below.

Transmitter Architecture

Referring to the block diagram in Figure 14.4, the baseband processing in the transmitter is segmented into three functional blocks:, payload/TB and control signaling generator, turbo encoder and modulator. These functional blocks are realized in four different modules shown in Figure 14.9: central controller, RF interfacing and control FPGA, Turbo encoder FPGA, and modulator FPGA with transmitter adapter module.

Based on the frame table, for each slot to be transmitted, the central controller performs certain functions.

- It generates payload bits and slot configuration and passes to the RF interfacing and control FPGA using two different communication mechanisms: DMA FIFO for the payload and registers for the slot configuration. The peak throughput required for the DMA transfer is equal to the peak coded throughput at 2.317 Gbps, which is well within the DMA throughput limits. Also, the peak throughput is required only when all the slots within a frame are to be transmitted with the highest MCS, i.e., 7/8 16 QAM.
- It transfers the corresponding slot type (UL/DL), slot index, and slot modulation type to the modulator FPGA using registers. The information is packed into a 32-bit number per slot. Thus register-based control transfer requires a throughput of only 312.5 kbps.

The RF interfacing and control FPGA acts as an interface between the central controller and multiple parallel FEC modules. Within the transmit path, it segments slots of payload bits into TBs that can be independently encoded. In addition, it generates control packets, which parameterize the encoder. The TBs are transferred to the encoder FPGA using P2P streaming. The peak throughput is 2.317 Gbps as before. The encoder control information is transmitted per TB at a rate of about 30 Mbps, allowing for very fast reconfiguration in real time.

Firstly, the encoder FPGA is configured through the encoder control information. Secondly, it uses parallel processing to encode the TBs received over the P2P link. The parallel encoder architecture is detailed in Section 14.4.4. The turbo encoder outputs CWs at a maximum rate of 2.5875 Gbps. These are transferred to the modulator FPGA through P2P streaming.

The modulator FPGA implements the remaining parts of the transmit baseband processing, as shown in Figure 14.4. SC-NCP modulation is performed according to the slot indices and modulation format parameters received from the central controller via DMA transfer and the CWs received from the encoder FPGA via P2P transfer. The pilot blocks are stored locally on the block RAM of the modulator FPGA. To keep track of the slot indices and frame

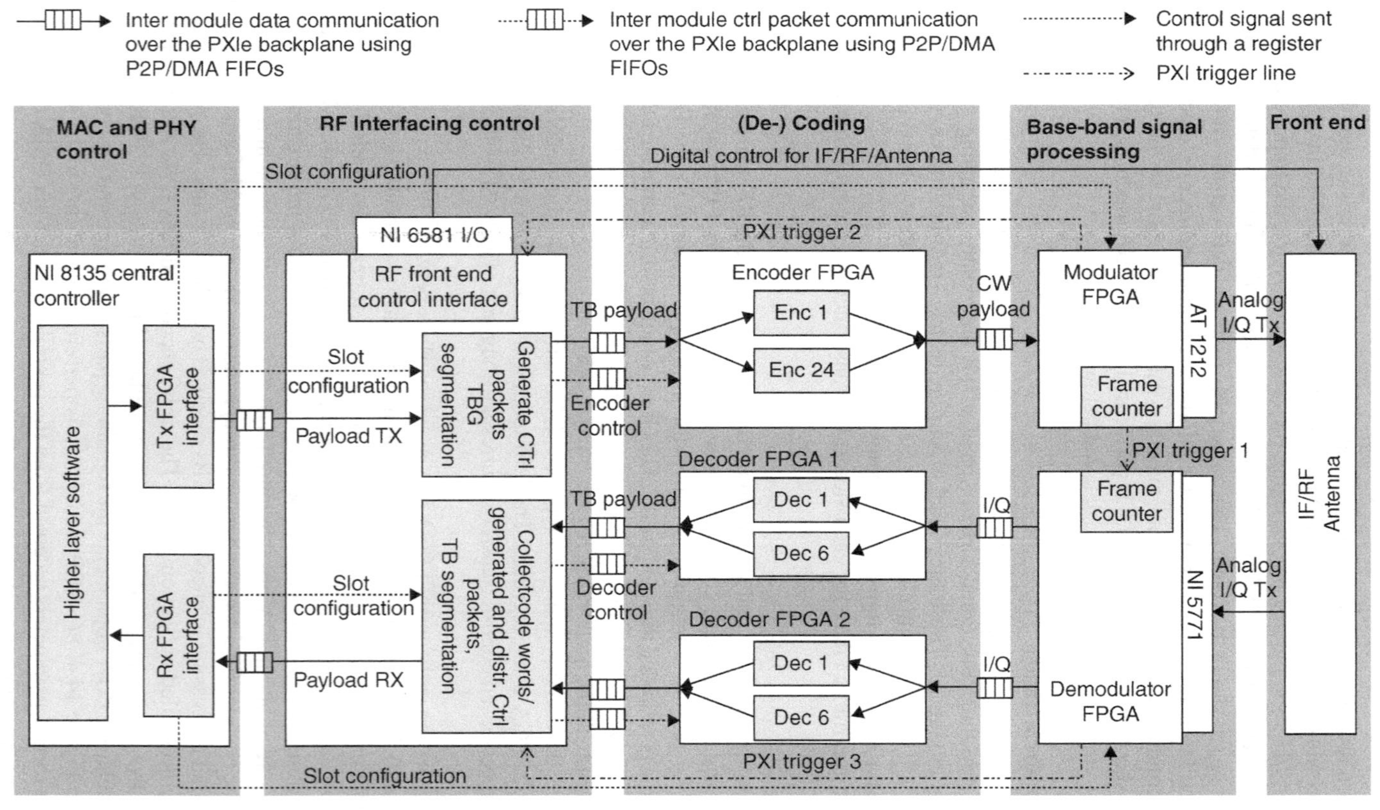

Figure 14.9 Mapping of baseband software to hardware

boundaries, the modulator FPGA keeps a sample counter. Using the sample counter, a state machine decides whether the current slot needs to be transmitted or not and correspondingly passes modulated symbols or zeros to the DAC. The transmitter adapter module (AT 1212), mated with the modulator FPGA as shown in Figure 14.8, is used to convert the digital IQ samples to baseband analog IQ signal. As the DAC input rate of the adapter module is 1.25 GS/s, the modulated symbols at 0.75 Gsym/s are upsampled by a factor of 5/3 before transfer to the DAC to match the DAC input rate. The clock required for driving the transmitter adapter module is generated with the NI 5791 adapter module. The clock signal is physically wired from the NI 5791 'clk out' pin to the 'clk in' pin of the transmitter adapter module as shown in Figures 14.7 and 14.8.

It is important to mention that the interface of the DAC with the adapter module is in the form of 16 parallel I/O ports each with 14-bit resolution, corresponding to 8 parallel I/Q samples. Thus the clock rate for the DAC data-transfer sub-process is $\frac{1.25 \times 10^9}{8} = 156.25$ MHz, and in each clock cycle the modulator FPGA needs to write an octet of output IQ samples to the DAC. Thus high-speed parallel signal processing is required to meet the throughput, as current FPGA technology cannot support clock rates in the gigahertz range to enable serial data transfer. Using parallelization of computational resources, the architecture of the modulator FPGA has been designed in a way that it can operate on arrays of samples in each clock cycle. An example of signal-processing blocks adapted for high-speed parallel signal processing on FPGA is given in Subsection 14.4.3.

The transmitter and receiver are implemented on separate modules in the mmWave PoC, but need to keep a common time base to agree on slot boundaries. Because of this, the starting instant of the sample counters on the transmitter and receiver must be synchronized. A PXI backplane trigger named "PXI trigger 1" in Figure 14.9 is used for this purpose. The modulator FPGA sends a trigger on one of the eight trigger lines of the backplane trigger bus, while both modulator and demodulator FPGA listen to the trigger. The sample counter is started when a rising edge is detected on the trigger line. As the trigger skew between different modules is about 1 ns, the sample counts get started in the same clock cycles.

The analog baseband IQ signal generated by the transmitter adapter module is transferred to the analog front end of the transceiver for upconversion and over-the-air transmission. To enable TDD communications, the front end needs to switch between transmit and receive modes based on the slot type. To turn on the transmitter circuitry and switch to transmit mode, the modulator FPGA generates another trigger (PXI trigger 2) at the start of each transmitted slot as shown in Figure 14.9. The RF interfacing and control FPGA listens to the corresponding trigger line and transfers the trigger to the analog front end via the I/O channel of the NI 6581 digital control module mated with the RF interfacing and control FPGA, as shown in Figure 14.8 and 14.9.

Receiver Architecture

Similar to the transmitter, the receiver processing is divided into three functional blocks: demodulator, turbo decoder and control signaling. These functional blocks are mapped to five hardware modules: demodulator FPGA and receiver adapter module, two turbo decoder FPGAs, RF interfacing and control FPGA, and the central controller. This is shown in Figure 14.9.

The central controller oversees the operation of the demodulator FPGA and also displays the key performance indicators as part of a user interface shown in Figure 14.10. Based on the

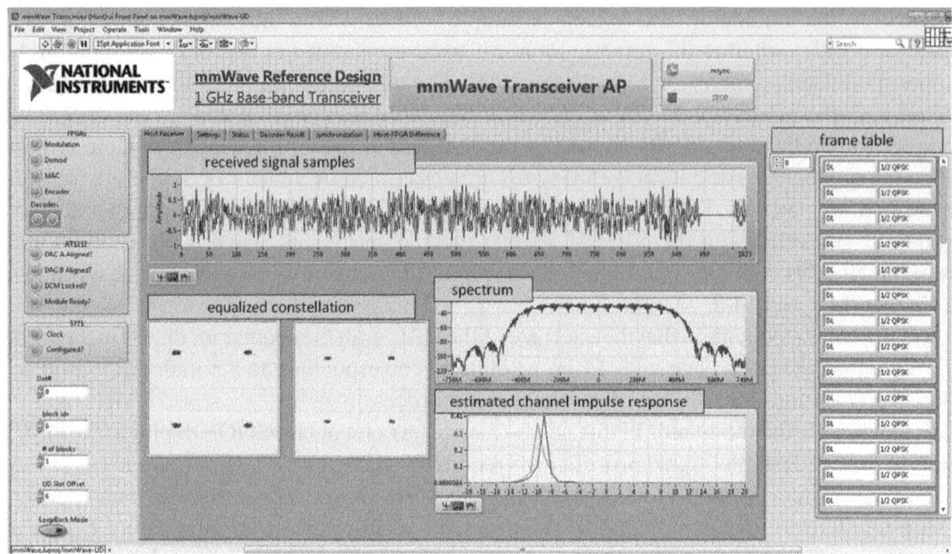

Figure 14.10 Snapshot of the mmWave PoC user interface

frame table, it transfers the indices of the slot to be demodulated to the demodulator FPGA, on a frame-by-frame basis.

The demodulator FPGA, mated with the receiver adapter module (NI 5771), performs the demodulation of the received signal. The subprocesses implemented in the demodulator FPGA are shown in Figure 14.5. The received signal is downconverted to analog baseband IQ format by the front end of the transceiver. The baseband signal is then sampled by the receiver adapter module at a rate of 1.5 GS/s. Thus the received signal is oversampled by a factor of two compared to the symbol rate. Similar to the DAC, the interface of the ADC to the demodulator FPGA contains 16 I/O pins of 8 bits each, providing eight IQ samples of the received signal for each clock cycle. Thus in the demodulator FPGA, octets of received IQ samples are obtained at a clock rate of $\frac{1.5\times10^9}{8} = 187.5$ MHz. With state-of-the-art FPGA technology, it is not possible to serialize the incoming data samples and implement the receiver on a sample-by-sample processing basis at a clock rate of 1.5 GHz. Thus, the baseband signal-processing architecture must be designed to operate on octets of received samples. We will present an example of parallel signal processing architecture, using the FFT algorithm implemented for FS-FDE, in Subsection 14.4.3.

The final output signal of the demodulator FPGA is the stream of equalized data symbols for each demodulated slot. Each symbol is represented by an IQ sample pair with 8-bit resolution for the real and imaginary parts. The equalized symbols are passed to two turbo decoder FPGAs using P2P streaming. The peak demodulated data rate out from the demodulator FPGA is

$$46\ (CWs\ per\ slot)\ \times\ 1440\ (symbols\ per\ CW)\ \times\ 2\ (bytes\ per\ symbol)$$
$$\times\ \frac{1}{102.4\ \times\ 10^6}\ (slot\ per\ sec) = 1.293\ GBps$$

The highly parallel turbo decoder implementation spans two FPGA modules. The rate computed above and the decoding task are shared evenly between these two FPGA modules. The turbo decoder architecture will be explained in the example in Subsection 14.4.4.

Decoded TBs are collected from two decoder FPGAs by the RF interfacing and control FPGA. The maximum data rate for this transfer is 2.317 Gbps (based on 16QAM, rate 7/8 turbo codes), similar to the encoding part at the transmitter side. The RF interfacing and control FPGA takes care of collecting the decoded data in the correct order. In addition, transport-block desegmentation is implemented in order to assemble TBs into slots. These operations are configured in real time according to the control information received from the central controller. The RF interfacing and control FPGA also computes control packets, which are distributed to the two decoder FPGAs at an aggregate data rate of about 9 Mbps. In a last step, the decoded and ordered information is transferred to the central controller at a data rate of 2.317 Gbps.

The demodulator FPGA also features an acquisition engine that can be triggered by the central controller to transfer one slot's worth of raw (before FS-FDE but after matched filtering) and demodulated data samples to the central controller via the DMA channel for debugging and display. As mentioned above, the central controller is used for higher-layer operations, longer-term statistics, and to control parts of the PHY layer. Thus, to display the results and debug information, data are captured in a non-real-time-buffered manner from the demodulator FPGA. For that purpose, the data to be transferred to the central controller are first stored on the DRAM of the demodulator FPGA and then transferred to the host via DMA at a lower throughput.

A snapshot of the mmWave PoC system GUI is shown in Figure 14.10. The demodulated signal status and derived parameters at different stages of the demodulation process are displayed periodically on a host PC that is connected to the chassis. This information is passed to the central controller from the FPGA modules via the controller DMA and register transfers. These messages are then passed to the host PC, which is connected to the chassis, for display. The displayed information includes the received signal samples, spectrum of the received signal, estimated CIR, throughput and status monitoring indicators for different FPGA modules.

14.4.3 Example: High Speed Parallel Signal Processing: FFT

FFT is one of the most common signal processing algorithms in modern-day communication systems. In the mmWave PoC system, FFT/IFFT is used to perform FS-FDE of the SC-NCP symbols as described in Subsection 14.4.1.5. As the sampling rate of the receiver is 1.5 GS/s, the interface of the ADC with the FPGA is in the form of 16 I/O pins of 8 bits each, as described in Subsection 14.4.2.2. Thus the ADC provides octets of time-interleaved I/Q samples at a clock rate of $\frac{1.5 \times 10^9}{8} = 187.5$MHz. In the case of FS-FDE, the incoming stream of octets is split into two parallel streams of quartets. For each stream, the FPGA needs to compute a 512-point FFT. However, it is not possible to serialize the incoming quartets to compute FFT, as the required clock rate for the computation will be at least 750 MHz, which is not possible using state-of-the-art FPGAs. Thus, the FFT algorithm must be adapted to the ordering of the incoming data symbols using a vector/parallel signal-processing architecture.

One such implementation of the FFT algorithm that can operate on multiple samples at a time is called in general the divide-and-conquer algorithm [17] for the discrete Fourier transform (DFT). It is also known as the Cooley–Tuckey FFT algorithm. The basic idea of the

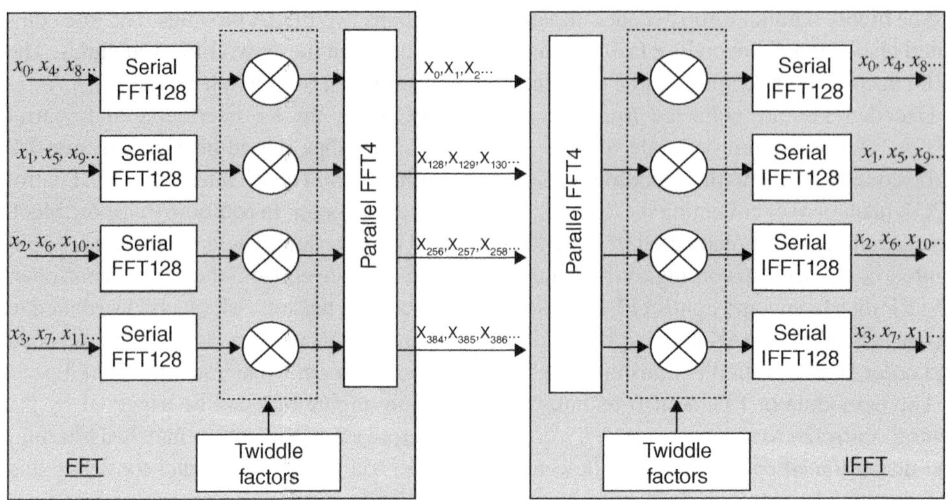

Figure 14.11 Divide and conquer approach to a parallel (I)FFT implementation

divide-and-conquer FFT algorithm is to divide the composite N = LM-point ($512 = 4 \times 128$ in this case) FFT into groups of smaller FFTs of size L and M. The algorithm is summarized in the following steps and the block diagram is shown in Figure 14.11. Details on the algorithm can be found in [17].

1. Arrange the 512 samples in a 4×128 matrix with column-major order. In the current implementation, step (1) is just a logical step as the incoming samples from the ADC are naturally in a column-major order. So no re-ordering of data is required and 128 quartet samples will naturally make a 4×128 matrix.
2. Compute 128-point FFT of each row of the matrix. This amounts to taking four 128-point *serial* FFTs, as shown in Figure 14.11.
3. Multiply the output of lth FFT $0 \leq l \leq 3$, with the twiddle factor $e^{j\frac{2\pi}{512}lm}$ $0 \leq m \leq 127$.
4. Store the output of the multiplication as a 4×128 matrix. This is again a logical step.
5. Take four-point DFT of each of the column of the matrix to compute the final FFT. This amounts to taking 128 four-point *parallel* DFTs. However, the result will be a 4×128 matrix in which the FFT entries are stored in row-major order as shown in Figure 14.11.

Thus, both the input and output of the overall divide-and-conquer FFT implementation is in quartets. At first sight, the column-major order of the output of FFT implementation seems to be a disadvantage as data may need to be rearranged to put it in the proper order. However, after frequency-domain equalization, IFFT of the equalized symbols is needed to bring the symbols back into the time domain, as discussed in Subsection 14.4.1.5. If the column-major order of the FFT output is kept as is, and equalization is performed in a column-major order, the divide-and-conquer IFFT implementation is a mirror-image of the FFT implementation as shown in Figure 14.11. Thus, the output of the IFFT will naturally come out to be in row-major order and re-arrangement of the data is avoided.

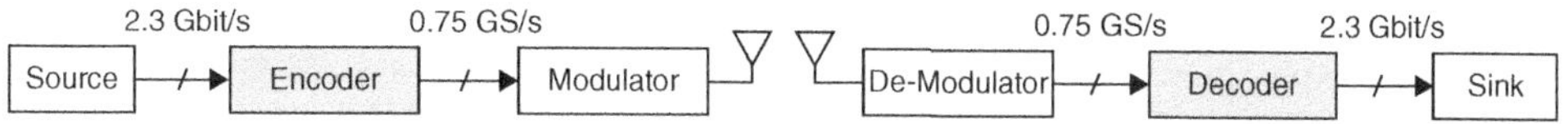

Figure 14.12 Transmission system supporting 2.3 Gbps data rate

14.4.4 Example: High Throughput Coding Architecture

14.4.4.1 Introduction

This subsection introduces an FPGA-based prototyping implementation and the respective design considerations for the FEC chain of the 1-GHz mmWave PoC system introduced in Subsection 14.4.1. The system is required to deliver 2.3 Gbps maximal throughput at a symbol rate of 750 MS/s, as shown in Figure 14.12.

The system is based on off-the shelf en-/decoding IP cores that have been designed for 3GPP-LTE applications and which are well understood and tested. These IP cores are designed to support throughputs of up to a few hundred megabits per second, as required by the 3GPP LTE standard. Hence a high-throughput mmWave application calls for massive parallelism, using a large number of IP cores in parallel, in a coordinated fashion. The design considerations that lead to a multi-FPGA design are outlined in the next subsection.

14.4.4.2 Parallel Forward Error Correction Coding Architecture

In order to derive a parallel coding architecture, consider the mapping between CWs and PHY signal structure first.

The PHY signal structure is divided into slots of 102.4 μs length, which is comparable to a transmission time interval in 3GPP LTE. These are further divided into 150 NCP blocks comprising 512 symbols each, as outlined in Section 14.4.1.1. Figure 14.13 sketches the mapping of CWs to NCP blocks.

A CW is mapped to two three-NCP blocks: 1440 adjacent complex symbols, neglecting the gap introduced by the NCP symbol portion. So a slot carries 46 CWs. Note that this mapping does not include any further interleaving on slot level, which reduces latency but comes at the cost of increased susceptibility to bursty signal distortions, such as interference. Note also that the CW-to-slot mapping in Figure 14.13 leads to an average maximal data rate of about 2.3 Gbps. The instantaneous maximal data rate during the used part of the slot is about 2.46 Gbps. The FEC design needs to account for the peak rates. Determining the number of

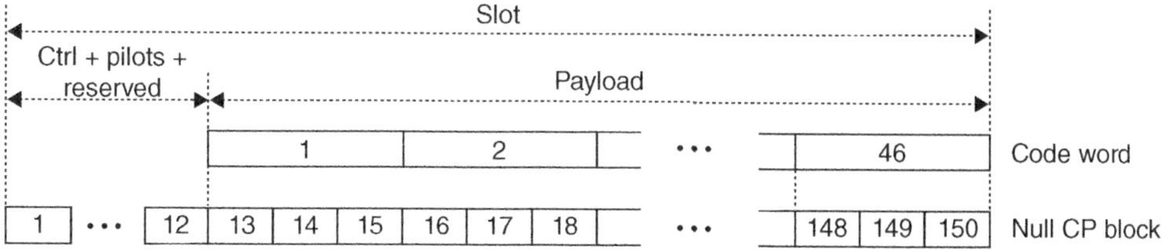

Figure 14.13 Mapping of CWs to Null CP blocks in a slot of 102.4 us length

parallel en-/decoder cores that will meet the data rate of 2.46 Gbps requires an experimental design-space exploration. This exploration should take into account:

- computational resources offered by the respective FPGA hardware, such as block RAM, DSP slices, registers and so on
- achievable clock rates of the FPGA implementation and the respective throughput
- settings to configure the decoder cores, such as any mathematical approximations or the number of iterations for iterative decoding, and their impact on the decoding performance and latency.

The first result of such an analysis delivers the achievable throughput, which is 220 Mbps per encoder core at a clock rate of 250 MHz, and 250 Mbps for each decoder core at a clock rate of 310 MHz, respectively. Choosing 12 parallel en-/decoder cores will be sufficient to achieve the desired target data rate. The FEC design is slightly over-provisioned, in particular at the decoder side. This margin helps to account for any overhead computation cycles in the FPGA design.

A round-robin-based mapping of uncoded TBs and encoded CWs within a slot to encoder and decoder cores is shown in Figure 14.14.

The mapping, in particular the (de)multiplexing operation, maintains the natural order of CWs and distributes the computational load evenly across cores.

The second result of the design space exploration is the footprint of an individual coding core with regard to FPGA resources at hand. For the FPGA hardware used in this example [15], it turns out that all 12 encoder cores do fit on a single FPGA, leaving a large portion of resources available for other purposes. Decoding is a more costly operation. Six cores fit on a single FPGA. This leads to the mapping of coding cores to FPGAs shown in Figure 14.15.

The general parallel architecture includes FPGAs with en-/decoding cores only, which are connected through the PXI backplane of the PoC system to an additional FPGA (RF interfacing and control FPGA), shown on the left-hand side of Figure 14.15, with two main tasks. Firstly, this FPGA takes care of distributing the TBs to the encoder FPGAs or collecting them in order from the decoder FPGAs respectively. Secondly, it takes care of (de)segmenting larger

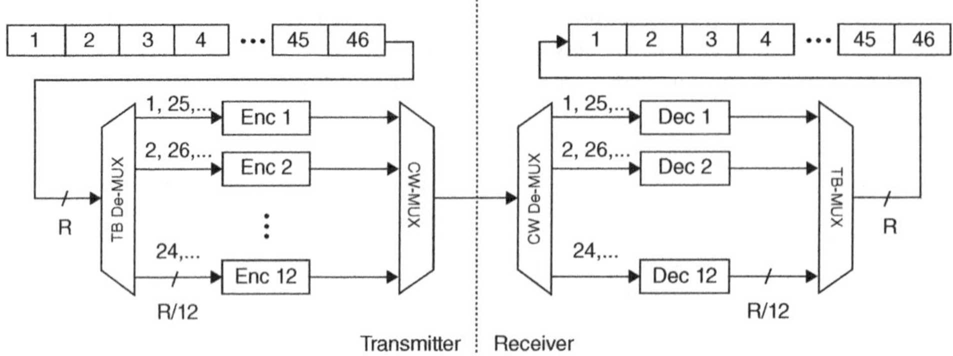

Figure 14.14 Round Robin based mapping of CWs to encoder and decoder cores

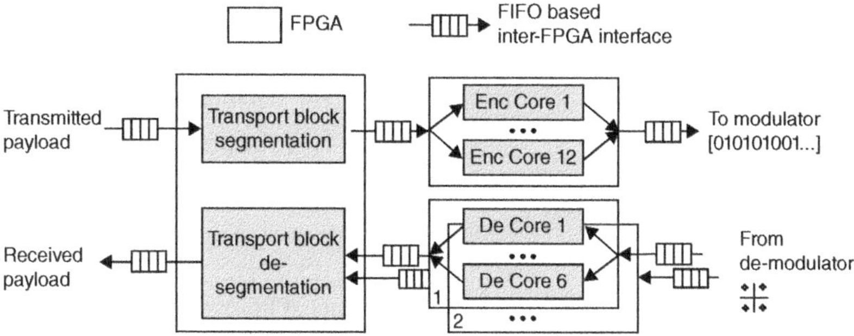

Figure 14.15 Mapping of forward error correction IP cores to FPGAs for a single mmWave transceiver node

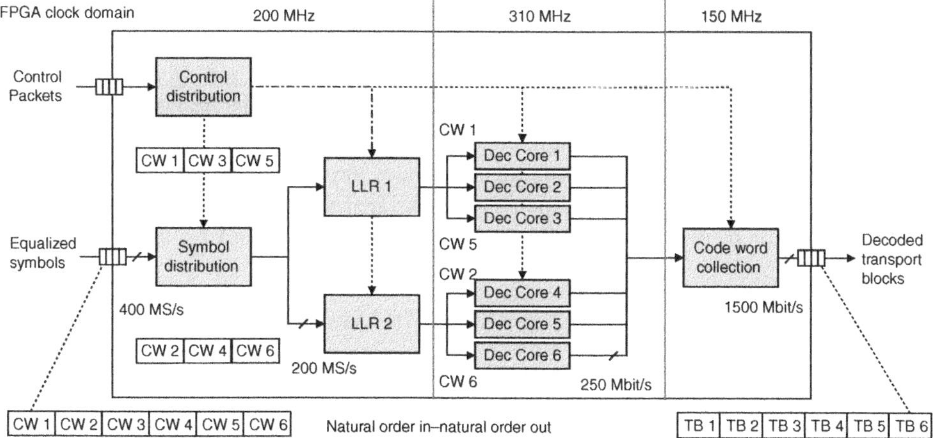

Figure 14.16 Decoding FPGA design: functional blocks executing in different clock domains

data units into TBs and interface to higher layers running on a real-time controller. Design considerations for a single-decoder FPGA are highlighted in the next section.

14.4.4.3 Parallel Decoder FPGA Design Example

Figure 14.16 shows the high-level block diagram of a single-decoder FPGA. The main functionality is to compute reliability information for each received encoded bit, followed by turbo decoding. The input to that FPGA comprises:

- control information per CW
- equalized symbols originating from another FPGA which implements equalization.

The output consists of decoded TBs.

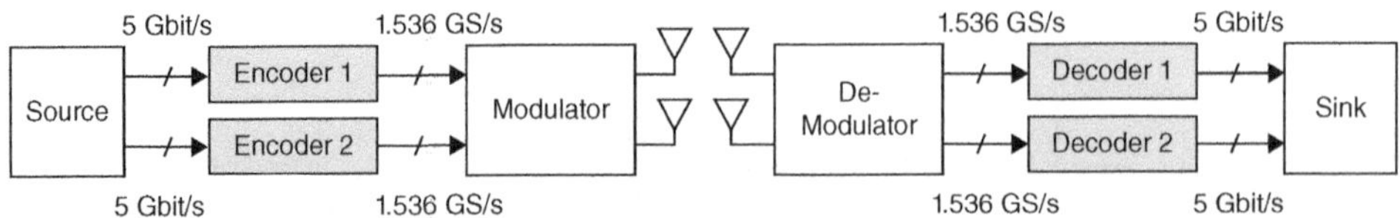

Figure 14.17 2×2 MIMO system supporting two independently encoded spatial streams to support 10-Gbps data rate

The first design decision to be made concerns the inter-FPGA connections. The inbound symbol rate at each decoder FPGA is 750 MS/s/4 = 375 MS/s. Computing reliability information per bit on the demodulator FPGA and forwarding these to the decoding FPGA will typically increase the inter-FPGA connection rate as compared to exchanging equalized data symbols, which is the preferred solution.

The second design decision to be made is driven by the high inbound symbol rate of 375 MS/s per decoding FPGA. Computing reliability information at this rate is a formidable task for the FPGA hardware. This problem can be circumvented by splitting the received symbol stream into two parallel streams and computing reliability information at half the rate in parallel. Each log-likelihood ratio stream is connected to three decoder cores, which are placed in a different, relatively high, clock domain in order to maximize throughput. A collection unit running at a lower clock domain cycles through the different decoder cores and collects the results in order. All functional blocks are reconfigured in real time based on control information received from the RF interfacing and control FPGA per code word.

14.4.5 MIMO Extension of the mmWave PoC System

Although the peak throughput of the 1-GHz mmWave PoC is 2.317 Gbps, the design of the system highlights the challenges, hardware/software architecture and the overall hardware system design. We now discuss the extension of the 1-GHz system to a 2-GHz 2 × 2 MIMO mmWave system based on work by Larew et al. [6] (see Figure 14.17). The most notable architectural change in the hardware setup is the use of a polarized antenna (horizontal and vertical polarization) for two MIMO streams and all the associated baseband ADC/DAC rate changes and mechanism to handle the data movements.

This system is capable of delivering some of the major 5G KPIs: peak throughput of 10 Gbps and latency lower than 1 ms. The system architecture represents a scaled up version of the 1-GHz SISO[1] [1] design presented earlier in this chapter. The system parameters are outlined in the first column of Table 14.1. In the following subsections, an overview of how the extension is achieved in the baseband receiver processing and FEC architecture will be described briefly.

14.4.5.1 Baseband Signal Processing

Compared to the 1-GHz air interface discussed in Section 14.4.1, the frame duration is shortened to 20 ms in order to represent a multiple of the 3GPP LTE radio frame size of 10 ms [18]. Consequently, the slot duration reduces slightly to 100 μs. The symbol rate is 1.536

[1] Single-input, single-output.

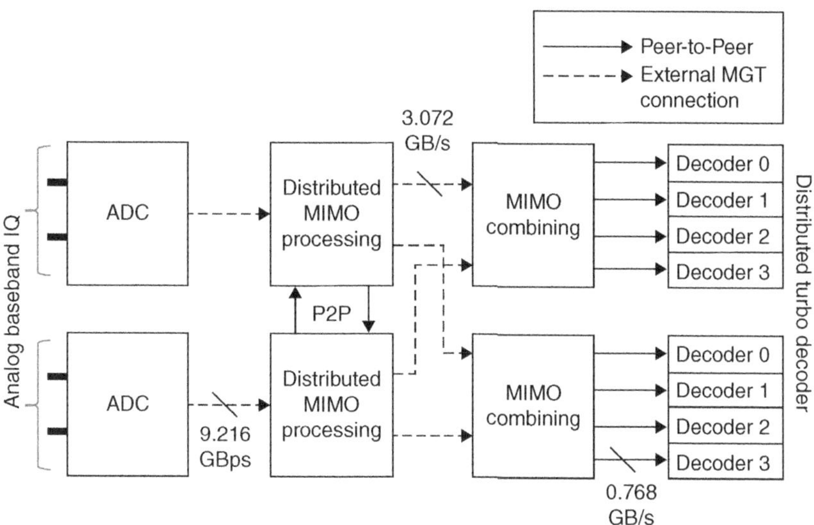

Figure 14.18 Block diagram of the 2 × 2 2 GHz mmWave baseband receiver

Gsym/s and with 2 times oversampling at the receiver, the sampling rate is 3.072 GS/s. Thus, the number of CWs per slot and the size of SC-NCP blocks get doubled: each SC-NCP block has 1024 symbols with 960 payload and 64 NCP symbols. With 7/8 16QAM, each spatial stream can deliver a throughput of $98 \times 5160 \times \frac{1}{100\mu} = 5.05$ Gbps, where 98 is the number of CWs per slot, 5160 is the TB size for 7/8 16QAM, and 100 μs is the slot duration. Thus, this system is capable of delivering 10.1 Gbps throughput.

Figure 14.18 shows an overview of the baseband receiver architecture. As each ADC samples at a rate of 3.072 GS/s with 12-bit resolution per I/Q component, the required throughput from the ADC to the demodulator FPGA amounts to 9.216 GBps. This throughput requirement is beyond the capability of the PXI chassis backplane. Alternatively, external multi-gigabit transceivers (MGT) are used to route data from the ADCs to the FPGAs. Each distributed MIMO-processing FPGA processes one of the two received signal streams. For equalization, MIMO channel estimates are shared between the MIMO processing FPGAs through P2P streaming. Each of the four streams of the FS-FDE subprocess coming out of the MIMO processing FPGAs, is at the symbol rate – 1.536 Gsym/s with 8 bits IQ samples – which amounts to 3.072 GBps per stream. The transfer of these streams to the MIMO combining FPGA is again achieved through external MGT connectors. The output of each MIMO combining FPGA is one spatial stream at 1.536 Gsym/s with 8-bit IQ samples, which amounts to an aggregate rate of 3.072 GBps. As this limit is within the one-way PXI backplane throughput limit, each demodulated spatial stream is distributed to a bank of four decoder FPGAs using P2P streaming. Thus the input data rate to each decoder FPGA is 0.768 GBps.

The respective FEC coding architecture is outlined in the next subsection.

14.4.5.2 Scaling the Coding Architecture from 2.3 to 10 Gbps

This subsection explains how the FEC coding architecture presented in Section 14.4.4 can be scaled to a 2 × 2 MIMO maximal throughput at a symbol rate of 1.536 GSym/s, using two

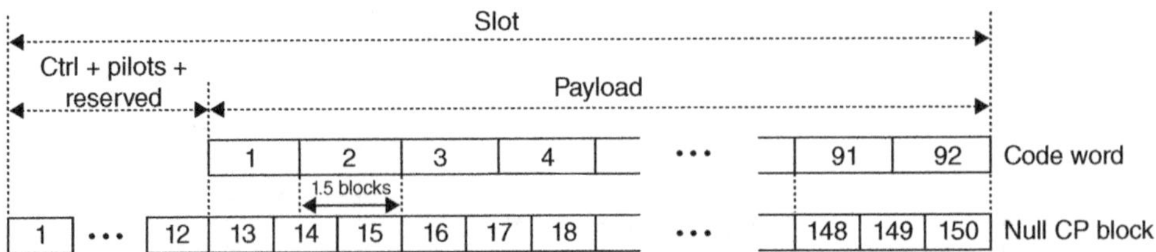

Figure 14.19 Mapping of CWs to null CP blocks in a slot of 100 us length

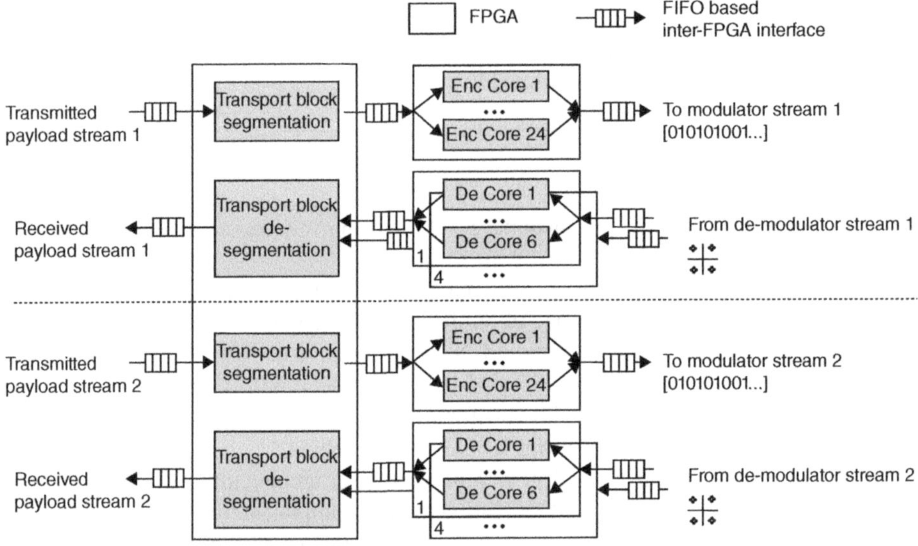

Figure 14.20 Mapping of FEC IP cores to FPGAs for a single 2 × 2, 2-GHz mmWave transceiver node

spatial streams. Each spatial stream, which employs its own FEC (per layer coding), is required to support a maximum rate of 5 Gbps as shown in Figure 14.17. At 2-GHz bandwidth, a NCP block comprises 1024 instead of 512 symbols. Therefore, the mapping of CWs to NCP blocks in Figure 14.13 naturally extends as shown in Figure 14.19.

A slot comprises twice the number of CWs per spatial stream: 92 instead of 46. Keeping in mind a possible use of reserved null CP blocks, enabling the transmission of 98 CWs per slot, requires support for 5 Gbps per spatial stream. Reusing the design space exploration analysis already done for the 1-GHz system shows that 24 en-/decoder cores suffice to maintain 5 Gbps. Figure 14.20 illustrates an updated mapping of coding functionality to FPGA modules. The additional encoder cores per spatial stream can fit into a single FPGA. On the decoder side, the data rate can be achieved by adding two FPGAs with six decoder cores each. A second spatial stream can be supported by duplicating the en-/decoder design for the first spatial stream. The RF interfacing and control FPGA orchestrates the en/de coding operation of the two spatial streams independently.

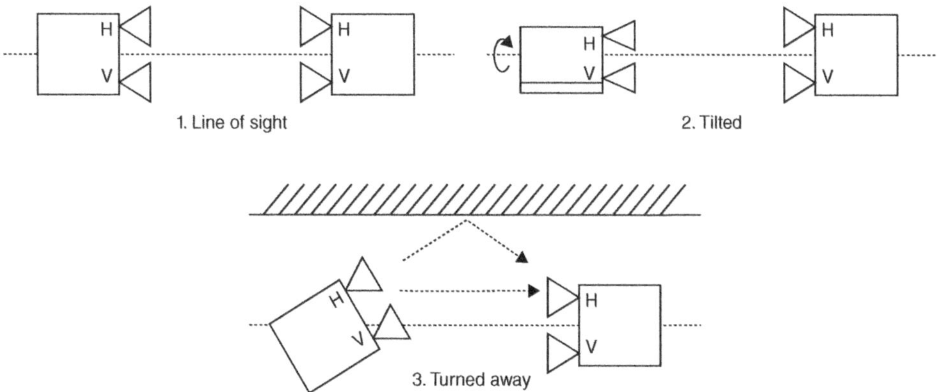

Figure 14.21 Test cases for the experimental results

14.4.6 Results

In this subsection, some of the results obtained from the testbed are presented. The following over-the-air results are obtained from the 2-GHz 2 × 2 MIMO testbed operating at 74 GHz. The symbol rate is 1.536 GSym/s per received channel, as discussed in Section 14.4.5, with horizontally and vertically polarized horn antennas for the two MIMO streams. The antenna has a gain of 23 dBi and a beam width of 14°. The main goal of these results is to show the relative variation of MIMO channel parameters for three different test cases as the receiver and transmitter are:

- in LoS
- tilted by 30–40°
- turned away by 30–40° relative to each other.

Since the relative changes are of interest, the system is not fully calibrated. These scenarios are depicted in Figure 14.21.

14.4.6.1 Test case 1: LoS

Figure 14.22 shows the frequency response of all the channels (H11, H12, H21, H22) in the 2 × 2 MIMO system. These measurements were taken with the transmitter and receiver placed about 20 feet apart, with the antennas almost pointing at each other and hence these measurements are labeled as LoS. Figure 14.23 represents the CIR for one of the components of the MIMO channel (H11). Figure 14.24, shows the condition number of the MIMO channel. The condition number of a matrix is defined as the ratio of the maximum and the minimum singular values. The channel matrix **H** is said to be "well-conditioned" if the condition number is close to 1 or 0 dB. In a practical system, a condition number equal to 0 dB is very difficult to achieve. However, a value as close to 0 dB as possible is desirable. The condition number provides a measure of how the angle between the two spatial signatures compares to the spatial resolution of the antenna array.

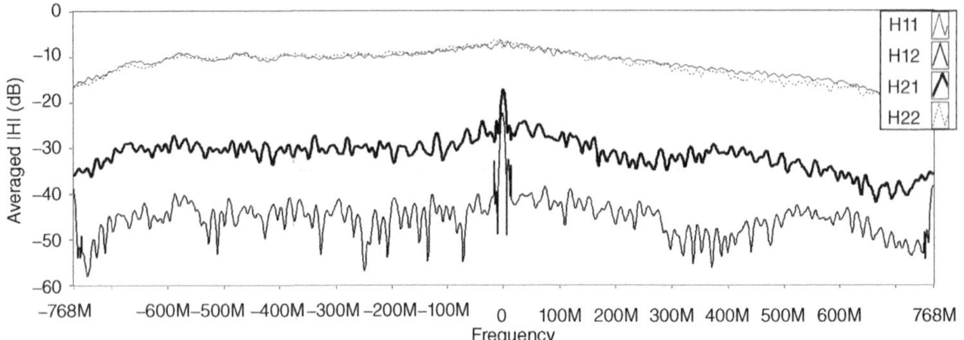

Figure 14.22 Over the air MIMO channel in frequency domain for the LoS setup

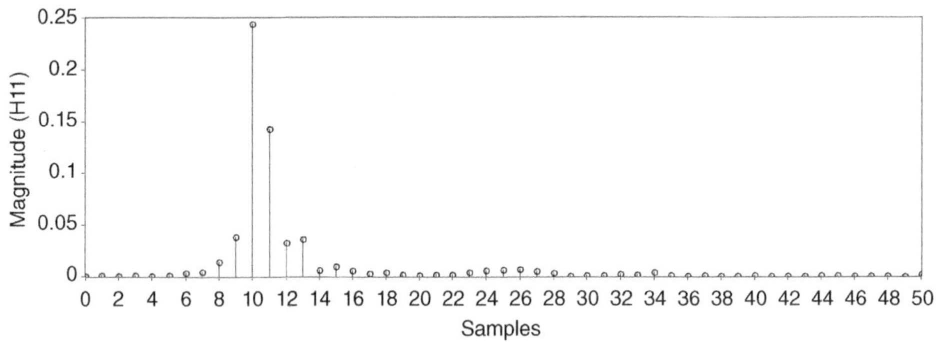

Figure 14.23 Channel impulse response for one component of the MIMO channel

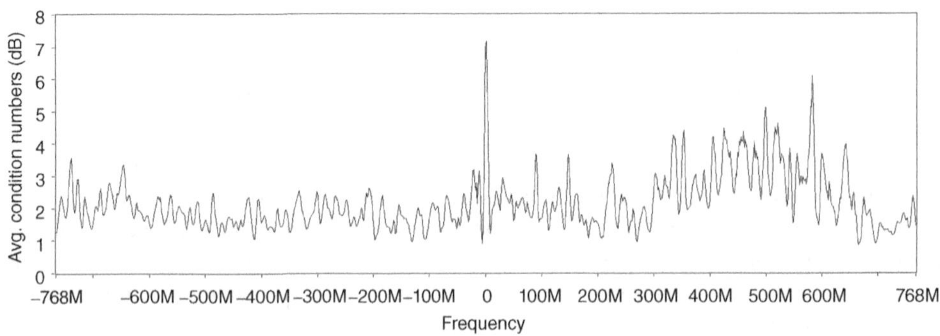

Figure 14.24 Condition number for the LoS MIMO channel

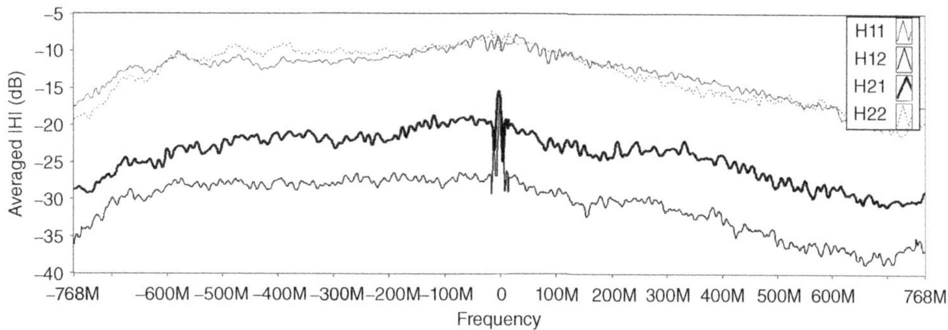

Figure 14.25 Over the air MIMO channel in frequency domain for the 30° tilted setup

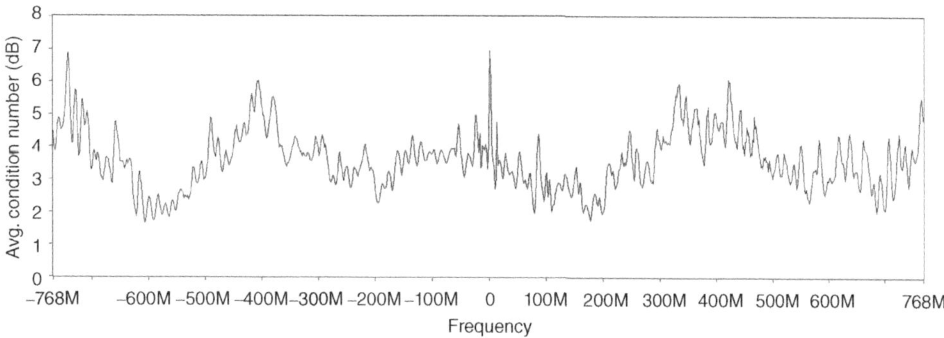

Figure 14.26 Condition number for the MIMO channel with 30° tilt

14.4.6.2 Test case 2: 30° tilt

Figure 14.25 shows the MIMO channel frequency response with 30° tilt, and Figure 14.26 shows the corresponding condition number for this channel. It is important to note the increase in condition number compared to the LoS channel, indicating the loss of spatial separation, which in turn will reflect the loss of system capacity.

14.4.6.3 Test case 3: System turned at 30°

Figure 14.27 shows the CIR as the same system is turned to one side by about 30°. Note the change in the power delay profile: new channel coefficients start to appear due to reflection.

Finally, Figure 14.28 shows the equalized constellation for the two MIMO streams.

It was also observed that the system can operate almost with no loss in performance despite some amount of tilt and rotation, indicating that 2×2 MIMO with beam tracking can potentially deliver the 5G mmWave system target peak data rate of 10 Gbps and can be fairly robust with careful system design and deployment. The results presented here are early and readily obtainable results from the PoC system. More controlled experiments will provide further

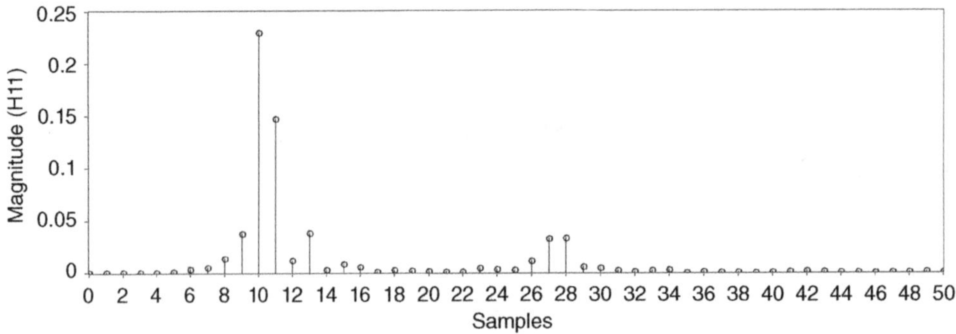

Figure 14.27 Channel impulse response for one of the MIMO channel with 30° rotation

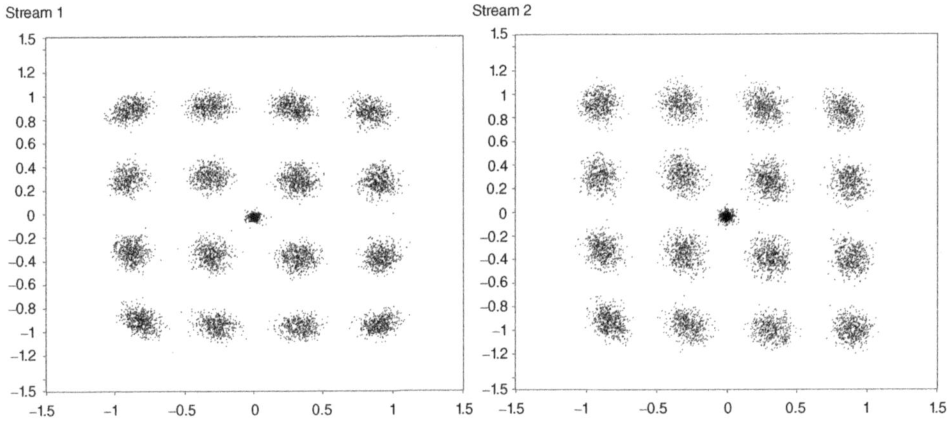

Figure 14.28 Equalized constellation for two MIMO streams

insight into the mmWave MIMO channel, air-interface design, system design in multiuser scenarios, handoffs, interoperability between different radio access technologies, and so on.

14.5 Conclusion

An overview of building a mmWave PoC system for 5G using a COTS platform is covered in this chapter. Some of the important features required for a flexible research platform were discussed, along with the software and hardware system architecture to enable high-throughput high-bandwidth applications such as mmWave radio access technology for 5G. It was also highlighted that movement of data and control across devices are as big a challenge as the signal processing itself. A case study was presented to show how some of these design challenges are handled in a PoC system.

While elegant theoretical research is providing the foundation for some of the core technologies for wireless research for 5G, the realization of 5G on hardware with existing technology will pave the way to its deployment within a target time frame. As simulations can capture only very limited operating scenarios, a PoC system is needed to validate even the simulation assumptions. PoC systems provide access to real-world data, interference, environmental challenges and so on. A PoC platform needs to operate in real-world scenarios and have the flexibility to be reprogrammed and reconfigured based on real world measurements and observations. In addition, the system needs to be extensible and maintainable throughout different phases of the research project. New tools and technologies are enabling researchers to achieve these objectives with ease. Significant improvements in the COTS platform for SDR, along with new software tools for designing wireless PoC systems, are redefining the wireless prototyping process and making it more accessible to the broader research community. This will hopefully allow us to build better and smarter solutions for 5G wireless systems.

References

[1] Nokia (2015). White paper: Network architecture for the 5G era. URL http://networks.nokia.com/file/40481/network-architecture-for-the-5g-era.

[2] NGMN (2015). *5G White paper*. URL https://www.ngmn.org/5g-white-paper.html

[3] GSMA, (2014) White paper: Understanding 5G: Perspectives on future technological advancements in mobile. URL http://www.gsma.com/newsroom/press-release/gsma-publishes-new-report-outlining-5g-future/

[4] Gosh, A., Thomas, T.A., Cudak, M.C., Ratasuk, R., Moorut, P., Vook, F.W., Rappaport, T.S., MacCartney, G.R., Sun, S., and Nie, S. (2014) Millimeter-wave enhanced local area systems: a high-data-rate approach for future wireless networks. *IEEE J. Select. Areas Commun.*, **32** (6), 1152–1163.

[5] Truchard, J. (2015) Platform drive innovation in RF and microwave system design. *Microwave J.*, **58** (3), 22–36.

[6] Larew, S.G., Thomas, T.A., Cudak, M., and Ghosh, A. (2013) Air interface design and ray tracing study for 5G millimeter wave communications. *IEEE Globecom Workshops (GC Wkshps)*, pp. 117–122.

[7] Roh, W., Ji-Yun, S., Jeongho, P., Byunghwan, L., Jaekon, L., Yungsoo, K., Jaeweon, C., Kyungwhoon, C., and Farshid, A. (2014) Millimeter-wave beamforming as an enabling technology for 5G cellular communications: theoretical feasibility and prototype results. *IEEE Commun. Mag.*, **52** (2), 106–113.

[8] Cudak, M., Kovarik, T., Thomas, T., Ghosh, A., Kishiyama, Y., and Nakamura, T. (2014) Experimental mm wave 5G cellular system, in *IEEE Globecom Workshops (GC Wkshps)*, pp. 377–381.

[9] Wang, Z. and Giannakis, G.B. (2000) Wireless multicarrier communications. *IEEE Signal Process. Mag.*, **17** (3), 29–48.

[10] Schmidl, T.M. and Cox, D.C. (1997) Robust frequency and timing synchronization for OFDM. *IEEE Trans. Commun.*, **45** (12), 1613–1621.

[11] Minn, H., Bhargava, V.K., and Letaief, K. (2003) A robust timing and frequency synchronization for OFDM systems , *IEEE Trans. Wireless Commun.*, **2** (4), 822–839.

[12] Gul, M., Ma, X., and Lee, S., (2014) Timing and frequency synchronization for OFDM downlink transmissions using Zadoff-Chu sequences. *IEEE Trans. Wireless Commun.*, **14** (3), 1716–1729.

[13] Sayed, A.H. (2003) *Appendix 3B in* Fundamentals of Adaptive Filtering. John Wiley.

[14] Ozdemir, M.K. and Arslan, H. (2007) Channel estimation for wireless OFDM systems. *IEEE Commun. Surveys Tut.*, **9** (2), 18–48.

[15] National Instruments (2014) NI PXIe-7976R specifications. URL http://www.ni.com/pdf/manuals/374546a.pdf.

[16] National Instruments Labview Communications webpage. URL http://www.ni.com/labview-communications/.

[17] Proakis, J.G. and Manolakis, D.G. (2007) *Digital Signal Processing*. Pearson Prentice Hall.

[18] 3GPP (2015) TS 36.211: Technical specification group radio access network; evolved universal terrestrial radio access (E-UTRA); physical channels and modulation (Release 12), v12.6.0.

15

*5G Millimeter-wave Communication Channel and Technology Overview

Qian (Clara) Li, Hyejung Jung, Pingping Zong and Geng Wu

15.1 Introduction

This chapter is devoted to addressing millimeter-wave (mmWave) channels and providing an overview on signal processing problems for mmWave communication in 5G. In the mmWave band, where frequency bands are above 30 GHz, a large number of spectrums are potentially available. In so far as a 5G wireless communication systems should be able to provide much

Signal Processing for 5G: Algorithms and Implementations, First Edition. Edited by Fa-Long Luo and Charlie Zhang.

higher area traffic capacity – downlink traffic density of 750 Gbps/km^2 [1], with a peak data rate of up to 10∼100 Gbps – a mobile network based on mmWave band radio access technology (RAT) is a key component.

The unique characteristics of the mmWave channel and the requirements for mmWave communications in 5G impose new requirements on mmWave channel modeling. New channel-modeling approaches therefore need to be developed for mmWave communication in 5G. This chapter presents two channel-modeling approaches to meet the requirements for a 5G mmWave channel model: an enhanced 3GPP/SCM model and a ray-propagation-based statistical model. Based on understanding of the mmWave channel characteristics, we then provide system-design considerations for 5G mmWave band RAT and signal-processing technologies related to 5G mmWave communications.

This chapter is organized into the following six sections. In Section 15.2, we will discuss the mmWave channel characteristics as compared to low-frequency-band channels. In Section 15.3, we will discuss the requirements on 5G mmWave channel modeling. In Section 15.4, we will present two mmWave channel models that meet the requirements of 5G. In Section 15.5, we discuss the key signal-processing techniques in mmWave communication. The final section of this chapter will give a summary and further discussion on the future work on modeling, estimation and other related signal-processing aspects of mmWave communication channels.

15.2 Millimeter-wave Channel Characteristics

Compared to centimeter-wave bands, more work is needed to understand the radio propagation characteristics of the mmWave band. Based on existing investigations [2]–[7], radio propagation in the mmWave bands is understood to have the following unique features:

- high path loss
- sensitivity to propagation environments
- vulnerability to geometry blockage
- non-stationarity in time and space.

According to Frii's equation, given isotropic Tx and Rx antennas of unit gain, the power loss of radio propagation in free space is inversely proportional to the square of the frequency of the radio signal:

$$P_r = P_t\left(\frac{c}{4\pi df}\right)^2 \tag{15.1}$$

where P_t is the transmit power, P_r is the receive power, d is the distance between transmitter and receiver, f is the radio signal frequency, and c is the speed of light. As a result, an increase in signal frequency by 10 times would result in a 20-dB decrease in received power. Higher atmospheric gaseous losses, rain losses and foliage losses are observed in mmWave radio propagation besides the free-space propagation loss [8]. Figure 15.1 shows the path-loss values generated from the path-loss model obtained on the basis of measurements at 28 GHz in the Manhattan city-canyon environment [3] and the path-loss values generated in the 3GPP UMi path loss model at 2 GHz. A gap of more than 20 dB can be observed across the evaluated Tx–Rx distances.

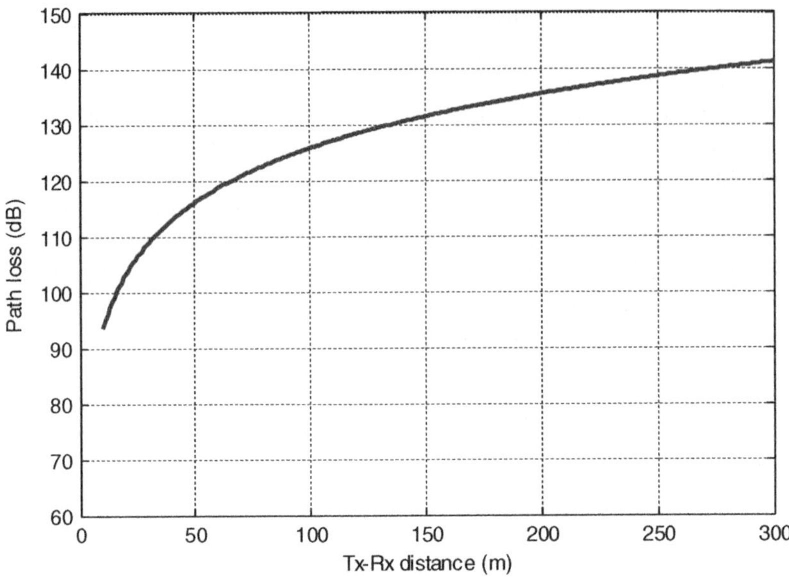

Figure 15.1 Path loss values with respect to Tx–Rx distance at 28 GHz and 2 GHz

In mmWave bands, because of the short wavelength, small objects in the propagation environment such as lamps, trees and small piece of furniture, almost invisible at lower frequency bands, become prominent. Also, irregularity or roughness of reflection surfaces, largely irrelevant at lower frequency bands [9], become important. The propagation environment becomes more detailed. Paths can be blocked due to small obstacles. Specular reflection can be undermined due to diffuse scattering. New paths can be created from reflections from small objects. As a result, in identical communication environments, the channel parameters – the number of paths, power delay profile and power angular spectrum – of a high frequency band channel can be different from those of a lower frequency band.

As the radio frequency increases, the propagation behaves more like optical propagation. The low diffraction probability of quasi-optical propagation leads to a high probability of blockage [10]. Receivers at locations behind a building or around a corner can be severely attenuated. We term such high shadow fading due to large objects in the map as geometry-induced blockage loss. High geometry-induced blockage loss is rarely observed in lower frequency bands due to the high diffraction probability. In cases where geometry-induced blockage losses brings down the channel gain to a level that falls below the receiver sensitivity, the receiver would be in an outage state. Figure 15.2 illustrates the effects of specular reflection, diffused scattering, blockage by small objects and diffraction.

In practical communication scenarios, the environment is dynamic: people and cars are moving around. Small moving objects, mostly invisible in the low frequency bands, will cause turbulence in high-frequency-band propagation. The channel in the high frequency band becomes non-stationary. Some of the paths can be temporarily blocked, while new paths could be created due to reflection from passing objects. The statistics of the transient blockage and the new paths depend on the traffic in the communication scenario. An example

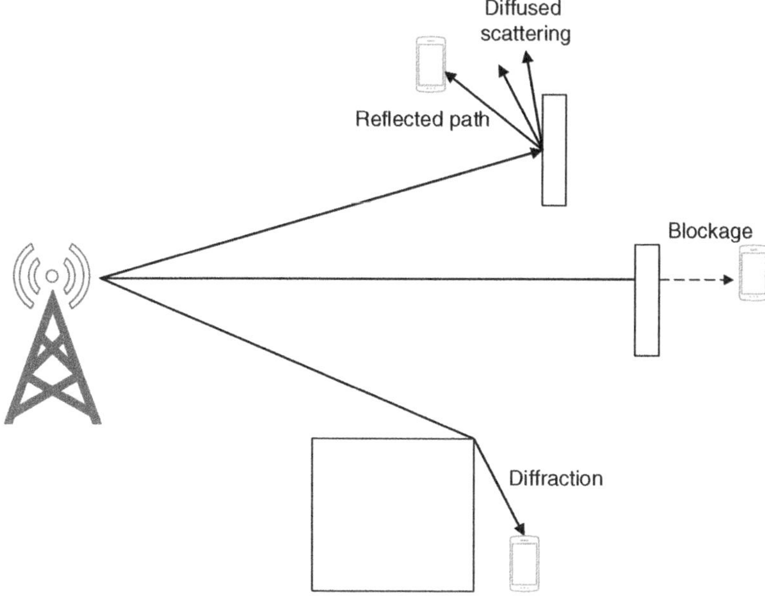

Figure 15.2 Effects of specular reflection, diffused scattering, blockage, and diffraction

from measurements of dynamic blockages is presented in Fig. 4.9 of the MiWEBA technical paper [7].

15.3 Requirements for a 5G mmWave Channel Model

The channel model abstracts the measured channel into a general format that can be used to regenerate it for system design and evaluation purposes. Channel model development must therefore fit with the needs of the system design. In a sense, channel measurement, channel modeling and system design are intertwined: channel measurement helps people understand the channel characteristics. Based on the channel characteristics and the targeted system deployment, technologies can be identified to enable communications under the channel condition. Based on the channel characteristics and the identified communication technology, the channel model can be developed to represent the channel characteristics to the level of detail that meets the needs of technology development and performance evaluation.

In a 5G cellular system, mmWave communication is expected to provide a mobile broadband service with diverse deployment scenarios: indoor offices, shopping malls, and urban/suburban micro and macro environments. The operation bandwidth can be up to 1 GHz or 2 GHz. High-gain antennas are needed to compensate for the path loss so as to achieve sufficient cellular coverage [11]. This need leads to the following requirements for mmWave channel modeling.

High spatial/temporal resolution At gigahertz bandwidth, the sampling rate would be at the sub-nanosecond level. As a result, propagation paths received nanoseconds apart can be differentiated. Therefore, the channel model needs to provide ns-level temporal resolution to match the signal bandwidth. In cluster-based channel models, where the component paths are modeled as clusters, ns-level temporal resolution requires detailed modeling on the intracluster structure, including of the delay and angle distributions of the intracluster rays. High antenna gain comes with high directivity and narrow beamwidth. For example, a 64-element critically sampled uniform linear array with proper beamforming can achieve an antenna gain of 18 dBi, while the corresponding half-power beamwidth is only 5.6°. To support the design and evaluation of highly directional transmission and reception, the channel model should have an angular resolution down to around 1°.

Geometry-induced blockage and temporal shadow fading Geometry-induced blockage creates coverage holes that need to be addressed in the system design to maintain connectivity. Therefore, the channel model needs to reflect the presence of geometry-induced blockages in terms of parameters such as blockage probability, blockage loss, blockage-area size and so on. As discussed in the previous section, the mmWave channel is sensitive to environmental dynamics. With highly directional transmission and reception, the impact of channel dynamics will be even more severe. For instance, if the path captured in the Tx and Rx narrow beams is blocked by a moving object, the link will be severely attenuated, which makes the transmitter and receiver need to switch to an alternative path. Where new paths are created due to reflections from moving objects, the transmitter and receiver may choose to steer the beams toward the new path during beam updating if the path gain is high. The effect of such temporal shadow fading (including temporal path blockage and temporal new path presence) will need to be modeled in the channel so that system design and evaluation take the channel dynamics into account.

Cluster-level spatial and temporal evolution Channel spatial and temporal evolution need to be modeled down to cluster level to provide consistent channel updates with directional narrow beam transmission and reception. This includes cluster gain and phase update and the cluster death and birth processes. The transition between line-of-sight (LoS) and non-line-of-sight (NLoS) states and the transition between a geometry-blockage state and a non-blockage state also need to be modeled.

Complexity The channel model should be representative of the channel characteristics. In the meantime, the complexity should be feasible for system-level performance evaluation: the model needs to achieve a balance between accuracy and complexity.

15.4 Millimeter-wave Channel Model for 5G

A wireless channel model can be generally expressed using the double-directional model [12]:

$$h(\tau, \phi_{AoD}, \phi_{AoA}, \theta_{ZoD}, \theta_{ZoA}) = \sum_{i=1}^{N}\sum_{j=1}^{M_i} H_{i,j} e^{j\Phi_{i,j}} \delta(\tau - \tau_{i,j})\delta(\phi_R - \phi_{AoA,i,j}) \times \delta(\phi_T - \phi_{AoD,i,j})\delta(\theta_R - \theta_{ZoA,i,j})\delta(\theta_T - \theta_{ZoD,i,j}) \tag{15.2}$$

The parameters that define the channel are the number of path clusters N of the link, the number of rays within the n-th cluster M_n, the path gain of each ray $H_{n,m}$, the phase of each ray $\Phi_{n,m}$, the time-of-arrival (ToA) of each ray $\tau_{n,m}$, the azimuth angle-of-arrival $\phi_{AoA,n,m}$

and azimuth angle-of-departure $\phi_{AoD,n,m}$ of each ray, and the zenith angle-of-arrival $\theta_{ZoA,n,m}$ and zenith angle-of-departure $\theta_{ZoD,n,m}$ of each ray. Based on the double-directional modeling approach, channel modeling activities at the link level have been focused on finding ways to generate the channel parameters for various deployment and communication scenarios. To this end, statistical models have been developed in the low-frequency bands. Examples include the WINNER model [13], which has been used in 3GPP as the channel model for the LTE system [14], and the COST models [15]–[18]. The statistical models developed are effective in generating the path delays, the path gains, the arrival/departure angles, the polarization and the Doppler. The complexity is feasible for system-level implementation.

In the high-frequency band, in order to meet the requirements of mmWave channel modeling, amendments to the existing models to account for low-frequency bands, or else new models, need to be developed. In this section, we introduce two statistical models for the mmWave channel. One model takes the 3GPP spatial channel model (SCM) as a baseline and adds new components to meet the mmWave channel model requirements. Another model is based on geometry statistics and ray propagation. Note that since the channel-model standardization work is on-going, the models introduced are intended to provide early implementation for system design and evaluation. The industry is still expecting a common channel model to be used for 5G system development above 6 GHz.

15.4.1 Enhanced SCM Model

The channel generation procedure in the 3GPP SCM is implemented in two steps: first, generate the large scale path loss, and then generate the small-scale fading. To accommodate the mmWave channel features, the 3GPP SCM model needs to be enhanced by adding geometry-induced blockages, temporal shadow fading, cluster evolution, and high spatial/temporal resolution. In this section, we focus on introducing these enhancements to the SCM model. Details of the SCM model can be found in textbooks and the 3GPP specification documents [14].

15.4.1.1 Large-scale Path Loss

Large-scale path loss is modeled as

$$PL(f,d) = c_{LOS} PL_{LOS}(f,d) + (1 - c_{LOS}) PL_{NLOS}(f,d) \tag{15.3}$$

where $c_{LOS} \in [0,1]$ is a coefficient modeling the transition between LoS and NLoS states.

For locations in LoS or NLoS states, the path loss is modeled as

$$\begin{aligned} PL(f,d) = {} & A(f,d) + B(f,d)\log_{10}(d) + C(f,d) + \mathbf{1}_{b0}(f,d)D(f,d) \\ & + (1 - \mathbf{1}_{b0}(f,d))(SF_0(f,d) + \mathbf{1}_{b1}(f,d)SF_1(f,d)) + E(f,d) \end{aligned} \tag{15.4}$$

where $A(f,d) + B(f,d)\log_{10}(d)$ models the expectation of a geometry-induced large-scale path loss in a static environment, $C(f,d)$ models the indoor–outdoor penetration loss, $D(f,d)$ models the geometry-induced static blockage loss, $SF_0(f,d)$ models the geometry-induced static shadow fading, $SF_1(f,d)$ models dynamic shadowing fading (including blockage loss

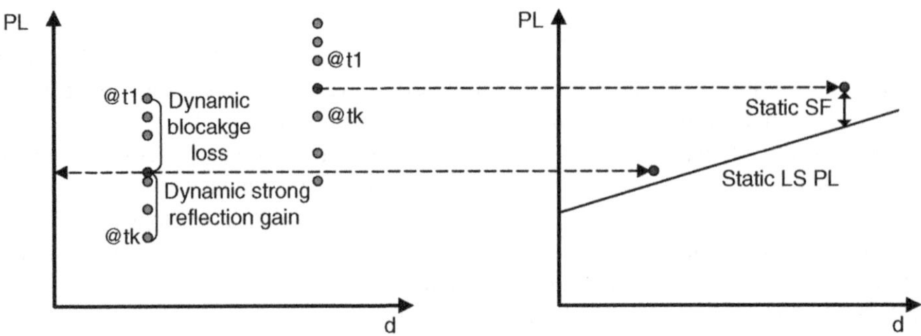

Figure 15.3 Illustration of the path-loss components. PL: path loss; SF: shadow fading

and strong reflection gain) due to environment dynamics, and $E(f, d)$ models atmosphere attenuation. $\mathbf{1}_{b0}(f, d) \in \{0, 1\}$ is an indicator function indicating the static blockage status. For example, $\mathbf{1}_{b0}(f, d) = 1$ indicates the link is experiencing geometry-induced static blockage, and $\mathbf{1}_{b0}(f, d) = 0$ otherwise. $\mathbf{1}_{b1}(f, d) \in \{0, 1\}$ is an indicator function indicating the dynamic shadow fading status. For example, $\mathbf{1}_{b1}(f, d) = 1$ indicates the presence of dynamic shadow fading. $\mathbf{1}_{b0}(f, d)$ and $\mathbf{1}_{b1}(f, d)$ are specified by static blockage probability p_{b0} and dynamic shadow fading probability p_{b1}, respectively.

Figure 15.3 provides a visualization of the different components. When we do measurements, multiple samples are collected at each measurement location at different time instances. The *static path loss* is obtained by taking the expectation over the multiple samples. The *dynamic shadow fading* is the difference between the static path loss and the measured path loss of the samples. Depending on how dynamic the environment is, dynamic shadow fading may or may not always be present. For instance, in a static environment, the path loss values at the different time instances could be the same. Therefore, the presence of dynamic shadow fading can be an opportunistic event specified by a probability value. By taking same measurement procedure across different network locations in the considered scenario, a set of static path loss values can be obtained. The static path loss values obtained can then be used to find a fitting on *static large-scale path loss*. In a NLoS scenario, no signals (in all time samples) can be detected in some locations. Those locations are regarded as in *static blockage*. A high static blockage loss can be added on to the static large-scale path loss. In a LoS scenario, an exceptionally high static path loss may be detected (in LoS scenario, the static shadow fading variance is small) due to (partial) blockage of the LoS path by a tree, street signs, furniture, people, etc. These locations can be regarded as being in static blockage and can be separately modeled in the same way as in the NLoS case.

15.4.1.2 Small-scale Fading

Based on the cluster generated by the 3GPP SCM model, we add the effects of dynamic shadow fading and cluster-level evolution. The cluster generation is done in four steps:

1. Generate clusters using the 3GPP SCM model for a static communication environment.
2. With a certain probability, model the effect of dynamic blockage on the clusters and additional new clusters due to strong reflection.

3. With a certain probability, model the effect of body blockage due to rotation.
4. In the case of moving transmitter or receiver, model the effect of cluster spatial evolution.

The blockage effect may result in additional shadow fading on (part of) the clusters, as shown in Figure 15.4 and Figure 15.5. This effect could naturally lead to a reduced number of clusters. In the case of a transient strong reflection path, an additional path needs to be added with the cluster power specified by the additional path gain applied in the large-scale path loss. Figure 15.6 demonstrates such an example. The probability of dynamic shadow fading and the distribution of the shadow fading value depend on the communication environment, and are subject to measurement to abstract the statistics in typical communication environments.

In order to model cluster evolution on short timescales (at the millisecond level), the cluster spatial/temporal evolution should mainly focus on the phase and delay evolution of the paths. A slight phase change would affect the beamforming coefficient adjustment. On longer timescales (say, seconds), a smooth cluster birth-and-death process should be modeled. The 3GPP SCM is a drop-based model. The spatial correlation is taken into account by means of correlated large-scale parameters. To enable cluster-level spatial correlation in the drop-based model, we divide the network area into a grid based on the cluster-coherence distance. The channel is generated at each cross-point in the grid as drop-based. Channels

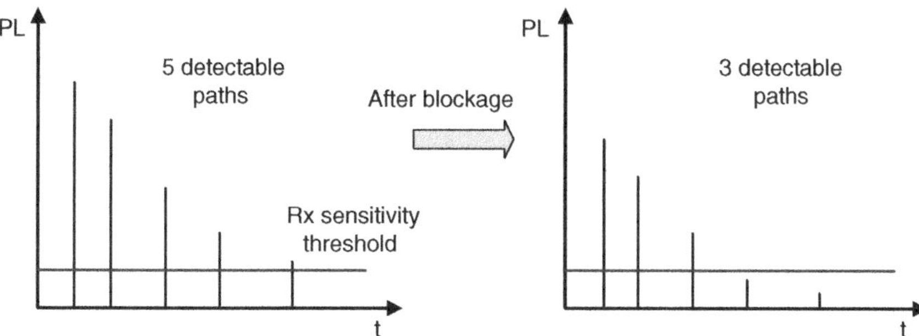

Figure 15.4 CIR before and after blockage – in the case of blockage loss uniformly applied to all clusters

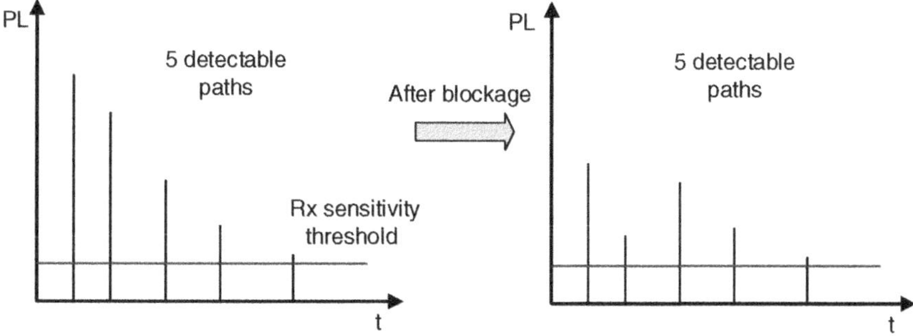

Figure 15.5 CIT before and after blockage – In the case of blocking the two strongest paths

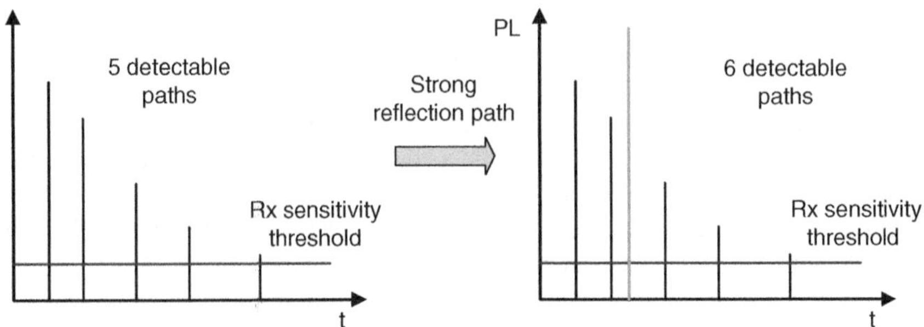

Figure 15.6 CIR before and after the presence of strong reflection path

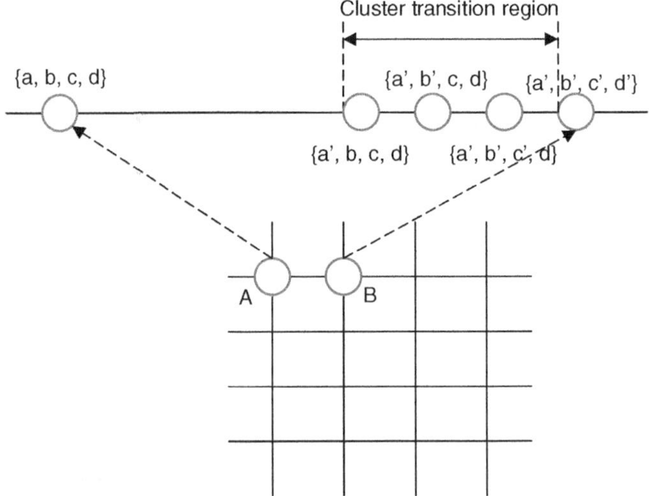

Figure 15.7 Cluster evolution in a drop-based grid

at locations between the grid points are generated by gradually replacing the clusters of one grid point with the clusters of the neighboring grid. In the example shown in Figure 15.7, as a user moves from grid point A to grid point B, the clusters of a channel generated at point A would be gradually replaced by clusters of a channel generated at point B. In each instance, one cluster is randomly selected and replaced with a cluster with similar excess delay. For example, if excess delay of cluster a is similar to excessive delay of cluster a', then cluster a will be replaced by cluster a'. In this way, we can ensure a consistent power-delay profile during the cluster evolution.

15.4.2 Ray-propagation-based Statistical Model

Signal propagation in the network area depends only on the location of the transmitter and the propagation interactions with the environment, regardless of the presence of receivers. As

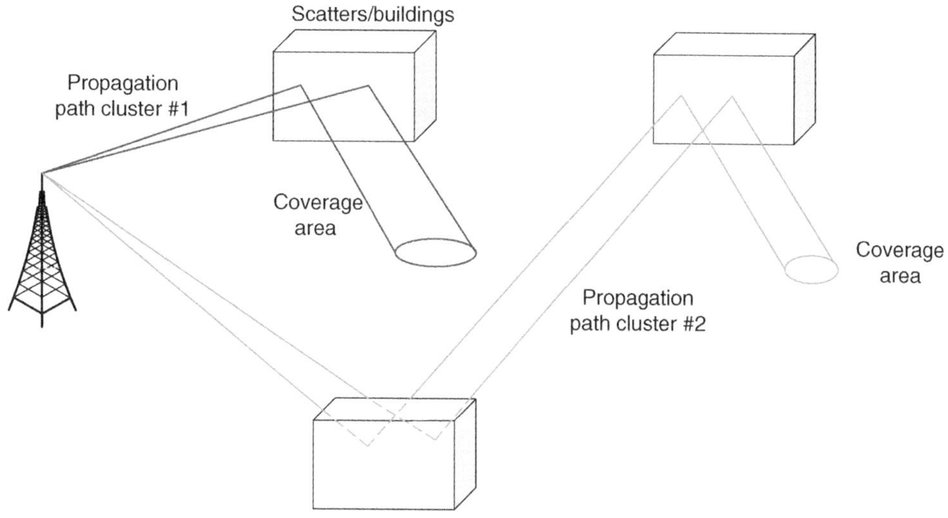

Figure 15.8 Illustration of propagation

illustrated in Figure 15.8, each propagation path cluster can be defined in the spatial and temporal domains by its coverage area, the departure/arrival angles, and the propagation time. The coverage area specifies the network area that the cluster reaches, while the propagation time defines the transmission time from the transmitter to the network area along the propagation path.

Based on this understanding, in the ray-propagation-based statistical model, channel generation is implemented in three steps:

- cluster deployment
- cluster parameter generation
- cluster–receiver association.

Cluster deployment generates the clusters in the network area by specifying the cluster center location, cluster coverage area, and the transition region. The transition region is used to model the ramping up and ramping down of the cluster. The cluster parameter generation step generates the cluster gain, delay, angle of arrival and angle of departure. Cluster–receiver association associates the specific receiver with the clusters that cover its location. In this way, nearby receivers that fall in the coverage area of the same cluster will be correlated naturally. Channel evolution due to mobility can also be naturally modeled as the associated clusters that evolve when the receiver location changes.

Figure 15.9 illustrates the cluster deployment associated with one BS in the x–y plane of the network area. In deploying the clusters, we first generate the cluster center and then generate the coverage range. The cluster-center dropping and the cluster-range values should be generated following certain distributions obtained from measurement and/or ray-tracing simulations. Clusters can be overlapping; in other words multiple clusters can be observed at each location. Different distributions for NLoS and LoS clusters are also expected.

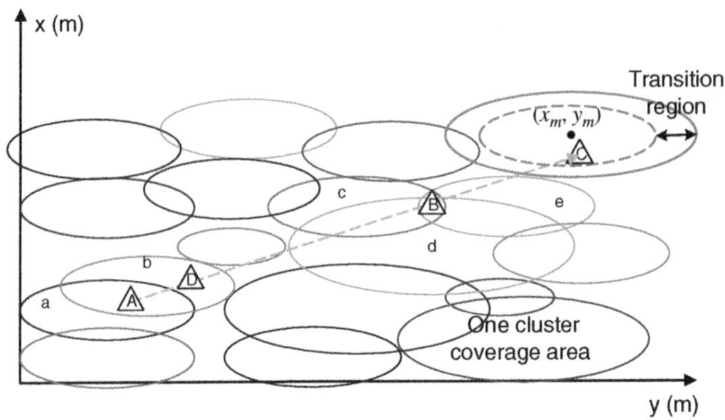

Figure 15.9 Illustration of cluster deployment

For each of the deployed clusters, we then generate the channel parameters as defined in the double-directional channel model. The cluster parameter generation takes a geometry-based approach that takes as input the statistical geometry distribution of the communication environment, the statistical distribution of the reflection/diffraction loss and the cluster ToA distribution. The path gain is calculated on the basis of geometry statistics: given a path ToA, we can calculate the traveling distance of the path, and together with the environmental geometry distribution and the reflection/diffraction loss statistics, we can also calculate the shadow fading of each path. The path gain can be calculated by adding the shadow fading with free-space path gain.

To be more specific, given the ToA of the n-th path t_n, the length of the path can be calculated as $d_n = ct_n$, where $c = 3 \times 10^8$ m/s is the speed of light in a vacuum. The probabilities of encountering k_r reflections and k_d diffractions along the trajectory followed by the path at hand are instead described by geometrical distribution of the environment. The shadow fading can thus be calculated as

$$SF(d) = \sum_{k_r} p_r(k_r|d)SF_r(k_r) + \sum_{k_d} p_d(k_d|d)SF_d(k_d), \tag{15.5}$$

where $p_r(k_r|d)$ and $p_d(k_d|d)$ are the probabilities of encountering k_r reflections and k_d diffractions at distance d, respectively.

As an example, if the environment geometry follows a Poisson distribution, $p_r(k_r|d)$ and $p_d(k_d|d)$ can be found as

$$p_r(k_r|d) = \frac{e^{-\lambda_r d}(\lambda_r d)^{k_r}}{k_r!} \tag{15.6}$$

and

$$p_d(k_d|d) = \frac{e^{-\lambda_d d}(\lambda_d d)^{k_d}}{k_d!} \tag{15.7}$$

with parameters λ_r and λ_d. $SF_r(k_r)$ and $SF_d(k_d)$ are the respective reflection and diffraction losses by the k_r and k_d reflections and diffractions. These can be randomly drawn from the reflection and diffraction loss distributions. The path gain can then be calculated as

$$H_n(d) = A(d)SF(d)\left(\frac{4\pi d}{\lambda_c}\right)^2 \quad (15.8)$$

where $A(d)$ is the atmosphere attenuation, which can be obtained from measurement.

With $H_n(d)$, the remaining paths in the cluster can be calculated by applying an intracluster power-delay profile (PDP). The intra-cluster PDP closely relates to the reflection surface and the material. Examples of such intracluster PDP in the 60 GHz bands [7],[19] and further details on cluster parameter generation [20] can be found in the literature.

With the generated clusters in the network area, the channel observed by each user equipment (UE) in the network area is the aggregation of clusters related to the UE location. The number of clusters can be accounted for accordingly. For the example shown in Figure 15.9, location A is associated with clusters *a* and *b*, and location B is associated with clusters *c*, *d*, *e*. Location A and location D can both observe cluster *b*. Cluster-level correlation between locations A and D is therefore naturally introduced.

Further extensions of the model include the association of each cluster with a coherence time, which is used to model the dynamic appearance of the cluster due to changes in the surrounding environment, and the modeling of the arrival time of each cluster as time-varying to account for double mobility at the transmitter and receiver.

The modeling approach can be verified using ray-tracing simulation. By ray-tracing, the input parameters for the model – the geometry statistics, the ToA distribution and the reflection/diffraction loss distribution – can be obtained. Then, by taking the input parameters, channel coefficients can be generated from the model. A comparison of the generated channel with the ray-tracing results leads to an effective evaluation of the channel model.

Here we use a commercial ray-tracing tool called Wireless InSite and maps of Rosslyn, Virginia, USA as an example to demonstrate channel-model verification. Figure 15.10 and Figure 15.11 show the ray-tracing simulation scenario.

We use the path loss and root-mean-square (RMS) delay spread (DS) as the metric for calibration. At 28 GHz carrier frequency, the path-loss exponent of the Rosslyn scenario is 3.22:

$$PathGain_{NLOS} = -61.38 - 32.2\log(d) + ShadowFading \quad (15.9)$$

Figure 15.12 shows the path gain generated by using the path-loss model obtained from ray tracing and the path gain generated from the ray-propagation-based statistical model. For clear comparison, we did not show shadow fading in the ray-tracking path loss. Figure 15.13 shows RMS DS generated by ray-tracing simulation and by the model. Comparison of Figure 15.12 and Figure 15.13 shows a good match between the model and ray-tracing simulation. From Figure 15.12, we can also observe that as transmitter–receiver distance increases, the probability of deep fading increases, too.

15.5 Signal Processing for mmWave Band 5G RAT

This section presents some of the system design challenges that must be overcome and the main technology components that will have to be developed to enable a mmWave band 5G RAT to become a reality. Signal-processing problems related to beam acquisition, beam tracking, channel estimation and cooperative transmission and reception will also be discussed.

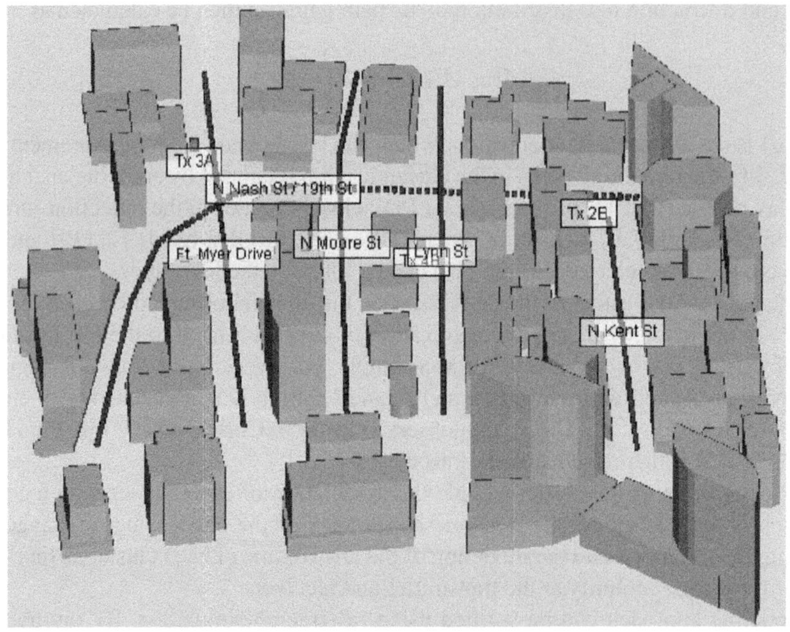

Figure 15.10 Rosslyn map – 3D

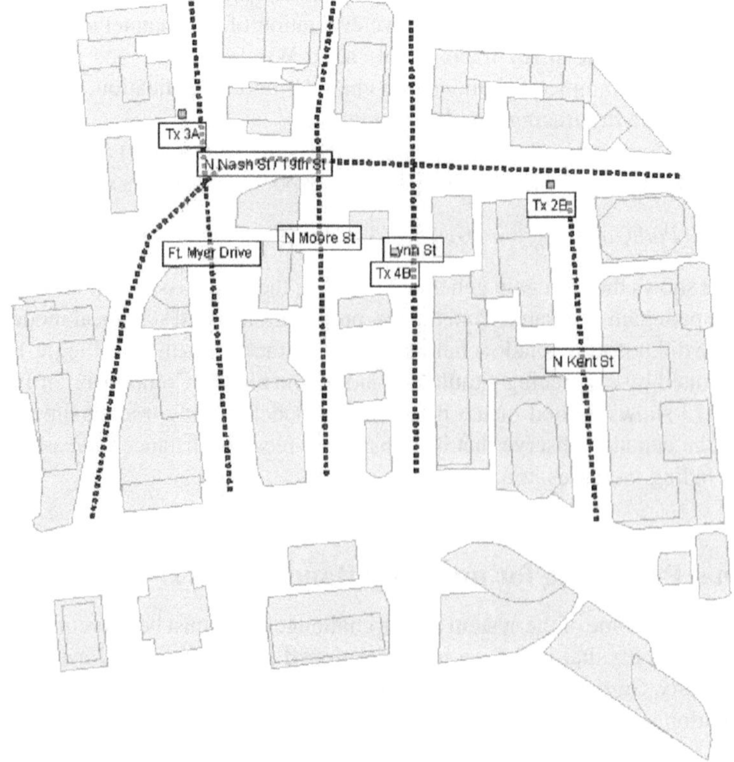

Figure 15.11 Rosslyn map – 2D

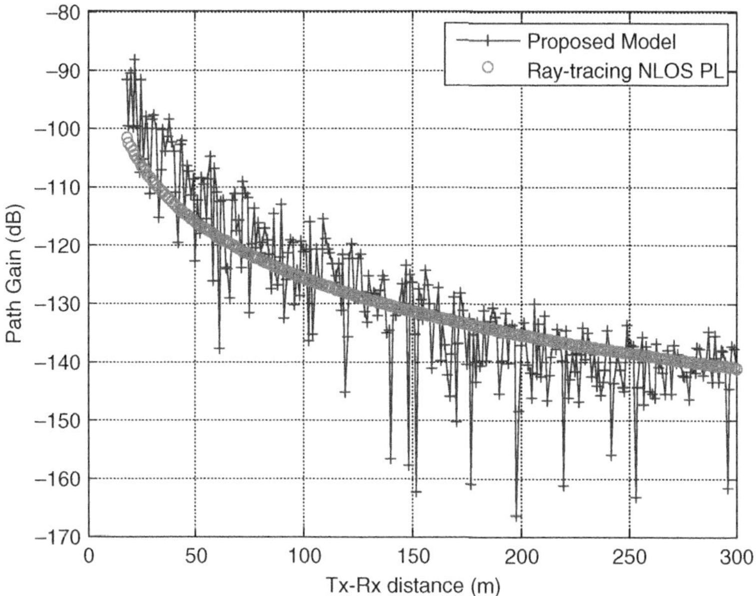

Figure 15.12 Path gains of the proposed model and ray-tracing model

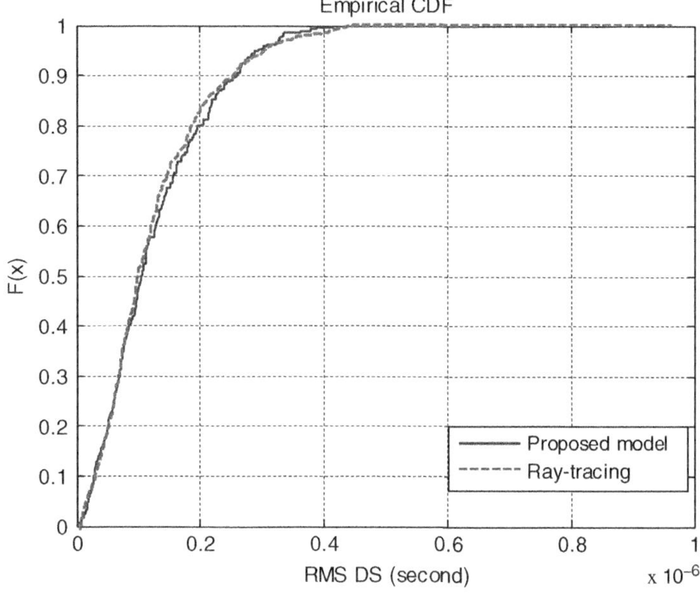

Figure 15.13 RMS DS generated by the proposed model and by ray-tracing

15.5.1 Beam Acquisition and Channel Estimation

For mmWave bands, directional beamforming both at BS and UE is essential to compensate for a large amount of path loss and, accordingly, to create a reliable radio link. Therefore, the mmWave 5G RAT is required to accommodate narrow beam-based system operation from an initial access to the data transmission, which is quite different from operation of the current 4G LTE-Advanced system in the low frequency bands. For example, cell identification and mobility measurements in the LTE or LTE-Advanced systems are based on broadcast signals such as primary/secondary synchronization or cell-specific reference signals [21], for which omnidirectional transmission or wide-beam transmission from the BS is mainly assumed. During a cell-search phase, optimal transmit (Tx) and receive (Rx) beam directions are known neither at BS nor at the UE. In order to fully realize coverage extension via Tx/Rx beamforming, transmission and reception of cell-discovery signals should properly incorporate a Tx/Rx beam-search operation, and UE and/or BS receivers may need to perform joint time, frequency and spatial synchronization. The beamforming-centric operation of the mmWave 5G RAT also makes the physical link quality very sensitive to user movements: a slight change in the propagation environments may result in misaligned Tx/Rx beam directions and significant link quality degradation. Thus, a mmWave 5G RAT design should strive to achieve fast beam acquisition and cell association, taking into account the different hardware capabilities between the BS and UE, and balancing the trade-off between beam-tracking accuracy and system overhead/power consumption.

Analog-digital hybrid beamforming [22, 23], which employs a two-step beamforming operation – radio frequency (RF) beamforming and baseband precoding – is a key enabler for the mmWave 5G RAT, as it allows multi-user multiplexing in the frequency and spatial domains and can achieve higher beamforming gains than analog-only beamforming by combining a few dominant paths, with manageable implementation complexity. Massive MIMO systems at mmWave bands are likely to have a limited number of RF chains due to hardware limitations such as the analog-to-digital converter (ADC) power consumption. Since a mmWave BS has small coverage, for example a 50–150-m cell radius, the number of users to be scheduled simultaneously by each mmWave BS may be smaller than one using massive MIMO macro-cell BSs in low-frequency bands. Thus, hybrid beamforming with the relatively smaller number of RF chains, say 4–8 RF chains for 128 antenna elements, is a proper design assumption for the mmWave BS. As to further hardware limitations at the UE side, analog-only or hybrid beamforming with four RF chains or fewer may be assumed.

The analog beamforming component in hybrid beamforming imposes unique challenges in obtaining channel state information, as well as some restrictions on multi-user scheduling and multiplexing of reference signals and data. Analog beamforming at a receiver requires repeated reference signal transmissions to allow the receiver to find the optimal analog beamforming weights, although multiple RF chains at the receiver can reduce the number of repeated transmissions. In addition, while the receiver performs Rx analog beam training, it may not be able to receive the data properly unless separate RF chains drive optimal Rx beams for data reception.

On the other hand, channel sounding with different Tx analog beams can extend the coverage. It does this through beamforming gains and higher Tx power for training symbols, exploiting multiple small power amplifiers instead of channel sounding of each antenna element. In addition, as shown in Figure 15.14, beamforming makes RMS delay spread of

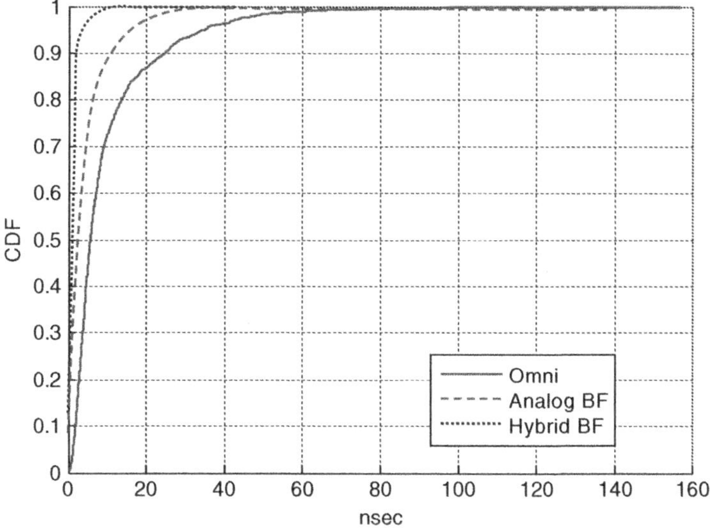

Figure 15.14 Impact of transmit beamforming on RMS delay spread with simulated 28 GHz dense urban channels

effective channels much smaller than for omni-directional channels. This can lead to lower cyclic prefix (CP) overhead in CP-based block transmission schemes and potentially to low training symbol density in the frequency-domain, if a block duration or a subcarrier spacing is properly selected. According to the results shown in Figure 15.15, multi-beam aggregation via hybrid beamforming may further reduce analog beam training overhead and latency by reducing a required analog beam codebook size. Figure 15.15 shows that similar SNRs are obtained from hybrid beamforming with four RF chains at the BS, for analog codebook sizes of 8, 16, and 64.

Among many studies related to mmWave channel estimation, Thomas and Vook present methods to estimate and provide feedback on full channel state information of 2D (elevation and azimuth) array antennas with low sounding resource overhead by transmitting 1D directive beams in the elevation domain and in the azimuth domain [24]. Alkhateeb *et al.* propose hierarchical beam-training-based channel estimation algorithms, taking into account hybrid beamforming in multi-path mmWave channels [25].

15.5.2 Cooperative Communication and Interference Handling

In addition to massive-MIMO-based beamforming, cooperative multi-point (CoMP) communications are expected to be important components of the mmWave 5G RAT. It has been shown that dense deployment of mmWave BSs can significantly reduce the outage probability and increase the fifth percentile UE throughput [26]. However, when a large number of small cells are densely deployed, frequent serving-cell changes may also occur with UE mobility. Ideal backhaul-based CoMP schemes or non-ideal backhaul-based multi-connectivity can reduce signaling overhead and service interruption due to frequent handovers, which leads

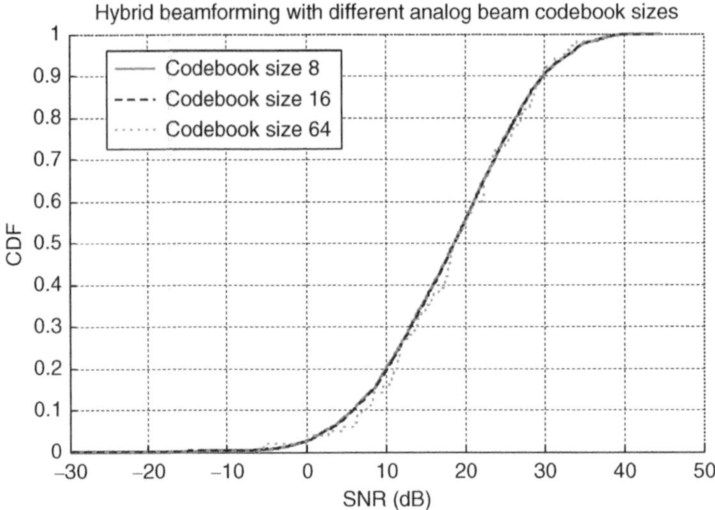

Figure 15.15 Downlink SNR CDF for single-user Tx beamforming. 28-GHz carrier frequency, 1-GHz system bandwidth, 30-dBm Tx power, 8×8 uniform planar array (UPA) and four RF chains for BS, 2×8 UPA and one RF chain for UE, BS–UE distance of 100 m

to consistent user experiences. Moreover, flexible transmission and reception-point switching and multi-point beam aggregation would help overcome intermittent connection problems caused by transient channel blockage in mmWave channels.

In order to support fast transmission- and reception-point switching, the UE may need to perform multi-link time and frequency synchronization and tracking. This suggests that reference signals suitable for multi-link tracking and corresponding advanced receiver algorithms may need to be developed. Furthermore, beam coordination among cooperative base stations based on efficient interference measurements becomes important in order to avoid strong interference from unwanted narrow beams.

15.6 Summary

This chapter described the mmWave channel for 5G and the related signal-processing techniques. It discussed the mmWave channel characteristics, the requirements for 5G mmWave channel modeling, and two mmWave channel models. Following the channel discussion, signal-processing techniques for 5G mmWave communication were discussed. When compared to centimeter-wave bands, the mmWave channel is still not well understood. More measurements need to be taken to better understand the propagation characteristics, build representative channel models and parametrize the channel model. In designing the radio access air-interface for the 5G mmWave system, challenges related to directional transmission/reception – beam acquisition, beam tracking, mobility management and blockage management – need to be addressed.

References

[1] The NGMN Alliance (2015) NGMN 5G White Paper v1.0.

[2] Rappaport, T.S., Heath, R.W. Jr. , Daniels, R., and Murdock, J. (2015) *Millimeter Wave Wireless Communications*. Pearson/Prentice-Hall.

[3] Rappaport, T.S., Sun, S., Mayzus, R., Zhao, H., Azar, Y., Wang, K., Wong, G.N., Schulz, J.K., Shamimi, M., and Gutierrez, F. (2013) Millimeter wave mobile communications for 5G cellular: It will work! *IEEE Access*, **1**, 335–345.

[4] Rappaport, T.S., Ben-Dor, E., Murdock, J.N., and Qiao, Y. (2012) 38 GHz and 60 GHz angle-dependent propagation for cellular and peer-to-peer wireless communications, in *Proc. IEEE Int. Conf. Commun.*, pp. 4568–4573.

[5] Samimi, M., Wang, K., Azar, Y., Wong, G.N., Mayzus, R., Zhao, H., Schulz, J.K., Sun, S., Gutierrez, F., and Rappaport, T.S. (2014) 28 GHz angle of arrival and angle of departure analysis for outdoor cellular communications using steerable-beam antennas in New York City, in *Proc. IEEE Veh. Technol. Conf.*, pp. 1–6.

[6] METIS Project (2014) Initial channel models based on measurements, ICT-317669-METIS/D1.2.

[7] MiWEBA Project (2014) D5.1: channel modeling and characterization, FP7-ICT 368721/D5.1.

[8] FCC (1997) Millimeter wave propagation: spectrum management implications, Bulletin Number 70, July.

[9] Degli-Esposti, V., Fuschini, F., Vitucci, E.M., and Falciasecca, G., (2007) Measurement and modeling of scattering from buildings", *IEEE Trans. Antennas Propag.*, **55** (1), pp.143–153.

[10] Tenerelli, P.A. and Bostian, C.W. (1998) Measurement of 28 GHz diffraction loss by building corners, in *Proc. IEEE PIMRC* **1998**, 3, pp. 1166–1169.

[11] Pi, Z. and Khan, F. (2011) An introduction to millimeter-wave mobile broadband systems. *IEEE Commun. Mag.*, **49** (6), 101–107.

[12] Steinbauer, M., Molisch, A.F., and Bonek, E. (2001) The double-directional radio channel. *IEEE Antennas Propag. Mag.*, **43** (4), 51–63..

[13] WINNER II, (2007) D1.1.2 V1.2: WINNER II channel models, Part I channel models, IST-4-027756/ D1.1.2v1.2.

[14] 3GPP (2014) TR36.873. Study on 3D channel model for LTE, v12.0.0.

[15] Molisch, A.F., Asplund, H., Heddergott, R., Steinbauer, M., and Zwick, T. (2006) The COST259 directional channel model - Part I: Overview and methodology, *IEEE Trans. Wireless Commun.*, **5** (12), 3421–3433.

[16] Asplund, H., Glazunov, A.A., Molisch, A.F., Pedersen, K.I., and Steinbauer, M. (2006) The COST 259 directional channel model - Part II: Macrocells. *IEEE Trans. Wireless Commun.*, **5** (12), 3434–3450.

[17] Liu, L., Oestges, C., Poutanen, J., Haneda, K., Vainikainen, P., Quitin, F., Tufvesson, F., Doncker, P.D., (2012) The COST 2100 MIMO channel model. *IEEE Wireless Commun. Mag.*, **19** (6), 92–99.

[18] Zhu, M., Eriksson, G., and Tufvesson, F. (2013) The COST 2100 channel model: Parameterization and validation based on outdoor MIMO measurements at 300 MHz, *IEEE Trans. Wireless Commun.*, **12**, 888–897.

[19] Maltsev, A. Maslennikov, R., and Sevastyanov, A. (2010) Channel models for 60 GHz WLAN Systems, IEEE document 802.11–09/0334r8. URL https://mentor.ieee.org/802.11/.

[20] Li, Q., Wu, G., and Rappaport, T.S. (2014) Channel model for millimeter-wave communications based on geometry statistics, in *Proc. IEEE GLOBECOM 2014* Workshop on Mobile Communications in Higher Frequency Bands, pp. 1–5.

[21] 3GPP (2015) TS 36.211, Physical channels and modulation (Release12).

[22] Ayach, O.E., Heath, R.W., Abu-Surra, S., Rajagopal, S., and Pi, Z. (2012) Low complexity precoding for large millimeter wave MIMO systems", in *Proc. 2012 IEEE International Conf. on Commun.*, pp. 3724–3729.

[23] Stirling-Gallacher, R.A. and Rahman, M.S. (2014) Linear MU-MIMO pre-coding algorithms for a millimeter wave communication system using hybrid beam-forming", in Proc. 2014 IEEE International Conf. on Commun., pp. 5449–5454.

[24] Thomas, T.A. and Vook, F.W., (2014) Method for obtaining full channel state information for RF beamforming, in Proc. IEEE GLOBECOM 2014 Wireless Communications Symposium, pp. 4333–4337..

[25] Alkhateeb, A., El Ayach, O., Leus, G., and Heath, R.W. Jr,, (2014) Channel estimation and hybrid precoding for millimeter wave cellular systems", *IEEE J. Select. Topics Signal Process.*, **8** (5), 831–846.

[26] Ghosh, A., Thomas, T.A., Cudak, M.C., Ratasuk, R., Moorut, P., Book, F.W., Rappaport, T.S., MacCartney, G.R., Jr.,, Sun, S., and Nie, S., (2014) Millimeter-wave enhanced local area systems: a high-data-rate approach for future wireless networks", *IEEE J. Select. Areas Commun.*, **32** (6), 1152–1163.

16

General Principles and Basic Algorithms for Full-duplex Transmission

Thomas Kaiser and Nidal Zarifeh

Signal Processing for 5G: Algorithms and Implementations, First Edition. Edited by Fa-Long Luo and Charlie Zhang.

16.1 Introduction

5G and other future wireless systems promise the user access to services that require considerably higher data rates in spite of the limited wireless spectrum. This demand for higher data rates entails achieving superior performance and efficiency in terms of wireless resources utilization. Previous generations of mobile communication mainly depended on half-duplex transmission schemes, in which the transmitted and received signals are separated either:

- in the time domain: time division duplexing (TDD), as in Figure 16.1(a)
- in the frequency domain: frequency division duplexing (FDD), as in Figure 16.1(b)
- in both: half-duplex FDD as in Figure 16.1(c).

The term "full-duplex" (FD) was traditionally used when the device had simultaneous bidirectional communication, in contrast to half-duplex" (HD), which assumed time-division duplexing. Previously, use of the term full-duplex assumed utilizing a pair of frequencies to transmit and receive simultaneously. However, in recent years the term has carried a new concept: the device can transmit and receive at the same time and over the same frequency, as Figure 16.1(d) depicts. Many papers use the term "in-band full duplex" (IBFD) to clarify this new concept [1, 2]; however, most refer to it by an abbreviated version: "full duplex", which we will adopt in this chapter as well.

It is intuitive that enabling wireless devices to use full-duplex offers the potential to double the spectral efficiency (bit/second/Hz), considering that traditional approaches for increasing spectral efficiency, such as adaptive coding and modulation, multiple-input, multiple-output (MIMO) and smart antennas have almost reached their maximum limits. An additional advantage of full duplex is the improvement it provides to the reliability and flexibility of dynamic spectrum allocation in wireless systems, such as cognitive radio networks, either with in-band full duplex or partial band-overlap FDD systems. This will provide cheaper unpaired spectrum, which is traditionally allocated for TDD operations, and also simplify spectrum management. Furthermore, full-duplex technology will enable the small cells in 5G to reuse radio resources simultaneously for access and backhaul. In addition, full duplex can be a potential solution for other wireless problems, which are expected to be solved in next generation systems:

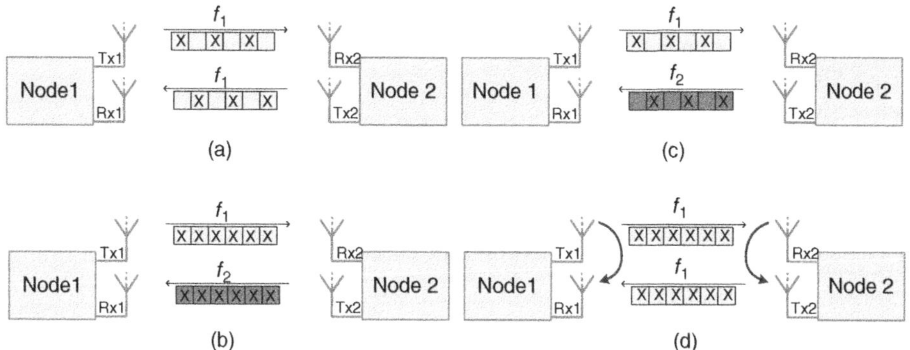

Figure 16.1 Duplex schemes: (a) time division duplex; (b) frequency division duplex; (c) half duplex FDD; (d) in-band full duplex or any division duplex with self-interference

- hidden terminals
- congestion
- collision
- excessive latency in sensitive applications and multi-hop networks.

For these reasons, full duplex presents an interesting and promising topic of research for next-generation mobile technology in different application scenarios: full duplex bidirectional, full duplex relaying (FDR) and backhauling. More discussions about full-duplex applications and possibilities in 5G can be found in the literature [3–6].

As shown in Figure 16.1(d), the major challenge to the implementation of full duplex is the self-interference (SI) signal: the part of transmitted signal that leaks into the receiver chain. With the desired signal from the remote node being weak, high-power SI is a major issue to the receiver. Nevertheless, the trend towards using short-range cells in new wireless networks provides an inherent capability to better manage the SI issue due to the lower cell-edge path loss, compared to the case in large cells [1].

A direct and simple question may be posed: in principle, the receiver knows the transmitted signal from its own transmitter, which is causing the self-interference, so wouldn't be easy to subtract the self-interference from the total received signal to cancel it out? The direct and simple answer to this question is: "No"; on the contrary it is challenging, because the assumption of receiver knowledge of the transmitted signal is inaccurate. As a matter of fact, although the transmitted digital samples are fully known, the signal that reaches the end of transmitter chain is quite different from its original baseband one, due to the multistage process it undergoes (digital-to-analog conversion, upconversion, amplification, and all other imperfect analog components), not to mention the effect of the SI channel between Tx and Rx chains. This situation has pushed researchers to explore new and advanced techniques to address it.

This chapter studies full-duplex technology from the perspective of signal-processing algorithms and implementation. It explains full-duplex system requirements, self-interference, and self-interference cancellation (SIC) techniques and the related algorithms and implementation challenges. It also shows the current achievements, from SISO, in order to treat the problem from the basics, to MIMO, which still needs further development.

The rest of this chapter is organized as follows. In Section 16.2, an analysis of the SI problem and a basic model of SI are shown. In Section 16.3, SIC techniques in a full-duplex SISO system are classified into four main categories:

- passive SI suppression in the propagation domain
- active SIC in the analog domain
- active SIC in the digital domain
- auxiliary chain SIC.

Hardware impairments and implementation challenges are covered in Section 16.4 and a full-duplex MIMO system is discussed in Section 16.5. Finally some conclusions are drawn and the outlook for full duplex in 5G networks are considered in Section 16.6.

16.2 Self-interference: Basic Analyses and Models

The main challenge for full duplex is self-interference and how to manage and suppress it. Self-interference was studied earlier in radar applications; the term "transmitter leakage" was

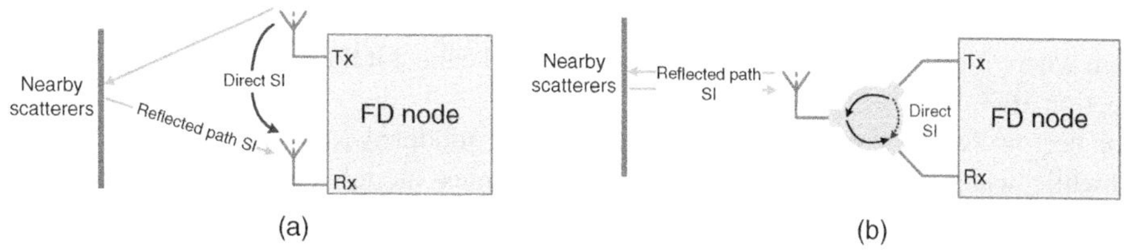

Figure 16.2 Two cases of antenna for Tx/Rx full-duplex node

used to describe the signal that leaks from the device transmitter to its own receiver. Generally, the transmitter signal is about 100 dB higher than the expected desired received signal. A considerable part of this transmitted signal leaks into the receiver chain, causing serious issues in decoding the desired signal, which could be considered noisy, with a dramatically affected signal-to-noise ratio (SNR). To achieve the best performance of a full-duplex system, the SI signal has to be suppressed to reach the receiver's noise floor. For example, when the transmitted downlink power from an LTE cell is 20 dBm, and the noise floor of the eNodeB receiver is −104 dBm, this means that about 124 dB of cancellation has to be achieved in order to completely cancel out the SI signal. This is of course the worst case; generally the difference between transmitted power and the receiver noise floor is between 90–110 dB. It is important to mention that these cancellation values cannot be achieved by one technique of cancellation, so hybrid methods are used in order to meet the cancellation requirement.

Figure 16.2 depicts the two cases of antenna setup in wireless communication transceiver: the first uses two separated antenna, one for transmission and another for reception, while the second uses one shared antenna for transmission and reception, with a circulator used to separate the signals. In both cases the SI signal consists of two components: the first is caused by the direct path/circulator leakage (and we may add here the mismatch between the transmission-line impedance and the antenna's input impedance), while the other component is caused by reflection paths. The direct and reflected paths of the SI channel may be considered as Rician and Rayleigh channels respectively [7].

Another classification of self-interference is presented in work by Bharadia *et al.* [8], which lists the components as follows:

Linear Components

This corresponds to the carrier itself, which is attenuated and reflected from the environment. The received distortion can be written as a linear combination of different delayed copies of the original carrier.

Non-linear Components

These components are created because the imperfect radio circuits take a signal x as an input and create outputs that contain nonlinear cubic and higher-order terms, such as x^3 and x^5. These higher-order signal terms have significant frequency content at frequencies close to the transmitted frequencies, which directly correspond to all the other harmonics [8].

Transmitter Noise

Transmitter noise is also referred to as broadband noise [9]. It appears as an increase – about 50 dB above receiver noise floor [8] – in the base signal level on the sides of the carrier at the receiver. It is by definition noise and is random, so the only way to cancel it is to get a copy of it from where it is generated in the analog domain and cancel it there [8]. This will be covered in Section 16.3.

For narrow band systems, the SI channel can be modeled as gain and delay functions, meanwhile wideband systems require a more complex model, because the reflected-path self-interference channel is often frequency-selective as a result of multipath propagation. In general we may build a basic equivalent baseband model in the digital domain that is valid for both the narrow- and wideband cases:

$$r(n) = r_d(n) + i(n) + w_r(n), \quad i(n) = r_{DSI}(n) + r_{SSI}(n) \tag{16.1}$$

where $r(n)$ is the total received complex baseband samples. $r_d(n)$ is the desired signal from the remote node. $r_{DSI}(n)$ are the complex samples caused by the direct self-interference component signal between the Tx and Rx antennas in case of two antennas, or leaked signal in the circulator in case of one antenna. $r_{SSI}(n)$ are the complex samples caused by scattered self-interference components, and $w_r(n)$ is additive white Gaussian noise.

Both direct and scattered SI can be represented in detail as a combination of linear and nonlinear components. The suppression in the propagation domain can mitigate both linear and nonlinear SI at the same time and with the same isolation value. Meanwhile, techniques in the analog and digital domains have different cancellation performances for the two components.

16.3 SIC Techniques and Algorithms

The simple model above shows clearly that the full-duplex system can reach maximum efficiency only when the SI signal $i(n)$ is suppressed to reach its own receiver's noise floor. The required cancellation has to meet certain requirements related to the system specifications, such as the full scale of the analog-to-digital converter (ADC) and noise-floor level. Generally, self-interference cancellation is implemented in three domains: propagation, analog and digital. None of these domains can meet the required cancellation value per se, so hybrid solutions are proposed in the literature. This section explains the requirements of SIC and how to achieve it in the three domains.

16.3.1 SIC Requirements

The primary role of SIC in the propagation and analog domains is to avoid the saturation of the receiver due to the high power of the SI signal; this power exceeds the ADC dynamic range and limits its precision after conversion because the desired signal is much weaker than the SI. Thus the required cancellation before the low-noise amplifier has to be sufficient to prevent such effects. Detailed analysis and calculations of the ADC and the linearity challenges of full-duplex system are found elsehwere [10]. To specify mathematically the SIC requirements for a full-duplex system, the dependencies among full-duplex transceiver specifications are illustrated in Figure 16.3.

We can classify the specifications of full-duplex system into three categories (see Table 16.1). While the residual SI power P_{RSI} is higher than the receiver noise floor level, the signal to self-interference plus noise ratio (SSINR) in a full-duplex system

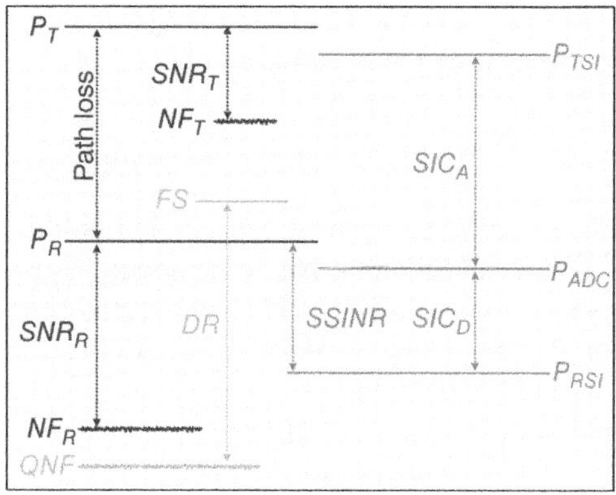

Figure 16.3 Dependencies of full-duplex system power levels

Table 16.1 Specifications of a full-duplex system

Main specifications	
P_T	Transmitter power level from the remote node
NF_T	Transmitter noise floor
SNR_T	Signal to noise ratio in the transmitter
P_R	Power level of the desired received signal from the remote node
NF_R	Receiver noise floor
SNR_R	Signal-to-noise ratio in the receiver
Analog-to-digital converter specifications	
FS	Full scale level of receiver ADC
QNF	Quantization noise floor; practically 6 dB (1 bit) below receiver noise floor [1].
DR	Dynamic range of ADC
Self-interference specifications	
P_{TSI}	Transmitted power from the own transmitter that causes SI
SIC_A	Self-interference cancellation capability in propagation and analog domains
P_{ADC}	Residual SI signal power after SIC before the ADC
SIC_D	SIC achieved in digital domain
P_{RSI}	Residual SI signal power after all cancellation operations

is lower than the SNR of a half-duplex system receiver. This means that the maximum efficiency of full duplex cannot be achieved. To clarify, a numeric example is now shown. Let $P_T = 0$ dBm for user equipment working in a Femto base station, with working frequency 2.6 GHz and line of sight (LoS) distance 50 m. The path loss is then about 75 dB in free space calculations. NF_R is -104 dBm (with bandwidth 10 MHz), then:

$$P_R = 0 \text{ dBm} - 75 \text{ dB} = -75 \text{ dBm}$$

$$\Rightarrow SNR_R = P_R - NF_R = -75 - (-104) = 29 \text{ dB}$$

Assuming: $P_{TSI} = 20$ dBm (as a Femto base station), $QNF = -110$ dBm, $DR = 60$ dB as a practical value for ADC based on the effective number of bits, then $FS = -110$ dBm + 60 dB = −50 dBm

$$P_{ADC} = P_{TSI} - SIC_A = 20 \text{ dBm} - SIC_A < -50 \text{ dBm} \Rightarrow SIC_A > 70 \text{ dBm}$$

This means that, in this example, the sum of cancellation before ADC (i.e. propagation domain suppression and analog cancellation) has to be about 70 dB or more to avoid ADC saturation. For instance, in the work by Jain *et al.* [11], with certain conditions, isolation of the antenna in the propagation domain achieves between 20 and 30 dB cancellation, and analog cancellation achieves between 20 and 45 dB.

Considering an example with $SIC_D = 30$ dBm, the residual SI power will be $P_{TSI} = P_{ADC} - SIC_D = -50 - 30 = -80$ dBm, so $SSINR = P_R - P_{RSI} = -75 - (-80) =$ 5 dB instead of $SNR_R = 29$ dB in a half-duplex system. The calculated minimum requirement $SIC_A = 70$ dBm means that the remaining cancellation to bring the residual SI signal below noise floor has to be performed in the digital domain SIC_D. SIC_D has to exceed $-50 - (-104) = 54$ dB in order to reach the ultimate performance of a full-duplex system, otherwise a comparison in performance between full-duplex system and half-duplex system should take place to evaluate the effectiveness of applying full duplex. Similar calculations have been done for WiFi systems [8]. The easier assumptions – receiver noise floor $NF_R = -90$ dBm and higher full-scale level of receiver ADC $FS = -30$ dBm – meant that the solution met the SIC requirement for WiFi systems.

16.3.2 SIC Technique Categories

Most SIC papers and the results of full-duplex testbeds are for SISO systems, and this chapter will similarly mainly analyze SIC and its properties for the SISO case, after which SIC techniques in MIMO systems will be discussed. Generally, SIC techniques can be classified into four main categories as illustrated in Figure 16.4:

- passive SI suppression in the propagation domain: conditional placement, directivity, polarization and shielding;
- active SIC in the analog domain: tapping the transmitted analog signal and feeding it with a negative sign to the receiver;
- active SIC in the digital domain: replication of the transmission samples and feeding it with a negative sign to the receiver;
- auxiliary chain SIC: a hybrid technique of the two previous methods, where replication and cancellation domains are different – one is analog and the other is digital – and using additional ADC/DAC as will be shown later.

In general SIC solutions are a combination of several techniques in order to meet the requirements of the system in question. Figure 16.5 gives an example, showing the average performance value in each domain.

16.3.3 Passive Suppression in the Propagation Domain

Passive SI suppression is defined as the signal power attenuation imposed by the path loss due to the physical separation between transmitting and receiving antennas of the same device [4].

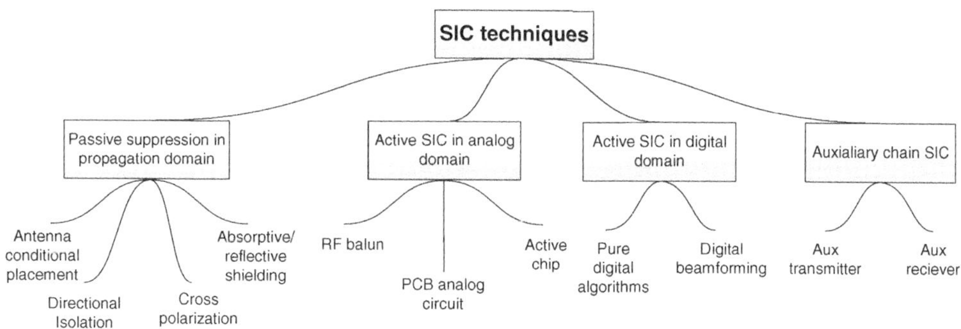

Figure 16.4 Categories of SIC technique

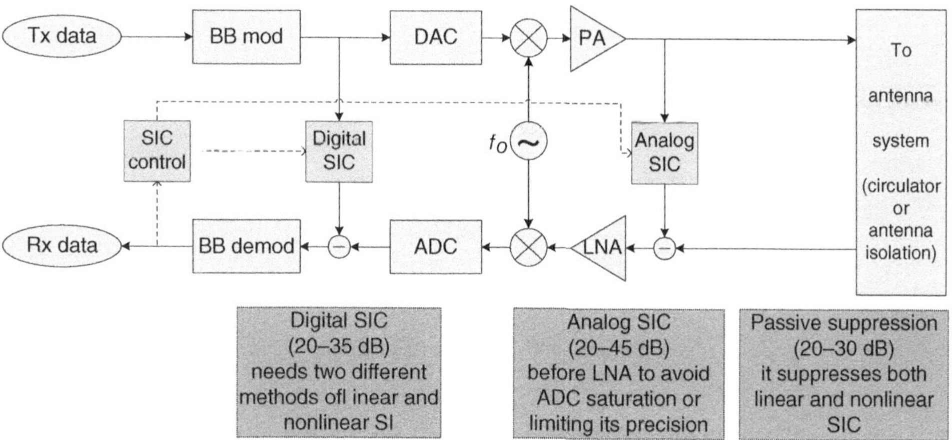

Figure 16.5 Example of hybrid SIC solution with performance average

In case of a shared antenna system, the suppression is done using a three-port RF circulator, as the ferrite within the device can be considered as a propagation domain [1]. Achievable isolation by the circulator is from 15 up to 30 dB [3, 8, 12–14], and in the case of wideband operation the maximum value would decrease [15–17].

In a separate-antenna system several SIC techniques can be used.

Antenna Conditional Placement [18–24]

As shown in Figure 16.6, the two transmit antennas are placed at distances d and $d + \lambda/2$ away from the receive antenna. Offsetting the two transmitters by half a wavelength causes their signals to cancel one another [22]. For narrowband signals, this technique is proved experimentally to be sufficiently robust, however, the suppression performance dramatically falls in case of wide band signals. Further details can be found in the literature.

Directional Isolation [25, 26]

Figure 16.7(a) illustrates how directionality isolates the receiving antenna from the interfering signals of the transmitting antenna. This technique is useful for the FDR scenario, in which

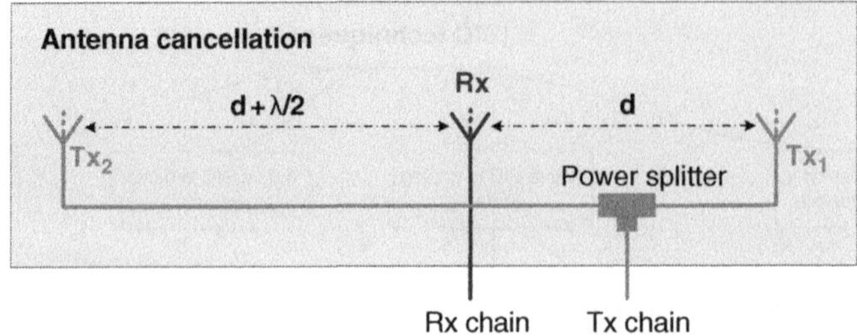

Figure 16.6 Example of antenna conditional placement

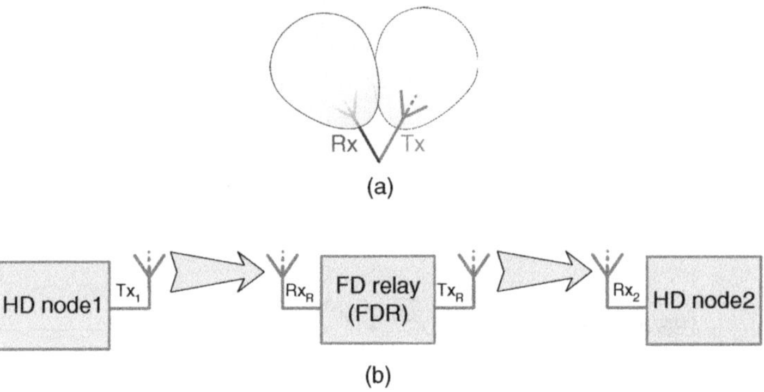

Figure 16.7 Directional isolation

the receiving and transmitting directions are generally separated from each other, as shown in Figure 16.7(b). Such an approach would not work for point-to-point full-duplex scenarios.

Absorptive/Reflective Shielding [19, 21, 25]

Electromagnetic shielding with copper or aluminum plates can enhance the isolation between antennas. However, one disadvantage is that the shielding affects the far-field coverage patterns because it prevents the antenna from transmitting to/receiving from the shielding direction. It is therefore relevant to the case of a directional antenna, as shown in Figure 16.8. Absorptive shielding is preferred on the reflective shielding plates, as the latter would couple with the transmit antenna and subsequently cause another component of self-interference. Experiments show that the absorption technique can achieve about 10 dB of isolation for 2.4 GHz [25]; intuitively, this value depends on specifications of the absorption plate, signal frequency and the surrounding environment.

Cross Polarization [23, 25–27]

Self-interference can be mitigated using orthogonal polarization, as shown in Figure 16.9. Achieving a low polarization match factor between the two antennas increases the isolation from about 10 to 20 dB in an anechoic chamber and 6 to 9 dB in a reflective room for 2.4 GHz [25].

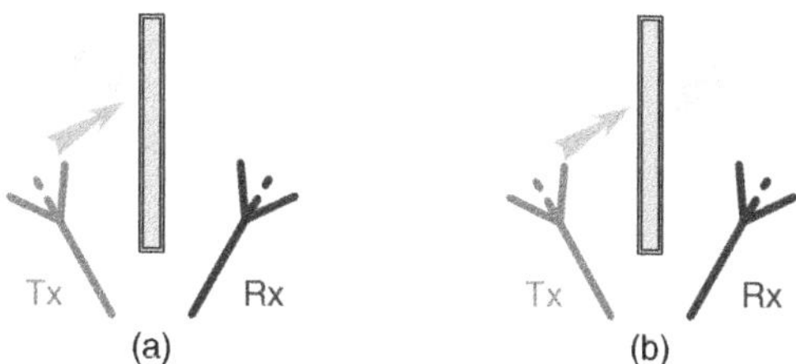

Figure 16.8 Antenna Shielding: (a) reflective; (b) absorptive

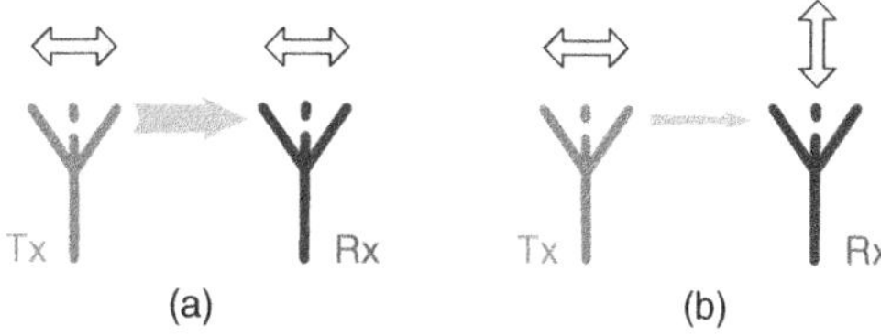

Figure 16.9 Cross polarization isolation: (a) co-polarization; (b) cross-polarization

Experiments in the propagation domain show, in optimal conditions, that up to 65 dB of SI can be suppressed with omnidirectional antennas [18, 26], and up to 72 dB with directional antennas when implementing multi-suppression techniques to achieve higher suppression performance [20, 25]. This suppression applies to the entire signal, including linear and nonlinear components as well as transmitter noise, since it is pure RF signal attenuation [8]. Although passive SI suppression techniques are appealing for reasons of their simplicity, they are highly sensitive to the wireless environment and its reflected paths, which cannot be known during the design. Moreoever, their effectiveness is greatly limited by the device form-factor: the smaller the device, the less room there is to implement such techniques [1].

16.3.4 Active SIC in the Analog Domain

Active SIC techniques in the analog domain adopt the following methodology: a replica of the transmission signal is created and then adjusted to match the SI channel, making the replica as similar to the SI signal as much as possible, in order to subtract it from the total received signal. This copy can be created either in the analog domain as in this subsection or in the digital domain before the DAC, as described in the next subsection. The SIC signal stays in the same domain from where it was copied; thus no additional ADC/DAC is required. Replication of the transmission signal in the analog domain can be achieved by tapping the Tx chain [8], using a power splitter [7], or using a balun (balanced–unbalanced) circuit in the case of two separate antennas [11]. Figure 16.10 illustrates the balun cancellation block diagram. Experiments show the practical benefits of the balun approach over using a phase shifter, notably the flatter response within a wide frequency band [11].

After creating an exact negative replication of the signal (RF reference signal) from the inverter, the replica is adjusted by delay and attenuation elements to match the self-interference. Adjusting the signal is often achieved using a noise-cancellation active chip Quellan QHx220 [11, 22, 28]; this is shown in Figure 16.11. The chip takes the input

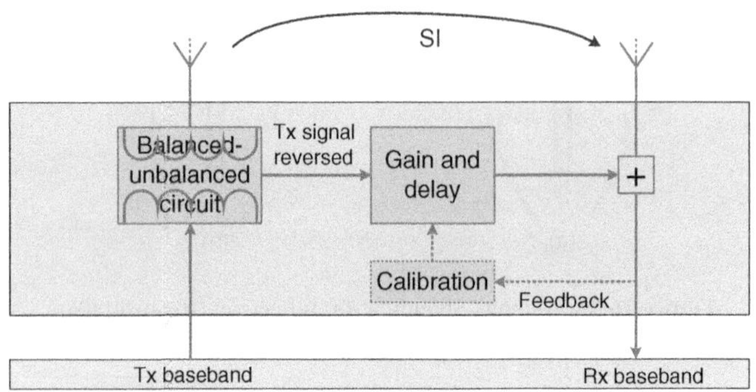

Figure 16.10 Balun cancellation block diagram [11]

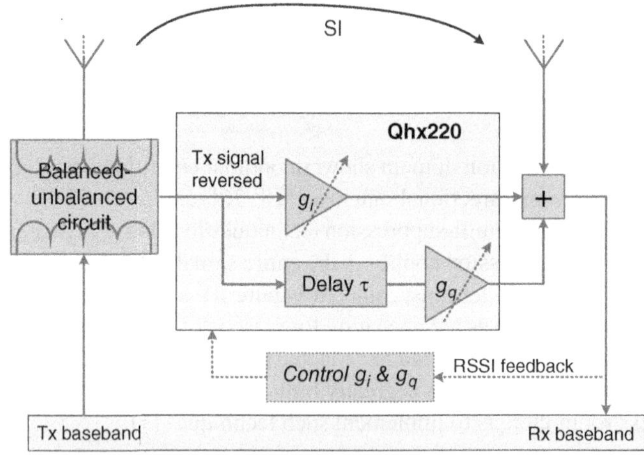

Figure 16.11 Block diagram of a full-duplex system with balun active cancellation. The RSSI values represent the energy remaining after cancellation [11]

signal from the balun circuit, and separates it into in-phase and quadrature components. A fixed delay is applied to the quadrature component; meanwhile any variable delay can be achieved by controlling the gains of in-phase and quadrature components. Adding this adjusted reference signal to the total received signal from the Rx antenna will partially cancel out the SI signal. The cancellation achieved by this method is limited to about 25 dB since it is very sensitive, and it requires a precise programmable delay with resolution as precise as 10 picoseconds, which is extremely challenging to build in practice [8]. One appealing aspect of method, however, is the inexpensive cost of the QuellanQHx220 and similar chips, although it cannot perform properly with wideband signals, where the SI channel cannot be simply modeled as a complex gain and delay between the Tx and Rx chains. The fixed delay and controlling steps of the gains show limitations in the case of wideband operations. Moreover, balun and QHx220 active chips produce nonlinear behavior that affects the performance of digital SIC, which assumes the SI channel is a linear time-invariant system [11].

Another method for adjusting the RF reference signal has been proposed [8]. While the previous method has limitations because of the fixed delay, a printed circuit board can be designed with several delay microstrip lines; each one has a different length and is connected to a tunable attenuator. A linear combination of eight or sixteen adjusted replicas of the transmission signal can approximately build one SIC signal that can mitigate a considerable portion of the self-interference [8]. The challenge here is to find an optimization algorithm for the variable attenuators. One example uses the Nyquist theorem to deal with SIC as a sampling and interpolation problem [8]; the cancellation achieved for a WiFi wideband signal with eight and sixteen delay tabs was about 30 and 48 dB respectively without considering the circulator isolation.

Nevertheless, it is worth mentioning that for a wideband signal the *direct path SI* component can be mitigated using the same analog techniques described above. This is due to the fact that the SIC direct path channel can be modeled as gain and delay functions with flat frequency behavior in the case that was considered during the antenna design. Meanwhile the *scattered SI* components are frequency selective, and therefore they require an adaptive analog filter to cancel them out [1]. This point forms one motivation toward duplicating the transmission baseband signal so that it can be cancelled in the digital domain with adaptive digital filters and further digital signal processing (DSP) algorithms that can mitigate both linear and nonlinear components as in [26, 29, 30].

16.3.5 Active SIC in the Digital Domain

Assuming that SIC in the propagation and analog domains ensures the minimum required cancellation that prevents ADC saturation, digital cancellation aims to cancel the residual SI. This opens the door to apply advanced DSP algorithms to process the SI signal and cancel both linear and nonlinear components. In this subsection linear SI cancellation is considered, and then nonlinear SI cancellation is examined when discussing hardware impairment and implementation challenges.

Earlier work on linear SIC in the digital domain proposed the same techniques that are used to digitally cancel the normal interference from an unwanted transmitter by solving the hidden node problem [22], when the desired packet is collided with another packet from the unwanted transmitter [31–33]. Firstly the receiver decodes the unwanted packet, reconstructs it and then subtracts it from the originally received collided signal. For SI, a correlation operation is performed without the need for decoding because the unwanted packet – its own transmitted samples – is already known to the receiver [22]. The correlation between the received signal and the clean transmitted signal is needed to detect the peaks that give the path delay of the SI channel. Experimentally this method cannot achieve more than 10 dB of cancellation, due to system nonlinearity, jitter and the hardware limitations that are discussed later in this chapter. More advanced techniques exist for SI channel estimation and cancellation [8, 11].

16.3.5.1 Baseband Equivalent Model

Full duplex with linear SI may be modeled in digital baseband as follows [8]. The SI channel $h_s(n)$ is considered a non-causal linear system that has the known preamble signal $x_{pr}(n)$. Any SI received sample $y(n)$ is modeled as a linear combination of transmitted samples x_{pr}

before and after the instant n. The non-causality assumption is possible, as all IQ baseband samples x_{pr} are known.

$$y(n) = x(n-k)h_s(k) + x(n-k+1)h_s(k-1) + \ldots + x(n+k-1)h_s(-k+1) + w(n) \quad (16.2)$$

where $w(n)$ is the receiver noise. To find $h_s(k)$, the above equation can be expressed as: $y = Ah_s + w$ where A is Toeplitz matrix of $x_{pr}(n)$:

$$A = \begin{bmatrix} x_{pr}(-k) & \cdots & x_{pr}(0) & \cdots & x_{pr}(k-1) \\ \vdots & & \vdots & & \vdots \\ x_{pr}(n-k) & \cdots & x_{pr}(n) & \cdots & x_{pr}(n+k-1) \end{bmatrix} \quad (16.3)$$

The problem is to find a maximum likelihood estimate of the vector h, which means $minimize\|y - Ah_s\|^2$ as y and A are already known. Using a convex optimization algorithm [34], computation of h_s coefficients is possible by multiplying each received sample of the preamble by the relevant column of the pseudo-inverse of A matrix $a_i^\dagger$

$$\hat{h}_s = \sum (y_i . a_i^\dagger) \quad (16.4)$$

The residual signal after subtraction of the SI signal $\hat{i}(n)$ from the total received signal $r(n)$ is:

$$r_{rs}(n) = r(n) - \hat{i}(n) = r_d(n) + \sum_{k=0}^{N-1} [h_s(k) - \hat{h}_s(k)]x(n-k) + z(n) \quad (16.5)$$

where $r_d(n)$ is the desired signal, and $z(n)$ is the noise after SIC.

This method is robust to noise as it includes the white noise in the channel estimation algorithm. However, an important consideration is the coherence time of the SI channel. Many other SIC algorithms in the digital domain have been proposed for MIMO systems; these will be tackled later on in this chapter.

16.3.6 Auxiliary Chain SIC

This method is one of the active SIC techniques [12, 18, 35]. It copies the baseband IQ samples of the transmitted signal in the digital domain, then it uses an additional transmitter chain (e.g. DAC, LPF, upconverter, PA) to generate the SIC signal and feed it back into the receiver in order to be subtracted from the total received signal. The SIC signal has to be adjusted (pre-distorted) in the digital domain before the DAC in order to match the transmitted signal through the SI channel.

Similar work proposes an auxiliary receiver chain instead of auxiliary transmitter chain [7]. The transmitted signal is tapped in the analog domain just before the antenna, and then fed back to the receiver digital domain as in Figure 16.12. However, like the auxiliary transmitter method, the SIC signal is also adjusted in the digital domain to exploit DSP algorithms. The auxiliary receiver chain method mitigates the effects on SIC of transmitter hardware impairments such as phase noise and nonlinearities. A common oscillator for the chains – ordinary and auxiliary – is used to mitigate the phase-noise effect on the SIC signal.

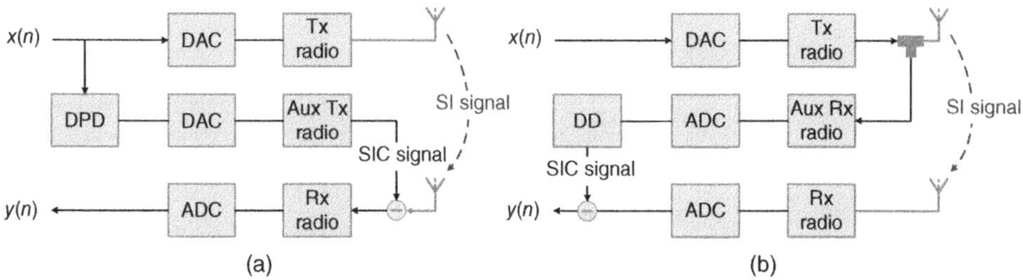

Figure 16.12 Auxiliary chain SIC: (a) auxiliary Tx; (b) auxiliary Rx

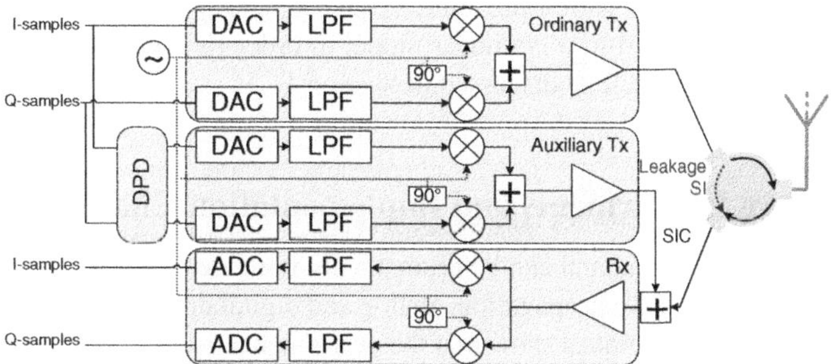

Figure 16.13 Structure of FD transceiver with auxiliary Tx SIC [12]

Figures 16.13 and 16.14 show the full-duplex transceiver structure with auxiliary transmitter [12, 36], and its system model respectively, with the following variables defined:

$h_{ord}(n)$	The equivalent baseband channel between the generated waveform from DAC and the RF power amplifier (PA) output at the ordinary transmission chain
$h_{aux}(n)$	The equivalent baseband channel between the generated waveform from DAC and the PA output at the auxiliary transmission chain
$h_{si}(n)$	The equivalent baseband channel of the SI radio channel
$h_{rx}(n)$	The equivalent baseband channel of the receiving chain
$s(n), r(n)$	The transmitted and the received complex baseband samples, respectively

The received baseband signal consists of the desired reception signal $y_d(n)$, the residual self-interference $i_{rsi}(n)$ and the noise $w_r(n)$. Assuming that the two transmitter chains are identical – in other words $h_{ord}(n) = h_{aux}(n)$ – then a digital predistortion function (DPD) has to emulate the estimated SI channel $\hat{h}_{si}(n)$. Then the residual SI in its baseband form is given as:

$$i_{rsi}(n) = [h_{si}(n) - \hat{h}_{si}(n)] * h_{rx}(n) * h_{ord}(n) * s(n) \tag{16.6}$$

This clearly demonstrates that the cancellation performance depends on minimizing the error in SI channel estimation, so a high cancellation value requires an accurate DPD model.

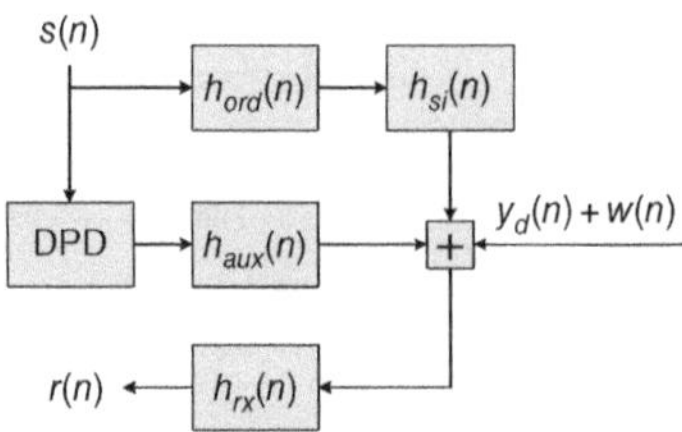

Figure 16.14 System model of FD transceiver with auxiliary Tx SIC

The DPD model is calculated with a preamble sequence, and it can be either linear as previously explained, or extended to a nonlinear model in order to handle the nonlinearities of hardware components. This will be discussed in Section 16.4.

16.4 Hardware Impairments and Implementation Challenges

In principle, a high-power SI signal can be dramatically mitigated by using a combination of the above SIC techniques in the propagation, analog and digital domains. Experiments during implementation of these techniques show that the performance is limited by hardware imperfections in the analog domain, for example nonlinearities, phase noise, and IQ mismatch (in case of direct conversion). These impairments have dissimilar effects of degradation on performance for different SIC techniques. The main impairments are discussed here, with one detailed example regarding IQ imbalance calibration to show how to deal with such imperfections in full-duplex systems.

16.4.1 Non-linear SIC

Experiments show that nonlinear components of SI can be 80 dB higher than the receiver noise floor [8, 30]. A major part may be eliminated along with the linear SI by SIC techniques in the analog and propagation domains, but the residual nonlinear SIC in the digital domain, which can be about 10 to 20 dB [8, 30], needs to be cancelled too. Normally, nonlinear SIC methods are added to the linear methods to achieve optimal performance. The general model to approximate the nonlinear function uses a Taylor series, so the output transmitted signal can be written as [8]:

$$y(n) = \sum_m a_m x_p^m(n) \tag{16.7}$$

where $x_p(n)$ is the ideal passband analog signal for the digital representation of $x(n)$. The Volterra series has been used in order to capture the memory effect [12].

It can be shown that for practical wireless systems only the odd orders of the polynomial contribute to the in-band distortion [37]. Furthermore, only a limited number of odd orders contribute to the distortion, and higher orders can be neglected [29]. In practical systems, the nonlinearity is typically characterized by the third-order intercept point, which is defined as the point at which the power of the third harmonic is equal to the power of the first harmonic [38].

Therefore, the above model can be simplified to [29]:

$$y = x + \alpha_3 x^3 \tag{16.8}$$

assuming a unity linear gain $\alpha_1 = 1$. The second term of the sum is presenting the nonlinear SI component that is transmitted, and it will cross the SI channel then suffer again from receiver nonlinearity. Total nonlinear SI can therefore be written as:

$$d = \underbrace{\alpha_3^t (x_{si})^3 * h_{si}}_{\text{Transmitter nonlinearity}} + \underbrace{\alpha_3^r (x_{si} * h_{si} + \alpha_3^t (x_{si})^3 * h_{si})^3}_{\text{Receiver nonlinearity}} \tag{16.9}$$

where α_3^t and α_3^r are the transmitter and receiver third-order nonlinearity coefficients. Expanding the above equation and neglecting the orders higher than 3 will give:

$$d = \alpha_3^t \underbrace{(x_{si})^3 * h_{si}}_{A} + \alpha_3^r \underbrace{(x_{si} * h_{si})^3}_{B} + 3\alpha_3^t \alpha_3^r \underbrace{(x_{si} * h_{si})^2 ((x_{si})^3 * h_{si})}_{C} \tag{16.10}$$

Accordingly, the main difference between transmitter and receiver nonlinearity is that the former affects only the signal, while the latter affects both the signal and the wireless channel function [29]. In the case of sending a preamble signal x_{pr}, the baseband representation of the received SI signal will be:

$$i = x_{pr} * h_{si} + d + z \tag{16.11}$$

where z is a random signal that sums up AWGN and the effects of all impairments – except nonlinearity – such as phase noise and quantization error. As i, x_{pr}, h_{si} are known to the receiver, the equation can be written:

$$i - x_{pr} * h_{si} = \hat{d} = d + z \tag{16.12}$$

and in matrix form:

$$\begin{bmatrix} \hat{d}_0 \\ \hat{d}_1 \\ \vdots \\ \hat{d}_N \end{bmatrix} = \underbrace{\begin{bmatrix} A_1 & B_1 & C_1 \\ A_2 & B_2 & C_2 \\ \vdots & \vdots & \vdots \\ A_N & B_N & C_N \end{bmatrix}}_{W} \begin{bmatrix} \alpha_3^t \\ \\ \alpha_3^r \\ \\ 3\alpha_3^t \alpha_3^r \end{bmatrix} + \begin{bmatrix} z_1 \\ z_2 \\ \vdots \\ z_N \end{bmatrix} \Rightarrow \hat{d} = W.\alpha + z \tag{16.13}$$

Using a least squares estimator, the matrix α, which represents the nonlinearity coefficient, can be found.

Simulations have shown that this scheme, with the simplifications above, can achieve a performance which is less than 0.5 dB off the performance of a linear OFDM full-duplex system [29]. However, this is only a simulation, and therefore needs to be experimentally verified, with expectation that there will be performance drop. Moreover, the complexity of the estimation algorithm has to be considered as a part of SIC process, as it may limit its performance with the shorter coherence time of the self-interference channel. Many other simulations and experiments have been conducted to study nonlinearity in a full-duplex system [2, 7, 8, 10, 12, 30, 39–41].

16.4.2 IQ Imbalance

IQ imbalance is caused by the gain and phase mismatches between I- and Q-branches of the transmitter and receiver chains. This imbalance results in the complex conjugate of the ideal signal being added on top of it, with some level of attenuation [2]. Thus the output of an imperfect IQ mixer is a transformation of an input signal $x(t)$ where both direct and conjugated signals are filtered and then summed together [39]. This is typically called widely-linear transformation [42, 43], and hence the IQ imbalance can be modeled as widely linear filters [39]. The output can be expressed as:

$$x_{IQ}(t) = g_1(t) * x(t) + g_2(t) * x^*(t) \tag{16.14}$$

where $g_1(t)$ and $g_2(t)$ are the responses for the direct and image components respectively [44]. The quality of the IQ mixer can be quantified using the image rejection ratio, which can be defined as: $IRR(f) = 10\log_{10}(\frac{|G_1(f)|^2}{|G_2(f)|^2})$. Taking the LTE-Advanced system for example, 3GPP specifications limit the minimum attenuation for the in-band image component in the user equipment transmitters to 25 or 28 dB [45]. Such image attenuation is sufficient in the transmission path, but when considering the full-duplex device self-interference problem, the IQ image of the SI signal represents an additional interference that leaks into the receiver [39].

The performance of full-duplex transceivers in the presence of IQ imbalance has been studied in detail [36, 39, 46, 47], and it has been shown that, with practical IQ image rejection ratio for a full-duplex transceiver, it is necessary to mitigate the IQ image of the SI signal. Otherwise the loss of SSINR might be of the order of tens of decibels [2], which negates the benefit of using full duplex.

Figure 16.15 shows the SIC performance in case of different gain and phase mismatch values in the auxiliary chain. Taking, for example, practical values for gain mismatch $g < 0.05$ and phase error $\phi < 1°$, the self-interference cancellation is 53 dB, and this means about 20 dB degradation of SIC performance from the best achieved cancellation. Therefore, the effect of IQ imbalance must be included in the design of the baseband cancellation signal [36, 47].

In the paper by Askar *et al.* [36], the IQ imbalance is calibrated by adjusting the cancellation signal in order to match the SI signal. This can be achieved by replicating the imbalance of the

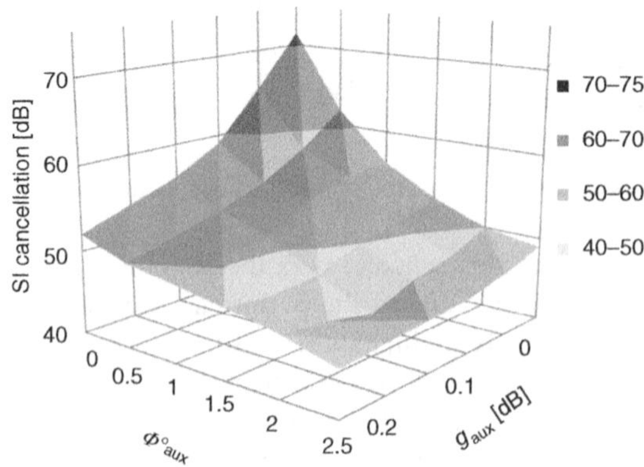

Figure 16.15 Self-interference suppression performance in presence of IQ imbalance [36]

ordinary transmitter and correcting the imbalance of the auxiliary transmitter. The direct and quadrature signals are:

$$I(t) = \cos(wt), \quad Q(t) = \sin(wt) \tag{16.15}$$

And after IQ imbalance with gain g and phase error ϕ

$$\acute{I}(t) = \cos(wt), \quad \acute{Q}(t) = g\sin(wt - \phi) \tag{16.16}$$

The above equations can be expressed in matrices:

$$\begin{bmatrix} \acute{I}(t) \\ \acute{Q}(t) \end{bmatrix} = \begin{bmatrix} 1 & 0 \\ -g\sin(\phi) & g\cos(\phi) \end{bmatrix} \begin{bmatrix} I(t) \\ Q(t) \end{bmatrix} \Rightarrow \begin{bmatrix} \acute{I}(t) \\ \acute{Q}(t) \end{bmatrix} = \begin{bmatrix} 1 & 0 \\ r_\beta & r_\alpha \end{bmatrix} \begin{bmatrix} I(t) \\ Q(t) \end{bmatrix} \tag{16.17}$$

where r_α and r_β are the replication parameters. These parameters can be used to replicate the IQ imbalance of the ordinary transmitter in the SIC signal. The correction parameters c_α and c_β can be extracted by inversion:

$$\begin{bmatrix} I_c(t) \\ Q_c(t) \end{bmatrix} = \begin{bmatrix} 1 & 0 \\ \tan(\phi) & (g\cos(\phi))^{-1} \end{bmatrix} \begin{bmatrix} \acute{I}(t) \\ \acute{Q}(t) \end{bmatrix} \Rightarrow \begin{bmatrix} I_c(t) \\ Q_c(t) \end{bmatrix} = \begin{bmatrix} 1 & 0 \\ c_\beta & c_\alpha \end{bmatrix} \begin{bmatrix} \acute{I}(t) \\ \acute{Q}(t) \end{bmatrix} \tag{16.18}$$

The parameters c_α and c_β can be used to correct the IQ imbalance of the auxiliary transmitter in the SIC signal. Before explaining further the calibration algorithm proposed by Askar *et al.* [36], let us check in Figures 16.16 and 16.17 the structure and equivalent baseband model of the full-duplex transceiver in the presence of IQ imbalance in the three chains (ignoring for now the calibration units PEU and PAU in both figures). The direct downconverter structure of the quadrature modulator is modeled as in [48].

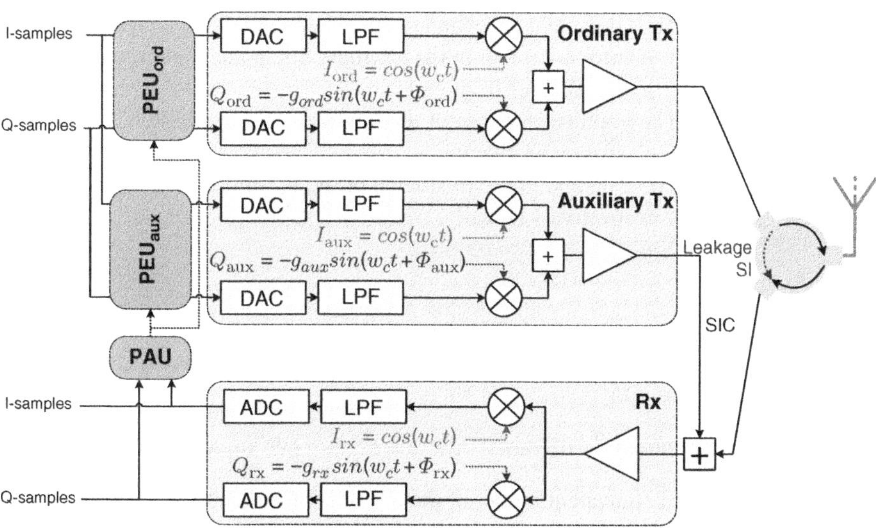

Figure 16.16 Structure of FD transceiver in presence of IQ imbalance [36]

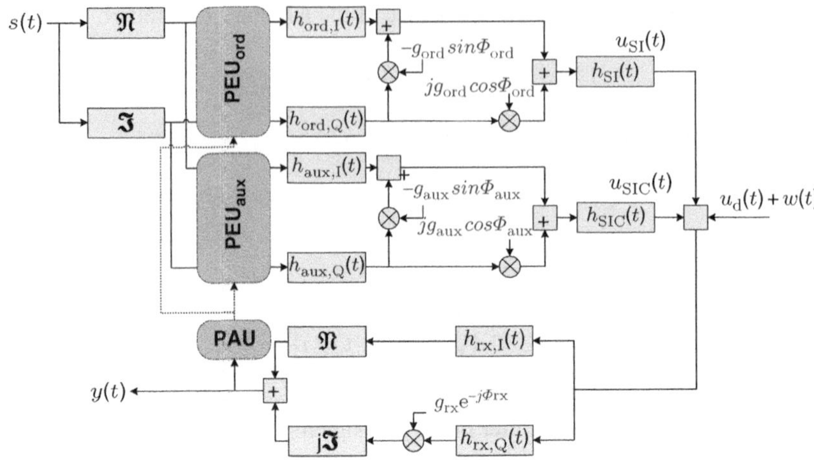

Figure 16.17 Equivalent baseband model of FD transceiver in presence of IQ imbalance. The model incorporates IQ imbalances in their equivalent baseband models [36]

The following definitions are used:

$h_{ord,I}(t)$, $h_{ord,Q}(t)$	The impulse response of the baseband filters in the ordinary transmitter at I- and Q-arm, respectively.
$h_{SI}(t)$	The equivalent baseband response of self-interference channel
$h_{aux,I}(t)$, $h_{aux,Q}(t)$	The impulse response of the baseband filters in the auxiliary transmitter at I- and Q-arm, respectively.
$h_{SIC}(t)$	The equivalent baseband channel between the upconverting quadrature mixer at the auxiliary chain and the directional coupler at the receiver front end.
$h_{rx,I}(t)$, $h_{rx,Q}(t)$	The impulse response of the low-pass baseband filters at I- and Q-arm respectively.
g, ϕ	The gain and phase mismatch of the IQ imbalance for each chain respectively (Ideally: $\phi_{ord} = \phi_{aux} = \phi_{rx} = 0$ and $g_{ord} = g_{aux} = g_{rx} = 1$).

The baseband equivalent model of the RF self-interference signal is

$$u_{SI}(t) = \Re\{s(t)\} * h_{ord,I}(t) * h_{SI}(t) \\ + g_{ord}(\mathcal{J} \cos\phi_{ord} - \sin\phi_{ord})\Im\{s(t)\} * h_{ord,Q} * h_{SI}(t) \tag{16.19}$$

Similarly, the baseband equivalent model of the RF self-interference cancellation signal is

$$u_{SIC}(t) = \Re\{s(t)\} * h_{aux,I}(t) * h_{SIC}(t) \\ + g_{aux}(\mathcal{J} \cos\phi_{aux} - \sin\phi_{aux})\Im\{s(t)\} * h_{aux,Q} * h_{SIC}(t) \tag{16.20}$$

Ideally, the self-interference signal $u_{SI}(t)$ and the self-interference cancellation one $u_{SIC}(t)$ have to be exactly identical in order to achieve perfect cancellation. Therefore, the IQ imbalance parameters – in addition to the SI channel – have to be estimated in order to compensate for their effect digitally, and the required SIC signal constructed. Adding the following units aims to estimate then calibrate IQ imbalance of SIC signal to match the SI signal:

PAU	Parameter acquisition unit
PEU_{ord}	Ordinary pre-equalization unit
PEU_{aux}	Auxiliary pre-equalization unit

Askar *et al.* [36] propose two algorithms for calibration: the replicator unit method and the distributed compensation units method.

Replicator Unit Method

In this case the PEU_{ord} will not be used; meanwhile the PEU_{aux} will perform the two required functions on SIC signal: to replicate the IQ imbalance of the ordinary transmitter and to correct the IQ imbalance of the auxiliary transmitter. This can be done by sending a pilot signal in order to estimate the parameters g_{ord}, ϕ_{ord}, g_{aux} and ϕ_{aux} by the PAU in the receiver, and then feed it back to the PEU_{aux}. Knowing these four parameters enables the PEU_{aux} to calculate the replication values $r_{ord,\alpha}$, $r_{ord,\beta}$ and the correction parameters $c_{aux,\alpha}$, and $c_{aux,\beta}$ as explained previously. In general, two types of IQ imbalance can be considered: frequency-independent IQ imbalance (FIIQ) and frequency-selective IQ imbalance (FSIQ). FIIQ imbalance occurs mainly due to local oscillator signal imperfections. Meanwhile FSIQ behavior is a result of baseband filter mismatches and differences in group delay between the I-arm and Q-arm. Figure 16.18 shows the replicator in both cases, where $g(t)$ is the adjustment function applied to the SIC signal to match the SI signal, and where its frequency-domain representation is $G(f) = \frac{H_{ord,I}(f)H_{SI}(f)}{H_{aux,I}(f)H_{SIC}(f)}$.

Distributed Compensation Units Method

In distributed compensation, each PEU corrects the IQ imbalance of its own branch (see Figure 16.19). As long as the IQ imbalance at the ordinary chain is completely compensated, there is no need to replicate its behavior at the auxiliary chain. Regardless of the full-duplex application, the main advantage of the distributed model is the fact that IQ imbalance of

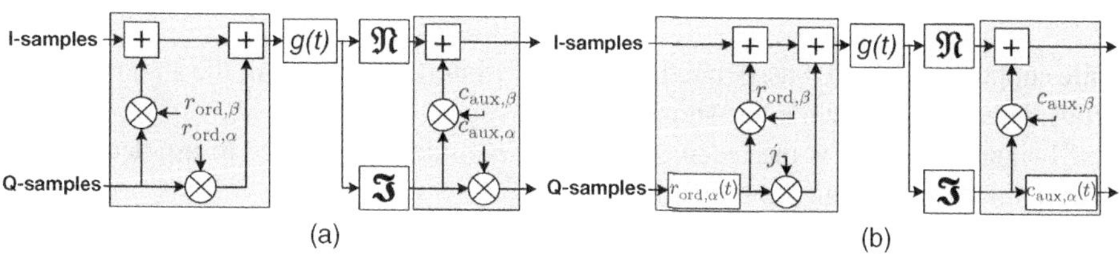

Figure 16.18 Structure of IQ imbalance replicators: (a) frequency independent; (b) frequency selective

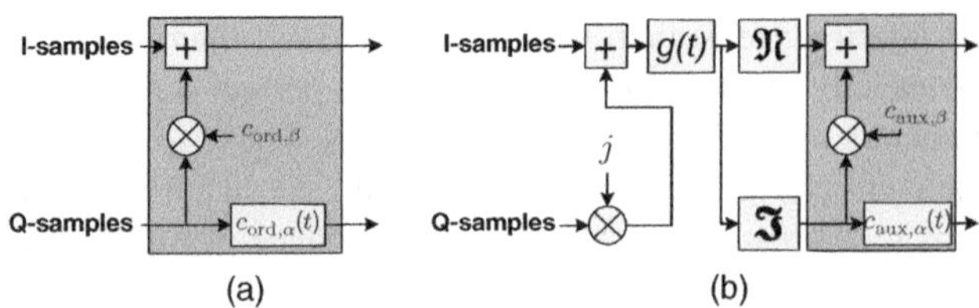

Figure 16.19 IQ imbalance compensation units: (a) in the ordinary Tx (b) in the auxiliary Tx

the ordinary transmission signal is digitally corrected, which is desirable for the remote reception node. The distributive property of this model might be considered as a disadvantage in some practical implementations where access to the digital domain of the ordinary chain is not recommended or more likely not possible [36]. Experiments on the above techniques show that the enhancement in cancellation performance is about 5 dB for the FIIQ method and about 12 dB for FSIQ and distributed compensation. The previous method – like many other works – is focusing on IQ imbalance in full-duplex systems, regardless of the impact of the other distortions. The concept of joint modeling of nonlinear distortions and IQ imbalance has been considered [49] and extended to a Volterra-series-based approach [50]. Many other studies have been conducted to study the IQ imbalance effect and calibration in full-duplex systems [39, 47, 51, 52].

16.4.3 Phase Noise

Analyses and experiments show that the oscillator phase noise is one of the main SIC challenges that limit the performance of full-duplex systems [53–56]. Earlier literature assumed that when transmitter and receiver use a common local oscillator, the level of phase noise would remain at a tolerable level [57, 58]. However, this consideration is not always valid, especially in the case of OFDM systems. The Wireless Open-access Research Platform (WARP) [58] shows that with a noise phase variance between 0.4° and 1.0°, the reduction in SIC performance is about 20–25 dB for OFDM systems (see Figure 16.20). This can be explained by the phase noise causing two effects: common phase error and intercarrier interference (ICI) [59, 60]. The former may have acceptable levels as previously assumed, but the latter stimulates an enhancement in SIC performance, which is achieved by consecutively estimating and suppressing the ICI signal.

The conventional half-duplex techniques for ICI suppression in the frequency domain [59–61] may be used in a full-duplex system with two considerations of full-duplex in mind [53]:

- While suppressing the ICI associated with the self-interference signal, the signal-of interest is considered an unknown noise signal.
- The SI signal is known at the receiver side, thus eliminating the need to use decision feedback techniques to obtain the transmitted signal.

Time-domain ICI estimation techniques [62, 63] can be modified and used in full-duplex systems [53]. A low-complexity least-squares (LS) algorithm plus filtering technique is used for

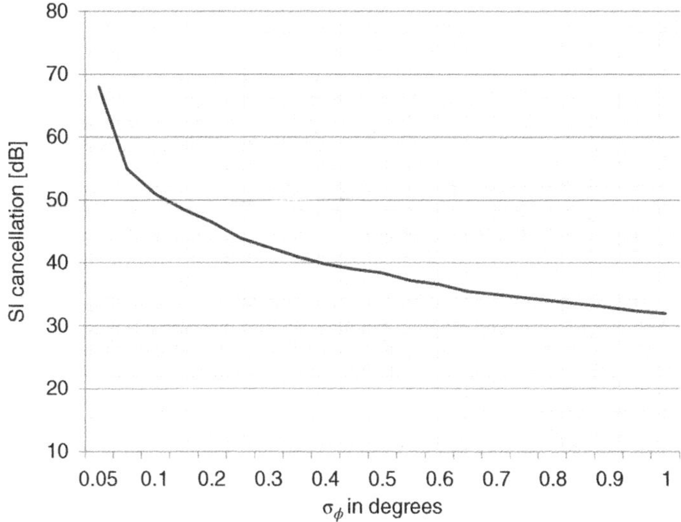

Figure 16.20 Amount of active analog cancellation as a function of the variance of phase noise [58]

ICI estimation in half-duplex systems. Despite its low complexity, using the LS algorithm in full-duplex systems has to be carefully considered, mainly due to the fact that the ICI has to be estimated in the presence of the signal of interest, which is typically higher than the ICI power in typical operating scenarios. This high power will negatively impact the LS estimator quality [53].

Other conventional estimation algorithms, such as minimum mean square error (MMSE), are used for phase-noise suppression in full-duplex systems. Assuming that the reader is familiar with such algorithms, and considering that an example of their use in the IQ imbalance problem has already been discussed, there is no need to further demonstrate their use here. The result of implementing these algorithms achieves enhancements of up to 10 dB beyond existing SIC schemes [53–56, 61].

16.5 Looking Toward Full-duplex MIMO Systems

Regarding MIMO, one may set forth the following argument: full duplex can double the capacity, but also two or more antenna (half-duplex MIMO) can do the same without all the challenges of full duplex. In fact, this would be true at first glance, and thus full-duplex MIMO may use one shared antenna for each Tx–Rx pair with a circulator, as shown in Figure 16.21, in order to strengthen its viability. However, when multiple circulators are in place, severe interference among the multiple shared antennas would occur, creating a bottleneck with respect to achieving full-duplex system feasibility [64]. Furthermore, multiple separate antennas have the advantage of exploiting the degree of freedom in the spatial domain. As more powerful SIC schemes are required to make full-duplex MIMO systems feasible, the increased degree of freedom is expected to provide full-duplex MIMO system with new solutions for SIC. However, so far convincing and practical full-duplex MIMO system designs, able to transmit and

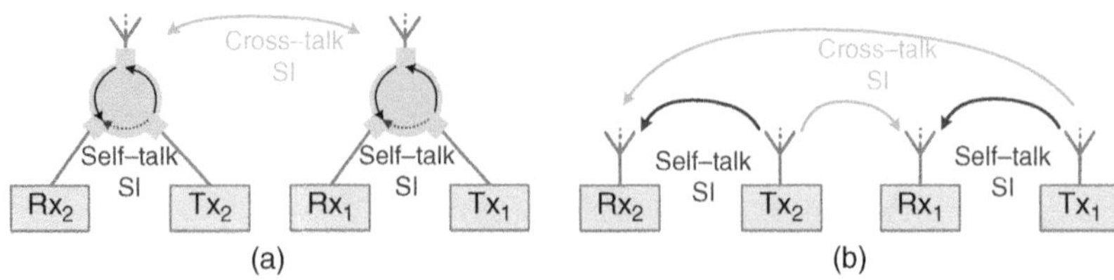

Figure 16.21 Self-interference in MIMO 2×2: (a) shared antenna; (b) separate antenna

receive from all antennas simultaneously with sufficient SIC, are yet to be realised. This section will therefore only address some of the promising proposals and have potential to evolve into dependable full-duplex MIMO systems for 5G.

16.5.1 Antenna Techniques

16.5.1.1 Antenna Conditional Placement

As shown in Section 16.3.3, this technique uses multiple antennas to deconstruct the transmitted SI signal. The same technique can be extended to the MIMO case, an example being the MIDU solution [20, 23]. MIDU performs a primary SIC in the propagation domain by employing antenna cancellation with symmetric placement of transmitting and receiving antennas (see Figure 16.22(a)). The achieved isolation is about 45 dB in open-space indoors, but the technique has limited performance of only 15 dB in indoor multipath rich environments, due to the fact that it can only mitigate the LoS SI component. The performance may be enhanced by using cross polarization as in Figure 16.22(b). In general, as in SISO, such techniques can be implemented only for narrowband systems, and they face major challenges regarding feasibility and scalability, not to mention the increasing cost of using $4N$ antennas to build an N-antenna MIMO system.

16.5.1.2 Directional Isolation

In case of full-duplex relaying, where the receiving and transmitting directions are generally separated from each other, it is possible to use MIMO to enhance the isolation (see Figure 16.7).

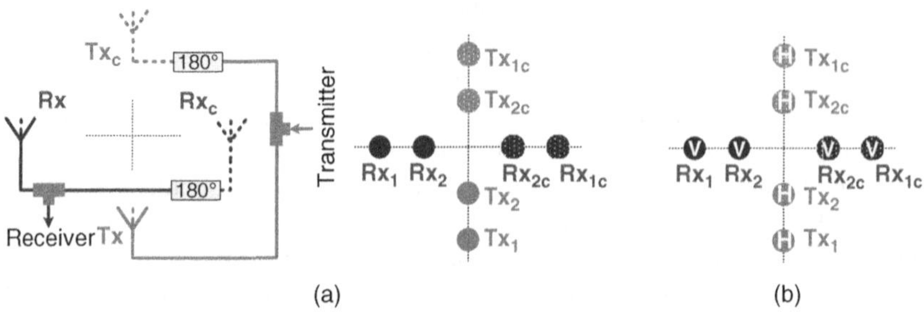

Figure 16.22 SIC in MIMO by antenna symmetric placement [23]: (a) multi cancellation level; (b) enhance with polarization

16.5.2 Cancellation in the Analog Domain

The same principle of analog SIC used in the SISO case is also used for the MIMO case [40, 65]; tapping the SI signals from N_T transmitters with N_R copies from each one and then adjusting each SIC signal in order to match the SI channel between every Tx–Rx pair ($N_T \times N_R$ pairs). The adjustment can be done by variable attenuators and phase shifters, or frequency-selective filters. In the case of 2×2 MIMO, practical implementation may seem achievable with four SI channels, but increasing numbers of antennas will make this solution impractical and expensive to implement.

16.5.3 Cancellation in the Digital Domain

The earlier literature of digital SIC in MIMO focused on full-uplex relaying applications, as the case of relay has the advantage of antenna directional isolation, which can be combined with digital beamforming. This combination can achieve considerable SIC value in case of decode-and-forward relays [66–71]. The later literature [46, 65, 72, 73] analyzes bidirectional full-duplex systems with digital schemes that depend on beamforming and phase rotation [74] to avoid SI.

Meanwhile the first mature 3×3 full-duplex MIMO system that cancels both components of the SI signal (self-talk and cross-talk) almost to the noise floor has been demonstrated [75]. The technique assumes that the co-located MIMO antennas share a similar environment since they will share the same reflectors in this environment, and the distances to these reflectors are almost the same from closely-spaced antennas. Thus it can be assumed that, for a certain transmit antenna, the cross-talk signals from other transmit antennas are similar to the self-talk signal with additional delays. Furthermore, cross-talk across chains is naturally reduced compared to the chain's own self-talk because of physical antenna separation. Such simplifying assumptions mean that cross-talk and self-talk transfer functions can be expressed as a function of each other, with a modifying factor to account for the antenna separation. This allows the system to be modeled as a cascade of transfer functions. Let $H_i(f)$ and $H_{ct}(f)$ be the transfer functions of the chain's own self-talk and cross-talk respectively. The overall relationship between these functions can be modeled as follows:

$$H_{ct}(f) = H_c(f)H_i(f) \tag{16.21}$$

where $H_c(f)$ is the cascade transfer function. The key observation is that $H_c(f)$, which cascaded with $H_i(f)$, results in the cross-talk transfer function, is a simple delay function. Further details about this work can be found in the original paper [75]. More analyses and experiments have to be conducted in order to verify the effect of such simplifying assumptions in different environments and systems.

16.5.4 Cancellation with a Auxiliary Transmitter

Scaling this technique up toward MIMO implies that the number of auxiliary transmitter chains is equal to number of receiving antennas N_R. A 2×1 MIMO OFDM system has been modeled and analyzed [26], as shown in Figure 16.23. For each receiver chain, the earlier SISO SIC technique is used, with SI channel estimation and predistortion of the signal digitally. For the

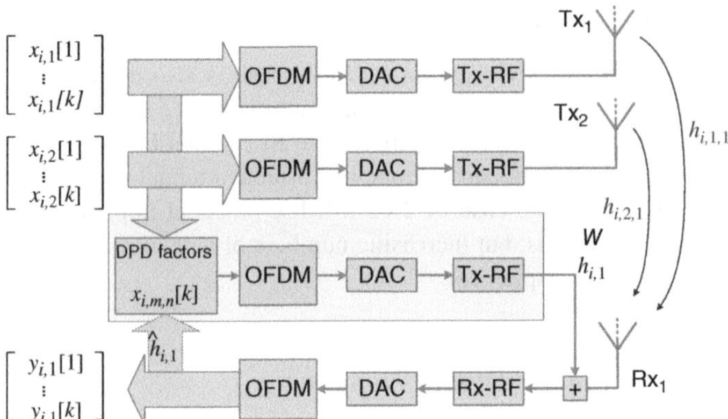

Figure 16.23 Block diagram of full-duplex MIMO 2 × 1 OFDM node using auxiliary Tx SIC. *Source:* Duarte 2014. Reproduced with permission of IEEE [26]

ordinary transmitter m, the pre-distortion factor $b_{i,m,n}$ is applied on the SI samples that leak to the receiver antenna n based on the following equation [26]:

$$b_{i,m,n}[k] = \frac{\hat{h}_{i,m,n}[k]}{\hat{h}^{W}_{i,n}[k]} \tag{16.22}$$

where i and k respectively are time and carrier indices of the time-frequency variable factor b. $h_{i,n}$ is the radio channel between the auxiliary transmitter n and receiver antenna n, meanwhile $h_{i,m,n}$ is the SI radio channel between the transmitter m and receiver antenna n.

Generally, in addition to scalability, all analog and digital SIC techniques still need further research and face a lot of challenges in MIMO full-duplex systems, for example:

- The length of the pilot that is used for SI channel estimation is proportional to number of antennas. Thus, channel estimation accuracy is limited by coherence time and correlation of the noise for long pilot sequences.
- The effects of hardware impairment, explained before for the SISO case, need to be analyzed and calibrated for MIMO. Apart from simplified models [40], this is yet to be achieved. The anticipation regarding this issue is that the complexity of impairment calibration in full-duplex systems may create a serious challenge as well.
- Cost efficiency and feasibility are important issues compared to half-duplex MIMO systems.

16.6 Conclusion and Outlook

This chapter demonstrated full-duplex technology from signal processing algorithm and implementation perspectives. It explained the self-interference cancellation requirements, the techniques to achieve them, the implementation challenges and recent achievements.

In general, one needs to define the characteristics of a full-duplex system (narrowband/wideband, SISO/MIMO, shared/separate antenna), and the self-interference cancellation specifications (linear/nonlinear, passive/active cancellation, channel aware/unaware, direct/scattered path), then develop a hybrid SIC solution and design related signal-processing algorithms to meet the system requirements. Optimal full-duplex technology faces many other challenges in the MAC and higher layers:

- standards modification to support both HD and FD nodes
- smart control schemes that switch between the two modes
- full-duplex networking
- optimization of resource management.

Such issues, which are not tackled here, are discussed in the duplo FP7 project [76].

Full-duplex technology offers the potential to participate in the evolution of 5G, and it leads the way toward new mobile generations that have higher capacity, more efficiency and optimal performance. Ever since Goldsmith said,

> It is generally not possible for radios to receive and transmit on the same frequency band because of the interference that results [77]

researchers have been trying to prove the opposite. The expectation is that within few years we shall see a mature full-duplex technology that will have a tremendous impact on 5G networks and beyond. It is a matter of when, rather than if, full duplex will become dependable. This is what science is all about, removing "not" from the quote above, turning "It is not possible" into "It *is* possible".

References

[1] Sabharwal, A., Schniter, P., Guo, D., Bliss, D., Rangarajan, S., and Wichman, R. (2014) In-band full-duplex wireless: Challenges and opportunities. *IEEE J. Select. Areas Commun.*, **32** (9), 1637–1652, doi: 10.1109/JSAC.2014.2330193.

[2] Korpi, D., Anttila, L., and Valkama, M. (2014) Feasibility of in-band full-duplex radio transceivers with imperfect RF components: Analysis and enhanced cancellation algorithms, in *Cognitive Radio Oriented Wireless Networks and Communications (CROWNCOM), 2014 9th International Conference on*, pp. 532–538.

[3] Hong, S., Brand, J., Choi, J., Jain, M., Mehlman, J., Katti, S., and Levis, P. (2014) Applications of self-interference cancellation in 5G and beyond. *IEEE Commun. Mag.*, **52** (2), 114–121, doi: 10.1109/MCOM.2014.6736751.

[4] Zhang, Z., Chai, X., Long, K., Vasilakos, A., and Hanzo, L. (2015) Full duplex techniques for 5G networks: self-interference cancellation, protocol design, and relay selection. *IEEE Commun. Mag.*, **53** (5), 128–137, doi: 10.1109/MCOM.2015.7105651.

[5] Liao, Y., Song, L., Han, Z., and Li, Y. (2015) Full duplex cognitive radio: a new design paradigm for enhancing spectrum usage. *IEEE Commun. Mag.*, **53** (5), 138–145, doi: 10.1109/MCOM.2015.7105652.

[6] Wang, L., Tian, F., Svensson, T., Feng, D., Song, M., and Li, S. (2015) Exploiting full duplex for device-to-device communications in heterogeneous networks. *IEEE Commun. Mag.*, **53** (5), 146–152, doi: 10.1109/MCOM.2015.7105653.

[7] Ahmed, E. and Eltawil, A. (2015) All-digital self-interference cancellation technique for full-duplex systems. *IEEE Trans. Wireless Commun.*, **14** (7), 3519–3532, doi: 10.1109/TWC.2015.2407876.

[8] Bharadia, D., McMilin, E., and Katti, S. (2013) Full duplex radios, in *Proceedings of the ACM SIGCOMM 2013 Conference*, ACM, pp. 375–386, doi: 10.1145/2486001.2486033.

[9] Lee, T. (2004) The design of CMOS radio-frequency integrated circuits. *Commun. Engineer*, **2** (4), 47–47.

[10] Korpi, D., Riihonen, T., Syrjala, V., Anttila, L., Valkama, M., and Wichman, R. (2014) Full-duplex transceiver system calculations: Analysis of ADC and linearity challenges. *IEEE Trans. Wireless Commun.*, **13** (7), 3821–3836, doi: 10.1109/TWC.2014.2315213.

[11] Jain, M., Choi, J.I., Kim, T., Bharadia, D., Seth, S., Srinivasan, K., Levis, P., Katti, S., and Sinha, P. (2011) Practical, real-time, full duplex wireless, in *Proceedings of the 17th Annual International Conference on Mobile Computing and Networking*, ACM, pp. 301–312, doi: 10.1145/2030613.2030647.

[12] Askar, R., Kaiser, T., Schubert, B., Haustein, T., and Keusgen, W. (2014) Active self-interference cancellation mechanism for full-duplex wireless transceivers, in *9th International Conference on Cognitive Radio Oriented Wireless Networks and Communications (CROWNCOM), 2014*, pp. 539–544.

[13] Chan, Y.K., Koo, V.C., Chung, B.K., and Chuah, H.T. (2009) A cancellation network for full-duplex front end circuit. *Prog. Electromag. Res. Lett.*, **7**, 139–148.

[14] Knox, M. (2012) Single antenna full duplex communications using a common carrier, in *IEEE 13th Annual Wireless and Microwave Technology Conference (WAMICON), 2012*, pp. 1–6, doi: 10.1109/WAMICON.2012.6208455.

[15] Miyoshi, T. and Miyauchi, S. (1980) The design of planar circulators for wide-band operation. *IEEE Trans. Microwave Theory Techn.*, **28** (3), 210–214, doi: 10.1109/TMTT.1980.1130042.

[16] Schloemann, E. (1989) Miniature circulators. *IEEE Trans. Magnetics*, **25** (5), 3236–3241, doi: 10.1109/20.42265.

[17] Katoh, H. (1975) Temperature-stabilized 1.7-GHz broad-band lumped-element circulator. *IEEE Trans. Microwave Theory Techn.*, **23** (8), 689–696, doi: 10.1109/TMTT.1975.1128649.

[18] Duarte, M. and Sabharwal, A. (2010) Full-duplex wireless communications using off-the-shelf radios: feasibility and first results, in *Conference Record of the Forty Fourth Asilomar Conference on Signals, Systems and Computers (ASILOMAR), 2010*, pp. 1558–1562, doi: 10.1109/ACSSC.2010.5757799.

[19] Anderson, C., Krishnamoorthy, S., Ranson, C., Lemon, T., Newhall, W., Kummetz, T., and Reed, J. (2004) Antenna isolation, wideband multipath propagation measurements, and interference mitigation for on-frequency repeaters, in *Proceedings. IEEE SoutheastCon, 2004.*, pp. 110–114, doi: 10.1109/SECON.2004.1287906.

[20] Khojastepour, M.A., Sundaresan, K., Rangarajan, S., Zhang, X., and Barghi, S. (2011) The case for antenna cancellation for scalable full-duplex wireless communications, in *Proceedings of the 10th ACM Workshop on Hot Topics in Networks*, ACM, pp. 17:1–17:6, doi: 10.1145/2070562.2070579.

[21] Sahai, A., Patel, G., and Sabharwal, A. (2011), Pushing the limits of full-duplex: Design and real-time implementation. Arxiv.org http://arxiv.org/abs/1107.0607.

[22] Choi, J.I., Jain, M., Srinivasan, K., Levis, P., and Katti, S. (2010) Achieving single channel, full duplex wireless communication, in *Proceedings of the Sixteenth Annual International Conference on Mobile Computing and Networking*, ACM, pp. 1–12, doi: 10.1145/1859995.1859997.

[23] Aryafar, E., Khojastepour, M.A., Sundaresan, K., Rangarajan, S., and Chiang, M. MIDU: enabling MIMO full duplex.

[24] Choi, J.I., Hong, S., Jain, M., Katti, S., Levis, P., and Mehlman, J. (2012) Beyond full duplex wireless, in *Conference Record of the Forty Sixth Asilomar Conference on Signals, Systems and Computers (ASILOMAR), 2012*, pp. 40–44, doi: 10.1109/ACSSC.2012.6488954.

[25] Everett, E., Sahai, A., and Sabharwal, A. (2014) Passive self-interference suppression for full-duplex infrastructure nodes. *IEEE Trans. Wireless Commun.*, **13** (2), 680–694, doi: 10.1109/TWC.2013.010214.130226.

[26] Duarte, M., Sabharwal, A., Aggarwal, V., Jana, R., Ramakrishnan, K., Rice, C., and Shankaranarayanan, N. (2014) Design and characterization of a full-duplex multiantenna system for WiFi networks. *IEEE Trans. Vehic.Tech.*, **63** (3), 1160–1177, doi: 10.1109/TVT.2013.2284712.

[27] Khandani, A. (2010), Methods for spatial multiplexing of wireless two-way channels, US Patent 7,817,641. URL http://www.google.com/patents/US7817641.

[28] Radunovic, B., Gunawardena, D., Key, P., Proutiere, A., Singh, N., Balan, V., and Dejean, G. (2010) Rethinking indoor wireless mesh design: Low power, low frequency, full-duplex, in *Fifth IEEE Workshop on Wireless Mesh Networks (WIMESH 2010)*, pp. 1–6, doi: 10.1109/WIMESH.2010.5507905.

[29] Ahmed, E., Eltawil, A., and Sabharwal, A. (2013) Self-interference cancellation with nonlinear distortion suppression for full-duplex systems, in *2013 Asilomar Conference on Signals, Systems and Computers*, pp. 1199–1203, doi: 10.1109/ACSSC.2013.6810483.

[30] Anttila, L., Korpi, D., Syrjala, V., and Valkama, M. (2013) Cancellation of power amplifier induced nonlinear self-interference in full-duplex transceivers, in *2013 Asilomar Conference on Signals, Systems and Computers*, pp. 1193–1198, doi: 10.1109/ACSSC.2013.6810482.

[31] Gollakota, S. and Katabi, D. (2008) Zigzag decoding: Combating hidden terminals in wireless networks. *SIGCOMM Comput. Commun. Rev.*, **38** (4), 159–170, doi: 10.1145/1402946.1402977. URL http://doi.acm.org/10.1145/1402946.1402977.

[32] Halperin, D., Anderson, T., and Wetherall, D. (2008) Taking the sting out of carrier sense: Interference cancellation for wireless LANs, in *Proceedings of the 14th ACM International Conference on Mobile Computing and Networking*, ACM, pp. 339–350, doi: 10.1145/1409944.1409983. URL http://doi.acm.org/10.1145/1409944.1409983.
[33] Katti, S., Gollakota, S., and Katabi, D. (2007) Embracing wireless interference: Analog network coding. *SIGCOMM Comput. Commun. Rev.*, **37** (4), 397–408, doi: 10.1145/1282427.1282425. URL http://doi.acm.org/10.1145/1282427.1282425.
[34] Boyd, S. and Vandenberghe, L. (2004) *Convex Optimization*, Cambridge University Press.
[35] Duarte, M., Dick, C., and Sabharwal, A. (2012) Experiment-driven characterization of full-duplex wireless systems. *IEEE Trans. Wireless Commun.*, **11** (12), 4296–4307, doi: 10.1109/TWC.2012.102612.111278.
[36] Askar, R., Zarifeh, N., Schubert, B., Keusgen, W., and Kaiser, T. (2014) I/Q imbalance calibration for higher self-interference cancellation levels in full-duplex wireless transceivers, in *1st International Conference on 5G for Ubiquitous Connectivity (5GU), 2014*, pp. 92–97, doi: 10.4108/icst.5gu.2014.258148.
[37] Schenk, T. (2008) *RF Imperfections in High-rate Wireless Systems: Impact and Digital Compensation*, Springer.
[38] Razavi, B. (2001) *Design of Analog CMOS Integrated Circuits*, McGraw-Hill.
[39] Korpi, D., Anttila, L., Syrjala, V., and Valkama, M. (2014) Widely linear digital self-interference cancellation in direct-conversion full-duplex transceiver. *IEEE J. Select. Areas Commun.*, **32** (9), 1674–1687, doi: 10.1109/JSAC.2014.2330093.
[40] Anttila, L., Korpi, D., Antonio-Rodriguez, E., Wichman, R., and Valkama, M. (2014) Modeling and efficient cancellation of nonlinear self-interference in MIMO full-duplex transceivers, in *Globecom Workshops (GC Wkshps), 2014*, pp. 777–783, doi: 10.1109/GLOCOMW.2014.7063527.
[41] Li, S. and Murch, R. (2011) Full-duplex wireless communication using transmitter output based echo cancellation, in *IEEE Global Telecommunications Conference (GLOBECOM 2011)*, pp. 1–5, doi: 10.1109/GLOCOM.2011.6133971.
[42] Schreier, P. and Scharf, L. (2010) *Statistical Signal Processing of Complex-Valued Data: The Theory of Improper and Noncircular Signals*, Cambridge University Press.
[43] Picinbono, B. and Chevalier, P. (1995) Widely linear estimation with complex data. *IEEE Trans. Signal Process.*, **43** (8), 2030–2033, doi: 10.1109/78.403373.
[44] Anttila, L. (2011) *Digital front-end signal processing with widely-linear signal models in radio devices*, Ph.D. thesis, Tampere University of Technology.
[45] 3GPP (2012) Evolved universal terrestrial radio access (E-UTRA); user equipment (UE) radio transmission and reception, *Tech. Rep. TS 36.101*.
[46] Hua, Y., Liang, P., Ma, Y., Cirik, A., and Gao, Q. (2012) A method for broadband full-duplex MIMO radio. *IEEE Signal Process. Lett.*, **19** (12), 793–796, doi: 10.1109/LSP.2012.2221710.
[47] Sakai, M., Lin, H., and Yamashita, K. (2014) Adaptive cancellation of self-interference in full-duplex wireless with transmitter IQ imbalance, in *2014 IEEE Global Communications Conference (GLOBECOM)*, pp. 3220–3224, doi: 10.1109/GLOCOM.2014.7037302.
[48] Rykaczewski, P., Brakensiek, J., and Jondral, F. (2004) Towards an analytical model of I/Q imbalance in OFDM based direct conversion receivers, in *2004 IEEE 59th Vehicular Technology Conference, VTC 2004-Spring.*, vol. 4, vol. 4, pp. 1831–1835, doi: 10.1109/VETECS.2004.1390589.
[49] Anttila, L., Handel, P., and Valkama, M. (2010) Joint mitigation of power amplifier and I/Q modulator impairments in broadband direct-conversion transmitters. *IEEE Trans. Microwave Theory Techn.*, **58** (4), 730–739, doi: 10.1109/TMTT.2010.2041579.
[50] Schubert, B., Gokceoglu, A., Anttila, L., and Valkama, M. (2013) Augmented Volterra predistortion for the joint mitigation of power amplifier and I/Q modulator impairments in wideband flexible radio, in *2013 IEEE Global Conference on Signal and Information Processing (GlobalSIP)*, pp. 1162–1165, doi: 10.1109/GlobalSIP.2013.6737113.
[51] Mokhtar, M., Al-Dhahir, N., and Hamila, R. (2014) On I/Q imbalance effects in full-duplex OFDM decode-and-forward relays, in *2014 IEEE Circuits and Systems Conference (DCAS)*, pp. 1–4, doi: 10.1109/DCAS.2014.6965343.
[52] Zhan, Z., Villemaud, G., Hutu, F., and Gorce, J.M. (2014) Digital estimation and compensation of I/Q imbalance for full-duplex dual-band OFDM radio, in *2014 IEEE 25th Annual International Symposium on Personal, Indoor, and Mobile Radio Communication (PIMRC)*, pp. 846–850, doi: 10.1109/PIMRC.2014.7136283.
[53] Ahmed, E. and Eltawil, A. (2015) On phase noise suppression in full-duplex systems. *IEEE Trans. Wireless Commun.*, **14** (3), 1237–1251, doi: 10.1109/TWC.2014.2365536.

[54] Ahmed, E., Eltawil, A., and Sabharwal, A. (2013) Self-interference cancellation with phase noise induced ICI suppression for full-duplex systems, in *2013 IEEE Global Communications Conference (GLOBECOM)*, pp. 3384–3388, doi: 10.1109/GLOCOM.2013.6831595.

[55] Ahmed, E., Eltawil, A., and Sabharwal, A. (2013) Rate gain region and design tradeoffs for full-duplex wireless communications. *IEEE Trans. Wireless Commun.*, **12** (7), 3556–3565, doi: 10.1109/TWC.2013.060413.121871.

[56] Sahai, A., Patel, G., Dick, C., and Sabharwal, A. (2013) On the impact of phase noise on active cancelation in wireless full-duplex. *IEEE Trans. Vehic.Tech.*, **62** (9), 4494–4510, doi: 10.1109/TVT.2013.2266359.

[57] Syrjala, V., Valkama, M., Anttila, L., Riihonen, T., and Korpi, D. (2014) Analysis of oscillator phase-noise effects on self-interference cancellation in full-duplex OFDM radio transceivers. *IEEE Trans. Wireless Commun.*, **13** (6), 2977–2990, doi: 10.1109/TWC.2014.041014.131171.

[58] Sahai, A., Patel, G., Dick, C., and Sabharwal, A. (2012) Understanding the impact of phase noise on active cancellation in wireless full-duplex, in *Conference Record of the Forty Sixth Asilomar Conference on Signals, Systems and Computers (ASILOMAR), 2012*, pp. 29–33, doi: 10.1109/ACSSC.2012.6488952.

[59] Petrovic, D., Rave, W., and Fettweis, G. (2007) Effects of phase noise on OFDM systems with and without PLL: characterization and compensation. *IEEE Trans Commun.*, **55** (8), 1607–1616, doi: 10.1109/TCOMM.2007.902593.

[60] Wu, S., Liu, P., and Bar-Ness, Y. (2006) Phase noise estimation and mitigation for OFDM systems. *IEEE Trans. Wireless Commun.*, **5** (12), 3616–3625, doi: 10.1109/TWC.2006.256984.

[61] Bittner, S., Rave, W., and Fettweis, G. (2007) Joint iterative transmitter and receiver phase noise correction using soft information, in *IEEE International Conference on Communications, 2007. ICC '07.*, pp. 2847–2852, doi: 10.1109/ICC.2007.473.

[62] Syrjälä, V. and Valkama, M. (2011) Receiver DSP for OFDM systems impaired by transmitter and receiver phase noise, in *Communications (ICC), 2011 IEEE International Conference on*, pp. 1–6, doi: 10.1109/icc.2011.5963413.

[63] Syrjälä, V. and Valkama, M. (2012) Iterative receiver signal processing for joint mitigation of transmitter and receiver phase noise in OFDM-based cognitive radio link, in *7th International Conference on Cognitive Radio Oriented Wireless Networks*, doi: 10.4108/icst.crowncom.2012.249438.

[64] Kim, D., Lee, H., and Hong, D. (2015) A survey of in-band full-duplex transmission from the perspective of PHY and MAC layers. *IEEE Commun. Surveys Tut.*, **17** (4), 2017–2046, doi: 10.1109/COMST.2015.2403614.

[65] Riihonen, T. and Wichman, R. (2012) Analog and digital self-interference cancellation in full-duplex MIMO-OFDM transceivers with limited resolution in a/d conversion, in *Conference Record of the Forty Sixth Asilomar Conference on Signals, Systems and Computers (ASILOMAR), 2012*, pp. 45–49, doi: 10.1109/ACSSC.2012.6488955.

[66] Hiep, P.T. and Kohno, R. (2014) Water-filling for full-duplex multiple-hop MIMO relay system. *EURASIP J. Wireless Commun. Network.*, **2014** (1), 1–10.

[67] Sangiamwong, J., Asai, T., Hagiwara, J., Okumura, Y., and Ohya, T. (2009) Joint multi-filter design for full-duplex MU-MIMO relaying, in *IEEE 69th Vehicular Technology Conference, 2009. VTC Spring 2009.*, pp. 1–5, doi: 10.1109/VETECS.2009.5073633.

[68] Riihonen, T., Werner, S., and Wichman, R. (2009) Spatial loop interference suppression in full-duplex MIMO relays, in *Conference Record of the Forty-Third Asilomar Conference on Signals, Systems and Computers, 2009*, pp. 1508–1512, doi: 10.1109/ACSSC.2009.5470111.

[69] Riihonen, T., Werner, S., and Wichman, R. (2010) Residual self-interference in full-duplex MIMO relays after null-space projection and cancellation, in *Conference Record of the Forty Fourth Asilomar Conference on Signals, Systems and Computers (ASILOMAR), 2010*, pp. 653–657, doi: 10.1109/ACSSC.2010.5757642.

[70] Lioliou, P., Viberg, M., Coldrey, M., and Athley, F. (2010) Self-interference suppression in full-duplex MIMO relays, in *Conference Record of the Forty Fourth Asilomar Conference on Signals, Systems and Computers (ASILOMAR), 2010*, pp. 658–662, doi: 10.1109/ACSSC.2010.5757643.

[71] Riihonen, T., Balakrishnan, A., Haneda, K., Wyne, S., Werner, S., and Wichman, R. (2011) Optimal eigenbeamforming for suppressing self-interference in full-duplex MIMO relays, in *Information Sciences and Systems (CISS), 2011 45th Annual Conference on*, pp. 1–6, doi: 10.1109/CISS.2011.5766241.

[72] Day, B., Margetts, A., Bliss, D., and Schniter, P. (2012) Full-duplex bidirectional MIMO: Achievable rates under limited dynamic range. *IEEE Trans Signal Process.*, **60** (7), 3702–3713, doi: 10.1109/TSP.2012.2192925.

[73] Riihonen, T., Vehkapera, M., and Wichman, R. (2013) Large-system analysis of rate regions in bidirectional full-duplex MIMO link: Suppression versus cancellation, in *Information Sciences and Systems (CISS), 2013 47th Annual Conference on*, pp. 1–6, doi: 10.1109/CISS.2013.6552337.

[74] Lee, J.H. (2013) Self-interference cancelation using phase rotation in full-duplex wireless. *IEEE Trans. Vehic.Tech.*, **62** (9), 4421–4429, doi: 10.1109/TVT.2013.2264064.

[75] Bharadia, D. and Katti, S. (2014) Full duplex MIMO radios, in *Proceedings of the 11th USENIX Conference on Networked Systems Design and Implementation*, USENIX Association, pp. 359–372. URL http://dl.acm.org/citation.cfm?id=2616448.2616482.

[76] DUPLO project, URL http://www.fp7-duplo.eu/.

[77] Goldsmith, A. (2005) *Wireless Communications*, Cambridge University Press. URL https://books.google.de/books?id=n-3ZZ9i0s-cC.

17

Design and Implementation of Full-duplex Transceivers

Katsuyuki Haneda, Mikko Valkama, Taneli Riihonen,
Emilio Antonio-Rodriguez and Dani Korpi

The upcoming 5G radio communication systems will be required to achieve as much as a thousand-fold throughput increase over current 4G systems. The increase is a huge challenge that needs to be addressed through different physical layer techniques such as the use of a massive number of antennas and of use of the centimeter- and millimeter-wave spectrum,

---

Signal Processing for 5G: Algorithms and Implementations, First Edition. Edited by Fa-Long Luo and Charlie Zhang.

among others. Still, it is likely that additional and more sophisticated transceiver technologies are required to increase the throughput further, and a technique that overlays these physical-layer techniques is required. In-band full-duplex technology can be complementary to other techniques and double the spectral efficiency [1], allowing the 5G networks to reach their full potential.

An in-band full-duplex-capable transceiver is able to transmit and receive simultaneously over the same center frequency [1]. With this definition, we can identify three basic communication scenarios of two or three nodes that cover any generic wireless network, such as the one shown at the bottom of Fig. 17.1, when they are freely combined together. In particular, the possible use cases benefiting from full-duplex operation include:

- bidirectional device-to-device transmission between two full-duplex user terminals
- full-duplex access point serving uplink and downlink (possibly half-duplex) users simultaneously in the same radio resource
- two-hop full-duplex wireless relaying as shown in the center panel of Fig. 17.1.

Actually, the latter two use cases are topologically the same, with only the conceptual difference that relays always forward their received information while the uplink user is not likely communicating with the downlink user but somebody else in a totally different cell. In addition to increasing spectral efficiency, full-duplex-capable nodes may decrease latency within multi-hop links and in time-division duplex (TDD) systems. In TDD, full-duplex transceivers may avoid the guard period that is necessary when switching the direction of the transmission from downlink to uplink. In addition, the latency in the user plane may decrease when access point and user equipment are able to transmit within the same slots. On control plane, full-duplex nodes may simultaneously process user-plane and control-plane signals so as to decrease the latency and boost the operation of the overall system.

In practice, the increase in throughput due to full-duplex operation is limited by the presence of unavoidable self-interference (SI) when the transmitted signal couples back to the receiver in the in-band full-duplex transceiver. Even when the transmitted signal is known in the digital baseband, it cannot be eliminated completely in the receiver because of radio frequency (RF) impairments [2] and a large power difference between the transmitted and received signals. On the other hand, RF cancellation is limited due to the simple processing capabilities of the electronics compared to digital signal processing. Nevertheless, 50–110 dB overall attenuation levels of SI due to antenna, RF and digital cancellation have been reported in the literature for varying bandwidth and surrounding environments [3–6]. Such attenuation levels are already so high that in local area networks, where transmit powers are limited and nodes may be close to each other, co-channel interference may already start to dominate over SI. It is more challenging to realize higher levels of SI cancellation as the required isolation bandwidth is greater, and as surrounding environments impose more multipath scattering. Holistic cancellation techniques are currently subject to intensive research by many groups around the world.

This chapter gives an overview of state-of-the-art techniques for SI mitigation and cancellation within multi-antenna in-band full-duplex radio transceivers. The techniques are described through concrete examples of an antenna array equipped with measures for passive isolation improvement and of an RF canceller dealing with time-varying SI signals. Signal models taking into account the non-idealities of transceiver components are formulated,

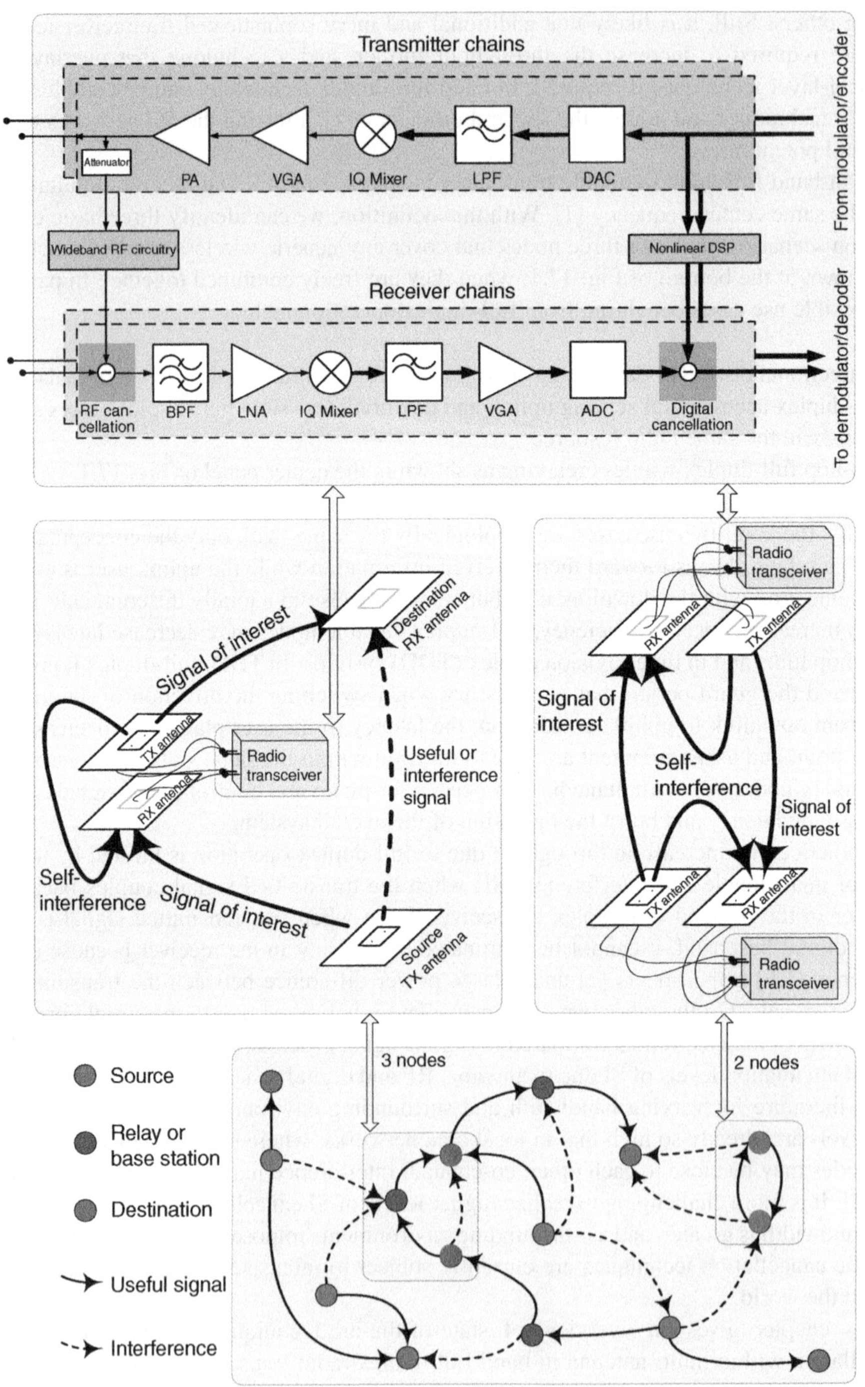

Figure 17.1 A multi-antenna radio transceiver and basic two- and three-node full-duplex scenarios occurring in wireless networks

making it possible to devise digital nonlinear adaptive SI cancellation algorithms. Finally, demonstration of the in-band full-duplex transceiver is given. The demonstration combines the antenna design with RF and digital cancellation in a relay case, showing overall SI suppression close to 100 dB – down to the noise level – even when using regular low-cost components.

The rest of this chapter is organized as six sections. Section 17.1 presents major factors in the in-band full-duplex transceiver that need to be addressed in SI mitigation and cancellation. Section 17.2 reviews methods for improving the natural and hence passive isolation between the transmitter (TX) and the receiver (RX) through advanced antenna design. Section 17.3 introduces SI mitigation techniques that can be implemented at the RF stage. Section 17.4 first provides mathematical formulation of signals in multiple-antenna transceivers, including the effect of non-idealities of RF components. The section then derives advanced digital algorithms for SI mitigation and cancellation, making use of multiple antennas and the non-ideality models of the RF components. Section 17.5 exemplifies the capability of the mentioned SI mitigation and cancellation techniques by integrating them and observing the residual SI level in a practical transceiver with off-the-shelf RF components. Finally, Section 17.6 summarizes this chapter with discussions on further challenges in in-band full-duplex transceivers.

17.1 Research Challenges

As already discussed, one of the most crucial issues in wireless single-channel full-duplex communications is the own transmit signal, or SI, that is coupled back to the RX and acts as a strong source of interference. The SI signal can be as much as 60–100 dB more powerful than the weak received signal of interest, and so it must be attenuated significantly to allow the detection of the actual received signal and enable in-band full-duplex communications.

In theory, mitigating the SI signal is an easier problem than mitigating co-channel interference. An interfering co-channel signal is unknown a priori and it must be estimated in the RX before it can be canceled. In contrast, the own transmitted signal is known within the device, and thus the SI can in principle be perfectly canceled merely by subtracting the original transmit signal from the received signal once the SI channel has been estimated. In practice, the transmitted signal is known perfectly only in digital baseband, while the transmitted RF signal is not completely known due to various distortion effects caused by RF components. The SI signal is typically much stronger than the received payload signal and consequently these distortion effects are significant as well. In conformance tests of wireless transceivers, distortion is typically expressed as an error-vector magnitude (EVM) and specifications define the upper limits for EVM depending on modulation. For example, in LTE, the maximum allowed EVM levels for base station TX are 17.5%, 12.5% and 8% for QPSK, 16QAM and 64QAM, respectively [7]. Therefore, achieving a sufficient amount of SI cancellation calls for advanced and elaborate measures since all the sources of distortion in the TX and RX must be accounted for when subtracting the cancellation signal from the received signal.

17.1.1 Passive Isolation and Active Cancellation

The passive isolation between the TX and RX should be maximized before the more complicated active cancellation techniques are utilized. This means that power from SI leaking to the RX is smaller in the first place, and thereby not as much active cancellation is needed to attenuate it to a tolerable level. When separate transmit and receive antennas are used, the electromagnetic isolation between the antennas can be improved by:

- increasing the spacing between the antennas
- having directional antennas pointing to different directions
- using different antenna polarizations [8, 9]
- having multiple TX antennas symmetrically or with a phase shift to create a null electromagnetic field at the RX [10–12].

However, one of the main limitations in increasing the spacing and having directional, electrically large, antennas is that the transceivers are usually space limited, and hence we do not have the luxury of accommodating them. Moreover, when multiple antennas are installed in the full-duplex transceivers to improve the link capacity, it is not advisable to devote the antenna polarization to isolation improvement. Finally, managing the high isolation between multiple TX and RX antennas through field nulling becomes a more difficult task as the number of antennas increases. Therefore advanced methods are required to improve the electromagnetic isolation between the antennas. These methods include, for instance, the use of band-gap structures as high-impedance surfaces to prevent surface waves between the antennas [13], inductive loops that produce counter-flowing magnetic fields [14], and resonant structures like wavetraps [15, 16] and slots on a ground plane [17] to alter ground plane currents to reduce the coupling between the antennas. Because these techniques are completely passive in nature, they require no tracking of the SI signal and its possible distortion, while still providing a potentially significant increase in SI attenuation.

In addition to these methods to improve the antenna isolation, the port-to-port isolation between the antenna feeds can be improved using different techniques. When the TX and RX share the same antenna, a circulator helps reduce the leakage of the TX RF signal to the RX RF chain [5, 18]. Electrically balanced hybrid transformers are also found to be effective in canceling the leakage of the TX signal to the RX [9, 19, 20]. The hybrid transformer is a four-port circuitry where the isolation between opposite ports is high, while adjacent ports are strongly coupled. When the TX and RX antennas can be separated in space, more options are available for antenna decoupling; a neutralization line [21], which uses a SI cancellation path between two antennas, and a lumped element network [22], which nullifies the mutual admittance between the antennas, are good examples.

It is, however, typically not possible to mitigate the SI signal perfectly with passive antenna-based techniques, and thus active SI cancellation is also required. A common solution is to do the active attenuation of the SI signal in two stages: first at the input of the RX chain, operating at RF frequencies, and then after the analog-to-digital conversion. These SI cancellation stages are usually referred to as RF cancellation and digital cancellation, respectively. RF cancellation is required in order to prevent the complete saturation of the RX components and the analog-to-digital converter (ADC). Finally, digital cancellation is performed to attenuate the remaining SI signal to below the noise floor. Both of these active cancellation methods rely on processing the known transmit signal to produce the cancellation signal.

A block diagram of an in-band full-duplex transceiver equipped with multiple antennas and employing the aforementioned active SI cancellation stages, is illustrated in the upper part of Figure 17.1. The TX and RX chains have been chosen to follow a direct-conversion architecture, which is a typical selection for modern wireless transceivers. This block diagram depicts the transceiver model that will be utilized in the following sections, thereby acting as a basis for the forthcoming analysis.

17.1.2 RF Imperfections

Typically, the performance of the active SI cancellation mechanisms based on linear processing is limited by the non-idealities occurring within the analog/RF circuit of the full-duplex transceiver [2]. This applies especially to the SI cancellation in the digital domain, where only the original transmit waveform is available. For this reason, the most prominent types of such analog/RF circuit non-idealities must be considered in this context, as they have to be regenerated in the cancellation processing to produce a sufficiently accurate cancellation signal. The most significant impairments include the nonlinear distortion of the SI signal, caused by the different amplifiers in the TX and RX, I/Q imbalance of the TX and RX, phase noise of the local oscillator, and ADC quantization noise. Some of these impairments can be modeled on the waveform level, with appropriate behavioral models, within digital cancellation, and thereby they do not pose an insurmountable obstacle for in-band full-duplex communications. There are many good examples of recent work in this field [2, 5, 14, 23–26]. In particular, the nonlinear distortion of the SI signal can be modeled in the digital domain using a memory polynomial or parallel Hammerstein type of model with carefully chosen coefficients [23, 25]. With such a model, the clean transmit data can be processed to create an accurate SI replica for efficient cancellation. Correspondingly, the effect of I/Q imbalance can be modeled digitally as an additional distortion component, which consists of the complex conjugate of the original baseband transmit signal with certain memory coefficients [24]. Nevertheless, there are also certain sources of distortion that cannot be dealt with afterwards, such as thermal noise and quantization errors produced by the ADC. These must be accounted for already in the design process of the transceiver [2].

To illustrate the need for modeling the most dominant impairments in the waveform level, Figure 17.2 shows the absolute power levels of the different signal components at the RX detector input, with respect to the transmit power. The power levels have been calculated for a SISO in-band full-duplex transceiver using simplified system calculations, similar to work by Korpi *et al.* [2], and using realistic RF component and ADC specifications. In this type of analysis, only the power levels of the different signal components are taken into account, which simplifies the calculations considerably while still providing useful insight into the system behavior. In this example, it is assumed that only the linear component of the SI signal can be perfectly attenuated by the different SI cancellation stages. To get the final numerical values for the power levels, typical component parameters from earlier literature and technical specifications have then been used, the most important of which are shown in the table in Table 17.1 for readers' convenience. Based on Table 17.1, it is evident that at least the TX power amplifier (PA) induced nonlinearities, alongside with the SI mirror image produced by the I/Q imbalances, must be attenuated by the digital cancellation algorithm to enable reliable in-band full-duplex operation and maintain an acceptable signal-to-interference-plus-noise ratio (SINR). Thus, simple linear modeling and processing in the digital domain does not suffice, and more advanced algorithms are required.

In addition to the analog circuit-level impairments, one additional aspect to consider is the time varying nature of the SI propagation channel. This is often overlooked when discussing

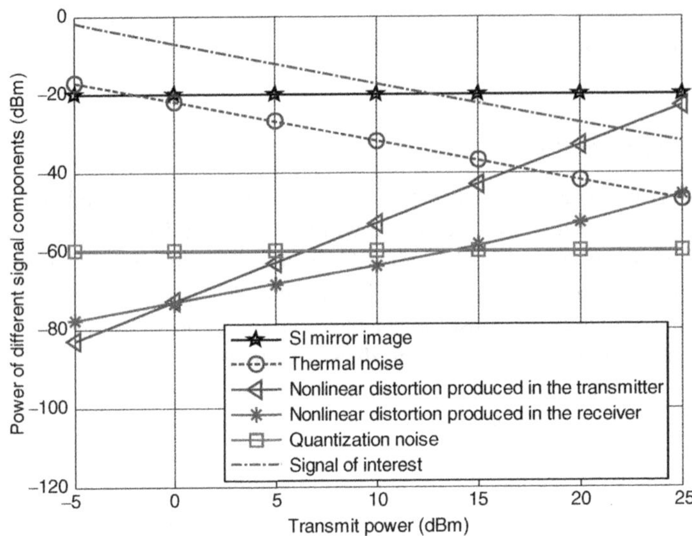

Figure 17.2 Example power levels of the different signal component at the input of the receiver detector. In this example, it is assumed that the linear SI component can be perfectly suppressed

Table 17.1 Example RF component parameters used in evaluating the power levels

Parameter	Value
SNR	15 dB
Bandwidth	12.5 MHz
RX noise figure	4 dB
Antenna isolation	40 dB
RF cancellation	30 dB
TX power amplifier IIP3	13 dBm
I/Q image rejection ratio	30 MHz
RX low-noise amplifier IIP3	−9 dBm
Number of bits at the ADC	12

SI cancellation in full-duplex devices but in practice it is obviously a key aspect in implementing an in-band full-duplex transceiver. The time varying nature of the SI channel becomes especially crucial in relaying applications where the transmit and receive antennas can be physically separated to achieve higher natural isolation between them. In such a scenario, the SI channel can vary significantly in time. However, even in a single-antenna full-duplex device, the reflections from nearby scatterers can change significantly from one moment to the next, thereby warranting the use of adaptive SI cancellation solutions.

When considering a complete and functional in-band full-duplex radio device, it is more than likely to involve all of the previously discussed methods to obtain a sufficient amount of

SI attenuation. First, the passive isolation between the TX and RX is maximized, after which the SI signal is attenuated actively. In the digital domain, the possible circuit-level impairments are then regenerated with advanced algorithms, and canceled efficiently. Furthermore, these algorithms also have the necessary adaptivity to constantly track the changes in the SI channel parameters. To demonstrate that such a device is quickly becoming a realistic prospect, the following sections provide an overview of SI reduction and cancellation methods, with a concrete example of in-band full-duplex relay implementation.

17.2 Antenna Designs

As reviewed in the previous section, the technique applied to achieve enhanced antenna isolation differs depending on the form factor of a radio device. When the form factor is extremely small, such as in a mobile phone, it is not desirable to have separate antennas for the TX and RX. In this case, it is necessary to resort to RF and digital cancellation entirely. When the space limitation is relaxed, as in for base stations or compact relays, the use of separate TX and RX antennas offers the ability to increase the electromagnetic isolation between the ports of those antennas before they are connected to the RF front-ends. In view of the goal of achieving multiple-antenna equipped full-duplex transceivers, the use of passive techniques without the need for redundant TX antennas or sacrificing the orthogonality of antenna polarization is of high relevance. One such solution is to reduce the ground-plane currents and electromagnetic fields that cause the TX signal to appear at the RX antenna. As mentioned in the previous section, high-impedance surfaces [15] and slots in the ground planes [27] have been considered. However, high-impedance surfaces usually need to be fairly large (larger than quarter-wavelength) and therefore are not feasible on compact devices, while slots in the ground plane usually mean prohibitive problems with meeting EMC requirements. Here, the use of resonant wavetraps is discussed as a possible method for improving the electromagnetic isolation in a compact device [16]. Wavetraps are resonant structures that behave similarly to baluns, consisting of a planar quarter-wavelength patch short-circuited at one end to the ground plane. The input impedance of a short-circuited quarter-wave transmission line is very high. This results in small currents at the open end of the patch, with large currents concentrated within the patch. When placed between two antennas in a suitable manner, wavetraps reduce the surface currents flowing from one to the other antenna and thereby effectively improve the electromagnetic isolation between them.

As an example, a back-to-back relay with TX and RX antennas on opposite sides of a box is shown in Figure 17.3. In Figure 17.3(a), a closed box of dimensions 180×150 mm with dual-polarized patch antennas (50×50 mm) on the top and on the bottom is shown. The closed box is a practical scenario with all the electronics contained inside the relay. The patch antenna on the bottom is co-located with the one on the top, but antenna feed locations are on opposite sides of the patch (see Figure 17.3(a)). The patch antennas are dual-polarized in order to support multiple-input multiple-output (MIMO) transmission. The relay can be used, for example, as an outdoor–indoor relay [14]. In order to enable in-band full-duplex operation, the electromagnetic isolation between the antennas on opposite sides of the box has to be improved. Figure 17.3(b) shows the schematic of the relay antenna with wavetraps on the ground plane.

The patch antennas of the relay are resonant at 2.56 GHz with -10 dB matching across a 139-MHz band, and -6 dB matching across 238 MHz. In the relay structure, ports 1 and 2

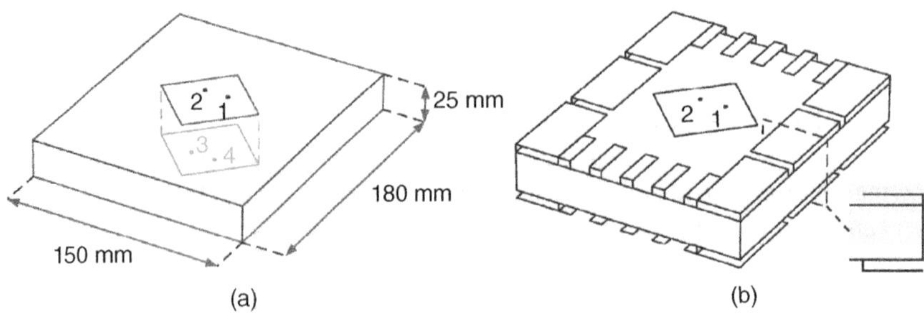

Figure 17.3 A compact in-band full-duplex relay antenna (a) without and (b) with wavetraps. In (b), the wavetraps are facing the opposite directions on either side of the relay

(TX) are on the top and ports 3 and 4 (RX) on the bottom, as shown in Figure 17.3(a). The wavetraps are quarter-wave patches, as shown in Figure 17.3(b). The dimensions and number of the wavetraps were optimized with EM-field simulations to obtain high isolation over a sufficient frequency band, and simultaneously between all the TX and all the RX antenna ports. The desired best minimum overall isolation is defined as the worst-case isolation among the four different combinations of all the RX and TX ports: S_{31}, S_{41}, S_{32} and S_{42}. The wavetraps on both sides of the relay are arranged in a similar order, but oriented in opposite directions; in other words the open ends of the wavetraps are located towards the center on the TX side and away from the center on the RX side as shown in Figure 17.3(b). This particular asymmetry provided better overall isolation between the antenna ports.

Figures 17.4(a) and 17.4(b) illustrate the simulated average electric field strength at the center cross-section parallel to the shorter edge of the box at 2.56 GHz with and without wavetraps, respectively, when port 1 of the antenna is excited. From Figure 17.4(a), it can be observed that the transmitted electromagnetic wave diffracts strongly around the corners of the relay enclosure and couples to the receiving-side antenna. With the wavetraps, as shown in Figure 17.4(b), the electric field strength is substantially lower on the receiving side of the relay than without the wavetraps, resulting in clearly improved isolation.

The designed relay antenna was manufactured, and then measured in an anechoic chamber. The isolation was evaluated by measuring the S-parameters of the relay antenna as a four-port network. The measured S-parameters are presented in Figure 17.4(c) and 17.4(d) where S_{11} denotes the antenna input matching and S_{21} and S_{43} the coupling between the antenna ports on the same side of the box. The remaining S-parameters show the isolation between the antenna ports on opposite sides of the box. In the relay without the wavetraps, the minimum isolation between RX and TX ports is 50 dB at the resonance frequency of the patch antenna as shown in Figure 17.4(c). After adding the wavetraps, the minimum isolation improved by 21 dB to 71 dB at the resonance frequency. In addition, the wavetraps offer also a wideband isolation improvement around the designed frequency, even though they are, by design, narrowband structures. The isolation across a 167-MHz band around the center frequency is better than 65 dB, which is an improvement of 15 dB over the 50-dB minimum isolation without wavetraps, the latter indicated by the dashed horizontal line in Figure 17.4(d). Within the given isolation bandwidth, the corresponding antenna port matching is better than −6 dB.

The gain of the relay antenna was also measured in the anechoic chamber to determine the impact of the wavetraps on the radiation pattern. The impact was minor, with a maximum gain

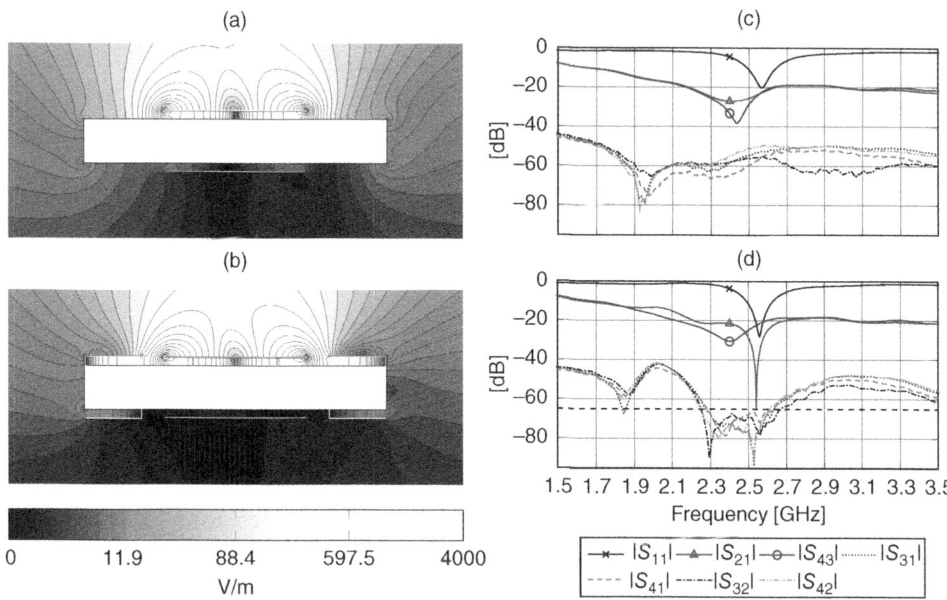

Figure 17.4 (a) Simulated electric field strength without wavetraps and (b) with wavetraps. (c) Measured S-parameters of the relay without wavetraps and (d) with wavetraps

of 10.2±0.1 dBi for all ports of the relay without wavetraps and 9.6 dBi for ports 1 and 2 and 10.5±0.1 dBi for ports 3 and 4 for the relay with wavetraps. In conclusion, the wavetraps efficiently concentrate the ground-plane currents and reduce the electromagnetic coupling to the other side of the relay box, thereby improving the antenna isolation.

17.3 RF Self-interference Cancellation Methods

Canceling SI in the analog domain is performed as follows: a replica of the SI is created and then subtracted from the received signal containing the unwanted SI. The presented passive isolation structures provide the first stage of SI attenuation but the major drawback is that they cannot account for changes in the SI channel. This calls for active electronic circuits that can cancel a time-varying SI-signal.

Two ways of creating the cancellation signal are used in the architectures presented in the literature. One is to couple part of the TX signal to a cancellation network [5, 9, 18, 28, 29], and another is to use a separate TX chain for cancellation signal creation [20]. The former approach is presented in the block diagram in Figure 17.5.

It has been found by simulation that the best place to couple the signal to a cancellation network is at the TX PA output [2]. This ensures that the cancellation signal contains all the impairments in the cancellation signal, including TX noise. In practice this means that those are canceled as well. A second coupler is then used to inject the cancellation signal into the received signal.

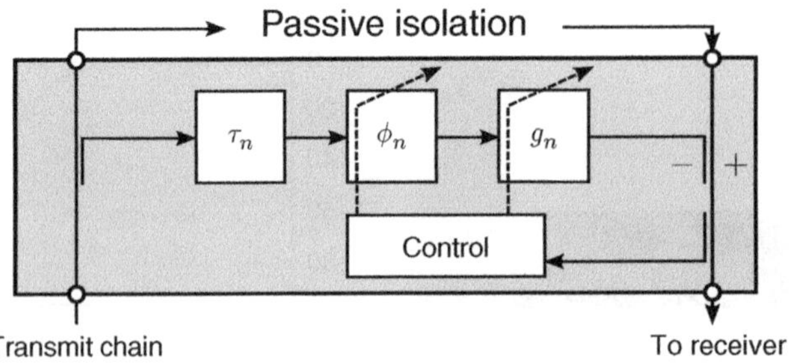

Figure 17.5 RF-canceller: overall diagram showing one RF-cancellation tap

The challenge of this design is the unavoidable trade-off between insertion losses and cancellation performance. First of all, the cancellation network must provide enough power to cancel the strongest-SI components. Maximizing passive isolation is therefore crucial. More power can be coupled to the cancellation network at the expense of increasing TX-chain insertion loss. The coupling ratio of the cancellation coupler can be adjusted at the expense of increasing the RX-noise figure as RX insertion loss is increased. Finally, the cancellation networks mostly do not provide any gain for the purpose of maximizing the linearity.

When using an additional signal generator to create the cancellation signal, only the cancellation coupler needs to be taken into account when trading off the RF noise figure and available cancellation power. Another benefit of using an additional generator is digitally controlled SI-cancellation signal generation, which can model all the required impairments in an adaptive manner. This option leaves the actual TX chain untouched, providing the lowest TX insertion loss. However, the drawbacks of this architecture are the second signal generator required and the presence of generator-induced noise at the RF input, which is coupled to the RX chain.

Overall, to ensure a high amount of SI cancellation, the amplitude, phase and delay of the cancellation signal need to match with the SI-signal well. Depending on the passive isolation method, one or two strong-SI components are present in the impulse response of the SI signal. In separate TX/RX antenna setups, typically only one strong component is present, whereas in circulator setups the leakage through the circulator and reflection from the antenna are the two main sources of SI.

In order to model a frequency-selective SI channel, the cancellation network needs to consist of multiple replicas with different delays [18]. Bharadia *et al.* used 16 analog taps to create the cancellation signal [5]; such a large number of taps was required as only the amplitude was controlled in the taps. In the design of Choi *et al.* [18, 30], both the amplitude and phase are controlled within the taps. Designs with a single RF tap have been reported [9, 28, 29]. When using only one tap in the RF canceller, the limitation is that a high amount of SI cancellation is achieved only when the attenuation of the SI signal does not show much frequency selectivity. In practice, this condition does not apply to all the operational conditions of the full-duplex transceiver.

The delays of the different taps can be implemented using transmission-line or delay-line components. The transmission lines, such as coaxial cables, are rather ideal delays but consume a significant amount of space. The delay-line components are physically small in size and can be slightly dispersive [31]. Amplitude and phase within the taps can be controlled by separate chips or using a vector modulator. Commonly, these components are analog devices but can be controlled digitally. They show reasonable linearity and noise properties, especially when amplitude control is implemented as a variable attenuator. The components used in the designs of Bharadia *et al.* [5] and Choi *et al.* [18] include a Peregrine PE43703 variable attenuator and Hittite HMC631LP3 vector modulator, respectively. The linearity of the components used should be as high as possible, since otherwise the already distorted signal from the PA output is further distorted. This results in double distortion, which can be very cumbersome to cancel in the digital domain.

In addition, the amplitude and phase control need to be self-adaptive in order to react to changes in the SI channel. Introducing a hand next to an antenna is enough to change the power reflected from the antenna such that new control settings are required. Otherwise the SI power will increase significantly. In the literature, two different solutions for adaptive control are reported. One approach is an analog least-mean-square-type filter [18], which has been shown to cancel SI channels and track their changes efficiently [30]. The other possibility is to monitor the power after the cancellation and then control the phases and amplitudes until the power is minimized [28, 32]. The ability to track changes depends on the available control speed, which translates into available computational speed.

From a digital-signal-processing point of view, the RF canceller can be viewed as another SI path, and the total SI path to the RX is a combination of the actual SI path and the SI coming from the canceller. Furthermore, if high powers are used, then the canceller can introduce additional distortion. For sensitive RXs this additional distortion must be canceled in the digital domain.

17.4 Digital Self-interference Cancellation Algorithms

Even though antenna isolation and RF cancellation provide a significant level of SI mitigation before digital processing is applied (~60–80 dB), the residual SI after analog-to-digital conversion may still be strong enough to desensitize the receiver [2, 23]. Therefore, the use of additional cancellation techniques in the digital domain is crucial in order to reduce the remaining interference below the noise level and guarantee the optimal performance of the system.

17.4.1 Signal Model

Consider the baseband model of a communication link with full-duplex transceivers in Figure 17.1. This generic case includes bidirectional communication (source and destination are the same node), the relay link (wherein the full-duplex transceiver acts as a relay) and the case in which the full-duplex transceiver is an access point simultaneously serving an uplink user and a downlink user (which are half-duplex devices) in the same frequency band. The source and destination have N_{S} transmit and N_{D} receive antennas, respectively, and the transceiver is equipped with N_{rx} receive and N_{tx} transmit antennas. We assume that transmit and receive antenna arrays in the full-duplex transceiver are spatially separated, although the model applies to the single-array design as well [33, 34].

Matrices $\mathbf{H}_{\mathrm{ST}}[n] \in \mathbb{C}^{N_{\mathrm{rx}} \times N_{\mathrm{S}}}$ and $\mathbf{H}_{\mathrm{TD}}[n] \in \mathbb{C}^{N_{\mathrm{D}} \times N_{\mathrm{tx}}}$ represent respective MIMO channels from the source to the transceiver, and from the transceiver to the destination. At time instant n, the source transmits signal vector $\mathbf{x}_{\mathrm{S}}[n] \in \mathbb{C}^{N_{\mathrm{S}}}$, and the transceiver transmits signal vector

$\mathbf{x}_\mathrm{T}[n] \in \mathbb{C}^{N_\mathrm{tx}}$ while it simultaneously receives signal vector $\mathbf{y}_\mathrm{T}[n] \in \mathbb{C}^{N_\mathrm{rx}}$. This creates an unavoidable SI feedback loop from the transceiver output to the transceiver input through channel $\mathbf{H}_\mathrm{TT}[n] \in \mathbb{C}^{N_\mathrm{rx}\times N_\mathrm{tx}}$. Spatial interference suppression is provided by precoding matrix $\mathbf{G}_\mathrm{tx}[n]$, and beamforming matrix $\mathbf{G}_\mathrm{rx}[n]$ and $\mathbf{A}[n]$ is used to subtract the contribution of the transmitted signal from the received signal for cancellation.

The corresponding received signals in the transceiver and in the destination node can be expressed as

$$\mathbf{y}_\mathrm{T}[n] = \mathbf{G}_\mathrm{rx}[n] \star (\mathbf{H}_\mathrm{ST}[n] \star \mathbf{x}_\mathrm{S}[n] + \mathbf{H}_\mathrm{TT}[n] \star \mathbf{G}_\mathrm{tx}[n] \star \mathbf{x}_\mathrm{T}[n]) + \mathbf{n}_\mathrm{T}[n] \tag{17.1a}$$

$$\mathbf{y}_\mathrm{D}[n] = \mathbf{H}_\mathrm{TD}[n] \star \mathbf{G}_\mathrm{tx}[n] \star \mathbf{x}_\mathrm{T}[n] + \mathbf{n}_\mathrm{D}[n] \tag{17.1b}$$

where $\mathbf{n}_\mathrm{T}[n] \in \mathbb{C}^{N_\mathrm{rx}}$ and $\mathbf{n}_\mathrm{D}[n] \in \mathbb{C}^{N_\mathrm{D}}$ are additive noise vectors in the transceiver and in the destination, respectively, and $\star$ refers to convolution. The signal model is capable of modeling multipath MIMO channels in the time domain assuming arbitrary modulation. In the case of OFDM, the model can be interpreted to describe transmit equations for one narrowband subcarrier in the frequency domain, and multiplication can be substituted for the convolution.

17.4.2 Basic Principles

Cancellation techniques in the digital domain work in a similar way to their analog counterparts: a digital replica of the SI signal is generated and then subtracted from the received signal. In addition to subtracting the replica signal, the full-duplex transceiver may also apply spatial suppression if transmitter and receiver antennas are spatially separated. This requires multiple antennas in both ends, such that the precoder at the transmitter side and the beamformer at the receiver are designed to simultaneously attenuate the SI signal and receive the payload signal. The RF beamformer is able to attenuate the effect of RF imperfections as well, because they pass through the same channel together with the known part of the SI signal.

In the case of subtracting the replica signal, its fidelity will determine the level of residual SI after cancellation and, therefore, will upper bound the system performance [35]. In order to achieve a nearly interference-free system, choosing the proper cancellation method for each application is of major importance. When both transmit and receive sides of the full-duplex device behave primarily like a linear filter, cancellation can be reduced to designing a filter that identifies the SI channel, since the transmitted signal is known at every time instant.

Unfortunately, any linear cancellation architecture cannot mitigate nonlinear behavior and noise sources, such as nonlinear distortion of the PA, I/Q imbalance during modulation/demodulation, phase noise, or quantization noise at the receiver [2, 5, 24, 25]. Some of these impairments can be mitigated by extending the linear architecture to a nonlinear architecture, as will be described later in this section.

Digital mitigation is typically performed either in the time domain or in the frequency domain. Mitigation in the frequency domain involves processing each subcarrier signal individually, which may result in a computationally demanding scheme if the number of subcarriers is large. On the other hand, mitigation in the time domain involves processing the signal samples independently of the number of subcarriers, but, due to the different interference paths between antennas, requires gauging the delay spread of the SI channel [36]. As shown at the top of Figure 17.1, mitigation takes place after baseband demodulation and digital conversion of the received signal, usually being the first operation within the digital pipeline. As a result, the employed signal must be sampled above the Nyquist rate, which demands the use of special techniques to deal with arbitrary signal spectra [36]. In this chapter, we focus entirely on time-domain mitigation techniques.

17.4.3 Spatial Suppression

Spatial-domain suppression [37] employs receive- and/or transmit-side feedforward filters, namely $\mathbf{G}_{\mathrm{rx}}[n]$ and $\mathbf{G}_{\mathrm{tx}}[n]$, which are matched to the SI channel alone without relying upon accurately knowing actual signals. This is in contrast to the feedback filters used in subtractive cancellation for generating an estimated copy of the SI signal from the imperfectly known transmitted signal. Nevertheless, spatial suppression can similarly mitigate SI; the pros and cons are weighed up next, when we compare the approach to time-domain cancellation.

In principle, suppression works by employing beamforming filters that direct reception and/or transmission of the signal of interest to the null space of the SI channel such that signal propagation is ideally blocked completely or, if the spatial degrees-of-freedom are scarce, to the weakest eigenmodes thereof such that the effective gain of the residual feedback channel is minimized. Mathematically this means solving

$$\min\|\mathbf{G}_{\mathrm{rx}}[n] \star \mathbf{H}_{\mathrm{TT}}[n] \star \mathbf{G}_{\mathrm{tx}}[n]\| \xrightarrow{\text{ideally}} \mathbf{G}_{\mathrm{rx}}[n] \star \mathbf{H}_{\mathrm{TT}}[n] \star \mathbf{G}_{\mathrm{tx}}[n] = \mathbf{0} \tag{17.2}$$

in terms of some suitable matrix norm. Such a mitigation approach is therefore applicable only with multi-antenna transceivers, and having low rank in the SI channel can boost suppression performance significantly. Likewise, beamforming always consumes the degrees-of-freedom, restricting the trade-off between spatial diversity and multiplexing in the signals of interest, which can be seen as reducing the effective number of antennas used for data transmission if suppression is implemented transparently around actual en/decoding blocks. Suppression is also obviously sensitive to estimation error in feedback channel-state information, not so unlike cancellation. However, the residual interference signal is not linearly proportional to the error term or its gain level because estimation error manifests itself as distorted beam patterns.

The main benefit of suppression over cancellation is the fact that it mitigates blindly all signal, distortion and noise components that pass through the loopback SI channel. Such a receive filter satisfies

$$\min\|\mathbf{G}_{\mathrm{rx}}[n] \star \mathbf{H}_{\mathrm{TT}}[n]\| \xrightarrow{\text{ideally}} \mathbf{G}_{\mathrm{rx}}[n] \star \mathbf{H}_{\mathrm{TT}}[n] = \mathbf{0} \tag{17.3}$$

This means that all the adverse transmit-side components caused by non-linear RF imperfections are suppressed at the receiver side together with the linear signal components, no matter how large the transmitter's EVM is. Yet suppression is achieved with simple linear digital signal processing without any need for modeling or implementing complex transceiver electronics, in contrast to analog or non-linear digital cancellation. Transmit-side beamforming will conversely suppress SI on-the-air before it even reaches the receiver front end such that problems related to limited ADC dynamic range and quantization noise are alleviated. In mathematical terms:

$$\min\|\mathbf{H}_{\mathrm{TT}}[n] \star \mathbf{G}_{\mathrm{tx}}[n]\| \xrightarrow{\text{ideally}} \mathbf{H}_{\mathrm{TT}}[n] \star \mathbf{G}_{\mathrm{tx}}[n] = \mathbf{0} \tag{17.4}$$

Thus it is also beneficial to employ spatial-domain suppression together with linear time-domain cancellation such that the latter efficiently eliminates the linear signal components and the former then takes care of the residual non-linear distortion components in an economic way.

17.4.4 Linear Cancellation

In general, the underlying architecture of the digital cancellation filter can be classified into two types: linear cancellation and nonlinear cancellation. Whereas linear cancellation assumes that the SI channel can be modeled as a linear filtering operation, nonlinear cancellation takes into account the presence of nonlinear components in the transmission/reception front ends. As it was seen in the system calculations example in Figure 17.2, impairments such as PA nonlinearity and I/Q imbalance may be significant. In this subsection, linear cancellation is elaborated.

The building blocks for digital baseband cancellation are depicted in Figure 17.6. The effects of transceiver impairments are modeled by noise vector $\mathbf{v}_{\mathrm{T}}[n]$ and quantization noise and limited dynamic range are captured by $\mathbf{w}[n]$. Let us first assume that $\mathbf{G}_{\mathrm{tx}}[n] = \mathbf{G}_{\mathrm{rx}}[n] = \mathbf{I}\delta[n]$ to simplify notation. Linear time-domain cancellation aims at estimating the SI channel $\mathbf{H}_{\mathrm{TT}}[n], n = 0, \cdots, L_T$, where L_T refers to the order of the channel. Then the canceller subtracts $\mathbf{A}[n] \star \mathbf{x}_{\mathrm{T}}[n] = \hat{\mathbf{H}}_{\mathrm{TT}}[n] \star \mathbf{x}_{\mathrm{T}}[n]$ from the received signal, where $\hat{\mathbf{H}}_{\mathrm{TT}}[n]$ refers to the estimate of the SI channel. It is important that $\mathbf{x}_{\mathrm{T}}[n]$ and $\mathbf{y}_{\mathrm{T}}[n]$ are uncorrelated so that the incoming signal-of-interest is not accidentally suppressed in the process. However, this condition is already satisfied by default in the case of bidirectional communication and when the full-duplex transceiver is an access point, because the uplink and downlink data are different. The condition is also true when the full-duplex transceiver is a decode-and-forward relay and when the output signal is delayed sufficiently with other protocols.

Let us arrange the $L_A + 1$ MIMO matrices $\mathbf{A}[n]$ of the interference canceller into the $N_{\mathrm{rx}} \times N_{\mathrm{tx}}(L_A + 1)$ matrix $\boldsymbol{\mathcal{A}}[n] = [\mathbf{A}[n], \cdots, \mathbf{A}[n + L_A]]$. In order to obtain a sufficient estimate of $\mathbf{H}_{\mathrm{TT}}[n]$, the order of $\mathbf{A}[n]$ should satisfy $L_A \geq L_T$. Thus we may write

$$\mathbf{y}_{\mathrm{T}}[n] = \mathbf{H}_{\mathrm{ST}}[n] \star \mathbf{x}_{\mathrm{S}}[n] + \mathbf{H}_{\mathrm{TT}}[n] \star \mathbf{x}_{\mathrm{T}}[n] - \boldsymbol{\mathcal{A}}[n]\S_{\mathrm{T}}[n] + \mathbf{n}_{\mathrm{T}}[n] \tag{17.5}$$

where $\S_{\mathrm{T}}[n] \in \mathbb{C}^{N_{\mathrm{tx}} \times (L_{\mathrm{A}}+1)}$ collects the samples $\mathbf{x}_{\mathrm{T}}[n], \cdots, \mathbf{x}_{\mathrm{T}}[n - L_A]$.

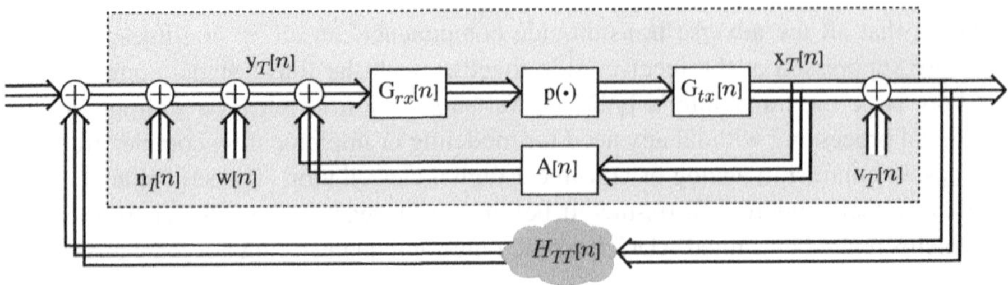

Figure 17.6 Baseband interference cancellation

If the source and the full-duplex transceiver employ orthogonal pilot sequences for channel estimation, and the transmit and receive signals are synchronized to maintain orthogonality, then the estimation process for $\boldsymbol{\mathcal{A}}[n]$ is similar to conventional multiuser channel estimation. Here we assume that the source and the full-duplex transceiver are continuously transmitting. Due to the decorrelation property of the transmit and receive signals, we may use a power-minimization algorithm without resorting to pilot signals for the estimation of the SI channel. Furthermore, by oversampling the received signal, the cancellation algorithm may operate without synchronizing the transmit and receive signals. This kind of architecture makes it possible to plug the digital cancellation block to an existing transceiver when, for example, more attenuation is needed after antenna isolation and RF cancellation.

To this end, we minimize $\mathcal{E}\{||\mathbf{y}_{\mathrm{T}}[n]||^2\}$ with respect to $\boldsymbol{\mathcal{A}}$ by the stochastic gradient descent (SGD) algorithm

$$\begin{aligned}\boldsymbol{\mathcal{A}}[i+1] &= \boldsymbol{\mathcal{A}}[n] - \mu\nabla_{\boldsymbol{\mathcal{A}}^*}\{||\mathbf{y}_{\mathrm{T}}[n]||^2\} \\ &= \boldsymbol{\mathcal{A}}[n] - \mu\mathbf{y}_{\mathrm{T}}[n]\S_{\mathrm{T}}[n]^\dagger \end{aligned} \tag{17.6}$$

where $\mu > 0$ is the adaptation step size controlling stability, convergence speed and misadjustment. Note that the algorithm operates in the time domain and is independent of modulation. In contrast, the operational domain of interference cancellation using channel estimation and pilot symbols is determined by the structure of the pilot symbols. For example, scattered pilot symbols in a time–frequency grid, as in LTE, would imply frequency-domain processing.

It can be shown that the stationary points of the algorithm provide perfect cancellation of the SI signal – in the sense that $\mathbf{A}[n]$ converges to $\mathbf{H}_{\mathrm{TT}}[n]$ – if the order of $\boldsymbol{\mathcal{A}}[n]$ is sufficient and $\mathbf{x}_{\mathrm{T}}[n]$ is persistently exciting [38]. Furthermore, the SGD algorithm can be combined with spatial suppression such that after the convergence of the algorithm, the spatial filters are alternatively optimized [36]. A recursive least squares (RLS) algorithm to minimize $\mathcal{E}\{||\mathbf{y}_{\mathrm{T}}[n]||^2\}$ has also been presented [38].

Figure 17.7 shows an example of the performance of the SGD and SGD-RLS within a MIMO-OFDM system. The figure depicts mean estimation error of the SI channel ($\hat{\mathbf{H}}_{\mathrm{TT}}[n]$) per iteration for SGD and SGD-RLS adaptation algorithms. Parameter κ defines the power ratio between information signal transmitted by the source node and SI; in other words, when $\kappa = 0$, the power of the SI and information signal are the same, and when $\kappa = \infty$ there is no SI. The antenna configuration of the relay is of size 2×2 for transmitting two data streams simultaneously. The SI channel has three taps and SNR is 15 dB. The number of subcarriers is 8192 and the length of the cyclic prefix is 8192/4 samples so the algorithms converge within few symbols depending on κ. The step size of SGD $\mu = 0.0005$ and the forgetting factor of SGD-RLS $\lambda = 0.9999$ and $\epsilon = 0.0001$ is used to initialize the autocorrelation matrix in SGD-RLS algorithm.

As mentioned before, linear cancellation by subtracting the contribution of the SI in the digital baseband domain is not able to remove the effects of transmitter imperfections. Instead of modeling the imperfections as in Section 17.4.5, another option is to combine the subtractive cancellation and spatial-domain suppression [36, 37, 39].

Let us model the total receiver noise as $\mathbf{n}_{\mathrm{T}}[n] = \mathbf{n}_{\mathrm{I}}[n] + \mathbf{w}[n] + \mathbf{H}_{\mathrm{TT}}[n] \star \mathbf{v}_{\mathrm{T}}[n]$, where $\mathbf{v}_{\mathrm{T}}[n]$ models the RF imperfections as Gaussian noise, whose power depends on the EVM value, $\mathbf{w}[n]$ models the quantization noise, and $\mathbf{n}_{\mathrm{I}}[i]$ denotes noise in the receiver. By modifying $\mathbf{w}[n]$ it is possible to model the effects of limited dynamic range of the

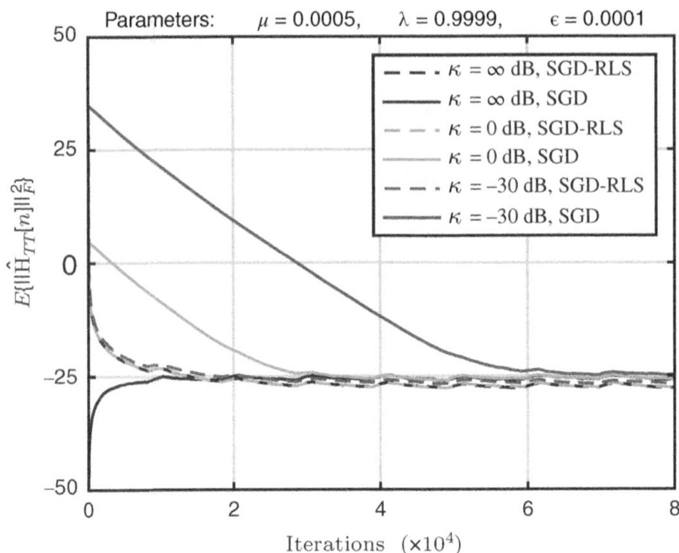

Figure 17.7 Convergence of LMS and RLS adaptation algorithms toward 25 dB suppression of self-interference

analog-to-digital converter. The received SINR is given by

$$\gamma = \frac{\mathcal{E}\{||\mathbf{G}_{\rm rx}[n] \star \mathbf{H}_{\rm ST}[n] \star \mathbf{x}_{\rm S}[n]||^2\}}{\mathcal{E}\{||\mathbf{G}_{\rm rx}[n] \star ((\mathbf{H}_{\rm TT}[n] - \mathbf{A}[n]) \star \mathbf{G}_{\rm tx}[n] \star \mathbf{x}_{\rm T}[n] + \mathbf{H}_{\rm TT}[n] \star \mathbf{v}_{\rm T}[n] + \mathbf{n}_{\rm I}[n])||^2\}} \tag{17.7}$$

Assuming that the SGD algorithm has converged, the term containing $\mathbf{x}_{\rm T}[n]$ vanishes. However, given the limited dynamic range of the receiver, it is nevertheless important to minimize $||\mathbf{H}_{\rm TT}[n] \star \mathbf{G}_{\rm tx}[n] \star \mathbf{x}_{\rm T}[n]||^2$. To this end, we can apply alternative optimization such that

$$\begin{aligned} \min_{\mathbf{G}_{\rm tx}[n]} \quad & ||\mathbf{H}_{\rm TT}[n] \star \mathbf{G}_{\rm tx}[n]||_F^2 \\ \text{subject to} \quad & \mathbf{H}_{\rm TD}[n] \star \mathbf{G}_{\rm tx}[n] = \mathbf{H}_{\rm TD}^{\rm eq}[n] \\ & \mathcal{E}\{||\mathbf{G}_{\rm tx}[n] \star \mathbf{x}_{\rm T}[n]||^2\} \leq P_{\max} \end{aligned}$$

where $||\cdot||_F$ refers to the Frobenious norm, and $P_{\max}$ is the maximum transmit power. Linear constraints can be set on the basis of an effective post-processing target channel $\mathbf{H}_{\rm TD}^{\rm eq}[n]$ between the transceiver and the destination. The target channel can be defined, for example by the maximum number of channel taps, so that $\mathbf{G}_{\rm tx}[n]$ performs channel shortening [39]. Once $\mathbf{G}_{\rm tx}[n]$ is determined $\mathbf{G}_{\rm rx}[n]$ can be solved by casting Eq. (17.7) as a generalized eigenvalue problem. This alternating optimization is repeated until convergence.

17.4.5 Nonlinear Cancellation

When assuming a practical radio transceiver, it is very unlikely that the accuracy of the linear digital canceller is enough to ensure sufficient performance. The different RF impairments will distort the observed SI signal such that it is no longer a linear transformation of the original TX signal. This means that nonlinear modeling is required to fully grasp the effects of the effective SI propagation channel, which includes various sources of nonlinear distortion. Nonlinear effects, such as amplifier distortion or mixer nonlinearities, can be accurately modeled using polynomial-based systems [5, 23, 25], whereas I/Q imbalance can be modeled using widely linear filters [24], which have been extensively studied in the literature.

As a starting point, the baseband signal of TX j ($j = 1, 2, \ldots, N_{\text{tx}}$) is denoted by $x_j(n)$. The first component distorting the TX signal is the I/Q modulator, which will inherently introduce some I/Q imbalance. The output signal of the I/Q modulator model (using a frequency-independent model for simplicity) is [25]:

$$x_j^{IQM}(n) = K_{1,j}x_j(n) + K_{2,j}x_j^*(n) \tag{17.8}$$

with $K_{1,j} = 1/2(1 + g_j \exp\,(j\varphi_j))$ and $K_{2,j} = 1/2(1 - g_j \exp\,(j\varphi_j))$, where g_j, φ_j are the gain and phase imbalance parameters of TX j. Notice that for any practical TX front end $|K_{1,j}| \gg |K_{2,j}|$. The strength of the induced I/Q image component, represented by the conjugated signal term in Eq. (17.8), is typically characterized with the image rejection ratio as $10\log_{10}(|K_{1,j}|^2/|K_{2,j}|^2)$.

The output signal of the I/Q modulator is then fed to the TX PA, which will further distort it. A common approach is to use polynomials to model the nonlinear distortion produced by the PA. In brief, polynomial-based systems model nonlinearities by processing higher-order terms of the input signal. Typically, a parallel Hammerstein (PH) model with polynomial branch nonlinearities and FIR branch filters is assumed for the PA. Using the PH model, the output signal of the PA can be written as follows.

$$x_j^{PA}(n) = \sum_{\substack{p=1 \\ p\ odd}}^{P} \sum_{m=0}^{M} h_{p,j}(m)\psi_p(x_j^{IQM}(n-m)) \tag{17.9}$$

where M and P denote the memory depth and nonlinearity order of the PH model, $h_{p,j}(n)$ denote the FIR filter impulse responses of the PH branches for TX j, and the basis functions are defined as

$$\psi_p(x(n)) = |x(n)|^{p-1}x(n) = x(n)^{\frac{p+1}{2}}x^*(n)^{\frac{p-1}{2}} \tag{17.10}$$

In general, the number of parameters of a Hammerstein model grows linearly with order P, while in the MIMO case, the increase is also relative to the number of TX and RX antennas [25]. The PH nonlinearity is a widely used nonlinear model for direct as well as inverse modeling of PAs and has been observed, through RF measurements, to characterize the operation of various PAs in an accurate manner [40].

Using Eqs. (17.8) and (17.9), the overall output signal of the TX can be expressed as

$$x_j^{PA}(n) = \sum_{\substack{p=1 \\ p\ odd}}^{P} \sum_{q=0}^{p} \sum_{m=0}^{M} h_{p,j}^{(q,p-q)}(m)x_j(n-m)^q x_j^*(n-m)^{p-q} \tag{17.11}$$

where $h_{p,j}^{(q,p-q)}$ are the coefficients for the basis function of the form $x^q x^{*p-q}$.

In fact, the RX(s) will also observe a signal model of the same form as Eq. (17.11) in the digital domain. In particular, assuming a MIMO propagation channel between the jth TX and

ith RX denoted by $c_{ij}(l)$, and a linear RF canceller whose coefficients are denoted by $h_{ij}^{RF}(l)$, the observed signal in the digital domain can be written as

$$\begin{aligned} r_i(n) &= \sum_{j=1}^{N_{\text{tx}}} \sum_{l=0}^{L} c_{ij}(l) x_j^{PA}(n-l) - \sum_{j=1}^{N_{\text{tx}}} \sum_{l=0}^{L'} h_{ij}^{RF}(l) x_j^{PA}(n-l) \\ &= \sum_{j=1}^{N_{\text{tx}}} \sum_{\substack{p=1 \\ p \text{ odd}}}^{P} \sum_{q=0}^{p} \sum_{m=0}^{M+\max(L,L')} \breve{h}_{p,ij}^{(q,p-q)}(m) x_j(n-m)^q x_j^*(n-m)^{p-q} \end{aligned} \tag{17.12}$$

with $\breve{h}_{p,ij}^{(q,p-q)}(m) = \sum_{l=0}^{m} c_{ij}^{RF}(l) h_{p,j}^{(q,p-q)}(m-l)$ and $c_{ij}^{RF}(l) = c_{ij}(l) - h_{ij}^{RF}(l)$. Here, L denotes the length of the MIMO propagation channels between the TX and RX, L' denotes the number of taps in the RF canceller and N_{tx} is the number of TX antennas. Hence, the structure of the model is still of the same form as in Eq. (17.11) but with modified impulse response coefficients and orders, taking into account the effects of the MIMO propagation channel and RF canceller.

The task of the nonlinear digital canceller is then to model as accurately as possible the received SI signal, whose signal model is characterized by Eq. (17.12). The principle is still the same as in the linear digital canceller but now the fundamental signal model assumed by the canceller is more complex. This will improve the performance of the digital canceller at a cost of higher computational complexity.

17.4.5.1 Widely Linear Digital Canceller

The simplest nonlinear canceller is the widely linear digital canceller, which only considers the I/Q imbalance [24]. The complexity of the SI signal model will slightly increase with respect to the linear canceller but the potential improvement in the cancellation capability is also significant. Referring to Eq. (17.12), the widely linear canceller merely means that the value for P is set to 1, which results in two basis functions: the linear SI component and its complex conjugate. Thus, the SI signal estimate is

$$\hat{r}_i(n) = \sum_{j=1}^{N_{\text{tx}}} \sum_{q=0}^{1} \sum_{m=0}^{M} \hat{h}_{1,ij}^{(q,1-q)}(m) x(n-m)^q x^*(n-m)^{1-q} \tag{17.13}$$

where $\hat{h}_{1,ij}^{(q,1-q)}(m)$ contains the channel estimates for the linear SI ($q = 1$) and its I/Q mirror image ($q = 0$) [24]. Assuming the same memory length for both estimates, the number of parameters is doubled with respect to the linear canceller since there are now two basis functions instead of only one.

However, the drawback of the widely linear digital canceller is that it completely ignores the effect of the TX PA, which is typically heavily distorting the SI signal. Thus the widely linear signal model can be expected to be accurate only with relatively low TX powers, when the PA is close to its linear operation region. Nevertheless, a significant gain in performance can still be expected when replacing a linear canceller with a widely linear canceller since most modern transceivers inherently suffer from I/Q imbalance.

17.4.5.2 Nonlinear Digital Canceller

The next step in increasing the accuracy and complexity of the SI signal model is the nonlinear digital canceller, which takes into account the nonlinear distortion produced by the TX PA but ignores the I/Q imbalance. Typically, it is not necessary to model the nonlinearities produced by the other components since usually the TX PA produces most of the nonlinear distortion [2]. The nonlinear signal model [23] can be obtained from Eq. (17.12) by setting $q = (p+1)/2$ and $p - q = (p-1)/2$. The corresponding estimate for the SI signal is then as follows:

$$\hat{r}_i(n) = \sum_{j=1}^{N_{\text{tx}}} \sum_{\substack{p=1 \\ p\ odd}}^{P} \sum_{m=0}^{M} \hat{h}_{p,ij}^{(\frac{p+1}{2},\frac{p-1}{2})}(m) x(n-m)^{\frac{p+1}{2}} x^*(n-m)^{\frac{p-1}{2}} \tag{17.14}$$

where, again, $\hat{h}_{p,ij}^{(\frac{p+1}{2},\frac{p-1}{2})}(m)$ contains the SI channel estimates for the different basis functions. It can be easily calculated that the number of basis functions, whose corresponding channel coefficients must be estimated, is now $(P+1)/2$, which means that the nonlinear canceller is bound to be somewhat more computationally demanding than the linear or widely linear canceller.

17.4.5.3 Augmented Nonlinear Digital Canceller

It has been shown that under typical levels of I/Q imbalance and PA nonlinearity, modeling only one of these impairments might not be sufficient to achieve reasonable SINR levels [25]. For this reason, to ensure sufficient performance, a full augmented nonlinear digital canceller should be used. This is capable of modeling TX and RX I/Q imbalance, alongside a heavily nonlinear TX PA, which means that it is resistant to the most significant impairments occurring in a typical radio transceiver. This type of solution can be expected to provide a considerable improvement in performance with the higher TX powers [24, 25]. In this case, the SI signal model used in the digital canceller is exactly the same as Eq. (17.12). The corresponding estimate of the SI signal is written as

$$\hat{r}_i(n) = \sum_{j=1}^{N_{\text{tx}}} \sum_{\substack{p=1 \\ p\ odd}}^{P} \sum_{q=0}^{p} \sum_{m=0}^{M} \hat{h}_{p,ij}^{(q,p-q)}(m) x(n-m)^q x^*(n-m)^{p-q} \tag{17.15}$$

with $\hat{h}_{p,ij}^{(q,p-q)}(m)$ again containing the SI channel estimates. Using the full augmented nonlinear SI signal model in the digital canceller will obviously result in an increase in the number of basis functions, for which the channel coefficients must be estimated. It can be shown that this particular signal model contains $\left(\frac{P+1}{2}\right)\left(\frac{P+1}{2}+1\right)$ basis functions. Thus, the computational complexity of the augmented nonlinear canceller can be rather high, but the cancellation performance is also likely to be better than the other solutions.

17.4.5.4 Simulation Example

Figure 17.8 shows a possible nonlinear canceller architecture for the full-duplex device in Figure 17.1, assuming two TX antennas and two RX antennas – in other words, $N = M =$

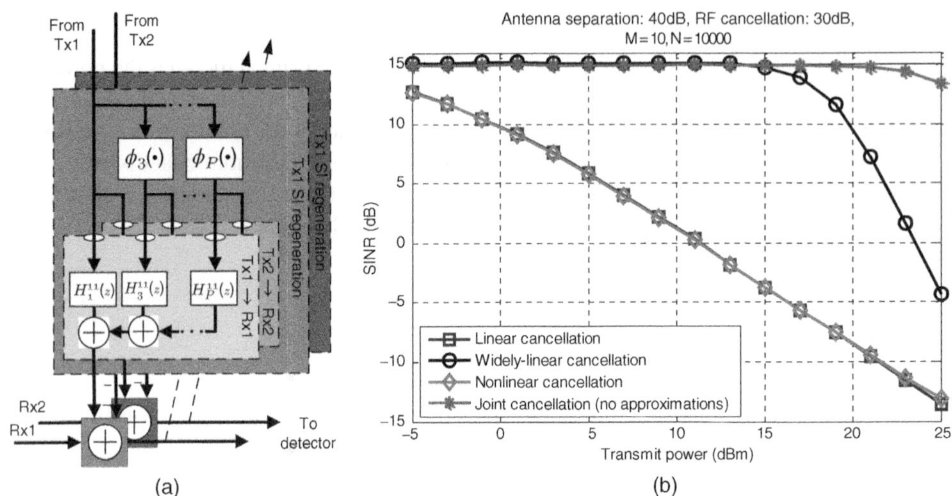

Figure 17.8 (a) Nonlinear adaptive self-interference canceller structure for a 2 ×2 full-duplex MIMO device. (b) Signal-to-interference-plus-noise ratio (SINR) at canceller output as a function of transmit power in a 2 ×2 MIMO full-duplex device. I/Q modulator mismatch and PA nonlinearity are present. SINR levels with linear, widely linear, and two different nonlinear cancellers are shown. Results from Anttila *et al.* [23], ©IEEE

2 – and a nonlinear architecture based on polynomial models. Function $\phi_p(x_n) = x_n|x_n|^{p-1}$ is the polynomial basis function of order p and $H_p(z)$ is the linear filter associated with $\phi_p(x_n)$.

Anttila *et al.* successfully used the parallel Hammerstein model to model the overall nonlinear SI channel comprising a nonlinear PA and a multipath SI channel [23]. While they treated the single-antenna case, in a later paper their model was extended to the MIMO case [25], where the I/Q imbalance at the TX side was also taken into account through the use of additional complex-valued basis functions. To illustrate the potential of digital cancellation solutions, we now present simulations with the techniques employed in the latter paper [25]. Figure 17.8(b) shows the performance results of a cancellation scheme for a full-duplex device where both TX and RX signals are 12.5 MHz OFDM signals, and the SNR at reception, per RX, is 15 dB. The SI propagation channel between the antennas is a multipath channel, and PA nonlinear distortion and I/Q imbalance during modulation are also included in the system model. Furthermore, an antenna isolation of 40 dB and analog RF cancellation of 30 dB are assumed, similar to the specifications in Figure 17.1. The exact description of the simulation setup can be found in the original paper [25].

Figure 17.8(b) shows the SINR after digital cancellation for different methods, such as basic linear cancellation [35], nonlinear cancellation with no I/Q imbalance [23], widely linear cancellation with no PA nonlinearity [24], and augmented nonlinear cancellation modelling all previous impairments [25], referred to as "joint cancellation" in the figure. Strictly speaking, the number of basis functions of the augmented nonlinear canceller deployed by Anttila *et al.* [25] is 12 times larger than that of the plain linear cancellation scheme, but they showed, only six of these are essential for the implementation. Furthermore, as shorter filters can be used for memory modeling with nonlinear terms, compared to the linear component, the total number of essential parameters to be estimated is still feasible. The augmented nonlinear canceller used is able to obtain SINR results within 1 dB of the optimal case when the TX power remains below +23 dBm [25].

The widely linear canceller [24] performs almost optimally when the TX power is under +15 dBm, while nonlinear cancellation [23] and plain linear cancellation fail to achieve good performance because of their inability to compensate for the I/Q mismatch, which is the major source of interference in this scenario. Actual RF measurement examples with the PH-based nonlinear digital canceller in the single-antenna case will be presented in the next section.

From the simulation results, we can conclude that digital cancellation techniques in combination with antenna isolation and RF cancellation can almost entirely remove the SI from the full-duplex device, whenever the cancellation architecture is able to accurately model the SI channel. In particular, a nonlinear cancellation architecture significantly improves the mitigation levels provided by its linear counterpart when operating with imperfect RF components.

17.4.6 Adaptive vs Block-based Processing

According to the operation mode of the cancellation scheme, we can distinguish between online and offline cancellation. Online cancellation uses an adaptive approach for obtaining the optimal cancellation filter, in which every new sample is employed to iteratively update the cancellation filter. Changes in the environment or in the internal state of the device, such as new coupling paths or reflections, as well as variations in the temperature or the TX power, can severely modify the SI channel and, consequently, an updated estimation of the SI channel is essential to sustain optimal interference levels.

An online cancellation scheme can track temporal variations of the device environment while maintaining a constant computational load, though the number of samples required to reach an optimal solution might be high [35]. On the other hand, offline cancellation obtains the optimal cancellation filter through a batch operation, after receiving several samples. This approach requires fewer samples to reach a solution, but the tracking capabilities are nonexistent and the computational load per sample is significantly higher than in online cancellation. A periodic estimation of the SI channel is necessary to maintain good performance levels when considering time-varying scenarios [23]. In general, the number of samples available for cancellation indicates which option to implement. An offline algorithm is preferable when few samples are available, whereas an online algorithm is preferable when a large number of samples is available.

Regardless of the operation mode, the optimal cancellation filter is obtained as the solution of a minimization problem. Assuming that the TX signal of the full-duplex device is uncorrelated with the incoming source signal, the optimal cancellation filter minimizes the signal power after cancellation or, equivalently, identifies the SI channel [35]. Consequently, the optimal filter can be obtained using common adaptive techniques such as gradient descent or recursive least-squares in the online cancellation case or conventional least-squares in the offline cancellation case. Detailed algorithm descriptions can be found in the literature [35, 36].

17.5 Demonstration

In order to evaluate the overall SI cancellation performance of an in-band full-duplex transceiver, a demonstrator for a relay antenna with analog/RF and digital SI cancellation

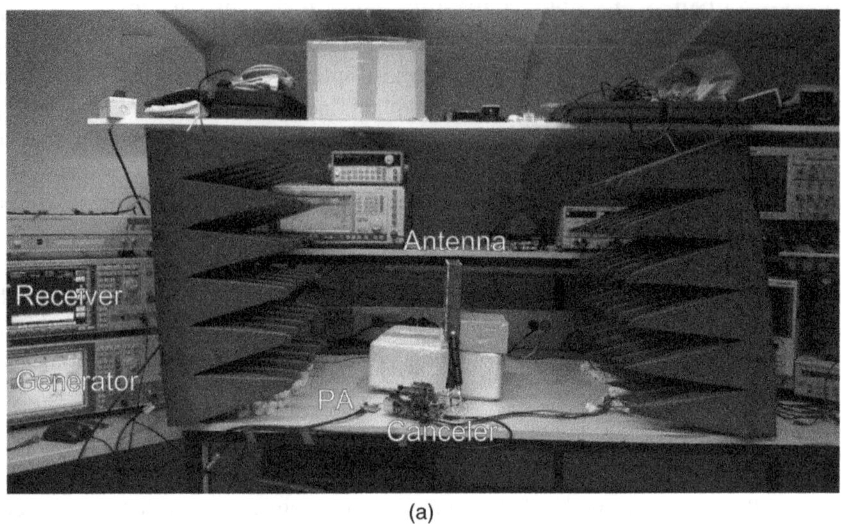

(a)

(b) PSD at different cancellation stages, transmit power +20 dBm

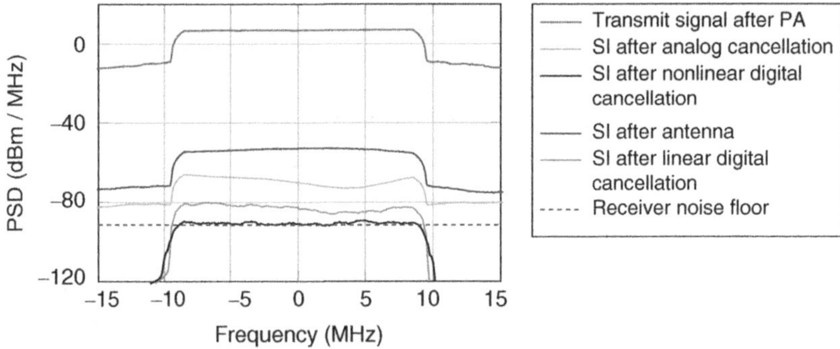

(c) PSD at different cancellation stages, transmit power +24 dBm

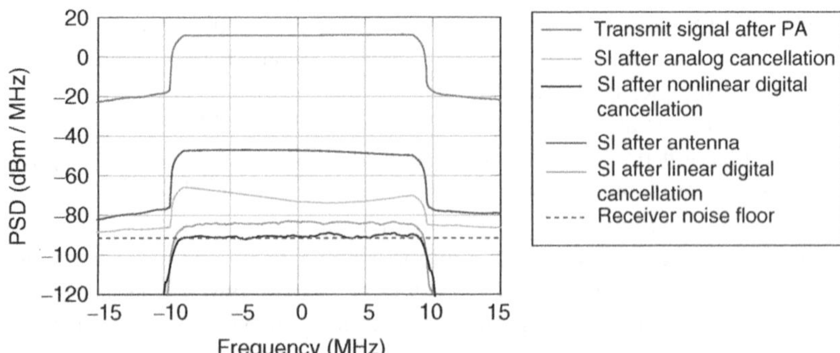

Figure 17.9 (a) Measurement setup and equipment. (b) Transmit signal and SI signal spectrum at the different cancellation stages for 20 dBm (Texas Instruments CC2595 power amplifier) and (c) 24 dBm (MiniCircuits ZVE-8G power amplifier) transmit powers. The RF center-frequency in the measurements is 2.47 GHz

was built (see Fig 17.9(a)). It consists of a vector signal generator, a PA, the relay antenna, an analog RF canceller, and a spectrum analyzer acting as an I/Q RX. Digital cancellation processing stages – linear and nonlinear – are implemented on the host computer processor. In the measurements, the relay antenna is used, but without the wavetraps, to test the performance limits of the active analog/RF and digital cancellers. The RF center frequency in the measurements is 2.47 GHz.

The analog/RF canceller is a novel design based on work by Choi *et al.* [18]. The canceller consists of two delay taps, such that the delay of the main SI component lies between the aforementioned canceller taps. The reference signal for these taps is taken from the PA output using a directional coupler. The phase and amplitude of these reference signals are adjusted manually using vector modulators such that a good cancellation level is observed from the spectrum analyzer, indicating that the cancellation signal matches well the actual SI signal. Cancellation is achieved simply by subtracting the cancellation signal from the RX signal.

The rest of the measurement equipment is standard laboratory equipment, except for the PA, a Texas Instruments CC2595, which is a low-cost chip intended for commercial use. It was deliberately chosen in order to demonstrate the performance of the digital nonlinear DSP algorithms under practical circumstances, where the PA distorts the TX signal heavily. As already discussed, this is inevitable in many applications due to restrictions on the size, cost and power consumption of the PA. Another PA, MiniCircuits ZVE-8G, is also deployed to demonstrate the operation with higher output power and enhanced linearity.

Figures 17.9(b) and 17.9(c) depict then the overall measurement results obtained using the setup of Figure 17.9(a). The test signal is configured to be a 20 MHz-wide OFDM signal at 2.47 GHz center frequency with an average TX power of +20 dBm (TI PA case) or +24 dBm (MiniCircuits PA case), respectively, to test challenging scenarios with fairly high power levels. In Figures 17.9(b) and 17.9(c), the uppermost spectrum is that of the actual TX signal, measured at the output of the PA. The high distortion levels at the PA output are also evident from the significant spectral regrowth visible outside the actual signal band. This is particularly clear in Figure 17.9(b), where the PA used is already operating in a highly nonlinear region. In Figure 17.9(c), the PA deployed is somewhat more linear, but clear nonlinear distortion is still created.

After the PA output, the TX signal propagates to the receiving antenna, turning into SI. From Figures 17.9(b) and 17.9(c) it can be observed that the SI signal is attenuated by approximately 60 dB, regardless of the TX power, when propagating to the RX, including 4 dB of losses from the setup. This result is closely in line with the observations presented in the antenna design (Figure 17.4(c)), where SI attenuations of approximately 55 dB were measured for the antenna without the wavetraps.

The following trace in the figure is the residual SI after analog RF cancellation, which was performed using the aforementioned RF canceller prototype. Depending on the TX power, the RF canceller is able to attenuate the SI signal by 15 or 22 dB. After this, the data is recorded to digital I and Q samples for offline postprocessing with the nonlinear digital-cancellation algorithm. In the digital canceller, nonlinear basis functions up to seventh order are considered, each of them having ten precursor and ten postcursor coefficients. The actual parameter estimation is carried out using blockwise least squares, as discussed earlier. From Figures 17.9(b) and 17.9(c) it can be observed that the residual SI power after digital nonlinear cancellation is at the IQ RX noise floor (−91.5 dBm/MHz) with both TX powers, meaning that the effective interference-plus-noise floor is suppressed by as much as 98 dB in total. This indicates that the total amount of SI cancellation from the antenna, RF canceller and digital canceller is sufficient to attenuate the SI signal below the RX noise floor, even with high TX powers. Also note that with ordinary linear digital cancellation the power of the residual SI is significantly higher, evidencing that nonlinear modeling is indeed required in the digital cancellation processing of a practical in-band full-duplex transceiver. This is particularly clear in Figure 17.9(b) where the PA is operating in a highly nonlinear region.

17.6 Summary

In this chapter, an overview of the state-of-the-art self-interference mitigation and cancellation techniques is given for multi-antenna in-band full-duplex communications, including bidirectional and relay transmissions. Several concrete examples of antenna, RF and digital canceller design are given; for example a novel antenna design utilizing resonant wavetraps assuming an on-frequency relay operation is first presented. The antenna is shown to provide substantially enhanced passive isolation, despite its compact size. Methodologies and implementation examples of the RF canceller are then provided. Having formulated interference-signal models, novel nonlinear and adaptive digital-cancellation algorithms are formulated. They enable enhanced digital SI cancellation levels when operating under practical nonlinear RF components. Furthermore, by combining the advanced antenna design with an active RF and digital canceller, we showed with actual RF measurements in a relay case that the SI can be suppressed below the RX noise floor, even when using regular low-cost components. The total measured aggregate SI suppression obtained in the measurements was close to 100 dB when operating at the TX power level of +24 dBm. Dealing with the SI problem in in-band full-duplex radio communications was shown to be technologically feasible, making it an essential element as an enabler of 5G radio networks. Further remaining challenges in reinforcing the in-band full-duplex technology as an indispensable technological element in 5G include wideband and multiband isolation improvement, application to multiple antenna devices and energy consumption, among many others. The insights and mathematical formulas presented in this chapter form bases for addressing the further challenges.

Acknowledgements

The authors of this chapter would like to thank their collaborators: M. Heino, S. Venkatasubramanian, Dr C. Icheln, Prof. R. Wichman from Aalto University and T. Huusari and Dr L. Anttila from Tampere University of Technology. The chapter is compiled with financial support of the Academy of Finland (Projects 258364 and 259915: “In-band full-duplex MIMO transmission: a breakthrough to high-speed low-latency mobile networks”) and the Finnish Funding Agency for Technology and Innovation (Tekes project “Full-duplex cognitive radio”).

References

[1] Sabharwal, A., Schniter, P., Guo, D., Bliss, D., Rangarajan, S., and Wichman, R. (2014) In-band full-duplex wireless: challenges and opportunities. *IEEE J. Select. Areas Commun.*, **32**, 1637–1652.

[2] Korpi, D., Riihonen, T., Syrjälä, V., Anttila, L., Valkama, M., and Wichman, R. (2014) Full-duplex transceiver system calculations: Analysis of ADC and linearity challenges. *IEEE Trans. Wireless Commun.*, **13** (7), 3821–3836.
[3] Heino, M., Korpi, D., and Huusari, T. *et al.* (2015) Recent advances in antenna design and interference cancellation algorithms for in-band full duplex relays. *IEEE Commun. Mag.*, **53** (5), 91–101.
[4] Chen, S., Beach, M., and McGeehan, J. (1998) Division-free duplex for wireless applications. *Electr. Lett.*, **34** (2), 147–148.
[5] Bharadia, D., McMilin, E., and Katti, S. (2013) Full duplex radios, in *Proc. ACM SIGCOMM '13*, pp. 375–386.
[6] Duarte, M., Dick, C., and Sabharwal, A. (2012) Experiment-driven characterization of full-duplex wireless systems. *IEEE Trans. Wireless Commun.*, **11** (12), 4296–4307.
[7] 3GPP (2014), Evolved universal terrestrial radio access (E-UTRA); base station (BS) radio transmission and reception.
[8] Debaillie, B., van den Broek, D.J., Lavin, C., van Liempd, B., Klumperink, E., Palacios, C., Craninckx, J., Nauta, B., and Parssinen, A. (2014) Analog/RF solutions enabling compact full-duplex radios. *IEEE J. Select. Areas Commun.*, **32** (9), 1662–1673.
[9] Debaillie, B., van den Broek, D.J., Lavin, C., van Liempd, B., Klumperink, E.A.M., Palacios, C., Craninckx, J., and Nauta, B. (2015) RF self-interference reduction techniques for compact full duplex radios, in *Proc. 81st Veh. Tech. Conf. (VTC2015-Spring)*, Glasgow, Scotland, pp. 1–5.
[10] Choi, J.I., Jain, M., Srinivasan, K., Levis, P., and Katti, S. (2010) in *Proc. 16th Annual Int. Conf. Mobile Computing and Netw.*, Chicago, IL, pp. 1–12.
[11] Aryafar, E., Khojastepour, M.A., Sundaresan, K., Rangarajan, S., and Chiang, M. (2012) MIDU: Enabling MIMO full duplex, in *Proc. 18th Annual Int. Conf. Mobile Computing and Netw.*, pp. 257–268.
[12] Alrabadi, O., Tatomirescu, A., Knudsen, M., Pelosi, M., and Pedersen, G. (2013) Breaking the transmitter-receiver isolation barrier in mobile handsets with spatial duplexing. *IEEE Trans. Ant. Propag.*, **61** (4), 2241–2251.
[13] Kang, H. and Lim, S. (2013) High isolation transmitter and receiver antennas using high-impedance surfaces for repeater applications. *J. Electromag. Waves Applic.*, **27** (18), 2281–2287.
[14] Korpi, D., Venkatasubramanian, S., Riihonen, T., and et al. (2013) Advanced self-interference cancellation and multiantenna techniques for full-duplex radios, in *Proc. 47th Asilomar Conf. Signals, Systems and Computers*, pp. 3–8.
[15] Lindberg, P. and Ojefors, E. (2006) A bandwidth enhancement technique for mobile handset antennas using wavetraps. *IEEE Trans. Antennas Propag.*, **54** (8), 2226–2233.
[16] Heino, M., Venkatasubramanian, S., Icheln, C., and Haneda, K. (2015) Design of wavetraps for isolation improvement in compact in-band full-duplex relay antennas. *IEEE Trans. Antennas Propag.*, **64**, 1061–1070.
[17] Karaboikis, M., Soras, C., Tsachtsiris, G., and Makios, V. (2004) Compact dual-printed inverted-F antenna diversity systems for portable wireless devices. *IEEE Ant. Wireless Propag. Lett.*, **3** (1), 9–14.
[18] Choi, Y.S. and Shirani-Mehr, H. (2014) Simultaneous transmission and reception: algorithm, design and system level performance. *IEEE Trans. Wireless Commun.*, **12** (12), 5992–6010.
[19] Laughlin, L., Beach, M., Morris, K., and Haine, J. (2014) Optimum single antenna full duplex using hybrid junctions. *IEEE J. Select. Areas Commun.*, **32** (9), 1653–1661.
[20] Laughlin, L., Beach, M., Morris, K., and Haine, J. (2015) Electrical balance duplexing for small form factor realization of in-band full duplex. *IEEE Commun. Mag.*, **53** (5), 102–110.
[21] Venkatasubramanian, S., Li, L., Icheln, C., Ferrero, F., Luxey, C., and Haneda, K. (2015) Impact of neutralization on isolation in co-planar and back-to-back antennas, in *Proc. 9th European conference on Antennas and Propagation (EUCAP 2015)*, Lisbon, Portugal, pp. 1347–1351.
[22] Venkatasubramanian, S., Lehtovuori, A., Icheln, C., and Haneda, K. (2015) On the constraints to isolation improvement in multi-antenna systems, in *Proc. 9th European conference on Antennas and Propagation (EUCAP 2015)*, Lisbon, Portugal, pp. 1130–1134.
[23] Anttila, L., Korpi, D., Syrjälä, V., and Valkama, M. (2013) Cancellation of power amplifier induced nonlinear self-interference in full-duplex transceivers, in *Proc. 47th Asilomar Conf. Signals, Systems and Computers*, pp. 1193–1198.
[24] Korpi, D., Anttila, L., Syrjälä, V., and Valkama, M. (2014) Widely-linear digital self-interference cancellation in direct-conversion full-duplex transceiver. *IEEE J. Select. Areas Commun.*, **32** (9), 1674–1687.
[25] Anttila, L., Korpi, D., Antonio-Rodriquez, E., Wichman, R., and Valkama, M. (2014) Modeling and efficient cancellation of nonlinear self-interference in MIMO full-duplex transceivers, in *Proc. IEEE Globecom Workshops*, pp. 862–868.

[26] Syrjälä, V., Valkama, M., Anttila, L., Riihonen, T., and Korpi, D. (2014) Analysis of oscillator phase-noise effects on self-interference cancellation in full-duplex OFDM radio transceivers. *IEEE Trans. Wireless Commun.*, **13** (6), 2977–2990.

[27] Knox, M. (2012) Single antenna full duplex communications using a common carrier, in *Proc. IEEE 13th Annual Wireless and Microwave Tech. Conf.*, pp. 1–6.

[28] Yang, B., Dong, Y., Yu, Z., and Zhou, J. (2013) An RF self-interference cancellation circuit for the full-duplex wireless communications, in *Antennas Propagation (ISAP), 2013 Proceedings of the International Symposium on*, pp. 1048–1051.

[29] Phungamngern, N., Uthansakul, P., and Uthansakul, M. (2013) Digital and RF interference cancellation for single-channel full-duplex transceiver using a single antenna, in *Electrical Engineering/Electronics, Computer, Telecommunications and Information Technology (ECTI-CON), 2013 10th International Conference on*, pp. 1–5, doi: 10.1109/ECTICon.2013.6559508.

[30] Huusari, T., Choi, Y.S., Liikkanen, P., Korpi, D., Talwar, S., and Valkama, M. (2015) Wideband self-adaptive RF cancellation circuit for full-duplex radio: Operating principle and measurements, in *Proc. 81st Veh. Tech. Conf. (VTC2015-Spring)*, Glasgow, Scotland, pp. 1–7.

[31] Anaren, Xinger XDL15-2-020S delay line, Data sheet revision B.

[32] Lavín, C. and Palacios, C. and Debaillie, B. and Hershberg, B. and van Liempd, B. and Klumperink, E. and van den Broek, D.-J. (2014) DUPLO Deliverable D2.2: full-duplex radios for local access, *Tech. Rep.*, DUPLO.

[33] Sohaib, S. and So, D. (2009) Asynchronous polarized cooperative MIMO communication, in *Proc. 69th Veh. Tech. Conf. (VTC Spring 2009)*, pp. 1–5.

[34] Sohaib, S. and So, D. (2010) Energy analysis of asynchronous polarized cooperative MIMO protocol, in *Proc. 21st Int. Symp. Personal Indoor and Mobile Radio Commun. (PIMRC 2010)*, pp. 1736–1740.

[35] Antonio-Rodriguez, E., Lopez-Valcarce, R., Riihonen, T., Werner, S., and Wichman, R. (2013) Adaptive self-interference cancellation in wideband full-duplex decode-and-forward MIMO relays, in *Proc. IEEE 14th Workshop on Signal Processing Advances in Wireless Commun.*, pp. 370–374.

[36] Antonio-Rodriguez, E., Lopez-Valcarce, R., Riihonen, T., Werner, S., and Wichman, R. (2014) SINR optimization in wideband full-duplex MIMO relays under limited dynamic range, in *Proc. IEEE 8th Workshop on Sensor Array and Multichannel Signal Processing*, pp. 177–180.

[37] Riihonen, T., Werner, S., and Wichman, R. (2011) Mitigation of loopback self-interference in full-duplex MIMO relays. *IEEE Trans. Signal Process.*, **59** (12), 5983–5993.

[38] Antonio-Rodriguez, E., Lopez-Valcarce, R., Riihonen, T., Werner, S., and Wichman, R. (2013) Wideband full-duplex MIMO relays with blind adaptive self-interference cancellation. *submitted to IEEE Trans. Signal Process.*

[39] Antonio-Rodriguez, E., Lopez-Valcarce, R., Riihonen, T., Werner, S., and Wichman, R. (2015) Subspace-constrained SINR optimization in MIMO full-duplex relays under limited dynamic range, in *Proc. IEEE 14th Workshop on Signal Processing Advances in Wireless Commun.*, pp. 370–374.

[40] Tehrani, A. Cao, H., Afsardoost, S., Eriksson, T., Isaksson, M., and Fager, C. (2010) A comparative analysis of the complexity/accuracy tradeoff in power amplifier behavioral models. *IEEE Trans. Microwave Theory Tech.*, **58**, 1510–1520.

Part Four

New System-Level Enabling Technologies for 5G

18

Cloud Radio Access Networks: Uplink Channel Estimation and Downlink Precoding

Osvaldo Simeone, Jinkyu Kang, Joonkhyuk Kang and Shlomo Shamai (Shitz)

Signal Processing for 5G: Algorithms and Implementations, First Edition. Edited by Fa-Long Luo and Charlie Zhang.

18.1 Introduction

The gains afforded by cloud radio access network (C-RAN) in terms of savings in capital and operating expenses, flexibility, interference management and network densification rely on the presence of high-capacity low-latency fronthaul connectivity between remote radio heads (RRHs) and a baseband unit (BBU). In light of the non-uniform and limited availability of fiber-optic cables, the bandwidth constraints on the fronthaul network call, on the one hand, for the development of advanced baseband compression strategies and, on the other hand, for a closer investigation of the optimal functional split between RRHs and BBU. In this chapter, after a brief introduction to the signal-processing challenges in C-RAN, this optimal functional split is studied at the physical (PHY) layer as it pertains to two key baseband signal-processing steps, namely channel estimation in the uplink and channel encoding/linear precoding in the downlink. Joint optimization of baseband fronthaul compression and of baseband signal processing is tackled under different PHY functional splits, whereby uplink channel estimation and downlink channel encoding/linear precoding are carried out either at the RRHs or at the BBU. The analysis, based on information-theoretical arguments and numerical results yields insight into the configurations of network architecture and fronthaul capacities in which different functional splits are advantageous. The treatment also emphasizes the versatility of deterministic and stochastic successive convex approximation strategies for the optimization of C-RANs.

18.2 Technology Background

In a C-RAN architecture, the base station (BS) functionalities, from the PHY layer to the higher layers, are implemented in a virtualized fashion on centralized general-purpose processors rather than on the local hardware of the BSs or access points. This results in a novel cellular architecture in which low-cost wireless access points–theRRHs–which retain only radio functionalities, are centrally managed by a reconfigurable centralized "cloud", the BBU. At a high level, the C-RAN concept can be seen as an instance of network function virtualization and hence as the RAN counterpart of the separation of control and data planes proposed for the core network in software-defined networking [1].

The C-RAN architecture has the following advantages, which make it a key contender for inclusion in a 5G standard:

- reduced capital expense due to the possibility to substitute fully fledged BSs with RRHs, which have reduced space and energy requirements;
- statistical multiplexing gain thanks to the flexible allocation of radio and computing resources across all the connected RRHs;
- easier implementation of coordinated and cooperative transmission/reception strategies, such as enhanced intercell interference coordination (eICIC) and coordinated multipoint (CoMP) in Long-term Evolution Advanced (LTE-A), to mitigate multicell interference;
- simplified network upgrades and maintenance owing to the centralization of RAN functionalities.

The C-RAN architecture depends on a network of so-called fronthaul links to enable the virtualization of BS functionalities at a BBU. This is because in the uplink the RRHs are required to convey their respective received signals, either in analog format or in the form of digitized baseband samples, to the BBU for processing. Moreover, in a dual fashion, in a C-RAN downlink, each RRH needs to receive from the BBU either directly the analog radio signal to be transmitted on the radio interface or a digitized version of the corresponding baseband samples. The RRH–BBU bidirectional links that carry such information are referred to as *fronthaul* links, in contrast to the backhaul links connecting the BBU to the core network.

The analog transport solution is typically implemented on fronthaul links by means of radio-over-fiber [2]. Instead, the digital transmission of baseband, or IQ, samples is currently carried out by following the Common Public Radio Interface (CPRI) standard [3], which most commonly requires fiber optic fronthaul links as well. The digital approach appears to be favored due to the traditional advantages of digital solutions, including resilience to noise and hardware impairment and flexibility in the transport options [4].

18.2.1 Signal Processing Challenges in C-RAN

The main roadblock to the realization of the promise of C-RAN hinges on the inherent restrictions on bandwidth and latency of the fronthaul links, which may limit the advantages of centralized processing at the BBU.

18.2.1.1 Fronthaul Capacity Limitations

Implementing the CPRI standard, the bit rate required for BSs that serve multiple cell sectors with carrier aggregation and with multiple antennas exceeds the 10 Gbit/s provided by standard fiber-optic links [4, 5]. This problem is even more pronounced for networks in which fiber-optic links are not available due to the large expense required for their deployment or lease, as for heterogeneous networks with smaller RRHs [6]. The capacity limitations of the fronthaul link call for the development of compression strategies that reduce the fronthaul rate with minor or no degradation in the quality of the quantized baseband signal. Typical solutions are based on filtering, per-block scaling, lossless compression and predictive quantization [7–12].

When quantization and compression are not sufficient, as reported in [13, 14], the bottleneck on the performance of C-RANs due to the capacity limitations of the fronthaul links can be alleviated by implementing a more flexible separation of functionalities between RRHs and BBU, rather than performing all baseband processing at the BBU. Examples of baseband operations that can be carried out at the RRH include fast Fourier transform and inverse fast Fourier transform (FFT and IFFT), demapping, synchronization, channel estimation, precoding and channel encoding. Note that the possibility of implementing functions at higher layers, such as error detection, at the RRHs has also been investigated [13]. We will elaborate on important aspects of the functional split between RRH and BBU below.

18.2.1.2 Fronthaul Latency Limitations

Two of the communication protocols that are most affected by fronthaul delays are uplink hybrid automatic repeat request (HARQ) and random access [13]. For HARQ, the problem is that the outcome of decoding at the BBU may only become available at the RRH after the time required for:

- the transfer of the baseband signals from the RRH to the BBU
- the processing at the BBU
- the transmission of the decoding outcome from the BBU to the RRH.

This delay may seriously affect the throughput achievable by uplink HARQ. For example, in LTE with frequency division multiplexing, the feedback latency should be less than 8 ms in order not to disrupt the operation of the system [13]. Similar issues impair the implementation of random access.

18.2.2 Chapter Overview

In this chapter, we explore the problem of optimal functional split between RRHs and BBU at the PHY layer by focusing on the two key baseband operations of channel estimation and channel encoding/precoding. We recall that alternative functional splits are envisaged to be potentially advantageous in the presence of significant fronthaul capacity constraints.

For the uplink, we compare the standard implementation in which all baseband processing, including channel estimation, is performed at the BBU, with an alternative architecture in which channel estimation, along with the necessary frame synchronization and resource demapping, is instead implemented at the RRHs. This is discussed in Section 18.3.

The downlink is discussed in Section 18.4, where we contrast the standard C-RAN implementation with an alternative one in which channel encoding and precoding are applied at the RRHs, while the BBU retains the function of designing the precoding matrices based on the available channel state information.

Throughout, we take an information-theoretic approach in order to evaluate analytical expressions for the achievable performance that illuminates the impact of different design choices. The analysis is corroborated by extensive numerical results that provide insight into the performance comparisons highlighted above. The chapter is concluded in Section 18.5.

18.3 Uplink: Where to Perform Channel Estimation?

In this section, we study the uplink and address the potential advantages that could be accrued by performing channel estimation at the RRHs rather than at the BBU. The rationale for the exploration of this functional split is that communicating the digitized signal received within the training portion of the received signal, as done in the conventional implementation, may impose a more significant burden on the fronthaul network than communicating directly the estimated channel state information (CSI). This split is also supported by the known information-theoretic optimality of separate estimation and compression [15]. In particular, we compare two different approaches:

- the conventional approach, in which the RRHs quantize the training signals and CSI estimation takes place at the BBU;
- channel estimation at the RRHs, in which the RRHs perform CSI estimation and forward a quantized version of the CSI to the BBU.

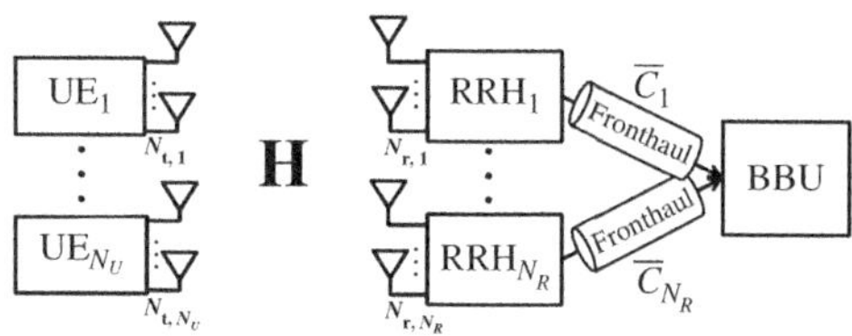

Figure 18.1 Uplink of a C-RAN system consisting of N_U UEs and N_R RRHs. Each jth RRH is connected to the BBU with a fronthaul link of capacity $\bar{C}_j$

Note that the conventional approach was the subject of an earlier study [16] and that this section is adapted from our earlier work [17], to which we refer for proofs and additional considerations.

We start by discussing the system model in Section 18.3.1 and then elaborate on the two approaches in Section 18.3.2 and Section 18.3.3. Finally, we present numerical results in Section 18.3.4.

18.3.1 System Model

We study the uplink of a cellular system consisting of N_U user equipments (UEs), N_R RRHs and a BBU, as shown in Figure 18.1. We denote the set of all UEs, or mobile users, $\mathcal{N}_U = \{1, \cdots, N_U\}$ and the set of all RRHs $\mathcal{N}_R = \{1, \cdots, N_R\}$. Each ith UE has $N_{t,i}$ transmit antennas, while each jth RRH is equipped with $N_{r,j}$ receive antennas. We define the number of total transmit antennas as $N_t = \sum_{i=1}^{N_U} N_{t,i}$. Each jth RRH is connected to the BBU via a fronthaul link of capacity $\bar{C}_j$. All rates, including $\bar{C}_j$, are normalized to the bandwidth available on the uplink channel from the UEs to the RRHs and are measured in bits/s/Hz. We assume that coding is performed across a large number of channel coherence blocks, for example over many resource blocks of an LTE system operating on a channel with significant time-frequency diversity. This implies that the ergodic capacity describes the system performance in terms of achievable rates (see, e.g., [18]).

Each channel coherence block, of length T channel uses, is split into a phase for channel training of length T_p channel uses and a phase for data transmission of length T_d channel uses, with

$$T_p + T_d = T \tag{18.1}$$

The signal transmitted by the ith UE is given by a $N_{t,i} \times T$ complex matrix $\mathbf{X}_i$, where each column corresponds to the signal transmitted by the $N_{t,i}$ antennas in a channel use. This signal is divided into the $N_{t,i} \times T_p$ pilot signal $\mathbf{X}_{p,i}$ and the $N_{t,i} \times T_d$ data signal $\mathbf{X}_{d,i}$. We assume that the transmit signal $\mathbf{X}_i$ has a total per-block power constraint $T^{-1}E[\|\mathbf{X}_i\|^2] = \bar{P}_i$, and we define $T_p{}^{-1}E[\|\mathbf{X}_{p,i}\|^2] = P_{p,i}$ and $T_d{}^{-1}E[\|\mathbf{X}_{d,i}\|^2] = P_{d,i}$ as the powers used for training and data, respectively, by the ith UE. Note that $E[\cdot]$ refers throughout to the expectation operator. In terms of pilot and data signal powers, the power constraint is hence expressed as

$$\frac{T_p}{T} P_{p,i} + \frac{T_d}{T} P_{d,i} = \bar{P}_i \tag{18.2}$$

For simplicity, we assume equal transmit power allocation for all UEs, and hence we have $\bar{P}_i = \bar{P}$, $P_{d,i} = P_d$ and $P_{p,i} = P_p$ for all $i \in \mathcal{N}_U$. Finally, we collect in matrices

$\mathbf{X}_p$ and $\mathbf{X}_d$ all the pilot signals and the data signals transmitted by all UEs, respectively: $\mathbf{X}_p = [\mathbf{X}_{p,1}^T, \cdots, \mathbf{X}_{p,N_U}^T]^T$ and $\mathbf{X}_d = [\mathbf{X}_{d,1}^T, \cdots, \mathbf{X}_{d,N_U}^T]^T$.

The training signal is $\mathbf{X}_p = \sqrt{P_p/N_t}\mathbf{S}_p$, where $\mathbf{S}_p$ is a $N_t \times T_p$ matrix with orthogonal rows and unitary power entries corresponding to the orthogonal training sequences transmitted from each antenna by all UEs [16]. Note that this implies that each training sequence is transmitted with power P_p/N_t and that the condition $T_p \geq N_t$ holds. During the data phase, the UEs transmit independent space-time codewords without precoding. Using random coding arguments, we write $\mathbf{X}_d = \sqrt{P_d/N_t}\mathbf{S}_d$, where $\mathbf{S}_d$ is a $N_t \times T_d$ matrix of independent and identically distributed (i.i.d.) $\mathcal{CN}(0,1)$ variables.

The $N_{r,j} \times T$ signal $\mathbf{Y}_j$ received by the jth RRH in a given coherence block, where each column corresponds to the signal received by the $N_{r,j}$ antennas in a channel use, can be split into the $N_{r,j} \times T_p$ received pilot signal $\mathbf{Y}_{p,j}$ and the $N_{r,j} \times T_d$ data signal $\mathbf{Y}_{d,j}$. The signal received at the jth RRH is then given by

$$\mathbf{Y}_{p,j} = \sqrt{\frac{P_p}{N_t}}\mathbf{H}_j\mathbf{S}_p + \mathbf{Z}_{p,j} \tag{18.3a}$$

$$\text{and } \mathbf{Y}_{d,j} = \sqrt{\frac{P_d}{N_t}}\mathbf{H}_j\mathbf{S}_d + \mathbf{Z}_{d,j} \tag{18.3b}$$

where $\mathbf{Z}_{p,j}$ and $\mathbf{Z}_{d,j}$ are respectively the $N_{r,j} \times T_p$ and $N_{r,j} \times T_d$ matrices of i.i.d. complex Gaussian noise variables with zero-mean and unit variance; in other words, $\mathcal{CN}(0,1)$. The $N_{r,j} \times N_t$ channel matrix $\mathbf{H}_j$ collects all the $N_{r,j} \times N_{t,i}$ channel matrices $\mathbf{H}_{ji}$ from the ith UE to the jth RRH as $\mathbf{H}_j = [\mathbf{H}_{j1}, \cdots, \mathbf{H}_{jN_U}]$.

The channel matrix $\mathbf{H}_{ji}$ is modeled as having i.i.d. $\mathcal{CN}(0, \alpha_{ji})$ entries, where α_{ji} is the path-loss coefficient between the ith UE and the jth RRH being given as

$$\alpha_{ji} = \frac{1}{1 + \left(\frac{d_{ji}}{d_0}\right)^{\eta}} \tag{18.4}$$

where d_{ji} is the distance between the ith UE and the jth RRH, d_0 is a reference distance, and η is the path-loss exponent. The channel matrices are assumed to be constant during each channel-coherence block and to change according to an ergodic process from block to block.

18.3.2 Conventional Approach

With the conventional approach, the RRH quantizes and compresses both its received pilot signal in Eq. (18.3a) and its received data signal in Eq. (18.3b), and forwards the compressed signals to the BBU on the fronthaul link. The BBU then estimates the CSI on the basis of the received quantized pilot signals and performs coherent decoding of the data signal. In the rest of Section 18.3, for simplicity of presentation, we limit the analytical treatment to the case of a single UE and a single RRH; in other words $N_U = 1$ and $N_R = 1$. We henceforth remove the subscripts indicating UE and RRH indices. A more general discussion can be found elsewhere [17].

18.3.2.1 Training Phase

During the training phase, the vector of received training signals $\mathbf{Y}_p$ in Eq. (18.3a) across all coherence times is quantized. In order to account for quantization and compression, throughout this chapter, we use the standard additive quantization noise model that follows conventional information-theoretical arguments based on random coding [19]. Accordingly, the quantized pilot signal can be written as

$$\widehat{\mathbf{Y}}_p = \mathbf{Y}_p + \mathbf{Q}_p \tag{18.5}$$

where the compression noise matrix $\mathbf{Q}_p$ is assumed to have i.i.d. $\mathcal{CN}(0, \sigma_p^2)$ entries. Note that the assumption of Gaussian i.i.d. quantization noise is made here for simplicity of analysis without claim of optimality. On a practical note, Gaussian quantization noise can be realized by high-dimensional vector quantizers such as trellis-coded quantization [20]. The quantization noise variance σ_p^2 dictates the accuracy of the quantization and depends on the fronthaul capacity via standard information-theoretic identities [19], as further discussed below.

Based on Eq. (18.5), the channel matrix $\mathbf{H}$ from the UE to the RRH is estimated at the BBU by the minimum mean-square error (MMSE) method. Hence, it can be expressed as

$$\mathbf{H} = \widehat{\mathbf{H}} + \mathbf{E} \tag{18.6}$$

where the estimated channel $\widehat{\mathbf{H}}$ is a complex Gaussian matrix with i.i.d. $\mathcal{CN}(0, \sigma_{\widehat{h}}^2)$ entries, and the estimation error $\mathbf{E}$ has i.i.d. $\mathcal{CN}(0, \sigma_e^2)$ entries. With $\sigma_{\widehat{h}}^2 = \alpha - \sigma_e^2$ and $\sigma_e^2 = \alpha N_t (1 + \sigma_p^2)/(T_p P_p + N_t(1 + \sigma_p^2))$, respectively [18, 21], where we recall that α is the power gain for the channel between UE and RRH.

18.3.2.2 Data Phase

The quantized data signal received at the BBU can be similarly expressed as $\widehat{\mathbf{Y}}_d = \mathbf{Y}_d + \mathbf{Q}_d$, where the quantization noise $\mathbf{Q}_d$ is assumed to have i.i.d. $\mathcal{CN}(0, \sigma_d^2)$ entries. Moreover, it can be written as the sum of a useful term $\widehat{\mathbf{H}}\mathbf{X}_d$ and of the equivalent noise $\mathbf{N}_d = \mathbf{E}\mathbf{X}_d + \mathbf{Z}_d + \mathbf{Q}_d$, namely

$$\widehat{\mathbf{Y}}_d = \widehat{\mathbf{H}}\mathbf{X}_d + \mathbf{N}_d \tag{18.7}$$

where the equivalent noise $\mathbf{N}_d$ has i.i.d. entries with zero mean and power $1 + \sigma_d^2 + P_d \sigma_e^2$. We observe that $\mathbf{N}_d$ is not Gaussian distributed and is not independent of $\mathbf{X}_d$. Further discussion can be found in the literature [17, 18].

18.3.2.3 Ergodic Rate

As mentioned above, we adopt as the performance criterion of interest the ergodic rate, which, under the assumption of Gaussian codebooks, is given by the mutual information $T^{-1} I(\mathbf{X}_d; \widehat{\mathbf{Y}}_d | \widehat{\mathbf{H}})$ [bits/s/Hz] (see, for example Chapter 3 of the book by Gamal and Kim [19]). This quantity can be lower-bounded by the following expression [17]:

$$R = \frac{T_d}{T} E[\log_2 \det(\mathbf{I}_{N_r} + \rho_{\text{eff}} \widehat{\mathbf{H}}\widehat{\mathbf{H}}^\dagger)] \tag{18.8}$$

with $\rho_{\text{eff}} = P_d/(N_t(1+\sigma_d^2+P_d\sigma_e^2))$ being the effective signal to noise ratio (SNR), which accounts for the effects of quantization and channel estimation, and $\widehat{\mathbf{H}}$ being distributed as in Eq. (18.6). The rate in Eq. (18.8) is hence an achievable ergodic rate [17]. Moreover, let us define C_p, the fronthaul rate allocated to transmit information about the pilot signals and C_d, the fronthaul rate for the data with $C_p + C_d = \bar{C}$. Then, if the conditions

$$C_p = \frac{T_p N_r}{T}\log_2\left(1+\frac{P_p\alpha+1}{\sigma_p^2}\right) \tag{18.9a}$$

$$\text{and } C_d = \frac{T_d N_r}{T}\log_2\left(1+\frac{P_d\alpha+1}{\sigma_d^2}\right) \tag{18.9b}$$

are satisfied, a quantization (and compression) scheme exists that guarantees the desired quantization errors (σ_d^2, σ_p^2) [17].

The ergodic achievable rate in Eq. (18.8) can now be optimized over the fronthaul allocation (C_p, C_d) under the fronthaul constraint $\bar{C} = C_p + C_d$, with C_p and C_d in Eq. (18.9), by maximizing the effective SNR ρ_{eff} in Eq. (18.8). This non-convex problem can be tackled using a line-search method [22] in a bounded interval (for examples, over C_p in the interval $[0, \bar{C}]$).

18.3.3 Channel Estimation at the RRHs

With the alternative functional split, each RRH estimates the CSI on the basis of its received pilot signal in Eq. (18.3a), and then quantizes and compresses both its estimated CSI and its received data signal in Eq. (18.3b) for transmission on the fronthaul.

18.3.3.1 Training Phase

The RRH performs the MMSE estimate of the channel $\mathbf{H}$ given the observation $\mathbf{Y}_p$ in Eq. (18.3a). As a result, similar to Eq. (18.6), we can decompose the channel matrix $\mathbf{H}$ into the MMSE estimate $\widetilde{\mathbf{H}}$ and the independent estimation error $\mathbf{E}$, as

$$\mathbf{H} = \widetilde{\mathbf{H}} + \mathbf{E} \tag{18.10}$$

where the error $\mathbf{E}$ has i.i.d. $\mathcal{CN}(0,\sigma_e^2)$ entries with $\sigma_e^2 = \alpha N_t/(T_pP_p+N_t)$ and $\widetilde{\mathbf{H}}$ has i.i.d. $\mathcal{CN}(0,\sigma_{\tilde{h}}^2)$ entries with $\sigma_{\tilde{h}}^2 = \alpha - \sigma_e^2$.

The sequence of channel estimates $\widetilde{\mathbf{H}}$ for all coherence times in the coding block is compressed by the RRH and forwarded to the BBU on the fronthaul link. The compressed channel $\widehat{\mathbf{H}}$ is related to the estimate $\widetilde{\mathbf{H}}$ as

$$\widetilde{\mathbf{H}} = \widehat{\mathbf{H}} + \mathbf{Q}_p \tag{18.11}$$

where the $N_r \times N_t$ quantization noise matrix $\mathbf{Q}_p$ has i.i.d. $\mathcal{CN}(0,\sigma_p^2)$ entries.

18.3.3.2 Data Phase

During the data phase, the RRH quantizes the signal $\mathbf{Y}_d$ in Eq. (18.3b) and sends it to the BBU on the fronthaul link. The signal obtained at the BBU is related to $\mathbf{Y}_d$ as

$$\widehat{\mathbf{Y}}_d = \mathbf{Y}_d + \mathbf{Q}_d \tag{18.12}$$

where $\mathbf{Q}_d$ is independent of $\mathbf{Y}_d$ and represents the quantization noise matrix with i.i.d. $\mathcal{CN}(0, \sigma_d^2)$ entries. Separating the desired signal and the noise in Eq. (18.12), the quantized signal $\widehat{\mathbf{Y}}_d$ can be expressed as

$$\widehat{\mathbf{Y}}_d = \widehat{\mathbf{H}}\mathbf{X}_d + \mathbf{N}_d \quad (18.13)$$

where $\mathbf{N}_d$ denotes the equivalent noise $\mathbf{N}_d = (\mathbf{Q}_p + \mathbf{E})\mathbf{X}_d + \mathbf{Z}_d + \mathbf{Q}_d$, which has i.i.d. zero-mean entries with power

$$\sigma_n^2 = P_d(\sigma_p^2 + \sigma_e^2)1 + \sigma_d^2 \quad (18.14)$$

We observe that, as in Eq. (18.7), $\mathbf{N}_d$ is not Gaussian distributed and is not independent of $\mathbf{X}_d$.

18.3.3.3 Ergodic Rate

Let C_p and C_d denote respectively the fronthaul rates allocated for the transmission of the quantized channel estimates in Eq. (18.11) and of the quantized received signals in Eq. (18.12) on the fronthaul link from the RRH to the BBU. An achievable ergodic rate is given as [17]:

$$R = \frac{T_d}{T}E[\log_2 \det(\mathbf{I}_{N_r} + \rho_{\text{eff}}\widehat{\mathbf{H}}\widehat{\mathbf{H}}^\dagger)] \quad (18.15)$$

with the effective SNR

$$\rho_{\text{eff}} = \frac{P_d}{N_t\sigma_n^2} = \frac{P_d}{N_t(1 + \sigma_d^2 + P_d(\sigma_p^2 + \sigma_e^2))} \quad (18.16)$$

$\widehat{\mathbf{H}}$ being distributed as in Eq. (18.11), and with σ_e^2 in Eq. (18.10). Moreover, if the conditions

$$C_p = \frac{N_r N_t}{T}\log_2\left(\frac{\alpha - \sigma_e^2}{\sigma_p^2}\right) \quad (18.17a)$$

$$\text{and } C_d = \frac{N_r T_d}{T}\log_2\left(1 + \left(\frac{\alpha P_d + 1}{\sigma_d^2}\right)\right) \quad (18.17b)$$

are satisfied, then a quantization scheme exists that guarantees the desired quantization error (σ_p^2, σ_d^2) [17]. The ergodic achievable rate in Eq. (18.15) can now be optimized over the fronthaul allocation (C_p, C_d) under the fronthaul constraint $\bar{C} = C_p + C_d$, with C_p and C_d in Eq. (18.17), by maximizing the effective SNR ρ_{eff} in Eq. (18.16) using a line search [22] in a bounded interval.

18.3.3.4 Adaptive Quantization

The alternative functional split studied here enables the RRHs to perform adaptive quantization of the data as a function of the estimated CSI in each coherence block. Specifically, rather than performing separate quantization of CSI and data, the data is quantized in each coherence period with a different accuracy depending on the corresponding CSI: a better channel quality calls for a more accurate quantization of the data field, and vice versa for worse CSI. We note that this is not possible in the conventional approach, in which CSI is not estimated at the RRHs. Further details can be found elsewhere [17].

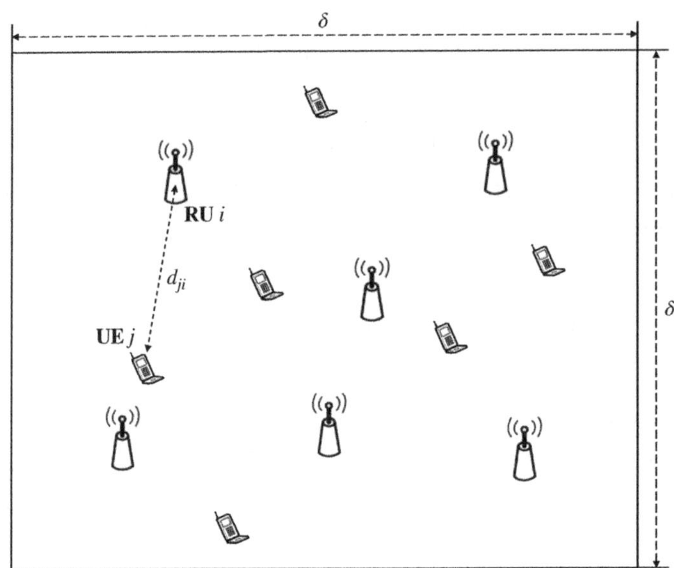

Figure 18.2 Set-up under consideration for the numerical results, where RRHs and UEs are located in a square with side δ. All RRHs are connected to the same BBU

18.3.4 Numerical Results

In this section, we evaluate the performance of the conventional and alternative strategies for the uplink. For the latter, we consider both the basic and adaptive implementations mentioned in the previous section. To this end, we consider a system with $N_R = N_U = 2$ RRHs and UEs with $N_t = N_r = 4$ antennas. The positions of the RRHs and the UEs are fixed[1] in the area with side $\delta = 500$ m as in Figure 18.2. In the path-loss formula of Eq. (18.4), we set the reference distance to $d_0 = 50$ m and the path loss exponent to $\eta = 3$. Throughout, we assume that each RRH has the same fronthaul capacity $\bar{C}$; that is, $\bar{C}_j = \bar{C}$ for $j \in \mathcal{N}_R$. We optimize over the power allocation (P_p, P_d) and we set $T_p = N_t$, which has been shown to be optimal in for a point-to-point link with no fronthaul limitation [18].

The effect of an increase of the coherence time on the ergodic achievable sum-rate is investigated in Figure 18.3 with fronthaul capacity $\bar{C} = 6$ bits/s/Hz, and power $\bar{P} = 10$ dB. As expected from information-theoretic considerations, Figure 18.3 demonstrates that the alternative approach is advantageous, although most of the gains are accrued by means of adaptive quantization. Moreover, it is observed that the performance of the conventional approach without adaptive quantization approaches that of the alternative approach as the coherence time T increases. This is because, for large coherence time T, the fraction of fronthaul capacity devoted to training becomes negligible and hence accurate CSI can be obtained at the BBU.

[1] The positions of RRHs are set as $\boldsymbol{p}_{R,1} = [307.50\ 233.18]^T$ and $\boldsymbol{p}_{R,2} = [430.3\ 192.64]^T$, where $\boldsymbol{p}_{R,i}$ is the position of ith RRH with coordinate origin at the lower left corner, and the positions of the UEs as $\boldsymbol{p}_{U,1} = [363.7\ 316.66]^T$ and $\boldsymbol{p}_{U,2} = [438.17\ 107.09]^T$, where $\boldsymbol{p}_{U,j}$ is the position of jth UE.

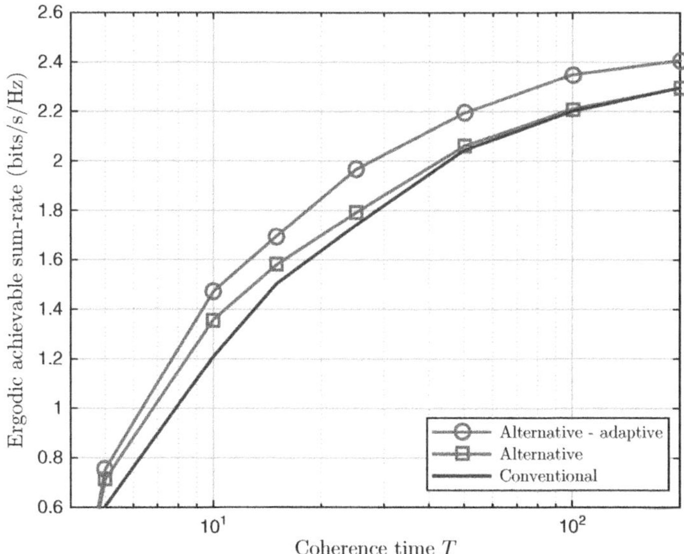

Figure 18.3 Ergodic achievable sum-rate vs coherence time ($N_R = N_U = 2$, $N_t = N_r = 4$, $\bar{C} = 6$ bits/s/Hz, and $\bar{P} = 10$ dB)

In Figure 18.4, we set the power as $\bar{P} = 10$ dB and the coherence time as $T = 10$, and we plot the ergodic achievable sum-rate versus the fronthaul capacity $\bar{C}$. The main conclusions are consistent with those discussed above for Figure 18.3. Moreover, it is seen that the performance gain of the alternative functional split is relevant as long as $\bar{C}$ is not too large, in which case the performance is limited by the uplink SNR and not by the limited fronthaul capacity.

18.4 Downlink: Where to Perform Channel Encoding and Precoding?

In this section, we turn to the downlink and address the issue of whether it is more advantageous to implement channel encoding and precoding at the RRHs rather than at the BBU as in the conventional implementation. Specifically, we compare the following two approaches:

- the conventional approach, in which the BBU performs channel coding and precoding and then quantizes and forwards the resulting baseband signals on the fronthaul links to the RRHs;
- channel encoding and precoding at the RRHs, in which the BBU does not perform precoding but rather forwards separately the information messages of a subset of UEs, along with the quantized precoding matrices to the all RRHs, which then perform channel encoding and precoding.

The conventional approach has been studied under a simplified quasi-static, rather than ergodic, channel model [23, 24], while the alternative functional split was investigated by Park *et al.* [25]. This section is adapted from our earlier paper [26], to which we refer for further

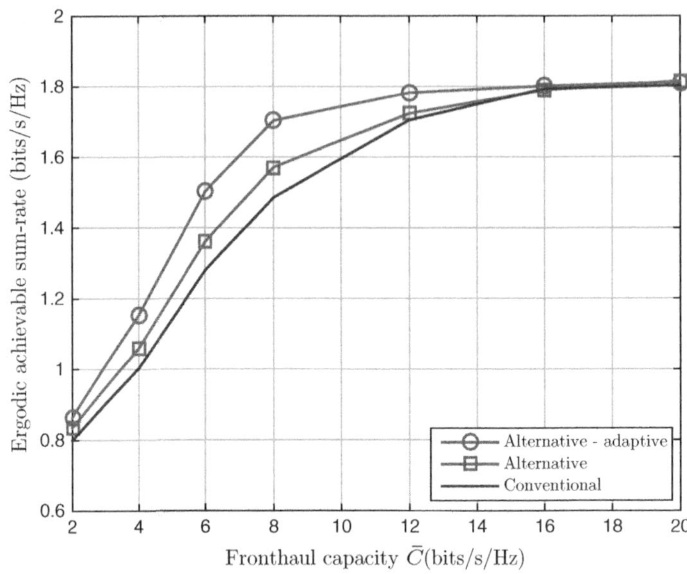

Figure 18.4 Ergodic achievable sum-rate vs fronthaul capacity ($N_R = N_U = 2$, $N_t = N_r = 4$, $\bar{P} = 10$ dB, and $T = 10$)

details and proofs. We also note that we focus here on linear precoding, or beamforming, and separate quantization for each RRH, and that related discussion on non-linear precoding and joint fronthaul quantization can be found in the literature [24].

We start by detailing the system model in Section 18.4.1. In Section 18.4.2, we study the conventional approach, while the alternative functional split mentioned above is studied in 18.4.3. In Section 18.4.4, numerical results are presented.

18.4.1 System Model

We consider the counterpart downlink C-RAN model of the uplink set-up studied in Sec 18.3, in which a cluster of N_R RRHs provides wireless service to N_U UEs as illustrated in Figure 18.5. Most of the baseband processing for all the RRHs in the cluster is carried out at a BBU that is connected to each ith RRH via a fronthaul link of finite capacity $\bar{C}_i$. Each ith RRH has $N_{t,i}$ transmit antennas and each jth UE has $N_{r,j}$ receive antennas. We denote the set of all RRHs as $\mathcal{N}_R = \{1, \cdots, N_R\}$ and the set of all UEs as $\mathcal{N}_U = \{1, \cdots, N_U\}$, and we define the number of total transmit antennas as $N_t = \sum_{i=1}^{N_R} N_{t,i}$ and of total receive antennas as $N_r = \sum_{j=1}^{N_U} N_{r,j}$. Moreover, we adopt a block-ergodic channel model in which the fading channels are constant within a coherence period but vary in an ergodic fashion across a large number of coherence periods.

Within each channel coherence period of duration T channel uses, the baseband signal transmitted by the ith RRH is given by a $N_{t,i} \times T$ complex matrix $\mathbf{X}_i$, where each column corresponds to the signal transmitted from the $N_{t,i}$ antennas in a channel use. The $N_{r,j} \times T$ signal $\mathbf{Y}_j$ received by the jth UE in a given channel coherence period, where each column

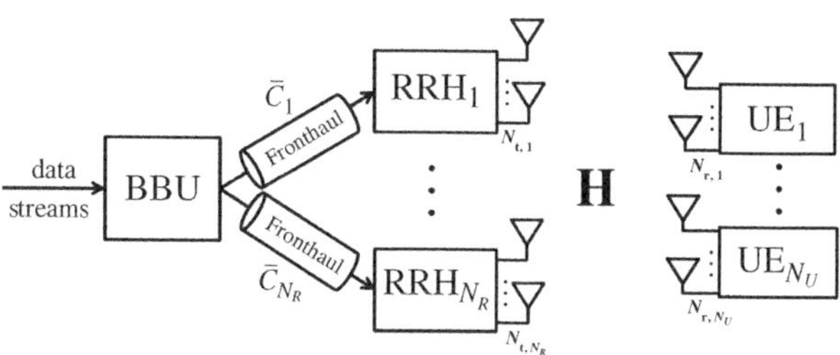

Figure 18.5 Downlink of a C-RAN system consisting of N_R RRHs and N_U UEs. The BBU is connected to each ith RRH with a fronthaul link of capacity $\bar{C}_i$

corresponds to the signal received by the $N_{r,j}$ antennas in a channel use, is given by

$$\mathbf{Y}_j = \mathbf{H}_j\mathbf{X} + \mathbf{Z}_j \tag{18.18}$$

where $\mathbf{Z}_j$ is the $N_{r,j} \times T$ noise matrix, which consists of i.i.d. $\mathcal{CN}(0,1)$ entries; $\mathbf{H}_j = [\mathbf{H}_{j1}, \cdots, \mathbf{H}_{jN_R}]$ denotes the $N_{r,j} \times N_t$ channel matrix for jth UE, where $\mathbf{H}_{ji}$ is the $N_{r,j} \times N_{t,i}$ channel matrix from the ith RRH to the jth UE; and $\mathbf{X}$ is the collection of the signals transmitted by all the RRHs, in other words $\mathbf{X} = [\mathbf{X}_1^T, \cdots, \mathbf{X}_{N_R}^T]^T$.

We consider the scenario in which the BBU has instantaneous information about the channel matrix $\mathbf{H}$ as well as the case in which the BBU is only aware of the distribution of the channel matrix $\mathbf{H}$; that is, it has *stochastic CSI*. Instead, the UEs always have full CSI about their corresponding channel matrices, as we will state more precisely in the next sections. The transmit signal $\mathbf{X}_i$ has a power constraint given as $T^{-1}E[\|\mathbf{X}_i\|^2] \leq \bar{P}_i$.

While the analysis applies more generally, in order to elaborate on the CSI requirements of the BBU, we consider as a specific channel model of interest the standard Kronecker model, in which the channel matrix $\mathbf{H}_{ji}$ is written as

$$\mathbf{H}_{ji} = \mathbf{\Sigma}_{R,ji}^{1/2}\widetilde{\mathbf{H}}_{ji}\mathbf{\Sigma}_{T,ji}^{1/2} \tag{18.19}$$

where the $N_{t,i} \times N_{t,i}$ matrix $\mathbf{\Sigma}_{T,ji}$ and the $N_{r,j} \times N_{r,j}$ matrix $\mathbf{\Sigma}_{R,ji}$ are the transmit-side and receiver-side spatial correlation matrices, respectively, and the $N_{r,j} \times N_{t,i}$ random matrix $\widetilde{\mathbf{H}}_{ji}$ has i.i.d. $\mathcal{CN}(0,1)$ variables and accounts for the small-scale multipath fading [27]. With this model, stochastic CSI entails that the BBU is only aware of the correlation matrices $\mathbf{\Sigma}_{T,ji}$ and $\mathbf{\Sigma}_{R,ji}$. Moreover, in the case that the RRHs are placed in a higher location than the UEs, one can assume that the receive-side fading is uncorrelated–$\mathbf{\Sigma}_{R,ji} = \mathbf{I}_{N_{r,j}}$–while the transmit-side covariance matrix $\mathbf{\Sigma}_{T,ji}$ is determined by the one-ring scattering model (see the paper by Adhikari *et al.*[27] and references therein). In particular, if the RRHs are equipped with $\lambda/2$-spaced uniform linear arrays, we have $\mathbf{\Sigma}_{T,ji} = \mathbf{\Sigma}_T(\theta_{ji}, \Delta_{ji})$ for the jth UE and the ith RRH located at a relative angle of arrival θ_{ji} and having angular spread Δ_{ji}, where the element (m,n) of matrix $\mathbf{\Sigma}_T(\theta_{ji}, \Delta_{ji})$ is given by

$$[\mathbf{\Sigma}_T(\theta_{ji}, \Delta_{ji})]_{m,n} = \frac{\alpha_{ji}}{2\Delta_{ji}} \int_{\theta_{ji}-\Delta_{ji}}^{\theta_{ji}+\Delta_{ji}} \exp^{-j\pi(m-n)\sin(\phi)} d\phi \tag{18.20}$$

with the path-loss coefficient α_{ji} between the jth UE and the ith RRH being given as Eq. (18.4).

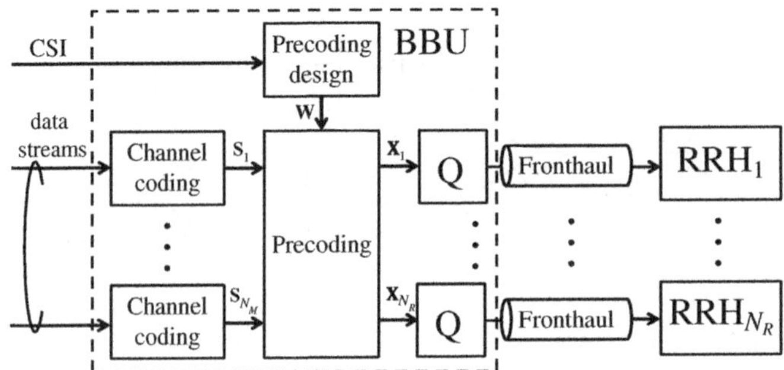

Figure 18.6 Downlink: Conventional approach ("Q" represents fronthaul compression)

18.4.2 Conventional Approach

We first describe the conventional approach in Section 18.4.2.1. Then, we discuss the joint optimization of fronthaul quantization and precoding with perfect instantaneous channel knowledge at the BBU in Section 18.4.2.2 and under the assumption of stochastic CSI at the BBU in Section 18.4.2.3.

18.4.2.1 Problem Formulation

With the conventional scheme as illustrated in Figure 18.6, the BBU performs channel coding and precoding, and then quantizes the resulting baseband signals so that they can be forwarded on the fronthaul links to the corresponding RRHs. Specifically, channel coding is performed separately for the information stream intended for each UE. This step produces the data signal $\mathbf{S} = [\mathbf{S}_1^\dagger, \cdots, \mathbf{S}_{N_U}^\dagger]^\dagger$ for each coherence block, where $\mathbf{S}_j$ is the $M_j \times T$ matrix containing, as rows, the $M_j \leq N_{r,j}$ encoded data streams for the jth UE. We define the number of total data streams as $M = \sum_{j=1}^{N_U} M_j$ and assume the condition $M \leq N_t$. Following standard random coding arguments, we take all the entries of matrix $\mathbf{S}$ to be i.i.d. as $\mathcal{CN}(0, 1)$. The encoded data $\mathbf{S}$ is further processed to obtain the transmitted signals $\mathbf{X}$ as detailed below.

The precoded data signal computed by the BBU for any given coherence time can be written as $\widetilde{\mathbf{X}} = \mathbf{WS}$, where $\mathbf{W}$ is the $N_t \times M$ precoding matrix. With instantaneous CSI, a different precoding matrix $\mathbf{W}$ is used for different coherence times in the coding block, while, with stochastic CSI, the same precoding matrix $\mathbf{W}$ is used for all coherence times.

In both cases, the precoded data signal $\widetilde{\mathbf{X}}$ can be divided into the $N_{t,i} \times T$ signals $\widetilde{\mathbf{X}}_i$ corresponding to the ith RRH for all $i \in \mathcal{N}_R$ as $\widetilde{\mathbf{X}} = [\widetilde{\mathbf{X}}_1^T, \cdots, \widetilde{\mathbf{X}}_{N_R}^T]^T$, with $\widetilde{\mathbf{X}}_i = \mathbf{W}_i^r\mathbf{S}$, where $\mathbf{W}_i^r$ is the $N_{t,i} \times N_r$ precoding matrix for the ith RRH, which is obtained by properly selecting the rows of matrix $\mathbf{W}$ (as indicated by the superscript r for "rows"): the matrix $\mathbf{W}_i^r$ is given as $\mathbf{W}_i^r = \mathbf{D}_i^{rT}\mathbf{W}$, with the $N_t \times N_{t,i}$ matrix $\mathbf{D}_i^r$ having all zero elements, except for the rows from $\sum_{k=1}^{i-1} N_{t,k} + 1$ to $\sum_{k=1}^{i} N_{t,k}$, which contain an $N_{t,i} \times N_{t,i}$ identity matrix.

The BBU quantizes each sequence of baseband signal $\widetilde{\mathbf{X}}_i$ for transmission on the ith fronthaul link to the ith RRH independently. We write the compressed signals $\mathbf{X}_i$ for the ith

RRH as

$$\mathbf{X}_i = \widetilde{\mathbf{X}}_i + \mathbf{Q}_{x,i} \tag{18.21}$$

where the quantization noise matrix $\mathbf{Q}_{x,i}$ is assumed to have i.i.d. $\mathcal{CN}(0, \sigma_{x,i}^2)$ entries. Note that the advantages of joint quantization across multiple RRHs are explored for static channels in a paper by Park *et al.* [24]. Based on Eq. (18.21), the design of the fronthaul compression reduces to the optimization of the quantization noise variances $\sigma_{x,1}^2, \cdots, \sigma_{x,N_R}^2$. The power transmitted by ith RRH is computed as

$$P_i(\mathbf{W}, \sigma_{x,i}^2) = \frac{1}{T} E[||\mathbf{X}_i||^2] = \mathrm{tr}(\mathbf{D}_i^{rT} \mathbf{W}\mathbf{W}^\dagger \mathbf{D}_i^r + \sigma_{x,i}^2 \mathbf{I}) \tag{18.22}$$

where we have emphasized the dependence of the power $P_i(\mathbf{W}, \sigma_{x,i}^2)$ on the precoding matrix $\mathbf{W}$ and quantization noise variances $\sigma_{x,i}^2$. Moreover, using standard rate-distortion arguments, the rate required on the fronthaul between the BBU and ith RRH in a given coherence interval can be quantified by $I(\widetilde{\mathbf{X}}_i; \mathbf{X}_i)/T$ (see, for example Chapter 3 of Gamal's book [19]), yielding, per Kang *et al.* [26]:

$$C_i(\mathbf{W}, \sigma_{x,i}^2) = \log\det(\mathbf{D}_i^{rT} \mathbf{W}\mathbf{W}^\dagger \mathbf{D}_i^r + \sigma_{x,i}^2 \mathbf{I}) - N_{t,i} \log(\sigma_{x,i}^2) \tag{18.23}$$

so that the fronthaul capacity constraint is $C_i(\mathbf{W}, \sigma_{x,i}^2) \leq \bar{C}_i$.

We assume that each jth UE is aware of the effective receive channel matrices $\widetilde{\mathbf{H}}_{jk} = \mathbf{H}_j \mathbf{W}_k^c$ for all $k \in \mathcal{N}_U$ at all coherence times, where $\mathbf{W}_k^c$ is the $N_t \times N_{r,j}$ precoding matrix corresponding to kth UE, which is obtained from the precoding matrix $\mathbf{W}$ by properly selecting the columns as $\mathbf{W} = [\mathbf{W}_1^c, \cdots, \mathbf{W}_{N_U}^c]$. We collect the effective channels in the matrix $\widetilde{\mathbf{H}}_j = [\widetilde{\mathbf{H}}_{j1}, \cdots, \widetilde{\mathbf{H}}_{jN_U}] = \mathbf{H}_j \mathbf{W}$. The effective channel $\widetilde{\mathbf{H}}_j$ can be estimated at the UEs via downlink training.

Under these assumptions, the ergodic achievable rate for the jth UE is computed as $E[R_j^{conv}(\mathbf{H}, \mathbf{W}, \boldsymbol{\sigma}_x^2)]$, with $R_j^{conv}(\mathbf{H}, \mathbf{W}, \boldsymbol{\sigma}_x^2) = I_{\mathbf{H}}(\mathbf{S}_j; \mathbf{Y}_j)/T$, where $I_{\mathbf{H}}(\widetilde{\mathbf{S}}_j; \mathbf{Y}_j)$ represents the mutual information for a fixed realization of the channel matrix $\mathbf{H}$, the expectation is taken with respect to $\mathbf{H}$ and

$$\begin{aligned} R_j^{conv}(\mathbf{H}, \mathbf{W}, \boldsymbol{\sigma}_x^2) &= \log\det(\mathbf{I} + \mathbf{H}_j(\mathbf{W}\mathbf{W}^\dagger + \boldsymbol{\Omega}_x)\mathbf{H}_j^\dagger) \\ &\quad - \log\det\left(\mathbf{I} + \mathbf{H}_j\left(\sum_{k \in \mathcal{N}_U \setminus j} \mathbf{W}_k^c \mathbf{W}_k^{c\dagger} + \boldsymbol{\Omega}_x\right)\mathbf{H}_j^\dagger\right) \end{aligned} \tag{18.24}$$

In Eq. (18.24), the covariance matrix $\boldsymbol{\Omega}_x$ is a diagonal with diagonal blocks given as $\mathrm{diag}([\sigma_{x,1}^2 \mathbf{I}, \cdots, \sigma_{x,N_R}^2 \mathbf{I}])$ and $\boldsymbol{\sigma}_x^2 = [\sigma_{x,1}^2, \cdots, \sigma_{x,N_R}^2]^T$.

The ergodic achievable weighted sum-rate can be optimized over the precoding matrix $\mathbf{W}$ and the compression noise variances $\boldsymbol{\sigma}_x^2$ under fronthaul capacity and power constraints. In the next subsections, we consider separately the cases with instantaneous and stochastic CSI.

18.4.2.2 Instantaneous CSI

In the case of instantaneous channel knowledge at the BBU, the design of the precoding matrix $\mathbf{W}$ and the compression noise variances $\boldsymbol{\sigma}_x^2$, is adapted to the channel realization $\mathbf{H}$ for each coherence block. To emphasize this fact, we use the notation $\mathbf{W}(\mathbf{H})$ and $\boldsymbol{\sigma}_x^2(\mathbf{H})$. The problem of optimizing the ergodic weighted achievable sum-rate with given weights $\mu_j \geq 0$ for $j \in \mathcal{N}_M$ is then formulated as

$$\underset{\mathbf{W}(\mathbf{H}),\boldsymbol{\sigma}_x^2(\mathbf{H})}{\text{maximize}} \sum_{j\in\mathcal{N}_U} \mu_j E[R_j^{conv}(\mathbf{H},\mathbf{W}(\mathbf{H}),\boldsymbol{\sigma}_x^2(\mathbf{H}))] \tag{18.25a}$$

$$\text{s.t.} \quad C_i(\mathbf{W},\sigma_{x,i}^2(\mathbf{H})) \leq \bar{C}_i, \tag{18.25b}$$

$$P_i(\mathbf{W}(\mathbf{H}),\sigma_{x,i}^2(\mathbf{H})) \leq \bar{P}_i \tag{18.25c}$$

where Eq. (18.25b)–(18.25c) apply for all $i \in \mathcal{N}_R$ and all channel realizations $\mathbf{H}$. Due to the separability of the fronthaul and power constraints across the channel realizations $\mathbf{H}$, the problem in Eq. (18.25) can be solved for each $\mathbf{H}$ independently. Note that the achievable rate in Eq. (18.25a) and the fronthaul constraint in Eq. (18.25b) are non-convex. However, the functions $R_j^{conv}(\mathbf{H},\mathbf{W}(\mathbf{H}),\boldsymbol{\sigma}_x^2(\mathbf{H}))$ and $C_i(\mathbf{W}(\mathbf{H}),\sigma_{x,i}^2(\mathbf{H}))$ are difference of convex (DC) functions of the covariance matrices $\widetilde{\mathbf{V}}_j(\mathbf{H}) = \widetilde{\mathbf{W}}_j^c(\mathbf{H})\widetilde{\mathbf{W}}_j^{c\dagger}(\mathbf{H})$ for all $j \in \mathcal{N}_U$ and the variance $\boldsymbol{\sigma}_x^2(\mathbf{H})$. The resulting rank-relaxed problem can be tackled via the majorization–minimization (MM) algorithm, as detailed in Park *et al.*'s paper [24], from which a feasible solution of problem in Eq. (18.25) can be obtained. We refer to [24] for details.

18.4.2.3 Stochastic CSI

With only stochastic CSI at the BBU, in contrast to the case with instantaneous CSI, the same precoding matrix $\mathbf{W}$ and compression noise variances $\boldsymbol{\sigma}_x^2$ are used for all the coherence blocks. Accordingly, the problem of optimizing the ergodic weighted achievable sum-rate can be reformulated as

$$\underset{\mathbf{W},\boldsymbol{\sigma}_x^2}{\text{maximize}} \sum_{j\in\mathcal{N}_U} \mu_j E[R_j^{conv}(\mathbf{H},\mathbf{W},\boldsymbol{\sigma}_x^2)] \tag{18.26a}$$

$$\text{s.t.} \quad C_i(\mathbf{W},\sigma_{x,i}^2) \leq \bar{C}_i \tag{18.26b}$$

$$P_i(\mathbf{W},\sigma_{x,i}^2) \leq \bar{P}_i \tag{18.26c}$$

where Eq. (18.26b)-(18.26c) apply to all $i \in \mathcal{N}_R$. In order to tackle this problem, we adopt the stochastic successive upper-bound minimization (SSUM) method [28], whereby, at each step, a stochastic lower bound of the objective function is maximized around the current iterate.[2] To this end, similar to Park et al.'s paper [24], we can recast the optimization over the covariance matrices $\mathbf{V}_j = \mathbf{W}_j^c\mathbf{W}_j^{c\dagger}$ for all $j \in \mathcal{N}_U$, instead of the precoding matrices $\mathbf{W}_j^c$ for all $j \in \mathcal{N}_U$. We observe that, with this choice, the objective function is expressed as the average of DC functions, while the constraint in Eq. (18.26b) is also a DC function, with respect to the

[2] We mention here that an alternative method to attack the problem is the strategy introduced in Yang *et al.* [29].

covariance $\mathbf{V} = [\mathbf{V}_1 \cdots \mathbf{V}_{N_U}]$ and the quantization noise variances $\boldsymbol{\sigma}_x^2$. Due to the DC structure, locally tight (stochastic) convex lower bounds can be calculated for objective function in Eq. (18.26a) and the constraint in Eq. (18.26b) (see, for example, Hunter and Lange [30]).

The algorithm proposed in Kang *et al.* [26] is based on SSUM [28] and contains two nested loops. At each outer iteration n, a new channel matrix realization $\mathbf{H}^{(n)} = [\mathbf{H}_1^{T\ (n)}, \cdots, \mathbf{H}_{N_U}^{T\ (n)}]$ is drawn based on the availability of stochastic CSI at the BBU. For example, with the model in Eq. (18.19), the channel matrices are generated based on the knowledge of the spatial correlation matrices. Following the SSUM scheme, the outer loop aims at maximizing a stochastic lower bound on the objective function, given as

$$\frac{1}{n}\sum_{l=1}^{n} \widetilde{R}_j^{conv}(\mathbf{H}^{(l)}, \mathbf{V}, \boldsymbol{\sigma}_x^2 | \mathbf{V}^{(l-1)}, \boldsymbol{\sigma}_x^{2\ (l-1)}) \tag{18.27}$$

where $\widetilde{R}_j^{conv}(\mathbf{H}^{(l)}, \mathbf{V}, \boldsymbol{\sigma}_x^2 | \mathbf{V}^{(l-1)}, \boldsymbol{\sigma}_x^{2\ (l-1)})$ is a locally tight convex lower bound on $R_j^{conv}(\mathbf{H}, \mathbf{W}, \boldsymbol{\sigma}_x^2)$ around solution $\mathbf{V}^{(l-1)}$, $\boldsymbol{\sigma}_x^{2\ (l-1)}$ obtained at the $(l-1)$ the outer iteration when the channel realization is $\mathbf{H}^{(l)}$. This can be calculated as (see, for example, [28])

$$\begin{aligned}\widetilde{R}_j^{conv}(\mathbf{H}^{(l)}, \mathbf{V}, \boldsymbol{\sigma}_x^2 | \mathbf{V}^{(l-1)}, \boldsymbol{\sigma}_x^{2\ (l-1)}) &\triangleq \log\det\left(\mathbf{I} + \mathbf{H}_j^{(l)}\left(\sum_{k=1}^{N_U}\mathbf{V}_k + \boldsymbol{\Omega}_x\right)\mathbf{H}_j^{(l)\ \dagger}\right)\\ &-f(\mathbf{I} + \mathbf{H}_j^{(l)}\boldsymbol{\Lambda}_j^{(l-1)}\mathbf{H}_j^{(l)\ \dagger}, \mathbf{I} + \mathbf{H}_j^{(l)}\boldsymbol{\Lambda}_j\mathbf{H}_j^{(l)\ \dagger})\end{aligned} \tag{18.28}$$

where $\boldsymbol{\Lambda}_j = \sum_{k=1,k\neq j}^{N_U}\mathbf{V}_k + \boldsymbol{\Omega}_x$, $\boldsymbol{\Lambda}_j^{(l-1)} = \sum_{k=1,k\neq j}^{N_U}\mathbf{V}_k^{(l-1)} + \boldsymbol{\Omega}_x$, the covariance matrix $\boldsymbol{\Omega}_x^{(l)}$ is a diagonal matrix with diagonal blocks given as $\mathrm{diag}([\sigma_{x,1}^{2\ (l)}\mathbf{I}, \cdots, \sigma_{x,N_R}^{2\ (l)}\mathbf{I}])$ and the linearized function $f(\mathbf{A}, \mathbf{B})$ is obtained from the first-order Taylor expansion of the log det function as

$$f(\mathbf{A}, \mathbf{B}) \triangleq \log\det(\mathbf{A}) + \frac{1}{\ln 2}\mathrm{tr}(\mathbf{A}^{-1}(\mathbf{B} - \mathbf{A})) \tag{18.29}$$

Since the maximization of Eq. (18.27) is subject to the non-convex DC constraint in Eq. (18.26b), the inner loop tackles the problem via the MM algorithm: by applying successive locally tight convex lower bounds to the left-hand side of the constraint in Eq. (18.26b) [31]. Specifically, given the solution $\mathbf{V}^{(n,r-1)}$ and $\boldsymbol{\sigma}_x^{2\ (n,r-1)}$ at $(r-1)$th inner iteration of the nth outer iteration, the fronthaul constraint in Eq. (18.26b) at the rth inner iteration can be locally approximated as

$$\begin{aligned}&\widetilde{C}_i(\mathbf{V}, \sigma_{x,i}^2 | \mathbf{V}^{(n,r-1)}, \sigma_{x,i}^{2\ (n,r-1)}) \triangleq\\ &f\left(\sum_{k=1}^{N_U}\mathbf{D}_i^{rT}\mathbf{V}_k^{(n,r-1)}\mathbf{D}_i^r + \sigma_{x,i}^{2\ (n,r-1)}\mathbf{I}, \sum_{k=1}^{N_U}\mathbf{D}_i^{rT}\mathbf{V}_k\mathbf{D}_i^r + \sigma_{x,i}^2\mathbf{I}\right) - N_{t,i}\log(\sigma_{x,i}^2)\end{aligned} \tag{18.30}$$

The resulting combination of SSUM and MM for the solution of the problem in Eq. (18.26) is summarized in Table Algorithm 18.1. The algorithm is completed by calculating, from the obtained solution $\mathbf{V}^*$ of the relaxed problem, the precoding matrix $\mathbf{W}$ by using the standard rank-reduction approach [32], which is given as $\mathbf{W}_j^* = \gamma_j \boldsymbol{\nu}_{\max}^{(M_j)}(\mathbf{V}_j^*)$ with the normalization factor γ_j, selected so as to satisfy the power constraint with equality, namely $P_i(\mathbf{W}, \sigma_{x,i}^2) = \bar{P}_i$.

Table 18.1 Design of fronthaul compression and precoding: conventional approach with stochastic CSI

Initialization: Initialize the covariance matrices $\mathbf{V}^{(0)}$ and the quantization noise variances $\boldsymbol{\sigma}_x^{2\,(0)}$, and set $n = 0$.
repeat (outer loop)
  $n \leftarrow n + 1$
  Generate a channel matrix realization $\mathbf{H}^{(n)}$ using the available stochastic CSI.
  Initialization: Initialize $\mathbf{V}^{(n,0)} = \mathbf{V}^{(n-1)}$ and $\boldsymbol{\sigma}_x^{2\,(n,0)} = \boldsymbol{\sigma}_x^{2\,(n-1)}$, and set $r = 0$.
  repeat (inner loop)
    $r \leftarrow r + 1$

$$\max_{\mathbf{V},\boldsymbol{\sigma}_x^2} \ \frac{1}{n}\sum_{l=1}^{n}\sum_{j\in\mathcal{N}_U} \mu_j \widetilde{R}_j^{conv}\left(\mathbf{H}^{(l)},\mathbf{V},\boldsymbol{\sigma}_x^2|\mathbf{V}^{(l-1)},\boldsymbol{\sigma}_x^{2\,(l-1)}\right)$$

$$\text{s.t.} \quad \widetilde{C}_i\left(\mathbf{V},\sigma_{x,i}^2|\mathbf{V}^{(n,r-1)},\sigma_{x,i}^{2\,(n,r-1)}\right) \le \bar{C}_i,$$

$$P_i\left(\mathbf{V},\sigma_{x,i}^2\right) \le \bar{P}_i, \quad \text{for all } i \in \mathcal{N}_R.$$

    Update $\mathbf{V}^{(n,r)} \leftarrow \mathbf{V}$ and $\boldsymbol{\sigma}_x^{2\,(n,r)} \leftarrow \boldsymbol{\sigma}_x^2$.
  until a convergence criterion is satisfied.
  Update $\mathbf{V}^{(n)} \leftarrow \mathbf{V}^{(n,r)}$ and $\boldsymbol{\sigma}_x^{2\,(n)} \leftarrow \boldsymbol{\sigma}_x^{2\,(n,r)}$.
until a convergence criterion is satisfied.
Solution: Calculate the precoding matrix $\mathbf{W}$ from the covariance matrices $\mathbf{V}^{(n)}$ via rank reduction as $\mathbf{W}_j = \gamma_j \nu_{\max}^{(M_j)}(\mathbf{V}_j^{(n)})$ for all $j \in \mathcal{N}_U$, where γ_j is obtained by imposing $P_i\left(\mathbf{W},\sigma_{x,i}^2\right) = \bar{P}_i$ using Eq. (18.22).

We finally note that, since the approximated functions in Eq. (18.28) and Eq. (18.30) are local lower bounds, the algorithm provides a feasible solution of the relaxed problem at each inner and outer iteration (see, for example, [28]).

18.4.3 Channel Encoding and Precoding at the RRHs

With this alternative functional split, the BBU calculates the precoding matrices, but does not perform precoding. Instead, as illustrated in Figure 18.7, it uses the fronthaul links to communicate the information messages of a given subset of UEs to each RRH, along with the corresponding compressed precoding matrices. Each RRH can then encode and precode the messages of the given UEs based on the information received from the fronthaul link. As will be discussed below, with this approach, a preliminary clustering step, whereby each UE is assigned to a subset of RRHs, is generally advantageous. In the following, we first describe the strategy in Section 18.4.3.1. Then we discuss the design problem; for fronthaul quantization and precoding under instantaneous CSI in Section 18.4.3.2 and with stochastic CSI in Section 18.4.3.3.

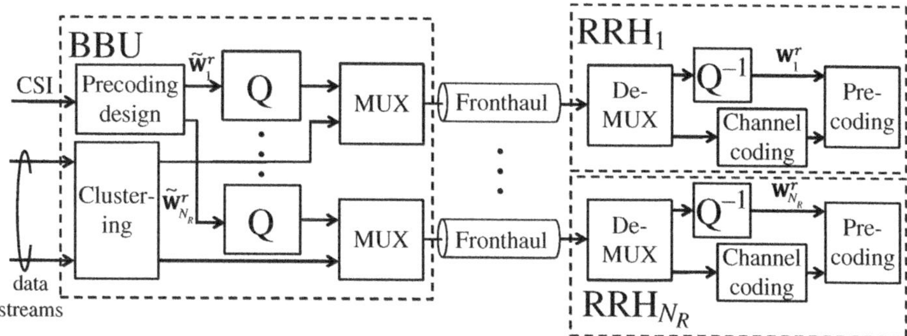

Figure 18.7 Downlink: alternative functional split. Q: fronthaul compression; Q^{-1}: fronthaul decompression

18.4.3.1 Problem Formulation

As shown in Figure 18.7, the precoding matrix $\widetilde{\mathbf{W}}$ and the information streams are separately transmitted from the BBU to the RRHs, and the received information bits are encoded and precoded at each RRH using the received precoding matrix. Note that, with this scheme, the transmission overhead over the fronthaul depends on the number of UEs supported by a RRH, since the RRHs should receive all the corresponding information streams.

Given the above, we allow for a preliminary clustering step at the BBU whereby each RRH is assigned by a subset of the UEs. We denote the set of UEs assigned by the ith RRH as $\mathcal{M}_i \subseteq \mathcal{N}_U$ for all $i \in \mathcal{N}_R$. This implies that the ith RRH only needs the information streams intended for the UEs in the set $\mathcal{M}_i$. We also denote the set of RRHs that serve the jth UE, as $\mathcal{B}_j = \{i | j \in \mathcal{M}_i\} \subseteq \mathcal{N}_R$ for all $j \in \mathcal{N}_U$. We use the notation $\mathcal{M}_i[k]$ and $\mathcal{B}_j[m]$ to respectively denote the kth UE and mth RRH in the sets $\mathcal{M}_i$ and $\mathcal{B}_j$, respectively. We define the number of all transmit antennas for the RRHs, which serve the jth UE, as $N_{t,\mathcal{B}_j}$. We assume here that the sets of UEs assigned by ith RRH are given and not subject to optimization (see Section 18.4.4 for further details).

The precoding matrix $\widetilde{\mathbf{W}}$ is constrained to have zeros in the positions that correspond to RRH–UE pairs such that the UE is not served by the given RRH. This constraint can be represented as

$$\widetilde{\mathbf{W}} = [\mathbf{E}_1^c \widetilde{\mathbf{W}}_1^c, \cdots, \mathbf{E}_{N_U}^c \widetilde{\mathbf{W}}_{N_U}^c] \tag{18.31}$$

where $\widetilde{\mathbf{W}}_j^c$ is the $N_{t,\mathcal{B}_j} \times N_{r,j}$ precoding matrix intended for the jth UE and RRHs in the cluster $\mathcal{B}_j$, and the $N_t \times N_{t,\mathcal{B}_j}$ constant matrix $\mathbf{E}_j^c$ ($\mathbf{E}_j^c$ only has either 0 or 1 entries) defines the association between the RRHs and the UEs as $\mathbf{E}_j^c = [\mathbf{D}_{\mathcal{B}_j[1]}^c, \cdots, \mathbf{D}_{\mathcal{B}_j[|\mathcal{B}_j|]}^c]$, with the $N_r \times N_{r,j}$ matrix $\mathbf{D}_j^c$ having all zero elements except for the rows from $\sum_{k=1}^{j-1} N_{r,k} + 1$ to $\sum_{k=1}^{j} N_{r,j}$, which contain an $N_{r,j} \times N_{r,j}$ identity matrix.

The sequence of the $N_{t,i} \times N_{r,\mathcal{M}_i}$ precoding matrices $\widetilde{\mathbf{W}}_i^r$ intended for each ith RRH for all coherence times in the coding block is compressed by the BBU and forwarded over the fronthaul link to the ith RRH. The compressed precoding matrix $\mathbf{W}_i^r$ for ith RRH is given by

$$\mathbf{W}_i^r = \widetilde{\mathbf{W}}_i^r + \mathbf{Q}_{w,i} \tag{18.32}$$

where the $N_{t,i} \times N_{r,\mathcal{M}_i}$ quantization noise matrix $\mathbf{Q}_{w,i}$ is assumed to have zero-mean i.i.d. $\mathcal{CN}(0, \sigma^2_{w,i})$ entries and to be independent across the index i. Overall, the $N_t \times N_r$ compressed precoding matrix $\mathbf{W}$ for all RRHs is represented as

$$\mathbf{W} = \widetilde{\mathbf{W}} + \mathbf{Q}_w \tag{18.33}$$

where $\mathbf{W} = [\mathbf{E}_1^{r\dagger}\mathbf{W}_{w,1}^{\dagger}, \cdots, \mathbf{E}_{N_R}^{r\dagger}\mathbf{W}_{w,N_R}^{\dagger}]^{\dagger}$, $\widetilde{\mathbf{W}}$ and $\mathbf{Q}_w$ are similarly defined.

Similar to Eq. (18.24), an ergodic rate achievable for the jth UE can be written as $E[R_j^{alt}(\mathbf{H}, \widetilde{\mathbf{W}}, \boldsymbol{\sigma}_w^2)]$, where

$$\begin{aligned} R_j^{alt}(\mathbf{H}, \widetilde{\mathbf{W}}, \boldsymbol{\sigma}_w^2) &= \frac{1}{T} I_{\mathbf{H}}(\mathbf{S}_j; \mathbf{Y}_j) = \log\det(\mathbf{I} + \mathbf{H}_j(\widetilde{\mathbf{W}}\widetilde{\mathbf{W}}^{\dagger} + \boldsymbol{\Omega}_w)\mathbf{H}_j^{\dagger}) \\ &\quad - \log\det\left(\mathbf{I} + \mathbf{H}_j\left(\sum_{k \in \mathcal{N}_U \setminus j} \widetilde{\mathbf{W}}_k^c \widetilde{\mathbf{W}}_k^{c\dagger} + \boldsymbol{\Omega}_w\right)\mathbf{H}_j^{\dagger}\right) \end{aligned} \tag{18.34}$$

18.4.3.2 Instantaneous CSI

With perfect CSI at the BBU, as discussed in Section 18.4.2.2, one can adapt the precoding matrix $\widetilde{\mathbf{W}}(\mathbf{H})$, the user rates $\{R_j(\mathbf{H})\}$ and the quantization noise variances $\boldsymbol{\sigma}_w^2(\mathbf{H})$ to the current channel realization at each coherence block. The rate required to transmit precoding information on the ith fronthaul in a given channel realization $\mathbf{H}$ is given by $C_i(\mathbf{H}, \widetilde{\mathbf{W}}_i^r, \sigma^2_{w,i})/T$, with

$$\begin{aligned} \frac{1}{T} C_i(\mathbf{H}, \widetilde{\mathbf{W}}_i^r, \sigma^2_{w,i}) &= \frac{1}{T} I_{\mathbf{H}}(\widetilde{\mathbf{W}}_i^r; \mathbf{W}_i^r) \\ &= \frac{1}{T}\{\log\det(\mathbf{D}_i^{rT}\widetilde{\mathbf{W}}\widetilde{\mathbf{W}}^{\dagger}\mathbf{D}_i^r + \sigma^2_{w,i}\mathbf{I}) - N_{t,i}\log(\sigma^2_{w,i})\} \end{aligned} \tag{18.35}$$

where the rate $C_i(\widetilde{\mathbf{W}}_i^r, \sigma^2_{w,i})$ required on i-fronthaul link is defined in Eq. (18.23). Note that the normalization by T is needed since only a single precoding matrix is needed for each channel coherence interval. Then, under the fronthaul capacity constraint, the remaining fronthaul capacity that can be used to convey precoding information corresponding to the ith RRH is $\bar{C}_i - \sum_{j \in \mathcal{M}_i} R_j$. As a result, the optimization problem of interest can be formulated as

$$\underset{\widetilde{\mathbf{W}}(\mathbf{H}), \boldsymbol{\sigma}^2_{w,i}(\mathbf{H}), \{R_j(\mathbf{H})\}}{\text{maximize}} \quad \sum_{j \in \mathcal{N}_U} \mu_j R_j(\mathbf{H}) \tag{18.36a}$$

$$\text{s.t.} \quad R_j(\mathbf{H}) \leq R_j^{alt}(\mathbf{H}, \widetilde{\mathbf{W}}(\mathbf{H}), \boldsymbol{\sigma}_w^2(\mathbf{H})) \tag{18.36b}$$

$$\frac{1}{T} C_i(\mathbf{H}, \widetilde{\mathbf{W}}_i^r(\mathbf{H}), \sigma^2_{w,i}(\mathbf{H})) \leq \bar{C}_i - \sum_{j \in \mathcal{M}_i} R_j(\mathbf{H}), \tag{18.36c}$$

$$P_i(\widetilde{\mathbf{W}}_i^r(\mathbf{H}), \sigma^2_{w,i}(\mathbf{H})) \leq \bar{P}_i \tag{18.36d}$$

where the constraints apply to all channel realizations, Eq. (18.36b) applies to all $j \in \mathcal{N}_U$, Eq. (18.36c)–(18.36d) apply to all $i \in \mathcal{N}_R$ and the transmit power $P_i(\widetilde{\mathbf{W}}_i^r(\mathbf{H}), \sigma^2_{w,i}(\mathbf{H}))$ at the ith RRH is defined in Eq. (18.22). Similar to Section 18.4.2.2, the problem in Eq. (18.36) can be solved for each channel realization $\mathbf{H}$ independently. In addition, each subproblem can be tackled by using the MM algorithm [24].

18.4.3.3 Stochastic CSI

With stochastic CSI at the BBU, the same precoding matrix is used for all the coherence blocks and hence the rate required to convey the precoding matrix $\widetilde{\mathbf{W}}_i^r$ to each ith RRH becomes negligible. As a result, we can neglect the effect of the quantization noise and set $\sigma_{w,i}^2 = 0$ for all $i \in \mathcal{N}_R$. Accordingly, the fronthaul capacity can be used to transfer the information stream under the constraint $\sum_{j \in \mathcal{M}_i} R_j \leq \bar{C}_i$, for all $i \in \mathcal{N}_R$. Based on the above considerations, the optimization problem of interest is formulated as

$$\underset{\widetilde{\mathbf{W}}, \{R_j\}}{\text{maximize}} \sum_{j \in \mathcal{N}_U} \mu_j R_j \tag{18.37a}$$

$$s.t. \ R_j \leq E[R_j^{alt}(\mathbf{H}, \widetilde{\mathbf{W}}, \boldsymbol{0})] \tag{18.37b}$$

$$\sum_{j \in \mathcal{M}_i} R_j \leq \bar{C}_i \tag{18.37c}$$

$$P_i(\widetilde{\mathbf{W}}_i^r, 0) \leq \bar{P}_i \tag{18.37d}$$

where Eq. (18.37b) applies to all $j \in \mathcal{N}_U$, Eq. (18.37c)–(18.37d) apply to all $i \in \mathcal{N}_R$ and the transmit power $P_i(\widetilde{\mathbf{W}}_i^r, \sigma_{w,i}^2)$ at ith RRH is defined in Eq. (18.22). In Eq. (18.37), the constraint in Eq. (18.37b) is not only non-convex but also stochastic. Similar to Section 18.4.2.3, the functions $R_j^{alt}(\mathbf{H}, \widetilde{\mathbf{W}})$ are DC functions of the covariance matrices $\widetilde{\mathbf{V}}_j = \widetilde{\mathbf{W}}_j^c \widetilde{\mathbf{W}}_j^{c\dagger}$ for all $j \in \mathcal{N}_U$, hence opening up the possibility of developing a solution based on SSUM. We refer to Kang *et al.* [26] for details of the resulting algorithm.

18.4.4 *Numerical Results*

In this section, we compare the performance of the conventional approach and the alternative split. To this end, we consider the RRHs and UEs to be randomly located in a square area with side $\delta = 500$ m as in Figure 18.2. As in Section 18.3.4, in the path-loss formula of Eq. (18.4), we set the reference distance to $d_0 = 50$ m and the path loss exponent to $\eta = 3$. We assume the spatial correlation model in Eq. (18.20) with the angular spread $\Delta_{ji} = \arctan(r_s / d_{ji})$, with the scattering radius $r_s = 10$ m and with d_{ji} being the Euclidean distance between the ith RRH and the jth UE. Throughout, we consider that the every RRH is subject to the same power constraint $\bar{P}$ and has the same fronthaul capacity $\bar{C}$; that is, $\bar{P}_i = \bar{P}$ and $\bar{C}_i = \bar{C}$ for $i \in \mathcal{N}_R$. Moreover, in the alternative split scheme, the UE-to-RRH assignment is carried out by choosing, for each RRH, the N_c UEs that have the largest instantaneous channel norms for instantaneous CSI and the largest average channel matrix norms for stochastic CSI. Note that this assignment is done for each coherence block in the former case, while in the latter the same assignment holds for all coherence blocks. Note also that a given UE is generally assigned to multiple RRHs.

The effect of the fronthaul capacity limitations on the ergodic achievable sum-rate is investigated in Figure 18.8, where the number of RRHs and UEs is $N_R = N_U = 4$, the number of transmit antennas is $N_{t,i} = 2$ for all $i \in \mathcal{N}_R$, the number of receive antennas is $N_{r,j} = 1$ for all $j \in \mathcal{N}_U$, the power is $\bar{P} = 10$ dB, and the coherence time is $T = 20$. We first observe that, with instantaneous CSI, the conventional approach is uniformly better than the alternative split as long as the fronthaul capacity is sufficiently large (here $\bar{C} > 2$).

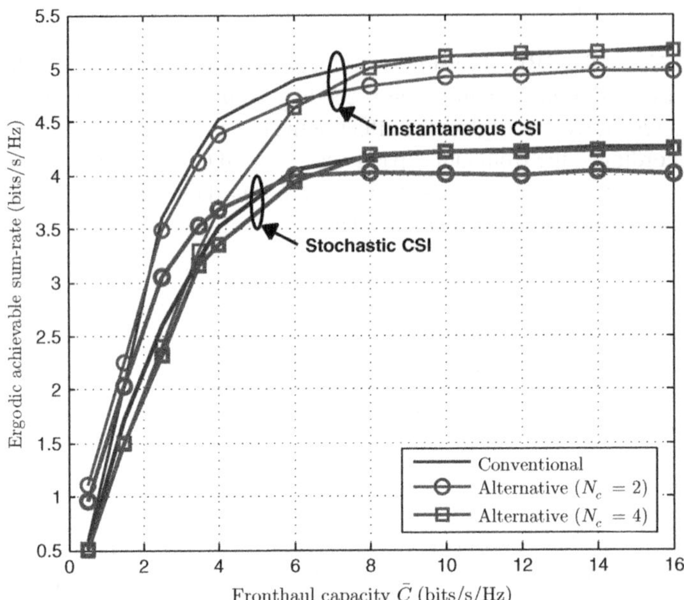

Figure 18.8 Ergodic achievable sum-rate vs the fronthaul capacity $\bar{C}$ ($N_R = N_U = 4$, $N_{t,i} = 2$, $N_{r,j} = 1$, $\bar{P} = 10$ dB, $T = 20$, and $\mu = 1$)

This is due to the enhanced interference mitigation capabilities of the conventional approach resulting from its ability to coordinate all the RRHs via joint baseband processing without requiring the transmission of all messages on all fronthaul links. Note in fact that, with the alternative split, only N_c UEs are served by each RRH, and that making N_c larger entails a significant increase in the fronthaul capacity requirements. We will later see that this advantage of the conventional approach is offset by the higher fronthaul efficiency of the alternative split in transmitting precoding information for large coherence periods T (see Figure 18.9). Instead, with stochastic CSI, in the low fronthaul capacity regime, here about $\bar{C} < 6$, the alternative split strategy is generally advantageous due to the additional gain that is accrued by amortizing the precoding overhead over the entire coding block. Another observation is that, for small $\bar{C}$, the alternative split schemes with progressively smaller N_c have better performance thanks to the reduced fronthaul overhead. Moreover, for large $\bar{C}$, the performance of the alternative split scheme with $N_c = N_U$, whereby each RRH serves all UEs, approaches that of the conventional scheme.

Figure 18.9 shows the ergodic achievable sum-rate as function of the coherence time T, with $N_R = N_U = 4$, $N_{t,i} = 2$, $N_{r,j} = 1$, $\bar{C} = 2$ bits/s/Hz, and $\bar{P} = 20$ dB. As anticipated, with instantaneous CSI, the alternative split is seen to benefit from a larger coherence time T, since the fronthaul overhead required to transmit precoding information gets amortized over a larger period. This is in contrast to the conventional approach for which such overhead scales proportionally to the coherence time T and hence the conventional scheme is not affected by the coherence time. As a result, the alternative split can outperform the conventional approach for sufficiently large T in the presence of instantaneous CSI. Instead, with stochastic CSI, the

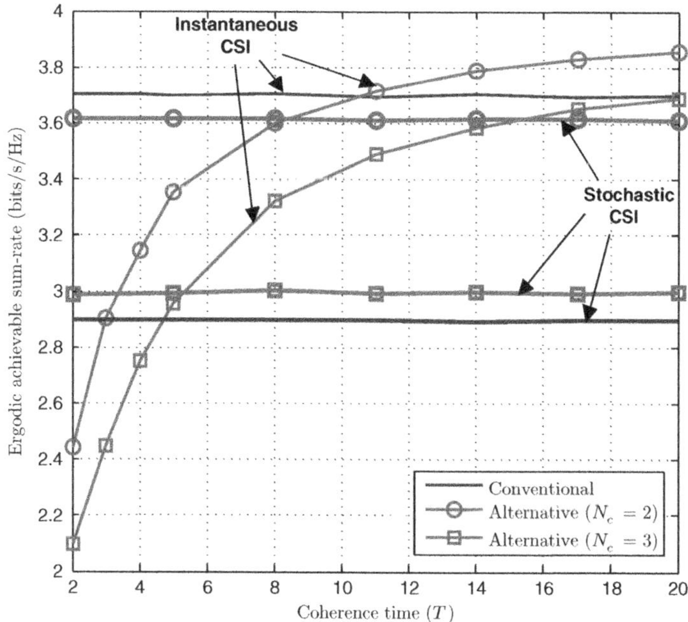

Figure 18.9 Ergodic achievable sum-rate vs the coherence time T ($N_R = N_U = 4$, $N_{t,i} = 2$, $N_{r,j} = 1$, $\bar{C} = 2$ bits/s/Hz, $\bar{P} = 20$ dB, and $\mu = 1$)

effect is even more pronounced due to the additional advantage that is accrued by amortizing the precoding overhead over the entire coding block.

Finally, in Figure 18.10, the ergodic achievable sum-rate is plotted versus the number of UEs N_U for $N_R = 4$, $N_{t,i} = 2$, $N_{r,j} = 1$, $\bar{C} = 4$, $\bar{P} = 10$ dB and $T = 10$. It is observed that the enhanced interference mitigation capabilities of the conventional approach without the overhead associated to the transmission of all messages on the fronthaul links yield performance gains for denser C-RANs; in other words for larger values of N_U. This remains true for both instantaneous and stochastic CSI cases.

18.5 Concluding Remarks

In this chapter, we have investigated two important aspects of the optimal functional split between RRH and BBU at the PHY layer, namely whether uplink channel estimation and downlink encoding/precoding should be implemented at the RRH or at the BBU. The analysis, based on information-theoretical arguments and numerical results built on proposed efficient design algorithms, yields insight into the configurations of network architecture, channel variability and fronthaul capacities in which different functional splits are advantageous. Among the main conclusions, we have argued that the alternative functional split in which uplink channel estimation is performed at the RRH is to be preferred for low or moderate values of the coherence period and fronthaul capacity, and mostly for its capability to enable adaptive quantization based on the channel conditions. Moreover, the alternative functional split in which downlink encoding and precoding are carried out at the RRH is beneficial for lightly loaded

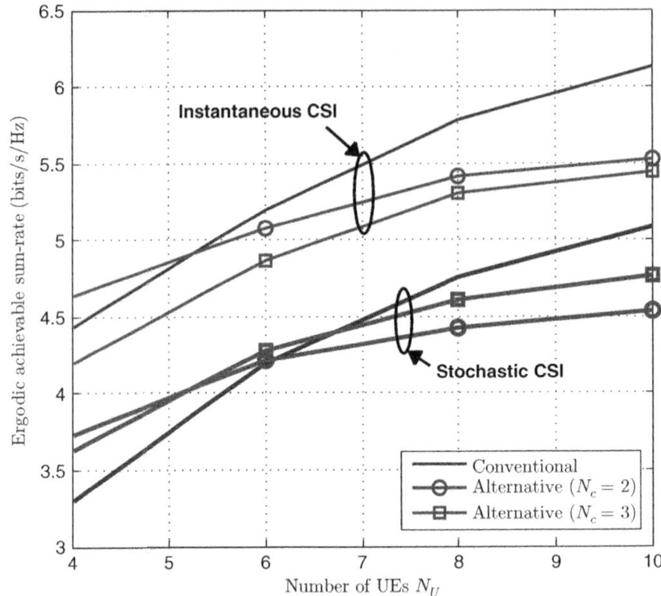

Figure 18.10 Ergodic achievable sum-rate vs the number of UEs N_U ($N_R = 4$, $N_{t,i} = 2$, $N_{r,j} = 1$, $\bar{C} = 4$ bits/s/Hz, $\bar{P} = 10$ dB, $T = 10$, and $\mu = 1$)

networks in the presence of slowly changing channels, particularly under the assumption of stochastic CSI, due to its reduced fronthaul overhead.

We close this chapter with some remark on related topics and open issues. For the uplink, an aspect that deserves further study is the integration of distributed source coding techniques (or Wyner–Ziv coding) with fronthaul processing for the joint transfer of CSI and data (see Park *et al.* [24] for some initial discussion). Analogously, for the downlink, the impact of joint, or multivariate, compression, as also proposed in Park *et al.* [24], on the optimal functional split in the presence of different degrees of CSI at the BBU is an interesting open problem. Finally, the analysis of alternative RRH–BBU functional splits in conjunction with structured coding, or compute-and-forward, techniques calls for further attention (see [33] and references therein).

References

[1] Bo, H., Gopalakrishnan, V., Ji, L., and Lee, S. (2015) Network function virtualization: Challenges and opportunities for innovations. *IEEE Commun. Mag.*, **53** (2), 90–97.

[2] Al-Raweshidy, H. and Komaki, S. (2002) *Radio over Fiber Technologies for Mobile Communications Networks*, Artech House.

[3] Ericsson AB, Huawei Technologies, NEC Corporation, Alcatel Lucent, and Nokia Siemens Networks (2011), Common public radio interface (CPRI); interface specification v5.0.

[4] Checko, A., Christiansen, H.L., Yan, Y., Scolari, L., Kardaras, G., Berger, M.S., and Dittmann, L. (2015) Cloud RAN for mobile networks–a technology overview. *IEEE Commun. Surveys Tut.*, **17** (1), 405–426.

[5] Integrated Device Technology Inc. (2013), White paper: Front-haul compression for emerging C-RAN and small cell networks.

[6] Fujitsu (2015), The benefits of cloud-RAN architecture in mobile network expansion.
[7] Samardzija, D., Pastalan, J., MacDonald, M., Walker, S., and Valenzuela, R. (2012) Compressed transport of baseband signals in radio access networks. *IEEE Trans. Wireless Commun.*, **11** (9), 3216–3225.
[8] Guo, B., Cao, W., Tao, A., and Samardzija, D. (2012) CPRI compression transport for LTE and LTE-A signal in C-RAN. *Proc. Int. ICST Conf. CHINACOM*, pp. 843–839.
[9] Nieman, K.F. and Evans, B.L. (2013) Time-domain compression of complex-baseband lte signals for cloud radio access networks. *Proc. IEEE Glob. Conf. on Sig. and Inf. Proc.*, pp. 1198–1201.
[10] Lorca, J. and Cucala, L. (2013) Lossless compression technique for the fronthaul of LTE/LTE-advanced cloud-RAN architectures. *Proc. IEEE Int. Symp. World of Wireless, Mobile and Multimedia Networks (WoWMoM)*, pp. 1–9.
[11] Grieger, S., Boob, S., and Fettweis, G. (2012) Large scale field trial results on frequency domain compression for uplink joint detection. *Proc. IEEE Glob. Comm. Conf.*, pp. 1128–1133.
[12] Vosoughi, A., Wu, M., and Cavallaro, J.R. (2012) Baseband signal compression in wireless base stations. *Proc. IEEE Glob. Comm. Conf.*, pp. 4505–4511.
[13] Dotsch, U., Doll, M., Mayer, H.P., Schaich, F., Segel, J., and Sehier, P. (2013) Quantitative analysis of split base station processing and determination of advantageous architectures for LTE. *Bell Labs Tech. J.*, **18** (1), 105–128.
[14] Wubben, D., Rost, P., Bartelt, J., Lalam, M., Savin, V., Gorgoglione, M., Dekorsy, A., and Fettweis, G. (2014) Benefits and impact of cloud computing on 5G signal processing: Flexible centralization through cloud-RAN. *IEEE Signal. Process. Mag.*, **31** (6), 35–44.
[15] Witsenhausen, H.S. (1980) Indirect rate distortion problems. *IEEE Trans. Info. Theory*, **26** (5), 518–521.
[16] Hoydis, J., Kobayashi, M., and Debbah, M. (2011) Optimal channel training in uplink network MIMO systems. *IEEE Trans. Signal Process.*, **59** (6), 2824–2833.
[17] Kang, J., Simeone, O., Kang, J., and Shamai, S. (2014) Joint signal and channel state information compression for the backhaul of uplink network MIMO systems. *IEEE Trans. Wireless Commun.*, **13** (3), 1555–1567.
[18] Hassibi, B. and Hochwald, B.M. (2003) How much training is needed in multiple-antenna wireless links? *IEEE Trans. Info. Theory*, **49** (4), 951–963.
[19] Gamal, A.E. and Kim, Y.H. (2011) *Network Information Theory*, Cambridge University Press.
[20] Zamir, R. and Feder, M. (1996) On lattice quantization noise. *IEEE Trans. Info. Theory*, **42** (4), 1152–1159.
[21] Bjornson, E. and Ottersten, B.E. (2010) A framework for training-based estimation in arbitrarily correlated Rician MIMO channels with Rician disturbance. *IEEE Trans. Signal Process.*, **58** (3), 1807–1820.
[22] S. Boyd and L. Vandenberghe (2004) *Convex Optimization*, Cambridge University Press.
[23] Simeone, O., Somekh, O., Poor, H.V., and Shamai, S. (2009) Downlink multicell processing with limited-backhaul capacity. *EURASIP J. Adv. Signal Process.*, 10.1155/2009/840814.
[24] Park, S.H., Simeone, O., Sahin, O., and Shamai, S. (2013) Joint precoding and multivariate backhaul compression for the downlink of cloud radio access networks. *IEEE Trans. Signal Process.*, **61** (22), 5646–5658.
[25] Park, S., Chae, C.B., and Bahk, S. (2013) Before/after precoded massive MIMO in cloud radio access networks. *Proc. IEEE Int. Conf. on Comm.*
[26] Kang, J., Simeone, O., Kang, J., and Shamai, S. (2014) Fronthaul compression and precoding design for C-RANs over ergodic fading channel. *arXiv:1412.7713.*
[27] Adhikary, A., Nam, J., Ahn, J.Y., and Caire, G. (2014) Joint spatial division and multiplexing: The large-scale array regime. *IEEE Trans. Info. Theory*, **59** (10), 6441–6463.
[28] Razaviyayn, M., Sanjabi, M., and Luo, Z.Q. (2013) A stochastic successive minimization method for nonsmooth nonconvex optimization with applications to transceiver design in wireless communication networks. *arXiv:1307.4457.*
[29] Yang, Y., Scutari, G., and Palomar, D.P. (2013) Parallel stochastic decomposition algorithms for multi-agent systems. *Proc. IEEE Workshop on Sign. Proc. Adv. in Wireless Comm.*, pp. 180–184.
[30] Hunter, D.R. and Lange, K. (2004) A tutorial on MM algorithms. *The American Statistician*, **58** (1), 30–37.
[31] Beck, A. and Teboulle, M. (2010) Gradient-based algorithms with applications to signal recovery problems, in *Convex Optimization in Signal Processing and Communications* (eds Y. Eldar and D. Palomar), Cambridge University Press, pp. 42–48.
[32] Vandenberghe, L. and Boyd, S. (1996) Semidefinite relaxation of quadratic optimization problems. *SIAM Rev.*, **38** (1), 49–95.
[33] Nazer, B., Cadambe, V., Ntranos, V., and Caire, G. (2015) Expanding the compute-and-forward framework: Unequal powers, signal levels, and multiple linear combinations. *arXiv:1504.01690.*

19

Energy-efficient Resource Allocation in 5G with Application to D2D

Alessio Zappone, Francesco Di Stasio, Stefano Buzzi and Eduard Jorswieck

Signal Processing for 5G: Algorithms and Implementations, First Edition. Edited by Fa-Long Luo and Charlie Zhang.

19.1 Introduction

The fifth generation (5G) of wireless communication systems will have to cope with an unprecedented number of connected devices, which is expected to reach 50 billion by 2020. On the one hand, this will require the data rates to grow by a factor of 1000 in order to serve such a massive number of devices, and to provide many new services, including e-health, e-banking, e-learning, and so on. On the other hand, such a data-rate increase cannot be achieved by simply scaling up the transmit powers, due to sustainable-growth, environmental and economic reasons. Instead, the thousand-fold data-rate increase will have to be achieved at a similar or lower level of energy consumption as present wireless networks [1]. It is therefore recognized that bit/Joule energy efficiency is a central design principle of 5G [2].

In the NGMN white paper for 5G [1], energy efficiency is identified as a key performance indicator of 5G, and is defined as the number of bits that can be transmitted per Joule of energy, where the energy is computed over the whole network, including potentially legacy cellular technologies, radio access and core networks, and data centers. This definition is in line with the physical meaning of efficiency, since the energy efficiency is defined as the system cost-benefit ratio in terms of the amount of data reliably transmitted divided by the energy that is required to do so. On the physical and medium-access control layers of a wireless network, the transmission-related energy is consumed for RF transceiver chains including power amplifiers, low noise amplifiers, pre- and decoding algorithms and signal processing, as well as channel coding and de-coding.

Before we arrive at a system-wide energy efficiency metric like the NGMN one [1], let us first focus on the energy efficiency of one single link. Consider a point-to-point link, with communication taking place over a time slot T, with transmit power p and signal-to-noise ratio (SNR) γ. Let the number of transmitted bits in the time slot T be proportional to $Tf(\gamma(p))$ and let the function $f(\cdot)$ be any measure of the amount of data that can be reliably sent to the destination per unit of time (i.e. the achievable rate, throughput, and so on). The energy to be consumed to transmit $Tf(\gamma(p))$ bits is $T(\alpha p + \theta)$. The constant α accounts for amplifier non-idealities and θ for the power dissipated in all other hardware components: the DA/AD converters, modulation filters, signal processing operation, and so on. Summing up, the energy efficiency for this single link is

$$\text{EE} = \frac{f(\gamma(p))}{\alpha p + \theta} \tag{19.1}$$

Two important properties of the EE in Eq. (19.1) are that it is neither monotonic increasing nor concave in the transmit power p. Therefore, it does not fulfill the typical properties of SNR or signal-to-interference-plus-noise ratio (SINR)-based utility functions, which are usually used to model resource allocation problems. Rather, it has the form of a fraction.

Now, if we look at a network consisting of multiple links, there are two conceptually different approaches available to generalize the concept of energy efficiency:

- Look at the overall system and define the global energy efficiency (GEE) as the ratio of the sum of all individual rates (the system throughput) to the overall consumed energy, as in the NGMN paper [1].
- Consider the design space consisting of single energy efficiency measures of individual links, services, users, cells, or subsystems.

There are different ways to formulate and de-construct the corresponding multi-objective programming problems into weighted arithmetic or geometric means of the energy efficiencies, or weighted minimum energy efficiency [3]. Regardless of which of the two approaches we follow, we are still faced with the problem of maximizing one or more fractions. As a consequence, fractional programming appears the most suitable optimization theory for energy-efficient resource allocation. This chapter is thus devoted to the illustration of fractional programming, and in particular of sequential fractional programming, for energy-efficient resource allocation in 5G wireless networks. A system-centric perspective will be taken, focusing on GEE maximization. In such a case, we have to deal with a single fraction, where the numerator is a term proportional to the system sum-rate, and the denominator is the overall network power consumption.

Interestingly, in a white paper published by Ericsson it is shown that in current LTE networks, the spectrum is under-utilized—below 20%—even in "extreme traffic" situations [4]. Additionally, the energy consumption varies only minimally with the traffic load. Therefore, in 5G wireless, we need better energy proportionality—from "always on" to "always available"—and better spectrum utilization. Therefore, this chapter proposes the device-to-device (D2D) communication paradigm to better utilize the spectrum. The energy-efficient resource allocation problem in D2D systems for energy proportionality is considered as a case study for the application of the outlined sequential fractional programming algorithms.

D2D communication is indeed being considered as one of the key ingredients of 5G wireless networks [5, 6]; D2D links permit replacement of long two-hop routes across base stations (BSs) with shorter one-hop routes, with benefits in terms of reduced latency, reduced power consumption and, possibly, increased spectral efficiency due to more aggressive reuse of the available spectrum. On the other hand, the use of D2D links brings new problems and challenges, due to the irregular deployment of the users and the increased level of interference, which call for advanced resource allocation and interference management schemes.

Recent studies in the literature have examined transmission schemes for D2D-enabled cellular systems based on geometric tools [7, 8, 9] and game theory [10, 11], and they have also considered the use of D2D transmission for cellular offloading [12]; there is also a survey of D2D communications [13]. In general, these papers show that the use of D2D links is a valid offloading technique that permits an increase in the overall network throughput provided, however, that suitable protection areas exist between D2D receivers and transmitting BSs.

In this chapter, we will consider the downlink of an orthogonal frequency-division multiple-access (OFDMA) cellular network, where infrastructure-to-device (I2D) and D2D communications coexist and are coordinated. The resource allocation policy that will be derived aims at maximizing the GEE of the coordinated I2D and D2D links. It will be also shown that the GEE cost function includes as a special case the weighted sum-rate (WSR), so that the procedure outlined for GEE-maximizing resource allocation can also be straightforwardly used for WSR-maximizing resource allocation. Thus, a comparison between the GEE-maximizing and WSR-maximizing resource allocation schemes will be given too.

The chapter is organized as follows. The next section contains a description of the system model and the received signal model for the case of a D2D-enabled cellular system. In Section 19.3 the two key performance measures considered in this chapter are formally presented, namely the GEE and the WSR. Sections 19.4–19.7 represent the mathematical part of the chapter, wherein the general theory on fractional programming (Section 19.4), on algorithms for solving fractional programs (Section 19.5), and on sequential fractional programming (Section 19.6) are surveyed. Section 19.7 focuses back on the case study and presents the application of sequential fractional programming algorithms to the problem of resource allocation for a 5G cellular system with D2D communications. Numerical results and their discussion are presented in Section 19.8, while concluding remarks are given in Section 19.9.

19.2 Signal Model

We consider a cluster of M coordinated BSs in the downlink of an OFDMA cellular network employing N subcarriers and universal frequency reuse. Devices and BSs are equipped with one antenna. There are K users served by the cellular infrastructure (also referred to as I2D users); each I2D user is connected to only one BS, which is selected according to long-term channel quality measurements. We denote by $\mathcal{B}_m$ the (non-empty) set of users assigned to BS m, and assume that each BS serves at most one user at a time on each subcarrier. Also, there are K_D D2D links between pairs of devices; D2D communication is network-controlled and takes place on the same frequency spectrum as I2D communication. Each D2D communication takes place on one subcarrier; we denote by $\mathcal{K}_D^{[n]}$ the set of D2D users transmitting on subcarrier n. Extension to the case in which D2D links use multiple carrier frequencies is straightforward and not considered here for the sake of simplicity. Channel coefficients are assumed to be perfectly estimated and the channel coherence time is such that resources may be allocated on the basis of their realizations.[1] Refer to Figure 19.1 for a pictorial view of the system and the notation used.

19.2.1 I2D Communication

Let $k(m,n) \in \mathcal{B}_m$ be the user served by BS m on subcarrier n; assuming perfect synchronization, the discrete-time baseband signal received by user $k(m,n)$ on subcarrier n is

$$\begin{aligned} r_{k(m,n)}^{[n]} = & \underbrace{H_{m,k(m,n)}^{[n]} x_m^{[n]}}_{\text{useful data}} + \underbrace{\sum_{\ell=1,\ell\neq m}^{M} H_{\ell,k(m,n)}^{[n]} x_\ell^{[n]}}_{\text{inter-cell interference}} + \\ & \underbrace{\sum_{j\in\mathcal{K}_D^{[n]}} \tilde{H}_{j,k(m,n)}^{[n]} \tilde{x}_j}_{\text{interference from D2D links}} + \underbrace{z_{k(m,n)}^{[n]}}_{\text{noise}} \end{aligned} \tag{19.2}$$

In Eq. (19.2), $H_{q,s}^{[n]}$ and $\tilde{H}_{j,s}^{[n]}$ are the complex channel responses between BS q and I2D user s, and between D2D transmitter j and I2D user s, respectively, at subcarrier n; these coefficients include small-scale fading, large-scale fading and path attenuation. The term $x_q^{[n]}$ is the data symbol transmitted by BS q on subcarrier n, while $\tilde{x}_j$ is the data symbol transmitted by D2D

[1] This is a customary hypothesis when dealing with resource allocation in cellular networks, which usually holds unless the mobile users are moving at very high speed.

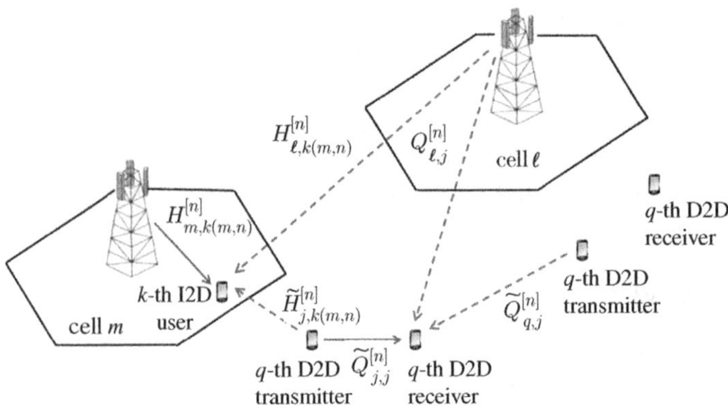

Figure 19.1 The scenario. Continuous lines represent the intended communication, while dashed lines represent interference

transmitter j. We model the transmitted symbols as independent random variables with zero mean and variance $\mathrm{E}\{|x^{[n]}_m|^2\} \triangleq p^{[n]}_m \geq 0$ and $\mathrm{E}\{|\tilde{x}_j|^2\} \triangleq \tilde{p}_j \geq 0$; we assume a per-subcarrier power constraint at each BS, in other words, $p^{[n]}_m \leq P_{m,\max}/N$ for $m = 1, \ldots, M$, and $n = 1, \ldots, N$, where $P_{m,\max}$ is the total power at BS m. We also assume $\tilde{p}_j \leq \tilde{P}_{j,\max}$ for $j = 1, \ldots, K_D$, where $\tilde{P}_{j,\max}$ is the power available at D2D transmitter j. Finally, $z^{[n]}_s$ is the additive noise received by I2D user s, which is modeled as a circularly-symmetric complex Gaussian random variable with variance $\mathcal{N}^{[n]}_s/2$ per real dimension. Different noise variances at each mobile may account for different levels of the (out-of-cluster) interference and for different noise figures of the receivers.

The SINR for BS m on subcarrier n when serving I2D user $s \in \mathcal{B}_m$ is written as

$$\gamma^{[n]}_{m,s} = \frac{p^{[n]}_m G^{[n]}_{m,s}}{1 + \sum\limits_{\ell=1,\ell\neq m}^{M} p^{[n]}_\ell G^{[n]}_{\ell,s} + \sum\limits_{j\in\mathcal{K}^{[n]}_D} \tilde{p}_j \tilde{G}^{[n]}_{j,s}} \tag{19.3}$$

where $G^{[n]}_{q,s} = |H^{[n]}_{m,s}|^2/\mathcal{N}^{[n]}_s$ and $\tilde{G}^{[n]}_{q,s} = |\tilde{H}^{[n]}_{q,s}|^2/\mathcal{N}^{[n]}_s$, while the corresponding achievable rate (in bit/s) is

$$\mathrm{R}^{[n]}_{m,s} = B\log_2[1 + \gamma^{[n]}_{m,s}]$$

where B is the bandwidth of each subcarrier. For notational convenience, we denote $I^{[n]}_{m,s}$ the denominator of $\gamma^{[n]}_{m,s}$ in Eq. (19.3), in other words the overall disturbance power affecting user s served by BS m on subcarrier n.

19.2.2 D2D Communication

We assume that each D2D link chooses for transmission the subcarrier with the largest channel gain.[2] Denoting $n(j) \in \{1, 2, \ldots, N\}$ the subcarrier used by the jth D2D link, the

[2] System optimization with respect to the choice of the subcarrier for the D2D links, although possible in principle, is not considered here.

discrete-time baseband signal at D2D receiver j is

$$y_j^{[n]} = \underbrace{\tilde{Q}_{j,j}^{[n]}\tilde{x}_j}_{\text{useful data}} + \underbrace{\sum_{q\in\mathcal{K}_D^{[n(j)]}\setminus j} \tilde{Q}_{q,j}^{[n]}\tilde{x}_q}_{\text{D2D interference}} +$$

$$\underbrace{\sum_{\ell=1}^{M} Q_{\ell,j}^{[n]}x_\ell^{[n]}}_{\text{interference from BSs}} + \underbrace{\tilde{z}_j^{[n]}}_{\text{noise}} \tag{19.4}$$

In Eq. (19.4), $\tilde{Q}_{q,j}^{[n]}$ and $Q_{\ell,j}^{[n]}$ are the complex channel responses between D2D transmitter q and D2D receiver j, and between BS ℓ and I2D receiver j, respectively, at subcarrier n; again, these coefficients include small-scale fading, large-scale fading and path attenuation. Moreover, $\tilde{z}_j^{[n]}$ is the additive noise received by D2D receiver j, which is again modeled as a circularly-symmetric complex Gaussian random variable with variance $\tilde{\mathcal{N}}_j/2$ per real dimension.

Hence, the SINR for the D2D receiver j can be expressed as

$$\tilde{\gamma}_j = \frac{\tilde{p}_j\tilde{M}_{j,j}}{1 + \sum_{q\in\mathcal{K}_D^{[n]}\setminus j} \tilde{p}_q\tilde{M}_{j,q} + \sum_{\ell=1}^{M}\sum_{n=1}^{N} p_\ell^{[n]}M_{\ell,j}^{[n]}} \tag{19.5}$$

where $\tilde{M}_{j,q} = |\tilde{Q}_{j,q}^{[n]}|^2/\tilde{\mathcal{N}}_j$, and $M_{\ell,j}^{[n]} = |Q_{\ell,j}^{[n]}|^2/\tilde{\mathcal{N}}_j$.

Accordingly, the corresponding achievable information rate (in bit/s) is written as

$$\tilde{\mathrm{R}}_j = B\log_2[1 + \tilde{\gamma}_j]$$

For notational convenience, we denote by $\tilde{I}_j$, the denominator of $\tilde{\gamma}_j$ in Eq. (19.5), in other words the overall disturbance power affecting the jth D2D receiver.

19.3 Resource Allocation

In this work, we aim at maximizing the network energy efficiency. We thus consider, as figure of merit, the GEE, which, as already discussed, is defined as the ratio between the network weighted sum-rate and the network power consumption:

$$\mathrm{GEE}(\mathbf{p}, \tilde{\mathbf{p}}, \mathbf{k}) = \frac{\sum_{m=1}^{M}\sum_{n=1}^{N} w_{m,k(m,n)}^{[n]}\mathrm{R}_{m,k(m,n)}^{[n]} + \sum_{j=1}^{K_D} \tilde{w}_j\tilde{\mathrm{R}}_j}{\sum_{m=1}^{M}\sum_{n=1}^{N}(\theta_m^{[n]} + \alpha_m^{[n]}p_m^{[n]}) + \sum_{j=1}^{K_D}(\tilde{\theta}_j + \tilde{\alpha}_j\tilde{p}_j)} \tag{19.6}$$

In Eq. (19.6), we have that

$$\mathbf{p} = [p_1^{[1]}, \ldots, p_1^{[N]}, p_2^{[1]}, \ldots, p_2^{[N]}, \ldots, p_M^{[1]}, \ldots, p_M^{[N]}]^T \tag{19.7}$$

where $(\cdot)^T$ denotes transpose, is the MN-dimensional vector of the powers transmitted by the BSs,

$$\tilde{\mathbf{p}} = [\tilde{p}_1, \ldots, \tilde{p}_{K_D}]^T \tag{19.8}$$

is the K_D-dimensional vectors of the powers transmitted by the D2D devices, while

$$\mathbf{k} = [k(1,1), \ldots, k(1,N), k(2,1), \ldots, k(2,N), \ldots, k(M,1), \ldots, k(M,N)]^T \tag{19.9}$$

is the MN-dimensional vector containing information of the scheduling allocation. The constants $w^{[n]}_{m,k(m,n)}$ and $\tilde{w}_j$ are the non-negative weights assigned to I2D and D2D links, respectively. Additionally, $\theta^{[n]}_m \geq 0$ is a parameter taking into account the power dissipated by the RF transmit chain of BS m on subcarrier n, while $\alpha^{[n]}_m \geq 1$ is a scaling coefficient modeling the amplifier and feeder losses; the accompanying parameters $\tilde{\theta}_j \geq 0$ and $\tilde{\alpha}_j \geq 1$ referring to D2D transmitter j have a similar meaning. For further details on the relevance of the GEE as a performance measure, we refer the reader to the literature [3, 14].

Note also that letting $\theta^{[n]}_m = 1/(MN)$, $\alpha^{[n]}_m = 0$, $\forall m, n$, and $\tilde{\theta}_j = 1/K_D$, $\tilde{\alpha}_j = 0$, $\forall j$, the GEE reduces to the well-known network WSR, which is the weighted sum of the achievable rates on the I2D and D2D links:

$$\mathrm{WSR}(\mathbf{p}, \tilde{\mathbf{p}}, \mathbf{k}) = \sum_{m=1}^{M} \sum_{n=1}^{N} w^{[n]}_{m,k(m,n)} R^{[n]}_{m,k(m,n)} + \sum_{j=1}^{K_D} \tilde{w}_j \tilde{\mathrm{R}}_j \tag{19.10}$$

WSR is a widely-accepted performance measure that has been used for decades in the area of wireless networks. We will also assume that the weights are given and fixed; typically, they are chosen based on the desired quality of service and, possibly, may be adjusted over successive scheduling intervals to ensure some form of fairness among the devices.

As can be seen, optimization of the GEE requires maximization of a fraction, and this is where fractional programming kicks in. In the next three sections, we therefore open an interlude devoted to the solution of optimization problems involving fractions; our treatment of the subject, although rigorous, will keep the mathematical content at its minimum, and will aim at providing the reader with algorithms suited for solving fractional programs. For a more comprehensive tutorial on fractional programming, we refer the reader to the paper by Zappone *et al.* [3].

19.4 Fractional Programming

Fractional programming is the branch of optimization theory that studies the optimization of fractional functions. A general fractional program has the form

$$\max_{\boldsymbol{x}} \frac{f(\boldsymbol{x})}{g(\boldsymbol{x})} \tag{19.11a}$$

$$\text{s.t. } \boldsymbol{x} \in \mathcal{X} \tag{19.11b}$$

with $f : \mathcal{C} \subseteq \mathbb{R}^n \to \mathbb{R}$, $g : \mathcal{C} \subseteq \mathbb{R}^n \to \mathbb{R}_+$ and $\mathcal{X} \subseteq \mathcal{C}$. In general, Problem (19.11a) is not guaranteed to be concave, even in the simple case in which both f and g are affine functions, which implies that standard convex optimization tools fail when applied to Problem (19.11). Instead, specific algorithms and solution methods are required.

The fundamental tools developed to deal with Problem (19.11) involved either a parametric approach [15, 16] or a parameter-free approach [17, 18]. The common denominator of these seminal papers is to establish that, while in regular optimization theory the complexity divide is between convex and non-convex problems, in fractional optimization it is between fractions with concave numerator and convex denominator, and fractions that do not possess this structure. In the rest of this section, this fundamental point will be illustrated in detail, and the most widely used fractional programming techniques will be described.

19.4.1 Generalized Concavity

As it has been observed, a fractional function in general does not enjoy any concavity property. The first step towards solving Problem (19.11) is therefore to extend the notion of concavity, searching for more general classes of functions that are not concave in general, but which nevertheless retain the useful properties of concave functions. Among the several possible generalizations of concavity, this chapter will be concerned with the following classes of functions, which are the most useful with regard to fractional programming:

- quasi-concave (QC) functions
- pseudo-concave (PC) functions.

For a more comprehensive overview of generalized concavity, we refer the reader to the literature [19].

19.4.1.1 Quasi-concavity

Definition 19.1 ***(Quasi-concavity)*** *Let $\mathcal{C} \subseteq \mathbb{R}^n$ be a convex set. Then $r : \mathcal{C} \to \mathbb{R}$ is QC if*

$$r(\lambda \boldsymbol{x}_1 + (1-\lambda)\boldsymbol{x}_2) \geq \min\{r(\boldsymbol{x}_1), r(\boldsymbol{x}_2)\} \tag{19.12}$$

for all $\boldsymbol{x}_1, \boldsymbol{x}_2 \in \mathcal{C}$ and $\lambda \in [0;1]$.

In words, quasi-concavity requires that the restriction of a function to a line joining two points of the domain should be above at least one of the endpoints of the line. By enforcing a strict inequality, we can define strict quasi-concavity.

Definition 19.2 ***(Strict quasi-concavity)*** *Let $\mathcal{C} \subseteq \mathbb{R}^n$ be a convex set. Then $r : \mathcal{C} \to \mathbb{R}$ is strictly QC (SQC) if*

$$r(\lambda \boldsymbol{x}_1 + (1-\lambda)\boldsymbol{x}_2) > \min\{r(\boldsymbol{x}_1), r(\boldsymbol{x}_2)\} \tag{19.13}$$

for all $\boldsymbol{x}_1, \boldsymbol{x}_2 \in \mathcal{C}$, $\boldsymbol{x}_1 \neq \boldsymbol{x}_2$ and $\lambda \in (0;1)$.

Similarly to concave functions, QC functions are such that the super-level set $\mathcal{S}_t = \{\boldsymbol{x} \in \mathcal{C} : r(\boldsymbol{x}) \geq t\}$ is convex for all $t \in \mathbb{R}$. This property follows directly from the definition of quasi-concavity, and is sometimes even taken as an alternative definition of quasi-concavity.

The interest for QC functions stems from the following result.

Proposition 19.1 *Let $r : \mathcal{C} \to \mathbb{R}$ be a QC function.*

(a) If $\boldsymbol{x}^*$ is a strict local maximum, then it is also global.
(b) If r is SQC, then a unique local maximizer exists, and it is also global.

It should be emphasized that Proposition 19.1 holds only for local maxima, and does not imply that any stationary point of a QC function is also a global maximum. For example, a QC function can have saddle points where the gradient vanishes, but which are not global optima. It should also be observed that so far no differentiability assumption has been made. For differentiable functions, quasi-concavity can be shown to admit the following characterization:

Proposition 19.2 *Let $\mathcal{C}$ be an open, convex set. Then, the function $r : \mathcal{C} \to \mathbb{R}$ is QC if and only if*

$$r(\boldsymbol{x}_2) \leq r(\boldsymbol{x}_1) \Rightarrow \nabla r(\boldsymbol{x}_2)^T(\boldsymbol{x}_1 - \boldsymbol{x}_2) \geq 0 \ , \forall\, \boldsymbol{x}_1, \boldsymbol{x}_2 \in \mathcal{C} \tag{19.14}$$

A close observation of Eq. (19.14) leads to the following two remarks.

Remark 19.1 *A condition equivalent to Eq. (19.14) can be written as:*[3]

$$\nabla r(\boldsymbol{x}_2)^T(\boldsymbol{x}_1 - \boldsymbol{x}_2) < 0 \Rightarrow r(\boldsymbol{x}_2) > r(\boldsymbol{x}_1) \ , \forall\, \boldsymbol{x}_1, \boldsymbol{x}_2 \in \mathcal{C} \tag{19.15}$$

Eq. (19.15) implies that any $\boldsymbol{x}_2$ such that $\nabla r(\boldsymbol{x}_2)^T(\boldsymbol{x}_1 - \boldsymbol{x}_2) < 0$ for all $\boldsymbol{x}_1, \boldsymbol{x}_2 \in \mathcal{C}$, is a global maximizer for r. However, no stationary point of r can fulfill Eq. (19.15), due to the strict inequality at the left-hand side. Therefore, Eq. (19.15) provides no insight as to the relation between stationary points and global maximizers of QC functions.

Remark 19.2 *It would be reasonable to think that a stronger condition than Eq. (19.14) holds for SQC functions, with a strict inequality at the right-hand side. Unfortunately this is not true. Indeed, if the right-hand side of Eq. (19.14) held with a strict inequality, it would be possible to obtain a similar condition as in Eq. (19.15), but with a non-strict inequality at the left-hand side, and this would imply that any stationary point of a SQC function is a global maximizer. However, this conclusion is wrong. As a counterexample, consider the SQC function $r : x \in \mathbb{R} \to r(x) = x^3$, which has a stationary point in $x = 0$, which is not a global maximizer. Indeed, taking $x_1 = 1$ and $x_2 = 0$, it follows that Eq. (19.14) is verified, but with an equality at the right-hand side.*

In order to strengthen the condition in Eq. (19.14) into a condition that allows one to show the equivalence between stationary points and global maximizers, the concept of quasi-concavity needs to be strengthened into pseudo-concavity.

[3] Recall that for any couple of propositions $\mathcal{P}_1$ and $\mathcal{P}_2$, if $\mathcal{P}_1$ implies $\mathcal{P}_2$, then $\bar{\mathcal{P}}_2$ implies $\bar{\mathcal{P}}_1$, with $\bar{\mathcal{P}}_1$ and $\bar{\mathcal{P}}_2$ denoting the negation of $\mathcal{P}_1$ and $\mathcal{P}_2$.

19.4.1.2 Pseudo-concavity

Definition 19.3 ***(Pseudo-concavity)*** *Let $C \subseteq \mathbb{R}^n$ be a convex set. Then $f : C \to \mathbb{R}$ is PC if and only if, for all $\boldsymbol{x}_1, \boldsymbol{x}_2 \in C$, it is differentiable and*

$$r(\boldsymbol{x}_2) < r(\boldsymbol{x}_1) \Rightarrow \nabla(r(\boldsymbol{x}_2))^T(\boldsymbol{x}_1 - \boldsymbol{x}_2) > 0 \tag{19.16}$$

Definition 19.4 ***(Strict pseudo-concavity)*** *Let $C \subseteq \mathbb{R}^n$ be a convex set. Then $f : C \to \mathbb{R}$ is strictly PC (SPC) if and only if, for all $\boldsymbol{x}_1 \neq \boldsymbol{x}_2 \in C$, it is differentiable and*

$$r(\boldsymbol{x}_2) \leq r(\boldsymbol{x}_1) \Rightarrow \nabla(r(\boldsymbol{x}_2))^T(\boldsymbol{x}_1 - \boldsymbol{x}_2) > 0 \tag{19.17}$$

From the given definitions, and recalling the definition of concavity, the following inclusion relations can be straightforwardly obtained.

Proposition 19.3 *Let $C \subseteq \mathbb{R}^n$ be a convex set and $r : C \to \mathbb{R}$.*

(a) If r is differentiable and concave, then r is PC.
(b) If r is differentiable and strictly concave, then r is SQC.
(c) If r is PC, then r is QC.
(d) If r is SPC, then r is SQC and PC.

So pseudo-concavity is stricter than quasi-concavity, but still more general than concavity. The next question is then about the stationarity properties of PC functions. By the same approach as in Remark 19.1, it follows that stationary points are indeed global maximizers of PC functions. Moreover, a unique maximizer exists for SPC functions. The following proposition shows an even stronger result.

Proposition 19.4 *Consider the optimization problem:*

$$\max_{\boldsymbol{x}} r(\boldsymbol{x}) \tag{19.18a}$$

$$\text{s.t. } c_i(\boldsymbol{x}) \geq 0\,, \forall i = 1, \dots, I \tag{19.18b}$$

with $r, c_i : C \subseteq \mathbb{R}^n \to \mathbb{R}$, differentiable functions. Assume r is PC while c_i is QC, for all $i = 1, \dots, I$. Then, assuming a constraint qualification holds, the Karush Kuhn Tucker (KKT) optimality conditions of Problem (19.18) are necessary and sufficient conditions for optimality.[4]

Roughly speaking, Proposition 19.4 makes pseudo-concavity and concavity equivalent as far as optimization is concerned.

Figure 19.2 summarizes the KKT properties and inclusion relations between quasi-concavity, pseudo-concavity, and concavity. It is worth stressing that no inclusion relation can be established between concavity and strict pseudo-concavity as well as between pseudo-concavity and strict quasi-concavity.

[4] The result can be straightforwardly extended to the case in which quasi-linear equality constraints are also included.

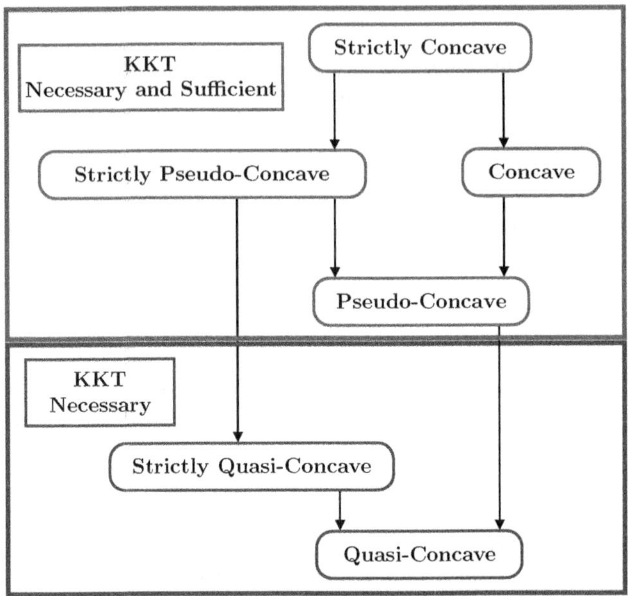

Figure 19.2 Inclusion relationships and optimality of KKT conditions for different classes of generalized concave functions

19.4.1.3 Generalized Concavity of Ratios

After introducing the framework of generalized concavity theory, we are finally ready to show the close connection between fractional functions and generalized concavity.

Proposition 19.5 *Let* $r(\boldsymbol{x}) = \frac{f(\boldsymbol{x})}{g(\boldsymbol{x})}$, *with* $f : \mathcal{C} \subseteq \mathbb{R}^n \to \mathbb{R}$ *and* $g : \mathcal{C} \subseteq \mathbb{R}^n \to \mathbb{R}_+$. *Then:*

(a) *If* f *is non-negative and concave, while* g *is convex, then* r *is QC.*
If g *is affine, the non-negativity of* f *can be relaxed.*

(b) *If* f *is non-negative, differentiable, and concave, while* g *is differentiable and convex, then* r *is PC.*
If g *is affine, the non-negativity of* f *can be relaxed.*

(c) *If* f *and* g *are affine, then* r *is pseudo-linear.*

19.5 Algorithms

It has been shown that a fractional function belongs to the class of QC or PC functions if the numerator is concave and the denominator is convex and that in this case any stationary point (local maximizer for QC functions) is a global maximizer. Bearing this in mind, it should come as no surprise that solution algorithms to solve fractional problems exhibit an affordable complexity precisely when the numerator and denominator follow the concavity–convexity structure. In this section, this point will be analyzed in detail, introducing the two most widely used algorithms to solve fractional problems, namely Dinkelbach's algorithm and the Charnes–Cooper transform.

19.5.1 Dinkelbach's Algorithm

Dinkelbach's algorithm [20, 15] is a parametric method that solves a fractional problem by solving a sequence of auxiliary, non-fractional maximizations. The fundamental result upon which it is based is the relation between the fractional Problem (19.11) and the function of real variable:

$$F(\lambda) = \max_{\boldsymbol{x} \in \mathcal{X}} \{f(\boldsymbol{x}) - \lambda g(\boldsymbol{x})\} \tag{19.19}$$

It is easy to see that F is continuous, provided f and g are continuous and that $\mathcal{X}$ is compact. Moreover, the following result holds:

Lemma 19.1 *Assume f and g are continuous, g is positive, and $\mathcal{X}$ is compact. Then:*

(a) F is convex for $\lambda \in \mathbb{R}$;
(b) F is strictly monotonic decreasing for $\lambda \in \mathbb{R}$;
(c) $F(\lambda)$ has a unique root λ_0;
(d) For any $\tilde{\boldsymbol{x}} \in \mathcal{X}$, $F(\lambda_{\tilde{x}}) \geq 0$, with $\lambda_{\tilde{x}} = \frac{f(\tilde{\boldsymbol{x}})}{g(\tilde{\boldsymbol{x}})}$, with equality when $\tilde{\boldsymbol{x}} = \arg\max_{\boldsymbol{x} \in \mathcal{X}} \{f(\boldsymbol{x}) - \lambda_{\tilde{x}} g(\boldsymbol{x})\}$.

Leveraging Lemma 19.1, the connection between Eq. (19.19) and Problem (19.11) can be proved as follows.

Proposition 19.6 *Consider $\boldsymbol{x}^* \in \mathcal{X}$ and $\lambda^* = \frac{f(\boldsymbol{x}^*)}{g(\boldsymbol{x}^*)}$. Then, $\boldsymbol{x}^*$ is a solution of Problem (19.11) if and only if*

$$\boldsymbol{x}^* = \arg\max_{\boldsymbol{x} \in \mathcal{X}} \{f(\boldsymbol{x}) - \lambda^* g(\boldsymbol{x})\} \tag{19.20}$$

Thus, solving Problem (19.11) is equivalent to finding the unique zero of the auxiliary function $F(\lambda)$. Dinkelbach's algorithm is able to determine the unique zero of $F(\lambda)$ with super-linear convergence rate. The formal pseudo-code is stated in Algorithm 19.1.

Proposition 19.7 *Algorithm 19.1 converges to the global solution of Problem (19.11).*

It should be emphasized that so far we have not actually exploited any concavity/convexity assumption on the numerator and denominator of the objective function. As a consequence, Dinkelbach's algorithm is guaranteed to converge to the global solution of Problem (19.11) under the very mild assumptions that f and g are continuous, real-valued functions, that $\mathcal{X}$ is a compact set, and that g is positive. However, in order to implement Dinkelbach's algorithm, the auxiliary problem $\max_{\boldsymbol{x} \in \mathcal{X}} \{f(\boldsymbol{x}) - \lambda_n g(\boldsymbol{x})\}$ must be globally solved in each iteration. Without making any further assumptions on f, g, and $\mathcal{X}$, the auxiliary problem is in general non-convex, and therefore it requires an unaffordable complexity to obtain the global solution.

Instead, if f is concave, g is convex, and $\mathcal{X}$ is convex, the auxiliary problem $\max_{\boldsymbol{x}\in\mathcal{X}}\{f(\boldsymbol{x}) - \lambda_n g(\boldsymbol{x})\}$ becomes convex,[5] and Dinkelbach's algorithm can be implemented with affordable complexity.

Algorithm 19.1 Dinkelbach's algorithm.

$\epsilon > 0$; $n = 0$; $\lambda_n = 0$;
while $F(\lambda_n) > \epsilon$ **do**
  $\boldsymbol{x}_n^* = \arg\max_{\boldsymbol{x}\in\mathcal{X}} \{f(\boldsymbol{x}) - \lambda_n g(\boldsymbol{x})\}$;
  $F(\lambda_n) = f(\boldsymbol{x}_n^*) - \lambda_n g(\boldsymbol{x}_n^*)$;
  $\lambda_{n+1} = \dfrac{f(\boldsymbol{x}_n^*)}{g(\boldsymbol{x}_n^*)}$;
  $n = n + 1$;
end while

19.5.1.1 Computational Complexity

Dinkelbach's algorithm converts the original fractional problem into a sequence of auxiliary problems, indexed by the parameter λ. The overall computational complexity then depends both on the convergence rate of the subproblem sequence and on the complexity of each individual subproblem.

In order to learn the convergence rate of the subproblem sequence, we proceed by rewriting the update rule for λ as

$$\lambda_{n+1} = \frac{f(\boldsymbol{x}_n^*)}{g(\boldsymbol{x}_n^*)} = \lambda_n - \frac{f(\boldsymbol{x}_n^*) - \lambda_n g(\boldsymbol{x}_n^*)}{-g(\boldsymbol{x}_n^*)} = \lambda_n - \frac{F(\lambda_n)}{F'(\lambda_n)} \tag{19.21}$$

thereby observing that Dinkelbach's algorithm follows Newton's method as far as updating λ is concerned. This means that Algorithm 19.1 can be interpreted as Newton's method applied to the convex function $F(\lambda)$. An immediate consequence of this fact is that Dinkelbach's algorithm exhibits a superlinear convergence rate in the subproblem sequence.

As for the complexity of each subproblem, it is difficult to make general statements, as it depends on the specific expression of the auxiliary function F and on the number of constraints. Focusing on the more practical case in which the auxiliary problem is convex, we can employ standard convex analysis results to state that the computational complexity of the auxiliary problem is polynomial in the number of variables and constraints [21, 22].

19.5.2 Charnes–Cooper Transform

Charnes–Cooper transform [17, 18], unlike Dinkelbach's algorithm, is a parameter-free method, which employs a variable transformation to reformulate Problem (19.11) into

[5] It should be mentioned that this holds provided that $\lambda_n > 0$ for all n. This is equivalent to requiring $\lambda_1 \geq 0$, which is ensured if $\max_{\boldsymbol{x}\in\mathcal{X}}\{f(\boldsymbol{x})\} \geq 0$. Notice that this additional assumption is not required if g is affine.

an equivalent, non-fractional problem. Specifically, let us denote by $c_i(\boldsymbol{x}) \leq 0$, with $i = 1, \ldots, I$, the inequalities defining the constraint set[6] $\mathcal{X}$ in Problem (19.11), and consider the transformation:

$$\boldsymbol{y} = \frac{\boldsymbol{x}}{g(\boldsymbol{x})}\,, t = \frac{1}{g(\boldsymbol{x})} \tag{19.22}$$

Problem (19.11) can then be equivalently recast as

$$\max_{t,\boldsymbol{y}/t} \; tf\left(\frac{\boldsymbol{y}}{t}\right) \tag{19.23a}$$

$$\text{s.t. } tg\left(\frac{\boldsymbol{y}}{t}\right) = 1 \tag{19.23b}$$

$$tc_i\left(\frac{\boldsymbol{y}}{t}\right) \leq 0\,, \forall\, i = 1, \ldots, I \tag{19.23c}$$

The objective in Eq. (19.23a) and the left-hand side of the constraints in Eq. (19.23c) are the perspective of the functions f and c_i, respectively. Since the perspective operation preserves concavity/convexity [22], the function in Eq. (19.23a) is concave and Eq. (19.23c) represent convex constraints, provided f is concave and c_i is convex for all $i = 1, \ldots, I$. Instead, the convexity of the function in Eq. (19.23b) is guaranteed only when g is affine, but not when it is convex. However, the following result proves that it is possible to relax the constraint in Eq. (19.23b) to an inequality constraint without loss of optimality, thus making Constraint (19.23b) convex, under the milder assumption that g is a convex function.

Proposition 19.8 *Assume f is non-negative and consider the relaxed version of Problem (19.23):*

$$\max_{t,\boldsymbol{y}/t} \; tf\left(\frac{\boldsymbol{y}}{t}\right) \tag{19.24a}$$

$$\text{s.t. } tg\left(\frac{\boldsymbol{y}}{t}\right) \leq 1 \tag{19.24b}$$

$$tc_i\left(\frac{\boldsymbol{y}}{t}\right) \leq 0\,, \forall\, i = 1, \ldots, I \tag{19.24c}$$

Then, Problems (19.23) and (19.24) are equivalent.

19.6 Sequential Fractional Programming

So far, a framework has been developed to maximize fractional functions with concave numerator and convex denominator, subject to convex constraints. It is natural to ask whether these assumptions are commonly met in the context of energy-efficiency maximization. Unfortunately, the answer is only partially affirmative. While the denominator of the energy efficiency is usually modeled by convex (very often even affine) functions, the numerator is concave only in simple, noise-limited scenarios. Instead, in interference-limited scenarios,

[6] Without loss of generality, we consider only inequality constraints. The extension to equality constraint can be easily obtained.

which are typical for 5G networks, the presence of multi-user interference makes the numerator of the energy efficiency non-concave. Moreover, it is common practice in cellular networks to consider quality-of-service constraints, which also turn out to be non-convex in interference-limited scenarios.

Thus in order to successfully apply fractional programming in 5G networks, the framework developed so far needs to be widened. To-date, no low-complexity algorithm exists to globally solve fractional problems in which the numerator of the objective is not concave. Typical approaches are to adopt (semi)-orthogonal transmission schemes and/or linear interference neutralization techniques to remove multi-user interference, thus falling back to the noise-limited scenario [23, 24]. While leading to simple resource allocation algorithms, this approach has the drawback of reducing the resource reuse factor or of suffering from noise-enhancement effects. Several other proposals that implement Dinkelbach's algorithm have been put forward, but in which a sub-optimal solution of the auxiliary problem is determined in each iteration [25, 26]. This approach can exhibit very limited complexity, but convergence of the resulting algorithm cannot be ensured, because Dinkelbach's algorithm is guaranteed to converge only if the global solution of each auxiliary problem is computed.

It is therefore desirable to develop a framework that allows one to obtain low-complexity solutions of fractional problems that do not possess the concave/convex structure, while at the same time guaranteeing some optimality claim. This can be accomplished by integrating fractional programming theory with sequential convex optimization theory, and in particular with the following result [27].

Proposition 19.9 *Let $\mathcal{P}$ be a maximization problem with continuous objective $r_0(\boldsymbol{x})$ and constraints $r_i(\boldsymbol{x}) \geq 0$, for $i = 1, \ldots, I$, defining a compact set. Let $\{\mathcal{P}_\ell\}_\ell$ be a sequence of maximization problems with objective $r_{0,\ell}(\boldsymbol{x})$, constraints $r_{i,\ell}(\boldsymbol{x}) \geq 0$, for $i = 1, \ldots, I$, and optimal solution $\boldsymbol{x}_\ell^*$ for all $\ell \geq 1$, and $\boldsymbol{x}_0^*$ any feasible point of $\mathcal{P}$. Assume that, for any ℓ and $i = 0, \ldots, I$, $r_{i,\ell}(\boldsymbol{x})$ enjoys the following properties.*

P1) $r_{i,\ell}(\boldsymbol{x}) \leq r_i(\boldsymbol{x})$, *for all* $\boldsymbol{x}$.
P2) $r_{i,\ell}(\boldsymbol{x}_{\ell-1}^*) = r_i(\boldsymbol{x}_{\ell-1}^*)$.
P3) $\nabla r_{i,\ell}(\boldsymbol{x}_{\ell-1}^*) = \nabla r_i(\boldsymbol{x}_{\ell-1}^*)$.

The sequence of the values of the original objective monotonically increases and converges. Moreover, upon convergence, the KKT conditions of the original problem are fulfilled.

Proof. Due to Properties P1 and P2, for all ℓ and $i = 1, \ldots, I$ we have

$$r_{i,\ell+1}(\boldsymbol{x}_\ell^*) = r_i(\boldsymbol{x}_\ell^*) \geq r_{i,\ell}(\boldsymbol{x}_\ell^*) \geq 0 \tag{19.25}$$

thus implying that $\boldsymbol{x}_\ell^*$ is feasible both for the original problem $\mathcal{P}$ and for next problem in the sequence $\mathcal{P}_{\ell+1}$. Next, again exploiting properties P1 and P2, we obtain

$$r_0(\boldsymbol{x}_\ell^*) \geq r_{0,\ell}(\boldsymbol{x}_\ell^*) \geq r_{0,\ell}(\boldsymbol{x}_{\ell-1}^*) = r_0(\boldsymbol{x}_{\ell-1}^*) \tag{19.26}$$

thereby showing that the sequence $\{r_0(\boldsymbol{x}_\ell^*)\}_\ell$ of the values achieved by the objective of the original problem $\mathcal{P}$ is non-decreasing. This shows the second part of the thesis, and also implies convergence because r_0 is continuous, and therefore admits a maximum on the compact feasible set. By a similar argument, also exploiting Property P3, it follows that upon convergence the KKT conditions of the original problem $\mathcal{P}$ are fulfilled. ■

Proposition 19.9 allows one to monotonically increase the objective of a difficult problem, eventually converging to a KKT-optimal point. However, in order to apply this tool to a complicated fractional problem $\mathcal{P}$, the critical issue is to find a sequence of problems $\mathcal{P}_\ell$ that fulfills Properties P1–P3, but that at the same time are easier to solve, for example because they enjoy the concave/convex structure. If the sequence $\mathcal{P}_\ell$ is available, this method provides solutions with affordable complexity, which at the same time enjoy the beneficial KKT optimality claim. Moreover, another advantage of this method over state-of-the-art approaches is that it is not tailored for a particular system model, and can be implemented in a variety of different settings. The first successful implementations of this sequential fractional programming framework were for:

- heterogeneous, multiple-antenna networks [28]
- multi-cell coordinated multi-point systems using the OFDMA transmission format [14]
- full-duplex systems[29]
- relay-assisted OFDMA coordinated multi-point systems and hardware-impaired massive MIMO systems, subject to minimum rate constraints [30].

In the rest of this chapter, the application of the sequential fractional programming approach will be illustrated in detail, considering the specific case of D2D communications.

19.7 System Optimization

We now revert to the original problem of GEE maximization. System optimization is carried out with respect to vectors **p**, **k**, $\tilde{\mathbf{p}}$. Hence, the problem that we consider is the following:

$$\begin{cases} \arg \max\limits_{\mathbf{p},\tilde{\mathbf{p}},\mathbf{k}} \text{GEE}(\mathbf{p}, \tilde{\mathbf{p}}, \mathbf{k}), \\ \text{s. to} \begin{matrix} 0 \le p_m^{[n]} \le P_{m,\max}/N, & \forall m, n \\ 0 \le \tilde{p}_j \le \tilde{P}_{j,\max}, & \forall j \end{matrix} \end{cases} \tag{19.27}$$

In general, Problem (19.27) is a mixed integer-continuous problem that is not convex. Following the steps outlined by others [31, 14], we now propose an heuristic iterative algorithm derived from the KKT conditions and that provably converges.

To solve Problem (19.27), first of all note that for given **p** and $\tilde{\mathbf{p}}$, the maximization over **k**, which in principle is a combinatorial problem, is actually separable across BSs and subcarriers, and the solution is given by:

$$\hat{k}(m,n) = \arg \max_{s \in \mathcal{B}_m} w_{m,s}^{[n]} \mathrm{R}_{m,s}^{[n]} \tag{19.28}$$

in other wordes, the mth BS must schedule on subcarrier n the I2D user in the set $\mathcal{B}_m$ maximizing the weighted rate on that subcarrier. For fixed **p** and $\tilde{\mathbf{p}}$, thus, user scheduling can be solved in an optimal way with a linear complexity in the system parameters M, N and K.

On the other hand, for any feasible $\mathbf{k}$, maximization of the GEE with respect to $\mathbf{p}$ and $\tilde{\mathbf{p}}$ is a fractional program. Due to the presence of co-channel interference in the SINR expressions of Eqs (19.3) and (19.5), the objective function in Problem (19.27) is not of the concave-over-convex type, and sequential fractional programming is needed.

In order to find suitable approximate problems, observe that for any $z \geq 0$ and $\bar{z} \geq 0$, the following inequality holds [14]:

$$\log_2(1+z) \geq \eta \log_2 z + \beta \tag{19.29}$$

where η and β are defined as follows

$$\eta = \frac{\bar{z}}{1+\bar{z}}, \quad \beta = \log_2(1+\bar{z}) - \frac{\bar{z}}{1+\bar{z}}\log_2 \bar{z} \tag{19.30}$$

and the bound is tight (i.e. holds with an equality sign) for $z = \bar{z}$. As a consequence, for a given feasible user selection $\mathbf{k}$, the following lower bound to the objective function is obtained:

$$\text{GEE} \geq \frac{\overbrace{\sum_{m=1}^{M}\sum_{n=1}^{N} B w_{m,k(m,n)}^{[n]} \left[\eta_m^{[n]} \log_2(\gamma_{m,k(m,n)}^{[n]}) + \beta_m^{[n]}\right] + \sum_{j=1}^{K_D} B\tilde{w}_j [\tilde{\eta}_j \log_2(\tilde{\gamma}_j) + \tilde{\beta}_j]}^{f(\mathbf{p},\tilde{\mathbf{p}},\mathbf{k})}}{\underbrace{\sum_{m=1}^{M}\sum_{n=1}^{N}(\theta_m^{[n]} + \alpha_m^{[n]} p_m^{[n]}) + \sum_{j=1}^{K_D}(\tilde{\theta}_j + \tilde{\alpha}_j \tilde{p}_j)}_{g(\mathbf{p},\tilde{\mathbf{p}})}} \tag{19.31}$$

where $\eta_m^{[n]}$, $\beta_m^{[n]}$, $\tilde{\eta}_j$ and $\tilde{\beta}_j$ are approximation constants computed as in Eq. (19.30) for some $\bar{z}_m^{[v]}$ and $\bar{\tilde{z}}_j$ to be specified in the following. Although the function $f(\mathbf{p}, \tilde{\mathbf{p}}, \mathbf{k})$ is not concave with respect to the vectors $(\mathbf{p}, \tilde{\mathbf{p}})$, using the transformations $p_m^{[n]} = e^{q_m^{[n]}}$, $\forall m, n$, and $\tilde{p}_j = e^{\tilde{q}_j}$, $\forall j$, and defining the NM-dimensional vector

$$\mathbf{q} = [q_1^{[1]}, \ldots, q_1^{[N]}, q_2^{[1]}, \ldots, q_2^{[N]}, \ldots, q_M^{[1]}, \ldots, q_M^{[N]}]^T \tag{19.32}$$

and the K_D-dimensional vector

$$\tilde{\mathbf{q}} = [\tilde{q}_1, \ldots, \tilde{q}_{K_D}]^T \tag{19.33}$$

it is easily seen that the function $f(e^{\mathbf{q}}, e^{\tilde{\mathbf{q}}}, \mathbf{k})$ is concave with respect to $(\mathbf{q}, \tilde{\mathbf{q}})$, while $g(e^{\mathbf{q}}, e^{\tilde{\mathbf{q}}})$ is convex with respect to $(\mathbf{q}, \tilde{\mathbf{q}})$. The ratio $\frac{f(e^{\mathbf{q}}, e^{\tilde{\mathbf{q}}}, \mathbf{k})}{g(e^{\mathbf{q}}, e^{\tilde{\mathbf{q}}})}$ is thus a lower bound for the GEE that is amenable to a maximization using Dinkelbach's algorithm and that can be thus exploited in order to apply sequential convex programming to maximize the GEE. More precisely, the optimization Problem (19.27) can be solved by iteratively optimizing the power allocation according to the lower bound in Eq. (19.31), computing the best user selection according to Eq. (19.28), and tightening the bound in Eq. (19.31), as summarized in Algorithm 19.2. The optimization Problem (19.34) in Algorithm 19.2 can in turn be solved using Dinkelbach's algorithm, as detailed in Algorithm 19.3.

Algorithm 19.2 The sequential fractional programming algorithm to solve Problem (19.27)

Initialize $I_{\max}$ and set $i = 0$;
Initialize $\mathbf{p}, \widetilde{\mathbf{p}}$ and compute $\mathbf{k}$ according to Eq. (19.28);
repeat
Set $\overline{z}_m^{[n]} = \gamma_{m,k(m,n)}^{[n]}$ and compute $\eta_m^{[n]}$ and $\beta_m^{[n]}$ as in Eq. (19.29), for $m = 1, \ldots, M$ and $n = 1, \ldots, N$;
Set $\overline{\widetilde{z}}_j = \widetilde{\gamma}_j$ and compute $\widetilde{\eta}_j$ and $\widetilde{\beta}_j$ as in Eq. (19.29), for $j = 1, \ldots, K_D$;
Update $\mathbf{p}$, and $\widetilde{\mathbf{p}}$ by solving the following problem

$$\begin{cases} \arg\max\limits_{\mathbf{q},\widetilde{\mathbf{q}}} \dfrac{f(e^{\mathbf{q}}, e^{\widetilde{\mathbf{q}}}, \mathbf{k})}{g(e^{\mathbf{q}}, e^{\widetilde{\mathbf{q}}})}, \\ \text{s. to } \begin{array}{ll} q_m^{[n]} \leq \ln(P_{m,\max}/N), & \forall m, n \\ \widetilde{q}_j \leq \ln(\widetilde{P}_{j,\max}), & \forall j \end{array} \end{cases} \tag{19.34}$$

Update $\mathbf{k}$ according to Eq. (19.28);
Set $i = i + 1$
until Convergence or $i = I_{\max}$

Algorithm 19.3 Dinkelbach's procedure to solve Problem (19.34)

Set $\epsilon > 0$, $\pi = 0$, and FLAG $= 0$
repeat
Update $\mathbf{q}$ and $\widetilde{\mathbf{q}}$ by solving the following concave maximization:

$$\begin{cases} \arg\max\limits_{\mathbf{q},\widetilde{\mathbf{q}}} f(e^{\mathbf{q}}, e^{\widetilde{\mathbf{q}}}, \mathbf{k}) - \pi g(e^{\mathbf{q}}, e^{\widetilde{\mathbf{q}}}), \\ \text{s. to } \begin{array}{ll} q_m^{[n]} \leq \ln(P_{m,\max}/N), & \forall m, n \\ \widetilde{q}_j \leq \ln(\widetilde{P}_{j,\max}), & \forall j \end{array} \end{cases} \tag{19.35}$$

if $f(e^{\mathbf{q}}, e^{\widetilde{\mathbf{q}}}, \mathbf{k}) - \pi g(e^{\mathbf{q}}, e^{\widetilde{\mathbf{q}}}) < \epsilon$ **then**
FLAG $= 1$
else
Set $\pi = f(e^{\mathbf{q}}, e^{\widetilde{\mathbf{q}}}, \mathbf{k}) / g(e^{\mathbf{q}}, e^{\widetilde{\mathbf{q}}})$
end if
until FLAG $= 1$

The optimization Problem (19.35), meanwhile, is a concave maximization problem with convex constraints, so it can be solved with standard techniques. Interestingly, note that the optimal solution to Problem (19.35) solves the following KKT conditions:

$$\frac{\mathrm{d}}{\mathrm{d}q_m^{[n]}}[f(e^{\mathbf{q}}, e^{\tilde{\mathbf{q}}}, \mathbf{k}) - \pi g(e^{\mathbf{q}}, e^{\tilde{\mathbf{q}}})] - \lambda_m^{[n]} e^{q_m^{[n]}} = 0, \forall\, m, n \tag{19.36a}$$

$$\frac{\mathrm{d}}{\mathrm{d}\tilde{q}_j}[f(e^{\mathbf{q}}, e^{\tilde{\mathbf{q}}}, \mathbf{k}) - \pi g(e^{\mathbf{q}}, e^{\tilde{\mathbf{q}}})] - \tilde{\lambda}_j e^{\tilde{q}_j} = 0, \forall\, j \tag{19.36b}$$

$$\lambda_m^{[n]} \geq 0, \forall\, m, n \tag{19.36c}$$

$$\tilde{\lambda}_j \geq 0, \forall\, j \tag{19.36d}$$

$$e^{q_m^{[n]}} \leq P_{m,\max}/N, \forall\, m, n,\ e^{\tilde{q}_j} \leq \tilde{P}_{j,\max}, \forall\, j \tag{19.36e}$$

$$\lambda_m^{[n]}(P_{m,\max}/N - e^{q_m^{[n]}}) = 0, \forall\, m, n \tag{19.36f}$$

$$\tilde{\lambda}_j(\tilde{P}_{j,\max} - e^{\tilde{q}_j}) = 0, \forall\, j \tag{19.36g}$$

where $\lambda_m^{[n]}$ is the Lagrange multiplier associated to the constraint on the maximum power radiated by BS m on subcarrier n, while $\tilde{\lambda}_j$ is the Lagrange multiplier associated to the constraint on the maximum power level radiated by D2D transmitter j.

$$\begin{aligned} q_m^{[n]} &= \ln(Bw_{m,k(m,n)}^{[n]}\eta_m^{[n]}/\ln 2) \\ &- \ln\left(\pi\alpha_m^{[n]} + \lambda_m^{[n]} + \sum_{j=1,j\neq m}^{M} \frac{Bw_{j,k(j,n)}^{[n]}\eta_j^{[n]}}{\ln 2}\frac{G_{m,k(j,n)}^{[n]}}{I_{j,k(j,n)}^{[n]}} + \sum_{j\in\mathcal{K}_D^{[n]}} \frac{B\tilde{w}_j\tilde{\eta}_j}{\ln 2}\frac{M_{m,j}^{n}}{\tilde{I}_j}\right) \end{aligned} \tag{19.37}$$

and

$$\begin{aligned} \tilde{q}_j &= \ln(\tilde{w}_j\tilde{\eta}_j/\ln 2) \\ &- \ln\left(\pi\tilde{\alpha}_j + \tilde{\lambda}_j + \sum_{m=1}^{M} \frac{Bw_{m,k(m,n(j))}^{[n(j)]}\eta_m^{[n(j)]}}{\ln 2}\frac{G_{j,k(m,n(j))}^{[n(j)]}}{I_{m,k(m,n(j))}^{[n(j)]}} + \sum_{\ell\in\mathcal{K}_D^{[n(j)]}\setminus j} \frac{B\tilde{w}_\ell\tilde{\eta}_\ell}{\ln 2}\frac{\tilde{M}_{\ell,j}}{\tilde{I}_\ell}\right) \end{aligned} \tag{19.38}$$

respectively. Note that Eq. (19.37), $\forall m$ and n, and Eq. (19.38), $\forall j$, represent fixed-point equations with respect to the vectors $(\mathbf{q}, \tilde{\mathbf{q}})$. Since it can be easily confirmed that the right-hand sides of these equations are standard, in the sense specified in the paper by Yates [32], the optimal vectors $(\mathbf{q}, \tilde{\mathbf{q}})$ solving Problem (19.35) can be practically obtained by starting from any feasible power allocation and by iteratively solving, until convergence, the following fixed point equations:

$$\begin{aligned} q_m^{[n]} = \min\Bigg\{ &\ln\left(\frac{P_{m,\max}}{N}\right), \ln(Bw_{m,k(m,n)}^{[n]}\eta_m^{[n]}/\ln 2) \\ &- \ln\left(\pi\alpha_m^{[n]} + \sum_{j=1,j\neq m}^{M} \frac{Bw_{j,k(j,n)}^{[n]}\eta_j^{[n]}}{\ln 2}\frac{G_{m,k(j,n)}^{[n]}}{I_{j,k(j,n)}^{[n]}} + \sum_{j\in\mathcal{K}_D^{[n]}} \frac{B\tilde{w}_j\tilde{\eta}_j}{\ln 2}\frac{M_{m,j}^{n}}{\tilde{I}_j}\right)\Bigg\} \end{aligned} \tag{19.39}$$

for $m = 1, \ldots, M$ and $n = 1, \ldots, N$, and

$$\tilde{q}_j = \min \left\{ \ln(P_{j,\max}), \ln(\tilde{w}_j \tilde{\eta}_j / \ln 2) \right. \\ \left. - \ln \left(\pi \tilde{\alpha}_j + \sum_{m=1}^{M} \frac{B w_{m,k(m,n(j))}^{[n(j)]} \eta_m^{[n(j)]}}{\ln 2} \frac{G_{j,k(m,n(j))}^{[n(j)]}}{I_{m,k(m,n(j))}^{[n(j)]}} + \sum_{\ell \in \mathcal{K}_D^{[n(j)]} \setminus j} \frac{B \tilde{w}_\ell \tilde{\eta}_\ell}{\ln 2} \frac{\tilde{M}_{\ell,j}}{\tilde{I}_\ell} \right) \right\} \tag{19.40}$$

for $j = 1, \ldots, K_D$. Summing up, the original GEE maximization Problem (19.27) can be solved by combining Iterations (19.39) and (19.40) with Algorithms 19.3 and 19.2. The overall procedure, for the reader's perusal, is reported in Algorithm 19.4.

Algorithm 19.4 Complete procedure to solve Problem (19.27)

Initialize $I_{\max}$ and set $i = 0$;
Initialize $\mathbf{p}, \widetilde{\mathbf{p}}$ and compute $\mathbf{k}$ according to Eq. (19.28);
repeat
  Set $\overline{z}_m^{[n]} = \gamma_{m,k(m,n)}^{[n]}$ and compute $\eta_m^{[n]}$ and $\beta_m^{[n]}$ as in Eq. (19.29), for $m = 1, \ldots, M$ and $n = 1, \ldots, N$;
  Set $\overline{\overline{z}}_j = \widetilde{\gamma}_j$ and compute $\widetilde{\eta}_j$ and $\widetilde{\beta}_j$ as in Eq. (19.29), for $j = 1, \ldots, K_D$;
  Set $\epsilon > 0$, $\pi = 0$, and FLAG $= 0$
  repeat
    Update $\mathbf{q}$ and $\widetilde{\mathbf{q}}$ by iterating, until convergence, fixed-point Eq. (19.39) for $m = 1, \ldots, M$ and $n = 1, \ldots, N$, and Eq. (19.40) for $j = 1, \ldots, K_D$;
    if $f(e^{\mathbf{q}}, e^{\widetilde{\mathbf{q}}}, \mathbf{k}) - \pi g(e^{\mathbf{q}}, e^{\widetilde{\mathbf{q}}}) < \epsilon$ **then**
      FLAG $= 1$
    else
      Set $\pi = f(e^{\mathbf{q}}, e^{\widetilde{\mathbf{q}}}, \mathbf{k}) / g(e^{\mathbf{q}}, e^{\widetilde{\mathbf{q}}})$
    end if
  until FLAG $= 1$
  Update $\mathbf{k}$ according to Eq. (19.28);
  Set $i = i + 1$
until Convergence or $i = I_{\max}$

As already discussed, letting $\theta_m^{[n]} = 1/(MN), \alpha_m^{[n]} = 0, \forall m, n$, and $\tilde{\theta}_j = 1/K_D, \tilde{\alpha}_j = 0, \forall j$ in Eq. (19.39) and (19.40) the GEE reduces to the WSR and Algorithm 19.4 actually maximizes the system WSR.

19.8 Numerical Results

We consider a cluster with $M = 3$ BSs and $N = 32$ subcarriers. There are 15 I2D users per cell and 5 D2D links per cell, so that the total number of I2D users and D2D links is $K = 45$ and $K_D = 15$, respectively. The distance between adjacent BSs is equal to 2 km (cell radius equal to 1000 m), while the distance covered by each D2D link is uniformly distributed in the range $[5, 50]$ m. The D2D transmitters have a maximum feasible transmit power $\tilde{P}_{j,\max} = 13\,\text{dBm}$, for all j. The channel coefficients from a generic BS to the mobile users (either I2D or D2D) and from each D2D transmitter to all its unintended receivers on each subcarrier have been generated accounting for both fast and slow fading. The fast-fading component has been generated as a realization of a zero-mean, complex Gaussian random variable with variance $\sigma^2 = \text{PL}_0\left(\frac{d_0}{d}\right)^3$, wherein PL_0 is the free-space attenuation at the reference distance $d_0 = 100$ m with a carrier frequency of 1800 MHz, while d is the distance between the considered transmitter and receiver pair. The slow-fading (shadowing) component has been generated according to a lognormal distribution with variance equal to 8 dB. With regard to the channel coefficients between each D2D transmitter and its intended receiver, the fast-fading component has been instead generated according to a Rice distribution where the ratio between the direct path strength and the scattered paths strengths is equal to -6 dB, and the path loss has been assumed to increase with the square of the length of the link. The thermal noise power at each receiver is generated as $F\mathcal{N}_0 W$, with $W = 180$kHz being the communication bandwidth, $F = 3$ dB the receiver noise figure and $\mathcal{N}_0 = -174$ dBm/Hz the receive power spectral density, which are typical values of LTE networks [33]. The amplifier efficiency and circuit power consumption have been set to $\alpha = 3.8$ and $\theta = 500$ mW for the links served by the BSs, while the values $\tilde{\alpha} = 1$ and $\tilde{\theta} = 100$ mW have been taken for the D2D transmitters.[7]

We compare the situation in which D2D links are active (referred to as "with D2D links" in the figure legends) with that in which all communications flow through the BSs and thus D2D receivers are included in the set of I2D users (referred to as "without D2D links" in the figures). We also compare the situation in which I2D users and D2D users are randomly deployed according to a uniform distribution (referred to as "random deployment" in the figures) with that in which, in order to protect D2D receivers from BS-originated interference, D2D links are at a minimum distance $d_{\min}$ from the BS.

In the following, we report on two performance measures. Firstly there is the system WSR, as reported in Eq. (19.10), with all the weights set equal to one, for both the cases in which resources are allocated in order to maximize WSR and GEE. Secondly there is the GEE of the BSs, namely the quantity in Eq. (19.6) without the sum over j (both at the numerator and denominator), and all the weights equal to one, again for both the cases in which resources are allocated in order to maximize WSR and GEE. Thus even though resources are optimized in order to maximize the WSR and the GEE in Eqs (19.10) and (19.6), respectively, which include the D2D transmitters, we show results on the GEE of the BSs only. In other words, we assume an operator-centric point of view and evaluate the impact that activation of D2D links has on the energy efficiency of the BSs.[8]

[7] These values are in agreement with the ones commonly found in the literature.

[8] By the same token, we instead include the rate of D2D links when plotting the WSR, in order to evaluate the gains in throughput granted by the activation of D2D links.

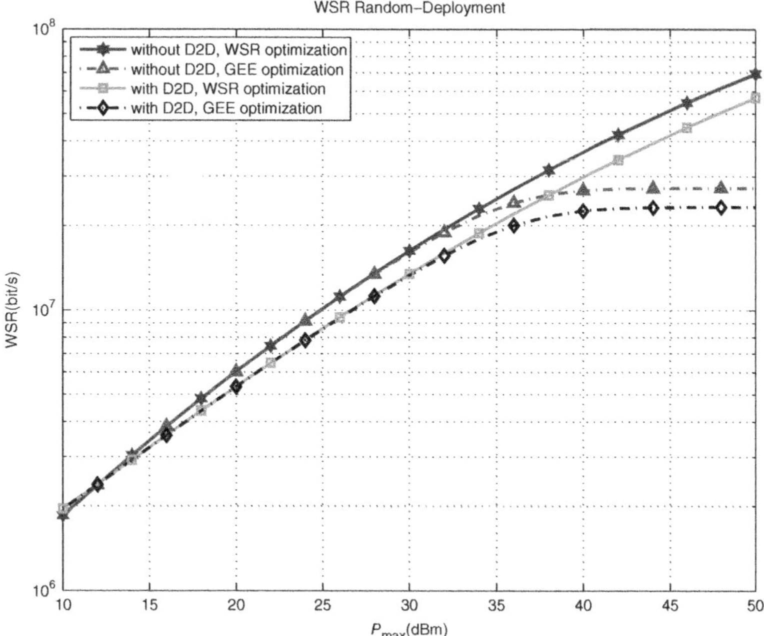

Figure 19.3 WSR versus $P_{\max}$ for the case of random deployment

The results that are reported versus the maximum transmit power available at each BS, come from an average over 500 independent trials. The maximum transmit power is assumed to be the same for all the BSs in the cluster: $P_{m,\max} = P_{\max}$ for all m.

Figures 19.3 and 19.4 refer to the "random deployment" scenario. In Figure 19.3 we report the WSR arising from the maximization of the WSR in Eq. (19.10) and of the GEE in Eq. (19.6), for both the "with D2D links" and "without D2D links" situations; in Figure 19.4, instead, we report the BSs' GEE arising from the maximization of the WSR in Eq. (19.10) and of the GEE in Eq. (19.6), for both the situations "with D2D links" and "without D2D links". Figures 19.5 and 19.6 report the same results as Figures 19.3 and 19.4, respectively, for the case in which D2D receivers are at a minimum distance $d_{\min} = 150$m from the BS.

Inspecting the figures, the following comments are in order. For the GEE-maximizing solution, the performance measures considered eventually saturate as $P_{\max}$ increases. A different behavior is observed for the WSR, which is monotonically increasing in $P_{\max}$. From a deeper inspection of, for instance, Figures 19.3 and 19.4 it is seen that allocating resources according to a maximum WSR criterion yields an increase in the data-rate at the price of a substantial reduction in the GEE for $P_{\max} > 33$dBm; instead, for $P_{\max} < 33$dBm, resource allocation according to WSR maximization and GEE maximization yields similar results and both the performance measures increase with $P_{\max}$. A similar conclusion may be drawn after inspecting Figures 19.5 and 19.6.

Moreover, it is seen that activation of D2D links has a moderately negative impact on the BS energy efficiency; this behavior is explained by noticing that D2D transmitters cause additional

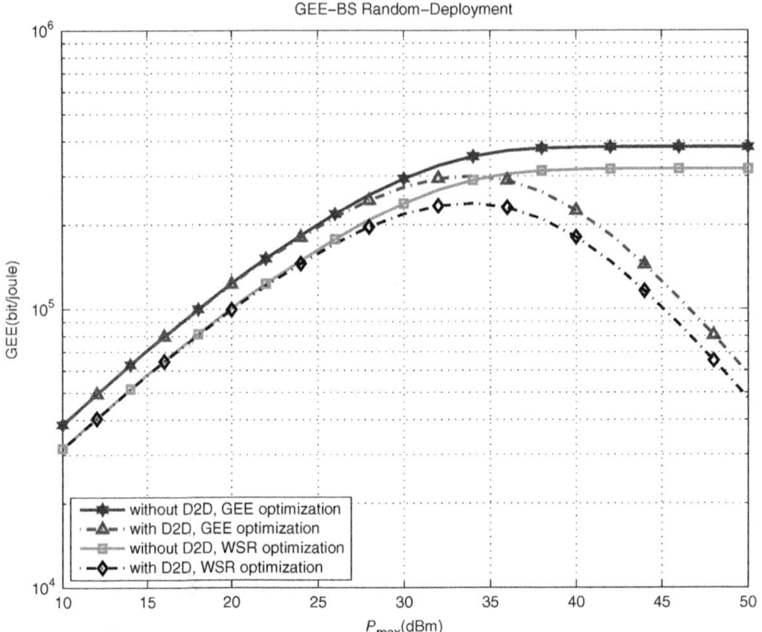

Figure 19.4 GEE of BSs versus $P_{\max}$ for the case of random deployment

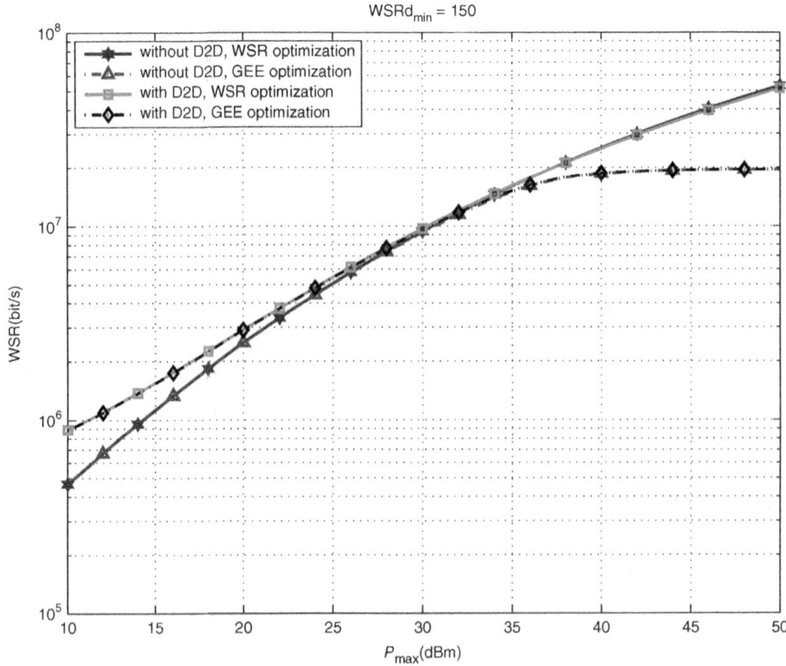

Figure 19.5 WSR versus $P_{\max}$ for the case in which a minimum distance of 150 m is observed between the BSs and the D2D receivers

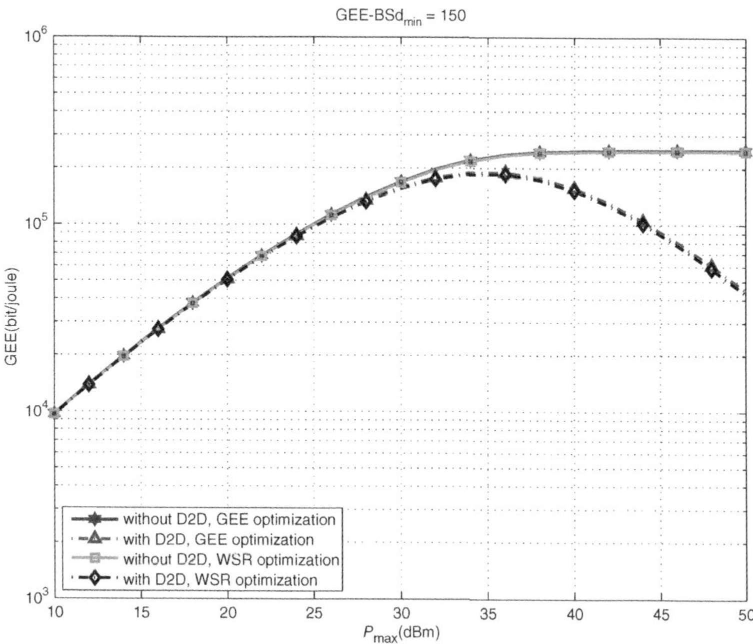

Figure 19.6 BSs WSEE versus $P_{\max}$ for the case in which a minimum distance of 150 m is observed between the BSs and the D2D receivers

interference to the I2D users served by the BSs, and this leads to a reduction of their energy efficiency. With regard to the system sum-rate, the activation of D2D links leads to an improvement of the WSR in the region of small values of $P_{\max}$, while for larger values of $P_{\max}$ the scenario "without D2D links" achieves a larger WSR than the scenario "with D2D links". This behavior can be explained by noticing that when D2D links are disabled, the D2D receivers become I2D users, thus competing for BS resources and increasing the multi-user diversity order of the system; this fact should not lead to the wrong conclusion that D2D links are not instrumental in the increase of the overall WSR. Indeed, comparing Figures 19.3 and 19.5 it is seen that the range of values of $P_{\max}$ wherein D2D links lead to an increase in the WSR is larger when a minimum distance between BSs and D2D receivers is introduced. This is confirmed by the results reported in Figure 19.7, wherein the WSR of the D2D receivers only resulting from WSR optimization for both the case "with D2D links" and "without D2D links" is reported, for $d_{\min} = 500$ m and $d_{\min} = 750$ m. It is clearly seen that the D2D receivers achieve a much larger rate when D2D links are active than when the D2D receivers are served by the BSs. Additionally, in the setup considered, D2D links use only one of the N available subcarriers; simulation results, not reported here for the sake of brevity, indicate that allowing D2D links to use two or three subcarriers leads to a further expansion of the operating zone wherein D2D links have a positive impact on the system sum-rate. A further expansion of this range may be also achieved by increasing the maximum allowed transmit power at the D2D transmitters.

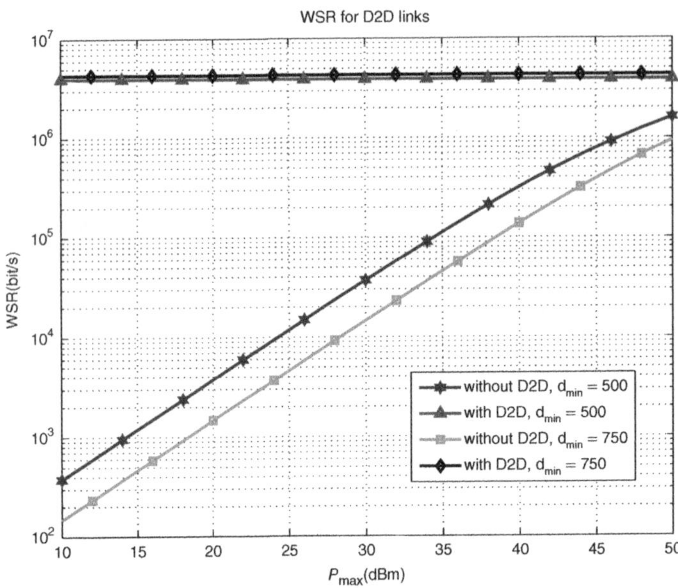

Figure 19.7 WSR of the D2D receivers versus P_{max} for two different values of the minimum distance between the BSs and the D2D receivers

19.9 Conclusion

Motivated by the consideration that energy efficiency is one of the drivers of 5G networks, this chapter has addressed the problem of power allocation for energy efficiency in wireless interference networks. The problem has been formulated as the maximization of the network GEE, with respect to all of the UEs' transmit powers. The resulting optimization problem turns out to be a fractional program, the solution of which is made particularly challenging by the interference-limited nature of the network. The presence of multi-user interference complicates the analysis and results in exponential computational complexity if standard fractional programming algorithms are used.

In order to overcome this issue, the framework of sequential fractional programming has been employed, thereby allowing one to obtain power allocations that trade off performance and complexity. In particular, the proposed optimization framework provides provably convergent algorithms, which require only the solution of a sequence of convex programs, while at the same time yielding a power allocation fulfilling the first-order optimality conditions of the power-control problem.

As a case study, the downlink of an OFDMA cellular network in which a group of BSs and of D2D transmitters coexist has been considered. The transmissions of all the transmitters in the network—both BSs and D2D transmitters—have been coordinated according to the proposed sequential fractional programming framework for GEE maximization.

The sequential fractional programming framework presented here is a powerful tool that can be used to perform energy-efficient resource allocation in a number of disparate scenarios. As an example, the consideration of the case in which the BSs and the mobile terminals are equipped with multiple antennas is a topic that is definitely worth being investigated. Here, beamforming can be used in order to concentrate transmitted signals towards the intended destination thus further reducing the interference in the system. Additionally, the proposed approach should be generalized to the case in which imperfect channel estimation is available and a finite-rate feedback channel is available. Finally, distributed coordination schemes, wherein each BS and each D2D transmitter independently allocate resources based on local information only is another key topic worth being investigated.

References

[1] NGMN Alliance (2015), 5G white paper. URL https://www.ngmn.org/uploads/media/NGMN_5G_White_Paper_V1_0.pdf.

[2] Infrastructure, G (2015), 5G Vision. URL https://5g-ppp.eu/wp-content/uploads/2015/02/5G-Vision-Brochure-v1.pdf.

[3] Zappone, A. and Jorswieck, E. (2015) Energy efficiency in wireless networks via fractional programming theory. *Found. Trends Commun. Info. Theory*, **11** (3–4), 185–396.

[4] Ericsson (2015), White paper: 5G energy performance. URL http://www.ericsson.com/res/docs/whitepapers/wp-5g-energy-performance.pdf.

[5] Andreev, S., Pyattaev, A., Johnsson, K., Galinina, O., and Koucheryavy, Y. (2014) Cellular traffic offloading onto network-assisted device-to-device connections. *IEEE Commun. Mag.*, **52** (4), 20–31, 10.1109/MCOM.2014.6807943.

[6] Tehrani, M., Uysal, M., and Yanikomeroglu, H. (2014) Device-to-device communication in 5G cellular networks: challenges, solutions, and future directions. *IEEE Commun. Mag.*, **52** (5), 86–92, 10.1109/MCOM.2014.6815897.

[7] George, G., Mungara, R.K., and Lozano, A. (2014), An analytical framework for device-to-device communication in cellular networks. URL http://arxiv.org/abs/1407.2201.

[8] Lee, N., Lin, X., Andrews, J.G., and Heath, Jr, R.W. (2015) Power control for D2D underlaid cellular networks: modeling, algorithms and analysis. *IEEE J. Select. Areas Commun.*, **33** (1), 1–13, 10.1109/JSAC.2014.2369612.

[9] Lin, X., Andrews, J.G., and Ghosh, A. (2014) Spectrum sharing for device-to-device communication in cellular networks. *IEEE Trans. Wireless Commun.*, **13** (12), 6727–6740, 10.1109/TWC.2014.2360202.

[10] Li, Y., Jin, D., Yuan, J., and Han, Z. (2014) Coalitional games for resource allocation in the device-to-device uplink underlaying cellular networks. *IEEE Trans. Wireless Commun.*, **13** (7), 3965–3977, 10.1109/TWC.2014.2325552.

[11] Wu, D., Wang, J., Hu, R., Cai, Y., and Zhou, L. (2014) Energy-efficient resource sharing for mobile device-to-device multimedia communications. *IEEE Trans. Vehic. Techn.*, **63** (5), 2093–2103, 10.1109/TVT.2014.2311580.

[12] Al-Kanj, L., Poor, H.V., and Dawy, Z. (2014) Optimal cellular offloading via device-to-device communication networks with fairness constraints. *IEEE Trans. Wireless Commun.*, **13** (8), 4628–4643, 10.1109/TWC.2014.2320492.

[13] Asadi, A., Wang, Q., and Mancuso, V. (2014) A survey on device-to-device communication in cellular networks. *IEEE Commun. Surveys Tut.*, **16** (4), 1801–1819, 10.1109/COMST.2014.2319555.

[14] Venturino, L., Zappone, A., Risi, C., and Buzzi, S. (2015) Energy-efficient scheduling and power allocation in downlink OFDMA networks with base station coordination. *IEEE Trans. Wireless Commun.*, **14** (1), 1–14.

[15] Dinkelbach, W. (1967) On nonlinear fractional programming. *Management Sci.*, **13** (7), 492–498.

[16] Schaible, S. (1976) Fractional programming. II, on Dinkelbach's algorithm. *Management Sci.*, **22** (8), 868–873.

[17] Charnes, A. and Cooper, W.W. (1962) Programming with linear fractional functionals. *Naval Res. Log. Quarterly*, **9**, 181–186.

[18] Schaible, S. (1974) Parameter-free convex equivalent and dual programs of fractional programming problems. *Zeitschrift für Operations Research*, **18** (5), 187–196.

[19] Cambini, A. and Martein, L. (2009) *Generalized Convexity and Optimization: Theory and Applications*, Springer-Verlag.

[20] Jagannathan, R. (1966) On some properties of programming problems in parametric form pertaining to fractional programming. *Management Sci.*, **12** (7).

[21] Bertsekas, D.P. (1999) *Nonlinear Programming*, Athena Scientific.

[22] Boyd, S.P. and Vandenberghe, L. (2004) *Convex Optimization*, Cambridge University Press.

[23] Ng, D.W.K., Lo, E.S., and Schober, R. (2012) Energy-efficient resource allocation in multi-cell OFDMA systems with limited backhaul capacity. *IEEE Trans. Wireless Commun.*, **11** (10), 3618–3631.

[24] Xiao, X., Tao, X., and Lu, J. (2014) Energy-efficient resource allocation in LTE-based MIMO-OFDMA systems with user rate constraints. *IEEE Trans. Vehic. Tech.*, **PP** (99), 1–1.

[25] He, S., Huang, Y., Jin, S., and Yang, L. (2013) Coordinated beamforming for energy efficient transmission in multicell multiuser systems. *IEEE Trans. Commun.*, **61** (12), 4961–4971.

[26] Du, B., Pan, C., Zhang, W., and Chen, M. (2014) Distributed energy-efficient power optimization for CoMP systems with max-min fairness. *IEEE Commun. Lett.*, **18** (6), 999–1002.

[27] Marks, B.R. and Wright, G.P. (1978) A general inner approximation algorithm for non-convex mathematical programs. *Operations Res.*, **26** (4), 681–683.

[28] Zappone, A., Jorswieck, E.A., and Buzzi, S. (2014) Energy efficiency and interference neutralization in two-hop MIMO interference channels. *IEEE Trans. Signal Process.*, **62** (24), 6481–6495.

[29] Dan, N., Tran, L.N., Pirinen, P., and Latva-aho, M. (2013) Precoding for full duplex multiuser MIMO systems: spectral and energy efficiency maximization. *IEEE Trans. Signal Process.*, **61** (16), 4038–4050.

[30] Zappone, A., Sanguinetti, L., Bacci, G., Jorswieck, E.A., and Debbah, M. (2015) Energy-efficient power control: a look at 5G wireless technologies. *IEEE Trans. Signal Process.*, *submitted*, http://arxiv.org/abs/1503.04609.

[31] Venturino, L., Prasad, N., and Wang, X. (2009) Coordinated scheduling and power allocation in downlink multicell OFDMA networks. *IEEE Trans. Vehic. Tech.*, **58** (6), 2835–2848.

[32] Yates, R.D. (1995) A framework for uplink power control in cellular radio systems. *IEEE J. Select. Areas Commun.*, **13** (7), 1341–1347.

[33] Auer, G., Giannini, V., Desset, C., Godor, I., Skillermark, P., Olsson, M., Imran, M.A., Sabella, D., Gonzalez, M.J., Blume, O. *et al.* (2011) How much energy is needed to run a wireless network? *IEEE Wireless Commun.*, **18** (5), 40–49.

20

Ultra Dense Networks: General Introduction and Design Overview

Jianchi Zhu, Xiaoming She and Peng Chen

Signal Processing for 5G: Algorithms and Implementations, First Edition. Edited by Fa-Long Luo and Charlie Zhang.

20.1 Introduction

The evolution of 3G into 4G is driven by the creation and development of new services for mobile devices, and is enabled by advancement of the technology available for mobile communication systems. Mobile communication has experienced explosive growth over the past decade, fueled by the popularity of smart phones and tablets, the number of which almost doubles every year. According to the Cisco Visual Networking Index [1], mobile data traffic volume is expected to grow at a compound annual growth rate of 61% from 2013 through 2018. Furthermore, a broad consensus in the wireless industry anticipates a continuation of this trend for several years to come. It is predicted that data traffic volume will reach 1000 times its 2010 value by 2020, referred to as the 1000× data challenge [2]. The main factors accounting for the significant growth in mobile data traffic are the emergence of mobile internet applications and the proliferation of Internet-of-Things (IoT) devices. Mobile internet and IoT will become the main forces driving the evolution of mobile communications into 5G [3].

To develop 5G, efforts have been made in three areas: spectrum expansion, spectrum efficiency enhancement and network densification. Network densification is considered to be the paramount and dominant approach to address the data challenge. It can be achieved by deploying a large number of small cells – microcells, picocells, femtocells, relay nodes and WiFi access points – which are low-powered radio access nodes and have smaller coverage areas than macrocells. Ultra-dense networks (UDNs), as shown in Figure 20.1, are the key technology to meet the requirements of ultra-high traffic volume density in year 2020 and beyond in hot-spot scenarios [4–9]. High spectral reuse factors can be achieved in UDNs via densely deployed wireless equipment. This results in capacity improvements by factors of 100 or more in hot-spot areas.

20.1.1 Application Scenarios

5G will need to fulfill many types of application in all aspects of life. The typical scenarios identified for UDNs include offices, dense residential areas, dense urban areas, campuses,

Figure 20.1 UDN with densely deployed small cells

Table 20.1 Application scenarios and services

Application scenarios	Services
Offices	Video telephony, desk cloud, data download and cloud storage.
Dense residential areas	Video telephony, HDTV, virtual reality for shopping, online gaming, data download, cloud storage, OTT messaging and intelligent household.
Dense urban areas	Virtual reality, Augmented reality, online gaming, data download, cloud storage, OTT message, vehicle-to-vehicle safe driving.
Campuses	Video telephony, HDTV, virtual reality for shopping, online gaming, data download, cloud storage, OTT.
Open-air gatherings	Video playing, Augmented reality/live video sharing, high-definition picture upload and OTT messaging
Stadiums	Video playing, enhanced reality/live video sharing, high-definition picture upload and OTT messaging
Shopping malls	Augmented reality, OTT message and video monitoring
Subways	Video playing, online gaming and OTT messaging

open-air gatherings, stadiums, subways and apartments [4–9]. In each scenario, there could be several types of services: high-definition TV programs (HDTV), video-telephony, video-conferencing, mobile online gaming, augmented reality, virtual reality, live video sharing, desk clouds, wireless data download, cloud storage, high-definition image upload, over-the-top (OTT) messaging, video monitoring, intelligent household control, and so on. The potential scenarios and services are listed in Table 20.1.

20.1.2 Challenges

UDNs will play a vital role in 5G, especially in hotspot scenarios. However, the widespread deployment of small cells will pose significant challenges, such as site acquisition and expenditure, network operation and management, interference management, mobility management, backhaul resources, and so on [4, 5].

20.1.2.1 Site Acquisition and Expenditure

In wireless networks, site rental and equipment cost accounts for most capital expenditure. Site acquisition is the critical problem for operators and a large number of sites is required for UDNs. Furthermore, as the network density increases, operation expenditure also increases. Thus UDNs, with their high cell density, will lead to extremely high capital and operational expenditure.

20.1.2.2 Network Operation and Management

With a large number of small cells, network operation and management becomes very complex as it needs lots of human effort to install, configure, monitor and maintain them. Additionally, the traffic may change frequently, depending on the location and time, so it is difficult for network operators to efficiently use network resources. In addition, there may be unplanned deployment and withdrawal of small cells located in residences and enterprises. In this scenario, small cells can be switched on/off or moved at any time by the users and they are beyond the reach of the network operators, making optimization and management of the network very challenging. The cells thus need to be capable of being configured, optimized and healed by themselves, so as not to cause any noticeable disruption to the network. Unplanned deployment without such autonomous processes may cause significant problems to both macro cell users and small cell users.

20.1.2.3 Interference Management

As the network density increases, intercell interference (ICI) becomes more severe, which significantly limits the gain achieved by the densely deployed small cells. ICI can arise in different ways. Strong interference between macrocells and co-channel small cells arises from the high imbalance in path loss and transmission power of these two types of cell. This may result in mobile users suffering significant degradation in the received signal quality. On top of macro–small cell interference, interference between densely deployed small cells further complicates the ICI landscape. The geographically random, unplanned small-cell deployment may generate out-of-cell interference to adjacent small cells especially in densely deployed areas [10]. Without mitigating interference, UDNs cannot be deployed successfully.

20.1.2.4 Mobility Management

In UDNs, mobile users may experience more frequent handover due to smaller cell coverage. Recent studies in LTE show that there are very frequent handover events with nearby picocells in the shared spectrum, but not even very dense [11]. The analysis indicates that handover performance in heterogeneous-network (HetNet) deployments is not as good as in pure macro. Numerous mobility enhancements and corresponding analyses have been studied in the context of HetNets [11–13]. As the network density increases, the handover frequency increases and this may result in higher handover-failure rates. Thus mobility management becomes more challenging in UDNs.

20.1.2.5 Backhaul Resources

As the network density increases, backhaul will rise in importance [7]. With densification, the goal of operators is to deliver additional capacity and coverage with sufficient backhaul capacity and low latency without recurring extra operational expenditure, with solutions that range from fiber and Ethernet to wireless. From the backhaul perspective, the challenges may include the following three aspects:

- With the increase in the number of deployed small cells, the deployed backhaul links should also be increased. Considering the cost of network development and maintenance, it is not possible to deploy high-speed wired backhaul for all the cells, for example, fiber.

- Usually, it is not easy to predict the deployment location of small cells and it is always hard to provide wired backhaul for locations where it is convenient to deploy small cells, such as kerbsides, on roofs or on street lamps.
- The ICI between base stations in UDNs is more severe than in traditional deployments. Therefore, fast, even real-time information exchange and coordination between base stations is needed to enable highly efficient interference coordination and suppression.

20.1.3 Key Technologies

In this chapter, promising technologies and solutions are discussed to address the challenges raised above. Interference management, mobility management, architecture and backhaul are presented and analyzed in Sections 20.2, 20.3 and 20.4, respectively. Other issues such as synchronization, energy efficiency, flexible duplex and promising solutions to address the site acquisition problem are discussed in Section 20.5. Section 20.6 is devoted to summarizing what has been presented in this chapter and is to discussion of future work.

20.2 Interference Management

Extensive research related to interference management has been performed in academia, industry, and standardization bodies such as the Third Generation Partnership Project (3GPP). The interference problem can be addressed from the network side or the user side. Table 20.2 shows an overview of the different mechanisms [28].

20.2.1 Network Coordination

Network coordination can be conducted in the spatial domain, time domain, frequency domain or power domain [23]. The spatial-domain techniques include coordinated multi-point techniques (CoMP) [15] and interference alignment techniques [20]. In the time domain, cross-tier interference between the macro layer and the small cell layer has been extensively investigated in the literature [22] for HetNet and ICI can be mitigated by switching off the small cells when there is no packet to be transmitted. Frequency-domain coordination can be realized by assigning different carriers to base stations, or by using different OFDMA subcarriers for transmission [14]. Finally, the adjustment of the base station transmit power is another network coordination technique that has often been applied to closed subscriber group femto cells with the goal of reducing cross-tier interference toward co-channel macro users [27].

20.2.1.1 Coordinated Multi-Point Transmission and Reception

CoMP transmission/reception is an effective method to improve both average and cell-edge throughput by mitigating the ICI [16]. Categorized by the different number of points for coordinated transmission, CoMP techniques include coordinated scheduling/beamforming (CS/CB) [17] and joint processing (JP) [18], corresponding to single-point and multiple-point coordinated transmission, respectively.

Table 20.2 Different mechanisms of interference management

	Mechanisms	Descriptions
Network coordination	Spatial-domain coordination	Use of spatial filtering techniques, including use of arrays of transmit antennas or active antennas with coordinated beamforming and joint transmission between cells.
	Time-domain coordination	Cells are time-synchronized and coordinate at which time-instances they transmit, such that there are time-instances where Cell A can serve its users without interference from Cell B. This is also known as coordinated muting. An example is enhanced ICI coordination (eICIC).
	Frequency-domain coordination	Include options such as using hard or soft frequency reuse between neighboring cells. The frequency-domain resource partitioning can be on physical resource block resolution, or on carrier resolution if there are networks with multiple carriers.
	Power-domain coordination	Adjustment of transmit power per cell to improve the interference conditions. Examples include the defined techniques for femto cell transmit power calibration to reduce interference toward co-channel macro-users.
Advanced receivers	Linear receiver with interference suppression	Interference suppression by means of linear combining of received signals at the users' antennas. Examples of such techniques are optimal combining and interference rejection combining.
	Non-linear receiver with interference cancellation	Interference cancellation with non-linear techniques where the user estimates one or multiple interfering signals and subtracts them from the received signal, followed by detection of the desired signal. Examples include successive or parallel interference cancellation schemes.

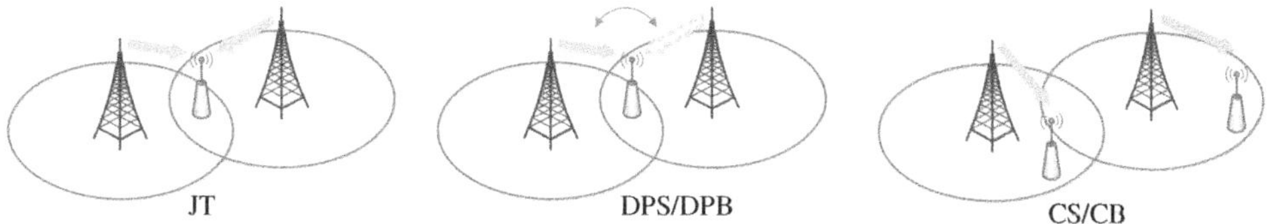

Figure 20.2 CoMP schemes

In CS/CB, by single-point transmission, coordination of scheduling decisions and transmit-beam selection among multiple cells is used to reduce ICI. In JP, data are shared among multiple points for joint transmission to one or more users simultaneously so as to achieve a performance gain by signal combining and interference nulling from multiple points. For JP, there are two schemes:

- joint transmission (JT)
- dynamic point selection (DPS) and dynamic point blanking (DPB) [19].

The two JP schemes and CS/CB are shown in Figure 20.2.

Among the above CoMP schemes, JT combined with multi-user (MU-) multiple-input, multiple-output (MIMO) achieved the best performance. For a JT MU-MIMO system, assuming that there are M transmit antennas at each remote radio head (RRH) and N receive antennas at each user, in the downlink, the L cooperative small cells and the K users can form a $LM \times KN$ virtual MIMO system, depicted in the Figure 20.3.

The channel matrix from the network to the user k on one subcarrier is denoted by $\mathbf{H}_k \in \mathbf{C}^{N \times LM}$. Therefore, the composite channel matrix of the cooperative $LM \times KN$

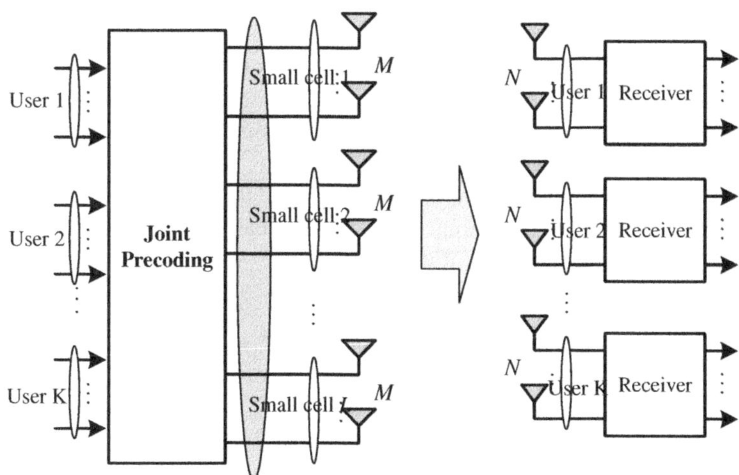

Figure 20.3 Structure of CoMP MU-MIMO

virtual MIMO system is given by

$$\mathbf{H} = [\mathbf{H}_1^T \ \mathbf{H}_2^T \cdots \mathbf{H}_K^T]^T \tag{20.1}$$

The data vector $\mathbf{x}_k$ intended for user k with power constraint $\mathbf{E}\{\mathbf{x}_k^H\mathbf{x}_k\} \leq \mathbf{P}_k$ is given by

$$\mathbf{x}_k = \begin{bmatrix} x_{k,1} & x_{k,2} & \cdots & x_{k,s_k} \end{bmatrix}^T \tag{20.2}$$

where s_k denotes the number of data streams for user k.

The joint precoding matrix for user k's data vector is denoted as $\mathbf{T}_k \in (LM \times s_k)$.

The receive signal $\mathbf{y}_k$ at user k can be written as

$$\mathbf{y}_k = \mathbf{H}_k\mathbf{T}_k\mathbf{x}_k + \sum_{j=1, j\neq k}^{K} \mathbf{H}_k\mathbf{T}_j\mathbf{x}_j + \mathbf{n}_k \tag{20.3}$$

where the second item on the right-hand side is the interference seen by user k from other users' signals and $\mathbf{n}_k$ denotes the additive Gaussian white noise (AWGN) vector for user k with variance $\mathbf{E}[\mathbf{n}_k\mathbf{n}_k^H] = \sigma^2\mathbf{I}$.

In UDNs, the employment of CoMP can be facilitated by new centralized network architectures such as cloud radio access network (C-RAN) [21]. In the architecture based on C-RAN, the signal processing of dozens of cells is concentrated at the building baseband unit (BBU) pool, which has super capability in computing and storage, while the radio transceiver parts are kept at the cell sites and connected to the BBU pool via fibers or high-speed wireless links. In this way, the coordination of adjacent cells can be performed conveniently since the data and channel-state information (CSI) of these cells are co-located in the BBU pool.

20.2.1.2 Interference Alignment

The fundamental concept of interference alignment (IA) is to align the interference signals in a particular subspace at each receiver so that an interference-free orthogonal subspace can be solely allocated for data transmission [20]. This is a promising technique to efficiently mitigate interference and to enhance the capacity of UDNs.

In an L-cell system with one user in each cell, assume that each base station is equipped with M antennas and each user is equipped with N antennas. Base station l transmits s_k data streams to user k. The received signal of user k is

$$\mathbf{y}_k = \mathbf{H}_{kk}\mathbf{V}_k\mathbf{x}_k + \sum_{j=1, j\neq k}^{K} \mathbf{H}_{kj}\mathbf{V}_j\mathbf{x}_j + \mathbf{n}_k \tag{20.4}$$

where $\mathbf{H}_{kj}$ denotes the channel matrix from base station j to user k, $\mathbf{V}_j$ is the precoding matrix of base station j, $\mathbf{x}_k = \begin{bmatrix} x_{k,1} & x_{k,2} & \cdots & x_{k,s_k} \end{bmatrix}^T$ is the transmitted signal for user k with power constraint $\mathbf{E}\{\mathbf{x}_k^H\mathbf{x}_k\} \leq \mathbf{P}_k$ and $\mathbf{n}_k$ denotes the AWGN vector for user k with variance $\mathbf{E}[\mathbf{n}_k\mathbf{n}_k^H] = \sigma^2\mathbf{I}$.

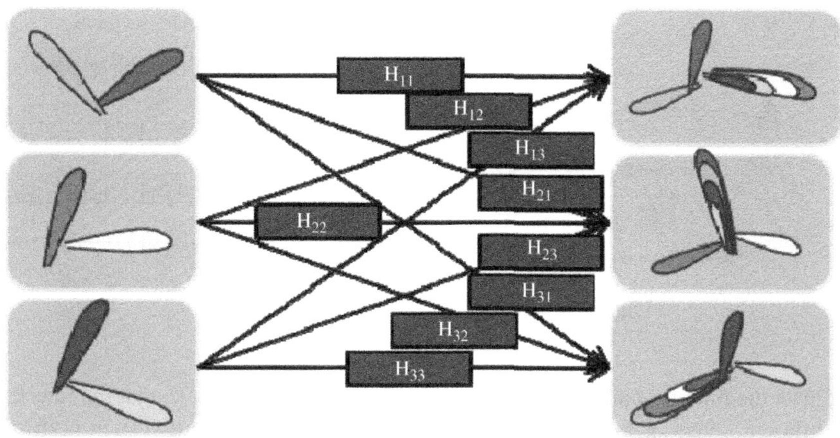

Figure 20.4 Example of interference alignment

As an example, take three cells with one user in each cell, as shown in Figure 20.4. To align the interference subspace at the receiver side, the following conditions should be satisfied.

$$\begin{aligned} \mathbf{span}(\mathbf{H}_{12}\mathbf{V}_2) &= \mathbf{span}(\mathbf{H}_{13}\mathbf{V}_3) \\ \mathbf{span}(\mathbf{H}_{21}\mathbf{V}_1) &= \mathbf{span}(\mathbf{H}_{23}\mathbf{V}_3) \\ \mathbf{span}(\mathbf{H}_{31}\mathbf{V}_1) &= \mathbf{span}(\mathbf{H}_{32}\mathbf{V}_2) \end{aligned} \tag{20.5}$$

Furthermore, the conditions can be constrained as

$$\begin{aligned} \mathbf{span}(\mathbf{H}_{12}\mathbf{V}_2) &= \mathbf{span}(\mathbf{H}_{13}\mathbf{V}_3) \\ \mathbf{H}_{21}\mathbf{V}_1 &= \mathbf{H}_{22}\mathbf{V}_2 \\ \mathbf{H}_{31}\mathbf{V}_1 &= \mathbf{H}_{32}\mathbf{V}_2 \end{aligned} \tag{20.6}$$

Then we can get the precoding matrix of interference alignment for each user.

$$\begin{aligned} \mathbf{span}(\mathbf{V}_1) &= \mathbf{span}(\mathbf{G}\mathbf{V}_1) \\ \mathbf{V}_2 &= (\mathbf{H}_{32})^{-1}\mathbf{H}_{31}\mathbf{V}_1 \\ \mathbf{V}_3 &= (\mathbf{H}_{23})^{-1}\mathbf{H}_{21}\mathbf{V}_1 \end{aligned} \tag{20.7}$$

where $\mathbf{G} = (\mathbf{H}_{31})^{-1}\mathbf{H}_{32}(\mathbf{H}_{12})^{-1}\mathbf{H}_{13}(\mathbf{H}_{23})^{-1}\mathbf{H}_{21}$.

20.2.1.3 Enhanced Inter-cell Interference Cancellation

An eICIC for co-channel deployment has been developed for LTE-A to protect both control channels and data channels in the HetNet scenario. Defining the interfering node as an aggressor and the interfered node as a victim, the basic principle of time-domain eICIC schemes is

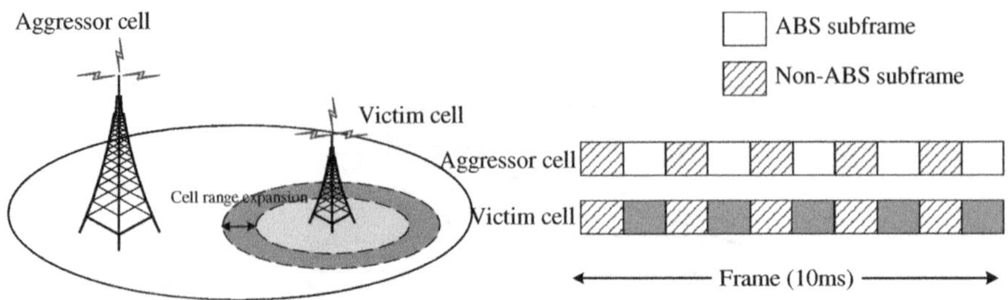

Figure 20.5 eICIC under macro-pico scenario

to coordinate the time-domain transmission of victims and aggressors so that the interference among different layers or in the same layer is reduced and the overall system performance is improved. Based on this idea, a time-domain eICIC approach, termed "almost blank subframe'' (ABS), is proposed and adopted in 3GPP LTE Release10 [22].

The basic idea of the ABS approach is to coordinate subframe utilization across different cells in the time domain. As shown in Figure 20.5, in the scenario with one aggressor cell (macro cell) and one victim cell (pico cell), a number of subframes are configured as ABSs in the macro cell, where no data signals are transmitted and only necessary control signals such as common reference signals are transmitted. In other words, with the ABS approach, the aggressors are almost muted in the configured almost blank subframes, and the users suffering strong ICI in the victim cells can be scheduled with high priority to transmit in the protected low-interference subframes.

20.2.1.4 Small Cell On/Off

One method to mitigate ICI is to exploit the uneven and dynamically fluctuating nature of wireless traffic by enabling a small cell on or off. The small cell can be signaled to be dormant when there is no packet to be transmitted and to be active again when there is a packet arrival. A dormant cell ceases overthe-air transmission of almost all its signals, hence unnecessary interference to its neighboring cells can be largely reduced. This method of turning small cells on or off opportunistically is referred to as "small cell on/off operation'' and it has been shown in 3GPP LTE Release 12 to be beneficial for improving user throughput. The mechanisms that allow different timescales of small cell on/off have also been investigated [23]. In particular, it was observed that the extent of performance gain in user throughput depends on the off-to-on and on-to-off transition time, in other words how fast cells can react to dynamic changes in traffic. For example, it was observed that with a transition time of less than 100 ms, a performance gain of 10–20% can be achieved [23].

20.2.1.5 Frequency-domain Coordination

Frequency-domain coordination can be realized by assigning different carriers to base stations, or by using different OFDMA subcarriers for transmission [14]. The simplest approach is hard frequency reuse, where nearby base stations use orthogonal frequency carriers. However, hard frequency reuse seldom results in the best performance. An alternative option is fractional frequency reuse (or soft frequency reuse), where some resources are reused by all base stations, while others are dedicated to only certain base stations. Furthermore, autonomous base station mechanisms for dynamically choosing the best carrier(s) have been widely investigated in the context of femto cell networks [24]. In all cases, the potential of time- and/or frequency-domain intercell partitioning methods is fully exploited when they are dynamically adjusted in step with the time-variant behavior of the system and the traffic fluctuations. The benefits of fast versus slow intercell coordination have been explored in the literature [25], while aspects of centralized versus distributed coordination have also been examined [26].

20.2.2 Advanced Receivers

An alternative to network coordination is to rely on advanced user receivers with interference mitigation or cancellation [7]. Users with multiple antennas can exploit linear interference suppression techniques such as interference rejection combining (IRC). Another variant of receiver-based interference mitigation is to apply non-linear interference cancellation, where the user reconstructs the interfering signal(s) followed by subtraction before decoding the desired signal. These techniques are especially attractive for canceling interference from semi-static signals such as common reference signals and broadcast and synchronization channels. However, applying non-linear interference cancellation to data-channel transmissions is much more challenging, as the scheduling and link adaptation (i.e. the selection of modulation and coding schemes) are highly dynamic, and conducted independently per cell. Getting the most out of non-linear interference cancellation therefore requires additional network assistance [30]. The idea is to simplify the processing at the user by providing a priori knowledge of the interfering signal characteristics such that the blind estimation of all their features can be removed.

20.2.2.1 Linear Receiver with Interference Suppression

The $\mathbf{N}_{RX}$-dimensional received signal vector $\mathbf{y}$ of the kth subcarrier and the l th OFDM symbol is assumed to be expressed as a sum of its own signal $\mathbf{H}_1(k,l)\mathbf{x}_1(k,l)$, interference signals $\mathbf{H}_j(k,l)\mathbf{x}_j(k,l)$ (j>1) and white noise :$\mathbf{n}(k,l)$:

$$\mathbf{y}(k,l) = \mathbf{H}_1(k,l)\mathbf{x}_1(k,l) + \sum_{j=2}^{N_{BS}} \mathbf{H}_j(k,l)\mathbf{x}_j(k,l) + \mathbf{n}(k,l) \tag{20.8}$$

where $\mathbf{x}_j(k,l)$ represents the $\mathbf{N}_{TX} \times \mathbf{1}$ transmitted signal vector. $\mathbf{H}_j(k,l), j = \{1,\dots,N_{BS}\}$ represents the ($\mathbf{N}_{RX} \times \mathbf{N}_{TX}$) channel matrix between the j th cell and the user containing the contribution from both receiver branches, with $\mathbf{H}_j = \left[\mathbf{H}_{j,1}^H \ \mathbf{H}_{j,2}^H\right]^T$ and $\mathbf{H}_{j,i}$ channel-matrix of size $\mathbf{N}_{TX} \times \mathbf{1}$ for the ith receiver antenna. The recovered $\mathbf{N}_{Stream} \times \mathbf{1}$ signal vector at the user, $\widehat{\mathbf{x}_1}(k,l)$, is detected by using the ($\mathbf{N}_{Stream} \times \mathbf{N}_{RX}$) receiver weight matrix $\mathbf{W}_{RX,1}(k,l)$ as follows.

$$\widehat{\mathbf{x}_1}(k,l) = \mathbf{W}_{RX,1}(k,l)\mathbf{y}(k,l) \tag{20.9}$$

IRC is a typical linear receiver with interference suppression. The IRC receiver can suppress not only interstream interference but also ICI when the degrees of freedom at the receiver are sufficient; in other words when the number of receiver antennas is higher than that of the number of desired data streams. For instance, one user equipped with M antennas has M degrees of freedom: one is used for the reception of its own stream; the remaining $M-1$ are available to exploit either diversity or interference suppression. The IRC receiver weight matrix is expressed as follows [29]:

$$\mathbf{W}_{RX,1}(k,l) = \widehat{\mathbf{H}}_1^H(k,l)\mathbf{R}^{-1} \tag{20.10}$$

where $\widehat{\mathbf{H}}_j(k,l)$ and $\mathbf{R}$ denote the estimated channel matrix and covariance matrix, respectively.

To obtain the IRC receiver weight matrix, the covariance matrix including the sources of ICI needs to be estimated.

20.2.2.2 Non-linear Receiver with Interference Cancellation

Nonlinear interference cancellation (IC) may involve estimating the interference signal at the modulation symbol level (SLIC) or at the codeword level (CLIC) [30], as shown in Figure 20.6. Error propagation issues associated with SLIC may be overcome by adopting a soft cancellation approach, incorporating the confidence level in estimated interference symbols. CLIC is mostly immune to error propagation effects, but requires that the spectral efficiency targeted by the interfering transmitter be consistent with the interference signal quality SINR at the victim receiver.

To successfully demodulate or decode the interference's signal, both approaches require knowledge of various transmission parameters of the interfering signal. For SLIC, interference parameters that can enable interferer channel estimation and interferer detection at symbol level are needed. For CLIC, interference parameters used for interference de-scrambling and

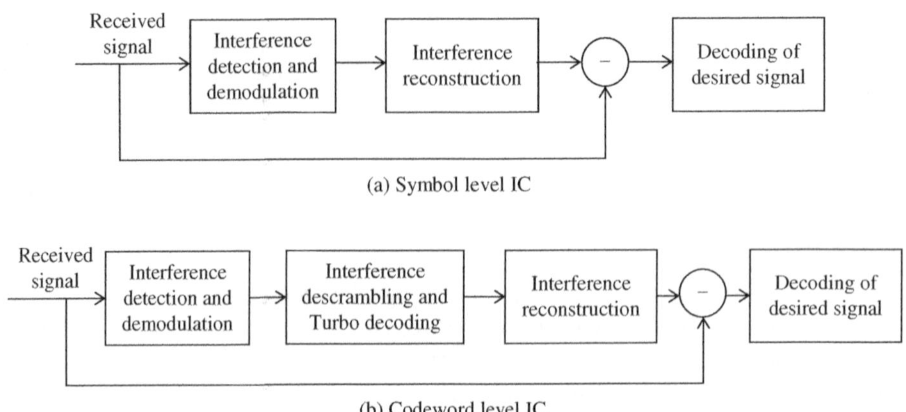

Figure 20.6 Non-linear receiver structure

turbo decoding are also needed, in addition to the required interference parameters for symbol level IC. In summary:

- Interference parameters required for SLIC for a cell-center user include the reference signal configuration, the number of data layers and the modulation scheme.
- Interference parameters required for CLIC for a cell-center user include the reference signal configuration, the number of data layers, the modulation scheme and code rate, the hybrid automatic repeat request redundancy version and the user radio network temporary identity.

20.3 Mobility Management

UDN will raise new challenges for mobility management.

Mobility performance Mobility performance will deteriorate. For instance, the rate of handover failure will increase due to severe co-channel interference between small cells located in one cluster; the ping-pong handover rate will increase as well.

Signaling load Mobile users may trigger frequent handovers when they move across the coverage areas of small cells, and the signaling load, including radio resource control (RRC) messages, X2 interface messages and signaling between core-network nodes, may be too heavy to bear.

Battery consumption To access the cell with better channel conditions, users have to take measurement of the signaling quality of a larger number of surrounding small cells, which will significantly increases their battery consumption.

In the following subsection, potential solutions are introduced for UDNs, including dual connectivity, virtual cells, virtual layers, mobility anchors and handover command diversity.

20.3.1 Dual Connectivity

Dual connectivity can be adopted in HetNet to improve per-user throughput and mobility robustness by allowing users to be connected simultaneously to a master cell group and a secondary cell group via the MeNB (master base station) and SeNB (secondary base station), respectively [31]. The architecture of dual connectivity is shown in Figure 20.7.

The control plane (C-plane) protocols and architectures for dual connectivity assume that there will be only one S1-MME connection per user. The MeNB is primarily responsible for handling the user's RRC state. In dual connectivity, the SeNB and MeNB could carry different bearers or a single bearer split into two streams. Different options for user-plane (U-plane) architectures have been considered for dual connectivity, as shown in Figure 20.8 [32].

20.3.2 Virtual Cell

A user-centric virtual cell is configured on the fly by a mobile user at the cell center and a set of cooperative small cells located in a circular area around them. The small cell in each virtual cell

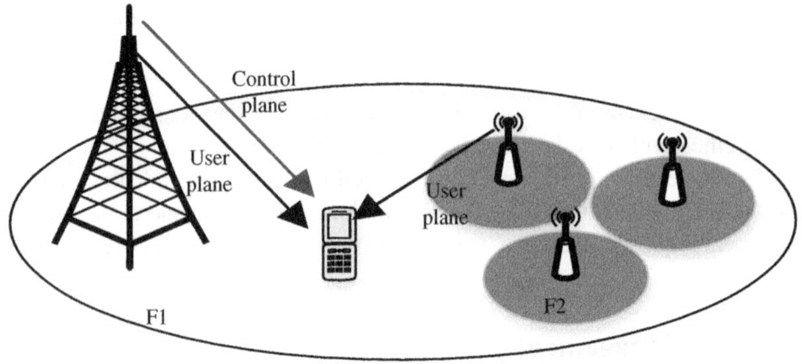

Figure 20.7 Dual connectivity

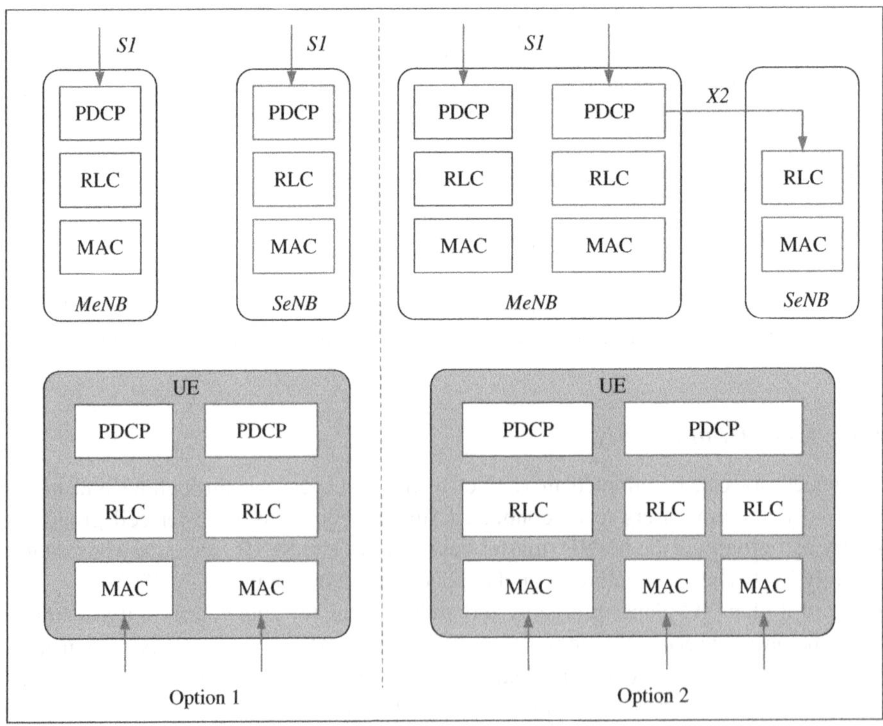

Figure 20.8 User-plane architectures for dual connectivity

may vary from time to time according to the movement of the user, or a change in the wireless environment [36]. The virtual cells, which are a cluster of cooperating hyper-dense small cells, may need a centralized node to manage them. In contrast to traditional cells, user-centric virtual cells enjoy advantages including:

- complete elimination of cell-edge users
- significant capacity gain from cooperative transmission or reception
- dynamic local cooperation, which reduces signaling overhead and computational complexity compared with static network-wide cooperation.

The physical cell included in the virtual cell changes when the user moves in the network, but the virtual cell ID remains the same. Therefore, no handover happens while the user is moving, and the user experience will be much enhanced. The diagram of virtual cell changes for a moving user is shown in Figure 20.9.

20.3.3 Virtual Layer

Virtual layer technology is a solution that relies on multi-layer networking: a virtual layer and a real layer (see Figure 20.10) [4]. The virtual layer is responsible for broadcasting, paging and mobility management, while data transmission is carried on the real layer. UDNs are divided into multiple clusters, each cluster corresponding to one virtual layer. UEs in idle mode camp on the virtual layer and have no need to recognize the real layer; no cell re-selection is needed

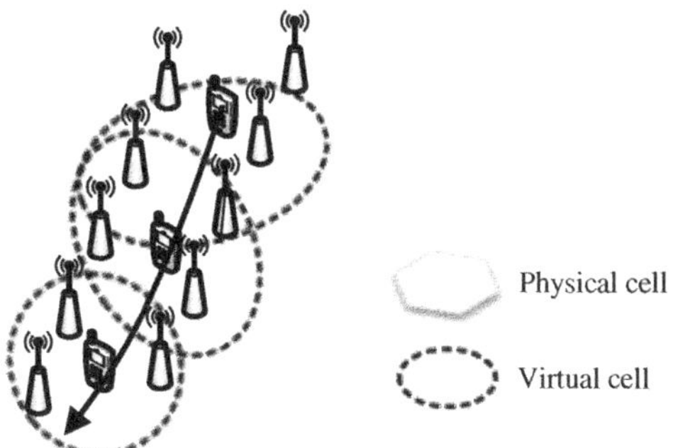

Figure 20.9 Virtual cell diagram

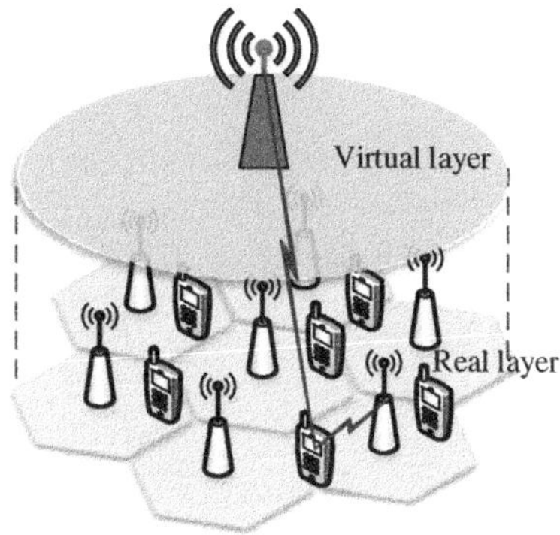

Figure 20.10 Virtual layer technology

either if UEs move within the same virtual layer. For UEs in connected mode, both the virtual layer and the real layer can be recognized; no handover is needed for the mobile users within the same virtual layer and so the user experience will be good.

Virtual layer technology can be realized by single-carrier or multiple-carrier solutions. For the single-carrier solution, the virtual layer and the real layer are constructed by different signals or channels respectively. For the multiple-carrier solution, the network can configure one carrier for each individual layer: one for the virtual layer and one for the real layer. For instance, to construct the virtual layer the network configures all cells with the same physical cell ID (PCI) on carrier 1 in the same cluster. It configures those with a different PCI for each cell in the same cluster to construct the real layer. Users in idle mode camp on carrier 1 and have no need to recognize carrier 2; no cell re-selection is needed either if users move within the same virtual layer. Users in connected mode can access both carrier 1 and carrier 2 with carrier aggregation; the network manages mobility via carrier 1 and no handover is needed for the mobile users within the same virtual layer since carrier 1 is not changed.

20.3.4 Mobility Anchor

The mobility anchor solution can reduce/hide signaling load towards the core network by hiding subsequent mobility involving SeNBs. Such a mobility anchor would be independent of the dual connectivity solution and could also be applied in case of limited user capability (single RX/TX) or high system load. In this solution a logical entity called a mobility anchor is introduced. The macro cell can be used as the anchor base station for the location of mobility anchor or the mobility anchor can be a new entity. The potential architectures for a mobility anchor are shown in Figure 20.11 [33].

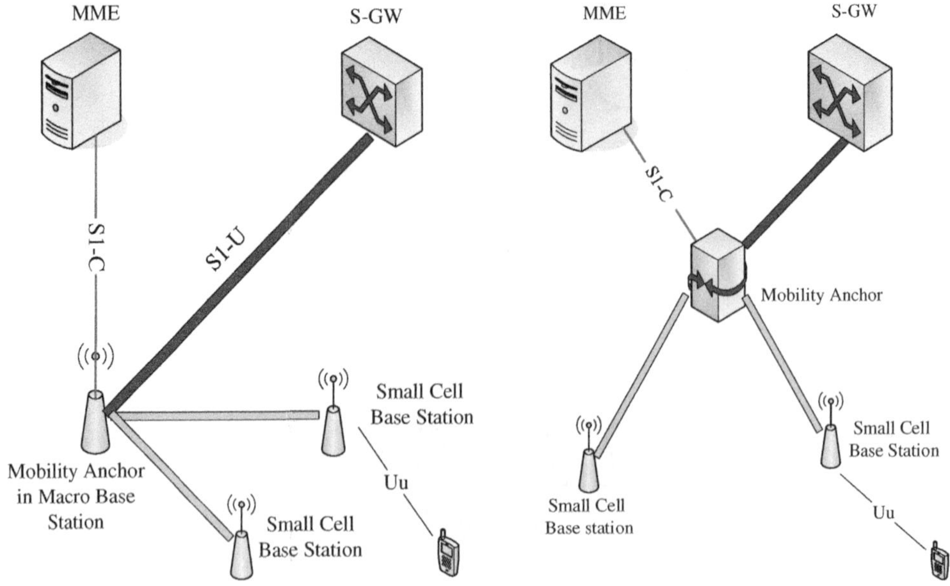

Figure 20.11 Architecture for mobility anchor

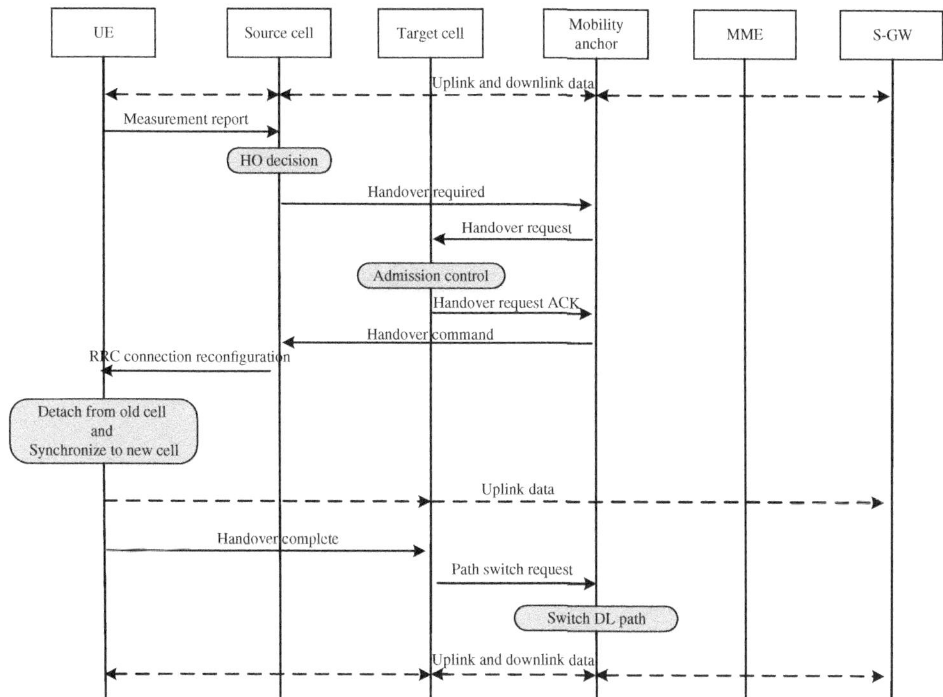

Figure 20.12 Handover procedure of mobility anchor solution

A mobility anchor is introduced as a centralized controller or a proxy for all small cells in one area; both S1-MME and S1-U terminate in the mobility anchor. Figure 20.12 gives an example of the handover process in the mobility anchor solution for the case of X2 interface absence between the source and target small cell [34]. It can be seen that messages relating to handover signaling can be disposed and generated by the mobility anchor, and the signaling overhead burden of mobility management entity (MME) is alleviated.

20.3.5 Handover Command Diversity

The major reason for handover failures is the failure of handover command transmission for users that are out-of-sync. Higher handover thresholds will further increase these failures. One possible solution is to enable the user to also receive the handover command from the target cell, so that a higher SINR can be achieved when compared to a late handover command transmission from the source cell [35]. This behavior is shown in Figure 20.13. Moreover, if the user is able to also receive the handover signal from the target cell, going out-of-sync in the source cell can be prevented.

20.4 Architecture and Backhaul

C-RAN [37, 38] is key for future mobile networks in order to meet the vast capacity demand of mobile traffic, and reduce the capital and operating expenditure burden faced by operators.

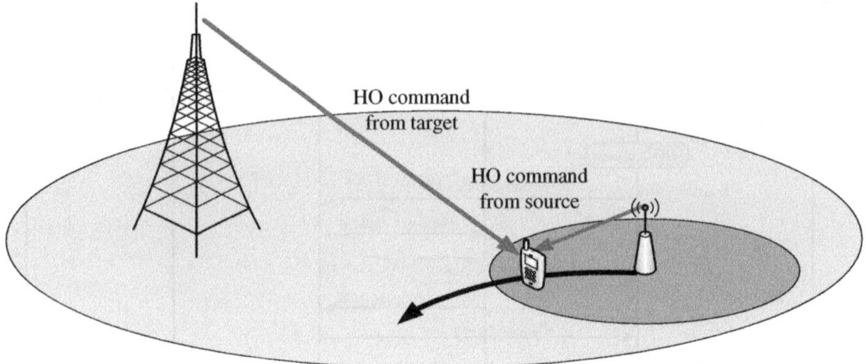

Figure 20.13 Handover command transmission diversity

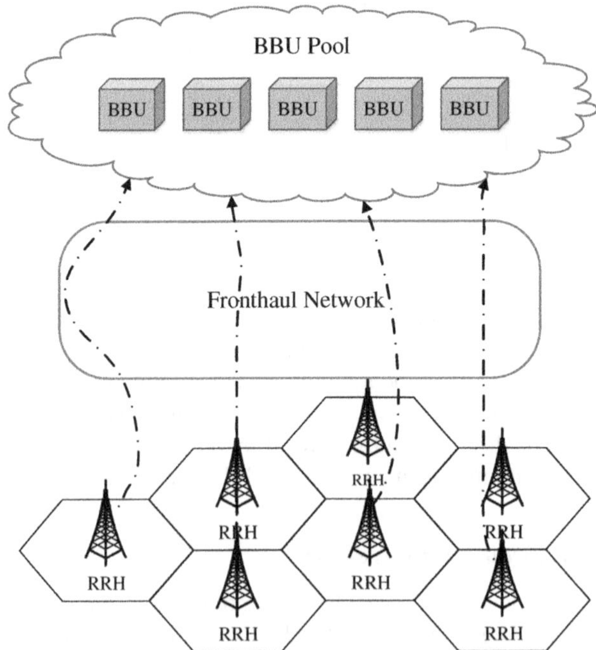

Figure 20.14 C-RAN architecture

As shown in Figure 20.14, C-RAN is a new centralized, cloud-computing-based cellular network architecture that has the ability to support current and future wireless communication standards. It does this by separating the BBUs from the radio access units, and migrating the BBUs to the cloud, forming a BBU pool for centralized processing. This gives better flexibility and scalability in terms of deployment of further RRHs compared to traditional radio access networks.

Table 20.3 The capability of different backhaul technologies

	Backhaul technology	Latency(one way)	Capacity
Ideal backhaul	Fiber	<2.5 us	>10 Gbps
Non-ideal backhaul	Fiber	10–30 ms	10M–10 Gbps
	Fiber	5–10 ms	100–1000 Mbps
	Fiber	2–5 ms	50M–10 Gbps
	xDSL	15–60 ms	10–100 Mbps
	Cable	25–35 ms	10–100 Mbps
	Wireless backhaul	5–35 ms	10M–100 Mbps, up to Gbps

As a significant scenario of a 5G system, UDN deployment based on C-RAN will play a very important role. More specifically, the design of UDN involves joint consideration of various issues, among them interference management/coordination, mobility management, control and user plan decoupling, backhauling, as well as multi-radio access technology (RAT) integration/interworking. In this case, C-RAN will come to play an important role in internal high-speed, low-latency coordination schemes and the central processing required to implement them.

The backhaul for UDN needs to have a significant role in aggregating and sending data traffic between the radio access link and the backbone network segment. Currently, the technologies of the backhaul network mainly include microwave, copper and fiber, and so on. Operators usually select a backhaul technology according to their needs in terms of system capacity, reliability, cost and deployment period.

Wired backhaul is a technology based on x-digital subscriber line (xDSL) or fiber. According to statistics, approximately 20% of all current backhaul deployments are copper-based xDSL and 30% are fiber-based [39]. Some types of fiber-based backhaul can offer nearly unlimited capacity over long distances as shown in Table 20.3; this can be called "ideal" backhaul. Capacity levels of up to 100 Mbps can be guaranteed only up to 500 ms with copper-based backhaul, which will most likely be superseded for this reason and also because of its inability to scale in a cost-efficient manner with a large number of base stations.

Wireless backhaul is a technology based on microwave, WiFi, and potentially other air-interface technologies. Nowadays, microwave represents nearly 50% of all current backhaul deployments. It is widely used in Europe, and it is expected that the same share will be maintained in the years to come. Wireless backhaul will be valuable for mobile networks, especially UDNs, due to its moderate installation cost and relatively short deployment times.

Moreover, although fiber-based backhaul is better on capacity and latency, when the backhaul challenges for UDNs mentioned in the Section 20.3 are considered, wireless backhaul has significant advantages over wired in terms of site acquisition. It is also beneficial to plug-and-play base stations. Therefore, a hybrid backhaul architecture, which includes wired and wireless options, should be reasonable for UDNs.

The wireless mesh network, which is a hybrid backhaul architecture, aims to construct a high-speed, highly efficient, self-optimizing and self-maintenance wireless transmission network between base stations so as to fulfill the demands for high data-transmission rates and traffic-volume density. The wireless mesh is also a crucial networking technology for UDN scenarios of the 5G network. Figure 20.15 shows 5G wireless mesh networks. From

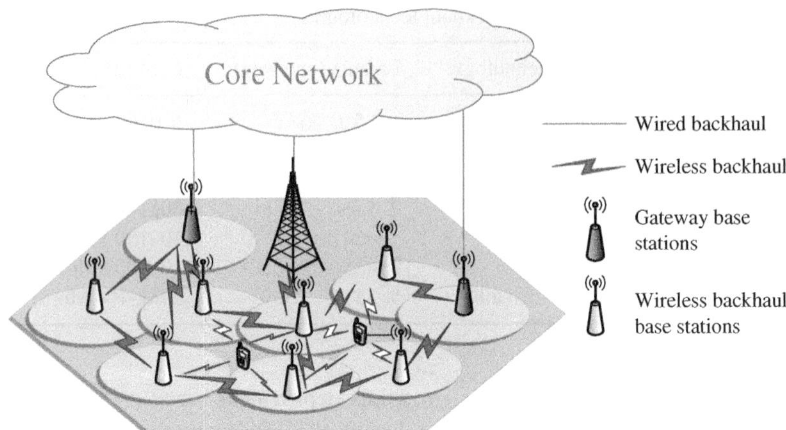

Figure 20.15 Wireless mesh network

the backhaul aspect, the basic backhaul network includes both wired backhaul and wireless backhaul. The base stations, acting as the gateways, are linked with the core network with wired backhaul; while the wireless backhaul base stations are linked with each other with wireless backhaul that makes up a wireless mesh network. The wireless backhaul base station can relay other base stations' backhaul data while transmitting its own data. The wireless backhaul base stations can quickly exchange the information of served users, transmission resources and so on via the wireless networks and make coordination decisions to ensure high performance and a consistent service for users.

For better deployment, a wireless mesh network can be constructed with a hierarchical backhaul architecture, as shown in Figure 20.16, where different base stations are marked by layers. The first backhaul layer includes macro cells and other small cells that have wired backhaul. The small cells belonging to the second backhaul layer are connected to the base stations in the first backhaul layer with one-hop wireless transmission. Likewise, the small cells belonging to the third and subsequent layers are connected to the upper layer with one-hop wireless transmission. This architecture combines wired and wireless backhaul together with a fixed or adaptive structure, which can provide an easy and plug-and-play networking mode.

From a backhaul transmission-structure perspective, wireless backhaul includes point-to-point (PTP) and point-to-multipoint (PTMP) approaches. The literature includes a wireless PTP scheme with an adaptive backhaul architecture [40], which can adapt to the load balance among all the small cells. In this architecture, there are three topologies considered:

- a tree topology with a single gateway node that connects the backhaul network to the backbone fiber network;
- multiple parallel gateway nodes;
- a mesh topology arrangement with redundant links.

In this type of architecture, different frequencies can be assigned to different parts of the backhaul network in order to satisfy changing traffic conditions and suppress interference. This adaptive PTP frequency division duplex (FDD) architecture for mobile small-cell backhaul not

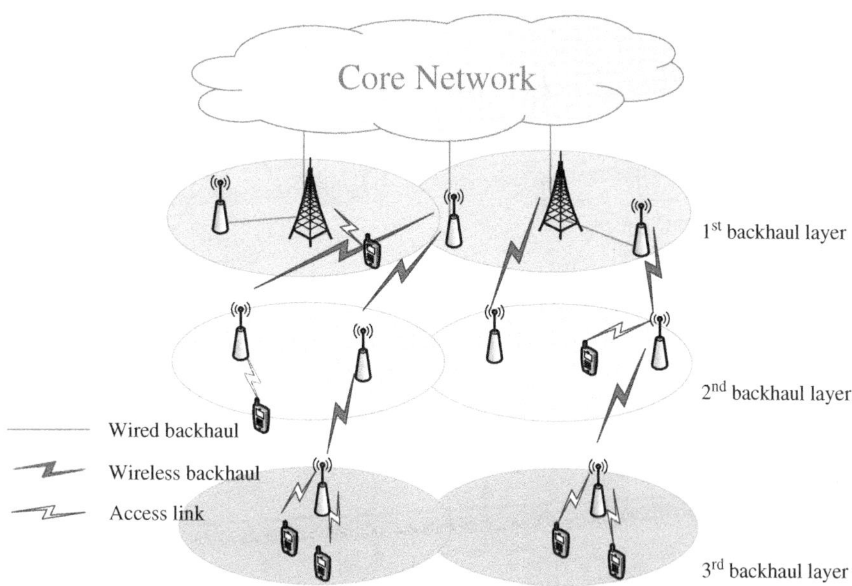

Figure 20.16 The hierarchical backhaul architecture

only allows changes to the overall backhaul topology, but also allows each individual backhaul link to vary its frequency to satisfy the traffic requirements and mitigate interference.

As for deployment expenditure, it would be valuable if the existing sites could be reused to deploy 5G base stations. Based on this design principle, wireless backhaul is used to connect 5G base stations at new and old sites, as illustrated in Figure 20.17(a). Assuming fiber backhaul to the original 4G sites is available and no additional fiber backhaul is deployed to the new sites, depending on how the cells are stacked, the resulting fiber density can be 1/3 – in other words one of every three 5G base stations is fibered – as shown in Figure 20.17(b). The literature includes a PTMP in-band microwave backhaul architecture in which two new 5G base stations connect to the fibered base station via wireless backhaul [41].

In order to realize highly efficient hybrid backhaul networks, the following points should be considered and need further research:

- The overall design and optimization of wireless backhaul links and wireless access links.
- Planning and management of the backhaul gateway.
- Management and optimization of the wireless backhaul topology.
- Resource management, protocols and interfaces of hybrid backhaul networks.

20.5 Other Issues in UDNs for 5G

Time and frequency synchronization in large networks of small cells like the UDNs in 5G enables efficient interference management, through techniques such as ICI coordination and CoMP, which are expected to boost the throughput performance in such dense scenarios. Existing mechanisms, including global positioning system (GPS) and IEEE precision time protocol

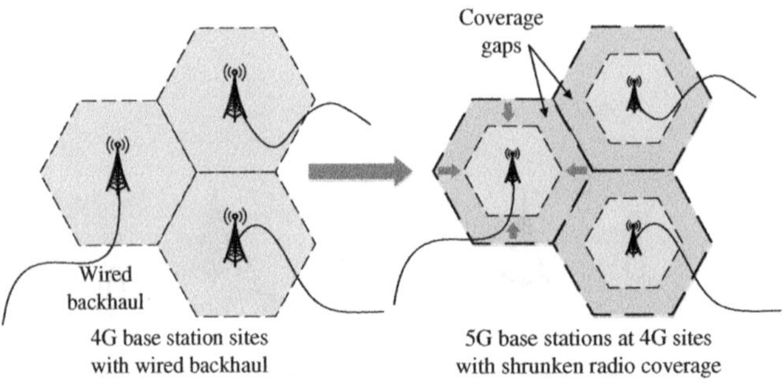

(a) 5G base stations deployed at backhauled 4G sites

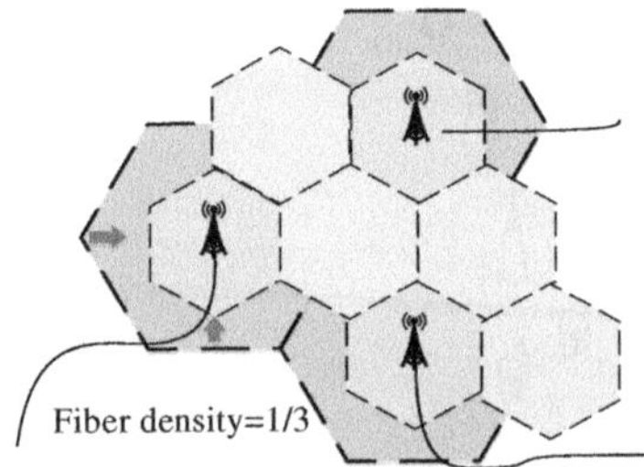

(b) 5G base stations deployed at new sites via wireless backhaul with fiber density of 1/3

Figure 20.17 PTMP backhaul structure

(PTP) are not sufficient considering the severe penetration losses and dependence on the traffic load in the backhaul network.

To solve this problem, the self-organizing solutions, in which multiple transmission points autonomously agree on a common timeline without centralized coordination, have been proposed [42]. These are based on autonomous timing corrections performed by each transmission point upon reception of beacon messages from its neighbors. The procedure would contain:

- initial synchronization, in which the initial synchronization and time alignment among multiple points is achieved
- runtime synchronization, in which synchronization is maintained despite non-idealities of the hardware clocks.

Energy efficiency is also a crucial issue for UDNs because deployment and expanding of dense small cells can drastically increase maintenance and operation costs. In order to achieve sustainability, 5G network needs to make improvements in energy efficiency; by a factor of the order of 100 [3].

Various energy efficiency schemes have been defined, for example discontinuous transmission. For UDNs, these schemes can be further extended to small cell on/off, in which the small cell can be dynamically turned off when there is no data to transmit. With small cell on/off, interference is less so that data transfer can be faster, resource utilization is reduced and energy consumption is reduced as well. Significant energy savings can be obtained by small

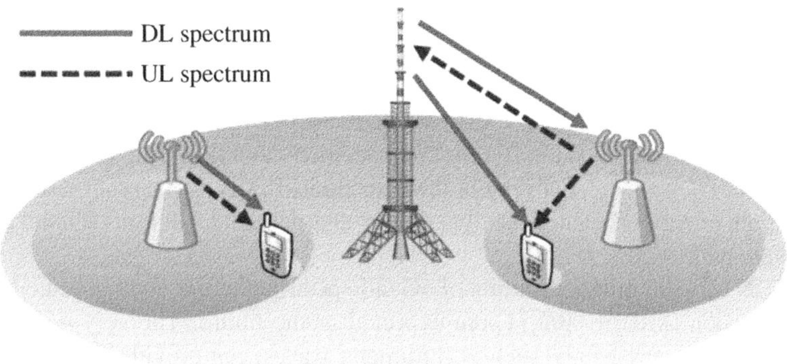

Figure 20.18 Use cases for flexible duplex

cell on/off, and the energy saving could be up to 35% or 48% depending on the parameters used [37].

In FDD systems, symmetric paired spectrum for downlink (DL) and uplink (UL) is required. This approach can make full use of the spectrum to accommodate services with symmetric UL and DL traffic, such as voice. However, with the boosting of mobile broadband services in recent years, more and more popular services asymmetric UL and DL traffic [43]. As such services increase, and the ratio between UL and DL traffic changes over the time, the static paired spectrum for UL and DL is not efficient for supporting dynamic asymmetric traffic, especially in UDNs. Flexible duplex can better adapt to dynamic asymmetric traffic. With flexible duplex, the UL spectrum defined in FDD systems can be re-allocated for DL transmission with high flexibility [44, 45]. Considering the potential crosslink interference – DL to UL or UL to DL – the transmission power for DL transmission on the UL spectrum should be constrained to a relatively low level. Flexible duplex can be applicable to small cells with low transmission power and a relay base station, as shown in Figure 20.18.

To solve the site-acquisition problem, many intelligent incentive schemes have been proposed and investigated. A practical realization of such intelligent incentive schemes is "Fon'' [46]. Fon motivates users to open up their resources for wider usage. This approach is essentially an instance of a cooperative game. A broader range of game-theoretic approaches, such as a coalition game and the Stackelberg game [47], is also being considered for intelligent incentive schemes. A utility-aware refunding framework has been proposed to motivate hybrid access in femtocells [47]. Under this framework, a unique Nash equilibrium is achieved and a hybrid access protocol is designed according to the analysis. Another promising solution is soft sector technology. For soft sector technology, multiple soft sectors are formed by means of multiple beams generated by the centralized entity, which can significantly reduce the cost of site rental, equipment and transmission. Soft sector technology can provide a unified management platform between soft sectors and real sectors, and reduce operation complexity.

20.6 Conclusions

In this chapter, we have provided an overview of UDNs from several viewpoints. In Subsection 20.1.1, we introduced the application scenarios identified for UDN, as well as the typical services for each scenario. Subsection 20.1.2 described the potential challenges for deployment and operation of UDNs, including site acquisition and expenditure, network operation and management, interference management, mobility management, and backhaul resources. Solutions to the potential problems were discussed in Sections 20.2–20.4 from several different perspectives. More specifically, interference management, mobility management, architecture and backhaul for UDNs were analyzed in these sections. Lastly, some other issues, including synchronization, energy efficiency, flexible duplex and potential solutions for site acquisition for UDNs were addressed.

To take UDN forward in the evolution of 5G, all-spectrum access and tight integration with other systems, such as IEEE 802.11, are expected in the future. The very wide bandwidth at much higher frequency bands, such as millimeter waves, can provide abundant resources. Also, operations at higher frequency bands can also provide opportunities to deploy novel air-interface designs, which can better solve the problems of UDNs. Licensed spectrum will remain a dominant spectrum-assignment approach in 5G, but there is clear value in the introduction of spectrum-sharing functionality in UDNs, since power levels are lower and deployment is more localized.

Acknowledgements

The authors would like to appreciate Yang Liu, Da Wang, Bei Yang and Shan Yang for their contributions to this chapter.

References

[1] Cisco (2015) White paper: Cisco visual networking index: Global mobile data traffic forecast update, 2014–2019.

[2] Qualcomm Inc. (2013) *The 1000× data challenge*. URL http://www.qualcomm.com/1000x/.

[3] IMT-2020(5G) Promotion Group (2014) 5G vision and requirements.

[4] IMT-2020(5G) Promotion Group (2015) 5G wireless technology architecture.

[5] I. Hwang, B. Song, and S. Soliman, (2013) A holistic view on hyper-dense heterogeneous and small cell networks. *IEEE Commun. Mag.*, **51** (6), 20–27.

[6] METIS (2013) The METIS view of 5G networks.

[7] Bhushan N., Li J., Malladi, D., Gilmore, R., Brenner D., Damnjanovic, A., Sukhavasi, R., Patel, C., and Geirhofer, S. (2014) Network densification: the dominant theme for wireless evolution into 5G. *IEEE Commun. Mag.*, **52** (2), 82–89.

[8] Boccardi, F., Heath, R., Lozano, A., Marzetta, T., and Popovski, P. (2014) Five disruptive technology directions for 5G. *IEEE Commun. Mag.*, **52**, 74–80.

[9] Yuan Y. and Zhu L. (2014) Application scenarios and enabling technologies of 5G. *China Commun.* **11**, 69–79.

[10] Andrews, J.G. et al., (2012) Femtocells: past, present, and future. *IEEE J. Select. Areas Commun.* **30** (3), 497–508.

[11] 3GPPTech. Spec. Group RAN (2013) TR 36.839: Mobility enhancements in heterogeneous networks (Release 11).

[12] Barbera S., Michaelsen P., Säily, M., and Pedersen, K. (2012) Improved mobility performance in LTE co-channel HetNets through speed differentiated enhancements, in *IEEE Proc. Globecom, Wksp. Heterogeneous, Multi-hop, Wireless, and Mobile Networks, December* **2012**, pp. 426–430.

[13] Prasad, A. Tirkkonen, O., Lundén P., Yilmaz, O., Dalsgaard, L., and Wijting, C. (2013) Energy-efficient inter-frequency small cell discovery techniques for LTE-Advanced heterogeneous network deployments. *IEEE Commun. Mag.,* **51** (5), 72–81.

[14] Boudreau, G., Panicker, J., Guo, N., Chang, R., Wang, N., and Vrzic, S. (2009) Interference coordination and cancellation for 4G networks. *IEEE Commun. Mag.*, **47** (4), 74–81.

[15] Lee J., Kim, Y., Lee, H., Ng, B., Mazzarese, D., Liu, J., Xiao, W., and Zhou Y. (2012) Coordinated multipoint transmission and reception in LTE-Advanced systems. *IEEE Commun. Mag.*, **50** (11), 44–50.
[16] 3GPP (2010) TR 36.814: Evolved universal terrestrial radio access (E-UTRA); Further advancements for E-UTRA physical layer aspects (Release 9).
[17] Karakayali, M., Foschini, G., and Valenzuela, R. (2006) Network coordination for spectrally efficient communications in cellular systems. *IEEE Wireless Commun.*, **13** (4), 56–61.
[18] Andrews, J., Choi, W., and Heath, R. (2007) Overcoming interference in spatial multiplexing MIMO cellular networks. *IEEE Wireless Commun.*, **14** (6), 95–104.
[19] 3GPP (2013) TR 36.819: Coordinated multi-point operation for LTE physical layer aspects (Release 11).
[20] Cadambe, V.R. and Jafar, S.A. (2008) Interference alignment and degrees of freedom of the K-User interference channel. *IEEE Trans. Info. Theory*, **54**, (8), 3425–3441.
[21] Zhai, G.W., Tian, L., Zhou Y., and Shi, J.L. (2014) Load diversity based optimal processing resource allocation for super base stations in centralized radio access networks. *Science China Info. Sciences*, **57** (4), 1–12.
[22] Pedersen. K.I., Wang, Y., Strzyz, S., and Frederiksen, F. (2013) Enhanced inter-cell interference coordination in co-channel multilayer LTE-Advanced networks. *IEEE Wireless Commun. Mag.*, **20**, (3), 120–27.
[23] 3GPP (2013) TR 36.872: Small cell enhancements for E-UTRA and E-UTRAN – Physical layer aspects (Release 12).
[24] Garcia, L.G., Kovacs, I.Z., Pedersen, K.I., Costa, G.W.O., and Mogensen, P.E. (2012) Autonomous component carrier selection for 4G femtocells – a fresh look at an old problem. *IEEE J. Select. Areas Commun.*, **30**, (3), 525–537.
[25] Soret, B., Pedersen, K.I., Kolding, T., and Kroener, H. (2013) Fast muting adaptation for LTE-A Het-Nets with remote radio heads, in *IEEE Proc. Globecom*, December 2013, pp. 3790–3795.
[26] Agrawal, R., Bedekar, A., Kalyanasundaram, S., Arulselvan, N., Kolding, T., and Kroener, H. (2014) Centralized and decentralized coordinated scheduling with muting, in *IEEE Proc. Vehic. Tech. Conf.*, May 2014, pp. 1–5.
[27] Lopez-Perez, D., Valcarce, A., de la Roche, G., and Zhang, J. (2009) OFDMA femtocells: a roadmap on interference avoidance. *IEEE Commun. Mag.*, **47**, (9), 41–48.
[28] Soret, B., Pedersen, K.I., Jørgensen, N.T.K., and Fernández-López, V. (2015) Interference coordination for dense wireless networks. *IEEE Commun. Mag.*, **53**, 102–109.
[29] 3GPP (2012) TR 36.829 v11.1.0: Enhanced performance requirement for LTE user equipment (UE) (Release 11).
[30] 3GPP (2014) TR 36.866 v12.0.1: Study on network-assisted interference cancellation and suppression (NAIC) for LTE (Release 12).
[31] Jha, S.C., Sivanesan, K., Vannithamby, R., and Koc, A.T. (2014) Dual connectivity in LTE small cell networks, in *Globecom Workshops*, pp. 1205–1210.
[32] 3GPP (2013) TR 36.842 v12.0.0: Study on small cell enhancements for E-UTRA and E-UTRAN; higher layer aspects (Release 12).
[33] Huawei, China Unicom, (2015) R3-150577: History discussion on mobility anchor, in *TSG-RAN3 Meeting #87bis*, Tenerife, Santa Cruz, Spain, 20–24 April 2015.
[34] China Telecom, Huawei, (2015) R3-150016: Consideration on mobility anchor solution, in *WG3 Meeting #87, Athens, Greece, 9–13* February 2015.
[35] Ericsson (2013) R2-130469: Heterogeneous networks mobility enhancements with handover signaling diversity, TSG-RAN WG2 #81, St Julian's, Malta, 28 January–1 February 2013.
[36] Zhang Y. and Zhang Y. (2014) User-centric virtual cell design for cloud radio access networks, in *2014 SPAWC workshop*, 22–25 June 2014, pp. 249–253.
[37] Li J. and Wu, H. (2014) Energy efficient small cell operation under ultra dense cloud radio access networks, in *Globecom Workshops*, pp. 1120–1125.
[38] Wang, R. and Hu, H. (2014) Potentials and challenges of C-RAN supporting multi-RATs toward 5G mobile networks, *IEEE Access*, **2**, 1187–1195.
[39] Mahloo, M., Monti, P., Chen, J., and Wosinska, L. (2014) Cost modeling of backhaul for mobile networks, in *Workshop on Fiber-Wireless Integrated Technologies, Systems and Networks*, pp. 397–402.
[40] Ni, W., Collings, I.B., Wang, X., and Liu, R.P. (2014) Multi-hop point-to-point FDD wireless backhaul for mobile small cells. *IEEE Wireless Commun.*, **21**, 88–96.
[41] Taori R. and Sridharan, A. (2015) Point-to-multipoint in-band mmWave backhaul for 5G networks. *IEEE Commun. Mag.*, **53**, 196–201.

[42] Berardinelli, G. Tavares, F., Tirkkonen, T. et al. (2014) Distributed initial synchronization for 5G small cells, in *IEEE 79th Vehicular Technology Conference (VTC Spring)*, pp. 1–5.
[43] Huawei, HiSilicon, and China Telecom (2014) RP-140062: Motivation of new SI proposal: evolving LTE with flexible duplex for traffic adaptation.
[44] Wan, L., Zhou, M., and Wen, R. (2013) Evolving LTE with flexible duplex, in *IEEE Globecom Workshops*, pp. 49–54.
[45] LG Electronics, China Telecom, Huawei, and HiSilicon (2015) RP-150470: New SID: Study on regulatory aspects for flexible duplex for E-UTRAN.
[46] Fon (2016) Webpage. http://corp.fon.com/en/this-is-fon/easy-tojoin/
[47] Chen, Y., Zhang, J. and Zhang, Q. (2012) Utility-aware refunding framework for hybrid access femtocell network, *IEEE Trans. Wireless Commun.*, **11**, (5), 1688–1697.

21

Radio-resource Management and Optimization in 5G Networks

Antonis Gotsis, Athanasios Panagopoulos, Stelios Stefanatos and Angeliki Alexiou

Signal Processing for 5G: Algorithms and Implementations, First Edition. Edited by Fa-Long Luo and Charlie Zhang.

21.1 Introduction

The introduction of previous cellular generations (2G to 4G) was marked by advances in link spectral efficiency (coding, MIMO) and user multiplexing (CDMA, OFDMA, SDMA). The 5G wireless network is expected to efficiently exploit new technologies, infrastructures and user devices, providing a transparent massive connectivity framework [1]. The main enablers of this approach will be the availability of hyperdense deployments of access/serving points and the exploitation of communications proximity. In this sense, ultradense networks (UDNs) are considered a cornerstone of 5G network evolution [2–6], building on:

- extreme resource (bandwidth) reuse capabilities, up to the point that each user is exclusively served by their own or even multiple access nodes
- improved signal conditions, by bringing the infrastructure closer to the user side.

Early studies on how to reach the thousand-fold capacity increase targets of 5G [7] have indicated that the vast majority of the capacity gains (up to 100 times) will come from network densification [8, 9]. Other prospective key 5G enablers, including the incorporation of new spectrum bands in the mmWave range and higher-order MIMO processing (3D/full-dimension and massive MIMO), may be built upon the UDN infrastructure and ultimately provide the anticipated 5G capacity levels [10, 11].

UDNs have their roots in the heterogeneous-networks ("HetNet") concept: adding smaller footprint cells within macrocell deployments as a means of offloading traffic in hotspot localized areas. When pushing densification to its limits, as in UDNs, traditional wireless network design principles should be rethought. First, the inherent highly irregular deployment of UDNs coupled with universal frequency reuse patterns, renders cellular planning practices irrelevant, giving rise to unpredictable and dynamic interference environments, and even questioning the cell notion. Secondly, due to the similar densities of serving and served nodes, the traffic load per serving/access node varies significantly within a network cluster, leading to the presence of a significant number of unloaded or lightly loaded access nodes, which drastically contribute to network interference levels while minimally contributing to overall network capacity. This is in contrast to conventional cellular networks, where the number of users is typically several orders of magnitude greater than the number of access nodes, leading to more or less balanced traffic loads. Thirdly, the increase in the number of active users in a given area and the number of access nodes in the vicinity of each user for a UDN dramatically impacts the signaling overhead (for example, for reporting channel-state information) and the backhauling requirements in terms of carried load, synchronization and operation dynamics; static preplanned backhaul in a dynamic UDN environment with optimized access node activation/de-activation becomes irrelevant.

To address the particular challenges and realize a future UDN-based system, major advances in several areas are of the utmost importance:

- the underlying technology
  - –the air-interface
  - –frame design
  - –flexible duplexing
  - –associated overhead signaling structures

- radio resources management
  –access
  –backhaul design.

The scope of this chapter is to provide a thorough analysis and discussion of the radio-resource management (RRM) aspects of homogeneous single-technology UDNs, with emphasis on centralized optimization problem modeling and solving. We will introduce a series of mathematical programming models and algorithms for making optimal RRM decisions, covering the majority of the UDN optimization dimensions. By applying the models and algorithms to potential UDN system setups, we will demonstrate the importance of optimal RRM as a key enabler for getting the most of the resource-reuse and proximity benefits offered by UDNs.

The rest of this chapter is organized as follows. In Section 21.2 we provide a brief overview of state-of-the-art RRM and the set of optimization tools used through the rest of the chapter. In Sections 21.3 and 21.4, the various RRM optimization models and solution algorithms are described, along with performance evaluation results, for single-antenna coordinated and multi-antenna coordinated/cooperative UDNs. Finally Section 21.5 concludes the chapter and briefly states potential research directions.

21.2 Background

In the current section we briefly provide the necessary background on how to manage and optimize radio-resource utilization in 5G networks. In particular, in Section 21.2.1 we introduce the challenges and new perspectives of UDN deployments in terms of RRM. Then, in Section 21.2.2 we provide a short list of the optimization tools and methods used in the rest of the chapter for modeling and solving RRM problems in the envisioned 5G-UDN system setups.

21.2.1 Radio-resource Management Challenges

The issue of quantifying the impact of network densification on 4G-LTE network capacity performance levels has attracted significant interest lately. The literature examines – from a system-level simulation point of view – the deployment densification levels required to scale current load levels so as to meet the projected network capacity demands for 2020 and beyond [8, 12–17]. However, the majority of these studies have considered traditional local – per cell – RRM techniques, ignoring the potential for network coordination through sophisticated RRM.

In this respect, the recent works by Hu *et al.* [18] and Hossain *et al.* [19] introduced two relevant RRM frameworks, focused on energy and spectral efficiency performance optimization of HetNets, with emphasis on mobile association, interference management and cooperative dynamic resource allocation. Although the above frameworks provide useful insights, their scope within UDNs is rather limited since they focused on traditional cellular setups.

To fill the gap in the current state of the art, this chapter will build on our previous work in the area [20–23] and deliver a comprehensive RRM optimization framework that is applicable to UDN setups, providing the basis for identifying the most critical optimization dimensions and quantifying their impact. In particular, we move beyond local, uncoordinated operation (which provides the baseline UDN performance levels), and consider empowering a future

5G-UDN deployment with intelligent network coordination capabilities facilitated by RRM mechanisms. RRM aims at maximizing system performance while guaranteeing any individual quality-of-service (QoS) requirement, by optimizing the mapping of UEs to the various system dimensions (space, frequency and time) taking into account the channel conditions. In our UDN context, RRM involves the following optimization dimensions:

User-association or Pairing

This is the association of each served user equipment (UE) node with one or more serving access nodes (ANs). In contrast to cellular networks, where typically each UE is set to be served by its single closest AN, in future UDNs, in addition to proximity, load balancing/shifting, energy and interference-control considerations will be accounted for in UE-to-AN association, given that that a multitude of ANs lie at the vicinity of each UE.

Scheduling or Partitioning

This involves the allocation of various AN–UE communicating pairs to different orthogonal resource blocks or groups of blocks (carriers); usually this is performed either in the time domain or in the frequency domain. Since the rate performance depends both on spectral and temporal resources usage (which is reduced through partitioning) and the achieved signal to interference and noise ratio, or SINR, (which in contrast increases through partitioning), the selection of AN–UE pairs to be orthogonalized should strike a balance between these factors.

Power

This is the control of the assigned power to each AN–UE pair; through network power coordination – tuning the power to each pair according to the proximity conditions and QoS requirements – interference can be controlled.

Precoding

This is the network-wise tuning of beamforming vectors and power-scaling factors, if multiple antennas are available at the ANs/UEs and/or ANs are able to cooperatively serve one or more UEs. In case of downlink, non-cooperative operation for example, the space dimension could be explored so as to increase the received signal strength to the intended UEs and/or reduce interference to neighbor UEs. For cooperative operation, multi-AN precoding could even totally eliminate interference, using zero-forcing for example.

21.2.2 Optimization Tools applied to Centralized Radio-resource Management

In the context of the proposed RRM framework, four mathematical programming tools will be leveraged.

Linear Programming

Linear programming (LP) is utilized when the objective and constraint functions involved are linear with respect to the design variables, where the latter are constrained to lay in the continuous domain. For LPs, solutions can be acquired under polynomial complexity and are associated with global optimality guarantees. However, their applicability is rare due to linearity and continuity requirements. An application example is the power-coordination problem for maximizing the achievable common rate across all UEs in a network [20].

Second-order Cone Programming

Second-order cone programming (SOCP) is utilized when the formulation involves conic (and possibly linear) expressions with respect to continuous variables. SOCP belongs to the class of convex programs, which, like LPs, can be exactly solved, providing global optimality guarantees with polynomial complexity. An example of an application is the precoding coordination problem for maximizing the achievable common rate across all UEs in a network [22].

Mixed-integer LP

The discrete version of LP, namely mixed-integer LP (MILP), is introduced when the design-variable set not only involves continuous variables (such as power allocation) but also integer ones (such as user-association or resource scheduling). Although the solution complexity for MILPs grows exponentially with the problem dimension, they constitute an attractive formulation since there exist powerful algorithms and software solver implementations [24, 25] for reaching optimal decisions with reasonable complexity. When MILPs involve continuous and binary variables the term BILP is used, and when they contain only integer variables (and no continuous ones) the term ILP is used. An example application is the user-association problem for maximizing the achievable common rate across all UEs in a network [20].

MI-SOCP

Along the same lines, the discrete mixed-integer version of SOCP (MI-SOCP) involves binary and continuous design variables interrelated through linear and conic expressions. An example application for MI-SOCPs is the user-association and precoding problem for cooperative UDNs with limited backhaul [21].

The fundamental formulations of the vast majority of RRM problems encountered in cellular and UDN setups belong to the class of mixed-integer non-linear and non-convex programs (MINLPs), which are hard to solve (and provide global optimal guarantees) even for small-scale setups. A common strategy for dealing with such problems is to look for opportunities for linearization reformulation [26]. In mixed binary-continuous variable problems, it is often the case that non-linearity stems from products involving multiple binary variables and/or products of a binary and a continuous variable [20, 23]. The particular non-linear expressions can be exactly reformulated by replacing each product with a single auxiliary variable, along with a set of linear-constraint expressions. The transformed formulation is then linear with respect to initial and auxiliary variables and may be solved using MILP. The drawback of this approach is the inevitable increase in the number of problem variables and constraint expressions, which impacts the computational complexity. Details and proofs for the validity of various linearization-reformulation methods are found in the literature [26–29]. Here we reproduce two methods utilized through the rest of the chapter.

Linearization Method 1 [28] A product of two binary variables x and y can be replaced by a new auxiliary binary variable $z = xy$ and a set of three linear constraint expressions: $z \leq x$, $z \leq y$, and $z \geq x + y - 1$ or equivalently by the more compact form: $2z \leq x + y \leq z + 1$.

Linearization Method 2 [29] A product of a binary variable x and a continuous positive variable y can be replaced by a new continuous auxiliary variable $z = xy$, along with a set of four linear constraint expressions: $y - z \leq K_y(1 - x)$, $z \leq y$, $z \leq K_y x$, and $z \geq 0$, where K_y is large number guaranteed to be greater than the maximum value that y could take.

21.3 Optimal Strategies for Single-antenna Coordinated Ultradense Networks

We now turn our attention on the RRM optimization opportunities encountered within random topology coordinated single-antenna access node clusters. First, in Section 21.3.1 we provide a detailed description of the system model considered. In Section 21.3.2 we study the problem of user-association and scheduling, while in Section 21.3.3 we extend the previous problem by considering power optimization. In Section 21.3.4 we examine the applicability of RRM optimization to a representative UDN system setup using system-level simulations, and demonstrate the associated performance benefits.

21.3.1 System Model and Scope

We consider a random-topology deployment comprising clusters of homogeneous (small-cell) ANs serving a set of UEs overlaid in a macrocell area, and operating in a distinct carrier frequency (e.g. 3.5 GHz as suggested by 3GPP [30]). The downlink (DL) of a single cluster is studied, although the model could be extended to multicluster and multi-macro-cell scenarios. Each network cluster serves a "hotspot" area (typically 50–70 m radius) and in contrast to typical macrocell setups, the number of serving and served nodes, M and K respectively, are of the same order. A maximum power budget per AN, $p_{\max}$, is considered, and the available bandwidth is split into N orthogonal groups of frequency resources, such as LTE carriers. Both ANs and UEs are equipped with single antennas and coordination among ANs is only performed in the signaling plane, hence neither single-AN (single-user/multi-user) nor multi-AN cooperative MIMO processing techniques are employed. Therefore, each UE k is associated with and receives data from a single serving AN m at a single carrier n.

An ideal (lossless, zero-delay and infinite capacity) backhaul network for relaying UE data to the respective ANs is also assumed. The wireless access network therefore constitutes the performance bottleneck. DL transmissions are impacted only by large-scale path loss, since the current study focuses on the optimization over longer timescales (say, several hundreds or thousands of frames) and UDNs are expected to be tailored to bandwidth-hungry but static/portable/low-mobility user conditions. In this respect, perfect knowledge of global path-loss network channel-state information is considered, and leveraged as the main tool for optimizing network performance. For an arbitrary mth-AN/kth-UE pair, we denote by g_{km} the corresponding power path gain. From a QoS point of view, this study focuses on how to provide high data rates uniformly across the network. Considering a set of UEs requesting service from the network, the respective network optimization design target is how to support the maximum achieved common rate (which is a priori not known) for all UEs. A slight variation of the particular QoS model targets the support of a set of known minimum rates per UE.

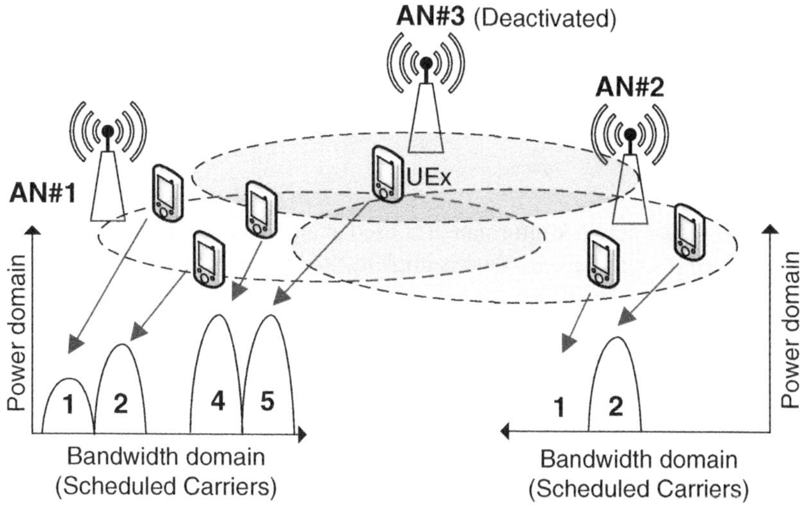

Figure 21.1 A cluster of three randomly deployed access nodes with overlapping areas serving six users. Notice the coordinated scheduling and power allocation of carriers between AN1 and AN2 for managing interference. Also notice that although UEx's closest serving node is AN3, it is actually associated with AN1, allowing AN3 to be deactivated, reducing interference to the other UEs. Sections 21.3.2 and 21.3.3 provide models and solutions for optimizing these decisions

In summary, the scope of this section is to propose a set of RRM strategies that select on a clusterwise basis the optimal:

- serving access node per user node (user-association)
- bandwidth partition or carrier assigned to each user node (scheduling)
- power level assigned to each AN–UE paired transmission (power allocation).

Figure 21.1 provides an example. By applying the above strategies one is able to explore the maximum performance limits of future UDN deployments as well as revealing, both in a qualitative and quantitative way, the most crucial factors affecting UDN optimization.

21.3.2 User Association and Scheduling

Initially we concentrate on the first two problem dimensions: user-association and scheduling. We assume that the power assigned to each pair is fixed and does not depend on channel conditions.

21.3.2.1 Baseline Uncoordinated Scheme

We define a baseline scheme operating in two steps. The first is a user-association criterion, which determines that each UE is associated with the AN providing the maximum path-gain conditions, without considering traffic load and/or network-interference conditions. This policy is followed in today's typical macrocell networks, which are more or less equally loaded. The second is a random scheduling approach, in which each UE is arbitrarily assigned to an exclusive orthogonal partition without considering network interference. This policy is similar to an LTE reuse approach, where no intercell interference coordination policy (ICIC) is applied.

This scheme is clearly:

- local, since no global network information is used at any point of decision making
- distributed, since resources assignment is implemented independently in each AN.

Beyond its practical applicability, implementation simplicity and minimal control-signaling requirements, such an uncoordinated approach will also let us evaluate the importance of the sophisticated centralized network-coordination approaches proposed below.

21.3.2.2 Joint Optimal User-association and Scheduling Scheme

Problem Formulation

We first consider the case of a common target rate level for all UEs, r_0, or equivalently an SINR level $\theta_0 = 2^{r_0} - 1$, and assume that each UE should be assigned to an exclusive carrier. We introduce the set of binary indicator variables $\{\rho_{kmn}\}, 1 \leq k \leq K, 1 \leq m \leq M, 1 \leq n \leq N$, for which each arbitrary element ρ_{kmn} is one if user k is associated with access node m and assigned a carrier n. We seek the optimal assignment decisions $\{\rho^*_{kmn}\}$. For an arbitrary UE/AN/carrier triplet, the achieved SINR γ_{kmn}, given a power level p (normalized over noise power), is stated in Eq. (21.1a), while Eq. (21.1b) guarantees that each UE enjoys at least the common required rate level r_0:

$$\gamma_{kmn} = \frac{p \cdot g_{km} \cdot \rho_{kmn}}{1 + p \cdot \sum\limits_{i \neq k} \sum\limits_{j \neq m} g_{kj} \rho_{ijn}}, \tag{21.1a}$$

$$\forall k : \prod_n (1 + \gamma_{kmn}) \geq 2^{r_0} \tag{21.1b}$$

To complete the problem formulation we introduce the constraint expression sets Eqs (21.2a)–(21.2b), where the former guarantees that each UE is served by a single AN and assigned to a single carrier, whereas the latter that each carrier is assigned exclusively to one user, so as to avoid inter-UE interference within each AN (as it is the case for typical OFDMA-based networks):

$$\forall k : \sum_m \sum_n \rho_{kmn} = 1, \tag{21.2a}$$

$$\forall m, n : 0 \leq \sum_k \rho_{kmn} \leq 1 \tag{21.2b}$$

Eq. (21.1a) and Eq. (21.1b) reveal that the problem of interest belongs to the challenging class of nonlinear integer programs (a subclass of MINLPs) which are extremely hard to solve for global optimality even for small dimensions [31]. To ensure global optimality one may resort to an exhaustive search over all possible combinations. However, the complexity of complete enumeration for the particular problem scales as $(MN)!$, which is prohibitive for network dimensions encountered in typical 3GPP-based small-cell cluster setups, involving one or even multiple ten(s) of access nodes, user nodes and carriers.

Exact Problem Linearization

To overcome the complexity barrier of an exhaustive search but still respect global optimality guarantees, we propose an approach for exactly reformulating the original problem to a linear form, by eliminating the nonlinearities stemming from binary variable products appearing in Eq. (21.1a) and Eq. (21.1b). Although the resulting class of ILPs is still NP-hard to solve, as reported in Section 21.2.2, such problems are tractable for large dimensions involving thousands of 0–1 variables and linear constraint expressions. In particular, it has been demonstrated that MILP is applicable to practical system setups encountered in practical ultradense small-cell cluster setups [32]. Note that not all nonlinear programs can be exactly linearized, and in this case, a series of approximations (e.g. a piecewise linear approximation) may be applied. However, in this case the solution of the transformed program is by no means guaranteed to match the solution of the original program.

We first rewrite the minimum rate/SINR constraint expressions in Eq. (21.1a) and Eq. (21.1b) using an alternative, more compact form as follows:

$$\forall k, m, n : \gamma_{kmn} = \frac{p \cdot g_{km}}{1 + p \cdot \sum\limits_{i \neq k} \sum\limits_{j \neq m} g_{kj} \cdot \rho_{ijn}} \geq \theta_0 \cdot \rho_{kmn} \tag{21.3}$$

Notice that the (nonlinear) product expression in Eq. (21.1b) is no longer needed, since for an arbitrary UE/AN/carrier triplet, the right-hand side guarantees that the SINR achieved is at least equal to the target level if it refers to an actual assignment ($\rho_{kmn} = 1 \rightarrow \gamma_{kmn} \geq \theta_0$), else it can take an arbitrary (and irrelevant) value ($\rho_{kmn} = 0 \rightarrow \gamma_{kmn} \geq 0$). In essence, since for one UE k, there is a single non-zero ρ_{kmn} element, as per Eq. (21.2a), only one γ_{kmn} element should be at least equal to the required SINR level θ_0. After a simple manipulation, Eq. (21.3) may be written as:

$$\forall k, m, n : p \cdot g_{km} \geq \theta_0 \cdot \rho_{kmn} + \theta_0 \cdot p \cdot \sum_{i \neq k} \sum_{j \neq m} g_{kj} \cdot (\rho_{kmn} \cdot \rho_{ijn}) \tag{21.4}$$

Still, Eq. (21.4) is a nonlinear expression due to the binary variable products appearing in the right-hand side double summation term. We are able to exactly linearize each of these products by leveraging Linearization Method 1, as introduced in Section 21.2.2. To this end, we define:

- a new set of $KM\,(K-1)(M-1)$ auxiliary binary variables, with $\{z_{kmijn}\} \in \{0,1\}$, each of which linearizes a single product term; that is $\forall k, m, n, i \neq k, j \neq m : z_{kmijn} = \rho_{kmn} \cdot \rho_{ijn}$, resorting to a new purely linear expression for minimum SINR constraint in Eq. (21.5);
- a new set of linear constraint expressions Eq. (21.6) for linking the auxiliary variables with the original ones.

$$\forall k, m, n, i \neq k, j \neq m : p \cdot g_{km} \geq \theta_0 \cdot \rho_{kmn} + \theta_0 \cdot p \sum_{i \neq k} \sum_{j \neq m} g_{kj} \cdot z_{kmijn} \tag{21.5}$$

$$\forall k, m, n, i \neq k, j \neq m : 2 \cdot z_{kmijn} \leq \rho_{kmn} + \rho_{ijn} \leq z_{kmijn} + 1 \tag{21.6}$$

In summary, the following feasibility binary integer linear program should be solved for obtaining a user-association and carrier scheduling assignment decision that provides the minimum required SINR level for all network UEs.

Problem 1: Joint user-association and scheduling for given SINR target
Input: $\{g_{km}\}_{k\in\mathcal{K},m\in\mathcal{M}}, p, \theta_0$
Output: $\{\rho^*_{kmn}\} \in \{0,1\}$

$$\text{find}\{\rho_{kmn}\}, \{z_{kmijn}\} \text{ subject to Eqs.}(21.2a), (21.2b), (21.5), (21.6).$$

The Case of Unknown Target Rates

Defining specific minimum rate/SINR requirements is typical for non-elastic traffic, whereas for best-effort traffic the medium access control (MAC) scheduler usually optimizes over this dimension as well. In this case, we seek the assignment of access nodes and carriers that provides the maximum common SINR across all UEs. We therefore optimize over both $\{\rho_{kmn}\}$ and θ_0. We provide two approaches for solving this particular problem.

A straightforward way to deal with the extra optimization dimension is to leverage bisection search: given a bounded interval for the achievable common SINR, namely $\theta_0 \in [\theta_0^{lb}, \theta_0^{ub}]$, we may progressively reduce this interval up to an arbitrary small tolerance ϵ (e.g. 10^{-4}) by solving $\lceil \log_2(\theta_0^{ub} - \theta_0^{lb})/\varepsilon \rceil$ feasibility BILPs as in Alg. 1 (see also the book by Boyd and Vandenberghe for a detailed description of the concept [33]):

Algorithm 21.1 Bisection search for unknown target SINR

1: Given $\theta_0^{lb} \le \theta_0 \le \theta_0^{ub}$, convergence tolerance ϵ
2: **repeat**
3: $\quad \theta_0^l \leftarrow (\theta_0^{lb} + \theta_0^{ub})/2$;
4: $\quad$ Solve the ILP feasibility Problem 1;
5: $\quad$ If a feasible solution is found $\theta_0^{ub} \leftarrow \theta_0^l$ else $\theta_0^{lb} \leftarrow \theta_0^l$;
6: **until** $|\theta_0^{ub} - \theta_0^{lb}| \le \varepsilon$.

The drawback of the presented approach is the need to solve multiple BILPs (typically 20–30) to obtain the optimal solution. We propose an alternative approach, which provides the optimal resource assignment in a single shot at the expense of introducing a new set of auxiliary binary variables and the associated linearization constraints. Indeed, we observe two types of nonlinear term in the minimum SINR constraint expression of Eq. (21.5), $\theta_0 \cdot \rho_{kmn}$ and $\theta_0 \cdot z_{kmijn}$. Both products involve a single continuous and a single binary variable, hence following the Linearization Method 2, as presented in Section 21.2.2. They are transformed as follows.

Linearization of $\theta_0 \cdot \rho_{kmn}$

We define $w_{kmn} = \theta_0 \cdot \rho_{kmn}, w_{kmn} \in [0, \theta_0^{ub}]^{KMN}$ where K_{θ_0} a large positive number greater than the maximum value that the common SINR can take, namely $K_{\theta_0} \ge \theta_0^{ub}$. We also introduce the set of linear inequality constraints:

$$\forall k, m, n : \theta_0 - w_{kmn} \le K_{\theta_0}(1 - \rho_{kmn}), w_{kmn} \le \theta_0, w_{kmn} \le K_{\theta_0}\rho_{kmn} \quad (21.7)$$

Linearization of $\theta_0 \cdot z_{kmijn}$
We define $v_{kmijn} = \theta_0 \cdot z_{kmijn}, v_{kmijn} \in [0, \theta_0^{ub}]^{KM(K-1)(M-1)N}$ and the set of linear inequality constraints:

$$\begin{aligned} \forall k, m, n, i \neq k, j \neq m : \theta_0 - v_{kmijn} \leq K_{\theta_o}(1 - v_{kmijn}) \\ v_{kmijn} \leq \theta_0, v_{kmijn} \leq K_{\theta_o} v_{kmijn} \end{aligned} \tag{21.8}$$

Replacing the binary-continuous variable products with the new auxiliary variables, we ultimately get the following linear expressions set for the minimum SINR constraint per UE:

$$\forall k : p \cdot g_{km} \geq w_{kmn} + p \cdot \sum_{i \neq k} \sum_{j \neq m} g_{kj} \cdot v_{kmijn} \tag{21.9}$$

Summing up, we provide the complete linearized formulation of the joint user-association and carrier-scheduling problem when the common SINR target is not known.

Problem 2: Joint user-association and scheduling for unknown SINR target
Input: $\{g_{km}\}_{k \in \mathcal{K}, m \in \mathcal{M}}, p$
Output: $\{\rho^*_{kmn}\} \in \{0, 1\}, \theta^*_0$

$$\max_{\{\rho_{kmn}, \theta_0\}} \theta_0 \text{ subject to Eqs.(21.2a), (21.2b), (21.6), (21.7), (21.8), (21.9)}$$

21.3.3 User Association, Scheduling and Power Allocation

In Section 21.3.2 the power allocated to each AN–UE pair was considered fixed. However, by coordinating the power of different transmissions on a cluster basis, leveraging the received signal and interference-conditions information, the rate performance levels could be further enhanced. In the current section we extend the previous analysis to account for power coordination. First we show how we can obtain the optimal power allocation given a known user-association and scheduling assignment. As a second step, we examine the joint user-association, scheduling and power allocation problem, which is more complex than the previous decoupled problem but provides the optimal performance bounds.

21.3.3.1 Power Allocation for a given User-association and Scheduling Assignment

We assume that UEs have already been assigned to ANs and carriers. Let L denote the number of AN–UE pairs (where $L = K$ since each UE is granted access to a single carrier). For an arbitrary pair l, let h_{ll} be the experienced path power gain and $\mathcal{I}_l$ the set of AN–UE interferer pairs. Notice that not all complementary pairs impact the pair of interest, but only the ones that are assigned to the same carrier. In this respect let $h_{ll'}, l' \in \mathcal{I}_l$, be the power path-gain involving the UE of the pair of interest l and the AN of the potential interfering pair l' (since we consider the DL direction only). To guarantee a minimum SINR performance level θ_0 the following expression must hold for every pair:

$$\forall l : \gamma_l = \frac{h_{ll}p_l}{1 + \sum\limits_{l' \in \mathcal{I}_l} h_{ll'}p_{l'}} \geq \theta_0 \Rightarrow g_{ll}p_l \geq \theta_0 + \theta_0 \cdot \sum_{l' \in \mathcal{I}_l} g_{ll'}p_{l'} \tag{21.10}$$

where $\{p_l\} \in [0, p_{\max}]^{L \times L}$ is the set of L continuous bounded variables, which we need to optimize.

For a known target SINR level, it is clear that one needs to solve a feasibility LP, an operation performed optimally and globally with linear complexity. For an unknown target SINR level, Algorithm 1 should be run, where at each iteration the above LP is solved. Below, we summarize the power-optimization subproblem for the generalized case of unknown target SINR.

Problem 3: Optimal power allocation for a given user-association and scheduling
Input: $\{h_{ll}\}_{l \in \mathcal{L}}, p_{\max}$
Output:$\{p_l^*\} \in [0, p_{\max}]^{L \times L}$
Run Alg. 1 where at each step solve: find$\{p_l\}$ subject to Eq. (21.10)

21.3.3.2 Joint User Association, Scheduling and Power Allocation

In addition to the binary user-association and scheduling variables ρ_{kmn} already defined, we introduce the corresponding continuous and upper-bound power allocation variables p_{kmn}. In fact, for numerical reasons we introduce the normalized power variables δ_{kmn}, where $\delta_{kmn} = p_{kmn}/p_{\max}$, $0 \leq \delta_{kmn} \leq 1$, and $p_{\max}$ is the maximum allowed DL power. Constraints formulated previously in Eq. (21.2a),Eq. (21.2b) are exactly applicable to this setup as well. Furthermore, for an arbitrary $\langle k, m, n \rangle$ triplet, it must be guaranteed that nonzero power is allocated if and only if it corresponds to an actual user-association and carrier assignment (mathematically speaking, if $\rho_{kmn} = 1 \rightarrow \delta_{kmn} > 0$, else $\delta_{kmn} = 0$). It is easily deduced that it suffices to introduce the following linear constraints set:

$$\forall k, m, n : 0 \leq \delta_{kmn} \leq \rho_{kmn} \tag{21.11}$$

Given a target minimum SINR level θ_0, similar to the analysis followed for the user-association and scheduling problem, we introduce the following expressions set:

$$\forall k, m, n : \gamma_{kmn} = \frac{p_{\max} \cdot g_{km} \cdot \delta_{kmn}}{1 + p_{\max} \cdot \sum\limits_{i \neq k} \sum\limits_{j \neq m} g_{kj} \cdot \rho_{ijn} \cdot \delta_{ijn}} \geq \theta_0 \cdot \rho_{kmn} \tag{21.12}$$

which can be straightforwardly rewritten as:

$$\forall k, m, n : p_{\max} \cdot g_{km} \cdot \delta_{kmn} \geq$$
$$\theta_0 \cdot \rho_{kmn} + \theta_0 \cdot p_{\max} \cdot \sum_{i \neq k} \sum_{j \neq m} g_{kj} \cdot \rho_{kmn} \cdot \rho_{ijn} \cdot \delta_{ijn} \tag{21.13}$$

By simple inspection it is deduced that the non-linear terms in Eq. (21.13) involve products of two binary (ρ_{kmn}, ρ_{ijn}), and a single upper-bounded continuous variable (δ_{ijn}). Hence these product expressions can be exactly linearized in two steps:

1. Introduce the binary variables set $\{z_{kmijn}\}, \forall k, m, n, i \neq k, j \neq m$, where $z_{kmijn} = \rho_{kmn} \cdot \rho_{ijn}$, along with the linear inequality constraints set:

$$\forall k, m, n, i \neq k, j \neq m : 2 \cdot z_{kmijn} \leq \rho_{kmn} + \rho_{ijn} \leq z_{kmijn} + 1 \tag{21.14}$$

2. Introduce the positive continuous variables set w_{kmijn}, where $w_{kmijn} = z_{kmijn} \cdot \delta_{ijn}$; K_δ, a large number greater than the upper bound of δ_{kmn} variables; and the set of linear inequality constraints:

$$\begin{aligned} \forall k, m, n, i \neq k, j \neq m : \delta_{ijn} - w_{kmijn} \leq K_\delta \cdot (1 - z_{kmijn}) \\ w_{kmijn} \leq \delta_{ijn}, w_{kmijn} \leq K_\delta \cdot \rho_{km}, w_{kmijn} \geq 0 \end{aligned} \tag{21.15}$$

Finally Eq. (21.13) is rewritten as:

$$\forall k, m, n : p_{\max} g_{km} \cdot \delta_{kmn} \geq \theta_0 \cdot \rho_{kmn} + \theta_0 \cdot p_{\max} \cdot \sum_{i \neq k} \sum_{j \neq m} g_{kj} \cdot w_{kmijn} \tag{21.16}$$

which by simple inspection is purely linear with respect to the original and auxiliary variable sets.

Generalizing for an unknown target SINR level, we may leverage bisection search, where at each step we need to solve an MILP, as summarized below:

Problem 4: Joint optimal user-association, scheduling and power allocation
Input: $\{g_{km}\}_{k \in \mathcal{K}, m \in \mathcal{M}}, p_{\max}$
Output: $\{\rho^*_{kmn}\} \in \{0, 1\}, \{\delta^*_{kmn}\} \in [0, 1]^{K \times M \times N}, \theta^*_0$
Run Alg. 1 where at each step solve the MILP feasibility problem:

$$\text{find}\{\rho_{kmn}\}, \{\delta_{kmn}\} \text{s.t.} (21.2a), (21.2b), (21.11)(21.14), (21.15), (21.16)$$

21.3.4 *Numerical Results*

In this subsection we apply the optimization models we have developed in typical dense small-cell setups, so as to acquire the optimal rate limits and identify the impact of each RRM optimization dimension on the overall performance. Towards this end, we loosely follow the 3GPP outdoor dense small-cell performance evaluation modeling assumptions, and in particular Scenario 2a [30]. We consider non-interfering clusters, each one dedicated to serve a 140-m radius hotspot area, populated with randomly located access nodes and user nodes, as in Figure 21.2. A minimum internode distance (AN–UE, AN–AN, and UE–UE) of 5 m is also considered. Each AN operates on a common 3.5-GHz spectrum band, and the available

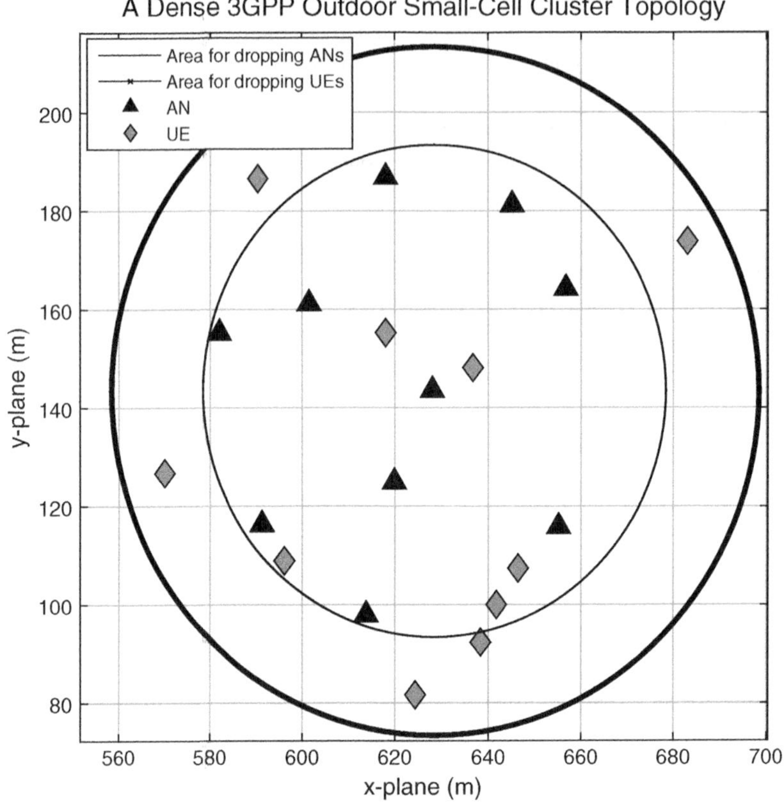

Figure 21.2 A random topology dense small-cell cluster snapshot

bandwidth is divided into $N = 5$ carriers. Large-scale channel effects are modeled using the log-distance path-loss model with propagation exponent $\alpha = 3.2$. A maximum DL power per AN p is considered, such that the transmit SNR (p/σ^2, where σ^2 is the noise power) is set to 120 dB. The number of users K within each cluster is kept constant at 10, while the number of ANs varies from 2 to 12, thus capturing the UDN region, where the densities of ANs and UEs are comparable. For each network scenario we evaluate the worst UE rate achieved by different RRM strategies, and we report the corresponding average values over 250 independent topology realizations. Optimization problems are modeled in MATLAB using the CVX tool [34] and solved using the Gurobi optimizer [24].

As for the RRM strategies, we examine five schemes, as analyzed in Sections 21.3.2 and 21.3.3. These schemes are elaborated below and summarized in Table 21.1:

- **The baseline scheme** reflects the uncoordinated network operation and harnesses only the infrastructure densification benefits
- **Strategy I** builds on the baseline user-association and carrier scheduling assignment, and employs only power optimization. This is the lightest network coordination strategy from a

Table 21.1 Overview of radio-resources management strategies for single-antenna coordinated UDNs

RRM strategy	Description	Cross-reference Section
Baseline	1) User-association based on closest AN	
	2) Random carrier scheduling	21.3.2.1
I	1) User-association and carrier scheduling as in Baseline	21.3.2.1
	2) Power optimization given previous assignment	21.3.3.1, Prob. 3
II	Joint user-association and scheduling	21.3.2.2, Prob. 2
III	1) Joint user-association and scheduling as in (II)	
	2) Power optimization given previous assignment	21.3.3.1, Prob. 3
IV	Joint user-association, scheduling and power optimization	21.3.3.2, Prob. 4

computational complexity point of view, since it requires the solution of a series of linear programs.

- **Strategy II** jointly optimizes user-association and carrier scheduling, requiring the solution of a series of integer linear programs.
- **Strategy III** builds on the previous strategy by employing power optimization over the user-association and carrier-scheduling assignment determined by Strategy II. This extra step involves the solution of a series of linear programs.
- **Strategy IV** jointly optimizes user-association, carrier scheduling and power allocation, requiring the solution of a series of integer linear programs, and providing the maximum achievable rate performance levels.

Figure 21.3a illustrates the rate performance results for the various RRM strategies as a function of infrastructure densification. Compared to the baseline uncoordinated network operation, it is observed that by exploiting the various RRM dimensions, namely user-association, carrier scheduling and power allocation, significant rate performance benefits can be achieved. In particular, by optimally coordinating the power allocated to each user/carrier pair following Strategy (I), gains of 1.6–2 times are achieved. Alternatively by optimally assigning users to access nodes and carriers to UE–AN pairs, following Strategy (II), performance is enhanced by 1.2–2.7 times compared to the baseline. Gains depend on network density conditions, and seem to be more prominent for the UDN region, where the number of access and user nodes becomes comparable. Notice that for the low-density region (2–4 ANs), Strategy I (optimal power allocation) outperforms Strategy II (optimal user-association and carrier scheduling). This is attributed to the fact that in this region there is limited flexibility in assigning the carriers to users in order to minimize interference, due to the shortage in the number of available degrees of freedom; recall that there are 10 UEs and in total $5M$ carriers. On the other hand, for the high-density region, Strategy II becomes significantly more beneficial; since it relies on fixed power allocation, further performance enhancements can be acquired by a network-wise tuning of the power levels per user/carrier pair. To this end, Strategy III employs optimal power coordination over the user/carrier pairs provided by Strategy II, resulting to excess rate enhancement levels of 1.05–2 times. These gains are prominent in the low-density region, and seem to vanish in the high-density region. Higher gains can be achieved by jointly optimizing user-association, carrier scheduling and power allocation, as in Strategy IV. Compared to

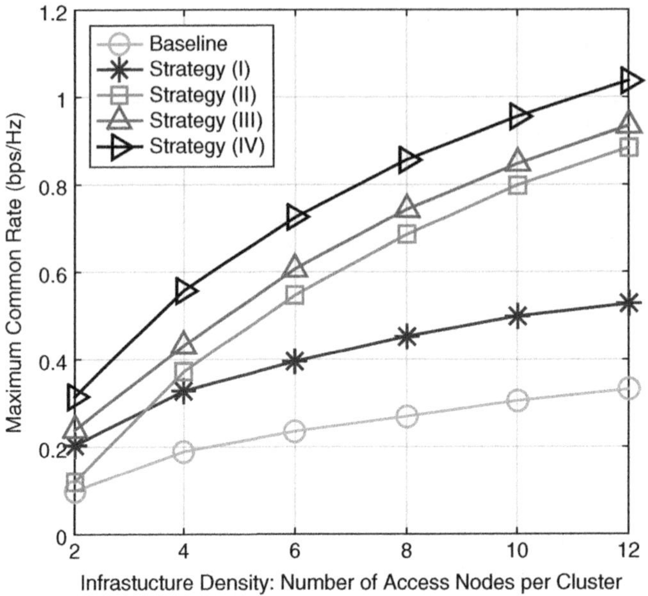

(a) Maximum Common Rate Performance vs Network Densification

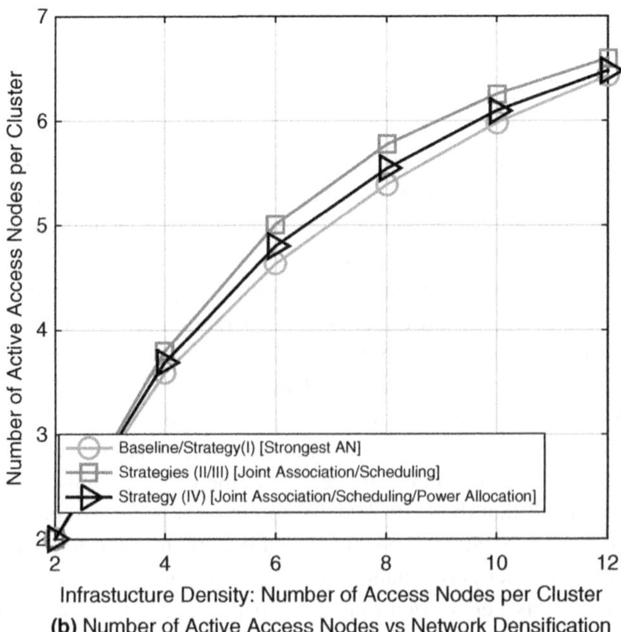

(b) Number of Active Access Nodes vs Network Densification

Figure 21.3 Comparative performance evaluation of various RRM strategies in coordinated UDNs, as a function of infrastructure densification

Strategy III, 1.1-1.3 times better performance is achieved. Strategy IV provides the ultimate performance levels achieved when global optimal network coordination is employed. In particular, in the system model considered, an approximate triple rate enhancement is observed, compared to the baseline uncoordinated operation.

In addition, Figure 21.3b illustrates the number of active ANs for three possible user-association/carrier scheduling approaches:

1. each UE served by its closest AN (baseline)
2. association is dictated by solving the joint optimal user-association and carrier-scheduling problem (strategies II and III)
3. association is dictated by solving the joint optimal user-association, carrier-scheduling and power-allocation problem (strategy IV).

As a first remark, no major difference in the number of utilized ANs is observed among the various approaches. Compared to the baseline user-association, both coordination schemes seem to (slightly) increase the access node utilization levels. This is justified by the fact that these policies tend to distribute user nodes to more ANs, such that the orthogonality among different AN carriers is exploited towards reducing inter-UE interference.

21.4 Optimal Strategies for Multi-antenna Coordinated and Cooperative Ultradense Networks

Following the analysis of single-antenna coordinated UDN system setups, in this section we examine the RRM opportunities when:

1. the spatial dimension is also considered, through equipping the access nodes with multiple antenna elements
2. access nodes are able to cooperatively serve multiple UEs, through an ideal or limited backhaul network.

In Section 21.4.2 we study the spatial resource management (precoder) design problem, while in Section 21.4.3 we consider the precoder and user-association problem for various backhaul conditions. Finally, indicative performance evaluation results for representative network setups are presented in Section 21.4.4.

21.4.1 System Model and Scope

We build on the system model of single-antenna coordinated UDNs presented in Section 21.3.1, but:

- each access node is equipped with L antennas while user nodes still have a single antenna
- two or more ANs could serve a UE node collaboratively
- all UEs are served over a single orthogonal transmission unit, namely a carrier n, thus we drop the carrier index.

On a per-AN basis, served UEs are multiplexed in the spatial dimension, leveraging up to L spatial transmission layers. For the arbitrary kth UE served by the mth AN, the complete channel state information (CSI) for the lth spatial dimension is reflected into $h_{km}^{l} = \sqrt{g_{km}} \cdot z_{km}^{l}$, where g_{km} is the power path-gain due to large-scale propagation effects and z_{km}^{l} is the complex channel amplitude due to small-scale fading effects. For a single AN–UE pair, we define the L-length CSI vector as $\mathbf{h}_{km} \in \mathbf{C}^{L\times 1}$, and for a specific UE node k, we define the LM-length CSI vector by stacking all the per-AN vectors as $\mathbf{h}_k = [\mathbf{h}_{k1}^T \cdots \mathbf{h}_{km}^T \cdots \mathbf{h}_{kM}^T]^T \in \mathbf{C}^{LM\times 1}$. Unless otherwise stated, perfect knowledge of all user CSI vectors $\{\mathbf{h}_k\}_{1\leq k\leq K}$ is assumed.

In the context of the new system model the "user-association" concept studied throughout Section 21.3 is re-defined, since each UE node can be served not only by a single AN but also collaboratively by multiple ANs. For the former network scenario (termed "coordinated" hereafter) exchange of information is allowed only in the signaling plane, since for each UE, payload should be present on a single serving AN; for the latter (termed "cooperative" hereafter) information exchange occurs on both signaling and data planes, since payload could be present on more than one AN. These ANs form the per-UE collaborative AN set. Note that for the coordinated operation, the collaborative AN set degenerates to a single-element set as in non-cooperative UDNs.

For cooperative serving, the major performance limitation is the data-sharing overhead, which depends on the backhaul capabilities. Within our system model, wireless access is optimized for given backhaul limitations, and no steps for joint access and backhaul optimization are taken. Regarding backhaul modeling, we assume a single "backhaul transmission unit" is needed for forwarding a UE's payload to an AN. Denoting by b the overall number of utilized backhaul units across the network, and by $b_{\max}$ the backhaul constraint, the network optimizer should form the collaborative AN sets for all UEs such that $b \leq b_{\max}$. From a radio-access performance point of view, optimal rates are achieved for unlimited backhaul conditions, such that all UEs' data are shared across all ANs, forming an ideal network MIMO. For a cluster of K UEs and M collaborative ANs, unlimited backhaul corresponds to having $b_{\max}^{unlim} = K \cdot M$ backhaul units available.

In multi-antenna uncoordinated, coordinated or cooperative network scenarios, significant performance gains can be obtained by adaptively designing the precoding vectors for each user node, based on network channel conditions. In this respect, the available spatial degrees of freedom can be leveraged towards either forming directive beams towards intended users or canceling interference towards co-cell or other-cell users. We denote by $\mathbf{w}_{km} \in \mathbf{C}^{L\times 1}$ the precoding vector for the arbitrary kth-UE/mth-AN pair, and by stacking up individual vectors we get the row-vector precoder for UE k, $\mathbf{w}_k = [\mathbf{w}_{k1}^T \cdots \mathbf{w}_{km}^T \cdots \mathbf{w}_{kM}^T] \in \mathbf{C}^{1\times LM}$. Then, for a given selection of precoding vectors $\{\mathbf{w}_k\}_{1\leq k\leq K}$, under known channel condition vector set $\{\mathbf{h}_k\}_{1\leq k\leq K}$, the achieved rate per UE r_k is written as:

$$\forall k : r_k = \log_2(1+\gamma_k), \text{ where } \gamma_k = \frac{|\mathbf{h}_k^H \mathbf{w}_k|^2}{1 + \sum\limits_{i\neq k} |\mathbf{h}_k^H \mathbf{w}_i|^2} \tag{21.17}$$

In summary, given:

- ideal network CSI knowledge ($\{\mathbf{h}_k\}$)
- network backhaul capability ($b_{\max}$, upper bounded by $b_{\max}^{unlim}$)
- maximum power budget per AN ($p_{\max}$)
- a common (known or unknown) target rate(r_0) or SINR (θ_0) per UE,

the scope of this section is to propose RRM optimizers that adaptively select for all UEs:

- the potentially collaborating AN serving sets;
- the DL precoding vectors (including beamforming and power scaling).

In Figure 21.4 an example is provided. Notice that in Eq. (21.17) the information on the selection of cooperative AN sets per UE is incorporated into the precoding vector information; in particular, for UE k, the AN m does not belong to the collaborating serving set if and only if $\mathbf{w}_{km}$ is an all-zero vector.

21.4.2 Precoder Design for given User Association

We begin by considering three precoder design approaches reflecting typical network operation "modes", based on fixed user-association.

- **Local/Uncoordinated Operation:** Each UE is associated with the strongest AN, considering only large-scale channel conditions ($\{g_{km}\}$), and is served exclusively by it. Precoders are optimized on a per-AN basis, without considering interference from the other ANs. This scheme provides the baseline rate performance levels.
- **Coordinated Operation:** This is based on the same user-association principle, but precoding vectors are optimized jointly over all UEs. This scheme provides the optimal

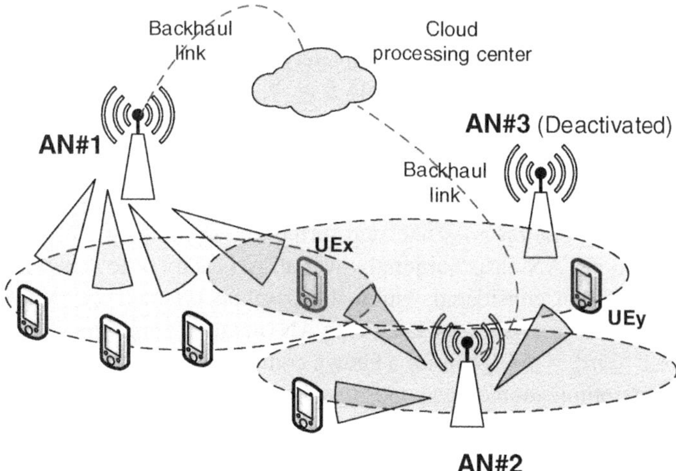

Figure 21.4 A cluster of 3 randomly deployed access nodes with overlapping areas serving 6 users, supported by a centralized processing centre and backhaul links. Notice the coordinated beam allocation (power scaling not shown) between AN1 and AN2. Notice how UEx is collaboratively served by AN1 and AN2 towards improving achieved SINR conditions given that its data is available to both ANs through backhaul. Also notice that although UEy's closest serving node is AN3, instead it is associated with AN2, allowing for AN3 to be deactivated, leading to interference and backhaul load reduction. Sections 21.4.2 and 21.4.3 will provide models and solutions for optimizing these decisions

performance limits for networks employing global optimal coordination over the signaling plane (CSI is exchanged among ANs).

- **Unlimited Cooperative Operation:** Each UE is associated with all the ANs, allowing the exchange of complete CSI and UE payload. This scheme enables the global optimal tuning of precoding vectors leveraging every possible spatial degree of freedom available, providing the ultimate performance limits.

The schemes are elaborated in what follows.

21.4.2.1 Baseline Scheme: Local/Uncoordinated Network Operation

The scheme requires the minimum amount of overhead signaling and backhaul, since precoding vectors are designed independently per AN. There is a variety of nonlinear and linear precoding design approaches, such as dirty paper coding and zero-forcing (ZF), trading off complexity with performance. Here, we consider a localized ZF approach, which completely removes interuser interference within each AN but take no steps for managing intercell interference.

We assume an AN m serving a UEs subset $\mathcal{U}_m$, comprising $N_m = |\mathcal{U}_m| \leq K$ UEs, and we form the local channel matrix $\mathbf{h}_m^{local} \in \mathbf{C}^{N_m \times L}$ by stacking the corresponding user vectors $\mathbf{h}_{im}^T, i \in \mathcal{U}_m$. The per-user ZF precoding vectors are extracted straightforwardly by taking the pseudo-inverse of the local channel matrix and employing appropriate normalization factors as follows:

$$(\mathbf{w}_m^{local})_{ZF} = \sqrt{\frac{p_{\max}}{N_m}} \cdot (\mathbf{h}_m^{local})^{-1} \in \mathbf{C}^{L \times N_m} \tag{21.18}$$

Interuser interference within each AN is completely eliminated as long as there are enough spatial degrees of freedom to leverage, namely $L \geq N_m$.

21.4.2.2 Coordinated Network Operation

Coordinated network operation goes one step further, by optimizing the per-UE precoding vectors not only on a per-AN basis but jointly over all ANs across the network.[1]

Following the notation considered within the baseline scheme, we denote by $\mathcal{U}_m$ the nonempty subset of network UEs associated with AN m, and by m_k^* the serving AN index for the UE k, where $1 \leq m_k^* \leq M$. Then for a known common target SINR level θ_0, coordinated network precoding optimization corresponds to solving the following feasibility program:

$$\text{find}\{\mathbf{w}_k\}_{1 \leq k \leq K} \ \text{subject to}$$

$$\forall k : \left|\mathbf{h}_k^H \mathbf{w}_k\right|^2 \Bigg/ \left(1 + \sum_{i \neq k} |\mathbf{h}_k^H \mathbf{w}_i|^2\right) \geq \theta_0 \tag{21.19a}$$

[1] In what follows, when we refer to the ANs belonging to the network it is implied that this set corresponds to the union of the active ANs, where an AN is considered active if and only if at least a single UE is associated with it. In this sense, information regarding inactive ANs is not considered within the channel and precoder representations.

$$\forall m: \sum_{k\in\mathcal{U}_m} \|\mathbf{w}_k\|_2 \leq \sqrt{p_{\max}} \tag{21.19b}$$

$$\forall k: \sum_{m\notin m_k^*} \|\mathbf{w}_{km}\|_0 = 0 \tag{21.19c}$$

Equation (21.19a) imposes that the minimum required SINR level per UE is achieved, Eq. (21.19b) that the maximum power budget per AN is not exceeded, and Eq. (21.19c) that each precoding sub-vector involving a UE and a non-serving AN is an all-zero vector (since collaboratively serving a UE is not allowed within the current scenario). Clearly, formulation Eq. (21.19) is a non-convex program, which cannot be straightforwardly solved for global optimality. However, following the analyses of Karakayali *et al.* [35] and Tolli *et al.* [36] we may acquire a convexified problem formulation. In particular the minimum SINR constrains set of Eq. (21.19a) could be identically rewritten as:

$$\left\| \begin{matrix} 1 \\ \left[(\mathbf{h}_k^H\mathbf{w}_1)\cdots(\mathbf{h}_k^H\mathbf{w}_K)\right]^H \end{matrix} \right\|_2 \leq \sqrt{1+\frac{1}{\theta_0}}\Re\{\mathbf{h}_k^H\mathbf{w}_k\} \tag{21.20a}$$

$$\Im\{\mathbf{h}_k^H\mathbf{w}_k\} = 0, \forall k \tag{21.20b}$$

We note that the constraint subset of Eq. (21.20a) is conic, since the left-hand side involves a norm and the right-hand side an affine expression [37]. In addition, the maximum power budget constraint set in Eq. (21.19b) is also conic and the constraints in Eq. (21.19c) and Eq. (21.20b) are linear. Therefore the coordinated precoding optimization problem belongs to the class of (convex) SOCPs [33]. For a given target SINR level the user precoding vectors can be found by solving a feasibility SOCP, whereas for common SINR optimization, a series of feasibility SOCPs should be solved following the bisection search approach (see Algorithm 1 in Section 21.3.2.2). The procedure is summarized below:

Problem 5: Optimal coordinated precoding
Input: $\{\mathbf{h}_k\}_{1\leq k\leq K}, p_{\max}, \{\mathcal{U}_m\}_{1\leq m\leq M}, \{m_k^*\}_{1\leq k\leq K}$
Output: $\{\mathbf{w}_k^*\}_{1\leq k\leq K}, \theta_0^*$
Run Alg. 1, at each step solving the feasibility SOCP
find $\{\mathbf{w}_k\}_{1\leq k\leq K}$ subject to Eqs. (21.20a), (21.20b), (21.19b), (21.19c)

21.4.2.3 Cooperative Network Operation with Unconstrained Backhaul

Cooperative network operation with unlimited backhaul not only tunes the user precoding vectors in a global fashion as above, but also allows the per-UE payload transmission to be facilitated by multiple ANs collaboratively. Mathematically speaking, the formulation is identical to Problem 5, without the need to impose the connectivity constraints subset of Eq. (21.19c). By removing the particular constraints, the feasible space is extended and the precoder design becomes significantly more flexible. In essence, by having at least as many spatial degrees of freedom as the number of served UEs, inter-UE interference can be completely eliminated, similar to multiuser MIMO (MU-MIMO) operation on a cell basis without intercell interference. From a theoretical performance point of view, such a distributed MU-MIMO approach (sometimes simply called network MIMO) is scalable with respect to user-node increases as long as there are enough spatial degrees of freedom to exploit in order to cancel interference [35].

21.4.3 Joint Precoder Design and User Association for Constrained Backhaul

As previously discussed, cooperative precoding provides the ultimate performance limits, but strongly depends on the backhaul infrastructure availability. It is typical to consider backhaul as the performance bottleneck in such scenarios, and then study the impact of limited backhaul on the rate performance. Based on the system model described in Section 21.4.1 the limited backhaul problem corresponds to solving the joint precoding and user-association design problem, where user-association dictates how each UE's payload is shared among different ANs.

Obviously this is a combinatorial problem, which can be solved through exhaustively searching over all possible user-association solutions. In particular, one must examine all possible data-sharing combinations and for each one acquire the optimal precoding design through solving a convex program (as in Section 21.4.2). Then the combination providing the maximum achieved performance level corresponds to the optimal joint precoder and user-association decision. The complexity of this procedure is dominated by the complete enumeration of user-association decisions, which scales as K^M with respect to the network size.

Alternatively, we could extend the mathematical programming formulation of the precoding design subproblem given in Section 21.4.2. In particular, given a target SINR level per UE θ_0, Eq. (21.20) guarantees that the target level is achievable for all UEs, while Eq. (21.19b) imposes a maximum transmission power level per AN ($p_{\max}$). Next, we introduce the binary user-association variables $\{\alpha_{km}\}_{1\leq k\leq K,1\leq m\leq M} \in \{0,1\}^{K\times M}$, for which $\alpha_{km}=1$ if the kth UE's payload is available at the mth AN. Then for every possible AN–UE pair, it must be guaranteed that:

- if the UE's data are not available at the particular AN ($\alpha_{km}=0$), the corresponding precoding sub-vector ($\mathbf{w}_{km}$) should be an all-zero vector;
- if the UE's data are available at the particular AN ($\alpha_{km}=1$), the norm of the corresponding precoding subvector is upper bounded by the maximum power budget.

It is straightforward to deduce that both conditions are met by the following constraint expressions subset:

$$\forall k,m : \|\mathbf{w}_{km}\|_2 \leq \alpha_{km}\cdot\sqrt{p_{\max}} \tag{21.21}$$

Note that Eq. (21.21) is a conic constraint. Finally, given a network backhaul availability $b_{\max}$, the following linear constraint fulfills this requirement:

$$\sum_k\sum_m \alpha_{km} \leq b_{\max} \tag{21.22}$$

Below we summarize the formulation for the generalized case of unknown common target SINR levels.

Problem 6: Joint precoder design and user-association for constrained backhaul
Input: $\{\mathbf{h}_k\}_{1\leq k\leq K}, p_{\max}, b_{\max}$
Output: $\{\mathbf{w}_k^*\}_{1\leq k\leq K}, \{\alpha_{km}\}_{1\leq k\leq K, 1\leq m\leq M}, \theta_0^*$
Run Alg. 1, at each step solving the feasibility MI-SOCP
find $\{\mathbf{w}_k\}_{1\leq k\leq K}, \{\alpha_{km}\}_{1\leq k\leq K, 1\leq m\leq M}$ s.t. (21.20), (21.19b), (21.21), (21.22)

To sum up, we have introduced to the original SOCP formulation two new constraint subsets, the first one being conic and the second one being linear, hence preserving its SOCP nature. However, due to the incorporation of the binary variables α_{km} we have destroyed its convexity. The problem thus belongs to the class of MI-SOCPs. Such problems are NP-hard, but at least for small-scale problem the dimensions are tractable using powerful methods and software tools within a reasonable time.

We finally report two special cases of Problem 6:

1. For $b_{\max} = KM$, the problem is identical to the cooperative precoding optimization problem with unlimited backhaul (ideal network MIMO, Section 21.4.2.3)
2. For $b_{\max} = K$, the problem is identical to the coordinated precoding optimization problem. However differently from Section 21.4.2.2 it performs a joint selection of the serving AN and the precoding vector for each UE, whereas in Section 21.4.2.2, each UE is forced to associate with its closest AN.

21.4.4 Numerical Results

We follow the evaluation assumptions for coordinated single-antenna UDNs, and in addition we consider:

- each AN equipped with $L = 4$ antennas
- on top of large-scale propagation effects, small-scale independent Rayleigh fading.

Each network scenario corresponds to a single cluster comprising a fixed number of $K = 8$ user nodes, while the number of ANs M varies from 2 to 10. For each scenario we test 50 instances of random AN and UE drops, where each instance comprises 20 small-scale fading realizations, and we record the average minimum rate achieved by all network users. We use CVX for modeling the problems in MATLAB [34], and MOSEK for solving the respective SOCP and MI-SOCP problem instances [25].

We first perform a comparative performance evaluation study for the three precoding strategies presented in Section 21.4.2, namely the baseline local/uncoordinated scheme ("Baseline"), the global coordination network scheme ("Coordinated") and the unconstrained global cooperation network scheme or ideal network MIMO ("Cooperative"). Figure 21.5a illustrates

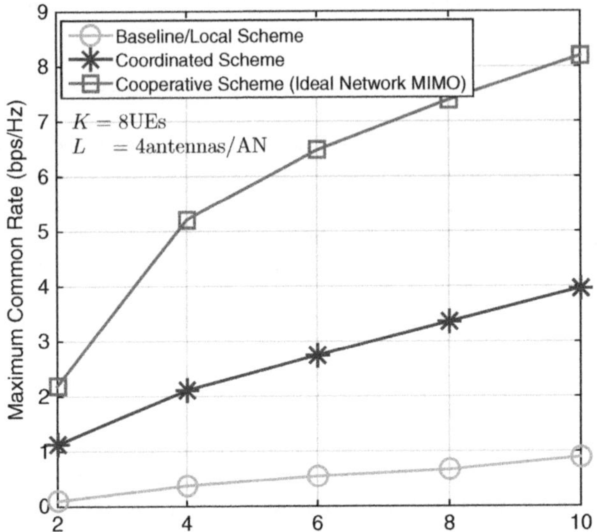

(a) Rate Performance of Local, Coordinated and Cooperative Precoding Optimization as a function of Network Densification

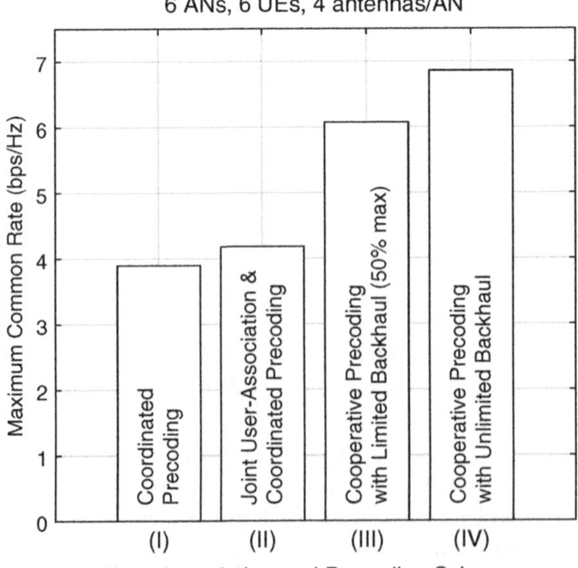

(b) Rate Performance of User-Association and Precoding Optimization with limited/unlimited backhaul conditions

Figure 21.5 Comparative performance evaluation of RRM optimization strategies for cooperative multi-antenna UDNs

the achieved rates for all schemes as a function of network infrastructure densification. As a first remark, global network coordination enhances the worst UE rate by almost an order of magnitude (5–11 times). This is attributed to the interference avoidance offered by the intelligent coordination of the spatial system dimensions. This benefit is maximized at low-density conditions (two ANs per cluster), and reduces with increased infrastructure densification. Secondly, global network cooperation with unlimited backhaul availability adds up a further 2–2.5 times performance boost, due to its capability to exploit (rather than avoid) interference, at the expense of a backhaul-load increase.

In the second set of experiments, we consider a network scenario consisting of six ANs, six UEs, and four antennas per AN, and we compare the maximum common rate achieved for a set of four schemes (Figure 21.5b). We assume that the global coordinated precoding optimization approach, corresponding to Problem 5 in Section 21.4.2, provides the the baseline rate-performance level. We then consider the joint user-association and coordinated precoding optimization approach, solving Problem 6 in Section 21.4.3 for the minimum backhaul overhead ($b_{\max} = 6$). It can be observed that by jointly optimizing user-association and coordinated precoding, instead of following a decoupled optimization approach as in the baseline scheme, within the considered system model, a limited performance enhancement ($\sim$7%) is acquired. Next we consider the cooperative precoding scheme with limited backhaul, which is set to 50% of its maximum value, the latter corresponding to ideal network MIMO where all users' data are available to all the cooperative access nodes. We report a significant performance boost ($\sim$56%), which is attributed to cooperative network operation, despite the fact that data sharing is not fully exploited. Finally for the sake of completeness, the performance of cooperative precoding with unlimited backhaul (solving Problem 6 in Section 21.4.3), for $b_{\max} = 36$ is also shown, providing the ultimate performance limits for the considered setup.

21.5 Summary and Future Research Directions

In this chapter we provided a detailed analysis of the RRM optimization potential within an envisioned 5G ultradense wireless access network environment. We first identified the various RRM dimensions encountered in UDNs, namely user-to-cell association, scheduling of orthogonal carriers, tuning of transmission power and precoding allocation, for single- and multi-antenna coordinated and cooperative network setups. In this context, we provided a set of mathematical programming formulations accompanied with approaches for globally and centrally optimizing the above dimensions, leveraging optimization tools, such as integer linear programming and second order cone programming. We applied the various optimization models into representative dense small-cell setups and explored the rate-performance trends as a function of infrastructure densification as well as the impact of individual RRM dimension optimization into the overall performance.

The proposed approaches are applicable to one-way (DL) UDN setups, hence they can be straightforwardly utilized for optimizing the DL of current and future frequency division duplexing (FDD) small-cell network deployments. Recently, and within the 5G evolution landscape, new trends have emerged focusing on flexible duplex operations, targeting two-way (both DL and UL) traffic optimization. Flexible duplex involves the concepts of:

- dynamic time division duplexing (TDD) [38], where each time-slot/resource-unit within each cell may be dynamically assigned to DL or UL directions, depending on the required load;
- full-duplex (FD) [39], which is a generalization of dynamic TDD/FDD by allowing the same slot to be assigned simultaneously to DL and UL directions;
- dynamic UL/DL decoupling [40], where a user node could associated with different access nodes for accommodating its DL/UL traffic.

The consideration of the above duplex approaches introduces new RRM system dimensions and calls for the development of novel problem formulations and corresponding solutions. In addition, it is commonly agreed that the massive MIMO technology will play a key role in the future 5G system [11]. Massive MIMO exploits the availability of massive spatial degrees of freedom per access node. Along these lines, the topic of RRM optimization should be revisited when massive MIMO meets homogeneous or heterogeneous UDNs [23, 41].

Acknowledgments

This work was supported by the Operational Program "Education and Lifelong Learning" under the project ARISTEIA II -FLAME, GSRT No. 3770, in National Technical University of Athens.

References

[1] Alexiou, A. (2014) Wireless world 2020: Radio interface challenges and technology enablers. *IEEE Vehic. Tech. Mag.*, **9** (1), 46–53, doi: 10.1109/MVT.2013.2295067.

[2] Andrews, J., Buzzi, S., Choi, W., Hanly, S., Lozano, A., Soong, A., and Zhang, J. (2014) What will 5G be? *IEEE J. Select. Areas Commun.*, **32** (6), 1065–1082, doi: 10.1109/JSAC.2014.2328098.

[3] Bhushan, N., Li, J., Malladi, D., Gilmore, R., Brenner, D., Damnjanovic, A., Sukhavasi, R., Patel, C., and Geirhofer, S. (2014) Network densification: The dominant theme for wireless evolution into 5G. *IEEE Commun. Mag.*, **52** (2), 82–89.

[4] Tullberg, H., Li, Z., Hoglund, A., Fertl, P., Gozalvez-Serrano, D., Pawlak, K., Popovski, P., Mange, G., and Bulakci, O. (2014) Towards the METIS 5G concept: First view on horizontal topics concepts, in *Networks and Communications (EuCNC), 2014 European Conference on*, doi: 10.1109/EuCNC.2014.6882694.

[5] ITU-R (2014), Report M.2320: Future technology trends of terrestrial IMT systems, URL `http://www.itu.int/pub/R-REP-M.2320-2014`.

[6] NGMN Alliance (2015), 5G White Paper, URL `http://www.ngmn.org/`.

[7] Osseiran, A., Boccardi, F., Braun, V., Kusume, K., Marsch, P., Maternia, M., Queseth, O., Schellmann, M., Schotten, H., Taoka, H., Tullberg, H., Uusitalo, M., Timus, B., and Fallgren, M. (2014) Scenarios for 5G mobile and wireless communications: the vision of the METIS project. *IEEE Commun. Mag.*, **52** (5), 26–35, doi: 10.1109/MCOM.2014.6815890.

[8] Hwang, I., Song, B., and Soliman, S. (2013) A holistic view on hyper-dense heterogeneous and small cell networks. *IEEE Commun. Mag.*, **51** (6), 20–27, doi: 10.1109/MCOM.2013.6525591.

[9] Li, Q., Niu, H., Papathanassiou, A., and Wu, G. (2014) 5G network capacity: Key elements and technologies. *IEEE Vehic. Tech. Mag.*, **9** (1), 71–78, doi: 10.1109/MVT.2013.2295070.

[10] Baldemair, R., Irnich, T., Balachandran, K., Dahlman, E., Mildh, G., Selen, Y., Parkvall, S., Meyer, M., and Osseiran, A. (2015) Ultra-dense networks in millimeter-wave frequencies. *IEEE Commun. Mag.*, **53** (1), 202–208, doi: 10.1109/MCOM.2015.7010535.

[11] Larsson, E., Edfors, O., Tufvesson, F., and Marzetta, T. (2014) Massive MIMO for next generation wireless systems. *IEEE Commun. Mag.*, **52** (2), 186–195, doi: 10.1109/MCOM.2014.6736761.

[12] Hiltunen, K. (2011) Comparison of different network densification alternatives from the LTE downlink performance point of view, in *Vehicular Technology Conference (VTC Fall), 2011 IEEE*, pp. 1112–1116.

[13] Gelabert, X., Legg, P., and Qvarfordt, C. (2013) Small cell densification requirements in high capacity future cellular networks, in *Communications Workshops (ICC), 2013 IEEE International Conference on*, pp. 1112–1116, doi: 10.1109/ICCW.2013.6649403.

[14] Hu, L., Luque Sanchez, L., Maternia, M., Kovacs, I., Vejlgaard, B., Mogensen, P., and Taoka, H. (2013) Heterogeneous LTE-Advanced network expansion for 1000x capacity, in *Vehicular Technology Conference (VTC Spring), 2013 IEEE 77th*, pp. 1–5, doi: 10.1109/VTCSpring.2013.6692578.

[15] Polignano, M., Mogensen, P.E., Fotiadis, P., Gimenez, L.C., Viering, I., and Zanier, P. (2014) The inter-cell interference dilemma in dense outdoor small cell deployment, in *Vehicular Technology Conference (VTC Spring), 2014 IEEE 79th*, Seoul, Republic of Korea, pp. 1–5.

[16] Koudouridis, G. (2014) On the capacity and energy efficiency of network scheduling in future ultra-dense networks, in *Computers and Communication (ISCC), 2014 IEEE Symposium on*, pp. 1–6.

[17] Yunas, S., Valkama, M., and Niemela, J. (2015) Spectral and energy efficiency of ultra-dense networks under different deployment strategies. *IEEE Commun. Mag.*, **53** (1), 90–100, doi: 10.1109/MCOM.2015.7010521.

[18] Hu, R. and Qian, Y. (2014) An energy efficient and spectrum efficient wireless heterogeneous network framework for 5G systems. *IEEE Commun. Mag.*, **52** (5), 94–101, doi: 10.1109/MCOM.2014.6815898.

[19] Hossain, E., Rasti, M., Tabassum, H., and Abdelnasser, A. (2014) Evolution toward 5G multi-tier cellular wireless networks: An interference management perspective. *IEEE Wireless Commun.*, **21** (3), 118–127, doi: 10.1109/MWC.2014.6845056.

[20] Gotsis, A.G. and Alexiou, A. (2013) Global Network Coordination in Densified Wireless Access Networks through Integer Linear Programming, in *Personal Indoor and Mobile Radio Communications (PIMRC), 2013 IEEE 24th International Symposium on*, London, UK, pp. 1548–1553, doi: 10.1109/PIMRC.2013.6666388.

[21] Gotsis, A. and Alexiou, A. (2013) Spatial resources optimization in distributed MIMO networks with limited data sharing, in *Globecom Workshops (GC Wkshps), 2013 IEEE, The First International Workshop on Cloud-Processing in Heterogeneous Mobile Communication Networks*, Atlanta, USA, pp. 789–794, doi: 10.1109/GLOCOMW.2013.6825085.

[22] Gotsis, A., Stefanatos, S., and Alexiou, A. (2014) Spatial coordination strategies in future ultra-dense wireless networks, in *Wireless Communications Systems (ISWCS), 2014 11th International Symposium on*, Barcelona, Spain, pp. 801–807.

[23] Gotsis, A., Stefanatos, S., and Alexiou, A. (2015) Optimal user association for massive MIMO empowered Ultra-Dense wireless networks, in *IEEE ICC 2015 - Workshop on Advanced PHY and MAC Techniques for Super Dense Wireless Networks (ICC'15 - Workshops 13)*, London, United Kingdom, pp. 2238–2244.

[24] GUROBI Optimization (2014), State-of-the-art mathematical programming solver, v6.0, URL `http://www.gurobi.com/`.

[25] MOSEK Aps (2014), MOSEK: Software for large-scale mathematical optimization problems, version 7.0.0.111, URL `http://www.mosek.com/`.

[26] Sherali, H.D. and Liberti, L. (2009) Reformulation-linearization technique for global optimization, in *Encyclopedia of Optimization* (eds C.A. Floudas and P.M. Pardalos), Springer, pp. 3263–3268, doi: 10.1007/978-0-387-74759-0_559.

[27] Hou, Y.T., Shi, Y., and Sherali, H.D. (2014) *Applied Optimization Methods for Wireless Networks*, Cambridge University Press.

[28] Chen, D.S., Batson, R., and Dang, Y. (2010) *Applied Integer Programming-Modeling and Solution*, John Wiley & Sons.

[29] Wu, T.H. (1997) A note on a global approach for general 0–1 fractional programming. *Europ J. Operational Res.*, **101** (1), 220–223.

[30] 3GPP (2013), TR 36.872 V12.1.0: Small cell enhancements for E-UTRA and E-UTRAN-Physical layer aspects (Release 12).

[31] Lee, J. and Leyffer, S. (eds) (2012) *Mixed Integer Nonlinear Programming*, The IMA Volumes in Mathematics and its Applications, Vol. 154, Springer.

[32] 3GPP (2012), TR 36.932 V12.0.0: Scenarios and requirements for small cell enhancements for E-UTRA and E-UTRAN (Release 12).

[33] Boyd, S. and Vandenberghe, L. (2004) *Convex Optimization*, Cambridge University Press.

[34] Grant, M. and Boyd, S. (2014), CVX: Matlab software for disciplined convex programming, version 2.1, URL `http://cvxr.com/cvx`.

[35] Karakayali, M., Foschini, G., and Valenzuela, R. (2006) Network coordination for spectrally efficient communications in cellular systems. *IEEE Wireless Commun.*, **13** (4), 56–61, doi: 10.1109/MWC.2006.1678166.

[36] Tolli, A., Codreanu, M., and Juntti, M. (2008) Cooperative MIMO–OFDM cellular system with soft handover between distributed base station antennas. *IEEE Trans. Wireless Commun.*, **7** (4), 1428–1440, doi: 10.1109/TWC.2008.061124.

[37] Lobo, M.S., Vandenberghe, L., Boyd, S., and Lebret, H. (1998) Applications of Second-Order Cone Programming. *Linear Alg. Applic.*, **284** (1-3), 193–228, doi: 10.1016/S0024-3795(98)10032-0.

[38] Shen, Z., Khoryaev, A., Eriksson, E., and Pan, X. (2012) Dynamic uplink-downlink configuration and interference management in TD-LTE. *IEEE Commun. Mag.*, **50** (11), 51–59.

[39] Goyal, S., Liu, P., Panwar, S., Difazio, R., Yang, R., and Bala, E. (2015) Full duplex cellular systems: will doubling interference prevent doubling capacity? *IEEE Commun. Mag.*, **53** (5), 121–127.

[40] Elshaer, H., Boccardi, F., Dohler, M., and Irmer, R. (2014) Downlink and uplink decoupling: A disruptive architectural design for 5G networks, in *Global Communications Conference (GLOBECOM), 2014 IEEE*, pp. 1798–1803.

[41] Bethanabhotla, D., Bursalioglu, O., Papadopoulos, H., and Caire, G. (2014) User association and load balancing for cellular massive MIMO, in *Information Theory and Applications Workshop (ITA), 2014*, pp. 1–10, doi: 10.1109/ITA.2014.6804284.

Part Five

Reference Design and 5G Standard Development

22

Full-duplex Radios in 5G: Fundamentals, Design and Prototyping

Jaeweon Kim, Min Soo Sim, MinKeun Chung, Dong Ku Kim and Chan-Byoung Chae

Signal Processing for 5G: Algorithms and Implementations, First Edition. Edited by Fa-Long Luo and Charlie Zhang.

22.1 Introduction

The proliferation of smartphones, tablet PCs and laptops is continually boosting demand for higher throughput of mobile devices with advanced wireless network capabilities. Between 2014 and 2019, global mobile data traffic is expected to increase nearly tenfold – expanding at a rate of 57% each year [1]. To support such demand, fifth generation (5G) wireless communications are expected to provide 1000-fold greater throughput than 4G. The wireless spectrum used to carry this traffic, however, is physically limited. This fact could give rise to a spectrum crunch. To increase spectral efficiency, researchers have developed several promising technologies. One of them, which could potentially double spectral efficiency, is known as full duplex.

Full-duplex radios simultaneously transmit and receive in the same frequency band. One of the obstacles to wireless communications – going back to 1895 when Guglielmo Marconi developed the wireless telegraph – has been self-interference. Self-interference is the phenomenon of a signal, transmitted from a transmitter, being received by its own receiver while that receiver is attempting to receive a signal sent from another device (the signal of interest). To avoid self-interference, wireless networks have operated in half duplex or out-of-band full duplex.[1] For example, Long Term Evolution (LTE) frequency-division duplex uses different frequency bands for downlink and uplink transmission, while LTE time-division duplex schedules downlink and uplink transmissions for different times.

Moreover, a sufficient amount of guard band (empty band) separating the two bands will prevent interference between the uplink and downlink signals (see Figure 22.1) [2]. It is commonly known that full duplex doubles the bandwidth efficiency but it can increase the spectral utilization by about 4.17 times if the guard band for data transmission is used for this frequency allocation.

To overcome the self-interference problem, researchers have investigated self-interference cancellation (SIC). Consider, as depicted in Figure 22.2, typical WiFi or LTE signals, which are transmitted at an average power of 23 dBm (200 mW). The thermal noise level of 20 MHz bandwidth is about -101 dBm ($= -174$ dBm/Hz $+73$ dB). Considering all radio frequency (RF) circuit components, the noise level can be increased to around -90 dBm. Therefore, to be used in a practical system, self-interference has to be canceled by at least 113 dB ($=$ 23 dBm $-$ (-90 dBm)). If self-interference is not suppressed to the noise level after SIC, the residual interference reduces the signal-to-interference-plus-noise ratio (SINR). In such a case, full duplex might be outperformed by half duplex in terms of throughput and bit error rate. For instance, assume that the SINR of a full-duplex link is -5 dB, while the SNR of the half-duplex link is 5 dB, due to insufficient SIC. By a simple calculation, the Shannon capacities of a full-duplex link and a half-duplex link are obtained as 0.79 and 2.06 bit/sec/Hz, respectively. Even though the capacity of the full-duplex link is multiplied by two, it is smaller than that of the half-duplex link. Thus SIC plays a crucial role in full-duplex systems.

Several SIC schemes have been investigated in prior work [3–18]. A simple design called antenna cancellation has been proposed [3], and this has been improved by a symmetric design [19]. To suppress self-interference more in the analog domain, an auxiliary transmit chain has been employed to generate the RF canceling signal [7–9]. A tunable single-path circuit that uses a copy of the transmitted signal as input has been used to mitigate self-interference [4]. A single-antenna system has been proposed using circulator-based isolation and a 16-path circuit [5]. Also, by applying a nonlinear model, not only linear components but also nonlinear

[1] In this chapter, full duplex means in-band full duplex.

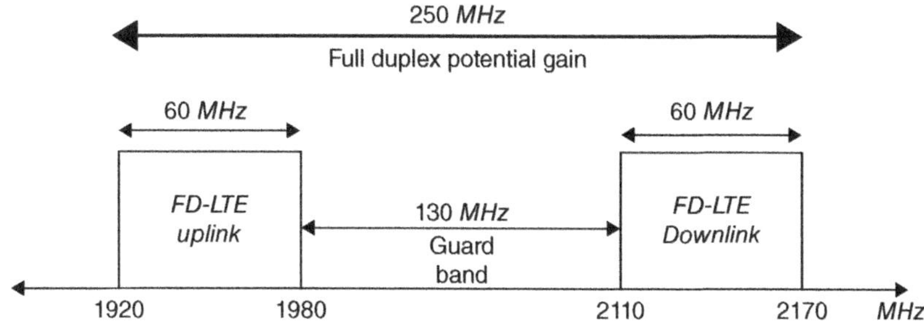

Figure 22.1 An example of LTE FDD spectrum allocation

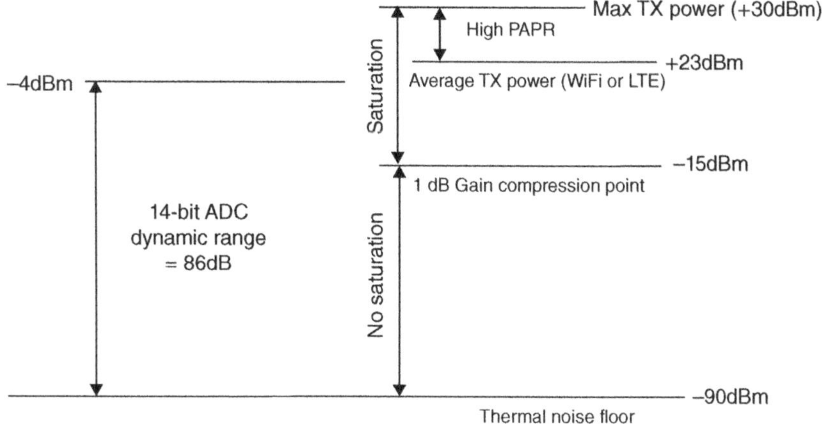

Figure 22.2 The required self-interference cancellation level

components are canceled in the digital domain. Including the nonlinearity of power amplifiers, the impact of hardware imperfections such as in-phase and quadrature (I/Q) imbalances and phase noise are considered in [10, 11, 16, 17].

This chapter is devoted to designing and prototyping of full-duplex communication systems, with emphasis on various SIC schemes. The rest of this chapter is organized as follows. Section 22.2 deals with the characterization of the major kinds of self-interference encountered in a full-duplex system. The next two sections introduce SIC schemes. These are categorized as analog SIC and digital SIC, and are explained in Sections 22.3 and 22.4, respectively. Section 22.5 reveals the details of the real-time full-duplex LTE prototyping performed by the authors. In Section 22.6, previous full-duplex radio researches are evaluated. Section 22.7 presents the summary of this chapter.

22.2 Self-interference

Self-interference has several components with different characteristics depending on the specifics of the full-duplex system implementation, such as the number of antennas used, the specifications of the RF circuit components and the surroundings. Roughly speaking, the components of the self-interference can be classified by linearity.

Linear components involve multipath propagation between a transmit and a receive antenna. For a single-antenna system, linear components include the leakage of a circulator or the reflections from impedance mismatch. Components of RF circuits, such as attenuators and delays, are also modeled as linear systems for analog SIC. The linear components of self-interference can be easily handled by existing channel estimation methods, as in most conventional wireless communication systems.

Nonlinear components are usually created by power amplifiers in transmitters and low-noise amplifiers in receivers. The nonlinearity of the power amplifier is generated because the power of the output signal is saturated for the high-power input signal, which gets worse for systems with high peak-to-average power ratios (PAPR) such as orthogonal frequency-division multiplexing (OFDM) and wideband code division multiple access. Intermodulation distortion, caused by the nonlinearity, interferes with the linear model of self-interference. Theoretically, intermodulation can be calculated by a Volterra series, while the usual approximation of it is obtained by generating a Taylor series. Since the even-ordered terms are out of band, the Taylor series includes odd-order terms only [20, 21].

Other RF imperfections – I/Q imbalances, phase noise and transmitter noise-occur in the transmitter too. The I/Q imbalance occurs when mismatches exist between the gain and phase of the two sinusoidal signals, and corrupts the baseband transmit signal. The imperfection of the local oscillator also degrades the linearity of the transmitted signal. In general, most of the influence of the oscillator imperfection is noticeable in random deviations in the output frequency, which can be modeled as phase noise. Transmitter noise also includes all kinds of thermal noise generated from RF components.

22.3 Analog Self-interference Cancellation

Analog SIC is used to suppress self-interference before it passes through an analog-to-digital converter (ADC) so that the residual self-interference can be digitized by a given ADC without saturation. The dynamic range, or the ratio between the largest and smallest values of a changeable quantity of a q-bit ADC, is given as

$$6.02 \times q + 1.76 \ [\text{dB}] \tag{22.1}$$

Recalling the previous example with the average transmit power of 23 dBm, the maximum transmit power is about 30 dBm due to the high PAPR of OFDM signals. Therefore, when a 14-bit ADC is used, the analog SIC must suppress self-interference by at least 34 dB ($=$ 30 dBm $-$ ($-$90 dBm) $-$ 86 dB). Furthermore, since an 1-dB gain compression point of a typical low-noise amplifier is -15 dBm, in order to ensure that receiver saturation is prevented, analog SIC is required to suppress up to 45 dB, as illustrated in Figure 22.2. Analog SIC can be implemented in two ways: passively and actively.

22.3.1 Analog Passive Self-interference Cancellation: Isolation

The simplest way to mitigate self-interference is to isolate the antennas. Isolation is also called passive analog SIC because no calculation or active component is used. The best advantage of

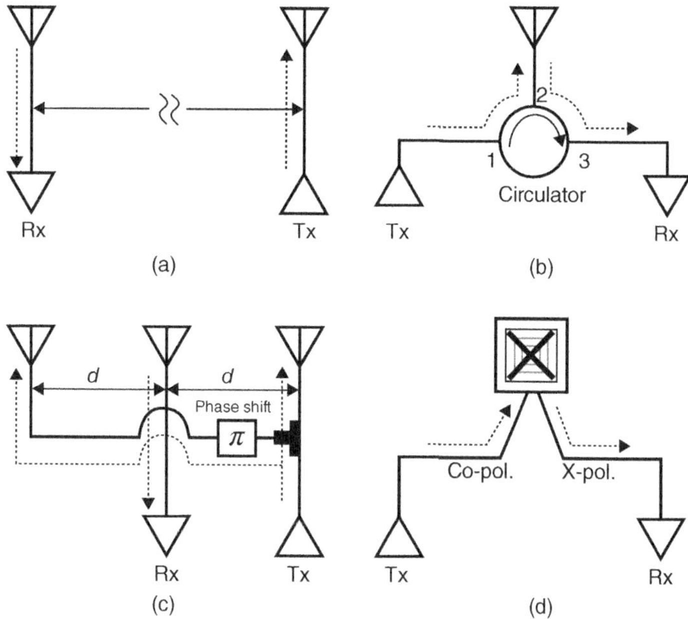

Figure 22.3 Antenna configurations of isolation methods: (a) antenna separation; (b) a circulator; (c) antenna cancellation; (d) a dual-polarized antenna

isolation is that it is able to reduce self-interference regardless the characteristics of the interference. It mitigates in a similar manner both linear and nonlinear self-interference and also reduces the thermal noise of a transmitter. Four widely used isolation methods are illustrated in Figure 22.3.

Antenna separation

is a straightforward method in which full-duplex systems use different antennas for transmitting and receiving. The distance between a transmit and receive antenna causes path loss, and the self-interference can be attenuated accordingly. The path loss can be calculated as

$$L = 10n \log_{10}(d) + C \tag{22.2}$$

where L is the path loss in decibels, n is the path loss exponent, d is the distance between the transmitter and the receiver, and C is a system-dependent constant. The SIC performance of antenna separation is limited by the physical size of the transceiver.

Circulator

A circulator is a passive 3-port device in which an RF signal entering any port is only transmitted to the next port in rotation [5]. For example, the signal that enters Port 1 is transmitted to Port 2, but is isolated to Port 3, as in Figure 22.3(b). Using this property, a circulator can be adopted as an isolator for a single-antenna full-duplex system, which uses an antenna for both the transmitter and receiver. When the signal is transmitted through Port 1 to Port 2 where the antenna is connected, it is not transmitted to the receiver at Port 3. A typical circulator provides about 15–20 dB of isolation.

Antenna cancellation

Antenna cancellation uses destructive interference to cancel the self-interference. When a signal and the π-phase rotated one is added, they cancel out each other. The simplest concept of antenna cancellation was proposed by Choi *et al.* [3], who placed two transmit antennas at distances d and $d + \lambda/2$ away from a receive antenna, where d is the distance between one transmit antenna and the receive antenna, and λ is the wavelength of the transmitted signal. By the placing them $\lambda/2$ apart, the phase of the signal is rotated by π. In this method, however, the received powers of the two signals are different due to asymmetric path loss. As a solution, a symmetric placement of antennas was proposed [19]. To obtain destructive interference, a π-phase shifter is employed on one of the transmit antennas, which are symmetrically placed at a distance d from the receive antenna as in Figure 22.3(c). After passing through the same distance, the two transmitted signals are assumed to be attenuated to the same degree. Furthermore, antenna cancellation is applied not only using two transmit antennas, but also two receive antennas. Combining two antenna cancellations in the transmitter and receiver, additional isolation can be obtained.

Other methods

Isolation is also obtained using directional antennas, dual-polarized antennas, or absorptive shielding [18]. By suppressing the line-of-sight component, these methods, compared to others, can provide a large amount of isolation. Directional isolation uses the directional antennas, which are arranged so that the gain of the transmit antenna is low in the direction of the receive antenna, and vice versa. Isolation by dual-polarized antennas is to transmit and receive in orthogonal polarization states. Using lossy materials to attenuate self-interference is also a simple way to create isolation.

The main limiting factor of isolation performance is environmental reflection [15, 18]. As can be inferred from its name, isolation operates passively, so it cannot suppress reflection and multipath components of self-interference, which vary relatively quickly. Isolation aims at the line-of-sight components of multi-antenna systems or the inter-port components of single-antenna systems. Other multipaths, therefore, become residual self-interference, and should be suppressed by analog active SIC and/or digital SIC.

22.3.2 Analog Active Self-interference Cancellation

The concept of active cancellation is to regenerate the self-interference signal and subtract it in the analog domain; in other words, before ADC. There are two main methods to recreate the signal: using tunable RF circuits and using an auxiliary transmit chain.

22.3.2.1 Tunable RF Circuits

Figure 22.4 shows a typical design of tunable RF circuits for a full-duplex system. The circuits obtain a small copy of the transmitted signal as input and generate a signal that replicates, as

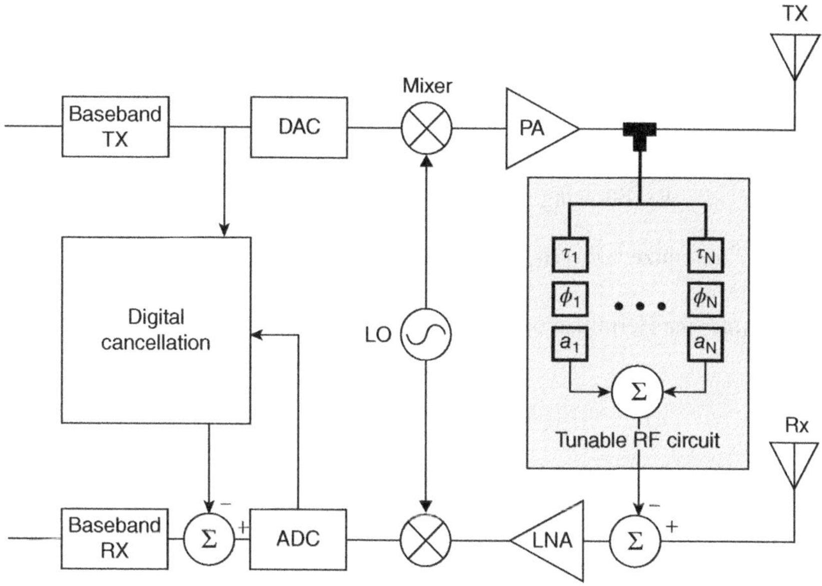

Figure 22.4 A full-duplex node with tunable RF circuits

far as possible, the self-interference. The circuits consist of multiple signal paths, where each path has attenuators (a_n), delays (τ_n) or phase shifters (ϕ_n). Depending on the full-duplex system design, various RF components are required. For example, Debaillie *et al.* proposed an analog SIC circuit, composed of a single path with a tunable attenuator and phase shifter [15]. Bharadia *et al.* proposed a 16-tap circuit with tunable attenuators and fixed delays [5].

To design an effective analog SIC circuit, the components of self-interference to be targeted should be decided. For example, the circuit design introduced by Bharadia *et al.* aims at the two primary components: the leakage of the circulator and the reflection due to impedance mismatch [5]. The average delays of the two signals are measured as 400 picoseconds and 1.4 nanoseconds. Thus, the first eight lines have fixed delays around 400 picoseconds, and the other eight lines around 1.4 nanoseconds. With these delays, the cancellation circuit can generate a signal similar to the target self-interference just by tuning the attenuators.

An optimal tuning of the cancellation circuit is obtained by convex optimization that minimizes the mean-square error between the frequency response of the circuit and the negative self-interference channel [13]. Assume that the circuit is composed of N paths, where each path contains a variable attenuator, phase shifter and fixed delay. Then, the circuit frequency response of the kth frequency component $H_{\text{cir}}[k]$ is expressed as

$$H_{\text{cir}}[k] = \sum_{n=1}^{N}(a_n e^{j\phi_n})e^{-j\Delta_\omega k\tau_n} \tag{22.3}$$

where a_n, ϕ_n and τ_n are attenuation, phase shift and delay of the nth path, respectively, and Δ_ω is the sampling interval over the bandwidth of interest. By defining

$\boldsymbol{X}_k = [e^{-j\Delta_\omega k\tau_1}, \cdots, e^{-j\Delta_\omega k\tau_N}]^T$ as a constant and $\boldsymbol{W} = [a_1 e^{j\phi_1}, \cdot, a_N e^{j\phi_N}]^T$ as a variable, $H_{\text{cir}}[k]$ is written as

$$H_{\text{cir}}[k] = \boldsymbol{X}_k^T \boldsymbol{W} \tag{22.4}$$

To minimize the average of the difference between the channel response of the kth subcarrier $H_{\text{chan}}[k]$ and $H_{\text{cir}}[k]$, the following minimization problem should be solved:

$$\underset{\boldsymbol{W}}{\text{minimize}} \quad \mathbb{E}\left[(H_{\text{chan}}[k] - \boldsymbol{X}_k^T \boldsymbol{W})(H_{\text{chan}}[k] - \boldsymbol{X}_k^T \boldsymbol{W})^H\right] \tag{22.5}$$

where $(\cdot)^H$ denotes the Hermitian operation. By convex optimization, the solution is obtained as

$$\boldsymbol{W} = (\mathbb{E}[(\boldsymbol{X}_k^T)^H \boldsymbol{X}_k^T])^{-1} \mathbb{E}[H_{\text{chan}}[k](\boldsymbol{X}_k^T)^H] \tag{22.6}$$

In practice, however, it is hard to assume that the frequency response of the circuit can be expressed in a polynomial form. Even though an RF circuit can be modeled theoretically, in practice RF components do not act ideally and are combined on several paths, resulting in frequency responses quite different from the ideal model.

To design a real-time full-duplex system, Bharadia *et al.* proposed a two-step tuning algorithm [5]: a closed-form coarse tuning followed by an additional algorithm. The frequency response of each path is premeasured and stored as a datasheet. The frequency response of the circuit is then calculated by summing individual frequency responses. Using the datasheet, the attenuators are tuned first, then more accurate tuning is done by random rounding.

22.3.2.2 Auxiliary Transmit Chain

An auxiliary transmit chain generates a copy of the self-interference and, as illustrated in Figure 22.5, subtracts it from the received signal. Since the subtraction, which is the actual canceling part, is performed in the analog domain, this is one of the analog SIC methods. It is also known, however, as "mixed-signal" cancellation because the baseband transmitted signal is processed in the digital domain and converted to an analog signal. Duarte and Sabharwal simply modeled self interference as [7]:

$$Y[k] = H_{\text{SI}}[k] X[k] \tag{22.7}$$

where $X[k]$, $Y[k]$ and $H_{\text{SI}}[k]$ are the transmitted signal, the received signal and the frequency response of kth subcarrier, respectively. Channel $\hat{H}_{\text{SI}}[k]$ is estimated in the digital domain by using $X[k]$ and $Y[k]$, and the cancellation signal $\hat{Y}[k]$ is generated by the auxiliary chain as:

$$\hat{Y}[k] = \frac{\hat{H}_{\text{SI}}[k]}{H_{\text{AUX}}} X[k] \tag{22.8}$$

where H_{AUX} is the pre-estimated frequency response of the wired path. Note that H_{AUX} is frequency-flat through all subcarriers. Since this model is linear, it cannot suppress the transmitter noise. Therefore, the overall SIC amount of this method is bounded.

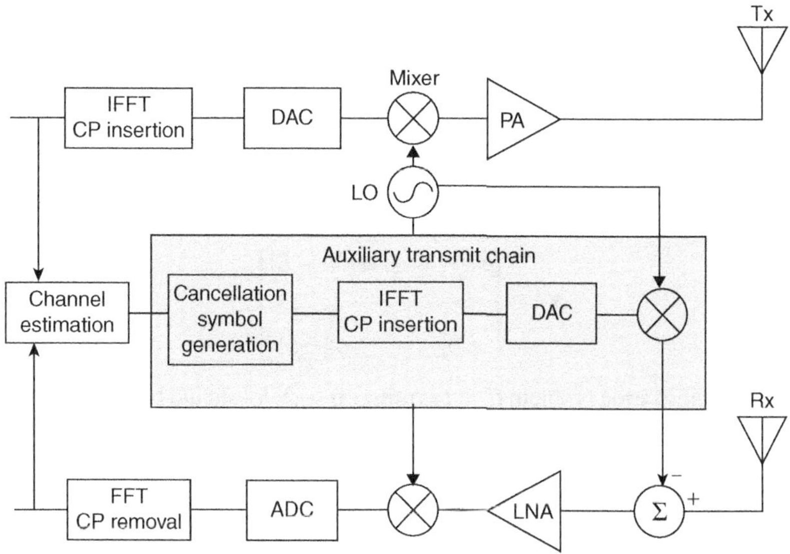

Figure 22.5 A full-duplex node with an auxiliary transmit chain

22.4 Digital Self-interference Cancellation

The goal of digital SIC is to cancel out residual self-interference after analog SIC, especially that originated from the non-line-of-sight reflections. All kinds of signal processing and calculations performed after ADC for the purpose of SIC are classified as digital SIC. Since they operate in the digital domain, the baseband equivalent models of self-interference should be determined before calculation. The models include linear and nonlinear self-interference models. Once the self-interference signal is modeled as a function of the transmitted signal, it can be estimated from the transmitted and received signals. The residual self-interference is then reconstructed as the output of the function, and subtracted from the received signal.

22.4.1 Linear Components

A simple and basic way to model a wireless link is with a linear model. It assumes that the transmitted signal that passes through a digital-to-analog converter (DAC) goes through a linear system, and returns to the ADC. The linear model can be written in a discrete-time form as:

$$y[n] = h[0]x[n] + h[1]x[n-1] + \cdots + h[L]x[n-L] + z[n] \tag{22.9}$$

where $x[n]$, $y[n]$ and $z[n]$ are, respectively, the transmitted and the received signals and the noise on time n, and $h[0], \cdots, h[L]$ are the linear coefficients, which represent the channel response. L is the maximum delay spread of the channel. To estimate the linear coefficients, a known preamble-type signal is usually transmitted. At the start of the packet, for example, WiFi uses a preamble of two known OFDM symbols. Let the samples $x_p[n]$ represent the preamble or known signal, and let the corresponding received samples be $y[n]$. Then the linear channel model can be written in matrix form as:

$$\boldsymbol{y} = \boldsymbol{X}\boldsymbol{h} + \boldsymbol{z} \tag{22.10}$$

where $(N+1)$ is the number of samples used for estimation, $\boldsymbol{y} = [y[N] \cdots y[0]]^T$, $\boldsymbol{h} = [h[L] \cdots h[0]]^T$ and $\boldsymbol{z} = [z[N] \cdots z[0]]^T$. An $(N+1) \times (L+1)$ Toeplitz matrix $\boldsymbol{X}$ is constructed as:

$$\boldsymbol{X} = \begin{bmatrix} x[N] & \cdots & x[N-L] \\ \vdots & \ddots & \vdots \\ x[0] & \cdots & x[-L] \end{bmatrix} \tag{22.11}$$

To obtain $\boldsymbol{h}$, a minimization problem $\boldsymbol{h} = \text{argmin}_{\boldsymbol{h}} \|\boldsymbol{y} - \boldsymbol{X}\boldsymbol{h}\|$ should be solved. For a column vector $\boldsymbol{v}$, $\|\boldsymbol{v}\| = \sqrt{\boldsymbol{v}^H \boldsymbol{v}}$. It is well known that the solution of this problem is obtained as $\hat{\boldsymbol{h}} = \boldsymbol{X}^{\dagger}\boldsymbol{y}$, where $(\cdot)^{\dagger}$ denotes the Moore–Penrose pseudo-inverse. Note that the pseudo-inverse of $\boldsymbol{X}$ can be pre-computed exploiting $\boldsymbol{X}$. This lightens the computational complexity of real-time digital SIC implementation.

The linear model can also be written in the frequency domain. For OFDM-based systems, the wireless system is modeled as:

$$Y[m] = H[m]X[m] + Z[m] \tag{22.12}$$

where $X[m]$, $Y[m]$, $Z[m]$ and $H[m]$ are the transmitted and received signals, the noise, and the frequency response of the wireless channel of the mth subcarrier, respectively. $H[m]$ is estimated by the least-squares method using pilots and interpolation as

$$\hat{H}[m] = \frac{Y[m]}{X[m]} \tag{22.13}$$

Compared to time-domain estimation, this model can adopt pilot-type rather than preamble-type reference symbols. In other words, the known data for channel estimation can be allocated to only some of the subcarriers, but done so more frequently. Therefore, the digital SIC performance can be improved in a time-selective fading environment.

22.4.2 Nonlinear Components

The linear system is suitable for modeling wireless channels or RF circuits for analog cancellation. However, certain nonlinear components that cannot be suppressed by linear SIC are essential in wireless communication systems. The two most common nonlinearities are perhaps power amplifier nonlinearity and I/Q imbalance.

Researchers have proposed several models of power amplifiers including the memoryless, the Volterra, the Wiener and the Hammerstein models [21]. The parallel Hammerstein model is widely used in wireless communications, and is described as

$$y[n] = \sum_{k=0}^{K} \sum_{\ell=0}^{L} b_{2k+1,\ell} |x[n-\ell]|^{2k} x[n-\ell] \tag{22.14}$$

where $b_{2k+1,\ell}$ are nonlinear coefficients, and $2K + 1$ is the highest order of nonlinearity. Similar to linear components SIC, the nonlinear coefficients have to be estimated. The methodology for estimating the nonlinear coefficients is the same as that for estimating the linear coefficients: construct the Toeplitz matrix with $|x[n]|^{2k}x[n]$ and solve the minimization problem. The number of the coefficients, however, is significantly increased. Compared to Eq. (22.11), the number of the nonlinear coefficients is $(K + 1)$ times more than that of the linear coefficients. This can be a sticking point when trying to obtain a reliable solution, as $(K+ 1)$ $(L+ 1)$ unknowns have to be obtained from $N + 1$ equations. Consider an LTE system with 20 MHz bandwidth and an extended cyclic prefix (CP). One OFDM symbol consists of 2560 digital samples (FFT size of 2048 and CP length of 512), allowing 2560 equations for coefficient estimation. The required number of nonlinear coefficients we need to calculate is $6 \times 512 = 3072$, by assuming up to an 11th-order-terms effect. The number of samples required is 512, which is the same as the length of the CP-the highest number expected for the multipath. It is obvious that this underdetermined system cannot give the correct solution.

One way to avoid to this problem is to decrease the number of nonlinear coefficients (unknowns) [5]. In practice, many of the coefficients of high-order terms are negligible compared to the noise level. Only the dominant coefficients need to be estimated, and these can be predetermined by measurement. For instance, only 224 coefficients are needed out of 1536 in Bharadia *et al.* [5].

The transmitter I/Q imbalances can also be modeled [11]. The output signal of the I/Q modulator model is

$$x^{\mathrm{IQ}}[n] = K_1 x[n] + K_2 x^H[n] \tag{22.15}$$

with $K_1 = (1 + ge^{j\phi})/2$ and $K_2 = (1 - ge^{j\phi})/2$, where g and ϕ are the gain and phase imbalance parameters of the transmitter, respectively. By inserting the I/Q modulator model into the parallel Hammerstein model, the combined model is written as

$$y[n] = \sum_{k=0}^{K} \sum_{p=0}^{2k+1} \sum_{\ell=0}^{L} b_{2k+1,\ell}^{(p)} x[n-\ell]^p x^H[n-\ell]^{2k+1-p} \tag{22.16}$$

where $b_{2k+1,\ell}^{(p)}$ are the coefficients for the basis function of $x^p(x^H)^{2k+1-p}$. The form of the equation is similar to the parallel Hammerstein model but has more coefficients to estimate.

22.4.3 *Auxiliary Receive Chain*

To cancel the nonlinear components of self-interference, use of an auxiliary receive chain can substitute for estimating the coefficients of the parallel Hammerstein model [14]. Figure 22.6 shows the block diagram of the full-duplex transceiver with an auxiliary receive chain. Unlike the method explained in Section 22.4.2, the nonlinear characteristic of a power amplifier need not be estimated; it just can be used through an auxiliary receive chain. The wireless channel is estimated as

$$Y_{\mathrm{WL}} = H_{\mathrm{WL}} X + Z_{\mathrm{WL}} \tag{22.17}$$

$$Y_{\mathrm{AUX}} = H_{\mathrm{AUX}} X + Z_{\mathrm{AUX}} \tag{22.18}$$

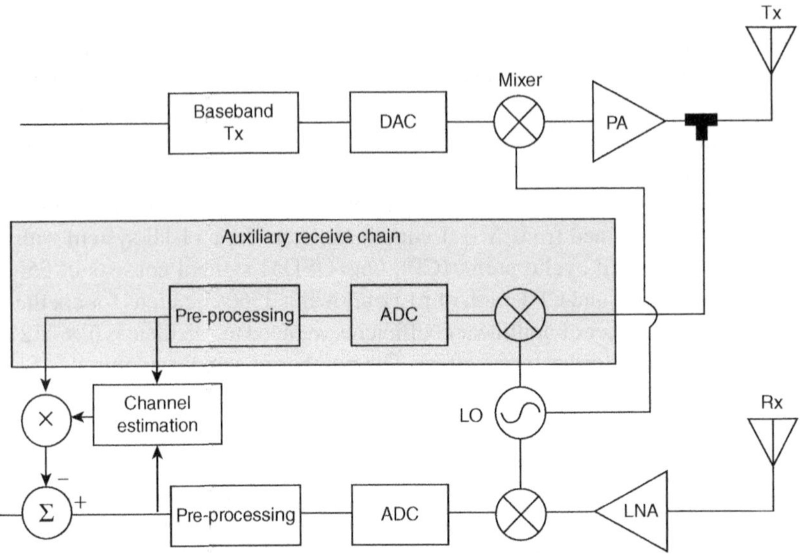

Figure 22.6 A full-duplex node with an auxiliary receive chain

where X, Y_{WL} and Y_{AUX} are the transmitted signal from the power amplifier, the received signal through the wireless channel, and the received signal through the wired channel from the power amplifier to the auxiliary receive chain, respectively. H_{WL} and H_{AUX} are the frequency responses of the wireless channel and the wired channel, respectively. The wireless self-interference channel, therefore, is obtained from the two received signals by

$$\hat{H}_{\mathrm{WL}} = \frac{Y_{\mathrm{WL}}}{Y_{\mathrm{AUX}}} H_{\mathrm{AUX}} \tag{22.19}$$

Note that the pre-estimated channel H_{AUX} is assumed to be fixed or slowly varying.

Using the estimated linear wireless channel, this method can mitigate transmitter impairments. Unlike other SIC methods, the distortion of the transmit signal by power amplifier nonlinearity or by I/Q imbalance is obtained directly by the auxiliary receive chain.

22.5 Prototyping Full-duplex Radios

Though it might seem simple to cancel self-interference in equation form, it is quite challenging to do it in real-world wireless systems: 100 dB of cancellation means suppressing interference signals that are billions of times stronger than the signal of interest. Conventional research on wireless communications has mostly depended on software simulations with simplified transceivers and channel models. This often ignores inevitable hardware impairments, such as power amplifier nonlinearity, I/Q imbalances and phase noise. These impairments can degrade the performance of wireless communications as well as the validity of the research. The importance of prototyping a real-time full-duplex system as a proof of concept cannot be overemphasized. Owing to the recent development of software-defined radio (SDR) platforms, prototyping has become a viable option for algorithm researchers. As an example for prototyping using SDR, this section lays out the details of Yonsei University's full-duplex system [22].

22.5.1 Hardware Architecture

The prototype was implemented using LabVIEW system design software and a state-of-the-art PXIe software-defined radio platform. It consists of four main components:

- a dual-polarized antenna
- a controller
- field-programmable gate array (FPGA) modules
- RF front ends.

The dual-polarized antenna suppresses self-interference by transmitting and receiving through orthogonal poles, as explained in Section 22.3.1. To guarantee a sufficient amount of isolation, a novel dual-polarized antenna, proposed by Oh *et al.* [23], was employed. Figure 22.7 shows the design of the antenna used. The isolation characteristic of a dual-polarized antenna can be expressed by its cross-polarization discrimination (XPD). The XPD is defined as the ratio of co-polarized average received power to cross-polarized average received power as

$$\mathrm{XPD} = \frac{E_{\mathrm{cross}}}{E_{\mathrm{co}}} \tag{22.20}$$

where E_{co} and E_{cross} are the average received power of the co-polarized and cross-polarized ports, respectively. Note that a dual-polarized antenna with high XPD in all directions could suppress self-interference properly. The one used by Chung *et al.* [22] has about 42 dB of XPD, which translates into that amount of analog passive SIC. Compared to the other analog passive SIC techniques described in Section 22.3.1, 42-dB isolation without huge antenna separation makes the dual-pole antenna a good choice for full-duplex radios.

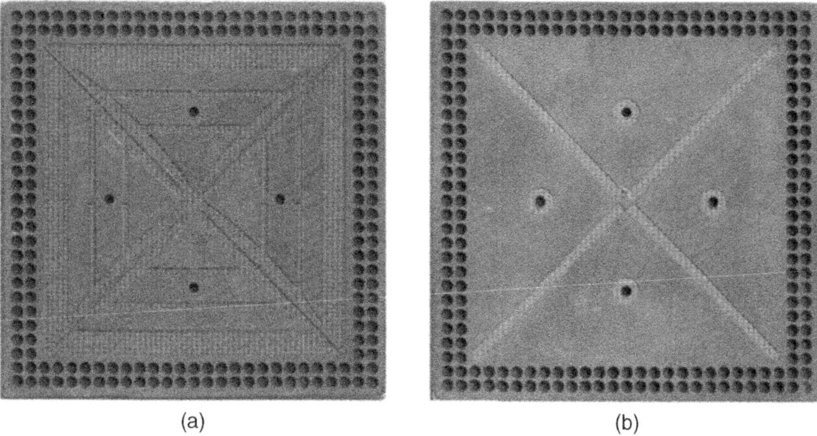

Figure 22.7 A dual-pole-based full-duplex antenna: (a) top view; (b) bottom view

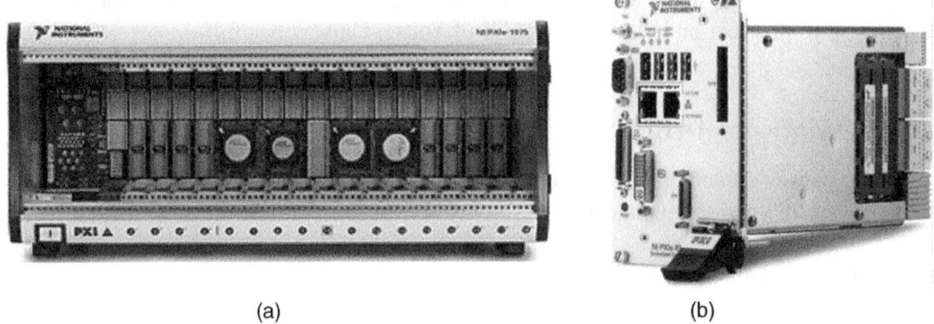

Figure 22.8 Hardware platform: (a)18-slot NI PXIe-1075 chassis and (b) NI PXIe-8133 controller

The PXIe-8133, shown in Figure 22.8(b), is a CPU-based controller that can run LabVIEW on either Windows 7 or a real-time OS [24]. It provides

- a user interface to input system parameters
- functionality to design and run a host-based algorithm and FPGA logic
- display of real-time data such as error rate, throughput and amount of digital SIC
- data bit sequences generation.

The RF transceiver module is a NI 5791R, which has dual 130 MS/s ADC with 14-bit accuracy, and a dual 130 MS/s DAC with 16-bit accuracy [25]. The transmit chain and receive chain share a common local oscillator, as shown in Figure 22.9. It provides 100 MHz of signal bandwidth for both transmitter and receiver.

The PXIe-7965R is a FlexRIO FPGA module with a Virtex-5 SX95T FPGA [26]. It interfaces with an NI FlexRIO adapter module (in this case NI 5791R) and a PXIe chassis. Figure 22.10 shows the PXIe-7965 and its architecture. Programmed with the NI LabVIEW

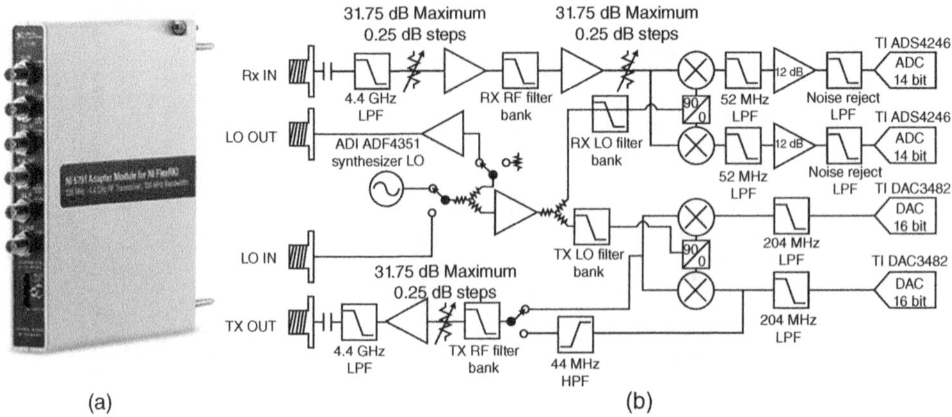

Figure 22.9 (a) NI 5791R RF transceiver adapter module and (b) block diagram

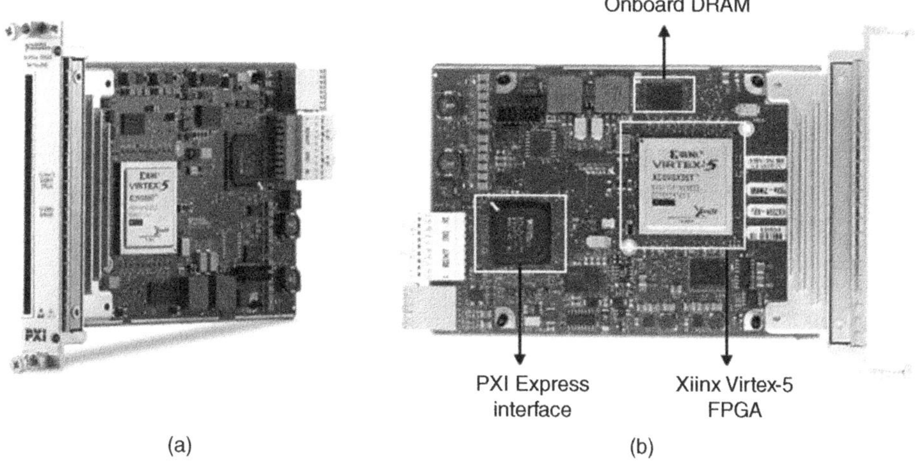

Figure 22.10 (a) NI PXIe-7965R FlexRIO FPGA module and (b) inner architecture

FPGA module [27], all the baseband signal processing, including transmitting and receiving the LTE signal and digital SIC, is executed on this module. All these modules sit in the NI PXIe-1075 chassis (see Figure 22.8(a)), which plays a role in data aggregation between FPGA processors and a real-time controller.

22.5.2 Implementing Full-duplex Radio using SDR

A real-time full-duplex system node is implemented with two NI 5791Rs and four PXIe 7965 FlexRIO FPGA modules, as depicted in Figure 22.11. This system is based on the standard LTE downlink [28]. Table 22.1 summarizes the system parameters used in the prototype.

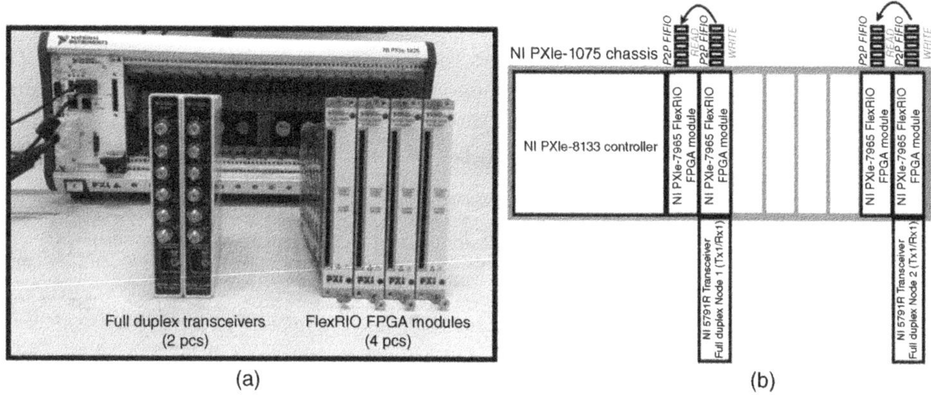

Figure 22.11 (a) SDR setup and (b) architecture in chassis

Table 22.1 Full-duplex LTE system parameters

Parameter	Notation	Value
Transmission bandwidth	W	20 MHz
Sampling rate	F_S	30.72 MS/s
Subcarrier spacing	Δf	15 kHz
FFT size	N_{FFT}	2048
CP length (extended)	N_{CP}	512
Modulation		4/16/64 QAM
Transmit power	P_T	5 dBm
Number of resource blocks	N_{RB}	100
Number of resource elements in a resource block	N_{RE}	7200
Slot duration	T_{slot}	0.5 ms

Figure 22.12 illustrates the prototype full-duplex system architecture. The transmission system follows the frame structure type 1 of the LTE downlink, where the frame duration is 10 ms, divided into 20 slots. Each slot consists of six OFDM symbols with a CP length (N_{CP}) of 512 (extended mode). The data bit is generated on the controller and modulated on the FPGA. The data symbols are interleaved with reference symbols stored in a look-up table and then zero-padded to form an array of 2048 samples, which are transformed into the time-domain waveform by an inverse FFT block with CP insertion. The following sections explain the details of implementation, mainly on the FPGA side.

22.5.2.1 Synchronization

Synchronization is one of the key blocks to implementation of real-time wireless communication systems, especially full-duplex systems. To facilitate timing synchronization, the LTE downlink standard specifies a primary synchronization signal (PSS). The PSS employs Zadoff–Chu (ZC) sequences, which have zero cyclic autocorrelation at all non-zero lags. Used as a PSS, a synchronization code can find the start of the symbol by searching for the greatest value of the correlation between the ideal sequence and the received sequence, as shown in Figure 22.13. Note that synchronization must be performed not only for the desired signal but also for the self-interference. Another property of ZC sequences is that ZC sequences with different root indices are orthogonal to each other. In other words, by using the ZC sequences with different root indices for uplink and downlink PSSs, synchronization of two links can be obtained simultaneously. For example, the PSS is modulated by a ZC sequence given as below:

$$P[k] = \begin{cases} e^{-j\frac{\pi}{N}uk(k+1)} & -31 \leq k \leq -1 \\ e^{-j\frac{\pi}{N}u(k+1)(k+2)} & 1 \leq k \leq 31 \end{cases} \tag{22.21}$$

where k is the subcarrier index, u is the root index and N is the sequence length ($N = 63$). Two different root indices relatively prime to N, $u_1 = 25$, $u_2 = 29$, are used.

The ZC sequence is allocated in a narrow band (1.4 MHz) around the DC-carrier in the last OFDM symbol of the first and eleventh slots of each frame. A low-pass filter is used prior to the synchronization block.

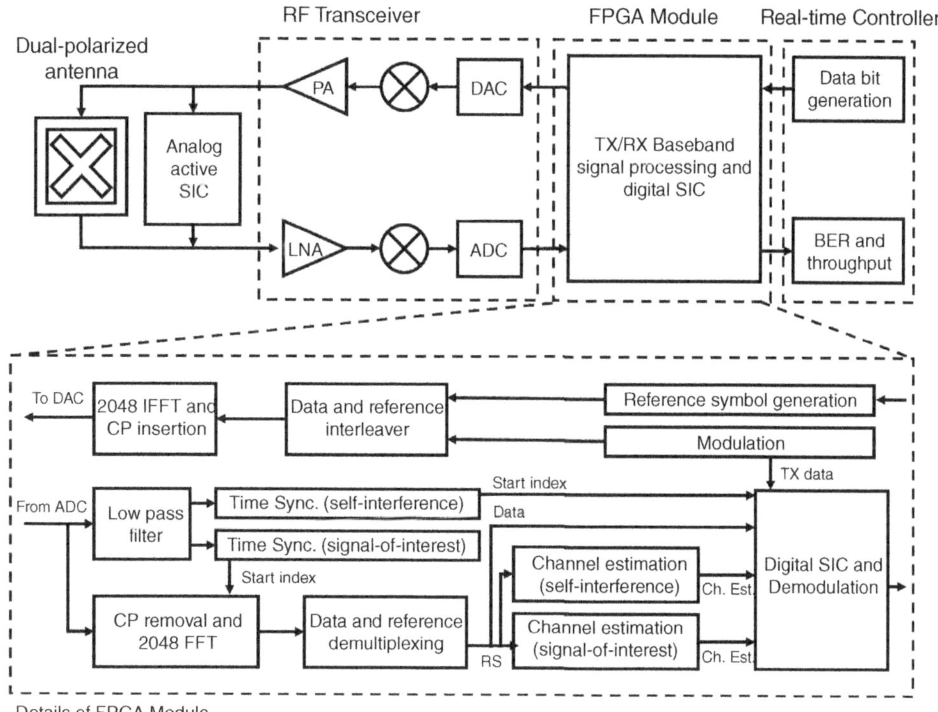

Figure 22.12 Full-duplex LTE system architecture

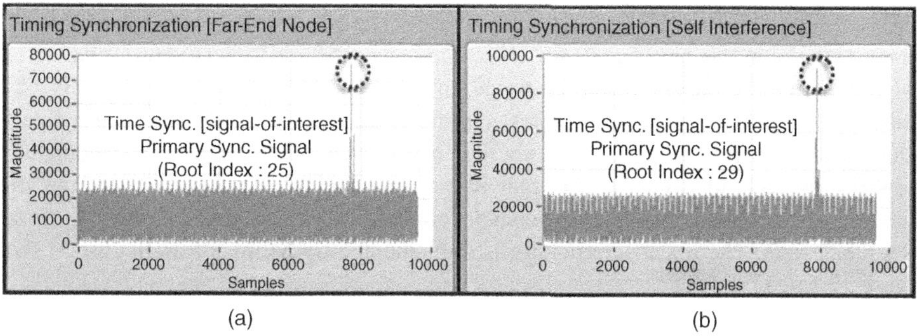

Figure 22.13 Peak detection for synchronization in LabVIEW front panel of full-duplex prototype: (a) signal-of-interest and (b) self-interference

22.5.2.2 Reference Symbols

The reference symbol (RS) is essential to estimating the channel for full-duplex radio. Unlike a conventional half-duplex LTE system, the uplink and downlink channels should be estimated simultaneously. Therefore, the reference symbols of the uplink signal are allocated where the

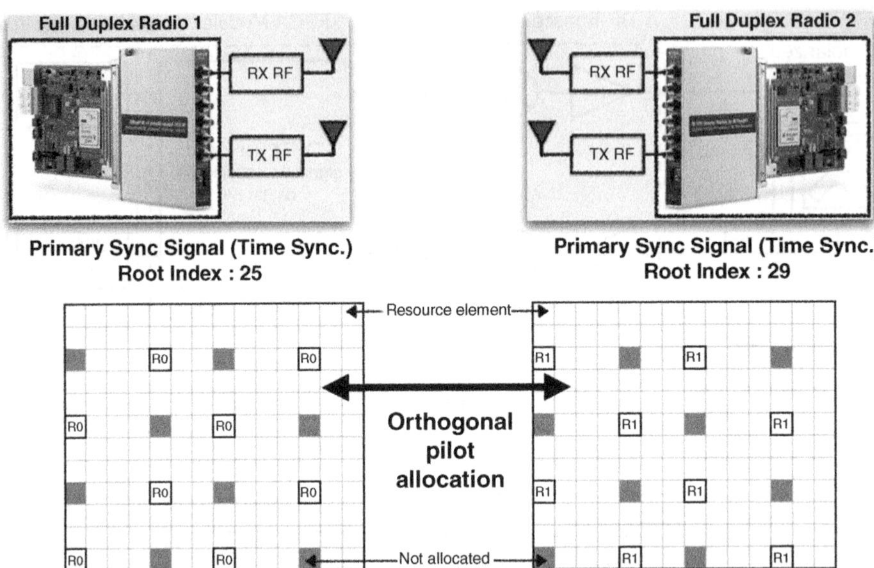

Figure 22.14 Reference symbol patterns for a full-duplex link

downlink signal is null, and vice versa; in other words reference symbols are allocated orthogonally. This RS pattern is similar to the two-antenna configuration of the cell-specific reference signal of the LTE standard, as illustrated in Figure 22.14. Note that the overhead of the reference symbol is double that of the half-duplex system.

22.5.2.3 Channel Estimation

Channel estimation is performed for two channels: the self-interference channel and the signal-of-interest channel. As illustrated in Figure 22.15(a) and (b), the RSs of the self-interference and the signal-of-interest can be extracted independently in the data and RS demultiplexing block. A channel coefficient of each RS subcarrier is calculated by a least-squares method using the original RSs stored in block memory. Then, the linear interpolator of each block estimates the channel coefficients of data subcarriers using those of the RS subcarriers. The linear interpolator is implemented by Xilinx's FIR IP core as shown in Figure 22.15(c).

22.5.2.4 Digital Self-interference Cancellation

A digital SIC block rebuilds the self-interference and subtracts the rebuilt self-interference from the received signal. To rebuild the self-interference, the digital SIC block receives the transmit data from the modulation block and the self-interference channel from the channel estimation block. When the rebuilt signal is subtracted from the received signal it is very important to align the rebuilt signal with the received signal in the resolution of a sample timing ($T_S = 1/F_S$). Accordingly, a counter is included in the digital SIC block. The counter starts counting when the starting index of the self-interference signal arrives at the digital SIC block

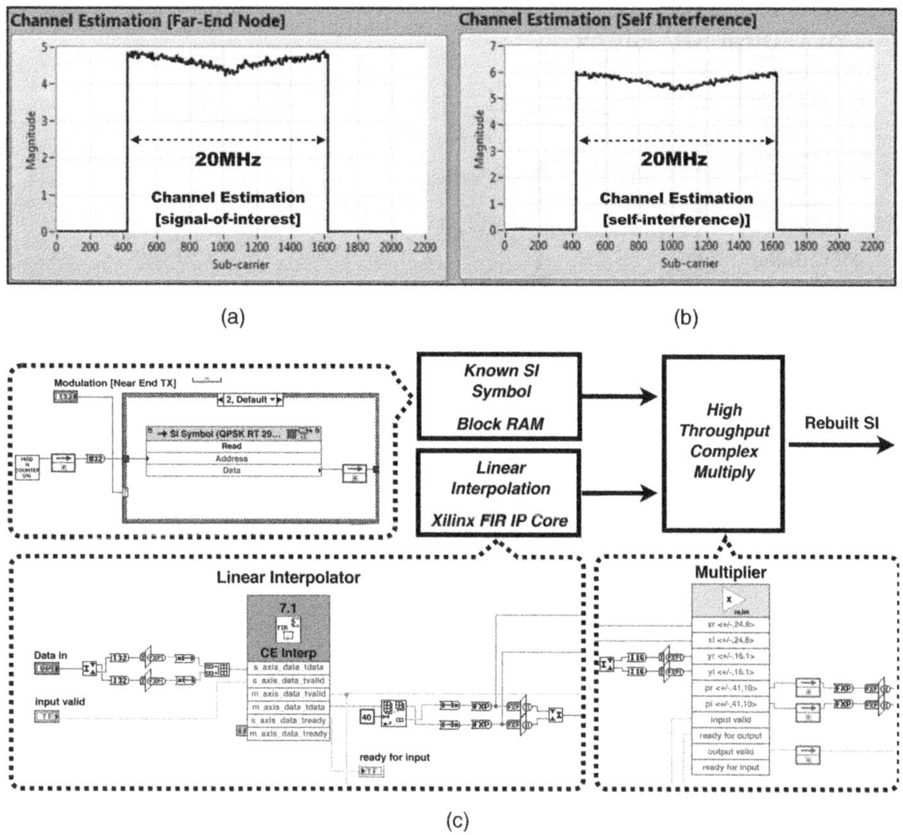

Figure 22.15 Channel estimation in LabVIEW front panel of full-duplex prototype: (a) signal-of-interest; (b) self-interference; (c) FPGA implementation

from the synchronization block. With the counted index, the transmitted symbol is multiplied by the corresponding channel coefficient and is subtracted from the received signal. After digital SIC, a demodulation block obtains the desired symbols of the signal-of-interest by using a zero-forcing channel equalizer, and gives the received data sequence to the real-time controller. Figure 22.16 shows the digital SIC implementation on the FPGA.

22.5.2.5 Prototype Validation: Cancellation and Throughput Measurements

The isolation amount of the dual-polarized antenna can be measured by a network analyzer. By connecting Port 1 to the transmitter and Port 2 to the receiver, the S_{21} parameter represents the amount of isolation. The antenna used by Chung *et al.* [22] achieves 42 dB of isolation in an indoor environment. In the digital domain, the error vector magnitude (EVM) for self-interference is calculated instantaneously, and the average level of digital SIC is measured. When the EVM is measured, analog SIC should be operating; in this case, the dual-polarized antenna should be connected. Chung *et al.* achieved 43 dB of digital SIC [22], which means the total SIC amount was 85 dB.

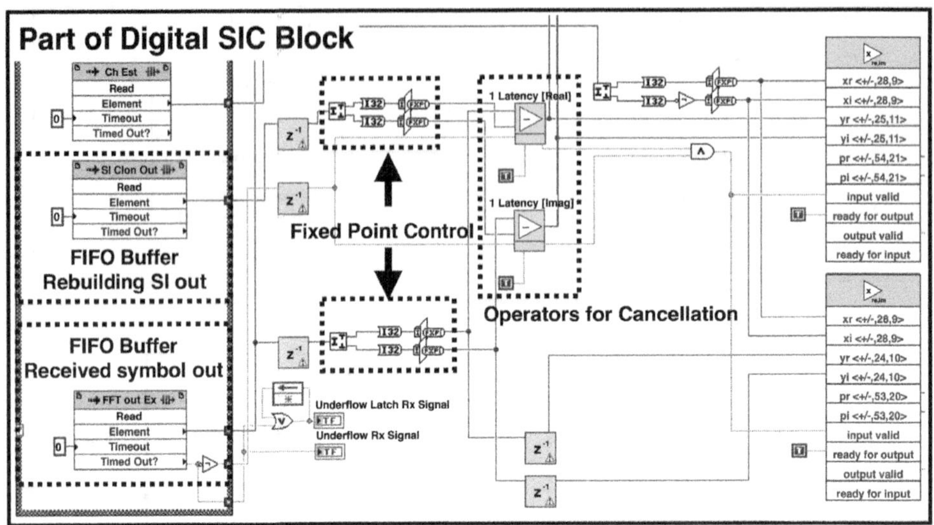

Figure 22.16 Digital SIC implementation

Table 22.2 Throughput of half/full-duplex prototype systems

	QPSK	16QAM	64QAM
Half duplex	21.57 Mbps	43.14 Mbps	64.71 Mbps
Full duplex	41.01 Mbps	82.01 Mbps	122.6 Mbps
Throughput gain	1.9 ×	1.9 ×	1.89 ×

To assess the throughput improvement, an LTE FDD prototype with the same system specifications and hardware architecture as used in half-duplex mode, was implemented. The throughput of the system was calculated as

$$T = \alpha_D \frac{N_{RE} \times N_{RB}}{T_{\text{slot}}} \times b \times \beta \times (1 - P_b) \tag{22.22}$$

where α_D is a duplex factor whose value equals 1 and 2 for half-duplex and full-duplex modes, respectively, b is a modulation order per resource element, β is an overhead factor and P_b is the bit error rate. The overhead factor (β) includes an RS overhead, a PSS overhead and a CP overhead. P_b is measured in the prototype system in real-time. The results are summarized in Table 22.2.

22.6 Overall Performance Evaluation

Good analog cancellation performance does not, however, always guarantee good overall cancellation in the full-duplex system. Passive SIC, in general, causes a decrease in the coherence bandwidth of the residual self-interference channel making it frequency-selective [18]. Active analog SIC also targets dominant components such as the line-of-sight components or circulator leakage. Since they focus on canceling the direct path between the transmitter and receiver, the non-line-of-sight components become dominant. In most cases, to estimate the frequency-selective channel needs greater effort, for example more pilot overhead. Otherwise, a high channel estimation error can be caused, which may degrade digital cancellation. Furthermore, as the amount of analog cancellation increases, residual self-interference decreases. This causes a low SNR for estimating self-interference in the digital domain [9].

Princeton University proposed a system called MIDU, which adopts antenna cancellation in the transmitter and receiver simultaneously [19]. After this isolation, simple linear digital cancellation reduces the residual self-interference. The total amount of SIC is 75 dB; the system was designed, however, as a single-subcarrier narrowband system.

Since both 4G wireless communications and the IEEE 802.11 (WiFi) family are based on OFDM, most studies adopt OFDM for transmit and receive signals. A prototype from Rice University employed separated antennas and is based on analog cancellation, as explained in Section 22.3.2.2 [7–9]. After analog cancellation, the residual self-interference is reduced by linear cancellation, as described in Section 22.4.1.

Stanford University proposed a full-duplex system that uses a single antenna for a SISO link [5]. Since the transmitter and receiver share the antenna, a circulator is employed for isolation. The system achieves 110 dB of overall cancellation, 15 dB of isolation, 45 dB of suppression by tunable circuit and 50 dB of digital nonlinear cancellation. The amount of cancellation is verified by a demonstration of a complete full-duplex communication link, which uses the full WiFi PHY.

Yonsei University applied rapid prototyping techniques to implement an LTE-based full-duplex system [22]. The system adopts a dual-polarized antenna for isolation, and allocates pilots for signal-of-interest and self-interference orthogonally, following the LTE standard. A dual-polarized antenna provides about 40 dB of isolation, and linear digital cancellation reduces the residual self-interference by about 43 dB.

22.7 Conclusion

Full-duplex is a key potential technology for increasing spectrum efficiency for the next generation of wireless communications. A major obstacle to full-duplex radios is the issue of strong self-interference, which hinders the receiving of a desired signal. In this chapter, we have investigated the characteristic of self-interference, which depends on hardware structure, and explored various techniques for self-interference suppression and cancellation. SIC schemes should be determined by inclusively considering target applications and system specifications. Also, we introduced one case of full-duplex prototypes implemented by software-defined radio. By implementing a practical system, the chapter has examined not only SIC techniques but also synchronization and reference symbol allocation for full-duplex systems. Several representative full-duplex systems have been explained along with their SIC performances.

References

[1] Cisco (2015), Cisco visual networking index: Global mobile data traffic forecast update, 2014–2019.

[2] 3GPP (2009) TS 36.104: Evolved universal terrestrial radio aAccess (E-UTRA); Base station (UE) radio transmission and reception (Release 8).

[3] Choi, J.I., Jain, M., Srinivasan, K., Levis, P., and Katti, S. (2010) Achieving single channel, full duplex wireless communication, in *Proc. of ACM Mobicom*, pp. 1–12.
[4] Jain, M., Choi, J.I., Kim, T.M., Bharadia, D., Seth, S., Srinivasan, K., Levis, P., Katti, S., and Sinha, P. (2011) Practical, real-time, full duplex wireless, in *Proc. of ACM Mobicom*, pp. 301–312.
[5] Bharadia, D., McMilin, E., and Katti, S. (2013) Full duplex radios, in *Proc. of ACM SIGCOMM*, pp. 375–386.
[6] Bharadia, D. and Katti, S. (2014) Full duplex MIMO radios, in *Proc. USENIX Symp. on Net. Sys. Design and Implementation*, pp. 359–372.
[7] Duarte, M. and Sabharwal, A. (2010) Full-duplex wireless communications using off-the-shelf radios: feasibility and first results, in *Proc. Asilomar Conf. on Sign., Syst. and Computers*, pp. 1558–1562.
[8] Duarte, M., Dick, C., and Sabharwal, A. (2012) Experiment-driven characterization of full-duplex wireless systems. *IEEE Trans. Wireless Commun.*, **11** (12), 4296–4307.
[9] Duarte, M., Sabharwal, A., Aggarwal, V., Jana, R., Ramakrishnan, K.K., Rice, C.W., and Shankaranarayanan, N.K. (2014) Design and characterization of a full-duplex multiantenna system for WiFi networks. *IEEE Trans. Vehic. Tech.*, **63** (3), 1160–1177.
[10] Anttila, L., Korpi, D., Syrjälä, V., and Valkama, M. (2013) Cancellation of power amplifier induced nonlinear self-interference in full-duplex transceivers, in *Proc. Asilomar Conf. on Sign., Syst. and Computers*, pp. 1193–1198.
[11] Anttila, L., Korpi, D., Antonio-Rodríguez, E., Wichman, R., and Valkama, M. (2014) Modeling and efficient cancellation of nonlinear self-interference in MIMO full-duplex transceivers, in *Proc. IEEE Glob. Telecom. Conf.*, pp. 777–783.
[12] Korpi, D., Venkatasubramanian, S., Riihonen, T., Anttila, L., Otewa, S., Icheln, C., Haneda, K., Tretyakov, S., Valkama, M., and Wichman, R. (2013) Advanced self-interference cancellation and multiantenna techniques for full-duplex radios, in *Proc. Asilomar Conf. on Sign., Syst. and Computers*, pp. 3–8.
[13] McMichael, J.G. and Kolodziej, K.E. (2012) Optimal tuning of analog self-interference cancellers for full-duplex wireless communication, in *Proc. Allerton Conf. on Comm. Control and Comp.*, pp. 246–251.
[14] Ahmed, E. and Eltawil, A.M. (2015) All-digital self-interference cancellation technique for full-duplex systems. *IEEE Trans. Wireless Commun.*, **14** (7), 3519–3532.
[15] Debaillie, B., van den Broek, D.J., Lavín, C., van Liempd, B., Klumperink, E.A.M., Palacios, C., Craninckx, J., Nauta, B., and Pärssinen, A. (2014) Analog/RF solutions enabling compact full-duplex radios. *IEEE J. Select. Areas Commun.*, **32** (9), 1662–1673.
[16] Sahai, A., Patel, G., Dick, C., and Sabharwal, A. (2013) On the impact of phase noise on active cancellation in wireless full-duplex. *IEEE Trans. Vehic. Tech.*, **62** (9), 4494–4510.
[17] Syrjälä, V., Valkama, M., Anttila, L., Riihonen, T., and Kropi, D. (2014) Passive self-interference suppression for full-duplex infrastructure nodes. *IEEE Trans. Wireless Commun.*, **13** (6), 2977–2990.
[18] Everett, E., Sahai, A., and Sabharwal, A. (2014) Passive self-interference suppression for full-duplex infrastructure nodes. *IEEE Trans. Wireless Commun.*, **13** (2), 680–694.
[19] Aryafar, E., Khojastepour, M.A., Sundaresan, K., Rangarajan, S., and Chiang, M. (2012) MIDU: enabling MIMO full duplex, in *Proc. of ACM Mobicom*, pp. 257–268.
[20] Ghosh, A., Zhang, J., Andrews, J.G., and Muhamed, R. (2010) *Fundamentals of LTE*, Prentice Hall.
[21] Cripps, S.C. (1999) *RF Power Amplifiers for Wireless Communications*, Artech House.
[22] Chung, M., Sim, M.S., Kim, J., Kim, D.K., and Chae, C.-B. (2015) Prototyping real-time full duplex radios. *IEEE Commun. Mag.*, **53** (9), 56–63.
[23] Oh, T., Lim, Y.G., Chae, C.-B., and Lee, Y. (2015) Dual-polarization slot antenna with high cross-polarization discrimination for indoor small-cell mimo systems. *IEEE Antennas Wireless Propagat. Lett.*, **14**, 374–377.
[24] National Instruments (2012), NI PXIe-8133 User manual. URL http://www.ni.com/pdf/manuals/372870d.pdf.
[25] National Instruments (2013), NI 5791R User manual and specifications. URL http://www.ni.com/pdf/manuals/373845c.pdf.
[26] National Instruments (2014), NI FlexRIO FPGA modules data sheet. URL http://www.ni.com/datasheet/pdf/en/ds-366.
[27] National Instruments (2015), LabVIEW FPGA module. URL http://www.ni.com/labview/fpga/.
[28] 3GPP (2011) TS 36.211: Evolved universal terrestrial radio sccess (E-UTRA); physical channels and modulation.

23

5G Standard Development: Technology and Roadmap

Juho Lee and Yongjun Kwak

23.1 Introduction

The wireless cellular network has been one of the most successful communications technologies of the last three decades. The advent of smartphones and tablets over the past several years has resulted in an explosive growth of data traffic. With the proliferation of more smart terminals communicating with servers and each other via broadband wireless networks, numerous new applications have also emerged to take advantage of wireless connectivity.

As 4G LTE-Advanced [1, 2] networks mature and become a global commercial success, the research community is now increasingly looking at future 5G technologies, both in standardization bodies such as 3GPP and in research projects such as the EU's FP7 METIS [3]. ITU-R has recently finalized their work on the vision for 5G systems, which includes support for an explosive growth of data traffic, support for a massive number of machine-type communication (MTC) devices, and support for ultra-reliable and low-latency communications [4]. While today's commercial 4G LTE-Advanced networks are mostly deployed in legacy cellular bands from 600 MHz to 3.5 GHz, recent technological advances will allow 5G to utilize any spectrum opportunities below 100 GHz, including existing cellular bands, new bands below 6 GHz, and

Signal Processing for 5G: Algorithms and Implementations, First Edition. Edited by Fa-Long Luo and Charlie Zhang.

new bands above 6 GHz including the so-called mmWave bands. There are coordinated efforts across the world to identify these new spectrum opportunities, with decisions expected for new frequency bands below 6 GHz at the World Radio-communication Conference (WRC)-2015, and for new frequency bands above 6 GHz at WRC-2019, respectively.

From the 5G technology roadmap perspective, we expect a dual-track approach to take place over the next few years in 3GPP. The first track is commonly known as the evolution track, where we expect that the evolution of LTE-Advanced will continue in Rel-13/14 and beyond in a backward-compatible manner with the goal of improving system performance in the bands below 6 GHz. It is also our expectation that at least a part of the 5G requirements could be met by the continued evolution of LTE-Advanced. For example, latency reduction with grant-less uplink access and shortened transmission time interval (TTI) could reduce over-the-air latency to less than 1 ms. The second track is commonly known as new radio access technology (RAT) track, which is not limited by backward-compatibility requirements and can integrate breakthrough technologies to achieve the best possible performance. The new RAT system should meet all 5G requirements as it would eventually need to replace the previous generation systems in the future. The new RAT track is also expected to have a scalable design that can seamlessly support communications in both above and below 6 GHz bands.

The rest of this chapter is organized into the following three sections, focusing on the technologies for the air interface of radio access networks. Section 23.2 is devoted to the standards roadmap from 4G to 5G. In Section 23.3, we will provide an overview of major enabling technologies and a more detailed roadmap of the 5G standard development and its deployment. Section 23.4 is the final section, which presents the summary of this chapter.

23.2 Standards Roadmap from 4G to 5G

Since the publication of the Rel-99 standards supporting wideband code division multiple access (WCDMA) – a representative 3G technology – 3GPP has been playing an important role in evolving cellular communication standards to 4G, namely LTE and LTE-Advanced. 3GPP is a partnership project between regional standardization bodies, or organizational partners (OPs). 3GPP was established in 1998 and has seven OPs as of 2015: ARIB and TTC from Japan, ATIS from USA, CCSA from China, ETSI from Europe, TSDSI from India, and TTA from Korea. After the success of 3G technologies, 3GPP introduced LTE as Rel-8 in 2009, and LTE-Advanced as Rel-10 in 2011, the latter declared by ITU-R "IMT-Advanced technology" and often called 4G. There has been continuing further evolution toward 5G. The 3GPP organization and its overall roadmap from LTE in Rel-8 to LTE-Advanced in Rel-13 are shown in Figures 23.1 and 23.2, respectively. The project coordination group coordinates the projects performed in 3GPP. Each technical specification group (TSG) decides whether specifications for a technology will be developed, typically taking into account the outcome of the related feasibility study in its working groups (WGs). WGs develop technical specifications, which are then formally approved by their TSG.

LTE in Rel-8 was the first standard in 3GPP that utilized frequency division multiplexing: orthogonal frequency division multiplexing (OFDM) in the downlink and single carrier frequency division multiple access (SC-FDMA) in the uplink as illustrated in Figure 23.3. SC-FDMA is a variant of OFDM, where discrete Fourier transform (DFT) processing is applied to the input signal before inverse DFT (IDFT) so that the output of IDFT mimics a single carrier signal. The peak-to-average power ratio of SC-FDMA is smaller than that of

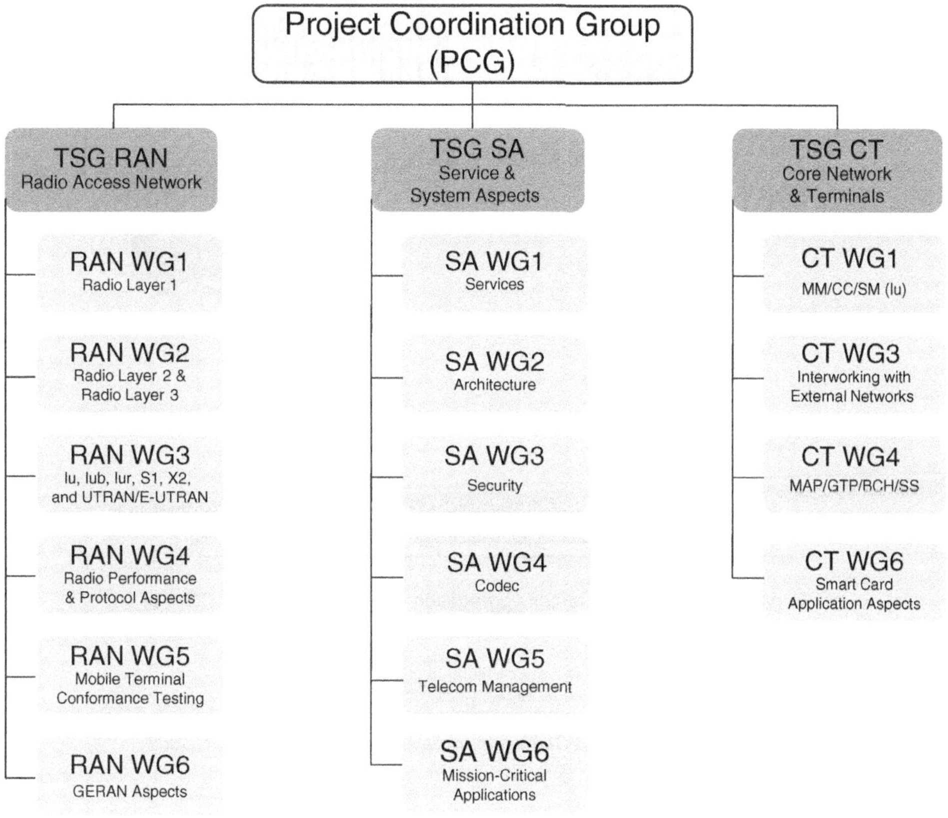

Figure 23.1 3GPP organization

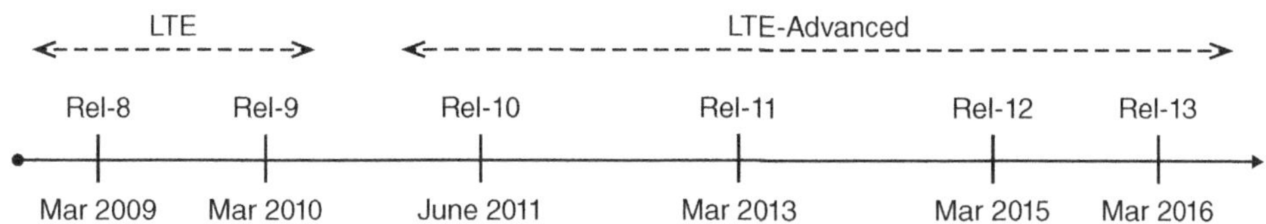

Figure 23.2 Overall roadmap from LTE in Rel-8 to LTE-Advanced in Rel-13

an OFDM signal and hence SC-FDMA helps increase the uplink coverage. The maximum bandwidth is 20 MHz, providing a 300 Mbps peak rate on the downlink with 4×4 MIMO and a 75 Mbps peak rate on the uplink. It is worth noting that the OFDM waveform on the downlink does not require a complicated equalization at the user equipment[1] (UE) receiver and hence helps reduce UE receiver complexity. This property motivated the specification of 4×4 MIMO on downlink from the very first release of LTE. LTE Rel-9 included a few

[1] 3GPP terminology for a mobile device.

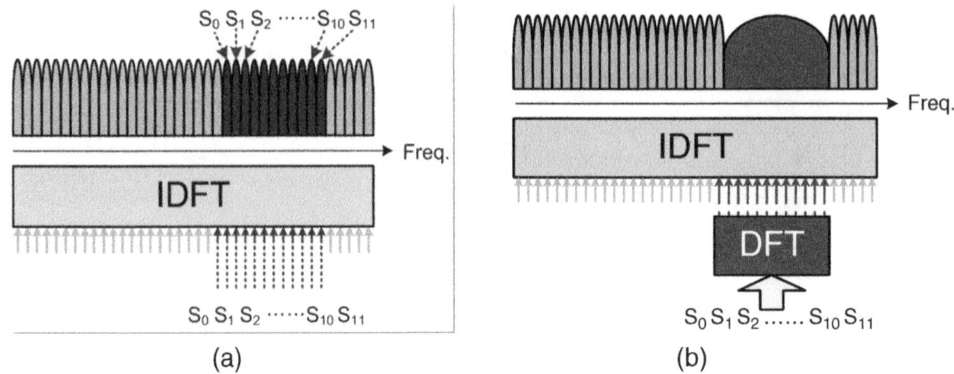

Figure 23.3 Signal generation for (a) OFDM on downlink and (b) SC-FDMA on uplink

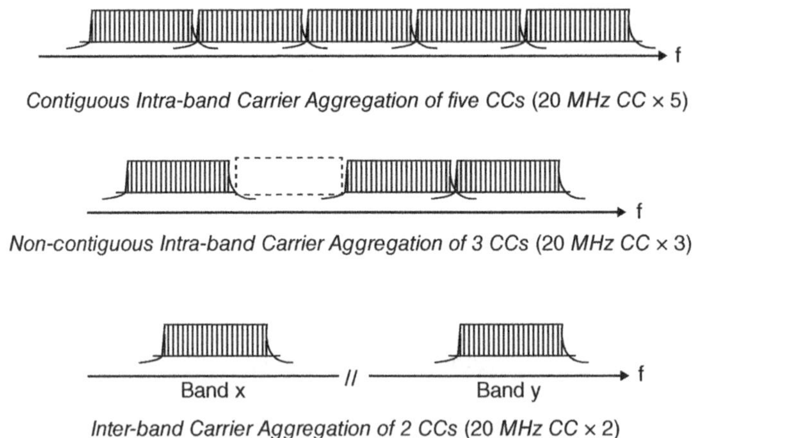

Figure 23.4 A few examples of carrier aggregation with up to five component carriers (CCs)

additional features, for example emergency call, that was required to support voice calls over LTE.

LTE-Advanced standard in Rel-10 was developed to meet not only IMT-Advanced requirements but also commercial requests for accommodating increased data traffic. As a natural approach for increasing the peak rate, aggregation of up to five component carriers was specified, resulting in the use of 100 MHz at maximum. A few examples of carrier aggregation are shown in Figure 23.4. It should be noted that carrier aggregation has been one of the most successful features of LTE-Advanced because it enables mobile network operators to provide higher peak rates and to improve the operational efficiency of radio access networks by utilizing scattered frequency resources. In addition, MIMO was further enhanced through the introduction of 8 × 8 MIMO and 4 × 4 MIMO on the downlink and uplink, respectively. A combination of carrier aggregation and the larger order of MIMO provided 3 Gbps and 1.5 Gbps peak rates on the downlink and uplink, respectively.

Support of MIMO with eight transmit antennas at the enhanced Node B[2] (eNB) necessitated introduction of a UE-specific demodulation reference signal (DM-RS) together with a channel status information reference signal (CSI-RS). This is because too much overhead results from increasing the number of common reference signal (CRS) ports, which are transmitted in a continuous manner (in every subframe). The DM-RS is transmitted, and introduces overhead, only for the UEs to which downlink transmissions are scheduled, and the CSI-RS overhead can be minimized by increasing its transmission period as shown in Figure 23.5, where RE, RB, and PRB denote resource element, resource block, and physical resource block, respectively.

While the carrier aggregation and MIMO are mainly aimed at increasing the peak rate, support of heterogeneous networks consisting of macro and pico cells on the same frequency layer by relying on time-domain interference coordination is a remarkable step that significantly improves the spectrum-utilization efficiency.

Following the improvements in peak data rates and system capacity provided by MIMO in LTE and LTE-Advanced, it was also shown to be possible to provide further performance improvement by having coordinated transmission and reception between multiple points [5], where a point can be treated as a set of geographically co-located transmit antennas, with the exception that sectors of the same site are considered different points. The Rel-11 standard introduced specification support for coordinated transmission in the downlink and coordinated reception in the uplink, which is commonly denoted coordinated multipoint (CoMP). The assumed deployment scenarios for CoMP illustrated in Figure 23.6 include homogeneous configurations, where the points are different cells, as well as heterogeneous configurations, where a set of low power points – for example remote radio heads (RRHs) or pico cells – are located in the geographical area served by a macro cell. For coordinated transmission in the downlink, the signals transmitted from multiple transmission points are coordinated to improve the received strength of the desired signal at the UE or to reduce the co-channel interference. The major purpose of coordinated reception in the uplink is to help ensure that the uplink signal from the UE is reliably received by the network while limiting uplink interference, taking into account the existence of multiple reception points.

While Rel-12 continued the evolution towards improving the peak rate by aggregating the available frequency resource and improving spectral efficiency, it also started specification of features that were required for support of new services such as machine-type communications and device-to-device communications (D2D) that had not been a major focus of previous releases. The features of the first category include the following:

- **Small cell enhancement:** Dual connectivity was introduced to enable a UE to connect to both the macro cell layer, which provides mobility support on lower frequencies, say 700 MHz, and the small cell layer, which provides a fat data pipe on higher frequencies, say 3.5 GHz.
- **Time division duplex (TDD)-frequency division duplex (FDD) joint operations:** Joint operations of both TDD and FDD carriers is important and useful for mobile network operators owning both TDD and FDD carriers. In order to enable this, the carrier aggregation between TDD and FDD carriers was specified in Rel-12.
- **Enhanced interference management and traffic adaptation for TDD:** This feature introduced specified support of dynamic adjustment of uplink and downlink resources in TDD

[2] 3GPP terminology for a base station.

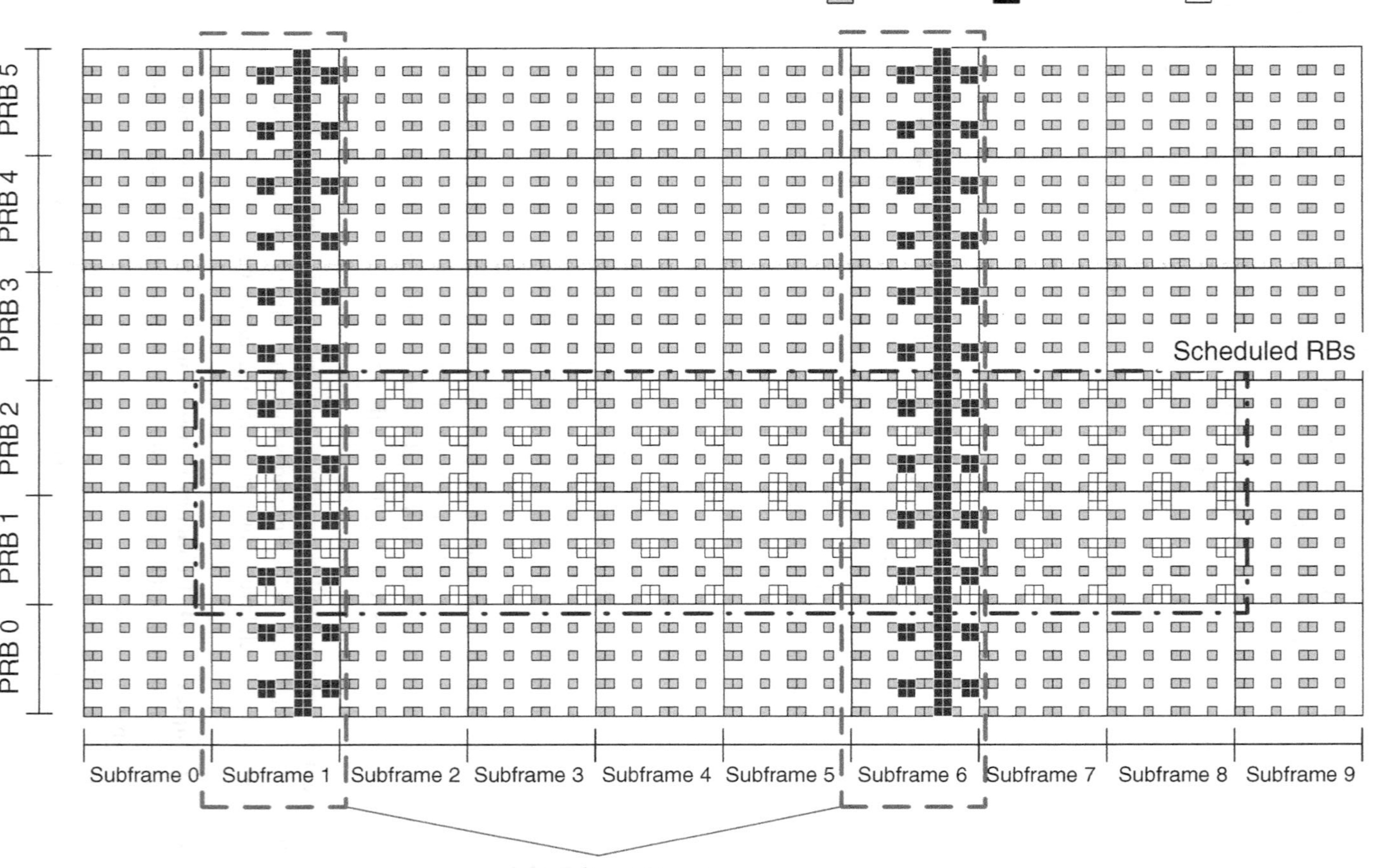

Figure 23.5 Downlink MIMO support in LTE-Advanced with DM-RS and CSI-RS

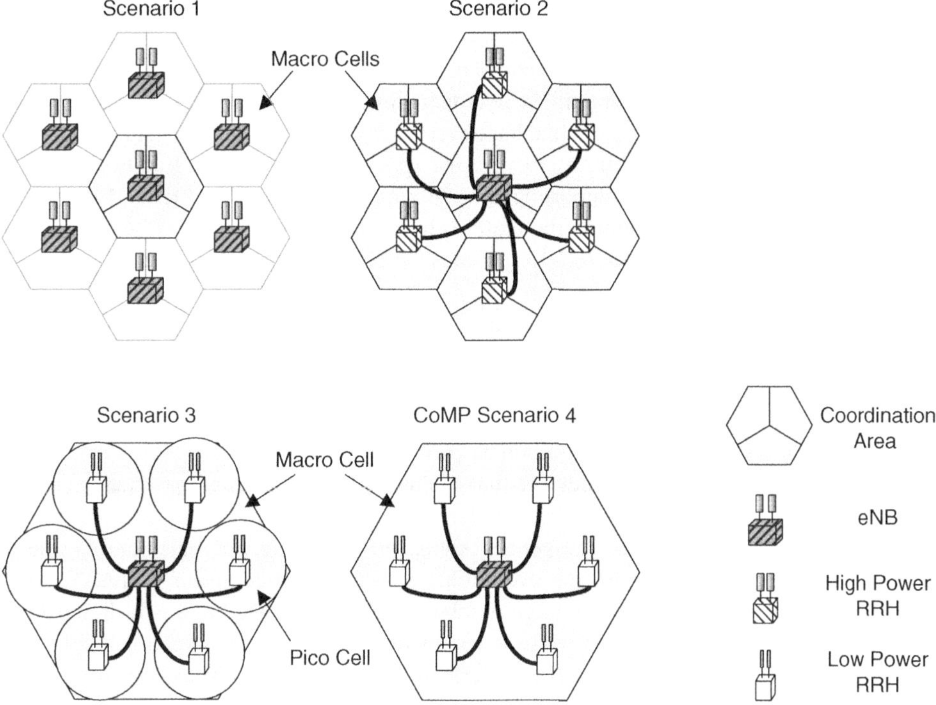

Figure 23.6 CoMP scenarios considered for the specification support of CoMP in LTE Rel-11

systems. While the dynamic adaptation of uplink and downlink resources according to the traffic situation has been understood as one of the biggest benefits of TDD systems, it was not really supported by standards in traditional macro-cell-based TDD systems due to the interference between uplink and downlink that would be caused if neighbor macro cells changed uplink–downlink ratios individually. The increasing need for small-cell operation justified the need for specification of this feature as it became feasible to assume the existence of isolated small cells or groups of small cells.

- **Network-assisted interference cancellation and suppression:** Specification support was introduced to reduce the search complexity for determining the transmission parameters of interference signals from neighboring cells in an advanced UE receiver; for example scrambling sequences used for the reference signals of neighboring cells.
- **Inter-eNB CoMP:** Network signaling between eNBs was introduced to enable coordinated downlink transmission from a set of eNBs connected with non-ideal backhaul. Note that CoMP in Rel-11 assumes an ideal backhaul between transmission points.

Rel-12 standard includes air-interface specifications for cost-efficient support of MTC by introducing UE category 0, where the maximum data rate is 1 Mbps and only one receive antenna is supported at the UE. For support of D2D operations, the Rel-12 standard specified peer discovery as well as direct communication between proximity UEs. Application of D2D includes not only commercial cases such as advertisements and social network service, but also public safety operations especially for the case when mobile networks have collapsed, for example due to earthquakes.

In the continued evolution of LTE-Advanced in Rel-13 and Rel-14, it is important to emphasize continuity and backward compatibility in order to leverage the massive economies of scale associated with the current ecosystem that has developed around the LTE/LTE-Advanced standards from Rel-8 to Rel-12. While improving the spectral efficiency has traditionally been emphasized, the importance of supporting diverse requests from mobile network operators was also well recognized in planning the features to be standardized in Rel-13 and upwards. As the result of this consideration, Rel-13 includes the following major features:

- full dimension MIMO (FD-MIMO) for a drastic increase of spectral efficiency via use of a large number of antennas at the base station,
- licensed assisted access (LAA) for utilizing unlicensed spectrum while guaranteeing coexistence with existing devices,
- carrier aggregation with up to 32 component carriers,
- further cost reductions for MTC devices that can also support extended coverage.

FD-MIMO heavily relies on advancement of signal processing technologies and is one of the key candidate technologies for the evolution from 4G to 5G cellular systems. The key idea behind FD-MIMO is to utilize a large number of antennas placed in a two-dimensional (2D) antenna-array panel for forming narrow beams in both the horizontal and vertical directions. Such beamforming allows an eNB to perform simultaneous transmissions to multiple UEs, realizing high-order multi-user spatial multiplexing. Figure 23.7 depicts an eNB with FD-MIMO implemented using a 2D antenna array panel, where every antenna is an active element allowing dynamic and adaptive precoding across all antennas. By utilizing such precoding, the eNB can simultaneously direct transmissions in the azimuth and elevation domains for multiple UEs. The key feature of FD-MIMO in improving the system performance is its ability to realize high-order multi-user multiplexing.

3GPP has conducted several studies since December 2012 in an effort to provide specification support for FD-MIMO. The first step was a study for developing a new channel model for future evaluation of antenna technologies based on 2D antenna-array panels [6]. The channel model provides the stochastic characteristics of a three-dimensional (3D) wireless channel. Based on the new channel model, a follow-up study item on FD-MIMO was initiated in September 2014 to evaluate the performance of FD-MIMO and identify key areas in LTE specifications that need to be enhanced in order to support 2D arrays with up to 64 antenna ports [7]. In June 2015, 3GPP started a work item to specify FD-MIMO operations for LTE-Advanced in Rel-13. FD-MIMO has two important differentiating factors compared to MIMO technologies from previous LTE releases. First, the number of antenna ports at the eNB transmitter is increased from 8 to 16. As a result, FD-MIMO significantly improves beamforming and spatial user multiplexing capabilities. Second, specification support for FD-MIMO is targeted for antennas placed on a 2D planar array. Using the 2D planar placement is also helpful to reduce the form factor of the antennas for practical applications.

LAA aims to use unlicensed spectrum as a complement for LTE systems operating in licensed spectrum to meet the sharply increased demand for wireless broadband data [8]. For tight integration of the unlicensed spectrum as a capacity boost with the licensed spectrum, LAA uses carrier aggregation. Considering the availability of the 5-GHz unlicensed band that can provide tens of 20-MHz carriers, carrier aggregation with up to 32 component carriers is specified to fully utilize the available frequency resource in both the licensed and unlicensed spectrums. Although the licensed spectrum affords operators exclusive control for providing guaranteed quality of service (QoS) and mobility, LTE operations in the unlicensed spectrum need to coexist with the operation of other radio access technologies such as WiFi. In order to meet the coexistence requirement, the listen-before-talk mechanism has been specified to govern when a wireless resource in the unlicensed spectrum can be utilized. It should be noted that the LAA specification in Rel-13 only supports downlink transmissions.

Machine-type communication through cellular networks is emerging as a significant opportunity for new applications in a networked world where devices communicate with humans and with each other. The Rel-12 specifications for MTC UEs achieved cost reductions of about 50% relative to the lowest category LTE UEs (category 1 LTE UEs) and Rel-13 MTC specifications are expected to achieve an additional 50% cost reduction primarily through restrictions in transmission/reception within only six resource blocks of a system bandwidth per transmission time interval and a lower power amplifier gain, where the RB bandwidth is 180 kHz [9]. The absence of receiver antenna diversity and thc rcduction in power amplifier gain can result in significant reductions in coverage, even for Rel-13 MTC UEs that do not experience large path-loss. A key design target is to provide up to 15 dB coverage enhancement while minimizing the impact on network spectral efficiency and MTC UE power consumption. The coverage enhancement is mainly achieved by repeated transmissions of same signals. In order to reduce to the required number of repetitions, other physical-layer techniques, such as use of multiple contiguous TTIs for a transmission to improve channel estimation accuracy at the receiver and frequency hopping across repetitions to increase the frequency diversity gain, are also specified. Narrowband Internet of Things (NB-IoT) is specified in Rel-13 as another approach

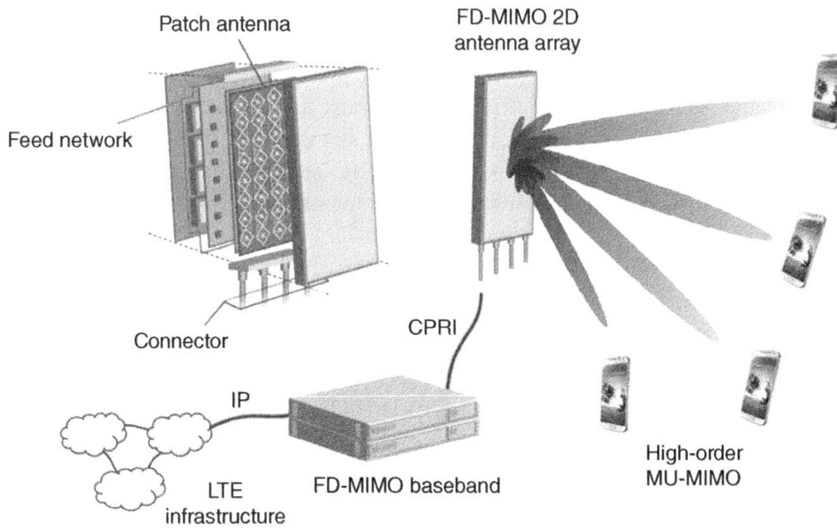

Figure 23.7 Conceptual diagram of a FD-MIMO system realizing high-order MU-MIMO through a 2D antenna array

for efficient support of the cellular Internet of Things (IoT) with low throughput up to about 50 kbps using a very narrow bandwidth of 180 kHz. The NB-IoT can be deployed by reusing a GSM carrier of 200 kHz bandwidth, using a single RB in LTE systems, or using a part of the guard band in LTE systems.

Discussion of the evolution of LTE-Advanced in Rel-14 has already started. FD-MIMO and LAA introduced in Rel-13 will naturally continue to be enhanced. In case of FD-MIMO, the number of eNB transmit antenna ports can be increased to 32. The LAA specification is expected to add support for uplink transmissions in the unlicensed band. In addition, it is expected that Rel-14 will introduce technologies for latency reduction, which is one of the most important aspects for improving the user experience but has not been improved much since the introduction of LTE. The uplink data transmission consists of a scheduling request, a resource grant and a data packet transmission. The request-grant procedure represents a big portion of the latency required for uplink data transmission, especially for the transmission of small payloads such as acknowledgement signaling in the file transfer protocol (FTP). By introducing a grantless procedure – in other words, removing the request-grant procedure – it becomes possible to significantly reduce the data download latency that is caused by the slow-start procedure of the transmission control protocol (TCP). Another approach gaining attention is to shorten the TTI length. In the current LTE standard, the TTI length is 1 ms and is equal to the duration of a subframe, which consists of 2 slots and corresponds to 14 OFDM symbols with normal cyclic prefix (CP) length. The TTI length will be reduced to for example 1 slot (0.5 ms) while guaranteeing backwards compatibility; in other words it will be possible for legacy UEs supporting the current TTI length of 1 ms to coexist with the new UEs supporting the reduced TTI length.

Technologies for vehicle-related services (V2X) such as vehicle-to-vehicle (V2V), vehicle-to-infra (V2I), and vehicle-to-pedestrian (V2P) have recently gained significant attention from the cellular industry as another opportunity for LTE-Advanced technologies to be extended to support vertical industries. These technologies are expected to be specified in Rel-14. Support for V2V and V2P over D2D communication links between UEs will be specified with the highest priority, including potential resource allocation and channel estimation enhancements to support efficient and robust transmissions with low latency. In addition, provisioning of V2X services over the link between the LTE network and the UE is also within the scope of the study. Considerations include the applicability of latency reduction and multi-cell multicast/broadcast enhancements to sufficiently meet industry and regulatory requirements for V2X.

23.3 Preparation of 5G Cellular Communication Standards

To provide a guideline for 5G technical work, ITU-R has been discussing its IMT-2020 vision in recent years and has recently finalized the IMT-2020 vision document [4]. Contrary to previous generations, which focused on enhancement of mobile broadband and improving voice or data capacity, 5G is expected to enhance three major usage scenarios, as shown in Figure 23.8(a): enhanced mobile broadband (eMBB), massive machine type communications (mMTC) and ultra-reliable and low-latency communications (URLLC). One example of the new services under the mMTC heading is the IoT, that will connect a very large number of objects: smart power meters, street lights, cars, home electronics such as refrigerators and

TVs, and surveillance cameras. The representative services of the URLLC category are factory automation, remote surgery and self-driving cars, which can be characterized by their requirement for very low latency and high reliability to prevent any accidents from happening. While mobile broadband has been enhanced quite a lot in previous generations, mobile network operators are emphasizing the importance of providing as uniform a user experience as possible regardless of where the mobile devices are located in a mobile network. It is now well understood that the energy consumption for operating mobile networks should also be reduced in order to reduce the operational cost as well as to be a good citizen to preserve the natural environment. These considerations resulted in defining the following eight parameters as key performance indicators of IMT-2020: peak data rate, user-experienced data rate, spectrum efficiency, mobility, latency, connection density, network energy efficiency, and area traffic capacity. Figure 23.8(b) shows target values for the identified parameters for IMT-2020 in comparison with those of IMT-Advanced.

Following the success of standardization and commercialization of 4G technologies, 3GPP has become the only place where technical work for standardizing 5G technologies will take place. The 5G standards from 3GPP will be brought to ITU-R for official publication in 2020 with the name of IMT-2020. ITU-R agreed on the timeline and process for IMT-2020 [10] as given in Figure 23.9. 3GPP has to finalize the standardization of 5G so that it satisfies all the requirements of IMT-2020 by the end of 2019. It must then submit the specifications of the 5G standard to ITU-R at the beginning of the "IMT-2020 specifications" process.

Even though the official IMT-2020 specifications will be published only in year 2020, there are commercial requests to deploy the first 5G system around the year 2020. It also should be noted that there are various activities in the mobile industry to demonstrate the expected 5G technologies, for example at the 2018 Winter Olympics in Pyeongchang, Korea and the 2020 Olympics in Tokyo, Japan. In order to meet such commercial requests, it is expected that 3GPP will take a phased approach for 5G standardization. The first phase of the 5G standard, which should satisfy a part of IMT-2020 requirements, will be completed in 2018 to enable early commercial deployment around 2020. The second phase, which should satisfy all of the IMT-2020 requirements, will be finalized in 2019 for submission to ITU-R as a candidate technology for IMT-2020.

To satisfy the 5G requirements, there are two candidate approaches in 3GPP: the first is to continue enhancing LTE-Advanced technologies and the second is the introduction of a new radio access technologies. LTE-Advanced was designed to satisfy IMT-Advanced requirements and its enhancement beyond Rel-13 may be able to meet a part of the IMT-2020 requirements. However, further enhancements based on LTE-Advanced to meet the more demanding IMT-2020 requirements could face significant difficulties because backwards compatibility for coexistence with legacy UEs needs to be maintained. Therefore, 3GPP will standardize a new radio access technology for IMT-2020.

One of the most promising technologies for the new RAT to satisfy the drastically increased data rate requirements of providing 20 Gbps is the utilization of higher frequency bands than used in traditional cellular communication bands, say around 30 GHz. Utilization of such higher frequency bands make it easier to use a very wide contiguous spectrum of bandwidth – more than 500 MHz – which would be difficult to operate by carrier aggregation as is done in LTE-Advanced. It is also a common understanding that there is not sufficient frequency spectrum available below 6 GHz even though WRC-15 discussed allocaiton of additional frequency bands below 6 GHz. Due to the above considerations, there is high

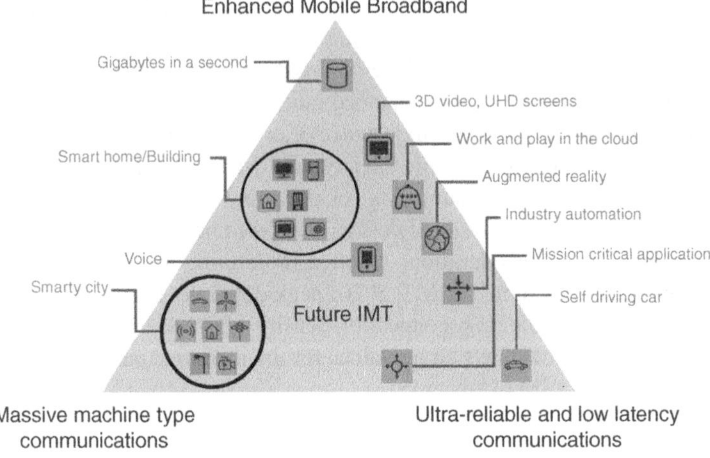

(a) Usage scenarios of IMT for 2020 and beyond

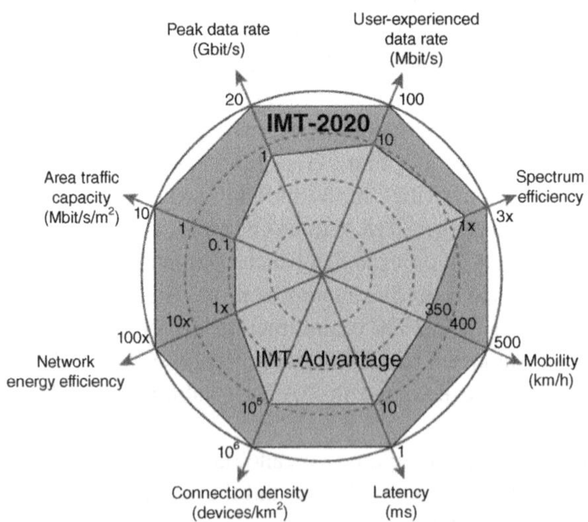

(b) Enhancement of key capabilities from IMT-Advanced to IMT-2020

Figure 23.8 Usage scenarios and key capabilities of IMT for 2020 and beyond [4]

level of interest in utilizing the new spectrum resource above 6 GHz, not only from mobile industries but also from governments and regional spectrum-related organizations. Therefore, it is expected that new frequency bands above 6 GHz will be allocated by WRC-19.

3GPP is expected to develop the new RAT standard to meet the IMT-2020 requirements, utilizing all frequency resources available in the traditional cellular frequency bands below 6 GHz as well as high-frequency bands above 6 GHz (up to 100 GHz). It is believed that the OFDM-based waveform will still be a baseline waveform, with potential variations, for example the application of additional filtering per subcarrier or per subband to reduce the

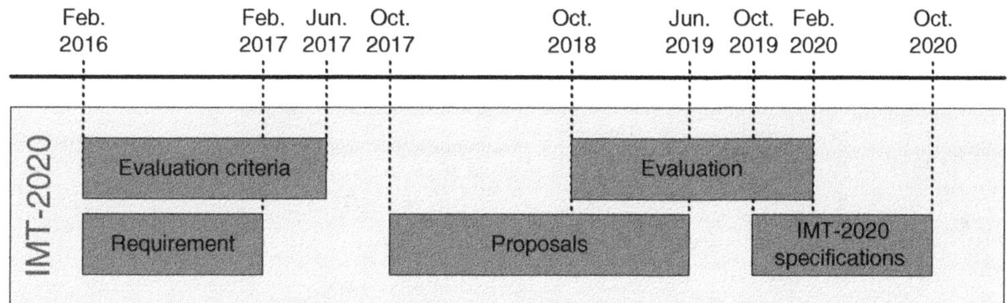

Figure 23.9 Timeline and process for IMT-2020

overhead caused by cyclic prefix and guard bands [11–16]. In order to support such a wide range of frequency bands, there will be a need to introduce multiple numerology sets defining OFDM-based waveforms. For example, the subcarrier spacing in the 2-GHz frequency band and the 30-GHz frequency band will have to be different, since the 15-kHz subcarrier spacing of the LTE standard is too narrow to be robust against RF phase noise in a frequency band around 30 GHz. The maximum bandwidth of a carrier and the supportable FFT size will also be reasons for defining different numerologies to support different frequency bands up to 100 GHz. Even though multiple numerologies will be introduced, it will be highly desirable to keep commonality and scalability for operations in different frequency bands so that the implementation complexity can be kept reasonable.

One of the main challenges for utilizing high-frequency bands around 30 GHz – known as the mmWave band – is the limited coverage due to large path loss. The conventional assumption is that the path loss is proportional to the frequency squared. However, the utilization of beamforming is quite useful to improve coverage. Furthermore, it is easy to have large beamforming gains in the mmWave band, since the wavelength becomes shorter as the frequency increases and there can be more antenna ports for the same antenna dimension, thereby allowing for sharper beams with higher beamforming gains. In the mmWave bands, conventional digital beamforming may not be feasible, since too many RF chains, each of which is used for each digital path, are required to support the massive number of antenna elements. In order to keep a reasonable RF complexity, a combination of analog beamforming and digital precoding – the hybrid beamforming illustrated in Figure 23.10 – is considered a practical approach for mmWave-based systems [17–19]. The analog part forms a set of beams to make sure that the terminals in the coverage area can be connected to the network and the digital part can be used to optimize performance of communication with scheduled terminals by combining analog beams.

It is expected that the new RAT may not provide full coverage in the early phases of 5G commercialization. In the case of mmWave band systems, even though the beamforming will help increase coverage, it may not be practical to assume that full coverage can be provided. This understanding motivates the utilization of LTE-Advanced and the idea that 4G base stations will give a coverage layer providing control-plane operations, while 5G base stations serve as the capacity layer – the user-plane operation – providing high data-rate services. This is shown in Figure 23.11, where a terminal is simultaneously connected to both 4G and 5G base stations when it is in the coverage area of a 5G base station. In order to guarantee proper

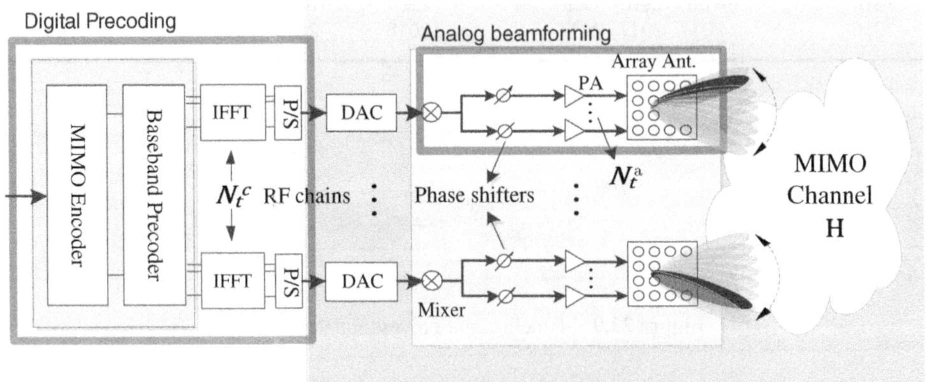

Figure 23.10 Hybrid beamforming

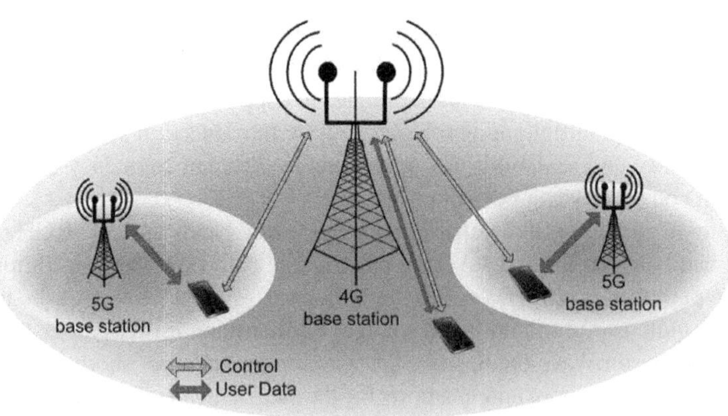

Figure 23.11 Tight integration between 4G and 5G

communication links between the terminals and the network with such split structure, it would be essential to have tight integration between 4G and 5G base stations. It is noted that if the 5G system supports full coverage, loose interworking with the 4G system may be enough.

A possible phased approach to support early commercial deployment around 2020 would be that eMBB is optimized in the first phase and the other usage scenarios are introduced or optimized in the second phase. In order to guarantee smooth migration from the first to the second phase, the first phase standard should guarantee easy and efficient introduction of new functions in the second phase and later. Provisioning of such forward compatibility is quite important, since it is becoming more difficult to predict what services will be needed in the future, as the information technology is evolving. A practical way to achieve this goal would be to leave as much air resource vacant as possible in the new RAT; in other words, signals and channels should be transmitted only when needed to serve for communication for a specific terminal(s). For example, transmission of periodic signals such as the CRS of LTE system,

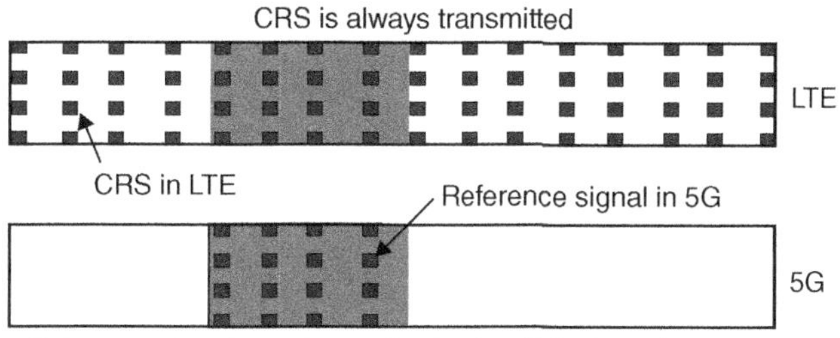

Figure 23.12 An example of forward-compatible design

which makes it difficult to introduce new features later, can be minimized in the new 5G RAT as illustrated in Figure 23.12.

23.4 Concluding Remarks

In this chapter, the standards roadmap from 4G to 5G was reviewed, and major enabling technologies and a more detailed roadmap of 5G standards were discussed. 5G technologies should be developed to enable efficient support of enhanced mobile broadband, which has been the major focus of previous generations, as well as new services such as massive machine-type communications and ultra-reliable and low-latency communications. In addition to the existing cellular frequency bands up to 3.5 GHz and new bands below 6 GHz, new frequency bands above 6 GHz (up to 100 GHz) are expected to be important in developing 5G technologies, especially for support of eMBB.

As can be seen recently, 4G technologies are having great commercial success globally. This became possible as a result of the industry momentum created through the standardization process in 3GPP, and the development of the LTE and LTE-Advanced standards, in which global manufacturers and major mobile network operators participated. Based on the industry experience with 4G, the standardization process for 5G is also expected to play a crucial role in leading research activity in 5G technologies to commercial success.

References

[1] Zhang, J.C, Ariyavisitakul, S., and Tao, M. (2012) LTE-advanced and 4G wireless communications. *IEEE Commun. Mag.*, **50** (2), 102–103,
[2] 3GPP (2010) TR 36.912, Feasibility study for Further Advancements for E-UTRA (LTE-Advanced)
[3] METIS (2015), ICT-317669-METIS/D8.4, METIS final project report.
[4] ITU-R WP-5D (2015) Document 5D/TEMP/625-E, IMT Vision – Framework and overall objectives of the future development of IMT for 2020 and beyond
[5] 3GPPTR 36.819 (2011) Coordinated multi-point operation for LTE physical layer aspects.
[6] 3GPP (2014) TR 36.873, Study on 3D channel model for LTE.
[7] 3GPP (2015) TR 36.897, Study on elevation beamforming/full-dimension (FD) MIMO for LTE.
[8] 3GPP (2015) TR 36.889, Study on licensed-assisted access using LTE.
[9] 3GPP (2013) TR 36.888, Study on provision of low-cost machine-type communications (MTC) user equipments (UEs) based on LTE.
[10] ITU-R (2015) WP-5D, Att. 2.12 to 5D/1042, ITU-R Working party 5D structure and work plan.

[11] Chang, R. (1966) High-speed multichannel data transmission with bandlimited orthogonal signals. *Bell Sys. Tech. J.*, **45**, 1775–1796.

[12] Saltzberg, B. (1967) Performance of an efficient parallel data transmission system. *IEEE Trans. Commun. Tech.*, **15** (6), pp. 805–811.

[13] Hirosaki, M.B. (1981) An orthogonally multiplexed QAM system using the discrete Fourier transform. *IEEE Trans. Commun.*, **29** (7), 982–989.

[14] Farhang-Boroujeny, B. (2011) OFDM versus filter bank multicarrier. *IEEE Signal Process.Mag.*, **28** (3), 92–112.

[15] Premnath S. Wasden, D., Kasera, S., Patwari, N., and Farhang-Boroujeny, B. (2013)Beyond OFDM: best-effort dynamic spectrum access using filterbank multicarrier. *IEEE/ACM Trans. Network.*, **21** (3), 869–882.

[16] Bogucka, H., Kryszkiewicz, P., and Kliks, A (2015) Dynamic spectrum aggregation for future 5G communications. *IEEE Commun. Mag.*, **53** (5), 35–43.

[17] Pi Z. and Khan F. (2011)An introduction to millimeter-wave mobile broadband systems. *IEEE Commun. Mag.*, **49** (6), 101–107.

[18] Kim C., Kim, T., and Seol, J.Y. (2013) Multi-beam transmission diversity with hybrid beamforming for MIMO-OFDM systems, in *Proc. IEEE GLOBECOM'13 Workshop*, pp. 61–65.

[19] Roh, W., Seol, J.Y., Park, J., Lee, B., Lee, J., Kim, Y., Cho, J., Cheun K., and Aryanfar, F. (2014) Millimeter-wave beamforming as an enabling technology for 5G cellular communications: theoretical feasibility and prototype results. *IEEE Commun. Mag.*, **52** (2), 106–113.

Index

Signal Processing for 5G: Algorithms and Implementations, First Edition. Edited by Fa-Long Luo and Charlie Zhang.

Printed and bound by CPI Group (UK) Ltd, Croydon, CR0 4YY

24/04/2024

14488604-0002